Handbook of Natural Computing

Main Editor
Grzegorz Rozenberg

Editors
Thomas Bäck
Joost N. Kok

Handbook of Natural Computing

Volume 1

With 734 Figures and 75 Tables

 Springer

Editors
Grzegorz Rozenberg
LIACS
Leiden University
Leiden, The Netherlands
and
Computer Science Department
University of Colorado
Boulder, USA

Joost N. Kok
LIACS
Leiden University
Leiden, The Netherlands

Thomas Bäck
LIACS
Leiden University
Leiden, The Netherlands

ISBN 978-3-540-92909-3   ISBN 978-3-540-92910-9 (eBook)
ISBN 978-3-540-92911-6 (print and electronic bundle)
DOI 10.1007/978-3-540-92910-9
Springer Heidelberg Dordrecht London New York

Library of Congress Control Number: 2010933716

Printed on acid-free paper

Springer is part of Springer Science+Business Media (www.springer.com)

Preface

Natural Computing is the field of research that investigates human-designed computing inspired by nature as well as computing taking place in nature, that is, it investigates models and computational techniques inspired by nature, and also it investigates, in terms of information processing, phenomena taking place in nature.

Examples of the first strand of research include neural computation inspired by the functioning of the brain; evolutionary computation inspired by Darwinian evolution of species; cellular automata inspired by intercellular communication; swarm intelligence inspired by the behavior of groups of organisms; artificial immune systems inspired by the natural immune system; artificial life systems inspired by the properties of natural life in general; membrane computing inspired by the compartmentalized ways in which cells process information; and amorphous computing inspired by morphogenesis. Other examples of natural-computing paradigms are quantum computing and molecular computing, where the goal is to replace traditional electronic hardware, by, for example, bioware in molecular computing. In quantum computing, one uses systems small enough to exploit quantum-mechanical phenomena to perform computations and to perform secure communications more efficiently than classical physics, and, hence, traditional hardware allows. In molecular computing, data are encoded as biomolecules and then tools of molecular biology are used to transform the data, thus performing computations.

The second strand of research, computation taking place in nature, is represented by investigations into, among others, the computational nature of self-assembly, which lies at the core of the nanosciences; the computational nature of developmental processes; the computational nature of biochemical reactions; the computational nature of bacterial communication; the computational nature of brain processes; and the systems biology approach to bionetworks where cellular processes are treated in terms of communication and interaction, and, hence, in terms of computation.

Research in natural computing is genuinely interdisciplinary and forms a bridge between the natural sciences and computer science. This bridge connects the two, both at the level of information technology and at the level of fundamental research. Because of its interdisciplinary character, research in natural computing covers a whole spectrum of research methodologies ranging from pure theoretical research, algorithms, and software applications to experimental laboratory research in biology, chemistry, and physics.

Computer Science and Natural Computing

A preponderance of research in natural computing is centered in computer science. The spectacular progress in Information and Communication Technology (ICT) is highly supported by the evolution of computer science, which designs and develops the instruments needed for this progress: computers, computer networks, software methodologies, etc. As ICT has such a tremendous impact on our everyday lives, so does computer science.

However, there is much more to computer science than ICT: it is the science of information processing and, as such, a fundamental science for other disciplines. On one hand, the only common denominator for research done in such diverse areas of computer science is investigating various aspects of information processing. On the other hand, the adoption of Information and Information Processing as central notions and thinking habit has been an important development in many disciplines, biology and physics being prime examples. For these scientific disciplines, computer science provides not only instruments but also a way of thinking.

We are now witnessing exciting interactions between computer science and the natural sciences. While the natural sciences are rapidly absorbing notions, techniques, and methodologies intrinsic to information processing, computer science is adapting and extending its traditional notion of computation, and computational techniques, to account for computation taking place in nature around us. Natural Computing is an important catalyst for this two-way interaction, and this handbook constitutes a significant record of this development.

The Structure of the Handbook

Natural Computing is both a well-established research field with a number of classical areas, and a very dynamic field with many more recent, novel research areas. The field is vast, and so it is quite usual that a researcher in a specific area does not have sufficient insight into other areas of Natural Computing. Also, because of its dynamic development and popularity, the field constantly attracts more and more scientists who either join the active research or actively follow research developments.

Therefore, the goal of this handbook is two-fold:

(i) to provide an authoritative reference for a significant and representative part of the research in Natural Computing, and
(ii) to provide a convenient gateway to Natural Computing for motivated newcomers to this field.

The implementation of this goal was a challenge because this field and its literature are vast — almost all of its research areas have an extensive scientific literature, including specialized journals, book series, and even handbooks. This implies that the coverage of the whole field in reasonable detail and within a reasonable number of pages/volumes is practically impossible.

Thus, we decided to divide the presented material into six areas. These areas are by no means disjoint, but this division is convenient for the purpose of providing a representative picture of the field — representative with respect to the covered research topics and with respect to a good balance between classical and emerging research trends.

Each area consists of individual chapters, each of which covers a specific research theme. They provide necessary technical details of the described research, however they are self-contained and of an expository character, which makes them accessible for a broader audience. They also provide a general perspective, which, together with given references, makes the chapters valuable entries into given research themes.

This handbook is a result of the joint effort of the handbook editors, area editors, chapter authors, and the Advisory Board. The choice of the six areas by the handbook editors in consultation with the Advisory Board, the expertise of the area editors in their respective areas, the choice by the area editors of well-known researchers as chapter writers, and the peer-review for individual chapters were all important factors in producing a representative and reliable picture of the field. Moreover, the facts that the Advisory Board consists of 68 eminent scientists from 20 countries and that there are 105 contributing authors from 21 countries provide genuine assurance for the reader that this handbook is an authoritative and up-to-date reference, with a high level of significance and accuracy.

Handbook Areas

The material presented in the handbook is organized into six areas: Cellular Automata, Neural Computation, Evolutionary Computation, Molecular Computation, Quantum Computation, and Broader Perspective.

Cellular Automata

Cellular automata are among the oldest models of computation, dating back over half a century. The first cellular automata studies by John von Neumann in the late 1940s were biologically motivated, related to self-replication in universal systems. Since then, cellular automata gained popularity in physics as discrete models of physical systems, in computer science as models of massively parallel computation, and in mathematics as discrete-time dynamical systems. Cellular automata are a natural choice to model real-world phenomena since they possess several fundamental properties of the physical world: they are massively parallel, homogeneous, and all interactions are local. Other important physical constraints such as reversibility and conservation laws can be added as needed, by properly choosing the local update rule. Computational universality is common in cellular automata, and even starkly simple automata are capable of performing arbitrary computation tasks. Because cellular automata have the advantage of parallelism while obeying natural constraints such as locality and uniformity, they provide a framework for investigating realistic computation in massively parallel systems. Computational power and the limitations of such systems are most naturally investigated by time- and space-constrained computations in cellular automata. In mathematics — in terms of symbolic dynamics — cellular automata are viewed as endomorphisms of the full shift, that is, transformations that are translation invariant and continuous in the product topology. Interesting questions on chaotic dynamics have been studied in this context.

Neural Computation

Artificial neural networks are computer programs, loosely modeled after the functioning of the human nervous system. There are neural networks that aim to gain understanding of biological neural systems, and those that solve problems in artificial intelligence without necessarily creating a model of a real biological system. The more biologically oriented neural networks model the real nervous system in increasing detail at all relevant levels of information processing: from synapses to neurons to interactions between modules of interconnected neurons. One of the major challenges is to build artificial brains. By reverse-engineering the mammalian brain in silicon, the aim is to better understand the functioning of the (human) brain through detailed simulations. Neural networks that are more application-oriented tend to drift further apart from real biological systems. They come in many different flavors, solving problems in regression analysis and time-series forecasting, classification, and pattern recognition, as well as clustering and compression. Good old multilayered perceptrons and self-organizing maps are still pertinent, but attention in research is shifting toward more recent developments, such as kernel methods (including support vector machines) and Bayesian techniques. Both approaches aim to incorporate domain knowledge in the learning process in order to improve prediction performance, e.g., through the construction of a proper kernel function or distance measure or the choice of an appropriate prior distribution over the parameters of the neural network. Considerable effort is devoted to making neural networks efficient so that large models can be learned from huge databases in a reasonable amount of time. Application areas include, among many others, system dynamics and control, finance, bioinformatics, and image analysis.

Evolutionary Computation

The field of evolutionary computation deals with algorithms gleaned from models of organic evolution. The general aim of evolutionary computation is to use the principles of nature's processes of natural selection and genotypic variation to derive computer algorithms for solving hard search and optimization tasks. A wide variety of instances of evolutionary algorithms have been derived during the past fifty years based on the initial algorithms, and we are now witnessing astounding successes in the application of these algorithms: their fundamental understanding in terms of theoretical results; understanding algorithmic principles of their construction; combination with other techniques; and understanding their working principles in terms of organic evolution. The key algorithmic variations (such as genetic algorithms, evolution strategies, evolutionary programming, and genetic programming) have undergone significant developments over recent decades, and have also resulted in very powerful variations and recombinations of these algorithms. Today, there is a sound understanding of how all of these algorithms are instances of the generic concept of an evolutionary search approach. Hence the generic term "evolutionary algorithm" is nowadays being used to describe the generic algorithm, and the term "evolutionary computation" is used for the field as a whole. Thus, we have observed over the past fifty years how the field has integrated the various independently developed initial algorithms into one common principle. Moreover, modern evolutionary algorithms benefit from their ability to adapt and self-adapt their strategy parameters (such as mutation rates, step sizes, and search distributions) to the needs of the task at hand. In this way, they are robust and flexible metaheuristics for problem-solving even without requiring too much special expertise from their users. The feature of self-adaptation illustrates the ability of evolutionary principles to work on different levels at the same time, and therefore provides a nice demonstration of the universality of the evolutionary principle for search and optimization tasks. The widespread use of evolutionary computation reflects these capabilities.

Molecular Computation

Molecular computing is an emergent interdisciplinary field concerned with programming molecules so that they perform a desired computation, or fabricate a desired object, or control the functioning of a specific molecular system. The central idea behind molecular computing is that data can be encoded as (bio)molecules, e.g., DNA strands, and tools of molecular science can be used to transform these data. In a nutshell, a molecular program is just a collection of molecules which, when placed in a suitable substrate, will perform a specific function (execute the program that this collection represents). The birth of molecular computing is often associated with the 1994 breakthrough experiment by Leonard Adleman, who solved a small instance of a hard computational problem solely by manipulating DNA strands in test tubes. Although initially the main effort of the area was focused on trying to obtain a breakthrough in the complexity of solving hard computational problems, this field has evolved enormously since then. Among the most significant achievements of molecular computing have been contributions to understanding some of the fundamental issues of the nanosciences. One notable example among them is the contribution to the understanding of self-assembly, a central concept of the nanosciences. The techniques of molecular programming were successfully applied in experimentally constructing all kinds of molecular-scale objects or devices with prescribed functionalities. Well-known examples here are self-assembly of Sierpinski triangles, cubes, octahedra, DNA-based logic circuits, DNA "walkers" that move along a track, and autonomous molecular motors. A complementary approach to understanding bioinformation and computation is through studying the information-processing capabilities of cellular organisms. Indeed, cells and nature "compute" by "reading" and "rewriting" DNA through processes that modify DNA (or RNA) sequences. Research into the computational abilities of cellular organisms has the potential to uncover the laws governing biological information, and to enable us to harness the computational power of cells.

Quantum Computation

Quantum computing has been discussed for almost thirty years. The theory of quantum computing and quantum information processing is simply the theory of information processing with a classical notion of information replaced by its quantum counterpart. Research in quantum computing is concerned with understanding the fundamentals of information processing on the level of physical systems that realize/implement the information. In fact, quantum computing can be seen as a quest to understand the fundamental limits of information processing set by nature itself. The mathematical description of quantum information is more complicated than that of classical information — it involves the structure of Hilbert spaces. When describing the structure behind known quantum algorithms, this reduces to linear algebra over complex numbers. The history of quantum algorithms spans the last fifteen years, and some of these algorithms are extremely interesting, and even groundbreaking — the most remarkable are Shor's factorization in polynomial time and Grover's search algorithm. The nature of quantum information has also led to the invention of novel cryptosystems, whose security is not based on the complexity of computing functions, but rather on the physical properties of quantum information. Quantum computing is now a well-established discipline, however implementation of a large-scale quantum computer continues to be extremely challenging, even though quantum information processing primitives, including those allowing secure cryptography, have been demonstrated to be practically realizable.

Broader Perspective

In contrast to the first five areas focusing on more-established themes of natural computing, this area encompasses a perspective that is broader in several ways. First, the reader will find here treatments of certain well-established and specific techniques inspired by nature (e.g., simulated annealing) not covered in the other five areas. Second, the reader will also find application-centered chapters (such as natural computing in finance), each covering, in one chapter, a collection of natural computing methods, thus capturing the impact of natural computing as a whole in various fields of science or industry. Third, some chapters are full treatments of several established research fields (such as artificial life, computational systems biology, evolvable hardware, and artificial immune systems), presenting alternative perspectives and cutting across some of the other areas of the handbook, while introducing much new material. Other elements of this area are fresh, emerging, and novel techniques or perspectives (such as collision-based computing, nonclassical computation), representing the leading edge of theories and technologies that are shaping possible futures for both natural computing and computing in general. The contents of this area naturally cluster into two kinds (sections), determined by the essential nature of the techniques involved. These are "Nature-Inspired Algorithms" and "Alternative Models of Computation". In the first section, "Nature-Inspired Algorithms", the focus is on algorithms inspired by natural processes realized either through software or hardware or both, as additions to the armory of existing tools we have for dealing with well-known practical problems. In this section, we therefore find application-centered chapters, as well as chapters focusing on particular techniques, not otherwise dealt with in other areas of the handbook, which have clear and proven applicability. In the second section, "Alternative Models of Computation", the emphasis changes, moving away from specific applications or application areas, toward more far-reaching ideas. These range from developing computational approaches and "computational thinking" as fundamental tools for the new science of systems biology to ideas that take inspiration from nature as a platform for suggesting entirely novel possibilities of computing.

Handbook Chapters

In the remainder of this preface we will briefly describe the contents of the individual chapters. These chapter descriptions are grouped according to the handbook areas where they belong and given in the order that they appear in the handbook. This section provides the reader with a better insight into the contents, allowing one to design a personal roadmap for using this handbook.

Cellular Automata

This area is covered by nine chapters.

The first chapter, "Basic Concepts of Cellular Automata", by Jarkko J. Kari, reviews some classical results from the theory of cellular automata, relations between various concepts of injectivity and surjectivity, and some basic dynamical system concepts related to chaos in cellular automata. The classical results discussed include the celebrated Garden-of-Eden and Curtis–Hedlund–Lyndon theorems, as well as the balance property of surjective cellular automata. All these theorems date back to the 1960s. The results are provided together with examples that illustrate proof ideas. Different variants of sensitivity to initial conditions and mixing properties are introduced and related to each other. Also undecidability results concerning cellular automata are briefly discussed.

A popular mathematical approach is to view cellular automata as dynamical systems in the context of symbolic dynamics. Several interesting results in this area were reported as early as 1969 in the seminal paper by G.A. Hedlund, and still today this research direction is among the most fruitful sources of theoretical problems and new results. The chapter "Cellular Automata Dynamical Systems", by Alberto Dennunzio, Enrico Formenti, and Petr Kůrka, reviews some recent developments in this field. Recent research directions considered here include subshifts attractors and signal subshifts, particle weight functions, and the slicing construction. The first two concern one-dimensional cellular automata and give precise descriptions of the limit behavior of large classes of automata. The third one allows one to view two-dimensional cellular automata as one-dimensional systems. In this way combinatorial complexity is decreased and new results can be proved.

Programming cellular automata for particular tasks requires special techniques. The chapter "Algorithmic Tools on Cellular Automata", by Marianne Delorme and Jacques Mazoyer, covers classical algorithmic tools based on signals. Linear signals as well as signals of nonlinear slope are discussed, and basic transformations of signals are addressed. The chapter provides results on using signals to construct functions in cellular automata and to implement arithmetic operations on segments. The methods of folding the space–time, freezing, and clipping are also introduced.

The time-complexity advantage gained from parallelism under the locality and uniformity constraints of cellular automata can be precisely analyzed in terms of language recognition. The chapter "Language Recognition by Cellular Automata", by Véronique Terrier, presents results and questions about cellular automata complexity classes and their relationships to other models of computations. Attention is mainly directed to real-time and linear-time complexity classes, because significant benefits over sequential computation may be obtained at these low time complexities. Both parallel and sequential input modes are considered. Separate complexity classes are given also for cellular automata with one-way communications and two-way communications.

The chapter "Computations on Cellular Automata", by Jacques Mazoyer and Jean-Baptiste Yunès, continues with the topic of algorithmic techniques in cellular automata. This chapter uses the basic tools, such as signals and grids, to build natural implementations of common algorithms in cellular automata. Examples of implementations include real-time multiplication of integers and the prime number sieve. Both parallel and sequential input and output modes are discussed, as well as composition of functions and recursion.

The chapter "Universalities in Cellular Automata", by Nicolas Ollinger, is concerned with computational universalities. Concepts of universality include Turing universality (the ability to compute any recursive function) and intrinsic universality (the ability to simulate any other cellular automaton). Simulations of Boolean circuits in the two-dimensional case are explained in detail in order to achieve both kinds of universality. The more difficult one-dimensional case is also discussed, and seminal universal cellular automata and encoding techniques are presented in both dimensions. A detailed chronology of important papers on universalities in cellular automata is also provided.

A cellular automaton is reversible if every configuration has only one previous configuration, and hence its evolution process can be traced backward uniquely. This naturally corresponds to the fundamental time-reversibility of the microscopic laws of physics. The chapter "Reversible Cellular Automata", by Kenichi Morita, discusses how reversible cellular automata are defined, as well as their properties, how they are designed, and their computing abilities. After providing the definitions, the chapter surveys basic properties of injectivity and surjectivity. Three design methods of reversible cellular automata are provided: block rules, partitioned, and second-order cellular automata. Then the computational power of reversible cellular automata is discussed. In particular, simulation methods of irreversible cellular automata, reversible Turing machines, and some other universal systems are given to clarify universality of reversible cellular automata. In spite of the strong constraint of reversibility, it is shown that reversible cellular automata possess rich information processing capabilities, and even very simple ones are computationally universal.

A conservation law in a cellular automaton is a statement of the invariance of a local and additive energy-like quantity. The chapter "Conservation Laws in Cellular Automata", by Siamak Taati, reviews the basic theory of conservation laws. A general mathematical framework for formulating conservation laws in cellular automata is presented and several characterizations are summarized. Computational problems regarding conservation laws (verification and existence problems) are discussed. Microscopic explanations of the dynamics of the conserved quantities in terms of flows and particle flows are explored. The related concept of dissipating energy-like quantities is also discussed.

The chapter "Cellular Automata and Lattice Boltzmann Modeling of Physical Systems", by Bastien Chopard, considers the use of cellular automata and related lattice Boltzmann methods as a natural modeling framework to describe and study many physical systems composed of interacting components. The theoretical basis of the approach is introduced and its potential is illustrated for several applications in physics, biophysics, environmental science, traffic models, and multiscale modeling. The success of the technique can be explained by the close relationship between these methods and a mesoscopic abstraction of many natural phenomena.

Neural Computation

This area is covered by ten chapters.

Spiking neural networks are inspired by recent advances in neuroscience. In contrast to classical neural network models, they take into account not just the neuron's firing rate, but also the time moment of spike firing. The chapter "Computing with Spiking Neuron Networks", by Hélène Paugam-Moisy and Sander Bohte, gives an overview of existing approaches to modeling spiking neural neurons and synaptic plasticity, and discusses their computational power and the challenge of deriving efficient learning procedures.

Image quality assessment aims to provide computational models to predict the perceptual quality of images. The chapter "Image Quality Assessment — A Multiscale Geometric Analysis-Based Framework and Examples", by Xinbo Gao, Wen Lu, Dacheng Tao, and Xuelong Li, introduces the fundamentals and describes the state of the art in image quality assessment. It further proposes a new model, which mimics the human visual system by incorporating concepts such as multiscale analysis, contrast sensitivity, and just-noticeable differences. Empirical results clearly demonstrate that this model resembles subjective perception values and reflects the visual quality of images.

Neurofuzzy networks have the important advantage that they are easy to interpret. When applied to control problems, insight about the process characteristics at different operating regions can be easily obtained. Furthermore, nonlinear model predictive controllers can be developed as a nonlinear combination of several local linear model predictive controllers that have analytical solutions. Through several applications, the chapter "Nonlinear Process Modelling and Control Using Neurofuzzy Networks", by Jie Zhang, demonstrates that neurofuzzy networks are very effective in the modeling and control of nonlinear processes.

Similar to principal component and factor analysis, independent component analysis is a computational method for separating a multivariate signal into additive subcomponents. Independent component analysis is more powerful: the latent variables corresponding to the subcomponents need not be Gaussian and the basis vectors are typically nonorthogonal. The chapter "Independent Component Analysis", by Seungjin Choi, explains the theoretical foundations and describes various algorithms based on those principles.

Neural networks has become an important method for modeling and forecasting time series. The chapter "Neural Networks for Time-Series Forecasting", by G. Peter Zhang, reviews some recent developments (including seasonal time-series modeling, multiperiod forecasting, and ensemble methods), explains when and why they are to be preferred over traditional forecasting models, and also discusses several practical data and modeling issues.

Support vector machines have been extensively studied and applied in many domains within the last decade. Through the so-called kernel trick, support vector machines can efficiently learn nonlinear functions. By maximizing the margin, they implement the principle of structural risk minimization, which typically leads to high generalization performance. The chapter "SVM Tutorial — Classification, Regression and Ranking", by Hwanjo Yu and Sungchul Kim, describes these underlying principles and discusses support vector machines for different learning tasks: classification, regression, and ranking.

It is well known that single-hidden-layer feedforward networks can approximate any continuous target function. This still holds when the hidden nodes are automatically and randomly generated, independent of the training data. This observation opened up many possibilities for easy construction of a broad class of single-hidden-layer neural networks. The chapter "Fast Construction of Single-Hidden-Layer Feedforward Networks", by Kang Li, Guang-Bin Huang, and Shuzhi Sam Ge, discusses new ideas that yield a more compact network architecture and reduce the overall computational complexity.

Many recent experimental studies demonstrate the remarkable efficiency of biological neural systems to encode, process, and learn from information. To better understand the experimentally observed phenomena, theoreticians are developing new mathematical approaches and tools to model biological neural networks. The chapter "Modeling Biological Neural Networks", by Joaquin J. Torres and Pablo Varona, reviews some of the most popular models of neurons and neural networks. These not only help to understand how living systems perform information processing, but may also lead to novel bioinspired paradigms of artificial intelligence and robotics.

The size and complexity of biological data, such as DNA/RNA sequences and protein sequences and structures, makes them suitable for advanced computational tools, such as neural networks. Computational analysis of such databases aims at exposing hidden information that provides insights that help in understanding the underlying biological principles. The chapter "Neural Networks in Bioinformatics", by Ke Chen and Lukasz A. Kurgan, focuses on proteins. In particular it discusses prediction of protein secondary structure, solvent accessibility, and binding residues.

Self-organizing maps is a prime example of an artificial neural network model that both relates to the actual (topological) organization within the mammalian brain and at the same time has many practical applications. Self-organizing maps go back to the seminal work of Teuvo Kohonen. The chapter "Self-organizing Maps", by Marc M. Van Hulle, describes the state of the art with a special emphasis on learning algorithms that aim to optimize a predefined criterion.

Evolutionary Computation

This area is covered by thirteen chapters.

The first chapter, "Generalized Evolutionary Algorithms", by Kenneth De Jong, describes the general concept of evolutionary algorithms. As a generic introduction to the field, this chapter facilitates an understanding of specific instances of evolutionary algorithms as instantiations of a generic evolutionary algorithm. For the instantiations, certain choices need to be made, such as representation, variation operators, and the selection operator, which then yield particular instances of evolutionary algorithms, such as genetic algorithms and evolution strategies, to name just a few.

The chapter "Genetic Algorithms — A Survey of Models and Methods", by Darrell Whitley and Andrew M. Sutton, introduces and discusses (including criticism) the standard genetic algorithm based on the classical binary representation of solution candidates and a theoretical interpretation based on the so-called schema theorem. Variations of genetic algorithms with respect to solution representations, mutation operators, recombination operators, and selection mechanisms are also explained and discussed, as well as theoretical models of genetic algorithms based on infinite and finite population size assumptions and Markov chain theory concepts. The authors also critically investigate genetic algorithms from the perspective of identifying their limitations and the differences between theory and practice when working with genetic algorithms. To illustrate this further, the authors also give a practical example of the application of genetic algorithms to resource scheduling problems.

The chapter "Evolutionary Strategies", by Günter Rudolph, describes a class of evolutionary algorithms which have often been associated with numerical function optimization and continuous variables, but can also be applied to binary and integer domains. Variations of evolutionary strategies, such as the $(\mu+\lambda)$-strategy and the (μ,λ)-strategy, are introduced and discussed within a common algorithmic framework. The fundamental idea of self-adaptation of strategy parameters (variances and covariances of the multivariate normal distribution used for mutation) is introduced and explained in detail, since this is a key differentiating property of evolutionary strategies.

The chapter "Evolutionary Programming", by Gary B. Fogel, discusses a historical branch of evolutionary computation. It gives a historical perspective on evolutionary programming by describing some of the original experiments using evolutionary programming to evolve finite state machines to serve as sequence predictors. Starting from this canonical evolutionary programming approach, the chapter also presents extensions of evolutionary programming into continuous domains, where an attempt towards self-adaptation of mutation step sizes has been introduced which is similar to the one considered in evolutionary strategies. Finally, an overview of some recent applications of evolutionary programming is given.

The chapter "Genetic Programming — Introduction, Applications, Theory and Open Issues", by Leonardo Vanneschi and Riccardo Poli, describes a branch of evolutionary algorithms derived by extending genetic algorithms to allow exploration of the space of computer programs. To make evolutionary search in the domain of computer programs possible, genetic programming is based on LISP S-expression represented by syntax trees, so that genetic programming extends evolutionary algorithms to tree-based representations. The chapter gives an overview of the corresponding representation, search operators, and technical details of genetic programming, as well as existing applications to real-world problems. In addition, it discusses theoretical approaches toward analyzing genetic programming, some of the open issues, as well as research trends in the field.

The subsequent three chapters are related to the theoretical analysis of evolutionary algorithms, giving a broad overview of the state of the art in our theoretical understanding. These chapters demonstrate that there is a sound theoretical understanding of capabilities and limitations of evolutionary algorithms. The approaches can be roughly split into convergence velocity or progress analysis, computational complexity investigations, and global convergence results.

The convergence velocity viewpoint is represented in the chapter "The Dynamical Systems Approach — Progress Measures and Convergence Properties", by Silja Meyer-Nieberg and Hans-Georg Beyer. It demonstrates how the dynamical systems approach can be used to analyze the behavior of evolutionary algorithms quantitatively with respect to their progress rate. It also provides a complete overview of results in the continuous domain, i.e., for all types of evolution strategies on certain objective functions (such as sphere, ridge, etc.). The chapter presents results for undisturbed as well as for noisy variants of these objective functions, and extends the approach to dynamical objective functions where the goal turns into optimum tracking. All results are presented by means of comparative tables, so the reader gets a complete overview of the key findings at a glance.

The chapter "Computational Complexity of Evolutionary Algorithms", by Thomas Jansen, deals with the question of optimization time (i.e., the first point in time during the run of an evolutionary algorithm when the global optimum is sampled) and an investigation of upper bounds, lower bounds, and the average time needed to hit the optimum. This chapter presents specific results for certain classes of objective functions, most of them defined over binary search spaces, as well as fundamental limitations of evolutionary search and related results on the "no free lunch" theorem and black box complexity. The chapter also discusses the corresponding techniques for analyses, such as drift analysis and the expected multiplicative distance decrease.

Concluding the set of theoretical chapters, the chapter "Stochastic Convergence", by Günter Rudolph, addresses theoretical results about the properties of evolutionary algorithms concerned with finding a globally optimal solution in the asymptotic limit. Such results exist for certain variants of evolutionary algorithms and under certain assumptions, and this chapter summarizes the existing results and integrates them into a common framework. This type of analysis is essential in qualifying evolutionary algorithms as global search algorithms and for understanding the algorithmic conditions for global convergence.

The remaining chapters in the area of evolutionary computation report some of the major current trends.

To start with, the chapter "Evolutionary Multiobjective Optimization", by Eckart Zitzler, focuses on the application of evolutionary algorithms to tasks that are characterized by multiple, conflicting objective functions. In this case, decision-making becomes a task of identifying good compromises between the conflicting criteria. This chapter introduces the concept and a variety of state-of-the-art algorithmic concepts to use evolutionary algorithms for approximating the so-called Pareto front of solutions which cannot be improved in one objective without compromising another. This contribution presents all of the required formal concepts, examples, and the algorithmic variations introduced into evolutionary computation to handle such types of problems and to generate good approximations of the Pareto front.

The term "memetic algorithms" is used to characterize hybridizations between evolutionary algorithms and more classical, local search methods (and agent-based systems). This is a general concept of broad scope, and in order to illustrate and characterize all possible instantiations, the chapter "Memetic Algorithms", by Natalio Krasnogor, presents an algorithmic engineering approach which allows one to describe these algorithms as instances of generic patterns. In addition to explaining some of the application areas, the chapter presents some theoretical remarks, various different ways to define memetic algorithms, and also an outlook into the future.

The chapter "Genetics-Based Machine Learning", by Tim Kovacs, extends the idea of evolutionary optimization to algorithmic concepts in machine learning and data mining, involving applications such as learning classifier systems, evolving neural networks, and genetic fuzzy systems, to mention just a few. Here, the application task is typically a data classification, data prediction, or nonlinear regression task — and the quality of solution candidates is evaluated by means of some model quality measure. The chapter covers a wide range of techniques for applying evolutionary computation to machine learning tasks, by interpreting them as optimization problems.

The chapter "Coevolutionary Principles", by Elena Popovici, Anthony Bucci, R. Paul Wiegand, and Edwin D. de Jong, deals with a concept modeled after biological evolution in which an explicit fitness function is not available, but solutions are evaluated by running them against each other. A solution is evaluated in the context of the other solutions, in the actual population or in another. Therefore, these algorithms develop their own dynamics, because the point of comparison is not stable, but coevolving with the actual population. The chapter provides a fundamental understanding of coevolutionary principles and highlights theoretical concepts, algorithms, and applications.

Finally, the chapter "Niching in Evolutionary Algorithms", by Ofer M. Shir, describes the biological principle of niching in nature as a concept for using a single population to find, occupy, and keep multiple local minima in a population. The motivation for this approach is to find alternative solutions within a single population and run of evolutionary algorithms, and this chapter discusses approaches for niching, and the application in the context of genetic algorithms as well as evolutionary strategies.

Molecular Computation

This area is covered by eight chapters.

The chapter "DNA Computing — Foundations and Implications", by Lila Kari, Shinnosuke Seki, and Petr Sosík, has a dual purpose. The first part outlines basic molecular biology notions necessary for understanding DNA computing, recounts the first experimental demonstration of DNA computing by Leonard Adleman in 1994, and recaps the 2001 milestone wet laboratory experiment that solved a 20-variable instance of 3-SAT and thus first demonstrated the potential of DNA computing to outperform the computational ability of an unaided human. The second part describes how the properties of DNA-based information, and in particular the Watson–Crick complementarity of DNA single strands, have influenced areas of theoretical computer science such as formal language theory, coding theory, automata theory, and combinatorics on words. More precisely, it explores several notions and results in formal language theory and coding theory that arose from the problem of the design of optimal encodings for DNA computing experiments (hairpin-free languages, bond-free languages), and more generally from the way information is encoded on DNA strands (sticker systems, Watson–Crick automata). Lastly, it describes the influence that properties of DNA-based information have had on research in combinatorics on words, by presenting several natural generalizations of classical concepts (pseudopalindromes, pseudoperiodicity, Watson–Crick conjugate and commutative words, involutively bordered words, pseudoknot bordered words), and outlining natural extensions in this context of two of the most fundamental results in combinatorics of words, namely the Fine and Wilf theorem and the Lyndon–Schützenberger result.

The chapter "Molecular Computing Machineries — Computing Models and Wet Implementations", by Masami Hagiya, Satoshi Kobayashi, Ken Komiya, Fumiaki Tanaka, and Takashi Yokomori, explores novel computing devices inspired by the biochemical properties of biomolecules. The theoretical results section describes a variety of molecular computing models for finite automata, as well as molecular computing models for Turing machines based on formal grammars, equality sets, Post systems, and logical formulae. It then presents molecular computing models that use structured molecules such as hairpins and tree structures. The section on wet implementations of molecular computing models, related issues, and applications includes: an enzyme-based DNA automaton and its applications to drug delivery, logic gates and circuits using DNAzymes and DNA tiles, reaction graphs for representing various dynamics of DNA assembly pathways, DNA whiplash machines implementing finite automata, and a hairpin-based implementation of a SAT engine for solving the 3-SAT problem.

The chapter "DNA Computing by Splicing and by Insertion–Deletion", by Gheorghe Păun, is devoted to two of the most developed computing models inspired by DNA biochemistry: computing by splicing, and computing by insertion and deletion. DNA computing by splicing was defined by Tom Head already in 1987 and is based on the so-called splicing operation. The splicing operation models the recombination of DNA molecules that results from cutting them with restriction enzymes and then pasting DNA molecules with compatible ends by ligase enzymes. This chapter explores the computational power of the splicing operation showing that, for example, extended splicing systems starting from a finite language and using finitely many splicing rules can generate only the family of regular languages, while extended splicing systems starting from a finite language and using a regular set of rules can generate all recursively enumerable languages. Ways in which to avoid the impractical notion of a regular infinite set of rules, while maintaining the maximum computational power, are presented. They include using multisets and adding restrictions on the use of rules such as permitting contexts, forbidding contexts, programmed splicing systems, target languages, and double splicing. The second model presented, the insertion–deletion system, is based on a finite set of axioms and a finite set of contextual insertion rules and contextual deletion rules. Computational power results described here include the fact that insertion–deletion systems with context-free insertion rules of words of length at most one and context-free deletion rules of words of unbounded length can generate only regular languages. In contrast, for example, the family of insertion–deletion systems where the insertion contexts, deletion contexts, and the words to be inserted/deleted are all of length at most one, equals the family of recursively enumerable languages.

The chapter "Bacterial Computing and Molecular Communication", by Yasubumi Sakakibara and Satoshi Hiyama, investigates attempts to create autonomous cell-based Turing machines, as well as novel communication paradigms that use molecules as communication media. The first part reports experimental research on constructing *in vivo* logic circuits as well as efforts towards building *in vitro* and *in vivo* automata in the framework of DNA computing. Also, a novel framework is presented to develop a programmable and autonomous *in vivo* computer in a bacterium. The first experiment in this direction uses DNA circular strands (plasmids) together with the cell's protein-synthesis mechanism to execute a finite state automaton in *E. coli*. Molecular communication is a new communication paradigm that proposes the use of molecules as the information medium, instead of the traditional electromagnetic waves. Other distinctive features of molecular communication include its stochastic nature, its low energy consumption, the use of an aqueous transmission medium, and its high compatibility with biological systems. A molecular communication system starts with a sender (e.g., a genetically modified or an artificial cell) that generates molecules, encodes information onto the molecules (called information molecules), and emits the information molecules into a propagation environment (e.g., aqueous solution within and between cells). A molecular propagation system (e.g., lipid bilayer vesicles encapsulating the information molecules) actively transports the information molecules to an appropriate receiver. A receiver (e.g., a genetically modified or an artificial cell) selectively receives the transported information molecules, and biochemically reacts to the received information molecules, thus "decoding" the information. The chapter describes detailed examples of molecular communication system designs, experimental results, and research trends.

The chapter "Computational Nature of Gene Assembly in Ciliates", by Robert Brijder, Mark Daley, Tero Harju, Nataša Jonoska, Ion Petre, and Grzegorz Rozenberg, reviews several approaches and results in the computational study of gene assembly in ciliates. Ciliated protozoa contain two functionally different types of nuclei, the macronucleus and the micronucleus. The macronucleus contains the functional genes, while the genes of the micronucleus are not functional due to the presence of many interspersing noncoding DNA segments. In addition, in some ciliates, the coding segments of the genes are present in a permuted order compared to their order in the functional macronuclear genes. During the sexual process of conjugation, when two ciliates exchange genetic micronuclear information and form two new micronuclei, each of the ciliates has to "decrypt" the information contained in its new micronucleus to form its new functional macronucleus. This process is called gene assembly and involves deleting the noncoding DNA segments, as well as rearranging the coding segments in the correct order. The chapter describes two models of gene assembly, the intermolecular model based on the operations of circular insertion and deletion, and the intramolecular model based on the three operations of "loop, direct-repeat excision", "hairpin, inverted-repeat excision", and "double-loop, alternating repeat excision". A discussion follows of the mathematical properties of these models, such as the Turing machine computational power of contextual circular insertions and deletions, and properties of the gene assembly process called invariants, which hold independently of the molecular model and assembling strategy. Finally, the template-based recombination model is described, offering a plausible hypothesis (supported already by some experimental data) about the "bioware" that implements the gene assembly.

The chapter "DNA Memory", by Masanori Arita, Masami Hagiya, Masahiro Takinoue, and Fumiaki Tanaka, summarizes the efforts that have been made towards realizing Eric Baum's dream of building a DNA memory with a storage capacity vastly larger than the brain. The chapter first describes the research into strategies for DNA sequence design, i.e., for finding DNA sequences that satisfy DNA computing constraints such as uniform melting temperature, avoidance of undesirable Watson–Crick bonding between sequences, preventing secondary structures, avoidance of base repeats, and absence of forbidden sequences. Various implementations of memory operations, such as access, read, and write, are described. For example, the "access" to a memory word in Baum's associative memory model, where a memory word consists of a single-stranded portion representing the address and a double-stranded portion representing the data, can be implemented by using the Watson–Crick complement of the address fixed to a solid support. In the Nested Primer Molecular Memory, where the double-stranded data is flanked on both sides by address sequences, the data can be retrieved by Polymerase Chain Reaction (PCR) using the addresses as primer pairs. In the multiple hairpins DNA memory, the address is a catenation of hairpins and the data can be accessed only if the hairpins are opened in the correct order by a process called DNA branch migration. After describing implementations of writable and erasable hairpin memories either in solution or immobilized on surfaces, the topic of *in vivo* DNA memory is explored. As an example, the chapter describes how representing the digit 0 by regular codons, and the digit 1 by wobbled codons, was used to encode a word into an essential gene of *Bacillus subtilis*.

The chapter "Engineering Natural Computation by Autonomous DNA-Based Biomolecular Devices", by John H. Reif and Thomas H. LaBean, overviews DNA-based biomolecular devices that are autonomous (execute steps with no external control after starting) and programmable (the tasks executed can be modified without entirely redesigning the DNA nanostructures). Special attention is given to DNA tiles, roughly square-shaped DNA nanostructures that have four "sticky-ends" (DNA single strands) that can specifically bind them to other tiles via Watson–Crick complementarity, and thus lead to the self-assembly of larger and more complex structures. Such tiles have been used to execute sequential Boolean computation via linear DNA self-assembly or to obtain patterned 2D DNA lattices and Sierpinski triangles. Issues such as error correction and self-repair of DNA tiling are also addressed. Other described methods include the implementation of a DNA-based finite automaton via disassembly of a double-stranded DNA nanostructure effected by an enzyme, and the technique of whiplash PCR. Whiplash PCR is a method that can achieve state transitions by encoding both transitions and the current state of the computation on the same DNA single strand: The free end of the strand (encoding the current state) sticks to the appropriate transition rule on the strand forming a hairpin, is then extended by PCR to a new state, and finally is detached from the strand, this time with the new state encoded at its end. The technique of DNA origami is also described, whereby a scaffold strand (a long single DNA strand, such as from the sequence of a virus) together with many specially designed staple strands (short single DNA strands) self-assemble by folding the scaffold strand — with the aid of the staples — in a raster pattern that can create given arbitrary planar DNA nanostructures. DNA-based molecular machines are then described such as autonomous DNA walkers and programmable DNA nanobots (programmable autonomous DNA walker devices). A restriction-enzyme-based DNA walker consists of a DNA helix with two sticky-ends ("feet") that moves stepwise along a "road" (a DNA nanostructure with protruding "steps", i.e., single DNA strands).

The chapter "Membrane Computing", by Gheorghe Păun, describes theoretical results and applications of membrane computing, a branch of natural computing inspired by the architecture and functioning of living cells, as well as from the organization of cells in tissues, organs, or other higher-order structures. The cell is a hierarchical structure of compartments, defined by membranes, that selectively communicate with each other. The computing model that abstracts this structure is a membrane system (or P system, from the name of its inventor, Gheorghe Păun) whose main components are: the membrane structure, the multisets of objects placed in the compartments enveloped by the membranes, and the rules for processing the objects and the membranes. The rules are used to modify the objects in the compartments, to transport objects from one compartment to another, to dissolve membranes, and to create new membranes. The rules in each region of a P system are used in a maximally parallel manner, nondeterministically choosing the applicable rules and the objects to which they apply. A computation consists in repeatedly applying rules to an initial configuration of the P system, until no rule can be applied anymore, in which case the objects in a priori specified regions are considered the output of the computation. Several variants of P systems are described, including P systems with symport/antiport rules, P systems with active membranes, splicing P systems, P systems with objects on membranes, tissue-like P systems, and spiking neural P systems. Many classes of P systems are able to simulate Turing machines, hence they are computationally universal. For example, P systems with symport/antiport rules using only three objects and three membranes are computationally universal. In addition, several types of P systems have been used to solve NP-complete problems in polynomial time, by a space–time trade-off. Applications of P systems include, among others, modeling in biology, computer graphics, linguistics, economics, and cryptography.

Quantum Computation

This area is covered by six chapters.

The chapter "Mathematics for Quantum Information Processing", by Mika Hirvensalo, contains the standard Hilbert space formulation of finite-level quantum systems. This is the language and notational system allowing us to speak, describe, and make predictions about the objects of quantum physics. The chapter introduces the notion of quantum states as unit-trace, self-adjoint, positive mappings, and the vector state formalism is presented as a special case. The physical observables are introduced as complete collections of mutually orthogonal projections, and then it is discussed how this leads to the traditional representation of observables as self-adjoint mappings. The minimal interpretation, which is the postulate connecting the mathematical objects to the physical world, is presented. The treatment of compound quantum systems is based mostly on operative grounds. To provide enough tools for considering the dynamics needed in quantum computing, the formalism of treating state transformations as completely positive mappings is also presented. The chapter concludes by explaining how quantum versions of finite automata, Turing machines, and Boolean circuits fit into the Hilbert space formalism.

The chapter "Bell's Inequalities — Foundations and Quantum Communication", by Časlav Brukner and Marek Żukowski, is concerned with the nature of quantum mechanics. It presents the evidence that excludes two types of hypothetical deterministic theories: neither a nonlocal nor a noncontextual theory can explain quantum mechanics. This helps to build a true picture of quantum mechanics, and is therefore essential from the philosophical point of view. The Bell inequalities show that nonlocal deterministic theories cannot explain the quantum mechanism, and the Kochen–Specker theorem shows that noncontextual theories are not possible as underlying theories either. The traditional Bell theorem and its variants, GHZ and CHSH among them, are presented, and the Kochen–Specker theorem is discussed. In this chapter, the communication complexity is also treated by showing how the violations of classical locality and noncontextuality can be used as a resource for communication protocols. Stronger-than quantum violations of the CHSH inequality are also discussed. They are interesting, since it has been shown that if the violation of CHSH inequality is strong enough, then the communication complexity collapses into one bit (hence the communication complexity of the true physical world seems to settle somewhere between classical and stronger-than quantum).

The chapter "Algorithms for Quantum Computers", by Jamie Smith and Michele Mosca, introduces the most remarkable known methods that utilize the special features of quantum physics in order to gain advantage over classical computing. The importance of these methods is that they form the core of designing discrete quantum algorithms. The methods presented and discussed here are the quantum Fourier transform, amplitude amplification, and quantum walks. Then, as specific examples, Shor's factoring algorithm (quantum Fourier transform), Grover search (amplitude amplification), and element distinctness algorithms (quantum random walks) are presented. The chapter not only involves traditional methods, but it also contains discussion of continuous-time quantum random walks and, more importantly, an extensive presentation of an important recent development in quantum algorithms, viz., tensor network evaluation algorithms. Then, as an example, the approximate evaluation of Tutte polynomials is presented.

The chapter "Physical Implementation of Large-Scale Quantum Computation", by Kalle-Antti Suominen, discusses the potential ways of physically implementing quantum computers. First, the DiVincenzo criteria (requirements for building a successful quantum computer) are presented, and then quantum error correction is discussed. The history, physical properties, potentials, and obstacles of various possible physical implementations of quantum computers are covered. They involve: cavity QED, trapped ions, neutral atoms and single electrons, liquid-form molecular spin, nuclear and electron spins in silicon, nitrogen vacancies in diamond, solid-state qubits with quantum dots, superconducting charge, flux and phase quantum bits, and optical quantum computing.

The chapter "Quantum Cryptography", by Takeshi Koshiba, is concerned with quantum cryptography, which will most likely play an important role in future when quantum computers make the current public-key cryptosystems unreliable. It gives an overview of classical cryptosystems, discusses classical cryptographic protocols, and then introduces the quantum key distribution protocols BB84, B92, and BBM92. Also protocol OTU00, not known to be vulnerable under Shor's algorithm, is presented. In future, when quantum computers are available, cryptography will most probably be based on quantum protocols. The chapter presents candidates for such quantum protocols: KKNY05 and GC01 (for digital signatures). It concludes with a discussion of quantum commitment, oblivious transfer, and quantum zero-knowledge proofs.

The complexity class BQP is the quantum counterpart of the classical class BPP. Intuitively, BQP can be described as the class of problems solvable in "reasonable" time, and, hence, from the application-oriented point of view, it will likely become the most important complexity class in future, when quantum computers are available. The chapter "BQP-Complete Problems", by Shengyu Zhang, introduces the computational problems that capture the full hardness of BQP. In the very fundamental sense, no BQP-complete problems are known, but the promise problems (the probability distribution of outputs is restricted by promise) bring us as close as possible to the "hardest" problems in BQP, known as BQP-complete promise problems. The chapter discusses known BQP-complete promise problems. In particular, it is shown how to establish the BQP-completeness of the Local Hamiltonian Eigenvalue

Sampling problem and the Local Unitary Phase Sampling problem. The chapter concludes with an extensive study showing that the Jones Polynomial Approximation problem is a BQP-complete promise problem.

Broader Perspective

This area consists of two sections, "Nature-Inspired Algorithms" and "Alternative Models of Computation".

Nature-Inspired Algorithms

This section is covered by six chapters.

The chapter "An Introduction to Artificial Immune Systems", by Mark Read, Paul S. Andrews, and Jon Timmis, provides a general introduction to the field. It discusses the major research issues relating to the field of Artificial Immune Systems (AIS), exploring the underlying immunology that has led to the development of immune-inspired algorithms, and focuses on the four main algorithms that have been developed in recent years: clonal selection, immune network, negative selection, and dendritic cell algorithms; their use in terms of applications is highlighted. The chapter also covers evaluation of current AIS technology, and details some new frameworks and methodologies that are being developed towards more principled AIS research. As a counterpoint to the focus on applications, the chapter also gives a brief outline of how AIS research is being employed to help further the understanding of immunology.

The chapter on "Swarm Intelligence", by David W. Corne, Alan Reynolds, and Eric Bonabeau, attempts to demystify the term Swarm Intelligence (SI), outlining the particular collections of natural phenomena that SI most often refers to and the specific classes of computational algorithms that come under its definition. The early parts of the chapter focus on the natural inspiration side, with discussion of social insects and stigmergy, foraging behavior, and natural flocking behavior. Then the chapter moves on to outline the most successful of the computational algorithms that have emerged from these natural inspirations, namely ant colony optimization methods and particle swarm optimization, with also some discussion of different and emerging such algorithms. The chapter concludes with a brief account of current research trends in the field.

The chapter "Simulated Annealing", by Kathryn A. Dowsland and Jonathan M. Thompson, provides an overview of Simulated Annealing (SA), emphasizing its practical use. The chapter explains its inspiration from the field of statistical thermodynamics, and then overviews the theory, with an emphasis again on those aspects that are important for practical applications. The chapter then covers some of the main ways in which the basic SA algorithm has been modified by various researchers, leading to improved performance for a variety of problems. The chapter briefly surveys application areas, and ends with several useful pointers to associated resources, including freely available code.

The chapter "Evolvable Hardware", by Lukáš Sekanina, surveys this growing field. Starting with a brief overview of the reconfigurable devices used in this field, the elementary principles and open problems are introduced, and then the chapter considers, in turn, three main areas: extrinsic evolution (evolving hardware using simulators), intrinsic evolution (where the evolution is conducted within FPGAs, FPTAs, and so forth), and adaptive hardware (in which real-world adaptive hardware systems are presented). The chapter finishes with an overview of major achievements in the field.

The first of two application-centered chapters, "Natural Computing in Finance — A Review", by Anthony Brabazon, Jing Dang, Ian Dempsey, Michael O'Neill, and David Edelman, provides a rather comprehensive account of natural computing applications in what is, at the time of writing (and undoubtedly beyond), one of the hottest topics of the day. This chapter introduces us to the wide range of different financial problems to which natural computing methods have been applied, including forecasting, trading, arbitrage, portfolio management, asset allocation, credit risk assessment, and more. The natural computing areas that feature in this chapter are largely evolutionary computing, neural computing, and also agent-based modeling, swarm intelligence, and immune-inspired methods. The chapter ends with a discussion of promising future directions.

Finally, the chapter "Selected Aspects of Natural Computing", by David W. Corne, Kalyanmoy Deb, Joshua Knowles, and Xin Yao, provides detailed accounts of a collection of example natural computing applications, each of which is remarkable or particularly interesting in some way. The thrust of this chapter is to provide, via such examples, an idea of both the significant impact that natural computing has already had, as well as its continuing significant promise for future applications in all areas of science and industry. While presenting this eclectic collection of marvels, the chapter also aims at clarity and demystification, providing much detail that helps see how the natural computing methods in question were applied to achieve the stated results. Applications covered include Blondie24 (the evolutionary neural network application that achieves master-level skill at the game of checkers), the design of novel antennas using evolutionary computation in conjunction with developmental computing, and the classic application of learning classifier systems that led to novel fighter-plane maneuvers for the USAF.

Alternative Models of Computation

This section is covered by seven chapters.

The chapter "Artificial Life", by Wolfgang Banzhaf and Barry McMullin, traces the roots, raises key questions, discusses the major methodological tools, and reviews the main applications of this exciting and maturing area of computing. The chapter starts with a historical overview, and presents the fundamental questions and issues that Artificial Life is concerned with. Thus the chapter surveys discussions and viewpoints about the very nature of the differences between living and nonliving systems, and goes on to consider issues such as hierarchical design, self-construction, and self-maintenance, and the emergence of complexity. This part of the chapter ends with a discussion of "Coreworld" experiments, in which a number of systems have been studied that allow spontaneous evolution of computer programs. The chapter moves on to survey the main theory and formalisms used in Artificial Life, including cellular automata and rewriting systems. The chapter concludes with a review and restatement of the main objectives of Artificial Life research, categorizing them respectively into questions about the origin and nature of life, the potential and limitations of living systems, and the relationships between life and intelligence, culture, and other human constructs.

The chapter "Algorithmic Systems Biology — Computer Science Propels Systems Biology", by Corrado Priami, takes the standpoint of computing as providing a philosophical foundation for systems biology, with at least the same importance as mathematics, chemistry, or physics. The chapter highlights the value of algorithmic approaches in modeling, simulation, and analysis of biological systems. It starts with a high-level view of how models and experiments can be tightly integrated within an algorithmic systems biology vision, and then deals in turn with modeling languages, simulations of models, and finally the postprocessing of results from biological models and how these lead to new hypotheses that can then re-enter the modeling/simulation cycle.

The chapter "Process Calculi, Systems Biology and Artificial Chemistry", by Pierpaolo Degano and Andrea Bracciali, concentrates on the use of process calculi and related techniques for systems-level modeling of biological phenomena. This chapter echoes the broad viewpoint of the previous chapter, but its focus takes us towards a much deeper understanding of the potential mappings between formal systems in computer science and systems interpretation of biological processes. It starts by surveying the basics of process calculi, setting out their obvious credentials for modeling concurrent, distributed systems of interacting parts, and mapping these onto a "cells as computers" view. After a process calculi treatment of systems biology, the chapter goes on to examine process calculi as a route towards artificial chemistry. After considering the formal properties of the models discussed, the chapter ends with notes on some case studies showing the value of process calculi in modeling biological phenomena; these include investigating the concept of a "minimal gene set" prokaryote, modeling the nitric oxide-cGMP pathway (central to many signal transduction mechanisms), and modeling the calyx of Held (a large synapse structure in the mammalian auditory central nervous system).

The chapter on "Reaction–Diffusion Computing", by Andrew Adamatzky and Benjamin De Lacy Costello, introduces the reader to the concept of a reaction–diffusion computer. This is a spatially extended chemical system, which processes information via transforming an input profile of ingredients (in terms of different concentrations of constituent ingredients) into an output profile of ingredients. The chapter takes us through the elements of this field via case studies, and it shows how selected tasks in computational geometry, robotics, and logic can be addressed by chemical implementations of reaction–diffusion computers. After introducing the field and providing a treatment of its origins and main achievements, a classical view of reaction–diffusion computers is then described. The chapter moves on to discuss varieties of reaction–diffusion processors and their chemical constituents, covering applications to the aforementioned tasks. The chapter ends with the authors' thoughts on future developments in this field.

The chapter "Rough–Fuzzy Computing", by Andrzej Skowron, shifts our context towards addressing a persistent area of immense difficulty for classical computing, which is the fact that real-world reasoning is usually done in the face of inaccurate, incomplete, and often inconsistent evidence. In essence, concepts in the real world are vague, and computation needs ways to address this. We are hence treated, in this chapter, to an overarching view of rough set theory, fuzzy set theory, their hybridization, and applications. Rough and fuzzy computing are broadly complementary approaches to handling vagueness, focusing respectively on capturing the level of distinction between separate objects and the level of membership of an object in a set. After presenting the basic concepts of rough computing and fuzzy computing in turn, in each case going into some detail on the main theoretical results and practical considerations, the chapter goes on to discuss how they can be, and have been, fruitfully combined. The chapter ends with an overview of the emerging field of "Wisdom Technology" (Wistech) as a paradigm for developing modern intelligent systems.

The chapter "Collision-Based Computing", by Andrew Adamatzky and Jérôme Durand-Lose, presents and discusses the computations performed as a result of spatial localizations in systems that exhibit dynamic spatial patterns over time. For example, a collision may be between two gliders in a cellular automaton, or two separate wave fragments within an excitable chemical system. This chapter introduces us to the basics of collision-based computing and overviews collision-based computing schemes in 1D and 2D cellular automata as well as continuous excitable media. Then, after some theoretical foundations relating to 1D cellular automata, the chapter presents a collision-based implementation for a 1D Turing machine and for cyclic tag systems. The chapter ends with discussion and presentation of "Abstract Geometrical Computation", which can be seen as collision-based computation in a medium that is the continuous counterpart of cellular automata.

The chapter "Nonclassical Computation — A Dynamical Systems Perspective", by Susan Stepney, takes a uniform view of computation, in which inspiration from a dynamical systems perspective provides a convenient way to consider, in one framework, both classical discrete systems and systems performing nonclassical computation. In particular, this viewpoint presents a way towards computational interpretation of physical embodied systems that exploit their natural dynamics. The chapter starts by discussing "closed" dynamical systems, those whose dynamics involve no inputs from an external environment, examining their computational abilities from a dynamical systems perspective. Then it discusses continuous dynamical systems and shows how these too can be interpreted computationally, indicating how material embodiment can give computation "for free", without the need to explicitly implement the dynamics. The outlook then broadens to consider open systems, where the dynamics are affected by external inputs. The chapter ends by looking at constructive, or developmental, dynamical systems, whose state spaces change during computation. These latter discussions approach the arena of biological and other natural systems, casting them as computational, open, developmental, dynamical systems.

Acknowledgements

This handbook resulted from a highly collaborative effort. The handbook and area editors are grateful to the chapter writers for their efforts in writing chapters and delivering them on time, and for their participation in the refereeing process.

We are indebted to the members of the Advisory Board for their valuable advice and fruitful interactions. Additionally, we want to acknowledge David Fogel, Pekka Lahti, Robert LaRue, Jason Lohn, Michael Main, David Prescott, Arto Salomaa, Kai Salomaa, Shinnosuke Seki, and Rob Smith, for their help and advice in various stages of production of this handbook. Last, but not least, we are thankful to Springer, especially to Ronan Nugent, for intense and constructive cooperation in bringing this project from its inception to its successful conclusion.

Leiden; Edinburgh; Nijmegen;   Grzegorz Rozenberg (Main Handbook Editor)
Turku; London, Ontario   Thomas Bäck (Handbook Editor and Area Editor)
October 2010   Joost N. Kok (Handbook Editor and Area Editor)
                                        David W. Corne (Area Editor)
                                        Tom Heskes (Area Editor)
                                        Mika Hirvensalo (Area Editor)
                                        Jarkko J. Kari (Area Editor)
                                        Lila Kari (Area Editor)

Editor Biographies

Prof. Dr. Grzegorz Rozenberg

Prof. Rozenberg was awarded his Ph.D. in mathematics from the Polish Academy of Sciences, Warsaw, and he has since held full-time positions at Utrecht University, the State University of New York at Buffalo, and the University of Antwerp. Since 1979 he has been a professor at the Department of Computer Science of Leiden University, and an adjunct professor at the Department of Computer Science of the University of Colorado at Boulder, USA. He is the founding director of the Leiden Center for Natural Computing.

Among key editorial responsibilities over the last 30 years, he is the founding Editor of the book series Texts in Theoretical Computer Science (Springer) and Monographs in Theoretical Computer Science (Springer), founding Editor of the book series Natural Computing (Springer), founding Editor-in-Chief of the journal Natural Computing (Springer), and founding Editor of Part C (Theory of Natural Computing) of the journal Theoretical Computer Science (Elsevier). Altogether he's on the Editorial Board of around 20 scientific journals.

He has authored more than 500 papers and 6 books, and coedited more than 90 books. He coedited the "Handbook of Formal Languages" (Springer), he was Managing Editor of the "Handbook of Graph Grammars and Computing by Graph Transformation" (World Scientific), he coedited "Current Trends in Theoretical Computer Science" (World Scientific), and he coedited "The Oxford Handbook of Membrane Computing" (Oxford University Press).

He is Past President of the European Association for Theoretical Computer Science (EATCS), and he received the Distinguished Achievements Award of the EATCS "in recognition of his outstanding scientific contributions to theoretical computer science". Also he is a cofounder and Past President of the International Society for Nanoscale Science, Computation, and Engineering (ISNSCE).

He has served as a program committee member for most major conferences on theoretical computer science in Europe, and among the events he has founded or helped to establish are the International Conference on Developments in Language Theory (DLT), the International

Conference on Graph Transformation (ICGT), the International Conference on Unconventional Computation (UC), the International Conference on Application and Theory of Petri Nets and Concurrency (ICATPN), and the DNA Computing and Molecular Programming Conference.

In recent years his research has focused on natural computing, including molecular computing, computation in living cells, self-assembly, and the theory of biochemical reactions. His other research areas include the theory of concurrent systems, the theory of graph transformations, formal languages and automata theory, and mathematical structures in computer science.

Prof. Rozenberg is a Foreign Member of the Finnish Academy of Sciences and Letters, a member of the Academia Europaea, and an honorary member of the World Innovation Foundation. He has been awarded honorary doctorates by the University of Turku, the Technical University of Berlin, and the University of Bologna. He is an ISI Highly Cited Researcher.

He is a performing magician, and a devoted student of and expert on the paintings of Hieronymus Bosch.

Prof. Dr. Thomas Bäck

Prof. Bäck was awarded his Ph.D. in Computer Science from Dortmund University in 1994, for which he received the Best Dissertation Award from the Gesellschaft für Informatik (GI). He has been at Leiden University since 1996, where he is currently full Professor for Natural Computing and the head of the Natural Computing Research Group at the Leiden Institute of Advanced Computer Science (LIACS).

He has authored more than 150 publications on natural computing technologies. He wrote a book on evolutionary algorithms, "Evolutionary Algorithms in Theory and Practice" (Oxford University Press), and he coedited the "Handbook of Evolutionary Computation" (IOP/Oxford University Press).

Prof. Bäck is an Editor of the book series Natural Computing (Springer), an Associate Editor of the journal Natural Computing (Springer), an Editor of the journal Theoretical Computer Science (Sect. C, Theory of Natural Computing; Elsevier), and an Advisory Board member of the journal Evolutionary Computation (MIT Press). He has served as program chair for all major conferences in evolutionary computation, and is an Elected Fellow of the International Society for Genetic and Evolutionary Computation for his contributions to the field.

His main research interests are the theory of evolutionary algorithms, cellular automata and data-driven modelling, and applications of these methods in medicinal chemistry, pharmacology and engineering.

Prof. Dr. Joost N. Kok

Prof. Kok was awarded his Ph.D. in Computer Science from the Free University in Amsterdam in 1989, and he has worked at the Centre for Mathematics and Computer Science in Amsterdam, at Utrecht University, and at the Åbo Akademi University in Finland. Since 1995 he has been a professor in computer science, and since 2005 also a professor in medicine at Leiden University. He is the Scientific Director of the Leiden Institute of Advanced Computer Science, and leads the research clusters Algorithms and Foundations of Software Technology.

He serves as a chair, member of the management team, member of the board, or member of the scientific committee of the Faculty of Mathematics and Natural Sciences of Leiden University, the ICT and Education Committee of Leiden University, the Dutch Theoretical Computer Science Association, the Netherlands Bioinformatics Centre, the Centre for Mathematics and Computer Science Amsterdam, the Netherlands Organisation for Scientific Research, the Research Foundation Flanders (Belgium), the European Educational Forum, and the International Federation for Information Processing (IFIP) Technical Committee 12 (Artificial Intelligence).

Prof. Kok is on the steering, scientific or advisory committees of the following events: the Mining and Learning with Graphs Conference, the Intelligent Data Analysis Conference, the Institute for Programming and Algorithms Research School, the Biotechnological Sciences Delft–Leiden Research School, and the European Conference on Machine Learning and Principles and Practice of Knowledge Discovery in Databases. And he has been a program committee member for more than 100 international conferences, workshops or summer schools on data mining, data analysis, and knowledge discovery; neural networks; artificial intelligence; machine learning; computational life science; evolutionary computing, natural computing and genetic algorithms; Web intelligence and intelligent agents; and software engineering.

He is an Editor of the book series Natural Computing (Springer), an Associate Editor of the journal Natural Computing (Springer), an Editor of the journal Theoretical Computer Science (Sect. C, Theory of Natural Computing; Elsevier), an Editor of the Journal of Universal

Computer Science, an Associate Editor of the journal Computational Intelligence (Wiley), and a Series Editor of the book series Frontiers in Artificial Intelligence and Applications (IOS Press).

His academic research is concentrated around the themes of scientific data management, data mining, bioinformatics, and algorithms, and he has collaborated with more than 20 industrial partners.

Advisory Board

Area Editors

Table of Contents

Volume 1

Volume 3

Volume 4

List of Contributors

Andrew Adamatzky
Department of Computer Science
University of the West of England
Bristol
UK
andrew.adamatzky@uwe.ac.uk

Paul S. Andrews
Department of Computer Science
University of York
UK
psa@cs.york.ac.uk

Masanori Arita
Department of Computational Biology,
Graduate School of Frontier Sciences
The University of Tokyo
Kashiwa
Japan
arita@k.u-tokyo.ac.jp

Wolfgang Banzhaf
Department of Computer Science
Memorial University of Newfoundland
St. John's, NL
Canada
banzhaf@cs.mun.ca

Hans-Georg Beyer
Department of Computer Science
Fachhochschule Vorarlberg
Dornbirn
Austria
hans-georg.beyer@fhv.at

Sander Bohte
Research Group Life Sciences
CWI
Amsterdam
The Netherlands
s.m.bohte@cwi.nl

Eric Bonabeau
Icosystem Corporation
Cambridge, MA
USA
eric@icosystem.com

Anthony Brabazon
Natural Computing Research and
Applications Group
University College Dublin
Ireland
anthony.brabazon@ucd.ie

Andrea Bracciali
Department of Computing Science and
Mathematics
University of Stirling
UK
braccia@cs.stir.ac.uk

Robert Brijder
Leiden Institute of Advanced Computer
Science
Universiteit Leiden
The Netherlands
robert.brijder@uhasselt.be

Časlav Brukner
Faculty of Physics
University of Vienna
Vienna
Austria
caslav.brukner@univie.ac.at

Anthony Bucci
Icosystem Corporation
Cambridge, MA
USA
anthony@icosystem.com

Ke Chen
Department of Electrical and Computer
Engineering
University of Alberta
Edmonton, AB
Canada
kchen1@ece.ualberta.ca

Seungjin Choi
Pohang University of Science and
Technology
Pohang
South Korea
seungjin@postech.ac.kr

Bastien Chopard
Scientific and Parallel Computing Group
University of Geneva
Switzerland
bastien.chopard@unige.ch

David W. Corne
School of Mathematical and Computer
Sciences
Heriot-Watt University
Edinburgh
UK
dwcorne@macs.hw.ac.uk

Mark Daley
Departments of Computer Science and
Biology
University of Western Ontario
London, Ontario
Canada
daley@csd.uwo.ca

Jing Dang
Natural Computing Research and
Applications Group
University College Dublin
Ireland
jing.dang@ucd.ie

Edwin D. de Jong
Institute of Information and Computing
Sciences
Utrecht University
The Netherlands
dejong@cs.uu.nl

Kenneth De Jong
Department of Computer Science
George Mason University
Fairfax, VA
USA
kdejong@gmu.edu

Benjamin De Lacy Costello
Centre for Research in Analytical, Material
and Sensor Sciences, Faculty of Applied
Sciences
University of the West of England
Bristol
UK
ben.delacycostello@uwe.ac.uk

Kalyanmoy Deb
Department of Mechanical Engineering
Indian Institute of Technology
Kanpur
India
deb@iitk.ac.in

Pierpaolo Degano
Dipartimento di Informatica
Università di Pisa
Italy
degano@di.unipi.it

Marianne Delorme
Laboratoire d'Informatique Fondamentale
de Marseille (LIF)
Aix-Marseille Université and CNRS
Marseille
France
delorme.marianne@orange.fr

Ian Dempsey
Pipeline Financial Group, Inc.
New York, NY
USA
ian.dempsey@pipelinefinancial.com

Alberto Dennunzio
Dipartimento di Informatica
Sistemistica e Comunicazione, Università
degli Studi di Milano-Bicocca
Italy
dennunzio@disco.unimib.it

Kathryn A. Dowsland
Gower Optimal Algorithms, Ltd.
Swansea
UK
k.a.dowsland@btconnect.com

Jérôme Durand-Lose
LIFO
Université d'Orléans
France
jerome.durand-lose@univ-orleans.fr

David Edelman
School of Business
UCD Michael Smurfit Graduate Business
School
Dublin
Ireland
david.edelman@ucd.ie

Gary B. Fogel
Natural Selection, Inc.
San Diego, CA
USA
gfogel@natural-selection.com

Enrico Formenti
Département d'Informatique
Université de Nice-Sophia Antipolis
France
enrico.formenti@unice.fr

Xinbo Gao
Video and Image Processing System Lab,
School of Electronic Engineering
Xidian University
China
xbgao@ieee.org

Shuzhi Sam Ge
Social Robotics Lab
Interactive Digital Media Institute, The
National University of Singapore
Singapore
elegesz@nus.edu.sg

Masami Hagiya
Department of Computer Science
Graduate School of Information Science
and Technology
The University of Tokyo
Tokyo
Japan
hagiya@is.s.u-tokyo.ac.jp

Tero Harju
Department of Mathematics
University of Turku
Finland
harju@utu.fi

Mika Hirvensalo
Department of Mathematics
University of Turku
Finland
mikhirve@utu.fi

Satoshi Hiyama
Research Laboratories
NTT DOCOMO, Inc.
Yokosuka
Japan
hiyama@nttdocomo.co.jp

Guang-Bin Huang
School of Electrical and Electronic
Engineering
Nanyang Technological University
Singapore
egbhuang@ntu.edu.sg

Thomas Jansen
Department of Computer Science
University College Cork
Ireland
t.jansen@cs.ucc.ie

Nataša Jonoska
Department of Mathematics
University of South Florida
Tampa, FL
USA
jonoska@math.usf.edu

Jarkko J. Kari
Department of Mathematics
University of Turku
Turku
Finland
jkari@utu.fi

Lila Kari
Department of Computer Science
University of Western Ontario
London
Canada
lila@csd.uwo.ca

Sungchul Kim
Data Mining Lab, Department of Computer
Science and Engineering
Pohang University of Science and
Technology
Pohang
South Korea
subright@postech.ac.kr

Joshua Knowles
School of Computer Science and
Manchester Interdisciplinary
Biocentre (MIB)
University of Manchester
UK
j.knowles@manchester.ac.uk

Satoshi Kobayashi
Department of Computer Science
University of Electro-Communications
Tokyo
Japan
satoshi@cs.uec.ac.jp

Ken Komiya
Interdisciplinary Graduate School of
Science and Engineering
Tokyo Institute of Technology
Yokohama
Japan
komiya@dis.titech.ac.jp

Takeshi Koshiba
Graduate School of Science and
Engineering
Saitama University
Japan
koshiba@mail.saitama-u.ac.jp

Tim Kovacs
Department of Computer Science
University of Bristol
UK
kovacs@cs.bris.ac.uk

Natalio Krasnogor
Interdisciplinary Optimisation Laboratory,
The Automated Scheduling, Optimisation
and Planning Research Group, School of
Computer Science
University of Nottingham
UK
natalio.krasnogor@nottingham.ac.uk

Lukasz A. Kurgan
Department of Electrical and Computer
Engineering
University of Alberta
Edmonton, AB
Canada
lkurgan@ece.ualberta.ca

Petr Kůrka
Center for Theoretical Studies
Academy of Sciences and Charles
University in Prague
Czechia
kurka@cts.cuni.cz

Thomas H. LaBean
Department of Computer Science and
Department of Chemistry and Department
of Biomedical Engineering
Duke University
Durham, NC
USA
thomas.labean@duke.edu

Kang Li
School of Electronics, Electrical Engineering
and Computer Science
Queen's University
Belfast
UK
k.li@ee.qub.ac.uk

Xuelong Li
Center for OPTical IMagery Analysis and
Learning (OPTIMAL), State Key Laboratory
of Transient Optics and Photonics
Xi'an Institute of Optics and Precision
Mechanics, Chinese Academy of Sciences
Xi'an, Shaanxi
China
xuelong_li@opt.ac.cn

Wen Lu
Video and Image Processing System Lab,
School of Electronic Engineering
Xidian University
China
luwen@mail.xidian.edu.cn

Jacques Mazoyer
Laboratoire d'Informatique Fondamentale
de Marseille (LIF)
Aix-Marseille Université and CNRS
Marseille
France
mazoyerj2@orange.fr

Barry McMullin
Artificial Life Lab, School of Electronic
Engineering
Dublin City University
Ireland
barry.mcmullin@dcu.ie

Silja Meyer-Nieberg
Fakultät für Informatik
Universität der Bundeswehr München
Neubiberg
Germany
silja.meyer-nieberg@unibw.de

Kenichi Morita
Department of Information Engineering,
Graduate School of Engineering
Hiroshima University
Japan
morita@iec.hiroshima-u.ac.jp

Michele Mosca
Institute for Quantum Computing and
Department of Combinatorics &
Optimization
University of Waterloo and St. Jerome's
University and Perimeter Institute for
Theoretical Physics
Waterloo
Canada
mmosca@iqc.ca

Nicolas Ollinger
Laboratoire d'informatique fondamentale
de Marseille (LIF)
Aix-Marseille Université, CNRS
Marseille
France
nicolas.ollinger@lif.univ-mrs.fr

Michael O'Neill
Natural Computing Research and
Applications Group
University College Dublin
Ireland
m.oneill@ucd.ie

Hélène Paugam-Moisy
Laboratoire LIRIS – CNRS
Université Lumière Lyon 2
Lyon
France
and
INRIA Saclay – Ile-de-France
Université Paris-Sud
Orsay
France
helene.paugam-moisy@univ-lyon2.fr
hpaugam@lri.fr

Gheorghe Păun
Institute of Mathematics of the Romanian
Academy
Bucharest
Romania
and
Department of Computer Science and
Artificial Intelligence
University of Seville
Spain
gpaun@us.es
george.paun@imar.ro

Ion Petre
Department of Information Technologies
Åbo Akademi University
Turku
Finland
ipetre@abo.fi

Riccardo Poli
Department of Computing and Electronic
Systems
University of Essex
Colchester
UK
rpoli@essex.ac.uk

Elena Popovici
Icosystem Corporation
Cambridge, MA
USA
elena@icosystem.com

Corrado Priami
Microsoft Research
University of Trento Centre for
Computational and Systems Biology
(CoSBi)
Trento
Italy
and
DISI
University of Trento
Trento
Italy
priami@cosbi.eu

Mark Read
Department of Computer Science
University of York
UK
markread@cs.york.ac.uk

John H. Reif
Department of Computer Science
Duke University
Durham, NC
USA
reif@cs.duke.edu

Alan Reynolds
School of Mathematical and Computer
Sciences
Heriot-Watt University
Edinburgh
UK
a.reynolds@hw.ac.uk

Grzegorz Rozenberg
Leiden Institute of Advanced Computer
Science
Universiteit Leiden
The Netherlands
and
Department of Computer Science
University of Colorado
Boulder, CO
USA
rozenber@liacs.nl

Günter Rudolph
Department of Computer Science
TU Dortmund
Dortmund
Germany
guenter.rudolph@tu-dortmund.de

Yasubumi Sakakibara
Department of Biosciences and Informatics
Keio University
Yokohama
Japan
yasu@bio.keio.ac.jp

Lukáš Sekanina
Faculty of Information Technology
Brno University of Technology
Brno
Czech Republic
sekanina@fit.vutbr.cz

Shinnosuke Seki
Department of Computer Science
University of Western Ontario
London
Canada
sseki@csd.uwo.ca

Ofer M. Shir
Department of Chemistry
Princeton University
NJ
USA
oshir@princeton.edu

Andrzej Skowron
Institute of Mathematics
Warsaw University
Poland
skowron@mimuw.edu.pl

Jamie Smith
Institute for Quantum Computing and
Department of Combinatorics &
Optimization
University of Waterloo
Canada
ja5smith@iqc.ca

Petr Sosík
Institute of Computer Science
Silesian University in Opava
Czech Republic
and
Departamento de Inteligencia Artificial
Universidad Politécnica de Madrid
Spain
petr.sosik@fpf.slu.cz

Susan Stepney
Department of Computer Science
University of York
UK
susan.stepney@cs.york.ac.uk

Kalle-Antti Suominen
Department of Physics and Astronomy
University of Turku
Finland
kalle-antti.suominen@utu.fi

Andrew M. Sutton
Department of Computer Science
Colorado State University
Fort Collins, CO
USA
sutton@cs.colostate.edu

Siamak Taati
Department of Mathematics
University of Turku
Finland
siamak.taati@gmail.com

Masahiro Takinoue
Department of Physics
Kyoto University
Kyoto
Japan
takinoue@chem.scphys.kyoto-u.ac.jp

Fumiaki Tanaka
Department of Computer Science
Graduate School of Information Science
and Technology
The University of Tokyo
Tokyo
Japan
fumi95@is.s.u-tokyo.ac.jp

Dacheng Tao
School of Computer Engineering
Nanyang Technological University
Singapore
dacheng.tao@gmail.com

Véronique Terrier
GREYC, UMR CNRS 6072
Université de Caen
France
veroniqu@info.unicaen.fr

Jonathan M. Thompson
School of Mathematics
Cardiff University
UK
thompsonjm1@cardiff.ac.uk

Jon Timmis
Department of Computer Science and
Department of Electronics
University of York
UK
jtimmis@cs.york.ac.uk

Joaquin J. Torres
Institute "Carlos I" for Theoretical and
Computational Physics and Department of
Electromagnetism and Matter Physics,
Facultad de Ciencias
Universidad de Granada
Spain
jtorres@ugr.es

Marc M. Van Hulle
Laboratorium voor Neurofysiologie
K.U. Leuven
Leuven
Belgium
marc@neuro.kuleuven.be

Leonardo Vanneschi
Department of Informatics, Systems and
Communication
University of Milano-Bicocca
Italy
vanneschi@disco.unimib.it

Pablo Varona
Departamento de Ingeniería Informática
Universidad Autónoma de Madrid
Spain
pablo.varona@uam.es

Darrell Whitley
Department of Computer Science
Colorado State University
Fort Collins, CO
USA
whitley@cs.colostate.edu

R. Paul Wiegand
Institute for Simulation and Training
University of Central Florida
Orlando, FL
USA
wiegand@ist.ucf.edu

Xin Yao
Natural Computation Group, School of
Computer Science
University of Birmingham
UK
x.yao@cs.bham.ac.uk

Takashi Yokomori
Department of Mathematics, Faculty of
Education and Integrated Arts and
Sciences
Waseda University
Tokyo
Japan
yokomori@waseda.jp

Hwanjo Yu
Data Mining Lab, Department of Computer
Science and Engineering
Pohang University of Science and
Technology
Pohang
South Korea
hwanjoyu@postech.ac.kr

Jean-Baptiste Yunès
Laboratoire LIAFA
Université Paris 7 (Diderot)
France
jean-baptiste.yunes@liafa.jussieu.fr

G. Peter Zhang
Department of Managerial Sciences
Georgia State University
Atlanta, GA
USA
gpzhang@gsu.edu

Jie Zhang
School of Chemical Engineering and
Advanced Materials
Newcastle University
Newcastle upon Tyne
UK
jie.zhang@newcastle.ac.uk

Shengyu Zhang
Department of Computer Science and
Engineering
The Chinese University of Hong Kong
Hong Kong S.A.R.
China
syzhang@cse.cuhk.edu.hk

Eckart Zitzler
PHBern – University of Teacher Education,
Institute for Continuing Professional
Education
Bern
Switzerland
eckart.zitzler@phbern.ch
eckart.zitzler@tik.ee.ethz.ch

Marek Żukowski
Institute of Theoretical Physics and
Astrophysics
University of Gdansk
Poland
marek.zukowski@univie.ac.at
fizmz@univ.gda.pl

Cellular Automata

Jarkko J. Kari

1 Basic Concepts of Cellular Automata

Jarkko J. Kari
Department of Mathematics, University of Turku, Turku, Finland
jkari@utu.fi

G. Rozenberg et al. (eds.), *Handbook of Natural Computing*, DOI 10.1007/978-3-540-92910-9_1,
© Springer-Verlag Berlin Heidelberg 2012

Abstract

This chapter reviews some basic concepts and results of the theory of cellular automata (CA). Topics discussed include classical results from the 1960s, relations between various concepts of injectivity and surjectivity, and dynamical system concepts related to chaos in CA. Most results are reported without full proofs but sometimes examples are provided that illustrate the idea of a proof. The classical results discussed include the Garden-of-Eden theorem and the Curtis–Hedlund–Lyndon theorem, as well as the balance property of surjective CA. Different variants of sensitivity to initial conditions and mixing properties are introduced and related to each other. Also, algorithmic aspects and undecidability results are mentioned.

1   Introduction

A cellular automaton (CA) is a discrete dynamical system that consists of an infinite array of cells. Each cell has a state from a finite state set. The cells change their states according to a local update rule that provides the new state based on the old states of the cell and its neighbors. All states use the same update rule, and the updating happens simultaneously at all cells. The updating is repeated over and over again at discrete time steps, leading to a time evolution of the system.

CA are among the oldest models of natural computing, dating back over half a century. The first CA studies by John von Neumann in the late 1940s were biologically motivated, related to self-replication in universal systems (Burks 1970; von Neumann 1966). Since then, CA have gained popularity as discrete models of physical systems. This is not surprising, considering that CA possess several fundamental properties of the physical world: they are massively parallel, homogeneous, and all interactions are local. Other important physical constraints such as reversibility and conservation laws can be added as needed by choosing the local update rule properly. See the chapter ❷ Cellular Automata and Lattice Boltzmann Modeling of Physical Systems by Chopard for more information on physical modeling using CA and related lattice Boltzmann methods, and the chapters ❷ Reversible Cellular Automata and ❷ Conservation Laws in Cellular Automata for details on the reversibility and conservation laws in CA.

CA have also been studied extensively as models of computation. Computational universality is common in CA, and even amazingly simple CA are capable of performing arbitrary computation tasks. Famous examples of this include Conway's Game-of-Life (Berlekamp et al. 1982; Gardner 1970) and the elementary rule 110 (Cook 2004; Wolfram 1986). The chapter ❷ Universalities in Cellular Automata by Ollinger contains other examples and more information on universality. As CA have the advantage of parallelism while obeying natural constraints such as locality and uniformity, CA provide a framework for investigating realistic computation in massively parallel systems. Programming such systems requires special techniques. In particular, the concept of signals turns out to be most useful, see the chapters ❷ Algorithmic Tools on Cellular Automata and ❷ Computations on Cellular Automata. The time complexity advantage gained from the parallelism under the locality/uniformity constraints of CA can be precisely analyzed in terms of language recognition. The time complexity classes of CA languages are among the classical research directions, see the chapter ❷ Language Recognition by Cellular Automata by Terrier.

A popular mathematical approach is to view CA as dynamical systems in the context of symbolic dynamics. Several interesting results in this area were reported already in 1969 in the seminal paper by Hedlund (1969), and until today this research direction has been among

the most fruitful sources of problems and new results in the theory of CA. The chapter
❯ Cellular Automata Dynamical Systems provides details on recent advances in this aspect.

The purpose of this chapter is to introduce the basic concepts and review some selected
results in the CA theory in order to make the more advanced chapters of the handbook more
accessible. In ❯ Sect. 2, the basic definitions and results are stated, including the Curtis–
Hedlund–Lyndon characterization of CA as continuous, translation-commuting maps
(❯ Theorem 1). In ❯ Sect. 3, other classical results from the 1960s are reviewed. These include
the balance property of surjective CA (❯ Theorem 2) as well as the Garden-of-Eden theorem
(❯ Theorems 3 and ❯ 4). In ❯ Sect. 4, different injectivity and surjectivity properties are
related to each other, and in ❯ Sect. 5, corresponding algorithmic questions are looked into.
❯ Sect. 6 introduces concepts related to topological dynamics and chaos, including variants of
sensitivity to initial conditions and mixing of the configuration space.

2 Preliminaries

Only the basic CA model is considered here: deterministic, synchronous CA, where the
underlying grid is infinite and rectangular. The cells are hence the squares of an infinite
d-dimensional checker board, addressed by $\mathbb{Z}^d$. The one-, two-, and three-dimensional cases
are most common and, as seen below, one-dimensional CA behave in many respects quite
differently from the higher dimensional ones.

2.1 Basic Definitions

The states of the automaton come from a finite *state set S*. At any given time, the *configuration*
of the automaton is a mapping $c : \mathbb{Z}^d \to S$ that specifies the states of all cells. The set of all
d-dimensional configurations over state set S is denoted by $S^{\mathbb{Z}^d}$. For any $s \in S$, the constant
configuration $c(\mathbf{x}) = s$ for all $\mathbf{x} \in \mathbb{Z}^d$ is called *s-homogeneous*.

The cells change their states synchronously at discrete time steps. The next state of each cell
depends on the current states of the neighboring cells according to an update rule. All cells use
the same rule, and the rule is applied to all cells at the same time. The neighboring cells may be
the nearest cells surrounding the cell, but more general neighborhoods can be specified by
giving the relative offsets of the neighbors. Let $N = (\mathbf{x}_1, \mathbf{x}_2, \ldots, \mathbf{x}_m)$ be a vector of m distinct
elements of $\mathbb{Z}^d$. Then the *neighbors* of a cell at location $\mathbf{x} \in \mathbb{Z}^d$ are the m cells at locations

$$\mathbf{x} + \mathbf{x}_i, \quad \text{for } i = 1, 2, \ldots, m$$

The *local rule* is a function $f : S^m \to S$ where m is the size of the neighborhood. State
$f(a_1, a_2, \ldots, a_m)$ is the state of a cell whose m neighbors were at states $a_1, a_2, \ldots, a_m$ one
time step before. This update rule then determines the global dynamics of the CA: Configuration c becomes in one time step the configuration e where, for all $\mathbf{x} \in \mathbb{Z}^d$,

$$e(\mathbf{x}) = f(c(\mathbf{x} + \mathbf{x}_1), c(\mathbf{x} + \mathbf{x}_2), \ldots, c(\mathbf{x} + \mathbf{x}_m))$$

We denote $e = G(c)$, and $G : S^{\mathbb{Z}^d} \to S^{\mathbb{Z}^d}$ is called the *global transition function* of the CA.

In summary, CA are dynamical systems that are homogeneous and discrete in both time
and space, and that are updated locally in space. A CA is specified by a quadruple (d, S, N, f)
where $d \in \mathbb{Z}_+$ is the dimension of the space, S is the state set, $N \in (\mathbb{Z}^d)^m$ is the neighborhood
vector, and $f : S^m \to S$ is the local update rule. One usually identifies a CA with its global

transition function G, and talks about CA function G, or simply CA G. In algorithmic questions, G is however always specified using the four finite items d, S, N, and f.

A CA is called *injective* if the global transition function G is one-to-one. It is *surjective* if G is onto. A CA is *bijective* if G is both onto and one-to-one.

Configuration c is *temporally periodic* for CA G, or *G-periodic*, if $G^k(c) = c$ for some $k \geq 1$. If $G(c) = c$ then c is a *fixed point*. Every CA has a temporally periodic configuration that is homogeneous: This follows from the facts that there are finitely many homogeneous configurations and that CA functions preserve homogeneity. Configuration c is *eventually periodic* if it evolves into a temporally periodic configuration, that is, if $G^m(c) = G^n(c)$ for some $m > n \geq 0$. This is equivalent to the property that the *(forward) orbit* $c, G(c), G^2(c), \ldots$ is finite.

CA G is called *nilpotent* if $G^n(S^{\mathbb{Z}^d})$ is a singleton set for sufficiently large n, that is, if there is a configuration c and number n such that $G^n(e) = c$ for all configurations e. Since homogeneous configurations remain homogeneous, one can immediately see that configuration c has to be a homogeneous fixed point.

Let G_1 and G_2 be two given CA functions with the same state set and the same dimension d. The *composition* $G_1 \circ G_2$ is also a CA function, and the composition can be formed effectively. If N_1 and N_2 are neighborhoods of G_1 and G_2, respectively, then a neighborhood of $G_1 \circ G_2$ consists of vectors $\mathbf{x} + \mathbf{y}$ for all $\mathbf{x}$ in N_1 and $\mathbf{y}$ in N_2.

Sometimes it happens that $G_1 \circ G_2 = G_2 \circ G_1 = id$ where id is the identity function. Then CA G_1 and G_2 are called *reversible*, and G_1 and G_2 are the *inverse automata* of each other. One can effectively decide whether two given CA are inverses of each other. This follows from the effectiveness of forming the composition and the decidability of the CA equivalence. Reversible CA are studied more in the chapter ❷ Reversible Cellular Automata in this handbook.

Examples of one-dimensional CA dynamics are often depicted as *space–time* diagrams. The horizontal rows of a space–time diagram are consecutive configurations. The top row is the initial configuration. More generally, the d-dimensional space–time diagram for initial configuration c is the mapping $sp \colon \mathbb{N} \times \mathbb{Z}^d \to S$ such that $sp(t, \mathbf{x}) = G^t(c)(\mathbf{x})$ for all $t \in \mathbb{N}$ and $\mathbf{x} \in \mathbb{Z}^d$.

2.2   Neighborhoods

Neighborhoods commonly used in CA are the *von Neumann* neighborhood N_{vN} and the *Moore* neighborhood N_M. The von Neumann neighborhood contains relative offsets $\mathbf{y}$ that satisfy $\|\mathbf{y}\|_1 \leq 1$, where

$$\|(y_1, y_2, \ldots, y_d)\|_1 = |y_1| + |y_2| + \cdots + |y_d|$$

This means that a cell in location $\mathbf{x}$ has $2d + 1$ neighbors: the cell itself and the cells at locations $\mathbf{x} \pm \mathbf{e}_i$ where $\mathbf{e}_i = (0, \ldots, 0, 1, 0, \ldots, 0)$ is the ith coordinate unit vector. The Moore neighborhood contains all vectors $\mathbf{y} = (y_1, y_2, \ldots, y_d)$ where each y_i is $-1, 0$ or 1, that is, all $\mathbf{y} \in \mathbb{Z}^d$ such that $\|\mathbf{y}\|_\infty \leq 1$, where

$$\|(y_1, y_2, \ldots, y_d)\|_\infty = \max\{|y_1|, |y_2|, \ldots, |y_d|\}.$$

Every cell has 3^d Moore neighbors. ❷ *Figure 1* shows the von Neumann and Moore neighborhoods in the case $d = 2$.

Generalizing the Moore neighborhood, one obtains *radius-r CA*, for any positive integer r. In radius-r CA, the relative neighborhood consists of vectors $\mathbf{y}$ such that $\|\mathbf{y}\|_\infty \leq r$. The Moore neighborhood is of radius 1.

■ Fig. 1

Two-dimensional (a) von Neumann and (b) Moore neighbors of cell *c*.

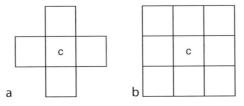

a   b

By *radius-$\frac{1}{2}$* neighborhood it is meant the neighborhood that contains the offsets $\mathbf{y} = (y_1, y_2, \ldots, y_d)$ where each y_i is 0 or 1. A one-dimensional, radius-$\frac{1}{2}$ CA is also called one-way, or OCA for short. Since the neighborhood $(0, 1)$ does not contain negative elements, information cannot flow to the positive direction. However, one can shift the cells as shown in ❯ *Fig.* 2 to obtain a symmetric trellis where information can flow in both directions.

2.3   Finite and Periodic Configurations

The *shift functions* are particularly simple CA that translate the configurations one cell down in one of the coordinate directions. More precisely, for each dimension $i = 1, 2, \ldots, d$ there is the corresponding shift function σ_i whose neighborhood contains only the unit coordinate vector $\mathbf{e}_i$ and whose local rule is the identity function. The one-dimensional shift function is the *left shift* $\sigma = \sigma_1$. *Translations* are compositions of shift functions and their inverses. Translation $\tau_{\mathbf{y}}$ by vector $\mathbf{y}$ is the CA with neighborhood $(-\mathbf{y})$ and the identity local rule. In $\tau_{\mathbf{y}}(c)$ cell $\mathbf{x} + \mathbf{y}$ has the state $c(\mathbf{x})$, for all $\mathbf{x} \in \mathbb{Z}^d$.

Sometimes, one state $q \in S$ is specified as a *quiescent state*. It should be *stable*, meaning that $f(q, q, \ldots, q) = q$. The quiescent configuration is the configuration where all cells are quiescent. A configuration $c \in S^{\mathbb{Z}^d}$ is called *q-finite* (or simply finite) if only a finite number of cells are non-quiescent, that is, the *q*-support

$$\mathrm{supp}_q(c) = \{\mathbf{x} \in \mathbb{Z}^d \mid c(\mathbf{x}) \neq q\}$$

is finite. Let us denote by $\mathfrak{F}_q(d, S)$, or briefly $\mathfrak{F}_q$, the subset of $S^{\mathbb{Z}^d}$ that contains only the *q*-finite configurations. Because of the stability of *q*, finite configurations remain finite in the evolution of the CA, so the restriction G_F of G on the finite configurations is a function $\mathfrak{F}_q \rightarrow \mathfrak{F}_q$.

A *periodic configuration*, or, more precisely, a *spatially periodic configuration* is a configuration that is invariant under *d* linearly independent translations. This is equivalent to the existence of *d* positive integers $t_1, t_2, \ldots, t_d$ such that $c = \sigma_i^{t_i}(c)$ for every $i = 1, 2, \ldots, d$, that is,

$$c(\mathbf{x}) = c(\mathbf{x} + t_i \mathbf{e}_i)$$

for every $\mathbf{x} \in \mathbb{Z}^d$ and every $i = 1, 2, \ldots, d$. Let us denote by $\mathfrak{P}(d, S)$, or briefly $\mathfrak{P}$, the set of spatially periodic configurations. CA are homogeneous in space and consequently they preserve the spatial periodicity of configurations. The restriction G_P of G on $\mathfrak{P}$ is hence a function $\mathfrak{P} \rightarrow \mathfrak{P}$.

Both $\mathfrak{F}_q$ and $\mathfrak{P}$ are dense in $S^{\mathbb{Z}^d}$, under the Cantor topology discussed in ❯ Sect. 2.5 below. Finite and periodic configurations are used in effective simulations of CA on computers.

■ **Fig. 2**
Dependencies in one-dimensional, radius-$\frac{1}{2}$ cellular automaton (CA).

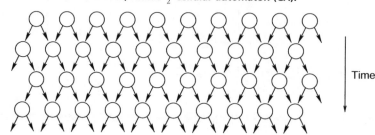

Time

Periodic configurations are often referred to as the periodic boundary conditions on a finite cellular array. For example, in the case $d = 2$, this is equivalent to running the CA on a torus that is obtained by "gluing" together the opposite sides of a rectangle. One should, however, keep in mind that the behavior of a CA can be quite different on finite, periodic, and general configurations, so experiments done with periodic boundary conditions may be misleading. Note that spatially periodic configurations are all temporally eventually periodic.

2.4 Elementary Cellular Automata

One-dimensional CA with state set $\{0, 1\}$ and the neighborhood $(-1, 0, 1)$ are called *elementary*. There are $2^3 = 8$ possible patterns inside the neighborhood of a cell, each of which may be mapped into 0 or 1. Hence, there are $2^8 = 256$ different elementary CA. Some of them are identical up to renaming the states or reversing right and left, so the number of essentially different elementary rules is smaller, only 88.

Elementary rules were extensively studied and empirically classified by Wolfram (1986). He introduced the following naming scheme: Each elementary rule is specified by an eight-bit sequence

$$f(111) \; f(110) \; f(101) \; f(100) \; f(011) \; f(010) \; f(001) \; f(000)$$

where f is the local update rule of the CA. The bit sequence is the binary expansion of an integer in the interval $0 \ldots 255$, called the *Wolfram number* of the CA (Wolfram 1986).

Example 1   Rule 110 is the elementary CA where

$$f(111) = 0, \quad f(110) = 1, \quad f(101) = 1, \quad f(100) = 0$$
$$f(011) = 1, \quad f(010) = 1, \quad f(001) = 1, \quad f(000) = 0$$

The name is obtained from the binary expansion $110 = (01101110)_2$. This rule will be used as a running example in the following sections.

2.5 Topology and CA Dynamics

A seminal paper in the dynamical systems approach to CA is by Hedlund (1969). In this paper, one-dimensional CA are studied as endomorphisms of the shift dynamical system, that is, as shift-commuting continuous transformations of the configuration space.

A metric on the configuration space $S^{\mathbb{Z}^d}$ is first defined. The distance between configurations c and e is

$$d(e, c) = \begin{cases} 0, & \text{if } e = c \\ 2^{-\min\{\|\mathbf{x}\|_\infty \mid c(\mathbf{x}) \neq e(\mathbf{x})\}}, & \text{if } e \neq c \end{cases}$$

where

$$\|(x_1, x_2, \ldots, x_d)\|_\infty = \max\{|x_1|, |x_2|, \ldots, |x_d|\}$$

It is easy to see that this function $d(\cdot, \cdot)$ is indeed a metric (even ultrametric) on $S^{\mathbb{Z}^d}$. Under this metric, two configurations that agree with each other on a large area around the origin are close to each other. The topology arising from metric $d(\cdot, \cdot)$ is the Cantor topology.

Next, a useful basis for the topology is built. Let $c \in S^{\mathbb{Z}^d}$ be an arbitrary configuration and let $D \subseteq \mathbb{Z}^d$ be a finite set of cells. The *cylinder*

$$\mathrm{Cyl}(c, D) = \{e \in S^{\mathbb{Z}^d} \mid e(\mathbf{x}) = c(\mathbf{x}) \text{ for all } \mathbf{x} \in D\}$$

determined by c and D contains all those configurations that agree with c inside domain D.

Note that all open balls under the metric are cylinders, and that every cylinder is a union of open balls. Consequently, the cylinders form a basis of the topology. The cylinders with the same domain D form a finite partitioning of $S^{\mathbb{Z}^d}$, so all cylinders are also closed. Hence cylinders are "clopen" (closed and opened), and the space $S^{\mathbb{Z}^d}$ has a clopen basis, that is, the space is zero-dimensional. Clopen sets are exactly the finite unions of cylinders.

Another way to understand the topology is to consider convergence of sequences. Under this topology, a sequence $c_1, c_2, \ldots$ of configurations converges to $c \in S^{\mathbb{Z}^d}$ iff for every $\mathbf{x} \in \mathbb{Z}^d$ there exists k such that $c_i(\mathbf{x}) = c(\mathbf{x})$ for all $i \geq k$. In other words, the state in every cell eventually stabilizes in the sequence. A standard argumentation (as in the proof of the weak König's lemma stating that every infinite binary tree has an infinite branch) shows that every sequence of configurations has a converging subsequence. But this simply means that $S^{\mathbb{Z}^d}$ is compact. Compactness further implies that $S^{\mathbb{Z}^d}$ is a Baire space and complete.

Let $G : S^{\mathbb{Z}^d} \to S^{\mathbb{Z}^d}$ be a CA function, and let $c_1, c_2, \ldots$ be a converging sequence of configurations, with limit c. Due to convergence, the states in the neighborhood of any cell $\mathbf{x} \in S^{\mathbb{Z}^d}$ stabilize, so in the sequence $G(c_1), G(c_2), \ldots$ the state of cell $\mathbf{x}$ eventually stabilizes to value $G(c)(\mathbf{x})$. The sequence $G(c_1), G(c_2), \ldots$ hence converges to $G(c)$. But this property simply states that G is continuous.

The following properties have been established:

Proposition 1 *The set $S^{\mathbb{Z}^d}$ is a compact metric space, and cylinders form a countable clopen basis. Every CA function $G : S^{\mathbb{Z}^d} \to S^{\mathbb{Z}^d}$ is continuous.*

Pairs (X, F) where X is compact and $F : X \to X$ is continuous are commonly called (topological) dynamical systems.

The continuity of G simply reflects the fact that all cells apply a local rule to obtain the new state. On the other hand, the fact that all CA functions G commute with translations τ (i.e., $G \circ \tau = \tau \circ G$) means that the local rule is the same at all cells. The Curtis–Hedlund–Lyndon theorem from 1969 shows that these two properties exactly characterize CA functions:

Theorem 1 (Hedlund 1969) *Function $G : S^{\mathbb{Z}^d} \to S^{\mathbb{Z}^d}$ is a CA function if and only if it is continuous and it commutes with translations.*

The proof is based on the fact that continuous functions on compact spaces are uniformly continuous. The following corollary of the theorem states that reversible CA are exactly the automorphisms of the full shift:

Corollary 1 (Hedlund 1969) *A cellular automaton function G is reversible if and only if it is a bijection.*

Proof By definition, a reversible CA function is bijective. Conversely, suppose that G is a bijective CA. The inverse of a continuous bijection between compact metric spaces is continuous, so G^{-1} is continuous. Trivially, G^{-1} commutes with translations, so, by ❷ Theorem 1, the inverse G^{-1} is a CA function.

3 Classical Results

In addition to ❷ Theorem 1, several significant combinatorial results were proved in the 1960s. These include the balance in surjective CA, and the Garden-of-Eden theorem by Moore and Myhill.

3.1 Balance in Surjective CA

A configuration c is a *Garden-of-Eden configuration* if it has no pre-images, that is, if $G^{-1}(c)$ is empty. A CA has Garden-of-Eden configurations iff the CA is not surjective.

Example 2 Consider the elementary CA number 110 from ❷ Example 1. Among eight possible neighborhood patterns there are three that are mapped to state 0 and five that are mapped to state 1. It is demonstrated below how this imbalance automatically implies that there are Garden-of-Eden configurations.

Let k be a positive integer, and consider a configuration c in which

$$c(3) = c(6) = \cdots = c(3k) = 0.$$

See ❷ *Fig. 3* for an illustration. There are

$$2^{2(k-1)} = 4^{k-1}$$

possible choices for the missing states between 0s in c (shown as "*" in ❷ *Fig. 3*).

In a pre-image e of c, the three state segments $e(3i-1)$, $e(3i)$, and $e(3i+1)$ are mapped into state 0 by the local rule f, for every $i = 1, 2, \ldots, k$. Since $|f^{-1}(0)| = 3$, there are exactly 3^k choices of these segments. If k is sufficiently large then $3^k < 4^{k-1}$. This means that some choice of c does not have a corresponding pre-image e. Therefore, the CA is not surjective.

Alternatively, one could show the non-surjectivity of rule 110 by directly verifying that any configuration containing pattern 01010 is a Garden-of-Eden configuration.

◼ **Fig. 3**

Illustration of the configuration c and its pre-image e in ❷ Example 2.

In this section, the previous example is generalized. The concept of a (finite) pattern is defined first. A *pattern*

$$p = (D, g)$$

is a partial configuration where $D \subseteq \mathbb{Z}^d$ is the *domain* of p and $g : D \rightarrow S$ is a mapping assigning a state to each cell in the domain. Pattern p is *finite* if D is a finite set. The cylinder determined by finite pattern $p = (D, g)$ is the set of configurations c such that $c(\mathbf{x}) = g(\mathbf{x})$ for all $\mathbf{x} \in D$.

If $N = (\mathbf{x}_1, \mathbf{x}_2, \dots, \mathbf{x}_m)$ is a neighborhood vector, the neighborhood of domain $D \subseteq \mathbb{Z}^d$ is defined as the set

$$N(D) = \{\mathbf{x} + \mathbf{x}_i \mid \mathbf{x} \in D, i = 1, 2, \dots, m\}$$

of the neighbors of elements in D.

Let G be a CA function specified by the quadruple (d, S, N, f). Let $q = (E, h)$ be a pattern, and let $D \subseteq \mathbb{Z}^d$ be a domain such that $N(D) \subseteq E$. An application of the local rule f on pattern q determines new states for all cells in domain D. A pattern $p = (D, g)$ is obtained where for all $\mathbf{x} \in D$

$$g(\mathbf{x}) = f[h(\mathbf{x} + \mathbf{x}_1), h(\mathbf{x} + \mathbf{x}_2), \dots, h(\mathbf{x} + \mathbf{x}_m)]$$

The mapping $q \mapsto p$ will be denoted by $G^{(E \rightarrow D)}$, or simply by G when the domains E and D are clear from the context. With this notation, the global transition function of the CA is $G^{(\mathbb{Z}^d \rightarrow \mathbb{Z}^d)}$.

A finite pattern $p = (D, g)$ is called an *orphan* if there is no finite pattern q with domain $N(D)$ such that $G^{(N(D) \rightarrow D)}(q) = p$. In other words, orphan is a pattern that cannot appear in any configuration after an application of the CA. For example, pattern 01010 is an orphan for rule 110.

Proposition 2 *A CA has an orphan if and only if it is non-surjective.*

Proof It is clear that if an orphan exists then the CA is not surjective: Every configuration that contains a copy of the orphan is a Garden-of-Eden. To prove the converse, the compactness of the configuration space is applied (see ❷ Proposition 1). If G is not surjective then $G(S^{\mathbb{Z}^d}) \subsetneq S^{\mathbb{Z}^d}$ is closed, so its complement (the set of Garden-of-Eden configurations) is open and nonempty. As cylinders form a basis of the topology, there is a cylinder all of whose elements are Garden-of-Eden configurations. The finite pattern that specifies the cylinder is then an orphan.

The following balance theorem generalizes ❷ Example 2. The result was proved in the one-dimensional case already in Hedlund (1969), but it holds in any dimension d (Maruoka and Kimura 1976). The theorem states that in surjective CA, all finite patterns of the same domain D must have the same number of pre-image patterns of domain $N(D)$. Any imbalance immediately implies non-surjectivity. As a special case, it can be seen that the local rule of a surjective CA must be balanced: each state appears in the rule table equally many times.

Theorem 2 *Let $A = (d, S, N, f)$ be a surjective CA, and let $E, D \subseteq \mathbb{Z}^d$ be finite domains such that $N(D) \subseteq E$. Then, for every pattern $p = (D, g)$ the number of patterns $q = (E, h)$ such that*

$$G^{(E \rightarrow D)}(q) = p$$

is $s^{|E| - |D|}$ where $s = |S|$ is the number of states.

Proof (sketch) The theorem can be proved analogously to ❷ Example 2, by considering imbalanced mappings from d-dimensional hypercube patterns of size n^d, into hypercubes of size $(n - 2r)^d$ where r is the neighborhood radius of the CA. Considering an arrangement of k^d such hypercubes (see ❷ Fig. 4) one can deduce an orphan as in ❷ Example 2, using the following technical Lemma.

Lemma 1 *For all $d, n, s, r \in \mathbb{Z}_+$ the inequality*

$$\left(s^{n^d} - 1\right)^{k^d} < s^{(kn - 2r)^d}$$

holds for all sufficiently large k.

Example 3 The balance condition of the local rule table is necessary but not sufficient for surjectivity. Consider, for example, the elementary CA number 232. It is the *majority* CA: $f(a, b, c) = 1$ if and only if $a + b + c \geq 2$. Its rule table is balanced because 000, 001, 010, 100 map to 0, and 111, 110, 101, 011 map to 1.

However, the majority CA is not balanced on longer patterns and hence it is not surjective: Any word of length 4 that contains at most one state 1 is mapped to 00, so 00 has at least five pre-images of length 4. Balanceness would require this number of pre-images to be four. Pattern 01001 is a particular example of an orphan.

❷ Theorem 2 actually states the fact that the uniform Bernoulli measure on the configuration space is invariant under applications of surjective CA: Randomly (under the uniform probability distribution) picked configurations remain random under iterations of surjective CA.

■ **Fig. 4**
Illustration for the proofs of ❷ Theorems 2, ❷ 3 and ❷ 4.

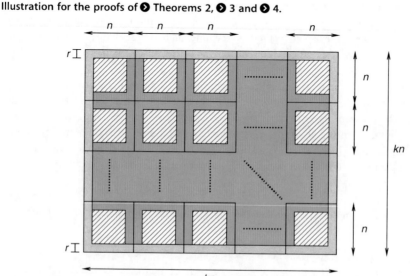

3.2   Garden-of-Eden Theorem

Two configurations c and d are *asymptotic* if they differ only in finitely many cells, that is, if

$$\mathrm{diff}(c, d) = \{\mathbf{x} \in \mathbb{Z}^d \mid c(\mathbf{x}) \neq d(\mathbf{x})\}$$

is finite. CA G is called *pre-injective* if $G(c) \neq G(d)$ for any asymptotic c, d such that $c \neq d$.

Let $s \in S$ be an arbitrary state. Note that CA C is pre-injective iff it is injective among s-finite configurations. Among the earliest discovered properties of CA is the Garden-of-Eden theorem by Moore and Myhill from 1962 and 1963, respectively, which proves that pre-injectivity is equivalent to surjectivity.

Example 4   As in ❷ Section 3.1, to start with, an illustration of the proof of the Garden-of-Eden theorem using elementary rule 110 is given. It demonstrates how non-surjectivity of rule 110 implies that it is not pre-injective.

By ❷ Example 2, rule 110 is not surjective. In fact, finite pattern 01010 is an orphan. It is shown in the following that there must exist different 0-finite configurations c and e such that $G(c) = G(e)$.

Let $k \in \mathbb{Z}_+$ be abitrary. Consider 0-finite configurations c whose supports are included in a fixed segment of length $5k - 2$, see ❷ *Fig. 5*. There are

$$2^{5k-2} = 32^k/4$$

such configurations. The support of $G(c)$ is included in a segment of length $5k$. Partition this segment in k subsegments of length 5. It is known that pattern 010101 cannot appear anywhere in $G(c)$, so there are at most $2^5 - 1 = 31$ different patterns that can appear in the length 5 subsegments. Hence, there are at most 31^k possible configurations $G(c)$. For all sufficiently large values of k, one has

$$32^k/4 > 31^k$$

so there must be two 0-finite configurations with the same image.

Theorem 3   (Myhill 1963) *If G is pre-injective then G is surjective.*

Proof   (sketch) The theorem can be proved along the lines of ❷ Example 4. If G is not surjective, there is an orphan whose domain is a d-dimensional hypercube of size n^d, for some n. Arranging k^d copies of the hypercube as in ❷ Fig. 4, one comes up with a domain in which at most $\left(s^{n^d} - 1\right)^{k^d}$ different non-orphans exist. Using ❷ Lemma 1 one can now easily conclude the existence of two different asymptotic configurations with the same image.

■ **Fig. 5**
Illustration of ❷ Example 4.

$$5k$$

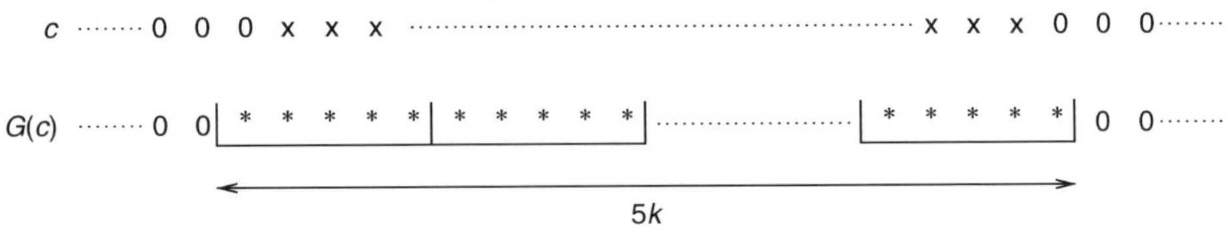

Injectivity trivially implies pre-injectivity, and therefore the following corollary is obtained.

Corollary 2 *Every injective CA is also surjective. Injectivity is hence equivalent to bijectivity and reversibility.*

The other direction of the Garden-of-Eden theorem is considered next. Again, one starts with a one-dimensional example that indicates the proof idea.

Example 5 Consider again rule 110. The asymptotic configurations

$$c_1 = \ldots 000011010000 \ldots$$
$$c_2 = \ldots 000010110000 \ldots$$

have the same image. In the following it is demonstrated how this implies that rule 110 is not surjective. (Of course, this fact is already known from prior examples.)

Extract patterns $p_1 = 011010$ and $p_2 = 010110$ of length 6 from c_1 and c_2, respectively. Both patterns are mapped into the same pattern 1111 of length 4. Moreover, p_1 and p_2 have a boundary of width 2 on both sides where they are identical with each other. Since rule 110 uses radius-1 neighborhood, one can replace in any configuration c pattern p_1 by p_2 or vice versa without affecting $G(c)$.

Let $k \in \mathbb{Z}_+$, and consider a segment of $6k$ cells. It consists of k segments of length 6. Any pattern of length $6k - 2$ that has a pre-image of length $6k$ also has a pre-image where none of the k subsegments of length 6 contains pattern p_2. Namely, all such p_2 can be replaced by p_1. This means that at most $(2^6 - 1)^k = 63^k$ patterns of length $6k - 2$ can have pre-images. On the other hand, there are $2^{6k-2} = 64^k/4$ such patterns, and for large values of k

$$64^k/4 > 63^k$$

so some patterns do not have a pre-image.

Theorem 4 (Moore 1962) *If G is surjective then G is pre-injective.*

Proof (sketch) The proof of the theorem is as in ❷ Example 5. If d-dimensional CA G is not pre-injective, there are two different patterns p_1 and p_2 with the same image such that the domain of the patterns is the same hypercube of size n^d, and the patterns are identical on the boundary of width r of the hypercube, where r is now twice the neighborhood radius of the CA. In any configuration, copies of pattern p_2 can be replaced by p_1 without affecting the image under G. Arranging k^d copies of the hypercubes as in ❷ Fig. 4, one comes up with a domain in which at most $\left(s^{n^d} - 1\right)^{k^d}$ patterns have different images. Using ❷ Lemma 1 again one sees that this number is not large enough to provide all possible interior patterns, and an orphan must exist.

4 Injectivity and Surjectivity Properties

The Garden-of-Eden theorem links surjectivity of CA to injectivity on finite configurations. The surjectivity and injectivity properties on $S^{\mathbb{Z}^d}$, on finite and on periodic configurations, are next related to each other, as reported in Durand (1998). On periodic configurations, the situation is slightly different in the one-dimensional case than in the higher-dimensional cases. The following implications are easily seen to be valid, however, regardless of the dimension:

- If G is injective then G_F and G_P are also injective (trivial).
- If G_F or G_P is surjective then G is also surjective (due to denseness of $\mathfrak{F}_q$ and $\mathfrak{P}$ in $S^{\mathbb{Z}^d}$).
- If G_P is injective then G_P is surjective (because the number of periodic configurations with given fixed periods is finite).
- If G is injective then G_F is surjective (due to reversibility and the stability of the quiescent state).

4.1   The One-Dimensional Case

In this subsection, we concentrate on one-dimensional CA. The balance theorem of surjective CA has the following interesting corollary that is valid in the one-dimensional case only.

Theorem 5   (Hedlund 1969) *For every one-dimensional surjective CA, there is a constant m such that every configuration has at most m pre-images.*

Proof   Let G be a one-dimensional surjective CA function defined using radius-r neighborhood. Let $s = |S|$ be the number of states. In the following, it is proved that every configuration has at most s^{2r} different pre-images.

Suppose the contrary: there is a configuration c with $s^{2r} + 1$ different pre-images

$$e_1, e_2, \ldots, e_{s^{2r}+1}$$

For some sufficiently large number $k > r$, each pair e_i and e_j of pre-images contains a difference inside the interval

$$E = \{-k, -k+1, \ldots, k-1, k\}$$

But this contradicts ❷ Theorem 2 since one now has a pattern with domain

$$D = \{-k+r, -k+r+1, \ldots, k-r\}$$

that has at least

$$s^{2r} + 1 > s^{|E|-|D|}$$

different pre-images in domain E.

Corollary 3   (Hedlund 1969) *Let G be a one-dimensional surjective CA function. The pre-images of spatially periodic configurations are all spatially periodic. In particular, G_P is surjective.*

Proof   Suppose $G(c)$ is spatially periodic, so that $\sigma^n(G(c)) = G(c)$ for some $n \in \mathbb{Z}_+$. Then for every $i \in \mathbb{Z}$

$$G(\sigma^{in}(c)) = \sigma^{in}(G(c)) = G(c)$$

so $\sigma^{in}(c)$ is a pre-image of $G(c)$. By ❷ Theorem 5, configuration $G(c)$ has a finite number of pre-images so

$$\sigma^{i_1 n}(c) = \sigma^{i_2 n}(c)$$

■ Fig. 6

Implications between injectivity and surjectivity properties in one-dimensional CA.

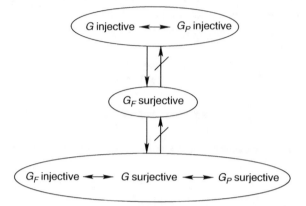

for some $i_1 < i_2$. But then, c is periodic with period $(i_2 - i_1)n$. It follows that all pre-images of periodic configurations are periodic.

In the one-dimensional setting, the injectivity of G_P implies that G is injective. This can be seen easily, for example, using the de Bruijn graph technique (Sutner 1991). Combining the results of this and the previous sections, the positive implications in ❷ *Fig. 6* are obtained. The following two examples confirm the two negative implications in the figure.

Example 6   The XOR automaton is the elementary CA rule 102. Every cell adds together (mod 2) its state and the state of its right neighbor. This CA is surjective, but it is not surjective on 0-finite configurations because the only two pre-images of the finite configuration . . . 001000 . . . are the non-finite configurations . . . 000111 . . . and . . . 111000 . . . .

Example 7   CONTROLLED-XOR is a one-dimensional radius-$\frac{1}{2}$ CA. It has four states 00, 01, 10, and 11. The first bit of each state is a control symbol that does not change. If the control symbol of a cell is 0, then the cell is inactive and does not change its state. If the control symbol is 1, then the cell is active and applies the XOR rule of ❷ Example 6 on the second bit.

State 00 is the quiescent state. CONTROLLED-XOR is easily seen to be surjective on finite configurations. It is not injective on unrestricted configurations as two configurations, all of whose cells are active, have the same image if their second bits are complements of each other.

4.2   The Higher-Dimensional Cases

In the two- and higher-dimensional cases, injectivity on periodic configurations is no longer equivalent to general injectivity, and surjectivity on periodic configurations is not known to be equivalent to surjectivity. ❷ *Figure 7* summarizes the currently known implications (Durand 1998). Note that in three cases the relation is unknown.

□ Fig. 7

Implications between injectivity and surjectivity properties in two- and higher-dimensional CA.

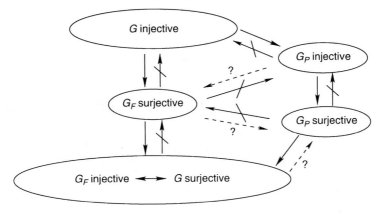

The difference in the one- and two-dimensional diagrams can be explained using tilings. Plane tilings are also useful in obtaining undecidability results for two-dimensional CA, as will be explained in ❥ Sect. 5. First, d-dimensional tilings are defined here using local matching rules. The definition closely resembles CA, the only difference being that tilings are static objects unlike dynamic CA. In symbolic dynamics terminology, tilings are d-dimensional subshifts of finite type.

A d-dimensional *tile set* $\mathfrak{T} = (d, T, N, R)$ consists of a finite set T whose elements are the *tiles*, a d-dimensional neighborhood vector N of size m (as defined in ❥ Sect. 2.1), and a *local matching rule* $R \subseteq T^m$ that gives a relation specifying which tilings are considered valid. *Tilings* are configurations $t \in T^{\mathbb{Z}^d}$. Configuration t is valid at cell $\mathbf{x} \in \mathbb{Z}^d$ iff

$$[t(\mathbf{x} + \mathbf{x}_1), t(\mathbf{x} + \mathbf{x}_2), \ldots, t(\mathbf{x} + \mathbf{x}_m)] \in R$$

that is, the neighborhood of $\mathbf{x}$ contains a matching combination of tiles. Configuration $t \in T^{\mathbb{Z}^d}$ is a *valid tiling* if it is valid at all positions $\mathbf{x} \in \mathbb{Z}^d$, and it is said that the tile set $\mathfrak{T}$ then *admits* tiling t.

There is an apparent similarity in the definitions of tile sets and CA. The only difference is that instead of a dynamic local rule f, tilings are based on a static matching relation R.

A fundamental property of CA is that they commute with translations. The tiling counterpart states the obvious fact that $\tau(t)$ is a valid tiling for every valid tiling t and every translation τ. The second fundamental property of CA is the continuity of CA functions. The tiling counterpart of this fact states that the set of valid tilings is closed in the Cantor topology, that is, the limit of a converging sequence of valid tilings is also valid. Proofs of these facts are straightforward. Compactness also directly implies that a tile set that can properly tile arbitrarily large hypercubes admits a tiling of the whole space.

A particularly interesting case is $d = 2$ because two-dimensional tilings can draw computations of Turing machines, leading to undecidability results (Wang 1961). A convenient way to describe a two-dimensional tile set is in terms of *Wang tiles*. Wang tiles use the von Neumann neighborhood. The tiles are viewed as unit squares whose edges are colored, and the local matching rule is given in terms of these colors: A tiling is valid at cell $\mathbf{x} \in \mathbb{Z}^2$ iff each of the four edges of the tile in position $\mathbf{x}$ has the same color as the abutting edge in the adjacent tiles.

Fig. 8

(a) Two Wang tiles, and (b) part of a valid tiling.

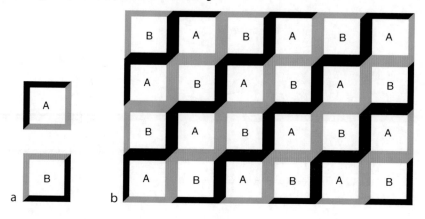

Example 8 Consider the two Wang-tiles A and B shown in ❷ *Fig. 8a*. Since all four neighbors of tile A have to be copies of B, and vice versa, the only valid tilings are infinite checkerboards where A's and B's alternate, as shown in ❷ *Fig. 8b*.

Tiling $t \in T^{\mathbb{Z}^d}$ is *periodic* if it is spatially periodic, as defined in ❷ Sect. 2.3. In the two-dimensional case, this simply means that the tiling consists of a horizontally and vertically repeating rectangular pattern. An interesting fact is that there exist two-dimensional tile sets that only allow nonperiodic tilings. A tile set is called *aperiodic* if

1. It admits some valid tilings, but
2. It does not admit any valid periodic tilings.

The existence of such aperiodic tile sets was demonstrated by Berger in 1966.

Theorem 6 (Berger 1966) *There exist aperiodic sets of Wang tiles.*

In the following, a particular aperiodic tile set SNAKES from Kari (1994) is used to construct an example of a non-injective two-dimensional CA that is injective on periodic configurations. The tiles SNAKES have an arrow printed on them. The arrow is horizontal or vertical and it points to one of the four neighbors of the tile. Given any tiling t, valid or invalid, the arrows determine paths, obtained by following the arrows printed on the tiles. The tile that follows tile (x, y) on a path is the von Neumann neighbor of (x, y) in the direction indicated by the arrow on $t(x, y)$.

The tile set SNAKES of Kari (1994) has the following plane-filling property: Consider a tiling t and a path P that indefinitely follows the arrows as discussed above. If the tiling is valid at all tiles that P visits, then the path necessarily covers arbitrarily large squares. In other words, for every $N \geq 1$ there is a square of $N \times N$ tiles on the plane, all of whose tiles are visited by path P. Note that the tiling may be invalid outside path P, yet the path is forced to snake through larger and larger squares. In fact, SNAKES forces the paths to follow the well-known Hilbert curve (Kari 1994).

Example 9 Using SNAKES, the two-dimensional SNAKE-XOR CA is constructed in the style of the one-dimensional CONTROLLED-XOR of ❷ Example 7. The states consist of two layers: a control layer and a xor layer. The control layer does not change: it only indicates which cells are active and which neighbor cell provides the bit to the XOR operation. In SNAKE-XOR, the control layer consists of SNAKES tiles discussed above. Only cells where the tiling on the control layer is valid are active. Active cells execute addition (mod 2) on their xor layer. The arrow of the tile tells which neighbor provides the second bit to the operation.

SNAKE-XOR is not injective: Two configurations c_0 and c_1 whose control layer consist of the same valid tiling have the same image if their xor layers are complementary to each other. However, SNAKE-XOR is injective on periodic configurations, as the plane-filling property ensures that on periodic configurations any infinite path that follows the arrows must contain nonactive cells.

5   Algorithmic Questions

Many relevant algorithmic questions can be formulated. It would be nice to have algorithms to determine, for example, whether a given CA is injective, surjective, nilpotent, etc. The one-dimensional case is usually algorithmically more accessible than higher-dimensional cases.

Theorem 7   (Amoroso and Patt 1972) *There exist algorithms to determine if a given one-dimensional CA is injective, or if it is surjective.*

Elegant polynomial-time decision algorithms can be based on de Bruijn graphs (Sutner 1991).

Already in the two-dimensional case, the injectivity and the surjectivity questions turn out to be undecidable. For injectivity, the proof is easily based on the SNAKES-tiles of ❷ Sect. 4.2, and the following classical result establishing the undecidability of the tiling problem:

Theorem 8   (Berger 1966) *It is undecidable if a given set of Wang tiles admits a valid tiling.*

The tiling problem can be reduced to the question concerning the injectivity of two-dimensional CA. Also, a proof in a similar style (but using a simpler variant of the tiling problem, known as the finite tiling problem) shows the undecidability of injectivity on finite configurations and, hence, by the Garden-of-Eden theorem, of surjectivity.

Theorem 9   (Kari 1994) *It is undecidable if a given two-dimensional cellular is injective. It is also undecidable whether it is surjective.*

Proof (sketch for injectivity) A reduction of the tiling problem to the injectivity question goes as follows. For a given set T of Wang tiles one effectively constructs a two-dimensional CA, similar to SNAKE-XOR of ❷ Example 9. The CA has a control layer and a xor layer. The control layer in turn consists of two layers: one with tiles T and one with tiles SNAKES. A cell is active if and only if the tiling is valid at the cell on both tile components. Active cells execute the addition (mod 2) on their xor layer, and the arrow on the SNAKES tile tells which neighbor provides the second bit to the sum. The plane-filling property of SNAKES guarantees that if two different configurations have the same successor then arbitrarily large squares must have a valid tiling. Conversely, if a valid tiling exists then two different configurations with identical control layers can have the same successor. Hence the CA that is constructed is injective iff T does not admit a valid tiling, and this completes the proof.

Note the following fundamental difference in the injectivity problems of G and G_F: A semi-algorithm exists for the injectivity of G (based on an exhaustive search for the inverse CA) and for the non-injectivity of G_F (based on looking for two finite configurations with the same image).

Even though ❯ Corollary 1 guarantees that the inverse function of every injective CA is a CA, ❯ Theorem 9 implies that the neighborhood of the inverse CA can be very large: there can be no computable upper bound, as otherwise all candidate inverses could be tested one by one. In contrast, in the one-dimensional case, the inverse automaton can only have a relatively small neighborhood. In one-dimensional radius-$\frac{1}{2}$ CA, the inverse neighborhood consists of at most $s - 1$ consecutive cells where s is the number of states (Czeizler and Kari 2005).

Analogously, ❯ Theorem 9 implies that the size of the smallest orphan of a non-surjective CA has no computable upper bound. In the one-dimensional case, however, a polynomial upper bound for the length of the shortest orphan was recently reported (Kari et al. 2009).

While injectivity and surjectivity are single time step properties, algorithmic questions on long-term dynamical properties have also been studied. Many such properties turn out to be undecidable already among one-dimensional CA. An explanation is that the space–time diagrams are two-dimensional and can be viewed as tilings.

Recall from ❯ Sect. 2.1 that a CA G is called nilpotent if it has trivial dynamics: for some n, the nth iterate $G^n(c)$ of every configuration is quiescent. It turns out that there is no algorithm to determine if a given one-dimensional CA is nilpotent Kari (1992). This is true even if the quiescent state q is *spreading*, that is, if $f(a_1, a_2, \ldots, a_m) \neq q$ implies that all $a_i \neq q$.

A reversible CA is called *periodic* if G^n is the identity function for some $n \geq 1$. Recently, it was shown that the periodicity of one-dimensional CA is undecidable (Kari and Ollinger 2008).

Theorem 10 (Kari 1992; Kari and Ollinger 2008) *It is undecidable if a given one-dimensional CA (with a spreading state) is nilpotent. It is also undecidable if a given reversible one-dimensional CA is periodic.*

Nilpotency is a simple property, and as such it can be reduced to many other dynamical properties. Some are indicated in ❯ Sect. 6 below.

6 Dynamical Systems Concepts

As advertised in ❯ Sect. 2.5, CA are rich examples of dynamical systems in the sense of topological dynamics. Chaotic behavior in dynamical systems means sensitivity to small perturbations in the initial state, as well as mixing of the configuration space. Sensitivity and mixing in CA under the Cantor topology have been investigated and related to each other.

Recall the metric defined in ❯ Sect. 2.5. Observing a configuration approximately under this metric simply means that states inside a finite window can be seen, but the observer has no information on the states outside the window. Better precision in the observation just means a larger observation window.

6.1 Sensitivity

Informally, a dynamical system is called sensitive to initial conditions if small changes in the state of the system get magnified during the evolution. This means that the long-term behavior of the system by simulation cannot be predicted unless the initial state is known precisely. In contrast, configurations whose orbits are arbitrarily well tracked by all sufficiently close configurations are called equicontinuous. Such orbits can be reliably simulated.

Precisely speaking, configuration $c \in S^{\mathbb{Z}^d}$ is an *equicontinuity point* for CA G if for every finite observation window $E \subseteq \mathbb{Z}^d$ there corresponds a finite domain $D \subseteq \mathbb{Z}^d$ such that for every $e \in \mathrm{Cyl}(c, D)$ one has $G^n(e) \in \mathrm{Cyl}(G^n(c), E)$ for all $n = 1, 2, 3, \ldots$. In other words, no difference is ever seen inside the observation window E if states of c are changed outside of D. Let $\mathscr{E}_G$ denote the set of equicontinuity points of G.

CA G is called *equicontinuous* if all configurations are equicontinuous, that is, if $\mathscr{E}_G = S^{\mathbb{Z}^d}$. It turns out that only eventually periodic CA are equicontinuous.

Theorem 11   (Blanchard and Tisseur; Kůrka 1997) *A CA G is equicontinuous if and only if $G^{m+p} = G^m$ for some $p \geq 1$ and $m \geq 0$. A surjective CA is equicontinuous if and only if it is periodic.*

It is undecidable to decide if a given one-dimensional CA is equicontinuous (Durand et al. 2003). In fact, equicontinuity is undecidable even among reversible CA (❯ Theorem 10).

CA G is *sensitive to initial conditions* (or simply *sensitive*) if there is a finite observation window $E \subseteq \mathbb{Z}^d$ such that for every configuration c and every domain D, there exists a configuration $e \in \mathrm{Cyl}(c, D)$ such that $G^n(e) \notin \mathrm{Cyl}(G^n(c), E)$, for some $n \geq 1$. In simple terms, any configuration can be changed arbitrarily far in such a way that the change propagates into the observation window E.

Example 10   Any nonzero translation τ is sensitive to initial conditions. The Xor-automaton of ❯ Example 6 is another example: For any configuration c and every positive integer n, a change in cell n propagates to cell 0 in n steps.

It is easy to see that a sensitive CA cannot have any equicontinuity points. In the one-dimensional case also the converse holds.

Theorem 12   (Kůrka 1997) *A one-dimensional CA G is sensitive if and only if $\mathscr{E}_G = \emptyset$. If G is not sensitive then $\mathscr{E}_G$ is a residual set (and hence dense).*

The previous theorem is not valid in the two-dimensional case (Sablik and Theyssier 2008). The difference stems from the fact that equicontinuity in the one-dimensional case is characterized by the presence of *blocking words* (words that do not let information pass between their left and right sides), while in the two-dimensional case information can circumvent any finite block.

A very strong form of sensitivity is *(positive) expansivity*. In positively expansive CA, every small change to any configuration propagates everywhere. More precisely, there exists a finite observation window $E \subseteq \mathbb{Z}^d$ such that for any distinct configurations c and e there exists time $n \geq 0$ such that configurations $G^n(c)$ and $G^n(e)$ differ in some cell in the window E.

Example 11   The Xor automaton of ❯ Example 6 is not positively expansive because differences only propagate to the left, but not to the right. The three neighbor Xor (elementary CA 150) is positively expansive since differences propagate to both directions.

Positively expansive CA are clearly surjective and sensitive but not injective. It turns out that in the two- and higher-dimensional spaces, there are no positively expansive CA (Shereshevsky 1993). It is not known whether there exists an algorithm to determine if a given CA is positively expansive, and this remains an interesting open problem. However, sensitivity is known to be undecidable (Durand et al. 2003), even among reversible one-dimensional CA (Lukkarila 2010).

6.2   Mixing Properties

Mixing of the configuration space is another property associated to chaos. Like sensitivity, mixing also comes in different variants. CA G is called *transitive* if for all cylinders U and V there exists $n \geq 0$ such that $G^n(U) \cap V \neq \emptyset$. Clearly transitive CA need to be surjective.

Example 12   The XOR automaton of ❷ Example 6 is easily seen to be transitive. For any fixed word w of length n, define the function $h_w : S^n \rightarrow S^n$ that maps $u \mapsto v$ iff $G^n(\ldots w\, u \ldots) = \ldots v \ldots$ where v is a word of length n starting in the same position as w, only n time steps later. Mapping h_w is injective, and therefore also surjective. It can be seen that for any two cylinders, U and V with domain $\{1,\ 2,\ \ldots,\ n\}$ holds $G^n(U) \cap V \neq \emptyset$. This is enough to prove transitivity.

The following theorem illustrates that transitivity is a stronger property than sensitivity.

Theorem 13   (Codenotti and Margara 1996) *A transitive CA (with at least two states) is sensitive to initial conditions.*

Example 13   As an example of a surjective CA that is sensitive but not transitive, consider the product $\tau \times I$ of a nonzero translation τ and the identity function I, both over the binary state set $\{0, 1\}$. The product is simply the four-state CA with two independent binary tracks, where the tracks compute τ and I, respectively. This CA is sensitive because τ is sensitive, but it is not transitive because I is not transitive.

Transitivity can equivalently be characterized in terms of dense orbits. Let us denote by

$$\mathscr{T}_G = \{c \in S^{\mathbb{Z}^d} \mid \text{the orbit } c,\ G(c),\ G^2(c),\ \ldots \text{ is dense}\}$$

the set of *transitive points* of G, that is, those points whose orbit visits every cylinder. It is not difficult to see that a CA is transitive iff it has transitive points.

Mixing is a more restrictive concept than transitivity: CA G is called *mixing* if for all cylinders U and V there exists positive integer n such that $G^k(U) \cap V \neq \emptyset$ for all $k \geq n$. It is immediate from the definition that every mixing CA is also transitive. Both transitivity and mixing are undecidable properties, even among reversible one-dimensional CA (Lukkarila 2010).

The following theorem relates positive expansivity to the mixing properties.

Theorem 14   (Blanchard and Maass 1997) *A positively expansive CA is mixing.*

We have the following implications between different variants of sensitivity and mixing in CA:

$$\textit{pos. expansive} \implies \textit{mixing} \implies \textit{transitive} \implies \textit{sensitive}$$

It is an open question whether every transitive CA is also mixing.

Notions of sensitivity and mixing of the space are related to chaotic behavior. Devaney (1986) defines a dynamical system to be chaotic if

1. It is sensitive to initial conditions.
2. It is transitive.
3. Temporally periodic points are dense.

As transitivity implies sensitivity, conditions 2 and 3 are sufficient. It remains a challenging open problem to show that periodic orbits are dense in every surjective CA, or even in every transitive CA. If this is the case, then Devaney's chaos is equivalent to transitivity.

7 Conclusions

In this chapter, some basic definitions, classical results, decidability issues, and dynamical system aspects of CA are covered. The goal was to provide background information that makes CA literature – and in particular other chapters in this handbook – more accessible. Full proofs were not provided, but in some cases a proof idea was given. For complete proofs, see the corresponding literature references.

References

Amoroso S, Patt Y (1972) Decision procedures for surjectivity and injectivity of parallel maps for tessellation structures. J Comput Syst Sci 6:448–464

Berger R (1966) The undecidability of the domino problem. Mem Amer Math Soc 66:1–72

Berlekamp ER, Conway JH, Guy RK (1982) Winning ways for your mathematical plays, vol II, chap 25. Academic, London

Blanchard F, Maass A (1997) Dynamical properties of expansive one-sided cellular automata. Israel J Math 99:149–174

Blanchard F, Tisseur P (2000) Some properties of cellular automata with equicontinuity points. Ann Inst Henri Poincaré, Probab Stat 36(5):569–582

Burks A (1970) Von Neumann's self-reproducing automata. In: Burks A (ed) Essays on cellular automata, University of Illinois Press, Champaign, pp 3–64

Codenotti B, Margara L (1996) Transitive cellular automata are sensitive. Am Math Mon 103(1):58–62

Cook M (2004) Universality in elementary cellular automata. Complex Syst 15:1–40

Czeizler E, Kari J (2005) A tight linear bound on the neighborhood of inverse cellular automata. In: Caires L, Italiano GF, Monteiro L, Palamidessi C, Yung M (eds) ICALP, Lecture notes in computer science, vol 3580, Springer, Berlin, pp 410–420

Devaney RL (1986) An introduction to chaotic dynamical systems. Benjamin/Cummings, Menlo Park, CA

Durand B (1998) Global properties of cellular automata. In: Goles E, Martinez S (eds) Cellular automata and complex systems. Kluwer, Dordrecht

Durand B, Formenti E, Varouchas G (2003) On undecidability of equicontinuity classification for cellular automata. In: Morvan M, Remila E (eds) Discrete models for complex systems, DMCS'03, DMTCS Conference Volume AB, Lyon, France, pp 117–128

Gardner M (1970) The fantastic combinations of John Conway's new solitaire game " Life ". Sci Am 223(4): 120–123

Hedlund GA (1969) Endomorphisms and automorphisms of the shift dynamical system. Math Syst Theor 3:320–375

Kari J (1992) The nilpotency problem of one-dimensional cellular automata. SIAM J Comput 21(3):571–586

Kari J (1994) Reversibility and surjectivity problems of cellular automata. J Comput Syst Sci 48(1): 149–182

Kari J, Ollinger N (2008) Periodicity and immortality in reversible computing. In: Ochmanski E, Tyszkiewicz J (eds) MFCS, Lecture notes in computer science, vol 5162, Springer, Berlin, pp 419–430

Kari J, Vanier P, Zeume T (2009) Bounds on non-surjective cellular automata. In: Královic R, Urzyczyn P (eds) MFCS, Lecture notes in computer Science, vol 5734, Springer, Berlin, pp 439–450

Kůrka P (1997) Languages, equicontinuity and attractors in cellular automata. Ergod Theor Dyn Syst 17:417–433

Lukkarila V (2010) Sensitivity and topological mixing are undecidable for reversible one-dimensional cellular automata. J Cell Autom 5:241–272

Maruoka A, Kimura M (1976) Condition for injectivity of global maps for tessellation automata. Inform Control 32(2):158–162. doi: 10.1016/S0019-9958 (76)90195-9, URL http://www.sciencedirect.com/ science/article/B7MFM-4DX4D7R-V/2/ 409f935b249fa44868bb569dffb5eb37

Moore EF (1962) Machine models of self-reproduction. In: Bellman RE (ed) Proceedings of symposia in applied mathematics XIV: "Mathematical problems in the biological sciences", AMS, pp 17–33

Myhill J (1963) The converse of Moore's Garden-of-Eden theorem. Proc Am Math Soc 14:685–686

von Neumann J (1966) Theory of self-reproducing automata. In: Burks AW (ed) University of Illinois Press, Urbana, London

Sablik M, Theyssier G (2008) Topological dynamics of 2D cellular automata. In: CiE '08: Proceedings of the 4th conference on computability in Europe, Springer, Berlin, Heidelberg, pp 523–532

Shereshevsky MA (1993) Expansiveness, entropy and polynomial growth for groups acting on subshifts by automorphisms. Indag Math 4(2):203–210

Sutner K (1991) De Bruijn graphs and linear cellular automata. Complex Syst 5:19–31

Wang H (1961) Proving theorems by pattern recognition - ii. Bell Syst Tech J 40:1–42

Wolfram S (ed) (1986) Theory and applications of cellular automata. World Scientific, Singapore

Wolfram S (2002) A new kind of science. Wolfram Media, Champaign, IL, USA

2 Cellular Automata Dynamical Systems

Alberto Dennunzio[1] · *Enrico Formenti*[2] · *Petr Kůrka*[3]
[1]Dipartimento di Informatica, Sistemistica e Comunicazione, Università degli Studi di Milano-Bicocca, Italy
dennunzio@disco.unimib.it
[2]Département d'Informatique, Université de Nice-Sophia Antipolis, France
enrico.formenti@unice.fr
[3]Center for Theoretical Studies, Academy of Sciences and Charles University in Prague, Czechia
kurka@cts.cuni.cz

G. Rozenberg et al. (eds.), *Handbook of Natural Computing*, DOI 10.1007/978-3-540-92910-9_2,
© Springer-Verlag Berlin Heidelberg 2012

Abstract

We present recent studies on cellular automata (CAs) viewed as discrete dynamical systems. In the first part, we illustrate the relations between two important notions: subshift attractors and signal subshifts, measure attractors and particle weight functions. The second part of the chapter considers some operations on the space of one-dimensional CA configurations, namely, shifting and lifting, showing that they conserve many dynamical properties while reducing complexity. The final part reports recent investigations on two-dimensional CA. In particular, we report a construction (slicing construction) that allows us to see a two-dimensional CA as a one-dimensional one and to lift some one-dimensional results to the two-dimensional case.

1   Introduction

Cellular automata (CAs) comprise a simple formal model for complex systems. They are used in many scientific fields (e.g., computer science, physics, mathematics, biology, chemistry, economics) with different purposes (e.g., simulation of natural phenomena, pseudo-random number generation, image processing, analysis of universal models of computation, and cryptography (Farina and Dennunzio 2009; Chaudhuri et al. 1997; Chopard 2012; Wolfram 1986)).

The huge variety of distinct dynamical behaviors has greatly determined the success of CA in applications. The study of CA as dynamical systems was first systematized by Hedlund in his famous paper (Hedlund 1969). Subsequent research introduced new details, unexpected phenomena, or links with aspects of dynamical systems theory. Presenting all these results would go beyond the purposes of this chapter, so the interested reader is directed to other surveys (Formenti and Kůrka 2009; Kůrka 2008; Cervelle 2008; Kari 2008; Pivato 2008; Di Lena and Margara 2008, 2009; Cattaneo et al. 2009; Di Lena 2006). This chapter aims to introduce basic results of CA viewed as dynamical systems and some recent developments.

2   Discrete Dynamical Systems

A **discrete dynamical system** (DDS) is a pair (X, F) where X is a set equipped with a metric d and $F: X \mapsto X$ is a map, which is continuous on X with respect to d. The nth **iteration** of F is denoted by F^n. If F is bijective (invertible), the negative iterations are defined by $F^{-n} = (F^{-1})^n$. The function F induces **deterministic dynamics** by its iterated application starting from a given element. Formally, the **orbit** of a point $x \in X$ is $\mathcal{O}_F(x) := \{F^n(x) : n > 0\}$. A point $x \in X$ is **periodic** with period $n > 0$, if $F^n(x) = x$, while it is **eventually periodic**, if $F^m(x)$ is periodic for some **preperiod** $m \geq 0$. A set $Y \subseteq X$ is **invariant**, if $F(Y) \subseteq Y$ and **strongly invariant** if $F(Y) = Y$.

A **homomorphism** $\varphi : (X, F) \to (Y, G)$ of DDS is a continuous map $\varphi : X \to Y$ such that $\varphi \circ F = G \circ \varphi$, that is, the following diagram commutes

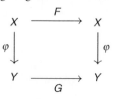

A **conjugacy** φ is a bijective homomorphism such that φ^{-1} is continuous. Two systems (X, F) and (Y, G) are topologically **conjugated**, if there exists a conjugacy between them. If a homomorphism φ is surjective, one can say that (Y, G) is a **factor** of (X, F), while if φ is injective, (X, F) is a **subsystem** of (Y, G). In that case $\varphi(X) \subseteq Y$ is a closed invariant set whenever X is compact. Conversely, if $Y \subseteq X$ is a closed invariant set, then (Y, F) is a subsystem of (X, F). The orbit $\mathbf{O}_F(x)$ of a point $x \in X$ is an invariant set, so its closure $(\overline{\mathbf{O}_F(x)}, F)$ is a subsystem of (X, F).

Denote by $B_\delta(x) = \{y \in X : d(y, x) < \delta\}$ the open ball centered in $x \in X$ and with radius $\delta > 0$.

A point $x \in X$ of a DDS (X, F) is **equicontinuous** (or Lyapunov stable), if

$$\forall \varepsilon > 0, \exists \delta > 0, \forall y \in B_\delta(x), \forall n \geq 0, d(F^n(y), F^n(x)) < \varepsilon$$

The existence of an equicontinuous point is a condition of local **stability** for the system. There are also notions of global stability based on the "size" of the set $\mathcal{E}_F$ of the equicontinuous points. A system is equicontinuous if

$$\forall \varepsilon > 0, \exists \delta > 0, \forall x, y \in X, (d(x, y) < \delta \Rightarrow \forall n \geq 0, d(F^n(x), F^n(y)) < \varepsilon)$$

If a system is equicontinuous then $\mathcal{E}_F = X$. The converse is also true in compact settings. A system is **almost equicontinuous** if $\mathcal{E}_F$ is a **residual** set, that is, it contains a countable intersection of dense open sets.

Conversely to stable behavior, sensitivity and positive expansivity are elements of instability for a system. Sensitivity to initial conditions recalls the well-known butterfly effects, in which small perturbations in the initial conditions can lead to strong differences in the evolution. More formally, a system is **sensitive** if

$$\exists \varepsilon > 0, \forall x \in X, \forall \delta > 0, \exists y \in B_\delta(x), \exists n \geq 0, d(F^n(y), F^n(x)) \geq \varepsilon$$

while it is **positively expansive** if

$$\exists \varepsilon > 0, \forall x, y \in X, x \neq y, \exists n \geq 0, d(F^n(y), F^n(x)) \geq \varepsilon$$

Note that in perfect spaces, positively expansive systems are sensitive.

Sensitivity and expansivity are often referred to as indicators of chaotic behavior. However, the most popular definitions of chaos also include other dynamical properties such as denseness of periodic orbits, transitivity, mixing, etc.

A DDS has the **DPO** property if its set of periodic points is dense. The set of **transitive points** is defined as $\mathcal{T}_F := \{x \in X : \overline{\mathbf{O}(x)} = X\}$. A system is **transitive** if for any nonempty open sets $U, V \subseteq X$ there exists $n > 0$ such that $F^n(U) \cap V \neq \emptyset$. In perfect spaces, if a system has a transitive point then it is transitive. On the other hand, in compact settings, any transitive system admits a transitive point and $\mathcal{T}_F$ is a residual set. A system (X, F) is **mixing** if for every nonempty open sets $U, V \subseteq X$, $F^n(U) \cap V \neq \emptyset$ for all sufficiently large n. A DDS is **strongly transitive** if for any nonempty open set U, $\bigcup_{n \in \mathbb{N}} F^n(U) = X$.

Using the notion of chain, the following weaker forms of transitivity and mixing can be introduced. A finite sequence $(x_i \in X)_{0 \leq i < n}$ is a $\boldsymbol{\delta}$-**chain** from x_0 to x_n if $d(F(x_i), x_{i+1}) < \delta$ for all $i < n$. A system is **chain-transitive** if for each x, y and for each $\varepsilon > 0$ there exists an ε-chain from x to y, and **chain-mixing**, if for any $x, y \in X$ and any $\varepsilon > 0$ there exist chains from x to y of arbitrary and sufficient length.

3 Symbolic Dynamical Systems

A **Cantor space** is any metric space that is **compact** (any sequence has a convergent subsequence), **totally disconnected** (distinct points are separated by disjoint **clopen**, that is, closed and open sets), and **perfect** (no point is isolated). Any two Cantor spaces are homeomorphic. A **symbolic space** is any compact, totally disconnected metric space, that is, any closed subspace of a Cantor space. A **symbolic dynamical system** (SDS) is any DDS (X, F) where X is a symbolic space.

The long-term behavior of a DDS can be conveniently described by the notions of limit set and attractor. Here, the definitions of these concepts adapted to SDS are given. Consider a SDS (X, F). The **limit set** of a clopen invariant set $V \subseteq X$ is $\Omega_F(V) := \bigcap_{n \geq 0} F^n(V)$. A set $Y \subseteq X$ is an **attractor**, if there exists a nonempty clopen invariant set V such that $Y = \Omega_F(V)$. There exists always the largest attractor $\Omega_F := \Omega_F(X)$. The number of attractors is at most countable. The union of two attractors is an attractor. If the intersection of two attractors is nonempty, it contains an attractor. The **basin** of an attractor $Y \subseteq X$ is the set $\mathscr{B}(Y) = \{x \in X; \lim_{n \to \infty} d(F^n(x), Y) = 0\}$. An attractor $Y \subseteq X$ is a **minimal attractor** if no proper subset of Y is an attractor. An attractor is a minimal attractor iff it is chain-transitive. A periodic point $x \in X$ is **attracting** if its orbit $\mathcal{O}(x)$ is an attractor. Any attracting periodic point is equicontinuous. A **quasi-attractor** is a nonempty set that is an intersection of a countable number of attractors.

3.1 Subshifts

For a finite alphabet A, denote by $A^* := \bigcup_{n \geq 0} A^n$ the set of words over A and by $A^+ := \bigcup_{n > 0} A^n$ the set of words of positive length. The length of a word $u = u_0, \ldots, u_{n-1} \in A^n$ is denoted by $|u| := n$ and the word of zero length is λ. One can say that $u \in A^*$ is a subword of $v \in A^*$ ($u \sqsubseteq v$) if there exists k such that $v_{k+i} = u_i$ for all $i < |u|$. One can denote by $u_{[i,j)} = u_i \ldots u_{j-1}$ and $u_{[i,j]} = u_i \ldots u_j$ subwords of u associated to intervals. One can denote by $A^{\mathbb{Z}}$ the space of A-**configurations**, or doubly infinite sequences of letters of A equipped with the metric

$$d(x, y) := 2^{-n}, \qquad \text{where } n = \min\{i \geq 0 : x_i \neq y_i \text{ or } x_{-i} \neq y_{-i}\}$$

The **shift map** $\sigma : A^{\mathbb{Z}} \to A^{\mathbb{Z}}$ is defined by $\sigma(x)_i := x_{i+1}$. For any $u \in A^+$ one has a σ-periodic configuration $u^\infty \in A^{\mathbb{Z}}$ defined by $(u^\infty)_i = u_{|i| \bmod |u|}$. A **subshift** is a nonempty subset $\Sigma \subseteq A^{\mathbb{Z}}$, which is closed and strongly σ-invariant, that is, $\sigma(\Sigma) = \Sigma$. For a given alphabet A define the bijective k-**block code** $\alpha_k : A^{\mathbb{Z}} \to (A^k)^{\mathbb{Z}}$ by $\alpha_k(x)_i = x_{[i,i+k)}$. For a subshift $\Sigma \subseteq A^{\mathbb{Z}}$ denote by $\Sigma^{[k]} = \alpha_k(\Sigma) \subseteq (A^k)^{\mathbb{Z}}$ its k-**block encoding**. Then $\alpha_k : \Sigma \to \Sigma^{[k]}$ is a conjugacy as it commutes with the shift maps $\alpha_k \sigma = \sigma \alpha_k$. For a subshift Σ there exists a set $D \subseteq A^*$ of forbidden words such that

$$\Sigma = \Sigma_D := \{x \in A^{\mathbb{Z}} : \forall u \sqsubseteq x, u \notin D\}$$

A subshift is uniquely determined by its **language**

$$\mathscr{L}(\Sigma) := \{u \in A^* : \exists x \in \Sigma, u \sqsubseteq x\}$$

We denote by $\mathscr{L}^n(\Sigma) := \mathscr{L}(\Sigma) \cap A^n$. The language of **first offenders** of Σ is

$$\mathscr{D}(\Sigma) := \{u \in A^+ \setminus \mathscr{L}(\Sigma) : u_{[0,|u|-1)}, u_{[1,|u|)} \in \mathscr{L}(\Sigma)\}$$

so $\Sigma = \Sigma_{\mathcal{D}(\Sigma)}$. The **extended language** of Σ is

$$\tilde{\mathcal{L}}(\Sigma) = \left\{ x_{|I} : x \in \Sigma, I \subseteq \mathbb{Z} \text{ is an interval} \right\}$$

A subshift $\Sigma \subseteq A^{\mathbb{Z}}$ is transitive, iff for any words $u, v \in \mathcal{L}(\Sigma)$ there exists $w \in A^*$ such that $uwv \in \mathcal{L}(\Sigma)$. A subshift $\Sigma \subseteq A^{\mathbb{Z}}$ is mixing if for any words $u, v \in \mathcal{L}(\Sigma)$ there exists $n > 0$ such that for all $m > n$ there exists $w \in A^m$ such that $uwv \in \mathcal{L}(\Sigma)$.

Definition 1 Given an integer $c \geq 0$, the c-**join** $\Sigma_0 \overset{c}{\vee} \Sigma_1$ of subshifts $\Sigma_0, \Sigma_1 \subseteq A^{\mathbb{Z}}$ consists of all configurations $x \in A^{\mathbb{Z}}$ such that either $x \in \Sigma_0 \cup \Sigma_1$, or there exist integers b, a such that $b - a \geq c$, $x_{(-\infty,b)} \in \tilde{\mathcal{L}}(\Sigma_0)$, and $x_{[a,\infty)} \in \tilde{\mathcal{L}}(\Sigma_1)$.

Proposition 1 (Formenti et al. 2010) *The c-join of two subshifts is a subshift and the operation of c-join is associative. A configuration $x \in A^{\mathbb{Z}}$ belongs to $\Sigma_1 \overset{c}{\vee} \cdots \overset{c}{\vee} \Sigma_n$ iff there exist integers $k > 0$, $1 \leq i_1 < i_2 < \cdots < i_k \leq n$, and intervals $I_1 = (a_1, b_1), I_2 = [a_2, b_2), \ldots, I_k = [a_k, b_k)$ such that $a_1 = -\infty$, $b_k = \infty$, $a_j < a_{j+1}$, $b_j < b_{j+1}$, $b_j - a_{j+1} \geq c$, and $x_{|I_j} \in \tilde{\mathcal{L}}(\Sigma_{i_j})$.*

3.2 Sofic Subshifts

A subshift $\Sigma \subseteq A^{\mathbb{Z}}$ is **sofic** if its language $\mathcal{L}(\Sigma)$ is regular. Sofic subshifts are usually described by finite automata or by labeled graphs.

A **labeled graph** over an alphabet A is a structure $G = (V, E, s, t, l)$, where V is a finite set of vertices, E is a finite set of edges, $s, t : E \to V$ are the **source** and **target maps**, and $l : E \to A$ is a **labeling function**. A finite or infinite word $w \in E^* \cup E^{\mathbb{Z}}$ is a **path** in G if $t(w_i) = s(w_{i+1})$ for all i. The source and target of a finite path $w \in E^n$ are $s(w) := s(w_0)$, $t(w) := t(w_{n-1})$. The **label** of a path is defined by $l(w)_i := l(w_i)$. A subshift Σ is sofic iff there exists a labeled graph G such that $\Sigma = \Sigma_G$ is the set of labels of all doubly infinite paths in G. In this case, it can be said that G is a **presentation** of Σ (see e.g., Lind and Marcus (1995) or Kitchens (1998)).

A subshift $\Sigma \subseteq A^{\mathbb{Z}}$ is **of finite type** (SFT) if $\Sigma = \Sigma_D$ for some finite set $D \subseteq A^+$ of forbidden words. The words of D can be assumed to be all of the same length, which is called the order $\mathfrak{o}(\Sigma)$ of Σ. A configuration $x \in A^{\mathbb{Z}}$ belongs to Σ iff $x_{[i,i+\mathfrak{o}(\Sigma))} \in \mathcal{L}(\Sigma)$ for all $i \in \mathbb{Z}$. Any SFT is sofic: if $p = \mathfrak{o}(\Sigma) - 1$, the **canonical graph** $G = (V, E, s, t, l)$ of Σ is given by $V = \mathcal{L}^p(\Sigma)$, $E = \mathcal{L}^{p+1}(\Sigma)$, $s(u) = u_{[0,p)}$, $t(u) = u_{[1,p]}$ and $l(u) = u_p$. If $\Sigma \subseteq A^{\mathbb{Z}}$ is a finite subshift, then it is of finite type and each its configuration is σ-periodic. The period $\mathfrak{p}(\Sigma)$ of Σ is then the smallest positive integer $\mathfrak{p}(\Sigma)$, such that $\sigma^{\mathfrak{p}(\Sigma)}(x) = x$ for all $x \in \Sigma$.

Example 1 The golden mean subshift is the SFT with binary alphabet $A = \{0, 1\}$ and forbidden word 11 ($\bullet$ *Fig. 1* left).

$\blacksquare$ **Fig. 1**
Golden (*left*) mean subshift and (*right*) even subshift.

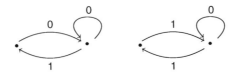

Example 2 The even subshift is the sofic subshift with binary alphabet and forbidden words $D = \{10^{2n+1}1 : n \geq 0\}$ (❷ *Fig. 1* right).

A labeled graph $G = (V, E, s, t, l)$ is **connected** if for any two vertices $q, q' \in V$ there exists a path $w \in E^*$ from q to q'. An **induced subgraph** of a graph G is a graph $G' = (V', E', s', t', l')$, such that $V' \subseteq V$, $E' = \{e \in E : s(e) \in V' \,\&\, t(e) \in V'\}$, and s', t', l' coincide respectively with s, t, l on E'. A **strongly connected component** of G is a subgraph of G, which is connected and maximal with this property. The subshift of a connected graph is transitive. Conversely, every transitive sofic subshift $\Sigma \subseteq A^{\mathbb{Z}}$ has a connected presentation. A labeled graph $G = (V, E, s, t, l)$ is **aperiodic** if the set of vertices cannot be partitioned into disjoint union $V = V_0 \cup \cdots \cup V_{p-1}$ such that $p \geq 2$ and if $s(e) \in V_i$, then $t(e) \in V_{(i+1) \bmod p}$. A sofic subshift is mixing iff it has a connected and aperiodic presentation.

For a labeled graph $G = (V, E, s, t, l)$ and $k > 1$ define the k-**block graph** $G^{[k]} := (V', E', s', t', l')$, where $V' \subseteq E^{k-1}$ is the set of paths of G of length $k - 1$, $E' \subseteq E^k$ is the set of paths of G of length k, s' and t' are the prefix and suffix maps and $l' : E' \to A^k$ is defined by $l'(u)_i = l(u_i)$. Then $G^{[k]}$ is a presentation of the k-block encoding $(\Sigma_G)^{[k]}$ of Σ_G. From $G^{[k]}$ one may in turn get a presentation of Σ_G, if l' is replaced by its composition with a projection $\pi_i : A^k \to A$ defined by $\pi(u)_i := u_i$, where $i < k$.

Given two sofic subshifts $\Sigma_0, \Sigma_1 \subseteq A^{\mathbb{Z}}$, their union and intersection (provided nonempty) are sofic subshifts. Moreover there exists an algorithm that constructs a presentation of $\Sigma_0 \cup \Sigma_1$ and $\Sigma_0 \cap \Sigma_1$ from those of Σ_0 and Σ_1. It is also decidable whether $\Sigma_0 \subseteq \Sigma_1$. Given a labeled graph G it is decidable whether Σ_G is an SFT (see Lind and Marcus 1995, p. 94). It is also decidable whether an SFT is mixing.

Proposition 2 *Let $\Sigma_0, \Sigma_1 \subseteq A^{\mathbb{Z}}$ be sofic subshifts and $c \geq 0$. Then $\Sigma_0 \overset{c}{\vee} \Sigma_1$ is a sofic subshift. There exists an algorithm that constructs a presentation G of $\Sigma_0 \overset{c}{\vee} \Sigma_1$ from the presentations G_i of Σ_i. Moreover, G has the same strongly connected components as the disjoint union $G_0 \cup G_1$.*

Proof Let $G_i = (V_i, E_i, s_i, t_i, l_i)$ be presentations of $\Sigma_i^{[c]}$, and assume that $V_0 \cap V_1 = \emptyset$ and $E_0 \cap E_1 = \emptyset$. Construct $G = (V, E, s, t, l)$, where $V = V_0 \cup V_1$

$$E = E_0 \cup E_1 \cup \{(e_0, e_1) \in E_0 \times E_1 : l_0(e_0) = l_1(e_1)\}$$

The source, target, and label maps extend s_i, t_i, l_i. For the new edges one has $s(e_0, e_1) = s_0(e_0)$, $t(e_0, e_1) = t_1(e_1)$, $l(e_0, e_1) = l_0(e_0) = l_1(e_1)$. Then $\Sigma_G = (\Sigma_0 \overset{c}{\vee} \Sigma_1)^{[c]}$, and one gets a presentation of $\Sigma_0 \overset{c}{\vee} \Sigma_1$, if one composes l with a projection $\pi_i : A^c \to A$. Clearly, the strongly connected components of $G_0 \cup G_1$ are not changed by the new edges.

Proposition 3 *The language $\mathscr{D}(\Sigma)$ of first offenders of a sofic subshift Σ is regular.*

4 One-Dimensional Cellular Automata (1D CAs)

In this section, the basic notions and results about 1D CA as SDS are reviewed. To introduce them, a more general definition is used whose relevance has been argued by Boyle and Kitchens (1999).

Definition 2 A cellular automaton is a pair (X, F), where $X \subseteq A^{\mathbb{Z}}$ is a mixing subshift of finite type, and $F : X \to X$ is a continuous mapping which commutes with the shift, i.e., $F\sigma = \sigma F$.

For a CA (X, F) there exists a **local rule** $f : \mathcal{L}^{a-m+1}(X) \to \mathcal{L}^1(X)$ such that $F(x)_i = f(x_{[i+m,i+a]})$. Here $m \le a$ are integers called **memory** and **anticipation**. The **diameter** of (X, F) is $d := a - m$. If $a = -m \ge 0$, then it can be said that a is the **radius** of (X, F). The local rule can be extended to a map $f : \mathcal{L}(X) \to \mathcal{L}(X)$ by $f(u)_i := f(u_{[i,i+d]})$ for $0 \le i < |u| - d$. Thus $|f(u)| = \max\{|u| - d, 0\}$. The **cylinder set** of a word $u \in \mathcal{L}(X)$ located at $l \in \mathbb{Z}$ is $[u]_l := \{x \in X : x_{[l,l+|u|)} = u\}$. The cylinder set of a finite set $U \subset A^*$ is $[U]_l = \bigcup_{u \in U}[u]_l$. We also write $[u] := [u]_0$ and $[U] := [U]_0$. Every clopen set of X is a cylinder set of a finite set $U \subset A^*$.

Proposition 4 (Formenti and Kůrka 2007) *Let (X, F) be a CA. If $\Sigma \subseteq X$ is a sofic subshift, then $F(\Sigma)$ and $F^{-1}(\Sigma)$ are sofic subshifts and there exists an algorithm that constructs their graphs from the local rule f, the graph of Σ and the set of forbidden words of X.*

A word $u \in \mathcal{L}(X)$ with $|u| \ge s \ge 0$ is s-**blocking** for a CA (X, F), if there exists an **offset** $k \in [0, |u| - s]$ such that

$$\forall x, y \in [u]_0, \forall n \ge 0, F^n(x)_{[k,k+s)} = F^n(y)_{[k,k+s)}$$

Theorem 1 (Kůrka 2008) *Let (X, F) be a CA with radius $r \ge 0$. The following conditions are equivalent.*

1. *(X, F) is not sensitive.*
2. *(X, F) has an r-blocking word.*
3. *$\mathcal{E}_F$ is **residual**, i.e., it contains a countable intersection of dense open sets.*
4. *$\mathcal{E}_F \ne \emptyset$.*

For a nonempty set $B \subseteq A^*$ define

$$\mathcal{T}_\sigma^n(B) := \{x \in A^{\mathbb{Z}} : (\exists j > i > n, x_{[i,j)} \in B) \,\&\, (\exists j < i < -n, x_{[j,i)} \in B)\}$$
$$\mathcal{T}_\sigma(B) := \bigcap_{n \ge 0} \mathcal{T}_\sigma^n(B)$$

Each $\mathcal{T}_\sigma^n(B)$ is open and dense, so the set $\mathcal{T}_\sigma(B)$ of B-**recurrent** configurations is residual. If B is the set of r-blocking words, then $\mathcal{E}_F \supseteq \mathcal{T}_\sigma(B)$.

Theorem 2 (Kůrka 2008) *Let (X, F) be a CA with radius $r \ge 0$. The following conditions are equivalent.*

1. *(X, F) is equicontinuous, i.e., $\mathcal{E}_F = X$.*
2. *There exists $k > 0$ such that any $u \in \mathcal{L}^k(X)$ is r-blocking.*
3. *There exists a preperiod $q \ge 0$ and a period $p > 0$, such that $F^{q+p} = F^q$.*

In particular, every CA with radius $r = 0$ is equicontinuous. A configuration is equicontinuous for F iff it is equicontinuous for F^n, that is, $\mathcal{E}_F = \mathcal{E}_{F^n}$.

Example 3 (a product rule ECA128) Consider the ECA given by the following local rule:
$F(x)_i = x_{i-1}x_ix_{i+1}$

000:0, 001:0, 010:0, 011:0, 100:0, 101:0, 110:0, 111:1

The ECA128 is almost equicontinuous and 0 is a 1-blocking word ($\bullet$ *Fig. 2*). It is not surjective since 101 has no preimage. The cylinder $[0]_0$ is a clopen invariant set whose omega-limit is the singleton $\Omega_F([0]_0) = \{0^\infty\}$ that contains stable fixed point 0∞. The maximal attractor is larger. We have $F(A^{\mathbb{Z}}) = \Sigma_{\{101,1001\}}$, $F^n(A^{\mathbb{Z}}) = \Sigma_{\{10^k1:\, 0<k\leq 2n\}}$ and

$$\Omega_F = \{u \in A^{\mathbb{Z}} :\ \forall n > 0,\, 10^n1 \not\sqsubseteq u\}$$

Example 4   (the majority rule ECA232) $F(x)_i = \left\lfloor \frac{x_{i-1}+x_i+x_{i+1}}{2} \right\rfloor$

000:0, 001:0, 010:0, 011:1, 100:0, 101:1, 110:1, 111:1

The majority rule has 2-blocking words 00 and 11, so it is almost equicontinuous ($\bullet$ *Fig. 3*). It is not surjective. The clopen sets $[00]_n$ and $[11]_n$ are invariant. More generally, let $E = \{u \in \mathbf{2}^* :\ |u| \geq 2,\, u_0 = u_1,\, u_{|u|-2} = u_{|u|-1},\, 010 \not\sqsubseteq u,\, 101 \not\sqsubseteq u\}$. Then for any $u \in E$ and for any $i \in \mathbb{Z}$, $[u]_i$ is a clopen invariant set, so its limit set $\Omega_F([u]_i)$ is an attractor. These attractors are not subshifts. There exists a subshift attractor $\Omega_F(U)$, where $U := \mathbf{2}^{\mathbb{Z}} \setminus ([010]_0 \cup [101]_0)$. The maximal attractor is

$$\Omega_F = \Sigma_{\{010^k1,10^k10,01^k01,101^k0,:\, k>1\}}$$

4.1   Combinatorial Properties for 1D CA

Some interesting dynamical properties can be conveniently expressed in a combinatorial manner allowing us to find decision algorithms. Such properties include surjectivity, injectivity and openness. A 1D CA is **surjective** (resp., **injective**, **open**) if its global function is surjective (resp., injective, open).

For the case of surjectivity, Hedlund gave the following characterization (Hedlund 1969), which will also be useful in what follows.

$\blacksquare$ **Fig. 2**

A space–time diagram for ECA128 (see $\bullet$ Example 3).

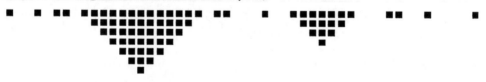

$\blacksquare$ **Fig. 3**

A space–time diagram for ECA232 (see $\bullet$ Example 4).

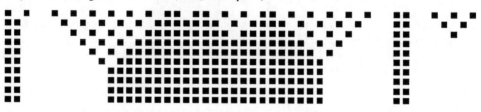

Theorem 3 *Let $(A^{\mathbb{Z}}, F)$ be a CA with local rule $f : A^{d+1} \to A$. The following conditions are equivalent.*

1. *$F : A^{\mathbb{Z}} \to A^{\mathbb{Z}}$ is surjective.*
2. *For each $x \in A^{\mathbb{Z}}$, $F^{-1}(x)$ is a finite set.*
3. *$f : A^* \to A^*$ is surjective.*
4. *For each $u \in A^+$, $|f^{-1}(u)| = |A|^d$.*

The above characterization (except for item (2)) has been extended to any dimension by Maruoka and Kimura (1976) via the notion of balance as shown in ❷ Sect. 5.2.

Closingness is another interesting combinatorial property. A CA $(A^{\mathbb{Z}}, F)$ is **right** (resp., **left) closing** iff $F(x) \neq F(y)$ for any pair $x, y \in A^{\mathbb{Z}}$ of distinct left (resp., right) asymptotic configurations, that is, $x_{(-\infty,n]} = y_{(-\infty,n]}$ (resp., $x_{[n,\infty)} = y_{[n,\infty)}$) for some $n \in \mathbb{Z}$, where $z_{(-\infty,n]}$ (resp., $z_{[n,\infty)}$) denotes the portion of a configuration z inside the infinite integer interval $(-\infty, n]$ (resp., $[n, \infty)$). A CA is said to be **closing** if it is either left closing or right closing.

In Boyle and Kitchens (1999), it is shown that a CA is open iff it is both left closing and right closing. Moreover, in the same paper, they prove that a closing CA has the joint denseness of periodic orbits (**JDPO**) property; that is, it has a dense set of periodic points, which are also periodic for the shift map. In what follows, it can be seen that this last property is closely related to the solution of a long-standing open conjecture.

Permutivity is a combinatorial property, which can be easily decided using the CA local rule. A 1D CA of radius r is permutive in the variable of index i ($|i| \leq r$) if its local rule f is injective with respect to the ith variable. A 1D CA is **leftmost permutive** (resp., rightmost permutive) if it is permutive in the variable of index $-r$ (resp., r).

Permutivity is closely related to chaotic behavior. Indeed, a 1D CA that is both leftmost and rightmost expansive is also positively expansive and open (Shereshevsky and Afraimovich 1992; Nasu 1995).

Moreover, the following two results will be also useful for the 2D CA case.

Proposition 5 (Cattaneo et al. 2004) *Let F be a 1D CA with local rule f on a possibly infinite alphabet A. If f is either rightmost or leftmost permutive, then F is topologically mixing.*

Proposition 6 (Dennunzio and Formenti 2008, 2009) *Let F be a 1D CA with local rule f on a possibly infinite alphabet A. If f is both rightmost and leftmost permutive, then F is strongly transitive.*

4.2 Attractors

The notion of attractor is essential to understand the long-term behavior of a DDS. The following results are a first characterization of attractors and quasi-attractors in CA.

Theorem 4 (Hurley 1990)

1. *If a CA has two disjoint attractors, then any attractor contains two disjoint attractors and an uncountably infinite number of quasi-attractors.*

2. *If a CA has a minimal (w.r.t. set inclusion) attractor, then it is a subshift, it is contained in any other attractor, and its basin of attraction is a dense open set.*
3. *If $x \in A^{\mathbb{Z}}$ is an attracting F-periodic configuration, then $\sigma(x) = x$ and $F(x) = x$.*

The ECA232 of ❷ Example 4 is of class (1) of ❷ Theorem 4. A trivial example of this class is the identity CA $(A^{\mathbb{Z}}, \mathrm{Id})$ in any alphabet. Each clopen set $U \subseteq A^{\mathbb{Z}}$ is invariant and its omega-limit set is itself: $\Omega_{\mathrm{Id}}(U) = U$. The ECA128 of ❷ Example 3 is of class (2) and satisfies also condition (3). A trivial example of class (2) is the zero CA in the binary alphabet $\{0, 1\}^{\mathbb{Z}}$ given by $F(x) = 0^{\infty}$.

Corollary 1 (Kůrka 2008) *For any CA, exactly one of the following statements holds.*

1. *There exist two disjoint attractors and a continuum of quasi-attractors.*
2. *There exists a unique quasi-attractor. It is a subshift and it is contained in any attractor.*
3. *There exists a unique minimal attractor contained in any other attractor.*

Both equicontinuity and surjectivity yield strong constraints on attractors.

Theorem 5 (Kůrka 2003b)

1. *A surjective CA has either a unique attractor or a pair of disjoint attractors.*
2. *An equicontinuous CA has either two disjoint attractors or a unique attractor that is an attracting fixed configuration.*
3. *If a CA has an attracting fixed configuration that is a unique attractor, then it is equicontinuous.*

4.3 Subshift Attractors

For non-surjective CA, attractors are often subshifts. In this section, this special type of attractor will be characterized. First, some definitions are needed.

Definition 3
1. A clopen F-invariant set $W \subseteq X$ is **spreading to the right** (or **left**) if $F^k(W) \subseteq \sigma^{-1}(W)$ (or $F^k(W) \subseteq \sigma(W)$) for some $k > 0$.
2. A clopen F-invariant set $W \subseteq X$ is **spreading** if it is spreading both to the right and to the left.

Proposition 7 (Formenti and Kůrka 2007) *Let (X, F) be a cellular automaton and $W \subseteq X$ a clopen F-invariant set. Then $\Omega_F(W)$ is a subshift attractor iff W is spreading.*

Theorem 6 (Formenti and Kůrka 2007) *Each subshift attractor is chain-mixing for the shift, contains a configuration which is both F-periodic and σ-periodic, and the complement of its language is recursively enumerable.*

It follows that every attractor that is a subshift of finite type is mixing.

Theorem 7 (Formenti and Kůrka 2007) *The only subshift attractor of a surjective CA* $(A^{\mathbb{Z}}, F)$ *is the full space* $A^{\mathbb{Z}}$.

Theorem 8 (Formenti and Kůrka 2007) *If* $\Sigma \subseteq A^{\mathbb{Z}}$ *is a mixing subshift of finite type, then there exists a cellular automaton* $(A^{\mathbb{Z}}, F)$ *such that* Σ *is an attractor of* $(A^{\mathbb{Z}}, F)$ *and* $F(x) = x$ *for every* $x \in \Sigma$.

Proposition 8 (Formenti and Kůrka 2007) *Let* G *be a labelled graph with two strongly connected components* G_0 *and* G_1 *such that* $\Sigma_G \neq \Sigma_{G_0} = \Sigma_{G_1} = \{\sigma^i(u^\infty) : i < |u|\}$ *for some* $u \in A^+$. *Then* Σ_G *is not a subshift attractor.*

The sofic subshifts in ❷ *Fig. 4* are not transitive. The left subshift is chain-transitive but not chain-mixing, so it is not subshift attractor by ❷ Theorem 6. The right subshift is chain-mixing, but it is not subshift attractor by ❷ Proposition 8. The union of two attractors is an attractor. The intersection of two attractors need not be an attractor, but if it is nonempty, it contains an attractor. The intersection of two subshift attractors is always nonempty but need not be an attractor either (see ❷ Example 5).

Example 5 (Kůrka 2007) The intersection of two subshift attractors need not to be an attractor.

Proof Take the alphabet $A = \{0, 1, 2, 3\}$ and local rule $f : A^3 \to A$ given by

$$x1v{:}0,\ z2x{:}1,\ x2z{:}1,\ y3x{:}1,\ x3y{:}1$$

where $x, v \in A$, $y \in \{0, 1, 2\}$, $z \in \{0, 1, 3\}$ and the first applicable production is used, otherwise the letter is left unchanged (see simulation in ❷ *Fig. 5*). Then $U_2 := [0] \cup [1] \cup [2]$, $U_3 := [0] \cup [1] \cup [3]$ are spreading sets whose attractors $\Sigma_2 := \Omega_F(U_2)$, $\Sigma_3 := \Omega_F(U_3)$ are sofic subshifts whose labeled graphs are in ❷ *Fig. 5*. The subshifts consist of all labels of all (doubly infinite) paths in their corresponding graphs. Then $\Sigma_0 := \Sigma_2 \wedge \Sigma_3 = \{0^\infty\}$ and $0^\infty 10^\infty \in (\Sigma_2 \cap \Sigma_3) \setminus \Sigma_0$. There is one more subshift attractor $\Omega_F = \Sigma_2 \cup \Sigma_3$.

Example 6 (Kůrka 2007) There exists a CA with an infinite sequence of subshift attractors $\Sigma_{n+1} \subset \Sigma_n$ indexed by positive integers.

A CA is first constructed with an infinite number of clopen invariant sets $U_{n+1} \subset U_n$ that spread to the right. One has the alphabet $A = \{0, 1, 2, 3\}$ and a CA $F(x)_i = f(x_{[i-1,i+1]})$, where $f : A^3 \to A$ is the local rule given by

◻ **Fig. 4**

A chain-transitive and a chain-mixing subshifts.

▣ Fig. 5

Intersection of attractors.

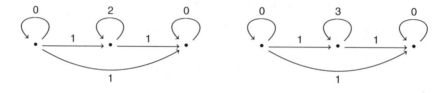

```
00022222211111333332222232323233333000
0001222210000013331122211111111133331000
00001221000000013100121000000013310000
0000011000000000100001000000000001100000
000000000000000000000000000000000000000000
```

▣ Fig. 6

Decreasing sequence of subshift attractors.

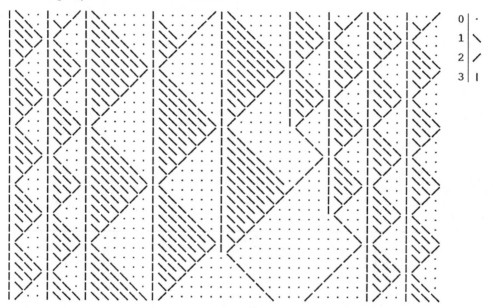

x33:0, 132:3, x32:0, xy2:2, x13:2

3xy:1, x2y:0, 10x:1, 11x:1, x1y:0

Here $x, y \in A$ and the first applicable production is used' otherwise the letter is left unchanged (see a simulation in ❷ *Fig. 6*). Letter 3 is a stationary particle, which generates right-going particles 1 via the production 3xx:1. When a 1 particle reaches another 3, it changes to a left-going particle 2 via the production x13:2 and erases all 1 particles which it encounters. When this 2 particle reaches a 3 particle simultaneously with a 1 particle, the 3 particle is preserved by the production 132:3. Otherwise 3 is destroyed by the production x32:0. Since the 3 particles are never created, they successively disappear unless they are distributed periodically. Set

$$U_n = \{x \in A^{\mathbb{Z}} : (0 \le i < j < 2n - 2) \ \& \ (x_i = x_j = 3) \Longrightarrow j - i \ge n\}$$

Each U_n is clopen and invariant, since the letter 3 is never created. We have $F^{4n-7}(U_n) \subseteq \sigma^{-1}(U_n)$ so U_n is spreading to the right. Define the mirror image CA G of F with local rule

$$33x{:}0, \ 231{:}3, \ 23x{:}0, \ 2xx{:}2, \ 31x{:}2$$
$$xx3{:}1, \ x2x{:}0, \ x01{:}1, \ x11{:}1, \ x1x{:}0$$

so that U_n is spreading to the left for G. Consider now the alphabet $B = \{(a, b) \in A^2 : a = 3 \Longleftrightarrow b = 3\}$ with 10 letters, and a CA on B defined by

$$H(x, y)_i = \begin{cases} (0, 0) & \text{if } x_i = y_i = 3 \text{ and} \\ & F(x)_i \ne 3 \text{ or } G(y)_i \ne 3 \\ (F(x)_i, G(y)_i) & \text{otherwise} \end{cases}$$

Thus the components F and G of H share the same particles 3 both F and G can destroy them. Then $V_n := B^{\mathbb{Z}} \cap (U_n \times U_n)$ is a spreading set for H and $\Sigma_n := \Omega_H(V_n)$ is a subshift attractor. Since $V_{n+1} \subset V_n$ we have $\Sigma_{n+1} \subseteq \Sigma_n$. For $n > 2$ there exists a σ-periodic configuration $x = (31^{n-1}320^{n-2})^{\infty} \in U_n \setminus U_{n+1}$ that is periodic for F, so $x \in \Omega_F(U_n) \setminus \Omega_F(U_{n+1})$. The mirror image configuration $y = (30^{n-2}231^{n-1})^{\infty}$ is periodic for G, so $(x, y) \in \Sigma_n \setminus \Sigma_{n+1}$. For each configuration $(x, y) \in \Sigma_n \setminus \Sigma_{n+1}$ we have $H^{2n}(x, y) = (x, y)$.

4.4 The Small Quasi-attractor

When a CA F is initialized on a random configuration x and if each word has positive probability, then this configuration contains with probability one each spreading set, not only of F, but also of all $F^q \sigma^p$. Therefore, the iterations of such a configuration converge to the intersection of all attractors of all $F^q \sigma^p$.

Definition 4 Let (Σ, F) be a cellular automaton. The intersection of all subshift attractors of all $F^q \sigma^p$, where $q > 0$ and $p \in \mathbb{Z}$, is called the **small quasi-attractor** of F and denoted by $\mathcal{Q}_F$.

Proposition 9 (Formenti and Kůrka 2007) *Let (Σ, F) be a cellular automaton. If $\Sigma_1, \ldots, \Sigma_k \subseteq \Sigma$ are subshifts such that each Σ_i is an attractor of some $F^{q_i} \sigma^{p_i}$, where $q_i > 0$ and $p_i \in \mathbb{Z}$, then the intersection $\Sigma_1 \cap \cdots \cap \Sigma_k$ is nonempty.*

Theorem 9 (Formenti and Kůrka 2007) *If (Σ, F) is a cellular automaton, then $\mathcal{Q}_F$ is a nonempty F-invariant subshift and $F : \mathcal{Q}_F \to \mathcal{Q}_F$ is surjective. Moreover, $(\mathcal{Q}_F, \sigma)$ is chain-mixing.*

The small quasi-attractor of ECA128 is $\mathcal{Q}_F = \{0^{\infty}\}$. The small quasi-attractor of ECA232 is
$$\mathcal{Q}_F = \Omega_F = \Sigma_{\{010^k 1, 10^k 10, 01^k 01, 101^k 0,: \ k>1\}}$$

Example 7 (A product rule ECA136) Consider the ECA given by the following local rule $F(x)_i = x_i x_{i+1}$.

The ECA136 is almost equicontinuous since 0 is a 1-blocking word (❷ *Fig.* 7). As in ❷ Example 3, we have $\Omega_F = \Sigma_{\{10^k 1: \ k>0\}}$. For any $m \in \mathbb{Z}$, $[0]_m$ is a clopen invariant

◼ **Fig. 7**
Example of space–time diagram for ECA136 (see ❷ Example 7).

set, which is spreading to the left but not to the right. Thus $Y_m = \Omega_F([0]_m) = \{x \in \Omega_F : \forall i \leq m, x_i = 0\}$ is an attractor but not a subshift. We have $Y_{m+1} \subset Y_m$ and $\bigcap_{m \geq 0} Y_m = \{0^\infty\}$ is the unique minimal quasi-attractor. Since $F^2 \sigma^{-1}(x)_i = x_{i-1} x_i x_{i+1}$ is the ECA128, which has a minimal subshift attractor $\{0^\infty\}$, F has the small quasi-attractor $\mathcal{Q}_F = \{0^\infty\}$.

4.5 Signal Subshifts

Definition 5 Let (X, F) be a cellular automaton. A configuration $x \in X$ is **weakly periodic** if $F^q \sigma^p(x) = x$ for some $q > 0$ and $p \in \mathbb{Z}$. We call (p, q) the **period** of x and p/q its speed. Let $\mathcal{S}_{(p,q)}(X, F) := \{x \in X : F^q \sigma^p(x) = x\}$ be the set of all weakly periodic configurations with period (p, q). A **signal subshift** is any non-empty $\mathcal{S}_{(p,q)}(X, F)$.

For fixed (X, F) we write $\mathcal{S}_{(p,q)} := \mathcal{S}_{(p,q)}(X, F)$. Note that $\mathcal{S}_{(p,q)}$ is closed and σ-invariant, so it is a subshift provided it is nonempty. Moreover, $\mathcal{S}_{(p,q)}$ is F-invariant and $F : \mathcal{S}_{(p,q)} \to \mathcal{S}_{(p,q)}$ is bijective, so $\mathcal{S}_{(p,q)} \subseteq \Omega_F(X)$. If $\mathcal{S}_{(p,q)}$ is finite, it consists only of σ-periodic configurations.
The ECA128 of ❷ Example 3 has nontransitive signal subshifts

$$\mathcal{S}_{(1,1)} = \{\sigma^n(0^\infty.1^\infty) : n \in \mathbb{Z}\} \cup \{0^\infty, 1^\infty\}$$
$$\mathcal{S}_{(-1,1)} = \{\sigma^n(1^\infty.0^\infty) : n \in \mathbb{Z}\} \cup \{0^\infty, 1^\infty\}$$

The ECA232 of ❷ Example 4 has transitive signal subshift $\mathcal{S}_{(0,1)} = \Sigma_{\{010,101\}}$.

Example 8 (The traffic rule ECA184, ❷ Fig. 8) $F(x)_i = 1$ if $x_{[i-1,i]} = 10$ or $x_{[i,i+1]} = 11$.

$$000{:}0\ 001{:}0\ 010{:}0\ 011{:}1\ 100{:}1\ 101{:}1\ 110{:}0\ 111{:}0$$

The ECA184 has three infinite signal subshifts

$$\mathcal{S}_{(1,1)}(F) = \Sigma_{\{11\}} \cup \{1^\infty\}, \quad \mathcal{S}_{(0,1)}(F) = \Sigma_{\{10\}}, \quad \mathcal{S}_{(-1,1)}(F) = \Sigma_{\{00\}} \cup \{0^\infty\}$$

and a unique F-transitive attractor $\Omega_F = \Sigma_{\{1(10)^n 0:\ n>0\}}$, which is sofic.

Theorem 10 *Let (X, F) be a cellular automaton with memory m and anticipation a, and let $d = a - m$ be the diameter.*

1. *If $\mathcal{S}_{(p,q)}$ is nonempty, then it is a subshift of finite type.*
2. *If $\mathcal{S}_{(p,q)}$ is infinite, then $\mathrm{o}(\mathcal{S}_{(p,q)}) \leq \max\{\mathrm{o}(X), dq + 1\}$ and $-a \leq p/q \leq -m$.*
3. *If $p_0/q_0 < p_1/q_1$, then $\mathcal{S}_{(p_0,q_0)} \cap \mathcal{S}_{(p_1,q_1)}$ is finite and $\mathrm{p}(\mathcal{S}_{(p_0,q_0)} \cap \mathcal{S}_{(p_1,q_1)})$ divides $\left(\frac{p_1}{q_1} - \frac{p_0}{q_0}\right)q$, where $q := \mathrm{lcm}(q_0, q_1)$ is the least common multiple.*

◻ Fig. 8
Example of space–time diagram for the traffic rule ECA184 (see ❯ Example 8).

Example 9 (Kůrka 2005) There exists a cellular automaton with infinitely many signal subshifts.

Consider a CA with alphabet $A = \{\text{b}, \text{s}, \text{r}, \text{l}\}$, where b is blank, s is a stationary particle, r is a right-going particle, and l is a left-going particle. When either of these two particles meets s, it shifts it to the right and changes to the other moving particle. If an r collides with an l, one of them annihilates. The transition table $f: A^4 \to A$ is given by

xxbl	xblx	xxry	xryx	xxsl	xslx	slxx	xxrs	xrsx	rsxx
l	b	b	r	b	s	r	b	l	s

where $x \in A$, $y \in \{\text{b}, \text{r}, \text{l}\}$, and the first applicable rule is used. We have $m = 2$ and $d = 3$, so $F(x)_i = f(x_{[i-2,i+1]})$. Then, for every $n \geq 0$, $\mathbf{S}_{(1,2n+3)}$ is a signal subshift. In ❯ *Fig. 9* one can see signals with speeds $1/3$, $1/5$, and $1/7$.

Theorem 11 (Formenti and Kůrka 2007) *Let (X, F) be a cellular automaton with diameter d and assume that $\mathbf{S}_{(p_1,q_1)}, \ldots, \mathbf{S}_{(p_n,q_n)}$ are signal subshifts with decreasing speeds, i.e., $p_i/q_i > p_j/q_j$ for $i < j$. Let $c > 0$ be an integer which satisfies the following inequalities:*

$$\mathfrak{o}(X) \leq c$$
$$\forall i \leq n, \mathfrak{o}(\mathbf{S}_{(p_i,q_i)}) \leq c$$
$$\forall i < j \leq n, (\mathbf{S}_{(p_i,q_i)} \cap \mathbf{S}_{(p_j,q_j)} \neq \emptyset \Longrightarrow \mathfrak{p}(\mathbf{S}_{(p_i,q_i)} \cap \mathbf{S}_{(p_j,q_j)}) \leq c)$$
$$\forall i < j \leq n, \left(\mathbf{S}_{(p_i,q_i)} \cap \mathbf{S}_{(p_j,q_j)} \neq \emptyset \Longrightarrow \left(d - \frac{p_i}{q_i} + \frac{p_j}{q_j}\right)q \leq c\right)$$

where $q = \text{lcm}\{q_1, \ldots, q_n\}$. Then for $\Sigma := \mathbf{S}_{(p_1,q_1)} \overset{c}{\vee} \cdots \overset{c}{\vee} \mathbf{S}_{(p_n,q_n)}$ the following assertions hold:

1. $\Sigma \subseteq F^q(\Sigma)$ *and therefore $\Sigma \subseteq \Omega_F(X)$.*
2. *If G is the graph of Σ constructed by ❯ Proposition 2 from graphs of $\mathbf{S}_{(p_i,q_i)}$, and if H is a strongly connected component of G, then Σ_H is F^q-invariant.*

⬛ **Fig. 9**
Infinitely many signal subshifts.

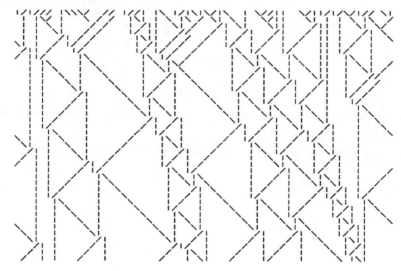

3. If G_k is the graph of $F^{kq}(\Sigma)$ constructed to ❷ Proposition 4 from G, then for each strongly connected component H_k of G_k there exists a strongly connected component H of G such that $\Sigma_{H_k} = \Sigma_H$.

Corollary 2 Let (X, F) be a CA. If $\Omega_F(X) = F^k(\mathbf{S}_{(p_1,q_1)} \overset{c}{\vee} \cdots \overset{c}{\vee} \mathbf{S}_{(p_n,q_n)})$ for some $(p_i, q_i) \in \mathbb{Z} \times \mathbb{N}^+$ and $k,c \geq 0$, then (X, F) has only a finite number of infinite transitive signal subshifts.

4.6   Decreasing Preimages

Notation
Given a language $L \subseteq A^*$ and $a, b \in \mathbb{N}$, $a < b$, define the following language:

$$L_{[a,b)} = \{v \in A^* : |v| \geq a + b, v_{[a,|v|-b)} \in L\}$$

Definition 6   Let $f : \mathscr{L}(X) \to \mathscr{L}(X)$ be a local rule of a cellular automaton (X, F). We say that a subshift $\Sigma \subseteq X$ has r-**decreasing preimages** if for each $u \in \mathscr{L}(X) \cap \mathscr{D}(\Sigma)$, each $v \in f^{-r}(u)$ contains as a subword a word $w \in \mathscr{D}(\Sigma)$ such that $|w| < |u|$. We say that Σ has **decreasing preimages** if it has r-decreasing preimages for some $r > 0$.

The condition of ❷ Definition 6 is satisfied trivially if $f^{-r}(u) = \emptyset$. Thus, for example, each $F^r(X)$ has r-decreasing preimages.

Theorem 12   (Formenti et al. 2010) If (X, F) is a CA and if a subshift $\Sigma \subseteq X$ has decreasing preimages, then $\Omega_F(X) \subseteq \Sigma$. If moreover $\Sigma \subseteq F^k(\Sigma)$ for some $k > 0$, then $\Omega_F(X) = \Sigma$.

Proposition 10 (Formenti and Kůrka 2007) *There exists an algorithm that decides whether for a given cellular automaton (X, F) and given $r > 0$, a given sofic subshift $\Sigma \subseteq X$ has r-decreasing preimages.*

The product CA (see ❯ Example 3) has two infinite nontransitive signal subshifts (see ❯ Fig. 10) $\mathbf{S}_{(1,1)} = \Sigma_{10}$, $\mathbf{S}_{(-1,1)} = \Sigma_{01}$, whose orders are $\mathfrak{o}(\mathbf{S}_{(1,1)}) = \mathfrak{o}(\mathbf{S}_{(-1,1)}) = 2$. Their intersection is $\{0^{\infty},\ 1^{\infty}\}$, so, by ❯ Theorem 11, $\mathfrak{p}(\mathbf{S}_{(1,1)}) \cap \mathbf{S}_{(-1,1)}) = 1$. Finally $\left(d - \frac{p_1}{q_1} + \frac{p_2}{q_2}\right)q = 0$, so, by ❯ Theorem 11, $c = 2$. The maximal attractor is constructed in ❯ Fig. 10. In the first row, labeled graphs for 2-block encodings of $\mathbf{S}_{(1,1)}$ and $\mathbf{S}_{(-1,1)}$ are given. Their join $\Sigma := \mathbf{S}_{(1,1)} \overset{2}{\vee} \mathbf{S}_{(-1,1)} = \Sigma_{\{10^n 1:\ n>0\} \cup \{010\}}$ is constructed in the second row. In the third row, there is the minimal presentation of (2-block encoding of) $F(\Sigma) = \Sigma_{\{10^n 1:\ n>0\}}$. We have $F^2(\Sigma) = F(\Sigma)$ and $F(\Sigma)$ has 1-decreasing preimages: If $u \in \mathcal{D}(F(\Sigma))$, then $u = 10^n 1$ for some $n > 0$. If $n \leq 2$, then u has no preimage. If $n > 2$, then the only preimage $1110^{n-2}111$ of u contains a shorter forbidden word $10^{n-2}1$. Thus $f^{-1}(\mathcal{D}(F(\Sigma)) \cap \mathcal{L}(F(\Sigma))_{[2,1)} = \emptyset$ as well as $f^{-1}(\mathcal{D}(F(\Sigma)) \cap \mathcal{L}(F(\Sigma))_{[1,2)} = \emptyset$. Thus $F(\Sigma)$ has 1-decreasing preimages and therefore $F(\Sigma) = \Omega_F(A^{\mathbb{Z}})$.

The majority CA (see ❯ Example 4) has a spreading set $U := A^3 \setminus \{010, 101\}$, whose subshift attractor is the signal subshift $\Omega_F(U) = \mathbf{S}_{(0,1)} = \Sigma_{\{010,101\}}$. There are two other infinite signal subshifts $\mathbf{S}_{(1,1)} = \Sigma_{\{011,100\}}$, $\mathbf{S}_{(-1,1)} = \Sigma_{\{001,110\}}$. The join Σ of all three signal subshifts is constructed in ❯ Fig. 11. We have $\mathfrak{o}(\mathbf{S}_{(1,1)}) = \mathfrak{o}(\mathbf{S}_{(0,1)}) = \mathfrak{o}(\mathbf{S}_{(-1,1)}) = 3$, $\mathbf{S}_{(1,1)} \cap \mathbf{S}_{(-1,1)} = \{0^{\infty}, 1^{\infty}, (01)^{\infty}, (10)^{\infty}\}$, $\mathbf{S}_{(1,1)} \cap \mathbf{S}_{(0,1)} = \mathbf{S}_{(0,1)} \cap \mathbf{S}_{(-1,1)} = \{0^{\infty}, 1^{\infty}\}$, and $\left(d - \frac{p_i}{q_i} + \frac{p_j}{q_j}\right)q \leq 2$, so $c = 3$. The arrows from $\mathbf{S}_{(1,1)}$ to $\mathbf{S}_{(0,1)}$ and from $\mathbf{S}_{(0,1)}$

◻ **Fig. 10**
The signal subshifts of ECA128.

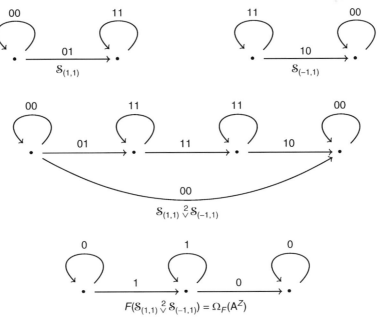

◘ **Fig. 11**

The join $S_{(0,1)} \overset{3}{\vee} S_{(1,1)} \overset{3}{\vee} S_{(-1,1)}$ **of the majority ECA232 (top) and the maximal attractor** $\Omega_F = S_{(0,1)} \cup (S_{(1,1)} \overset{3}{\vee} S_{(-1,1)})$ **(bottom).**

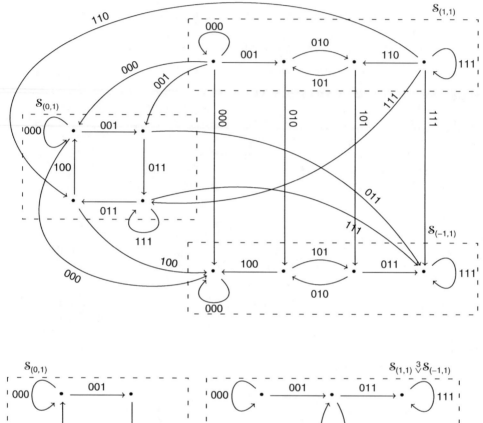

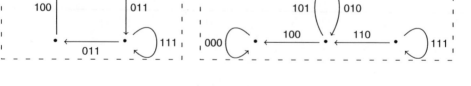

to $S_{(-1,1)}$ can be omitted without changing the subshift. Thus we get $\Sigma := S_{(1,1)} \overset{3}{\vee} S_{(0,1)} \overset{3}{\vee} S_{(-1,1)} = S_{(0,1)} \cup (S_{(1,1)} \overset{3}{\vee} S_{(-1,1)})$ (see ❷ *Fig. 3*). First offenders are $\mathscr{D}(\Sigma) = \{010^n1, 10^n10, 01^n01, 101^n0 : n > 1\}$. We have $F(\Sigma) = \Sigma$ and Σ has 1-decreasing preimages. To show it, note that 010, 101 have unique preimages $f^{-1}(010) = \{01010\}$, $f^{-1}(101) = \{10101\}$. For the first offenders we get $f^{-1}(01001) = \emptyset$, $f^{-1}(010001) = \{01010011\}$. If $n \geq 4$, then each preimage of 010^n1 has one of the form $01010u0011$ or $01010u0101$, where $|u| = n - 4$ and in each case it contains a shorter forbidden word. We get the same result for other forbidden words. Thus $f^{-1}(\mathscr{D}(\Sigma)) \cap \mathscr{L}(\Sigma)_{[2,1)} \cap \mathscr{L}(\Sigma)_{[1,2)} = \emptyset$, Σ has 1-decreasing preimages, and $\Omega_F(A^{\mathbb{Z}}) = \Sigma$.

◻ **Fig. 12**
Signal subshifts of ECA184.

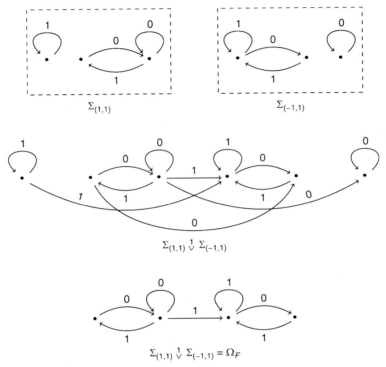

$$\Sigma_{(1,1)}$$

$$\Sigma_{(-1,1)}$$

$$\Sigma_{(1,1)} \overset{1}{\vee} \Sigma_{(-1,1)}$$

$$\Sigma_{(1,1)} \overset{1}{\vee} \Sigma_{(-1,1)} = \Omega_F$$

The traffic rule of ❯ Example 8 has two signal subshifts (see ❯ Fig. 12) representing holes and jams

$$\mathcal{S}_{(1,1)} = \{x \in A^{\mathbb{Z}} : 11 \nsubseteq x\} \cup \{1^{\mathbb{Z}}\}$$
$$\mathcal{S}_{(-1,1)} = \{x \in A^{\mathbb{Z}} : 00 \nsubseteq x\} \cup \{0^{\mathbb{Z}}\}$$

Their intersection is the finite subshift $\{0^{\mathbb{Z}}, 1^{\mathbb{Z}}, (01)^{\mathbb{Z}}, (10)^{\mathbb{Z}}\}$. There is one more (nontransitive) signal subshift $\Sigma_{(0,1)} = \{x \in A^{\mathbb{Z}} : \forall i < j, x_i \leq x_j\}$. The maximal attractor is constructed in ❯ Fig. 12. In the first row, 1-distinguishing presentations for $\Sigma_{(1,1)}$ and $\Sigma_{(-1,1)}$ are constructed. Their join is constructed in the second row. In the third row, the minimal presentation of $\Sigma_{(1,1)} \overset{1}{\vee} \Sigma_{(-1,1)}$ is given. As it has decreasing preimages, it equals Ω_F.

4.7   Measures

Given a metric space X, the family $\mathcal{B}(X) \subseteq \mathcal{P}(X)$ of Borel sets is the smallest family of sets, which contains open sets and is closed with respect to the operations of countable union and complementation. A **Borel probability measure** on a metric space X is a countably additive function $\mu : \mathscr{B}(X) \to [0, 1]$ that assigns 1 to the full space. The (topological) **support** of a measure μ is

$$\Sigma_\mu := \bigcap \{V \subseteq X : \mu(V) = 1 \text{ and } V \text{ is closed}\}$$

If $\Sigma_\mu = X$, it can be said that μ has **full support**. A measure on $A^{\mathbb{Z}}$ is determined by its values on centered cylinders $\mu(u) := \mu([u]_{-n})$ for $u \in A^{2n} \cup A^{2n+1}$. The map $\mu : A^* \to [0, 1]$ satisfies the **Kolmogorov compatibility conditions**

$$\mu(\lambda) = 1, \quad \sum_{a \in A} \mu(ua) = \mu(u) \text{ for } |u| \text{ even}, \quad \sum_{a \in A} \mu(au) = \mu(u) \text{ for } |u| \text{ odd}$$

Conversely, every function $\mu : A^* \to [0, 1]$ satisfying the Kolmogorov compatibility conditions determines a measure. Each point $x \in A^{\mathbb{Z}}$ determines the Dirac point measure δ_x defined by

$$\delta_x(U) = \begin{cases} 1 & \text{if } x \in U \\ 0 & \text{if } x \notin U \end{cases}$$

For each positive probability vector $p = (p_a)_{a \in A}$, we have the Bernoulli measure defined by

$$\mu(u) = p_{u_0} \cdots p_{u_{n-1}}, \quad u \in A^n$$

A stochastic matrix $R = (R_{ab})_{a,b \in A}$ is a matrix with nonnegative entries such that the sum of each row is one: $\sum_{b \in A} R_{a,b} = 1$. A stochastic matrix R together with a probability vector p determines a Markov measure

$$\mu(u) = p_{u_0} \cdot R_{u_0, u_1} \cdots R_{u_{n-2}, u_{n-1}}, \quad u \in A^n$$

Given two measures μ_0, μ_1, and positive real numbers a_0, a_1 with sum $a_0 + a_1 = 1$, then the convex combination $\mu(U) := a_0 \mu_0(U) + a_1 \mu_1(U)$ is a measure. If μ_0 and μ_1 are shift-invariant, then so is μ.

Denote by $\mathfrak{M}(A^{\mathbb{Z}})$ the set of all Borel probability measures on $A^{\mathbb{Z}}$. On $\mathfrak{M}(A^{\mathbb{Z}})$ we have the weak* topology. A sequence of measures μ_n converges to a measure μ if

$$\lim_{n \to \infty} \int f \, d\mu_n = \int f \, d\mu$$

for every continuous function $f : A^{\mathbb{Z}} \to \mathbb{R}$. The space of Borel probability measures with the weak* topology is compact and metrizable. A compatible metric is

$$\rho(\mu, v) := \sum_{n=0}^{\infty} \max\{|\mu(u) - v(u)| : u \in A^n\} \cdot 2^{-n}$$

Given two configurations $x, y \in A^{\mathbb{Z}}$ we have

$$d(x, y) = 2^{-n} \ \& \ x_{-n} \neq y_{-n} \implies \rho(\delta_x, \delta_y) = 2^{-2n+1}$$
$$d(x, y) = 2^{-n} \ \& \ x_{-n} = y_{-n} \implies \rho(\delta_x, \delta_y) = 2^{-2n}$$

Thus the map $\delta : (A^{\mathbb{Z}}, d) \to (\mathfrak{M}(A^{\mathbb{Z}}), \rho)$ is continuous. Given two probability vectors p, q, denote by $d := \max_{a \in A} |p_a - q_a|$, Then the distance of the corresponding Bernoulli measures is

$$\rho(\mu, v) = \sum_{n=1}^{\infty} d^n 2^{-n} = \frac{d}{2-d}$$

If $F : A^{\mathbb{Z}} \to A^{\mathbb{Z}}$ is a continuous map and $\mu \in \mathfrak{M}(A^{\mathbb{Z}})$, then $F\mu$ is a Borel probability measure defined by $(F\mu)(U) = \mu(F^{-1}(U))$ for any Borel set U. A measure μ is F-**invariant**, if $F\mu = \mu$. The map $\mu \mapsto F\mu$ is continuous and **affine**, that is, it preserves convex combinations. Thus, we obtain a dynamical system $F : \mathfrak{M}(A^{\mathbb{Z}}) \to \mathfrak{M}(A^{\mathbb{Z}})$.

If a measure μ is σ-invariant, then the measure of a cylinder does not depend on its location, so $\mu(u) = \mu([u]_i)$ for every $u \in A^*$ and $i \in \mathbb{Z}$. Thus a σ-invariant measure is determined by a map $\mu : A^* \to [0, 1]$ subject to **bilateral compatibility conditions**

$$\mu(\lambda) = 1, \quad \sum_{a \in A} \mu(ua) = \sum_{a \in A} \mu(au) = \mu(u), \quad u \in A^*$$

A point measure is not σ-invariant unless x is a σ-fixed point of the form $x = a^{\infty}$. On the other hand, each Bernoulli measure is σ-invariant. A Markov measure is σ-invariant iff the vector p is a stationary vector of R, that is, if $p \cdot R = p$.

Denote by $\mathfrak{M}_{\sigma}(A^{\mathbb{Z}}) := \{\mu \in \mathfrak{M}(A^{\mathbb{Z}}) : \sigma\mu = \mu\}$ the closed subspace of σ-invariant measures. The support $\Sigma_{\mu} := \{x \in A^{\mathbb{Z}} : \forall u \sqsubseteq x, \mu(u) > 0\}$ of a σ-invariant measure μ is a subshift whose forbidden words are the words of zero measure. If $\Sigma \subseteq A^{\mathbb{Z}}$ is a subshift, then

$$\mathfrak{M}_{\sigma}(\Sigma) := \{\mu \in \mathfrak{M}_{\sigma}(A^{\mathbb{Z}}) : \Sigma_{\mu} \subseteq \Sigma\} = \{\mu \in \mathfrak{M}_{\sigma}(A^{\mathbb{Z}}) : \forall u \in A^* \setminus \mathscr{L}(\Sigma), \mu(u) = 0\}$$

is a closed subspace of $\mathfrak{M}_{\sigma}(A^{\mathbb{Z}})$. A measure $\mu \in \mathfrak{M}_{\sigma}(\Sigma)$ can be identified with a map $\mu : \mathscr{L}(\Sigma) \to [0, 1]$, which satisfies the bilateral compatibility conditions. It can be said that $\mu \in \mathfrak{M}_{\sigma}(\Sigma)$ is σ-**ergodic**, if for every strongly σ-invariant set $Y \subseteq \Sigma$ either $\mu(Y) = 0$ or $\mu(Y) = 1$. If $\mu \in \mathfrak{M}_{\sigma}(\Sigma)$, then the support of μ is contained in the **densely recurrent subshift**

$$\mathscr{D}(\Sigma) := \mathrm{cl}\{x \in \Sigma : \forall k > 0, \limsup_{n \to \infty} \#\{|i| < n : x_{[i,i+k)} = x_{[0,k)}\}/n > 0\}$$

of Σ, that is, $\Sigma_{\mu} \subseteq \mathscr{D}(\Sigma)$ (see Akin (1991), Proposition 8.8, p. 164). On the other hand, there exists $\mu \in \mathfrak{M}_{\sigma}(\Sigma)$ such that $\Sigma_{\mu} = \mathscr{D}(\Sigma)$. If the support of a set $Y \subseteq \mathfrak{M}(A^{\mathbb{Z}})$ is defined as $\mathscr{S}(Y) := \mathrm{cl}(\bigcup_{\mu \in Y} \Sigma_{\mu})$, the following proposition can be arrived at.

Proposition 11 *Let $\Sigma \subseteq A^{\mathbb{Z}}$ be a subshift.*

1. $\mathscr{S}(\mathfrak{M}_{\sigma}(\Sigma)) = \mathscr{D}(\Sigma)$ *and therefore,* $\mathfrak{M}_{\sigma}(\Sigma) = \mathfrak{M}_{\sigma}(\mathscr{D}(\Sigma))$.
2. *Let G be a labelled graph and let $G_1, \ldots, G_k$ be all its strongly connected components. Then* $\mathscr{D}(\Sigma_G) = \Sigma_{G_1} \cup \cdots \cup \Sigma_{G_k}$.

If $\mu \in \mathfrak{M}_{\sigma}(A^{\mathbb{Z}})$, then $(F\mu)(u) = \sum_{v \in f^{-1}(u)} \mu(v)$ is also a σ-invariant measure. Thus we get a dynamical system $F : \mathfrak{M}_{\sigma}(A^{\mathbb{Z}}) \to \mathfrak{M}_{\sigma}(A^{\mathbb{Z}})$, which is a subsystem of $F : \mathfrak{M}(A^{\mathbb{Z}}) \to \mathfrak{M}(A^{\mathbb{Z}})$. The properties of this dynamical system do not depend on the memory constant m. CA F and $F\sigma^p$ determine the same dynamical system on $\mathfrak{M}_{\sigma}(A^{\mathbb{Z}})$. If (X, F) is a CA, and $\mu \in \mathfrak{M}_{\sigma}(X)$, then $F\mu \in \mathfrak{M}_{\sigma}(X)$, so we get a dynamical system $(\mathfrak{M}_{\sigma}(X), F)$.

In ECA128 of ❷ Example 3 if a measure μ gives positive probability to $\mu(0) > 0$, then it converges to the point measure $\delta_{0^{\infty}}$. This is, however, not an attractor in $\mathfrak{M}(A^{\mathbb{Z}})$, since any convex combination $a_0 \delta_{0^{\infty}} + a_1 \delta_{1^{\infty}}$ is a fixed point.

In ECA232 of ❷ Example 4, each measure with support included in the attractor $\Omega_F(A^{\mathbb{Z}} \setminus ([010]_0 \cup [101]_0)$ is a fixed point in $\mathfrak{M}(A^{\mathbb{Z}})$.

4.8   The Measure Attractor

Definition 7   Given a cellular automaton (X, F), we say that an F-invariant subshift $\Sigma \subseteq X$ **attracts ergodic measures** if there exists $\varepsilon > 0$, such that for every σ-ergodic measure $\mu \in \mathfrak{M}_\sigma(X)$ such that $\rho(\mu, \mathfrak{M}_\sigma(\Sigma)) < \varepsilon$, we have $\rho(F^n\mu, \mathfrak{M}_\sigma(\Sigma)) \to 0$ as $n \to \infty$.

Proposition 12   Let (X, F) be a cellular automaton and let $D \subseteq \mathcal{L}(X)^*$ be a set of forbidden words. Then for every $\mu \in \mathfrak{M}_\sigma(X)$ we have

$$\lim_{n \to \infty} \rho(F^n\mu, \mathfrak{M}_\sigma(\Sigma_D)) = 0 \iff \forall u \in D, \quad \lim_{n \to \infty}(F^n\mu)(u) = 0$$

Proposition 13   Let (X, F) be a cellular automaton, and $\Sigma \subseteq X$ a subshift attractor with basin $\mathcal{B}_F(\Sigma) = \{x \in X : \lim_{n \to \infty} d(x, \Sigma) = 0\}$. If $\mu \in \mathfrak{M}_\sigma(X)$ and $\mu(\mathcal{B}_F(\Sigma)) = 1$, then $\rho(F^n\mu, \mathfrak{M}_\sigma(\Sigma)) \to 0$ as $n \to \infty$. In particular, for all $\mu \in \mathfrak{M}_\sigma(X)$, $\rho(F^n\mu, \mathfrak{M}_\sigma(\Omega_F(X))) \to 0$ as $n \to \infty$.

Proposition 14   (Kůrka 2005) A subshift attractor for F attracts ergodic measures.

Definition 8   Given a CA (X, F) and a measure $\mu \in \mathfrak{M}_\sigma(X)$, the μ-limit $\Lambda_\mu(F)$ is defined by the condition

$$u \notin \mathcal{L}(\Lambda_\mu(F)) \iff \lim_{n \to \infty} F^n\mu([u]_0) = 0, \ \forall u \in \mathcal{L}(X)$$

Proposition 15   If $(A^\mathbb{Z}, F)$ is a CA, $\mu \in \mathfrak{M}(A^\mathbb{Z})$ a σ-ergodic measure, and $\Sigma \subseteq A^\mathbb{Z}$ a μ-attractor, then $\Lambda_\mu(F) \subseteq \Sigma$. In particular, $\Lambda_\mu(F) \subseteq \Omega_F(A^\mathbb{Z})$.

Proposition 16   (Kůrka 2005) If $\lim_{n \to \infty} F^n\mu = \nu$ in $\mathfrak{M}(X)$, then $\Lambda_\mu(F) = \Sigma_\nu$.

Proposition 17   (Kůrka 2005) If (X, F) is a cellular automaton, then $F : \mathfrak{M}_\sigma(X) \to \mathfrak{M}_\sigma(X)$ has a unique attractor $\Omega_F(\mathfrak{M}_\sigma(X)) = \mathfrak{M}_\sigma(\mathcal{D}(\Omega_F(X)))$.

Definition 9   Let (X, F) be a cellular automaton.

1. The **measure attractor** of F is $\mathcal{M}_F := \mathcal{S}(\Omega_F(\mathfrak{M}_\sigma(X))) = \mathcal{D}(\Omega_F(X))$, where the second equality is by ❷ Proposition 17.
2. The **measure quasi-attractor** of F is $\mathcal{M2}_F := \mathcal{D}(\mathcal{2}_F)$.

Proposition 18   (Pivato 2008) If (X, F) is a cellular automaton, then $\Lambda_\mu(F) \subseteq \mathcal{M}_F$ for every $\mu \in \mathfrak{M}_F(X)$,

$$\mathcal{L}(\mathcal{M}_F) = \bigcup\{\Lambda_\mu(F) : \ \mu \in \mathfrak{M}_\sigma(X)\}$$
$$\mathcal{M}_F = \overline{\bigcup\{\Lambda_\mu(F) : \ \mu \in \mathfrak{M}_\sigma(X)\}}$$

where $\mathfrak{M}_F(X)$ is the set of F-invariant measures in $\mathfrak{M}(X)$.

Corollary 3 (Kůrka 2003) *Let* (X, F) *be a cellular automaton.*

1. *If* $\mu \in \mathfrak{M}_\sigma(X)$, *then* $\rho(F^n\mu, \mathfrak{M}_\sigma(\mathscr{M}_F)) \to 0$ *as* $n \to \infty$.
2. $\mathscr{M2}_F$ *attracts ergodic measures.*

The maximal attractor of ECA128 is a nontransitive sofic subshift with strongly connected components for 0^∞ and 1^∞. The measure attractor of ECA128 is therefore $\mathscr{M}_F = \{0^\infty, 1^\infty\}$. The measure quasi-attractor is $\mathscr{M2}_F = \{0^\infty\}$. The measure attractor of ECA232 is $\mathscr{M}_F = \mathscr{M2}_F = \{0^\infty, 1^\infty\}$. The measure attractor and quasi-attractor of ECA184 are $\mathscr{M}_F = \mathscr{M2}_F = \mathbf{S}_{(-1,1)} \cup \mathbf{S}_{(1,1)}$.

4.9 Generic Space

The Besicovitch pseudometric on $A^{\mathbb{Z}}$ is defined by

$$d_B(x, y) = \limsup_{n \to \infty} \#\{i \in [-n, n] : x_i \neq y_i\}/2n$$

For a configuration $x \in A^{\mathbb{Z}}$ and words $u, v \in A^+$ set

$$\varphi_v(u) = \#\{i \in [0, |u| - |v|) : u_{[i,i+|v|)} = v\}$$
$$\varphi_v(x) = \#\{i \in \mathbb{Z} : x_{[i,i+|v|)} = v\}$$
$$\varphi_v^n(x) = \#\{i \in [-n, n] : x_{[i,i+|v|)} = v\}$$
$$\underline{\Phi}_v(x) = \liminf_{n \to \infty} \varphi_v^n(x)/2n$$
$$\overline{\Phi}_v(x) = \limsup_{n \to \infty} \varphi_v^n(x)/2n$$

For every $v \in A^*$, $\underline{\Phi}_v, \overline{\Phi}_v : A^{\mathbb{Z}} \to [0,1]$ are continuous in the Besicovitch topology. In fact we have

$$|\overline{\Phi}_v(x) - \overline{\Phi}_v(y)| \leq d_B(x, y)|v|, \quad |\underline{\Phi}_v(x) - \underline{\Phi}_v(y)| \leq d_B(x, y)|v|$$

Define the generic space (over the alphabet A) as

$$\mathscr{G}_A = \{x \in A^{\mathbb{Z}} : \forall v \in A^*, \underline{\Phi}_v(x) = \overline{\Phi}_v(x)\}$$

It is a closed subspace of $A^{\mathbb{Z}}$ (in the Besicovitch topology). For $v \in A^*$ denote by $\Phi_v : \mathscr{G}_A \to [0, 1]$ the common value of $\underline{\Phi}$ and $\overline{\Phi}$. Also for $x \in \mathscr{G}_A$ denote by $\Phi^x : A^* \to [0, 1]$ the function $\Phi^x(v) = \Phi_v(x)$. For every $x \in \mathscr{G}_A$, Φ^x is a shift-invariant Borel probability measure. The map $\Phi : \mathscr{G}_A \to \mathfrak{M}_\sigma(A^{\mathbb{Z}})$ is continuous with respect to the Besicovitch and weak topologies. In fact we have

$$d_M(\Phi^x, \Phi^y) \leq \sum_{n=1}^{\infty} d_B(x, y)n2^{-n} \leq 2d_B(x, y)$$

By the Kakutani theorem, Φ is surjective. Every shift-invariant Borel probability measure has a generic point. It follows from the Ergodic theorem that if μ is a σ-invariant measure, then $\mu(\mathscr{G}_A) = 1$ and for every $v \in A^*$, the measure of v is the integral of its density Φ_v,

$$\mu(v) = \int \Phi_v(x)\, d\mu.$$

If $F : A^{\mathbb{Z}} \to A^{\mathbb{Z}}$ is a CA and $\widetilde{F} : \mathfrak{M}_\sigma(A^{\mathbb{Z}}) \to \mathfrak{M}_\sigma(A^{\mathbb{Z}})$ its extension, then we have a commutative diagram $\Phi F = \widetilde{F}\Phi$.

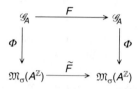

4.10   Particle Weight Functions

For a function $p : A^+ \to \mathbb{N}$ denote by

$$\mathscr{S}_p = \{u \in A^+ : p(u) \neq 0\}, \quad \Sigma_p = \{x \in A^{\mathbb{Z}} : \forall v \sqsubseteq x, p(v) = 0\}$$

its support and the subshift with forbidden set $\mathscr{S}_p$, respectively.

Definition 10   A particle weight function over an alphabet A is a map $p : A^* \to \mathbb{N}$ such that $p(\lambda) = 0$, $\mathscr{S}_p$ is finite and Σ_p is nonempty. The bound and order of p are defined by

$$b = \max\{p(u) : u \in A^*\}, \quad s = \max\{|u| : u \in \Sigma_p\}$$

The (total) weight of a word $u \in A^*$, or of a configuration $x \in A^{\mathbb{Z}}$ is defined by

$$\varphi_p(u) = \sum_{v \in A^+} \varphi_v(u) \cdot p(v) = \sum_{i=0}^{|u|-1} \sum_{j=i+1}^{|u|} p(u_{[i,j)})$$

$$\varphi_p(x) = \sum_{v \in A^+} \varphi_v(x) \cdot p(v) = \sum_{i=-\infty}^{\infty} \sum_{j=i+1}^{\infty} p(x_{[i,j)})$$

$$\varphi_p^n(x) = \sum_{v \in A^+} \varphi_v^n(x) \cdot p(v) = \sum_{i=-n}^{n} \sum_{j=i+1}^{\infty} p(x_{[i,j)})$$

The set of p-finite configurations is denoted by

$$\mathscr{F}_p = \{x \in A^{\mathbb{Z}} : \varphi_p(x) < \infty\}$$

The mean weight of a configuration $x \in \mathscr{G}_A$ and the weight of a measure $\mu \in \mathfrak{M}(A^{\mathbb{Z}})$ are defined by

$$\Phi_p(x) = \sum_{v \in A^+} \Phi_v(x) \cdot p(v) = \lim_{n \to \infty} \varphi_p^n(x)/(2n+1)$$

$$\Psi_p(\mu) = \sum_{v \in A^+} \mu(v) \cdot p(v)$$

In particular, if $x = u^\infty$ is a σ-periodic configuration, δ_u is the measure whose support is the σ-orbit of u, and if $k > 1$ is the least integer such that $(k-1)|u| \geq s$, then

$$\Psi_p(\delta_u) = \Phi_p(u^\infty) = \sum_{i=0}^{|u|-1} \sum_{j=i+1}^{i+s} p((u^k)_{[i,j)})/|u|$$

Lemma 1 *For $u,v \in A^+$ and $x \in A^{\mathbb{Z}}$ we have*

$$\varphi_p(u) \leq bs|u|, \quad \varphi_p(uv) \leq \varphi_p(u) + \varphi_p(v) + bs^2,$$

$$\varphi_p(x_{[-n,n]}) \leq \varphi_p^n(x) \leq \varphi_p(x_{[-n,n]}) + bs^2 \leq bs(2n+1+s)$$

If $0 \leq k < l \leq |u| - s$ and $u_{[k,k+s)} = u_{[l,l+s)}$, then

$$\varphi_p(u) = \varphi_p(u_{[0,k)} u_{[l,|u|)}) + (l - k)\Phi_p((u_{[k,l)})^{\infty})$$

The function $\Phi_p : \mathscr{G}_A \to [0, \infty)$ is continuous in the Besicovitch topology and the weight map $\Psi_p : \mathfrak{M}(A^{\mathbb{Z}}) \to [0, \infty)$ is bounded, affine, and continuous in the weak topology. We have a commutative diagram $\Psi_p\Phi = \Phi_p$. If $x \in \mathscr{G}_A$ and $\mu \in \mathfrak{M}(A^{\mathbb{Z}})$, then

$$\Phi_p(x) \leq bs, \quad \Psi_p(\mu) = \int \Phi_p(x)\, d\mu \leq bs$$

Theorem 13 *We say that $p : A^* \to \mathbb{N}$ is a Lyapunov particle weight function for a cellular automaton $F : A^{\mathbb{Z}} \to A^{\mathbb{Z}}$, if one of the following equivalent conditions is satisfied:*

1. *There exists a local rule f for F such that $\forall u \in A^*, \varphi_p(f(u)) \leq \varphi_p(u)$.*
2. *For any $x \in \mathscr{F}_p$, $\varphi_p(F(x)) \leq \varphi_p(x)$.*
3. *For any $x \in \mathscr{G}_A$, $\Phi_p(F(x)) \leq \Phi_p(x)$.*
4. *For any σ-periodic $x \in A^{\mathbb{Z}}$, $\Phi_p(F(x)) \leq \Phi_p(x)$.*
5. *For any $\mu \in \mathfrak{M}(A^{\mathbb{Z}})$, $\Psi_p(F\mu) \leq \Psi_p(\mu)$.*

Proof (1) $\to$ (2): Let $f : A^{d+1} \to A$ be a local rule for F and assume that the memory is $m = 0$. If $x \in \mathscr{F}_p$ then there exists n such that all particles of x occur in the interval $[-n, n+d]$ and all particles of $F(x)$ occur in the interval $[-n, n]$. Then, by ❷ Lemma 1, it holds $\varphi_p(F(x)) = \varphi_p(F(x)_{[-n,n]}) \leq \varphi_p(x_{[-n,n+d]}) = \varphi_p(x)$.

(2) $\to$ (3): For $n > 0$ let $y \in \mathscr{F}_p$ be a configuration such that $y_{[-n,n]} = x_{[-n,n]}$ and y contains no particles in $(-\infty, -n)$ and (n, ∞). Then

$$\varphi_p^{n-d}(F(x)) \leq \varphi_p(F(x)_{[-n+d,n-d]}) + bs^2 \leq \varphi_p(F(y)) + bs^2$$
$$\leq \varphi_p(y) + bs^2 \leq \varphi_p^n(x) + 3bs^2$$

(3) $\to$ (4) is trivial.
(4) $\to$ (2): Let $x \in \mathscr{F}_p$ and choose n so that all particles of x are within the interval $[-n+s+d, n-s-d]$. For the σ-periodic configuration $y = (x_{[-n,n]})^{\infty}$ we have

$$\varphi_p(F(x)) = (2n+1)\Phi_p(F(y)) \leq (2n+1)\Phi_p(y) = \varphi_p(x)$$

(2) & (4) $\to$ (1): Let $g : A^{e+1} \to A$ be a local rule for F. We show first that there exists $c \geq 0$ such that

$$\forall x \in \mathscr{F}_p, \forall j \leq |A|^{s+e}, \quad \varphi_p(F(x)_{[0,j)}) \leq \varphi_p(x_{[-c,j+c)})$$

Assume that the above condition does not hold. Then, for every $c > 0$ there exists $x^c \in \mathscr{F}_p$ and $j \leq |A|^{e+s}$ such that $\varphi_p(F(x^c)_{[0,j)}) > \varphi_p(x^c_{[-c,j+c)})$. Now, consider the sequence of x^c and extract a converging subsequence. Assume that this subsequence converges to $x \in A^{\mathbb{Z}}$. It is clear that $\varphi_p(F(x)_{[0,|A|^{e+s}})) > \varphi_p(x)$. Thus $x \in \mathscr{F}_p$ and $\varphi_p(F(x)) > \varphi_p(x)$ and this is a contradiction. Next we prove that for all $x \in \mathscr{F}_p$ and all $i < j$ we have

$$\varphi_p(F(x)_{[i,j)}) \leq \varphi_p(x_{[i-c,j+c)})$$

We already proved that the condition holds for $j - i \leq |A|^{e+s}$. If the condition does not hold for some $j - i > |A|^{e+s}$, then there exist $i \leq k < l \leq j$ such that $x_{[k,k+e+s)} = x_{[l,l+e+s)}$ and $F(x)_{[k,k+s)} = F(x)_{[l,l+s)}$. Set $y = x_{(-\infty,k)}x_{[l,\infty)}$. Then

$$\begin{aligned}
\varphi_p(F(y)_{[i,j-l+k)}) &= \varphi_p(F(x)_{[i,j)}) - (l - k)\Phi_p((F(x)_{[k,l)})^\infty) \\
&> \varphi_p(x_{[i-c,j+c)}) - (l - k)\Phi_p((x_{[k,l)})^\infty) \\
&\geq \varphi_p(y_{[i-c,j-l+k+c)})
\end{aligned}$$

and this is a contradiction. Set now $d = e + 2c$ and define $f \colon A^{d+1} \to A$ by $f(u) = g(u_{[c,|u|-c)})$. Then f is a local rule for F and $\varphi_p(f(u)) \leq \varphi_p(u)$.

$(3) \to (5)$: By the Ergodic theorem,

$$\Psi_p(F\mu) = \int \Phi_p(F(x))d\mu \leq \int \Phi_p(x)d\mu = \Psi_p(\mu)$$

$(5) \to (4)$: If $x = u^\infty$, then $\Phi_p(Fx) = \Psi_p(F\delta_u) \leq \Psi_p(\delta_u) = \Phi_p(x)$

Definition 11   Let $F \colon A^{\mathbb{Z}} \to A^{\mathbb{Z}}$ be a cellular automaton and let $p \colon A^* \to \mathbb{N}$ be a Lyapunov particle weight function for F. A word $v \in A^+$ is (F, p)-reducing, if there exists $r(v) > 0$, such that for every $x \in \mathscr{F}_p$, $\varphi_p(F(x)) \leq \varphi_p(x) - r(v) \cdot \varphi_v(x)$. Denote by $\Sigma^0_{(F,p)}$ the subshift whose forbidden words are (F, p)-reducing words and

$$\Sigma_{(F,p)} = \bigcap_{n=-\infty}^{\infty} F^n(\Sigma^0_{(F,p)})$$

Proposition 19   *Let v be a (F, p)-reducing word. Then for every $\mu \in \mathfrak{M}(A^{\mathbb{Z}})$ we have $\rho(F^n\mu, \mathfrak{M}(\Sigma_{\{v\}})) \to 0$ as $n \to \infty$.*

Proof   We have $\Psi_p(F^{j+1}\mu) \leq \Psi_p(F^j\mu) - r(v) \cdot (F^j\mu)(v)$. Adding these inequalities for $j = 0, \ldots, k$ we get

$$r(v) \cdot (\mu(v) + \cdots + (F^k\mu)(v)) \leq \Psi_p(\mu) - \Psi_p(F^{k+1}\mu) \leq \Psi_p(\mu)$$

Thus $\sum_{k=0}^{\infty}(F^k\mu)(v)$ converges and therefore $\lim_{k \to \infty}(F^k\mu)(v) = 0$. $\quad\square$

Corollary 4   *If p is a Lyapunov particle weight function for F then $\mathscr{M}_F \subseteq \Sigma_{(F,p)}$.*

Proof   If $f^n(w)$ is a reducing word for some $n \geq 0$, then $F^m\mu(v) \to 0$ as $m \to \infty$, so $\mathscr{M}_F(A^{\mathbb{Z}}) \subseteq \bigcap_{n=-\infty}^{0} F^n(\Sigma^0_{(F,p)})$. Since $\mathscr{M}_F(A^{\mathbb{Z}})$ is F-invariant, we get $\mathscr{M}_F(A^{\mathbb{Z}}) \subseteq \Sigma_{(F,p)}$. $\quad\square$

4.11   Particle Distribution

Assume that (I, λ), (J, λ) are finite linearly ordered sets and let $x = (x_i)_{i \in I}$, $y = (y_j)_{j \in J}$ be nonnegative vectors such that $\sum_{i \in I} x_i \leq \sum_{j \in J} y_j$. A distribution for x and y is a nonnegative matrix $z = (z_{ij})_{i \in I, \, j \in J}$ such that $\sum_{j \in J} z_{ij} = x_i, \sum_{i \in I} z_{ij} \le y_j$. Let λ be the lexicographic order on $I \times J$ defined by

$$(i,j)\lambda(i',j') \iff i\lambda i' \vee (i = i' \ \& \ j\lambda j')$$

On the set of distributions a linear order can be defined by

$$z\lambda w \iff \exists (i,j) \in I \times J, (z_{ij} > w_{ij} \ \& \ \forall (k,l)\lambda(i,j), z_{kl} = w_{kl})$$

Then there exists a unique minimal distribution. If $I = \{0,1\ldots,n-1\}$ and $J = \{0,\ldots,m-1\}$, then the minimal distribution is defined inductively by

$$z_{00} = \min\{x_0, y_0\}$$

and

$$z_{ij} = \min\{x_i - z_{i0} - \ldots - z_{i,j-1}, y_j - z_{0j} - \ldots - z_{j,i-1}\}$$

Assume now that p is a Lyapunov particle weight function for a local rule f, so $\varphi_p(f(u)) \le \varphi_p(u)$ for every $u \in A^*$. If $p(f(u)) > 0$, one can try to distribute $p(f(u))$ among the particles of u. Denote by $q_{ij}(u) \le p(u_{[i,j)})$ the share of a particle $u_{[i,j)}$. The sum of these shares is then $p(f(u))$. In a configuration $x \in A^{\mathbb{Z}}$, a particle $x_{[i,j)}$ may get shares from several particles of $F(x)$. If $k \le i < j \le l$, then a particle $F(x)_{[k-a,l-a-d)} = f(x_{[k,l)})$ requires the share $q_{i-k,j-k}(x_{[k,l)})$ from the particle $x_{[i,j)}$ (⊘ Fig. 13 right). The sum of these requirements cannot exceed the weight $p(x_{[i,j)})$.

Definition 12 A particle distribution for $f : A^{d+1} \to A$ and a particle weight function $p : A^* \to \mathbb{N}$ is a system $q = \{q_{ij}(u) : u \in A^+, 0 \le i < j \le |u|\}$ of integers such that for $u \in A^+$ and $x \in A^{\mathbb{Z}}$,

$$\sum_{i=0}^{|u|-1} \sum_{j=i+1}^{|u|} q_{ij}(u) = p(f(u)),$$

$$q(x,i,j) = \sum_{k=-\infty}^{i} \sum_{l=j}^{\infty} q_{i-k,j-k}(x_{[k,l)}) \le p(x_{[i,j)}), \quad i < j$$

For $k = 0$ and $u = x_{[k,l)}$ we get in particular $q_{ij}(u) \le p(u_{[i,j)})$.

Proposition 20 *If p is a Lyapunov pwf for a CA F, then there exists d, local rule $f : A^{d+1} \to A$ for F and a distribution for p and f.*

Proof By ⊘ Proposition 13 there exists d and a local rule $f : A^{d+1} \to A$ such that for every $u \in A^*, \varphi_p(f(u)) \le \varphi_p(u)$. We have the lexicographic order on the set of $I = \{[i,j) \in \mathbb{Z}^2 : i < j\}$ of intervals and also on the set I^2 of the pairs of intervals. Given $x \in \mathscr{F}_p$, we try to distribute

■ **Fig. 13**
Particle distribution.

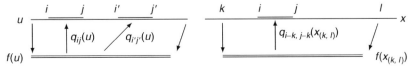

the weight of a particle $F(x)_{[k,l-d)} = f(x_{[k,l)})$ in $F(x)$ among particles $x_{[i,j)}$. Call a distribution for x any system of nonnegative numbers $g_{klij}(x)$ indexed by intervals $[k, l)$ and $[i, j)$ such that

$$\sum_{i=-\infty}^{\infty} \sum_{j=i+1}^{\infty} g_{klij}(x) = p(F(x)_{[k,l-d)})$$

$$\sum_{k=-\infty}^{\infty} \sum_{l=k+1}^{\infty} g_{klij}(x) \le p(x_{[i,j)})$$

This implies in particular $g_{klij}(x) \le p(x_{[i,j)})$, $g_{klij}(x) \le p(F(x)_{[k,l-d)}) = p(f(x_{[k,l)}))$. There is only a finite number of distribution for any $x \in \mathscr{F}_p$ and there exists a unique minimal distribution. It can now be shown that there exist distributions that are nonzero only on the index set

$$I_4 = \{(k, l, i, j) \in I^2 : k \le i < j \le l\}$$

This means that the particle $F(x)_{[k,l-d)}$ is distributed only between particles in $x_{[k, l)}$. Take the first interval $[k, l)$ for which $F(x)_{[k,l-d)}$ is a particle and construct its minimal distribution into particles of $x_{[k, l)}$. Take now the second particle, that is, the least $[k', l') \gg [k, l)$ such that $F(x)_{[k',l'-d)}$ is a particle. If a particle in $x_{[k',l')}$ has been charged in the first step by $F(x)_{[k,l-d)}$, we construct the minimal distribution for $F(x)_{[k,l'-d)}$ into $x_{[k,l')}$. If no particle in $x_{[k',l')}$ has been charged in the first step, then we construct the minimal distribution for $F(x)_{[k',l'-d)}$ in $x_{[k',l' + d)}$. In either case, we get the minimal distribution for the first two particles indexed by I_4. We continue and construct the minimal distribution of x indexed by I_4. Consider now a word u such that $p(f(u)) > 0$. Set $k = 0$, $l = |u|$ and let $Y_u = \mathscr{F}_p \cap [u]_0$. For each $x \in Y_u$, let

$$\{g_{klij}(x) : 0 \le i < j < |u|\}$$

be the restriction of the minimal distribution g of x to the interval $[k, l) = [0, |u|)$. The set of all these restricted distributions has a maximal element g and we set $q_{ij}(u) = g_{klij}(x)$. It can be now shown that q is a distribution for p and f. Given $x \in \mathscr{F}_p$, set

$$h_{klij}(x) = q_{i-k,l-k}(x_{[k,l+d)}).$$

Then h is a distribution for x so q satisfies the requirements.

Proposition 21 *Let F be a cellular automaton with a local rule $f : A^{d+1} \to A$. If a particle weight function $p : A^* \to \mathbb{N}$ has a distribution q, then p is a Lyapunov pwf for F.*

Proof Let $x \in \mathscr{F}_p$ be a p-finite configuration. For $k < l$ and $u = x_{[k,l)}$ we get

$$p(f(x_{[k,l)})) = \sum_{i=0}^{|u|-1} \sum_{j=i+1}^{|u|} q_{ij}(u) = \sum_{i=k}^{l-1} \sum_{j=i+1}^{l} q_{i-k,j-k}(x_{[k,l)})$$

Using the fact that $p(\lambda) = 0$ we get

$$\varphi_p(F(x)) = \sum_{k=-\infty}^{\infty} \sum_{l=k+1}^{\infty} p(f(x_{[k,l)})) = \sum_{k=-\infty}^{\infty} \sum_{l=k+1}^{\infty} \sum_{i=k}^{l-1} \sum_{j=i+1}^{l} q_{i-k,j-k}(x_{[k,l)})$$

$$= \sum_{i=-\infty}^{\infty} \sum_{j=i+1}^{\infty} \sum_{k=-\infty}^{i} \sum_{l=j}^{\infty} q_{i-k,j-k}(x_{[k,l)}) \le \sum_{i=-\infty}^{\infty} \sum_{j=i+1}^{\infty} p(x_{[i,j)}) = \varphi_p(x)$$

By ❿ Proposition 13, p is Lyapunov.

Proposition 22 Let $p : A^s \to \mathbb{N}$ be a particle weight function with order s, let $f : A^{d+1} \to A$ be a local rule and q a distribution for p. Then for every $x \in A^{\mathbb{Z}}$,

$$q(x, i, j) = \sum_{k=j-s-d}^{i} \sum_{l=j}^{i+s+d} q_{i-k,j-k}(x_{[k,l]}).$$

Proof If either $k < j - s - d \le l - s - d$ or $k + s + d \le i + s + d < l$, then $l - k > s + d$ and $q_{i-k,j-k}(x_{[k,l]}) \le p(f(x_{[k,l]})) = 0$.

Proposition 23 For $|v| \le 2(s + d)$ set $i_v = |v| - s - d$, $j_v = s + d$ and

$$r(v) = p(v_{[i_v,j_v)}) - \sum_{k=j-s-d}^{i_v} \sum_{l=j_v}^{i_v+s+d} q_{i_v-k,j_v-k}(v_{[k,l]}).$$

If $r(v) > 0$ then v is (F, p)-reducing and for every $x \in \mathscr{F}_p$, $\varphi_p(F(x)) \le \varphi_p(x) - r(v) \cdot \varphi_v(x)$.

Proof For $x \in \mathscr{F}_p$ set $\xi_{ij}(x) = 1$ if $x_{[j-s-d,i+s+d)} = v$, and $\xi_{ij}(x) = 0$ otherwise. If $\xi_{ij}(x) = v$, then $x_{[i,j)} = v_{[i_v,j_v)}$ and

$$r(v) \cdot \xi_{ij}(x) \le p(x_{[i,j)}) - \sum_{k=-\infty}^{i} \sum_{l=j}^{\infty} q_{i-k,j-k}(x_{[k,l]})$$

Similarly as in ❯ Proposition 22 we get

$$\varphi_p(F(x)) = \sum_{i=-\infty}^{\infty} \sum_{j=i+1}^{\infty} \sum_{k=-\infty}^{i} \sum_{l=j}^{\infty} q_{i-k,j-k}(x_{[k,l]})$$

$$\le \sum_{i=-\infty}^{\infty} \sum_{j=i+1}^{\infty} p(x_{[i,j)}) - r(v)\xi_{ij}(x) \le \varphi_p(x) - r(v)\varphi_v(x)$$

Definition 13 We say that a CA $(A^{\mathbb{Z}}, F)$ is p-attracting for a particle weight function $p : A^+ \to \mathbb{N}$ if there exists $s > 0$, such that each word $v \in A^+$ with $\varphi_p(v) > s$ is (F, p)-reducing.

Proposition 24 If a CA $(A^{\mathbb{Z}}, F)$ is p-attracting for a pwf $p : A^+ \to \mathbb{N}$, then $\mathscr{M}_F \subseteq \Sigma_p$.

Proof We get $\mathscr{M}_F \subseteq \Sigma_p^s := \{x \in A^{\mathbb{Z}} : \varphi_p(x) \le s\}$. The densely recurrent subshift of Σ_p^s is Σ_p.

Example 10 (ECA62) Consider the ECA given by the following local rule: $F(x)_i = 0$ iff $x_{[i-1,i+1]} \in \{000, 110, 111\}$.

There exists a Lyapunov particle weight function given by

$$p(00) = 1, \ p(111) = 1, \ p(010) = 2, \ p(u) = 0 \quad \text{for } u \notin \{00, 111, 010\}$$

A particle distribution for p is given in ❯ *Table 1*. The strokes in the preimages give positions of i, j such that $q_{ij}(u) = 1$.

The only preimages of 0^6 are 0^8, 01^7, and 1^8. However 1^7 is not in $\mathscr{L}(\Omega_F(A^{\mathbb{Z}}))$, so the only preimage of 0^6 in the omega-limit is 0^8. Thus there exists an attractor that does not contain 0^∞. Indeed, the set $V = [0^6]_2 \cup [1^7]_1 \cup \bigcup_{v \in f^{-1}(1^7)} [v]_0$ is inversely invariant, so $U = A^{\mathbb{Z}} \setminus V$ is a clopen invariant set and $\Omega_F(U)$ is a subshift attractor. It contains σ-transitive

⬛ **Table 1**

A particle distribution for ECA62

Image	Preimages						
00	0|00|0,	|111|0,	|111|1				
111	0|010|0,	0|010|1,	|010|01,	|010|11,	1|00|10,	1|00|11,	1|010|0, 1|010|1
010	11|00|0,	110|00|					

signal subshifts $\Sigma_{(1,2)}$ and $\Sigma_{(0,3)}$, whose intersection is $\mathfrak{o}((110)^\infty)$, as well as their join. Similarly as in ❷ Example 8 we show

$$\mathcal{Q}_F = \Omega_F(U) = F^2(\Sigma_{(1,2)} \vee \Sigma_{(0,3)}), \quad \mathcal{M}\mathcal{Q}_F = \Sigma_{(1,2)} \cup \Sigma_{(0,3)}$$

The only other signal subshifts are $\Sigma_{(4,4)}$ and $\Sigma_{(-1,1)}$. We get

$$\Omega_F(A^\mathbb{Z}) = F^2(\Sigma_{(4,4)} \vee \Sigma_{(1,2)} \vee \Sigma_{(0,3)} \vee \Sigma_{(-1,1)})$$

$$\mathcal{M}_F = \{0^\infty\} \cup \Sigma_{(1,2)} \cup \Sigma_{(0,3)}$$

Signal subshifs for ECA 62 are given in ❷ *Fig. 15*. In ❷ *Fig. 14*, a configuration whose left part belongs to $\Sigma_{(4,4)} \vee \Sigma_{(-1,1)}$ and whose right part belongs to $\Sigma_{(1,2)} \vee \Sigma_{(0,3)}$ can be seen. In the space–time diagram, the words 00, 111, and 010, which do not occur in $(110)^\infty$, are displayed in gray.

Example 11 (Gacs–Kurdyumov–Levin cellular automaton, ❷ *Fig. 16*) The alphabet $A = \{0, 1\}$ is binary, $d = 6$ and $a = -3$. The local rule is given by the table

0011	001	011	0010	1011	110	100	101	010	0	1
01	0	1	0	1	0	1	1	0	0	1

The first applicable rule is always used.

The original rule of Gacs et al. (1978), Gacs (2001), or de Sá and Maes (1992) was slightly modified, which reads

$$F(x)_i = \begin{cases} \mathbf{m}(x_{i-3}, x_{i-1}, x_i) & \text{if} \quad x_i = 0 \\ \mathbf{m}(x_i, x_{i+1}, x_{i+3}) & \text{if} \quad x_i = 1 \end{cases}$$

where $\mathbf{m}$ is the majority function. The modification also homogenizes short regions. Note that the iterates of any finite perturbation of 0^∞ eventually attain 0^∞ and likewise the iterates of any finite perturbation of 1^∞ eventually attain 1^∞.

This model has been considered to be an example of a computation that is stable with respect to random errors. Suppose that after each (deterministic) step, each site of the configuration is flipped with a given probability ε. This updating defines a dynamical system in the space $\mathfrak{M}(A^\mathbb{Z})$. It has been conjectured that for sufficiently small ε, the resulting dynamical system has two invariant measures, one close to $\mu_{\{0\}}$, the other close to $\mu_{\{1\}}$. The conjecture is still open but a much more complex CA with these properties has been constructed in Gacs (2001). Mitchell et al. (1994) consider more general class of CA, which

■ Fig. 14
ECA62.

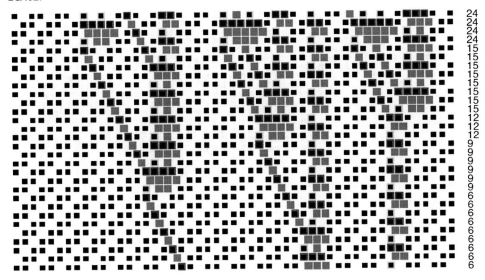

■ Fig. 15
Signal subshifts for ECA62.

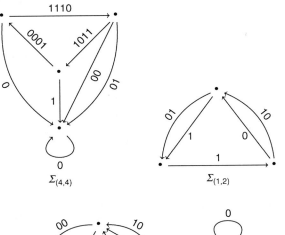

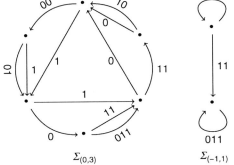

⬛ Fig. 16

Gacs–Kurdyumov–Levin CA.

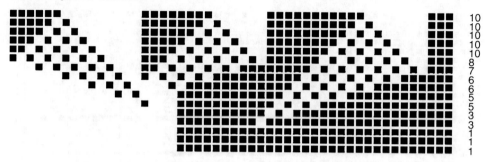

solve the density classification task, that is, decide whether a given configuration has more zeros or ones. We have nontransitive signal subshifts

$$S_{(3,1)} = \{\sigma^n(0^\infty.(10)^\infty) : \ n \in \mathbb{Z}\} \cup \{0^\infty, (01)^\infty, (10)^\infty\}$$
$$S_{(1,1)} = \{\sigma^n((01)^\infty.0^\infty) : \ n \in \mathbb{Z}\} \cup \{0^\infty, (01)^\infty, (10)^\infty\}$$
$$S_{(0,1)} = \{\sigma^n(0^\infty.1^\infty) : \ n \in \mathbb{Z}\} \cup \{0^\infty, 1^\infty\}$$
$$S_{(-1,1)} = \{\sigma^n(1^\infty.(01)^\infty) : \ n \in \mathbb{Z}\} \cup \{1^\infty, (01)^\infty, (10)^\infty\}$$
$$S_{(-3,1)} = \{\sigma^n((10)^\infty.1^\infty) : \ n \in \mathbb{Z}\} \cup \{1^\infty, (01)^\infty, (10)^\infty\}$$

There exists a particle weight function that charges these signal subshifts. It is given by $p(1100) = 2$,

$$p(1011) = 1, p(1101) = 1, p(0011) = 1, p(0100) = 1, p(0010) = 1$$
$$s(1011) = -3, s(1101) = -1, s(0011) = 0, s(0100) = 1, s(0010) = 3$$

where s gives the speeds of these particles. Particle 1100 has no speed as it immediately changes to a pair of particles 1101 and 0100. The algorithm of ❷ Proposition 22 shows that there exists no distribution for this particle weight function. However, there does exist a particle distribution restricted to the first image subshift with language $f(A^*) = \{u \in A^* : \ \forall v \in S, v \not\sqsubseteq u\}$, where $S = \{1100, 100100, 110110\}$ is the set of forbidden words. One such distribution is given in the following table (the strokes in a preimage u give positions i, j such that $q_{ij}(u) = 1$).

Image	Preimages
1011	xxx110|1011|, 101|0011|1xx, x11010|1011|, 101|0011|011, 101010|1011|
1101	xxx1|1101|0x, xx10|1101|0x
0011	xx|0010|11xx, x|0010|1011x, |0010|101011, u|0011|v
0100	x1|0100|0xxx, x1|0100|10xx
0010	|0010|100xxx, xx0|0011|010, |0010|10100x, 001|0011|010, |0010|101010

where $u \in \{$xx0, 001$\}$ and $v \in \{$1xx, 011$\}$. When two particles meet they annihilate or change as follows:

$$-1, -3 \to \lambda, \quad 0, -1 \to 3, \quad 3, -3 \to 0, \quad 1, 0 \to -3, \quad 3, 1 \to \lambda$$

It is also possible that three particles $1, 0, -1 \rightarrow \lambda$ meet simultaneously and disappear. These particles cannot occur in a configuration in an arbitrary order. Any possible sequence of their occurrence is represented by a path in the following graph. The particles are represented by their speeds and the edges between them represent one of the periodic words 0^∞, 1^∞, $(01)^\infty$, which separate them.

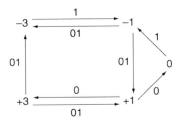

The path $-3 \rightarrow -1 \rightarrow 1 \rightarrow 3$ is the longest path with increasing speeds and represents a configuration $(01)^\infty 1^i (01)^j 0^k (01)^\infty$ in which no particle ever disappears. Any configuration that contains at least $s = 5$ particles contains a pair of neighboring particles, which approach one another and finally either disappear or are transformed into another particle. This means that the CA is p-attracting, so $\mathcal{M}_F \subseteq \Sigma_p$. The subshift $\Sigma_p = \{0^\infty, 1^\infty, (01)^\infty, (10)^\infty\}$ is finite and $\mathfrak{M}_\sigma(\Sigma_p)$ is a triangle consisting of convex combinations of $\mu_{\{0\}}$, $\mu_{\{1\}}$, and $\mu_{\{01\}}$. Every measure in this triangle is F-invariant. However, there are trajectories that go from an arbitrary neighborhood of any measure of $\mathfrak{M}(\Sigma_p)$ to any other measure in $\mathfrak{M}(\Sigma_p)$.

Example 12   (random walk) The alphabet is $\{0, 1\}^3$ and the rule is given by

$$F(x, y, z)_i = (1, y_{i-1}, z_{i+1}) \text{ if}$$
$$x_{i-1} = 1, y_{i-1} = z_{i-1}, x_i \cdot (y_i - z_i) = 0 \text{ or}$$
$$x_{i+1} = 1, y_{i+1} \neq z_{i+1} = 0, x_i \cdot (y_i + z_i - 1) = 0$$
$$F(x, y, z)_i = (0, y_{i-1}, z_{i+1}) \text{ otherwise}$$

Denote by $\pi_i : A^{\mathbb{Z}} \rightarrow \{0, 1\}^{\mathbb{Z}}$, $i = 1, 2, 3$ the projections. Then $\pi_2 : (A^{\mathbb{Z}}, F) \rightarrow (\{0, 1\}^{\mathbb{Z}}, \sigma^{-1})$ and $\pi_3 : (A^{\mathbb{Z}}, F) \rightarrow (\{0, 1\}^{\mathbb{Z}}, \sigma)$ are factor maps. Ones in the first coordinate are particles, which move to the right if the second and third coordinates are equal and to the left otherwise. When two neighboring particles cross, they annihilate. When they meet, they merge into one particle. Suppose that μ is a Bernoulli measure on $A^{\mathbb{Z}}$, then $\pi_2\mu$, $\pi_3\mu$ are Bernoulli measures; so neighboring particles perform independent random walk until they annihilate or merge. These random walk need not be symmetric, but the distance between two neighboring particles performs a symmetric random walk with absorbing states 0 and -1 (which represents annihilation). Thus the random walk CA implements the behavior that has been observed in ECA18 or ECA54. We have a particle weight function $p : A \rightarrow \mathbb{N}$ given by $p(x, y, z) = x$, which counts the number of particles. It is clearly nonincreasing for F (see ❷ *Fig. 17*) and Σ_p is the subshift of particle-free configurations. It can be proved (see Kůrka and Maass 2000 or Kůrka 2003a) that for every connected measure μ we have $\Lambda_\mu \subseteq \Sigma_p$.

4.12 Particle Weight Functions with Infinite Support

In Kůrka (2003a) particle weight functions with infinite support are considered. They are functions $p : A^* \rightarrow \mathbb{N}$, which satisfy the following conditions.

■ **Fig. 17**
Random walk.

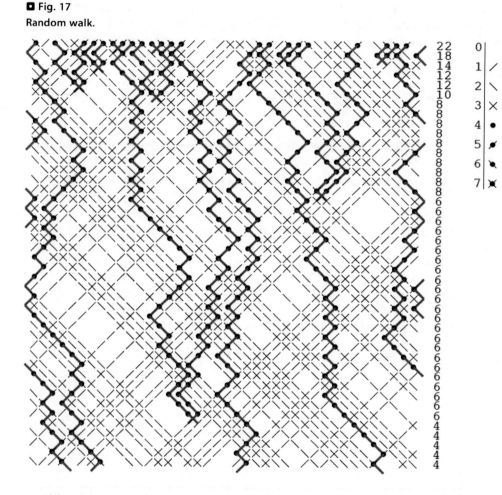

1. $p(\lambda) = 0$.
2. There exists $b_0 > 0$, such that for every $u \in A^*$, $p(u) \le b_0$.
3. There exists $b_1 > 0$, such that if $x \in A^{\mathbb{Z}}$, $p(x_{[i,j)}) > 0$, and $p(x_{[k,l)}) > 0$, then the interval $[i, j) \cap [k, l)$ has less than b_1 elements.
4. There exists $x \in A^{\mathbb{Z}}$ such that $p(x_{[i,j)}) = 0$ for all $i < j$.

Example 13 (ECA18, ❯ *Fig. 18*) $F(x)_i = 1$ iff $x_{[i-1,i+1]} \in \{001, 100\}$.

$$000{:}0,\ 001{:}1,\ 010{:}0,\ 011{:}0,\ 100{:}1,\ 101{:}0,\ 110{:}0,\ 111{:}0.$$

Consider a particle weight function given by

$$p(u) = \begin{cases} 1 & \text{if } u = 10^{2m}1 \quad \text{for some } m \ge 0 \\ 0 & \text{otherwise} \end{cases}$$

Then

$$\Sigma_p = \{x \in A^{\mathbb{Z}} : \forall i, x_{2i} = 0\} \cup \{x \in A^{\mathbb{Z}} : \forall i, x_{2i+1} = 0\}$$

☐ Fig. 18
Example of a space–time diagram for ECA18 (see ❯ Example 13). Particles are colored in red.

The dynamics on Σ_p is quite simple. If $x_{2i} = 0$ for all i, then $F(x)_{2i+1} = 0$ and $F(x)_{2i} = x_{2i-1} + x_{2i+1}$ mod 2. It can be proved that $\Lambda_\mu \subseteq \Sigma_p$ for every connected measure μ (see Kůrka 2003a).

There exist also particle weight functions, which are decreasing only in the long run, while they oscillate in the short run.

Example 14   (equalizing CA) The alphabet is $A = \{0, 1, 2, 3\}$ and the local rule is given by

xx33x:0 xx032:3 130xx:3 131xx:3 x132x:3 xx32x:0 x13xx:0

xxx2x:2 xx13x:2 x3xxx:1 xx2xx:0 x10xx:1 x11xx:1 xx1xx:0

Particles 1, 3 walk between two walls 3 similarly as in ❯ Example 6, but when they reach a wall they push it forward. The intervals between neighboring walls grow and shrink but the shorter intervals grow faster, so the lengths of all intervals approach the same value. If $(x_i)_{0 \le i \le n}$ are lengths of neighboring intervals in a periodic configuration $u \in A^{\mathbb{Z}}$, we define weight function $p(u) = \sum_{i<n} |x_{i+1} - x_i|$ (see ❯ Fig. 19), which decreases in the long run and then oscillates with small values.

4.13   Operations on CA

The local rule of a 1D CA can be represented by a look-up table. However, given a look-up table one cannot uniquely define the corresponding CA since the CA memory has to be fixed. Indeed, for each memory value, a different CA is obtained. It is therefore natural to wonder what properties are conserved by the CA when changing the memory but keeping the same look-up table for the local rule. This problem can be equivalently rephrased as follows. Let F be a CA. If a property holds for a CA F, does it hold for the CA $\sigma^n \circ F$? Along the same direction, a look-up table defines a CA both on $A^{\mathbb{Z}}$ and $A^{\mathbb{N}}$ (under the constraint that the memory $m \ge 0$), but what properties does these two CA share? In this section, attempts have been made to

◘ **Fig. 19**

Equalizing CA.

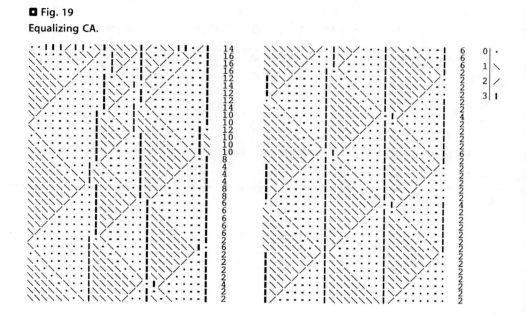

answer these questions. The obtained results help to shape a new scenario for the old-standing conjecture about the equivalence between surjectivity and DPO for CA: the study can be restricted to mixing CA (Acerbi et al. 2007, 2009).

4.13.1 Shifting

Throughout this section F_m stands for a CA F with memory $m \in \mathbb{Z}$. Given a CA $(A^{\mathbb{Z}}, F_m)$ and $h \in \mathbb{Z}$, consider the CA $(A^{\mathbb{Z}}, F_{m+h})$. Since $F_{m+h} = \sigma^h \circ F_m$, we say that the CA $(A^{\mathbb{Z}}, F_{m+h})$ is obtained by a *shifting operation*, which moves the memory of the originally given CA from m to $m+h$. In this section, properties that are preserved by the shifting operation are studied.

Proposition 25 (Acerbi et al. 2007, 2009) *Let $(A^{\mathbb{Z}}, F_m)$ be a CA. For any $h \in \mathbb{Z}$, $(A^{\mathbb{Z}}, F_{m+h})$ is surjective (resp., is injective, is right-closing, is left-closing, has JDPO) iff $(A^{\mathbb{Z}}, F_m)$ is surjective (resp., is injective, is right-closing, is left-closing, has JDPO).*

The following theorem establishes the behavior of the shift operation with respect to sensitivity, equicontinuity, and almost equicontinuity. Its proof is essentially contained in Sablik (2008).

Theorem 14 *For any CA $(A^{\mathbb{Z}}, F_m)$ one and only one of the following statements holds:*

$\mathscr{S}_0$: *the CA $(A^{\mathbb{Z}}, F_{m+h})$ is nilpotent (and then equicontinuous) for any $h \in \mathbb{Z}$;*

$\mathscr{S}_1$: *there exists an integer $\bar{h}$ with $\bar{h} + m \in [-d, d]$ such that the CA $(A^{\mathbb{Z}}, F_{m+h})$ is equicontinuous for $h = \bar{h}$ and it is sensitive for any $h \neq \bar{h}$;*

$\mathscr{S}_2$: *there is a finite interval* $I \subset \mathbb{Z}$, *with* $I + m \subseteq [-d, d]$, *such that the CA* $(A^{\mathbb{Z}}, F_{m+h})$ *is almost equicontinuous but not equicontinuous iff* $h \in I$ *(and then it is sensitive for any other* $h \in \mathbb{Z} \setminus I$);

$\mathscr{S}_3$: *the CA* $(A^{\mathbb{Z}}, F_{m+h})$ *is sensitive (ever-sensitivity) for any* $h \in \mathbb{Z}$.

In the case of surjective CA, ❯ Theorem 14 can be restated as follows.

Theorem 15 (Acerbi et al. 2007, 2009) *For any surjective CA* $(A^{\mathbb{Z}}, F_m)$ *one and only one of the following statements holds:*

$\mathscr{S}'_1$: *there exists an integer* h', *with* $h' + m \in [-d, d]$, *such that the CA* F_{m+h} *is equicontinuous for* $h = h'$ *and it is mixing for* $h \neq h'$;

$\mathscr{S}'_2$: *there exists an integer* h', *with* $h' + m \in [-d, d]$, *such that the CA* F_{m+h} *is almost equicontinuous but not equicontinuous for* $h = h'$ *and it is mixing for* $h \neq h'$;

$\mathscr{S}'_3$: *there is at most a finite set* $I \subset \mathbb{Z}$, *with* $I + m \subseteq [-d, d]$, *such that if* $h \in I$ *then the CA* F_{m+h} *is sensitive but not mixing, while it is mixing if* $h \in \mathbb{Z} \setminus I$.

The next proposition assures that positively expansive CA are in class $\mathscr{S}'_3$, in particular they are "ever-sensitive."

Proposition 26 (Acerbi et al. 2007, 2009) *If* $(A^{\mathbb{Z}}, F_m)$ *is a positively expansive CA, then for any* $h \in \mathbb{Z}$ *the CA* $(A^{\mathbb{Z}}, F_{m+h})$ *is sensitive.*

The following is a long-standing conjecture in CA theory, which dates back at least to Blanchard and Tisseur (2000).

Conjecture 1 Any Surjective CA has DPO.

❯ Proposition 25 states that the shift operation conserves surjectivity. Therefore, ❯ Conjecture 1 leads naturally to the following.

Conjecture 2 For any CA $(A^{\mathbb{Z}}, F_m)$ and any $h \in \mathbb{Z}$, $(A^{\mathbb{Z}}, F_{m+h})$ has DPO iff $(A^{\mathbb{Z}}, F_m)$ has DPO.

Recall that both surjective almost equicontinuous CA and closing CA have JDPO (Blanchard and Tisseur 2000; Boyle and Kitchens 1999). By ❯ Proposition 25, JDPO is preserved by the shifting operation, so all the CA in the classes $\mathscr{S}'_1$ and $\mathscr{S}'_2$ have JDPO. We conjecture that the same holds for (non closing) CA in $\mathscr{S}'_3$:

Conjecture 3 If a CA has DPO then it has JDPO.

Remark that ❯ Conjectures 1 and ❯ 3 are also true for the class of additive CA and the class of number-conserving CA (Cattaneo et al. 2004; Cervelle et al. 2008; Formenti and Grange 2003).

A notion that will be useful in what follows is now introduced. A CA $(A^{\mathbb{Z}}, F_m)$ is *strictly right* (resp., *strictly left*) if $m > 0$ (resp., $a < 0$) while it is *one-sided* if $m \geq 0$.

Proposition 27 (Acerbi et al. 2007, 2009) *A surjective strictly right (or strictly left) CA is mixing.*

Proposition 28 (Cattaneo et al. 2000) *Let* $(A^{\mathbb{Z}}, F_m)$ *be a strictly right CA. Any periodic configuration for the CA is also periodic for* σ.

The following corollary is a trivial consequence of the previous proposition.

Corollary 5   (Acerbi et al. 2007, 2009) *Consider a strictly right CA. If it has DPO then it has JDPO too.*

Proposition 29   (Acerbi et al. 2007, 2009) ❷ Conjecture 2 is equivalent to ❷ Conjecture 3.

As a by-product of the previous results, we have the following.

Theorem 16   (Acerbi et al. 2007, 2009) *In the CA settings, the following statements are equivalent*

1. *Surjectivity implies DPO.*
2. *Surjectivity implies JDPO.*
3. *For strictly right CA, topological mixing implies DPO.*

❷ Theorem 16 illustrates that in order to prove ❷ Conjecture 1 one can focus on mixing strictly right CA. Note that all known examples of topologically mixing CA have DPO. We now want to present a result that further supports the common feeling that ❷ Conjecture 1 is true.

Proposition 30   (Acerbi et al. 2007, 2009) *Any surjective CA (both on $A^{\mathbb{Z}}$ and on $A^{\mathbb{Z}}$) has an infinite set of points that are jointly periodic for the CA and σ.*

4.13.2 Lifting

Any CA with memory $m \geq 0$ is well defined also on $A^{\mathbb{N}}$. In that case, one can denote by $\Phi_m : A^{\mathbb{N}} \to A^{\mathbb{N}}$ its global rule. For a fixed local rule f and a memory $m \geq 0$, consider the two one-sided CA $(A^{\mathbb{Z}}, F_m)$ and $(A^{\mathbb{N}}, \Phi_m)$ on $A^{\mathbb{Z}}$ and $A^{\mathbb{N}}$, respectively. In this section, the properties that are conserved when passing from the CA on $A^{\mathbb{Z}}$ to the one on $A^{\mathbb{N}}$ and *vice-versa* are studied.

Consider the *projection* $P : A^{\mathbb{Z}} \to A^{\mathbb{N}}$ defined as follows: $\forall x \in A^{\mathbb{Z}}$, $\forall i \in \mathbb{N}, P(x)_i = x_i$. Then, P is a continuous, open, and surjective function. Moreover, the following diagram commutes

$$(1)$$

that is, $\Phi_m \circ P = P \circ F_m$. Therefore, the CA $(A^{\mathbb{N}}, \Phi_m)$ is a factor of $(A^{\mathbb{Z}}, F_m)$. For these reasons, we also say that the CA on $A^{\mathbb{Z}}$ is obtained by a *lifting (up) operation* from the CA on $A^{\mathbb{N}}$ (having the same rule and memory). As an immediate consequence of the fact that $\Phi_m \circ P = P \circ F_m$, the CA on $A^{\mathbb{N}}$ inherits from the CA on $A^{\mathbb{Z}}$ several properties such as surjectivity, transitivity, mixing, DPO, JDPO, left closingness, and openness.

The notion of left *Welch index* $L(f)$ for a local rule $f : A^{d+1} \mapsto A$ of a surjective CA can be recalled now. This index is an upper bound for the number of the left possible extensions of a block u, which are mapped by f in the same word (where with an abuse of notation, f represents the extended application of the rule f on any word of length greater than $d+1$). Let $n \geq d$ and $k > 0$ be two integers and let $u \in A^n$ a block. A left k-extension of u is a word vu where $v \in A^k$. A set of left k-extensions $v_1 u, v_2 u, \ldots, v_n u$ of u is said to be compatible with f if

$f(v_1 u) = f(v_2 u) = \ldots = f(v_n u)$. Set

$$L(u, k, f) = \max\{|W| : W \text{ set of left } k\text{-extensions of } u \text{ compatible with } f\}$$

and define $L(u, f) = \max_{k > 0}\{L(u, k, f)\}$. The left Welch index of the local rule f is $L(f) = L(u, f)$. Since the number $L(u, f)$ does not depend on the choice of u (see for instance Hedlund 1969), the left Welch index is well defined.

The following propositions reveal that the injectivity property is lifted down only under special conditions involving the Welch index, while the opposite case (lift up) is verified without additional hypothesis.

Proposition 31 (Acerbi et al. 2007, 2009) *Let $(A^{\mathbb{Z}}, F_m)$ be an injective one-sided CA. The CA $(A^{\mathbb{Z}}, \Phi_m)$ is injective if and only if $m = 0$ and the left Welch index $L(f) = 1$.*

Proposition 32 (Acerbi et al. 2007, 2009) *If $(A^{\mathbb{N}}, \Phi_m)$ is an injective (resp., surjective) CA, then the lifted CA $(A^{\mathbb{Z}}, F_m)$ is injective (resp., surjective).*

The following proposition directly follows from the definitions.

Proposition 33 *Left-closingness is conserved by the lifting up operation.*

The following results are immediately obtained from the fact that a word is blocking for a CA on $A^{\mathbb{N}}$ iff it is blocking for its lifted CA.

Proposition 34 *A CA $(A^{\mathbb{N}}, \Phi_m)$ is equicontinous (resp., almost equicontinuous) (resp., sensitive) iff the CA $(A^{\mathbb{Z}}, F_m)$ is equicontinous (resp., almost equicontinous) (resp., sensitive).*

Positive expansivity is not preserved by the lifting operation. Indeed, there are no positively expansive one-sided CA on $A^{\mathbb{Z}}$ Acerbi et al. (2007, 2009). For a proof of this result, see (Blanchard and Maass 1997).

Proposition 35 *No one-sided CA $(A^{\mathbb{Z}}, F_m)$ is positively expansive.*

Proposition 36 (Acerbi et al. 2007, 2009) *If $(A^{\mathbb{N}}, \Phi_m)$ is mixing (resp., transitive), then its lifted CA $(A^{\mathbb{Z}}, F_m)$ is mixing (resp., transitive).*

The lifting (up) of DPO remains an open problem even if on the basis of the results of ❯ Sect. 3 we conjecture that this should be true. This is also partially supported by the following proposition.

Proposition 37 (Acerbi et al. 2007, 2009) *If $(A^{\mathbb{N}}, \Phi_m)$ has JDPO, then $(A^{\mathbb{Z}}, F_m)$ has JDPO (i.e., JDPO is lifted up).*

The following result will be used later; it is the $A^{\mathbb{N}}$ version of ❯ Proposition 27.

Proposition 38 (Acerbi et al. 2007, 2009) *A surjective strictly right CA Φ_m on $A^{\mathbb{N}}$ is strongly transitive.*

❷ Proposition 28 and ❷ Corollary 5 also hold for CA on $A^{\mathbb{N}}$. As a consequence we have the following result.

Proposition 39 (Acerbi et al. 2007, 2009) *The following statements are equivalent for CA on $A^{\mathbb{N}}$:*

1. *Surjectivity implies DPO.*
2. *Surjectivity implies JDPO.*
3. *Strong transitivity implies DPO.*

As a by-product of the previous results another equivalent version of ❷ Conjecture 1 is obtained.

Proposition 40 (Acerbi et al. 2007, 2009) *The following statements are equivalent:*

1. *For CA on $A^{\mathbb{Z}}$, surjectivity implies (J)DPO.*
2. *For CA on $A^{\mathbb{N}}$, strong transitivity implies (J)DPO.*

5 From 1D CA to 2D CA

Von Neumann introduced CA as formal models for cells self-replication. These were two-dimensional objects. However, the study of the CA dynamical behavior focused essentially on the 1D case except for additive CA (Dennunzio et al. 2009a) and a few others (see, e.g., (Theyssier and Sablik 2008; Dennunzio et al. 2008, 2009c)). The reason of this gap is maybe twofold: from one hand, there is a common feeling that most of dynamical results are "automatically" transferred to higher dimensions; from the other hand, researchers mind the complexity gap. Indeed, many CA properties are dimension sensitive, that is, they are decidable in dimension 1 and undecidable in higher dimensions (Amoroso and Patt 1972; Durand 1993; Kari 1994; Durand 1998; Bernardi et al. 2005). In Dennunzio and Formenti (2008, 2009), in order to overcome this complexity gap, two deep constructions are introduced. They allow us to see a 2D CA as a 1D CA. In this way, well-known results of 1D CA can be lifted to the 2D case. The idea is to "cut" the space of configurations of a 2D CA into slices of dimension 1. Hence, the 2D CA can be seen as a new 1D CA operating on configurations made of slices. The only inconvenience is that this latter CA has an infinite alphabet. However, the constructions are refined so that slices are translation invariants along some fixed direction. This confers finiteness to the alphabet of the 1D CA allowing us to lift even more properties.

5.1 Notations and Basic Definitions

For a vector $\mathbf{i} \in \mathbb{Z}^2$, denote by $|\mathbf{i}|$ the infinite norm of $\mathbf{i}$. Let A be a finite alphabet. A *2D configuration* is a function from $\mathbb{Z}^2$ to A. The *2D configuration set* $A^{\mathbb{Z}^2}$ is equipped with the following metric, which is denoted for the sake of simplicity by the same symbol of the 1D case:

$$\forall x, y \in A^{\mathbb{Z}^2}, \quad d(x, y) = 2^{-k} \quad \text{where } k = \min\{|\mathbf{i}| : \mathbf{i} \in \mathbb{Z}^2, x_{\mathbf{i}} \neq y_{\mathbf{i}}\}$$

The 2D configuration set is a Cantor space.

For any $r \in \mathbb{N}$, let M_r be the set of all the two-dimensional matrices with values in A and entry vectors in the integer square $[-r, r]^2$. If $N \in M_r$, denote by $N(\mathbf{i}) \in A$ the element of the matrix N with entry vector $\mathbf{i}$.

A *2D CA* is a structure $\langle 2, A, r, f \rangle$, where A is the alphabet, $r \in \mathbb{N}$ is the *radius* and $f : M_r \to A$ is the *local rule* of the automaton. The local rule f induces a *global rule* $F : A^{\mathbb{Z}^2} \to A^{\mathbb{Z}^2}$ defined as follows,

$$\forall x \in A^{\mathbb{Z}^2}, \forall \mathbf{i} \in \mathbb{Z}^2, \qquad F(x)_{\mathbf{i}} = f\big(M_r^{\mathbf{i}}(x)\big)$$

where $M_r^{\mathbf{i}}(x) \in M_r$ is the *finite portion* of x of reference position $\mathbf{i} \in \mathbb{Z}^2$ and radius r defined by $\forall \mathbf{k} \in [-r, r]^2, M_r^{\mathbf{i}}(x)_{\mathbf{k}} = x_{\mathbf{i+k}}$. From now on, for the sake of simplicity, no distinction between a 2D CA and its global rule will be made.

For any $\mathbf{v} \in \mathbb{Z}^2$ the *shift map* $\sigma^{\mathbf{v}} : A^{\mathbb{Z}^2} \to A^{\mathbb{Z}^2}$ is defined by $\forall x \in A^{\mathbb{Z}^2}, \forall \mathbf{i} \in \mathbb{Z}^2$, $\sigma^{\mathbf{v}}(x)_{\mathbf{i}} = c_{\mathbf{i+v}}$. A function $F : A^{\mathbb{Z}^2} \to A^{\mathbb{Z}^2}$ is said to be *shift-commuting* if $\forall \mathbf{k} \in \mathbb{Z}^2$, $F \circ \sigma^{\mathbf{k}} = \sigma^{\mathbf{k}} \circ F$. As in 1D case, the 2D CA are exactly the class of all shift-commuting functions, which are (uniformly) continuous with respect to the metric d (Hedlund's (1990) theorem).

For any fixed vector $\mathbf{v}$, we denote by $S_{\mathbf{v}}$ the set of all configurations $x \in A^{\mathbb{Z}^2}$ such that $\sigma^{\mathbf{v}}(x) = x$. Note that, for any 2D CA global map F and for any $\mathbf{v}$, the set $S_{\mathbf{v}}$ is F-invariant, that is, $F(S_{\mathbf{v}}) \subseteq S_{\mathbf{v}}$.

A *pattern* P is a function from a finite domain $Dom(P) \subseteq \mathbb{Z}^2$ taking values in A. The notion of cylinder can be conveniently extended to general patterns as follows: for any pattern P, let $[P]$ be the set $\{x \in A^{\mathbb{Z}^2} \mid \forall \mathbf{i} \in Dom(P), x_{\mathbf{i}} = P(\mathbf{i})\}$. As in the 1D case, cylinders form a basis for the open sets. In what follows, with a little abuse of notation, for any pattern P, $F(P)$ is the pattern P' such that $Dom(P') = \{\mathbf{i} \in Dom(P), \mathcal{B}_r(\mathbf{i}) \subseteq Dom(P)\}$ and for all $\mathbf{i} \in Dom(P')$, $P'(\mathbf{i}) = f(\mathcal{B}_r(\mathbf{i}))$, where $\mathcal{B}_r(\mathbf{i}) = \{\mathbf{j} \in \mathbb{Z}^2, |\mathbf{i} - \mathbf{j}| \leq r\}$. We denote by $F^{-1}(P) = |\{P' : F(P') = P\}|$ the set of the pre-images of a given pattern P.

5.2 Combinatorial Properties

Some results about injectivity and surjectivity of 2D CA are reviewed.

A r-radius 2D CA F is said to be *k-balanced* if $|F^{-1}(P)| = |A|^{(k+2r)^2 - k^2}$ for any pattern P whose domain is a square of side k. A CA F is *balanced* if it is k-balanced for any $k > 0$.

Theorem 17 (Maruoka and Kimura 1976) *A 2D CA is surjective if and only if it is balanced.*

A 2D configuration x is *finite* if there are a symbol $a \in A$ and a natural n such that $x_{\mathbf{i}} = a$ for any $\mathbf{i}$ with $|\mathbf{i}| > n$. A symbol $a \in A$ is said to be a *quiescent state* if $f(N_a) = a$ where $N_a \in M_r$ is the matrix with all the elements equal to a. The following theorem was originally proved in Moore (1962) and Myhill (1963) for CA in any dimension admitting a quiescent state. As remarked in Durand (1998), it is valid for any CA.

Theorem 18 *A CA is surjective if and only if it is injective when restricted to finite configurations.*

As an immediate consequence of the previous result, we have that injectivity and bijectivity are equivalent notions for CA.

Proposition 41 (Durand 1998) *Bijectivity and surjectivity are equivalent for CA restricted to finite configurations.*

Proposition 42 (Durand 1998) *If a CA is bijective, then its restriction to finite configurations is bijective too.*

Call $\bigcup_{\mathbf{v}\in\mathbb{Z}^2} S_{\mathbf{v}}$ the set of the *spatially periodic configurations*.

Proposition 43 (Durand 1998) *Bijectivity and injectivity are equivalent for CA restricted to spatially periodic configurations.*

As to decidability the following results are arrived at.

Theorem 19 (Kari 1994) *It is undecidable to establish if a 2D CA is injective.*

Theorem 20 (Kari 1994) *It is undecidable to establish if a 2D CA is surjective.*

Proposition 44 (Durand 1998) *It is undecidable to establish if a 2D CA restricted to spatially periodic configurations is injective.*

5.3 Stability Classification of 2D CA

❯ Theorems 1 and ❯ 2 lead to the well-known classification of 1D CA partitioned by their stability degree as follows: equicontinuous CA, non-equicontinuous CA admitting an equicontinuity configuration, sensitive but not positively expansive CA, positively expansive CA. This classification is no more relevant in the context of 2D CA since the class of positively expansive CA is empty (Shereshevsky 1993). Moreover, the following recent result holds.

Theorem 21 (Theyssier and Sablik 2008) *There exist nonsensitive 2D CA without any equicontinuity point.*

Corollary 6 (Theyssier and Sablik 2008) *Each 2D CA falls exactly into one among the following classes:*

1. *Equicontinuous CA*
2. *Non-equicontinuous CA admitting an equicontinuity point*
3. *Nonsensitive CA without any equicontinuity point*
4. *Sensitive CA*

Theorem 22 (Theyssier and Sablik 2008) *Each of the above classes 2, 3, and 4 is neither recursively enumerable nor co-recursively enumerable.*

5.4 2D CA as 1D CA: Slicing Constructions

In this section, the slicing constructions that are fundamental to prove the results involving 2D closingness and permutivity are illustrated.

5.4.1 v-Slicing

Fix a vector $v \in \mathbb{Z}^2$ and let $\mathbf{d} \in \mathbb{Z}^2$ be a normalized integer vector (i.e., a vector in which the coordinates are co-prime) perpendicular to v. Consider the line L_0 generated by the vector $\mathbf{d}$ and the set $L_0^* = L_0 \cap \mathbb{Z}^2$ containing vectors of form $\mathbf{i} = t\mathbf{d}$ where $t \in \mathbb{Z}$. Denote by $\varphi : L_0^* \mapsto \mathbb{Z}$ the isomorphism associating any $\mathbf{i} \in L_0^*$ with the integer $\varphi(\mathbf{i}) = t$. Consider now the family $\mathscr{L}$ constituted by all the lines parallel to L_0 containing at least a point of integer coordinates. It is clear that $\mathscr{L}$ is in a one-to-one correspondence with $\mathbb{Z}$. Let l_a be the axis given by a direction $\mathbf{e}_a$, which is not contained in L_0. The lines are enumerated according to their intersection with the axis l_a. Formally, for any pair of lines L_i, L_j it holds that $i < j$ iff $p_i < p_j$ ($p_i, p_j \in \mathbb{Q}$), where $p_i\mathbf{e}_a$ and $p_j\mathbf{e}_a$ are the intersection points between the two lines and the axis l_a, respectively. Equivalently, L_i is the line expressed in parametric form by $\mathbf{i} = p_i\mathbf{e}_a + t\mathbf{d}$ ($\mathbf{i} \in \mathbb{R}^2, t \in \mathbb{R}$) and $p_i = ip_1$, where $p_1 = \min\{p_i, p_i > 0\}$. Remark that $\forall i, j \in \mathbb{Z}$, if $\mathbf{i} \in L_i$ and $\mathbf{j} \in L_j$, then $\mathbf{i} + \mathbf{j} \in L_{i+j}$. Let $\mathbf{j}_1 \in \mathbb{Z}^2$ be an arbitrary but fixed vector of L_1. For any $i \in \mathbb{Z}$, define the vector $\mathbf{j}_i = i\mathbf{j}_1$ that belongs to $L_i \cap \mathbb{Z}^2$. Then, each line L_i can be expressed in parametric form by $\mathbf{i} = \mathbf{j}_i + t\mathbf{d}$. Note that, for any $\mathbf{i} \in \mathbb{Z}^2$ there exist $i, t \in \mathbb{Z}$, such that $\mathbf{i} = \mathbf{j}_i + t\mathbf{d}$.

The construction can be summarized. We have a countable collection $\mathscr{L} = \{L_i : i \in \mathbb{Z}\}$ of lines parallel to L_0 inducing a partition of $\mathbb{Z}^2$. Indeed, defining $L_i^* = L_i \cap \mathbb{Z}^2$, it holds that $\mathbb{Z}^2 = \bigcup_{i \in \mathbb{Z}} L_i^*$ (see ❯ *Fig. 20*).

Once the plane has been sliced, any configuration $c \in A^{\mathbb{Z}^2}$ can be viewed as a mapping $c : \bigcup_{i \in \mathbb{Z}} L_i^* \mapsto \mathbb{Z}$. For every $i \in \mathbb{Z}$, the *slice* c_i of the configuration c over the line L_i is the

□ **Fig. 20**
Slicing of the plane according to the vector $v = (1, 1)$.

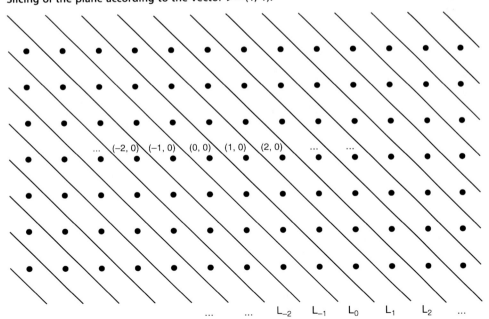

◻ **Fig. 21**

Slicing of a 2D configuration c according to the vector $v = (1, 1)$. The components of c viewed as a 1D configuration are not from the same alphabet.

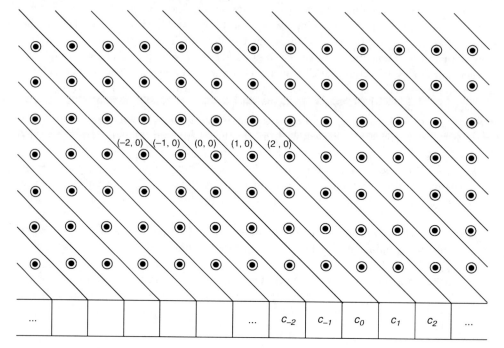

mapping $c_i : L_i^* \to A$. In other terms, c_i is the restriction of c to the set $L_i^* \subset \mathbb{Z}^2$. In this way, a configuration $c \in A^{\mathbb{Z}^2}$ can be expressed as the bi-infinite one-dimensional sequence $\prec c \succ = (\ldots, c_{-2}, c_{-1}, c_0, c_1, c_2, \ldots)$ of its slices $c_i \in A^{L_i^*}$ where the i-th component of the sequence $\prec c \succ$ is $\prec c \succ_i = c_i$ (see ❷ *Fig. 21*). One can stress that each slice c_i is defined only over the set L_i^*. Moreover, since $\forall \mathbf{i} \in \mathbb{Z}^2, \exists! i \in \mathbb{Z} : \mathbf{i} \in L_i^*$, for any configuration c and any vector $\mathbf{i}$ we write $c(\mathbf{i}) = c_i(\mathbf{i})$.

The identification of any configuration $c \in A^{\mathbb{Z}^2}$ with the corresponding bi-infinite sequence of slices $c \equiv \prec c \succ = (\ldots, c_{-2}, c_{-1}, c_0, c_1, c_2, \ldots)$ allows the introduction of a new one-dimensional bi-infinite CA over the alphabet $A^{\mathbb{Z}}$ expressed by a global transition mapping $F^* : (A^{\mathbb{Z}})^{\mathbb{Z}} \mapsto (A^{\mathbb{Z}})^{\mathbb{Z}}$, which associates any configuration $a : \mathbb{Z} \mapsto A^{\mathbb{Z}}$ with a new configuration $F^*(a) : \mathbb{Z} \to A^{\mathbb{Z}}$. The local rule f^* of this new CA that is about to be defined will take a certain number of configurations of $A^{\mathbb{Z}}$ as input and will produce a new configuration of $A^{\mathbb{Z}}$ as output.

For each $h \in \mathbb{Z}$, define the following bijective map $\mathcal{T}_h : A^{L_h^*} \mapsto A^{L_0^*}$, which associates any slice c_h over the line L_h with the slice $\mathcal{T}_h(c_h)$

$$\left(c_h : L_h^* \to A \right) \overset{\mathcal{T}_h}{\longrightarrow} \left(\mathcal{T}_h(c_h) : L_0^* \to A \right)$$

defined as $\forall \mathbf{i} \in L_0^*, \mathcal{T}_h(c_h)(\mathbf{i}) = c_h(\mathbf{i} + \mathbf{j}_h)$. Remark that the map $\mathcal{T}_h^{-1} : A^{L_0^*} \to A^{L_h^*}$ associates any slice c_0 over the line L_0 with the slice $\mathcal{T}_h^{-1}(c_0)$ over the line L_h such that

$\forall \mathbf{i} \in L_h^*$, $\mathcal{T}_h^{-1}(c_h)(\mathbf{i}) = c_0(\mathbf{i} - \mathbf{j}_h)$. Denote by $\Phi_0 : A^{L_0^*} \to A^{\mathbb{Z}}$ the bijective mapping putting in correspondence any $c_0 : L_0^* \to A$ with the configuration $\Phi_0(c_0) \in A^{\mathbb{Z}}$,

$$\left(c_0 : L_0^* \to A\right) \xrightarrow{\Phi_0} \left(\Phi_0(c_0) : \mathbb{Z}^2 \to A\right)$$

such that $\forall t \in \mathbb{Z}$, $\Phi_0(c_0)(t) := c_0(\varphi^{-1}(t))$. The map $\Phi_0^{-1} : A^{\mathbb{Z}} \to A^{L_0^*}$ associates any configuration $a \in A^{\mathbb{Z}}$ with the configuration $\Phi_0^{-1}(a) \in A^{L_0^*}$ in the following way: $\forall \mathbf{i} \in L_0^*$, $\Phi_0^{-1}(a)(\mathbf{i}) = a(\varphi(\mathbf{i}))$. Consider now the bijective map $\Psi : A^{\mathbb{Z}^2} \to (A^{\mathbb{Z}})^{\mathbb{Z}}$ defined as follows

$$\forall c \in A^{\mathbb{Z}^2}, \quad \Psi(c) = (\dots, \Phi_0(\mathcal{T}_{-1}(c_{-1})), \Phi_0(\mathcal{T}_0(c_0)), \Phi_0(\mathcal{T}_1(c_1)), \dots)$$

Its inverse map $\Psi^{-1} : (A^{\mathbb{Z}})^{\mathbb{Z}} \mapsto A^{\mathbb{Z}^2}$ is such that $\forall a \in (A^{\mathbb{Z}})^{\mathbb{Z}}$

$$\Psi^{-1}(a) = (\dots, \mathcal{T}_{-1}^{-1}(\Phi_0^{-1}(a_{-1})), \mathcal{T}_0^{-1}(\Phi_0^{-1}(a_0)), \mathcal{T}_1^{-1}(\Phi_0^{-1}(a_1)), \dots)$$

Starting from a configuration c, the isomorphism Ψ permits to obtain a one-dimensional configuration a whose components are all from the same alphabet (see ❷ Fig. 22).

Now we have all the necessary formalisms to correctly define the radius r^* local rule $f^* : (A^{\mathbb{Z}})^{2r^*+1} \to A^{\mathbb{Z}}$ starting from a radius r 2D CA F. Let r_1 and r_2 be the indexes of the lines passing for (r, r) and $(r, -r)$, respectively. The radius of the 1D CA is $r^* = \max\{r_1, r_2\}$. In other words, r^* is such that $L_{-r^*}, \dots L_{r^*}$ are all the lines that intersect the 2D r-radius Moore neighborhood. The local rule is defined as

$$\forall (a_{-r^*}, \dots, a_{r^*}) \in (A^{\mathbb{Z}})^{2r^*+1}, \quad f^*(a_{-r^*}, \dots, a_{r^*}) = \Phi_0(b)$$

where $b : L_0^* \to A$ is the slice obtained the simultaneous application of the local rule f of the original CA on the slices $c_{-r^*}, \dots, c_{r^*}$ of any configuration c such that $\forall i \in [-r^*, r^*], c_i = \mathcal{T}_i^{-1}(\Phi_0^{-1}(a_i))$ (see ❷ Fig. 23). The global map of this new CA is

❏ **Fig. 22**
All the components of the 1D configuration $a = \Phi(c)$ are from the same alphabet.

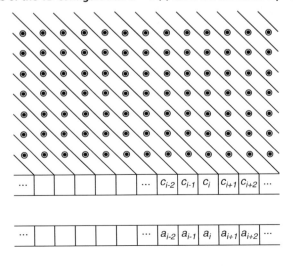

⬛ **Fig. 23**

Local rule f^* of the 1D CA as sliced version of the original 2D CA. Here $r=1$ and $r^*=2$.

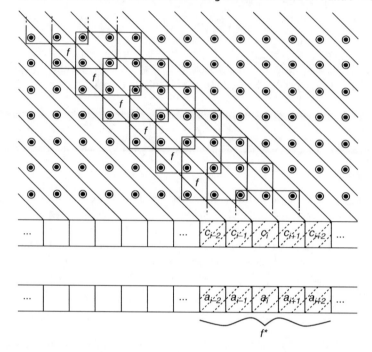

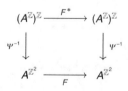

$F^* : A^{\mathbb{Z}^{\mathbb{Z}}} \to A^{\mathbb{Z}^{\mathbb{Z}}}$ and the link between F^* and f^* is given, as usual, by $(F^*(a))_i = f^*(a_{i-r^*}, \dots, a_{i+r^*})$ where $a = (\dots, a_{-1}, a_0, a_1, \dots) \in (A^{\mathbb{Z}})^{\mathbb{Z}}$ and $i \in \mathbb{Z}$.

The slicing construction can be summarized by the following theorem.

Theorem 23 (Dennunzio and Formenti 2008, 2009) *Let* $(A^{\mathbb{Z}^2}, F)$ *be a 2D CA and let* $((A^{\mathbb{Z}})^{\mathbb{Z}}, F^*)$ *be the 1D CA obtained by the v-slicing construction of it, where $v \in \mathbb{Z}^2$ is a fixed vector. The two CA are isomorphic by the bijective mapping* Ψ. *Moreover, the map* Ψ^{-1} *is continuous and then* $(A^{\mathbb{Z}^2}, F)$ *is a factor of* $((A^{\mathbb{Z}})^{\mathbb{Z}}, F^*)$.

$$
\begin{array}{ccc}
(A^{\mathbb{Z}})^{\mathbb{Z}} & \xrightarrow{\;\;F^*\;\;} & (A^{\mathbb{Z}})^{\mathbb{Z}} \\
{\scriptstyle \psi^{-1}}\downarrow & & \downarrow{\scriptstyle \psi^{-1}} \\
A^{\mathbb{Z}^2} & \xrightarrow[\;\;F\;\;]{} & A^{\mathbb{Z}^2}
\end{array}
$$

5.4.2 v-Slicing with Finite Alphabet

Fix a vector $v \in \mathbb{Z}^2$. For any 2D CA F, an associated sliced version F^* with finite alphabet can be built. In order to obtain one, it is sufficient to consider the v-slicing construction of the 2D CA restricted on the set $S_{\mathbf{v}}$, where $\mathbf{v}$ is any vector such that $\mathbf{v} \perp v$. This is possible since the set $S_{\mathbf{v}}$ is F-invariant and so $(S_{\mathbf{v}}, F)$ is a DDS. The obtained construction leads to the following theorem.

◻ **Fig. 24**

Example of v-slicing of a configuration $c \in S_v$ on the binary alphabet A where $v = (1, 1)$ and $v = (3, -3)$. The configuration $\Psi(c)$ is on the alphabet $B = A^3$.

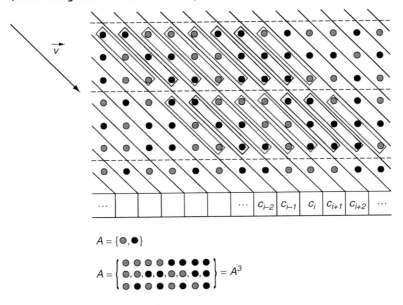

$$A = \{\bigcirc, \bullet\}$$

$$A = \left\{ \begin{matrix} \bigcirc\bigcirc\bigcirc\bullet & \bullet\bullet\bullet\bullet \\ \bigcirc,\bigcirc,\bullet,\bullet,\bigcirc,\bigcirc,\bullet,\bullet \\ \bullet\bullet\bullet\bigcirc & \bigcirc\bullet\bullet \end{matrix} \right\} = A^3$$

Theorem 24   (Dennunzio and Formenti 2008, 2009) *Let F be a 2D CA. Consider the v-slicing construction of F, where $v \in \mathbb{Z}^2$ is a fixed vector. For any vector $\mathbf{v} \in \mathbb{Z}^2$ with $\mathbf{v} \perp v$, the DDS (S_v, F) is topologically conjugated to the 1D CA $(B^{\mathbb{Z}}, F^*)$ on the finite alphabet $B = A^{|v|}$ obtained by the v-slicing construction of F restricted on S_v (❷ Fig. 24).*

$$
\begin{array}{ccc}
B^{\mathbb{Z}} & \xrightarrow{\ F^*\ } & B^{\mathbb{Z}} \\
{\scriptstyle \Psi^{-1}}\big\downarrow & & \big\downarrow{\scriptstyle \Psi^{-1}} \\
S_v & \xrightarrow[\ F\ ]{} & S_v
\end{array}
$$

The previous result is very useful since one can use all the well-known results about 1D CA and try to lift them to F.

5.5   2D Closingness

One can now generalize to 2D CA the notion of closingness.

　　Let $v \in \mathbb{Z}^2$ and denote $\bar{v} = -v$.

Definition 14   (v-asymptotic configurations) *Two configurations $x, y \in A^{\mathbb{Z}^2}$ are v-asymptotic if there exists $q \in \mathbb{Z}$ such that $\forall \mathbf{i} \in \mathbb{Z}^2$ with $v \cdot \mathbf{i} \geq q$ it holds that $x_{\mathbf{i}} = y_{\mathbf{i}}$.*

Definition 15   (v-closingness) *A 2D CA F is v-closing is for any pair of $\bar{v}$-asymptotic configurations $x, y \in A^{\mathbb{Z}^2}$ we have that $x \neq y$ implies $F(x) \neq F(y)$. A 2D CA is closing if it is v-closing for some v. A 2D CA is bi-closing if it is both v and $\bar{v}$-closing for some $v \in \mathbb{Z}^2$.*

Note that a 2D CA can be closing with respect to a certain direction but it cannot with respect to another one. For example, consider the radius $r = 1$ 2D CA on the binary alphabet whose local rule performs the xor operator on the four corners of the Moore neighborhood. It is easy to observe that this CA is $(1, 1)$-closing but it is not $(1, 0)$-closing.

Thanks to the v-slicing construction with finite alphabet, the following property holds.

Proposition 45 (Dennunzio and Formenti 2008, 2009) *Let F be a v-closing 2D CA. For any vector $\mathbf{v} \in \mathbb{Z}^2$ with $\mathbf{v} \perp v$, let $(B^{\mathbb{Z}}, F^*)$ be the 1D CA of ❷ Theorem 24 which is topologically conjugated to $(S_\mathbf{v}, F)$. Then F^* is either right or left closing.*

Using ❷ Proposition 45 and the fact that closing 1D CA have DPO, it is possible to prove the following theorem.

Theorem 25 (Dennunzio and Formenti 2008, 2009) *Any closing 2D CA has DPO.*

Corollary 7 *Any closing 2D CA is surjective.*

Regarding decidability, we have the following result whose proof is based on tiling and plane filling curves as in Kari (1994).

Proposition 46 (Dennunzio et al. 2009b) *For any $v \in \mathbb{Z}^2$, v-closingness is an undecidable property.*

Recalling that closingness is decidable in dimension 1, it is determined to be another dimension sensitive property (see Kari 1994; Bernardi et al. 2005 for other examples).

5.5.1 2D Openness

The relation between openness and closingness can be now dealt with. Recall that in 1D case, a CA is open iff it is both left and right closing. In 2D settings, we have the following properties.

Theorem 26 (Dennunzio et al. 2009b) *If a 2D CA F is bi-closing, then F is open.*

Proposition 47 (Dennunzio and Formenti 2008a, 2009) *Any open 2D CA is surjective.*

5.6 2D Permutivity

The notion of 1D permutive CA can be conveniently extended to 2D CA as follows.

Let $\tau \in \{(1, 1)(1, -1)(-1, -1)(-1, 1)\}$

Definition 16 (permutivity) A 2D CA of local rule f and radius r is τ-*permutive*, if for each pair of matrices $N, N' \in M_r$ with $N(\mathbf{i}) = N'(\mathbf{i})$ in all vectors $\mathbf{i} \neq r\tau$, it holds that $N(r\tau) \neq N'(r\tau)$ implies $f(N) \neq f(N')$. A 2D CA is *bi-permutive* iff it is both τ permutive and $\bar{\tau}$-permutive.

Several relationships among τ-permutivity, closingness, and dynamical properties are presented.

Firstly, as a consequence of the τ-slicing construction we have the following.

Proposition 48   (Dennunzio and Formenti 2008, 2009) *Consider a τ-permutive 2D CA F. For any v belonging either to the same quadrant or the opposite one as τ, the 1D CA $((A^{\mathbb{Z}})^{\mathbb{Z}}, F^*)$ obtained by the v-slicing construction is either rightmost or leftmost permutive.*

Theorem 27   (Dennunzio and Formenti 2008, 2009) *Any bi-permutive 2D CA is strongly transitive.*

Theorem 28   (Dennunzio and Formenti 2008, 2009) *Any τ-permutive 2D CA is topologically mixing.*

Theorem 29   (Dennunzio and Formenti 2008, 2009) *Consider a τ-permutive 2D CA F. For any v belonging to the same quadrant as τ, F is v-closing.*

Theorem 30   (Dennunzio and Formenti 2008, 2009) *Any τ-permutive 2D CA has DPO.*

Acknowledgments

The research was supported by the Research Program CTS MSM 0021620845, by the Interlink/MIUR project "Cellular Automata: Topological Properties, Chaos and Associated Formal Languages," by the ANR Blanc "Projet Sycomore" and by the PRIN/MIUR project "Mathematical aspects and forthcoming applications of automata and formal languages."

References

Acerbi L, Dennunzio A, Formenti E (2007) Shifting and lifting of cellular automata. In: Third conference on computability in europe (CiE 2007), Lecture notes in computer science, vol 4497. Springer, Siena, Italy, pp 1–10

Acerbi L, Dennunzio A, Formenti E (2009) Conservation of some dynamical properties for operations on cellular automata. Theor Comput Sci 410 (38–40):3685–3693. http://dx.doi.org/10.1016/j.tcs.2009.05.004; http://dblp.uni-trier.de

Akin E (1991) The general topology of dynamical systems. American Mathematical Society, Providence, RI

Amoroso S, Patt YN (1972) Decision procedures for surjectivity and injectivity of parallel maps for tessellation structures. J Comput Syst Sci 6:448–464

Bernardi V, Durand B, Formenti E, Kari J (2005) A new dimension sensitive property for cellular automata. Theor Comput Sci 345:235–247

Blanchard F, Maass A (1997) Dynamical properties of expansive one-sided cellular automata. Isr J Math 99:149–174

Blanchard F, Tisseur P (2000) Some properties of cellular automata with equicontinuity points. Ann Inst Henri Poincaré, Probabilité et Statistiques 36:569–582

Boyle M, Kitchens B (1999) Periodic points for cellular automata. Indagat Math 10:483–493

Cattaneo G, Finelli M, Margara L (2000) Investigating topological chaos by elementary cellular automata dynamics. Theor Comput Sci 244:219–241

Cattaneo G, Dennunzio A, Margara L (2002) Chaotic subshifts and related languages applications to one-dimensional cellular automata. Fundam Inf 52:39–80

Cattaneo G, Dennunzio A, Margara L (2004) Solution of some conjectures about topological properties of linear cellular automata. Theor Comput Sci 325:249–271

Cattaneo G, Dennunzio A, Formenti E, Provillard J (2009) Non-uniform cellular automata. In: Horia Dediu A, Armand-Mihai Ionescu, Martín-Vide C (eds) Proceedings of third international conference on language and automata theory and applications (LATA 2009), 2–8 April 2009. Lecture notes in computer science. Springer, vol 5457, pp 302–313. http://dx.doi.org/10.1007/978-3-642-00982-2_26; conf/lata/2009; http://dblp. uni-trier.de; http://dx. doi.org/10.1007/978-3-642-00982-2; http://dblp. uni-trier.de doi:978-3-642-00981-5

Cervelle J, Dennunzio A, Formenti E (2008) Chaotic behavior of cellular automata. In: Meyers B (ed) Mathematical basis of cellular automata, Encyclopedia of complexity and system science. Springer, Berlin, Germany

Chaudhuri P, Chowdhury D, Nandi S, Chattopadhyay S (1997) Additive cellular automata theory and applications, vol 1. IEEE Press, Mountain View, CA

Chopard B (2012) Cellular automata and lattice Boltzmann modeling of physical systems. Handbook of natural computing. Springer, Heidelberg, Germany

de Sá PG, Maes C (1992) The Gacs-Kurdyumov-Levin automaton revisited. J Stat Phys 67(3/4):507–522

Dennunzio A, Formenti E (2008) Decidable properties of 2D cellular automata. In: Twelfth conference on developments in language theory (DLT 2008), Lecture notes in computer science, vol 5257. Springer, New York, pp 264–275

Dennunzio A, Formenti E (2009) 2D cellular automata: new constructions and dynamics, 410(38–40): 3685–3693

Dennunzio A, Guillon P, Masson B (2008) Stable dynamics of sand automata. In: Fifth IFIP conference on theoretical computer science. TCS 2008, Milan, Italy, September, 8–10, 2008, vol 273, IFIP, Int. Fed. Inf. Process. Springer, Heidelberg, Germany, pp 157–179

Dennunzio A, Di Lena P, Formenti E, Margara L (2009a) On the directional dynamics of additive cellular automata. Theor Comput Sci 410(47–49):4823–4833. http://dx.doi.org/10.1016/j.tcs.2009.06.023; http://dblp.uni-trier.de

Dennunzio A, Formenti E, Weiss M (2009b) 2D cellular automata: expansivity and decidability issues. CoRR, abs/0906.0857 http://arxiv.org/abs/0906.0857; http://dblp.uni-trier.de

Dennunzio A, Guillon P, Masson B (2009c) Sand automata as cellular automata. Theor Comput Sci 410 (38–40):3962–3974. http://dx.doi.org/10.1016/j.tcs. 2009.06.016; http://dblp.uni-trier.de

Di Lena P (2006) Decidable properties for regular cellular automata. In: Fourth IFIP conference on theoretical computer science. TCS 2006, Santiago, Chile, August, 23–24, 2006, IFIP, Int. Fed. Inf. Process. vol 209. Springer, pp 185–196

Di Lena P, Margara L (2008) Computational complexity of dynamical systems: the case of cellular automata. Inf Comput 206:1104–1116

Di Lena P, Margara L (2009) Undecidable properties of limit set dynamics of cellular automata. In: Albers S, Marion J-Y (eds) Proceedings of 26th international symposium on theoretical aspects of computer science (STACS 2009), 26–28 February 2009, Freiburg, Germany, vol 3. pp 337–347. http://dx.doi.org/10.4230/LIPIcs.STACS.2009.1819; conf/stacs/2009; http://dblp.uni-trier.de

Durand B (1993) Global properties of 2D cellular automata: some complexity results. In: MFCS, Lecture notes in computer science, vol 711. Springer, Berlin, Germany, pp 433–441

Durand B (1998) Global properties of cellular automata. In: Goles E, Martinez S (eds) Cellular automata and complex systems. Kluwer, Dordrecht, The Netherlands

Farina F, Dennunzio A (2008) A predator-prey cellular automaton with parasitic interactions and environmental effects. Fundam Inf 83:337–353

Formenti E, Grange A (2003) Number conserving cellular automata II: dynamics. Theor Comput Sci 304(1–3):269–290

Formenti E, Kůrka P (2007) Subshift attractors of cellular automata. Nonlinearity 20:105–117

Formenti E, Kůrka P (2009) Dynamics of cellular automata in non-compact spaces. In: Robert A. Meyers (ed) Mathematical basis of cellular automata, Encyclopedia of complexity and system science. Springer, Heidelberg, Germany, pp 2232–2242. http://dx.doi. org/10.1007/978-0-387-30440-3_138; reference/ complexity/2009; http://dblp.uni-trier.de

Formenti E, Kůrka P, Zahradnik O (2010) A search algorithm for subshift attractors of cellular automata. Theory Comput Syst 46(3):479–498. http:// dx.doi.org/10.1007/s00224-009-9230-6; http://dblp. uni-trier.de

Gacs P (2001) Reliable cellular automata with self-organization. J Stat Phys 103(1/2):45–267

Gacs P, Kurdyumov GL, Levin LA (1978) One-dimensional uniform arrays that wash out finite islands. Peredachi Informatiki 14:92–98

Hedlund GA (1969) Endomorphisms and automorphisms of the shift dynamical system. Math Syst Theory 3:320–375

Hurley M (1990) Attractors in cellular automata. Ergod Th Dynam Syst 10:131–140

Kari J (1994) Reversibility and surjectivity problems of cellular automata. J Comput Syst Sci 48:149–182

Kari J (2008) Tiling problem and undecidability in cellular automata. In: Meyers B (ed) Mathematical basis of cellular automata, Encyclopedia of complexity and system science. Springer, Heidelberg, Germany

Kitchens BP (1998) Symbolic dynamics. Springer, Berlin, Germany

Kůrka P (1997) Languages, equicontinuity and attractors in cellular automata. Ergod Th Dynam Syst 17: 417–433

Kůrka P (2003a) Cellular automata with vanishing particles. Fundam Inf 58:1–19

Kůrka P (2003b) Topological and symbolic dynamics, Cours spécialisés, vol 11. Société Mathématique de France, Paris

Kůrka P (2005) On the measure attractor of a cellular automaton. Discrete Continuous Dyn Syst 2005 (suppl):524–535

Kůrka P (2007) Cellular automata with infinite number of subshift attractors. Complex Syst 17(3):219–230

Kůrka P (2008) Topological dynamics of one-dimensional cellular automata. In: Meyers B (ed) Mathematical basis of cellular automata, Encyclopedia of complexity and system science. Springer, Heidelberg, Germany

Kůrka P, Maass A (2000) Limit sets of cellular automata associated to probability measures. J Stat Phys 100 (5/6):1031–1047

Lind D, Marcus B (1995) An introduction to symbolic dynamics and coding. Cambridge University Press, Cambridge

Maruoka A, Kimura M (1976) Conditions for injectivity of global maps for tessellation automata. Inf Control 32:158–162

Mitchell M, Crutchfield JP, Hraber PT (1994) Evolving cellular automata to perform computations: mechanisms and impediments. Physica D 75:361–391

Moore EF (1962) Machine models of self-reproduction. Proc Symp Appl Math 14:13–33

Myhill J (1963) The converse to Moore's Garden-of-Eden theorem. Proc Am Math Soc 14:685–686

Nasu M (1995) Textile systems for endomorphisms and automorphisms of the shift, Memoires of the American Mathematical Society, vol 114. American Mathematical Society, Providence, RI

Pivato M (2008) The ergodic theory of cellular automata. In: Meyers B (ed) Mathematical basis of cellular automata, Encyclopedia of complexity and system science. Springer, Heidelberg, Germany

Sablik M (2008) Directional dynamics for cellular automata: a sensitivity to the initial conditions approach. Theor Comput Sci 400:1–18

Shereshevsky MA (1993) Expansiveness, entropy and polynomial growth for groups acting on subshifts by automorphisms. Indagat Math 4:203–210

Shereshevsky MA, Afraimovich VS (1992) Bipermutative cellular automata are topologically conjugate to the one-sided Bernoulli shift. Random Comput Dyn 1:91–98

Theyssier G, Sablik M (2008) Topological dynamics of 2D cellular automata. In: Computability in Europe (CIE'08), Lecture notes in computer science, vol 5028, pp 523–532

Wolfram S (1986) Theory and applications of cellular automata. World Scientific, Singapore

3 Algorithmic Tools on Cellular Automata

Marianne Delorme[1] · *Jacques Mazoyer*[2]
[1]Laboratoire d'Informatique Fondamentale de Marseille (LIF),
Aix-Marseille Université and CNRS, Marseille, France
delorme.marianne@orange.fr
[2]Laboratoire d'Informatique Fondamentale de Marseille (LIF),
Aix-Marseille Université and CNRS, Marseille, France
mazoyerj2@orange.fr

G. Rozenberg et al. (eds.), *Handbook of Natural Computing*, DOI 10.1007/978-3-540-92910-9_3,
© Springer-Verlag Berlin Heidelberg 2012

Abstract

This chapter is dedicated to classic tools and methods involved in cellular transformations and constructions of signals and of functions by means of signals, which will be used in subsequent chapters. The term "signal" is widely used in the field of cellular automata (CA). But, as it arises from different levels of understanding, a general definition is difficult to formalize. This chapter deals with a particular notion of signal, which is a basic and efficient tool in cellular algorithmics.

1 Introduction

The first significant works that lead to the notion of cellular automaton were those of von Neumann (1966) and Ulam (1957). Von Neumann adopted an engineering point of view: how to build an object with a wanted global behavior by means of local interactions between different components. S. Ulam observed that iterations of simple local rules may lead to global complex behaviors and interesting geometric constructions. While von Neumann's mechanical engineering project failed, the abstract cellular automaton (CA) object emerged and both processes or problematics – to look for bricks locally organized in order to obtain some wanted global result, or to discover global results from given local means – have since been at play in the history of cellular automata. Two powerful tools play a basic role in this history, especially in the emergence of "signals," namely *space–time* and *geometric diagrams*. The space–time diagrams are a way to represent the evolution of configurations by piling them up in succession, as an illustration of orbits of configurations. Geometric diagrams are a way to represent, when possible, a realization of an algorithm: a kind of abstraction of space–time diagrams. From these representations emerge patterns that seem significant and are diversely interpreted and dealt with, depending on the observation field, for example, as particles and backgrounds, signals, or information moves.

❯ *Figure 1a* shows one space–time diagram and ❯ *Fig. 1b* one geometric diagram (actually a decorated geometric diagram: to get a "true" (underlying) geometric diagram, let us replace discrete straight lines by continuous lines by removing the colors represented by white squares or gray triangles in black squares) of a solution of the FSSP (Balzer 1966). It gives an immediate intuition of "signals" as segments of discrete or real lines, carrying and transmitting information inside some "universe." Although this example may lead one to think there is some canonical correspondence between space–time and geometrical diagrams, it is not the case. Actually, a space–time diagram cannot necessarily be translated into a geometric one, and there are algorithms whose realizations cannot be drawn as geometric diagrams. Nevertheless, what follows will prove the power of these representations.

This chapter explains the classic tools and methods that lie at the foundations of cellular algorithmics in dimension 1 and that are essentially based on cellular constructions or transformations of signals. In the second section, after general definitions – in particular, the definition of a signal – an example of a construction of a signal is developed. This is done in order to introduce a more general notion of transformations by cellular automata, which is used in the following section. In the third section, the transformation formalism allows one to get the first negative result. The rest of the section is then dedicated to a special transformation: the translation of a finite right signal. The question about the uniformization of this translation is posed. A lot of techniques are used, such as extraction of signal moves, reconstruction of a signal, and data duplication. In the fourth section, other constructions are developed, concerning deformations of signals, and constructions of infinite families of

◻ Fig. 1

(a) An evolution of the automaton. (b) A corresponding geometric diagram.

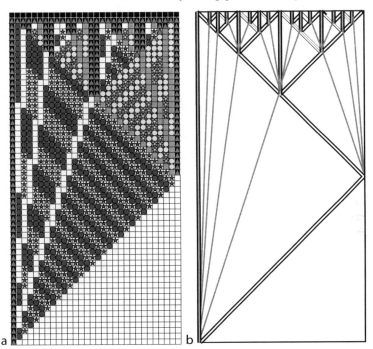

a   b

signals, discrete parabolas and exponentials. The fifth section is devoted to constructions of functions. Fischer's constructibility is explained and a limitation on signals slopes is presented and proved. In the sixth section, other tools and methods are presented on information transfers, arithmetic operations on segments, a technique called "riddle," the technique of freezing, as well as its generalization that is called "clipping." Finally, the seventh section is a conclusion, in which the question of signals in dimension 2 is also discussed.

Some definitions and notations are recalled here.

Definition 1   (one-dimensional cellular automata)

1. A 1D-CA is a 3-uplet $\mathcal{A} = (Q, V, \delta)$ where:
   - $Q = \{q_1, \ldots, q_{|Q|}\}$ is a finite set (the *states* set of $\mathcal{A}$).
   - $V = \{v_1, \ldots, v_{|V|}\}$ is a finite ordered subset of $\mathbb{Z}$, called the *neighborhood*, which from now on will be $\{-1, 0, +1\}$, unless clearly mentioned (so that $\mathcal{A}$ becomes $\mathcal{A} = (Q, \delta)$).
   - δ is a mapping from $Q^{|V|}$ into Q (the *states* (or *local*) *transition function* of $\mathcal{A}$).
2. A *configuration* of $\mathcal{A}$ is a mapping from $\mathbb{Z}$ into Q. The set $Q^{\mathbb{Z}}$ of $\mathcal{A}$-configurations is denoted $\mathcal{C}$.
   The *global transition function* of $\mathcal{A}$, $G : \mathcal{C} \rightarrow \mathcal{C}$, is defined by:
   $$\forall x \in \mathcal{C}, \forall z \in \mathbb{Z}, \; G(x)(z) = \delta(x(z + v_1), \ldots, x(z + v_{|V|}))$$
   $G(x)(z)$ will also be denoted $G(x)_z$.
   The *orbit of a configuration* x is the sequence $\mathcal{O}(x) = \{G^t(x) \mid t \in \mathbb{N}\}$.
3. A state q of Q is said to be *quiescent* when $\delta(q, \ldots, q) = q$.

Restricting the neighborhood of CAs to $\{-1, 0, +1\}$ is not a problem. Indeed, let $\mathcal{A} = (Q, \{-2, -1, 0, +1, +2\}, \delta_A)$ be considered, for example. Then, there exists an automaton $\mathcal{B} = (Q^2, \{-1, 0, +1\}, \delta_B)$ that has the behavior of $\mathcal{A}$ at the expense of some transformation. Here, δ_B operates as follows:

- Starting from three states of $\mathcal{B}$, $(q_{\ell,-1}, q_{r,-1})$, $(q_{\ell,0} \quad q_{r,0})$, $(q_{\ell,+1}, q_{r,+1})$, one gets $q_\ell = \delta_A(q_{\ell,-1}, q_{r,-1}, q_{\ell,0}, q_{r,0}, q_{\ell,+1})$ and $q_r = \delta_A(q_{r,-1}, q_{\ell,0}, q_{r,0}, q_{\ell,+1}, q_{r,+1})$. Then $\delta_B((q_{\ell,-1}, q_{r,-1}), (q_{\ell,0}, q_{r,0}), (q_{\ell,+1}, q_{r,+1})) = (q_\ell, q_r)$.

In fact, it is possible to build by grouping, for each cellular automaton $\mathcal{A}$ and all (m, p) in $\mathbb{N} \times \mathbb{N}$ (where m is the space factor and p the time factor), the automaton $\mathcal{A}^{(m,p)}$, which is algorithmically equivalent to $\mathcal{A}$ (Rapaport 1998; Ollinger 2002; Theyssier 2005). (See also the chapter ❷ Universalities in Cellular Automata of this volume.) The neighborhood of $\mathcal{A}^{(m,p)}$ is $\{-1, 0, +1\}$ for all sufficiently large m.

Orbits of random configurations of some cellular automata show regular arrangements of states, sometimes looking like threads, which can be diversely interpreted, but are usually considered to be significant and carrying "information". Here, we focus on some of them that we call signals. Understanding how pieces of information advance through a space–time diagram amounts to considering the states set Q as $Q_1 \cup Q_2$, Q_1 representing information which progresses through the states of Q_2 which can then be considered as equivalent relative to Q_1. This justifies the identification of Q_2 states as a unique state, L, which is latent for Q_1 states. Therefore, this chapter considers configurations made of a finite segment of Q_1 states inside lines of $L : {}^\infty L q_0 \ldots q_\ell L^\infty$. Using grouping, this can be seen as ${}^\infty L q L^\infty$. (In the following, we shall assume that the first non-quiescent state is on cell 0.)

2 Signals

2.1 Introduction

As the interest here is in cellular automata, the idea of signals is obtained by observing orbits; however, in reality it more generally concerns colorings of the discrete plane. Starting with a general idea, the attention here is restricted to signals that are constructible (or generated) by cellular automata.

2.2 Basic Definitions

Definition 2 (signal)

1. Let C be some finite set, $S \subset C$, and let $\mu : \mathbb{Z} \times \mathbb{N} \to C$ be some coloring of $\mathbb{Z} \times \mathbb{N}$. The coloring μ is said to present a *signal* S on S if S is a sequence of sites $(x_k, t_k)_{k < \alpha}$ (elements of $\mathbb{Z} \times \mathbb{N}$) with indices in an initial segment of $\mathbb{N}$ (possibly whole $\mathbb{N}$), such that:
   - $\mu(x_k, t_k)$ is in S.
   - (x_{k+1}, t_{k+1}) is $(x_k, t_k + 1)$ or $(x_k + 1, t_k)$ or $(x_k - 1, t_k)$.

- If (x_{k+1}, t_{k+1}) is $(x_k, t_k + 1)$, then $\mu(x_k + 1, t_k) \notin S$ or $(x_k + 1, t_k)$ is (x_{k-1}, t_{k-1}), and $\mu(x_k - 1, t_k) \notin S$ or $(x_k - 1, t_k)$ is (x_{k-1}, t_{k-1}).
- If (x_{k+1}, t_{k+1}) is $(x_k + 1, t_k)$, then $\mu(x_k - 1, t_k) \notin S$.
- If (x_{k+1}, t_{k+1}) is $(x_k - 1, t_k)$, then $\mu(x_k + 1, t_k) \notin S$.

Usually, because the context is clear, one speaks of a signal S without mentioning the coloring that presents it. The state (in the signal) of a site (x_k, t_k) is sometimes denoted by $\langle x_k, t_k \rangle$.

2. A signal S presented by some coloring μ is said to be CA-generated if there exists some cellular automaton, $\mathcal{A} = (Q, \delta)$, and some element $c_{\mathcal{A}}$ of $Q^{\mathbb{Z}}$ such that μ is the orbit of the configuration $c_{\mathcal{A}}$.

   A signal S presented by some coloring μ is said to be CA-constructible if there exists some cellular automaton $\mathcal{A} = (Q, \delta)$ with a quiescent state L, a distinguished state G and $S \subseteq Q \setminus \{L\}$ such that μ is the orbit of the configuration $^{\infty}LGL^{\infty}$. Such a configuration is called the *basic initial configuration*.

3. In what follows, sites (x_k, t_k) will represent both positions and states $\mu(x_k, t_k)$, and signals will be denoted by S_S or S.

4. A *signal* is said to be *right* (*left*) if, whatever k is, (x_{k+1}, t_{k+1}) is $(x_k, t_k + 1)$ or $(x_k + 1, t_k)$ $((x_k, t_k + 1)$ *or* $(x_k - 1, t_k)$, *respectively*).

5. Let S be a right signal with $(x_0, t_0) = (0, 0)$ and let ξ be in $\{x_k | k < \alpha\}$. Let η_ξ denote the smallest time t for S to reach the cell ξ, that is, $\eta_\xi = t_{k_0}$ where $k_0 < \alpha$ is such that
   - $x_{k_0} = \xi$ and $t_{k_0} = \min\{t \mid \exists k(x_k = \xi$ and $t = t_k)\}$
   - For each k, $k < k_0$, $x_k \neq \xi$

   The function $\Pi_S : \xi \mapsto \eta_\xi$ is called the *slope* of S. To the slope corresponds the *speed* of S that is the function $\mathcal{V}_S : t \mapsto (\max_{\eta_\xi \leq t} \xi)/t$.

   Most of time, signals are discrete straight lines. Their slopes are then simply identified to the slopes of these straight lines.

6. If a signal S is finite,
   - Its site (x_0, t_0) is said to be its *origin* if for all $t < t_0$, all x, $(x, t) \notin S$ and $(x_0, t_0 + 1) \in S$ (then $(x_0 - 1, t_0) \notin S$, $(x_0 + 1, t_0) \notin S$ or $(x_0 + 1, t_0) \in S$ and $(x_0 + 1, t_0 + 1) \in S$ or $(x_0 - 1, t_0) \in S$ and $(x_0 - 1, t_0 + 1) \in S$).
   - Its site (x_f, t_f) is said to be its *end* if for all $t > t_f$, all x, $(x, t) \notin S$ and $(x_f, t_f - 1) \in S$ or $(x_f - 1, t_f)$ and $(x_f - 1, t_f - 1)$ belong to S or $(x_f + 1, t_f)$ and $(x_f + 1, t_f - 1)$ belong to S.

Due to the choice of the initial configuration, if a signal S is CA-constructed by some cellular automaton (Q, δ), and is the only one to be constructed on $^{\infty}LGL^{\infty}$, then it is always an ultimately periodic sequence (of states in S, as well as in its slope and speed (Mazoyer 1989)). In **❯** *Fig. 2a*, for example, the signal is of period $(2, 3)$, the slope is $\frac{3}{2}$ and the speed $\frac{2}{3}$ ("it advances of two cells in three time steps").

Usually, several signals may appear in an orbit, as it is the case in **❯** *Fig. 2b*. Clearly, the choice to use 4-connected signals is arbitrary. One may consider 8-connected signals as well: in this case, there exists a uniform way to translate a 8-connected signal into a 4-connected one (e.g., if $(x_{k+1}, t_{k+1}) = (x_k + 1, t_k + 1)$, add $(x_k, t_k + 1)$ to the signal and renumber it).

Definition 3 (signals emissions and interactions) Let S_S be a signal generated by some cellular automaton $\mathcal{A} = (Q \cup \{L\}, \delta)$ (**❯** Definition 2) and let K be a subset of Q.

■ **Fig. 2**

(a) An example of a single signal. **(b)** Several signals inside an orbit. **(c)** Geometric representation of a signal. **(d)** Geometric representation of a signal with indications.

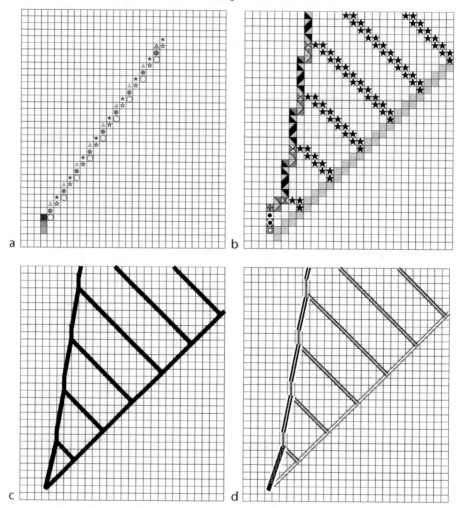

1. $\mathcal{S}_S$ is said to *send out* or *emit* a signal $\mathcal{S}_K$ if there is some k such that:
   - (x_k, t_k) belongs to $\mathcal{S}_S$
   - $\langle x_k - 1, t_k + 1 \rangle \in K$
   - $\langle x_k - 1, t_k + 2 \rangle \in K$

   or the symmetrical layout where $(x_k + 1, t_k + 1)$ and $(x_k + 1, t_k + 2)$ belong to K, or the

   case where $(x_k, t_k) \in S$, $(x_k, t_k + 1)$ and $(x_k, t_k + 2)$ belong to K: $\left(\begin{smallmatrix} \blacksquare \end{smallmatrix} \right)$, $\left(\begin{smallmatrix} \blacksquare \end{smallmatrix} \right)$,

   $\left(\begin{smallmatrix} \blacksquare \end{smallmatrix} \right)$.

2. A signal $\mathcal{S}_S$ is said to *interact* with some signal $\mathcal{S}_K$ if (ξ, η) and $(\xi + 1, \eta)$ belong to S and K, respectively (or K and S) (▪▪), or if (ξ, η) and $(\xi + 2, \eta)$ belong to S and K, respectively (or K and S) (▪ ▪).

In ❯ *Fig. 2b*, several signals may be distinguished. First, there is the signal on the only gray state, which goes to the right at speed 1. Second, there are the only star state signals, which are sent out to the left, at speed 1, by the gray signal. Finally, there is a right signal defined on all the other states that interacts with the star signals. But the previous figure can also be considered to be made of several finite signals. The site of the origin of the gray signal can be assumed to be $(1, 1)$. Then, one can distinguish a first finite signal $\mathcal{S}_0$ from the site $(0, 2)$ to $(2, 6)$, a second one $\mathcal{S}_1$, sent out by the first star signal, from $(2, 7)$ to $(2, 9)$, which sends out a new one $\mathcal{S}'_1$ from $(2, 10)$ up to $(3, 13)$, which interacts with the second star signal, which in turn sends out a new signal $\mathcal{S}_2$, and so on.

This example leads to at least two remarks about the notion of a signal. First, observing the star signals (as well as the $\mathcal{S}_i$ and $\mathcal{S}'_i$, $i \geq 1$) one would like to consider them as different instances of a single one. Second, it appears that significant information moves seem not to essentially depend on the states themselves, but on state sets and their geometrical displays. This leads to abstract signals (or instances of signals) as segments of straight lines as shown in ❯ *Fig. 2c, d*. One often says that ❯ *Fig. 2c* or ❯ *Fig. 2d* are *geometric diagrams* associated to the orbit in ❯ *Fig. 2b*. Historically, when one wants to build a cellular automaton realizing a given task, one tries to draw a geometrical representation of an algorithm instance to get a geometrical diagram; when it is possible, one assigns sets of states to each isolated segment in identifying instances of a same signal and managing junctions (simple meetings or creations) in order to obtain the corresponding orbit.

An important question arises: starting from a geometrical diagram, is it possible to define a cellular automaton whose orbit would have the former as a geometrical diagram? The question is not simple, and it is necessary to clearly define the notions. An answer can be found in a specific framework in Richard (2008). Nevertheless, one is able to assert the following.

Proposition 1 *Starting from a finite family of geometric diagrams such that*

- *The number of different lines is finite.*
- *The number of different junctions (inputs and outputs) is finite.*
- *All the meetings and all the emissions are of finite size.*

Then, there is some cellular automaton with a family of orbits corresponding to these geometric diagrams.

In what follows, geometrical representations will be extensively used. When necessary, we will verify that the conditions of ❯ Proposition 1 are satisfied. If they are not, proper justifications will be provided.

2.3 Translation of a Finite Right Signal: A First Example of Construction by Cellular Automaton

A first example of a construction of a signal is developed with the purpose of introducing the concept of (uniform) transformations of signals by a cellular automaton.

Let S be a finite right signal, with origin (x_0, t_0), and end (x_f, t_f), generated by some cellular automaton $\mathcal{B}$, starting from an initial configuration $c_\mathcal{B}$ (see ❷ Definition 2). One wants to design a cellular automaton $\mathcal{A}$, with $T \subset Q_\mathcal{A}$, such that:

- The states of T make up a finite right signal T in the $\mathcal{A}$-orbit of $c_\mathcal{B}$, beginning at $(x_f, t_f + 1)$, ending at $(2x_f - x_0, 2t_f - t_0 + 1)$, and such that $(\xi + x_f - x_0, \eta + t_f - t_0 + 1) \in T$ for each $(\xi, \eta) \in S$. In other words T is obtained from S by the translation $(x_f - x_0, t_f - t_0 + 1)$.
- $Q_\mathcal{A} = Q_\mathcal{B} \cup Q_{\mathcal{A}^*}$ with $Q_\mathcal{B} \cap Q_{\mathcal{A}^*} = S$ and $Q_\mathcal{A}$ contains a quiescent state $q_\mathcal{A}$. Moreover, if $\delta_\mathcal{A} \upharpoonright Q_\mathcal{B} = \delta_\mathcal{B}$, $\delta_\mathcal{A}$ is stable on $Q_{\mathcal{A}^*}$, the actions of all the states of $Q_\mathcal{B} \setminus S$ on $Q_{\mathcal{A}^*}$ are identical and the states in $Q_{\mathcal{A}^*}$ take into account only the states of S in $Q_\mathcal{B}$. So the only effective knowledge that $\mathcal{A}_{\upharpoonright Q_{\mathcal{A}^*}}$ has from $\mathcal{B}$ is S.

One may observe that the previous conditions are satisfied if $\mathcal{A}$ maintains in $Q_{\mathcal{A}^*}$ a "joke copy" of $Q_\mathcal{B}$.

The result is obtained by building a family of signals S_i. Because S is a right signal, it is a sequence of stationary and rightward moves. Let St_i and Sd_i, for $i \geq 0$, be the successions of sites along vertical and diagonal segments, respectively.

- S_0 : This signal (see ❷ Fig. 3a) counts the sites of St_0. Let st_0 be the result. If $st_0 \neq 0$, S_0 brings it along S up to (x_f, t_f), where it emits a signal which marks the sites $(x_f, t_f + 1)$, ..., $(x_f, t_f + st_0 + 1)$ (giving Tst_0) and then goes to the right at speed 1 along a diagonal DT_0. If $st_0 = 0$, S_0 brings information up to (x_f, t_f) and generates DT_0 starting from $(x_f, t_f + 1)$.
- S_1 : First, this signal (see ❷ Fig. 3b) counts the moves to the right of Sd_0. The result is assumed to be sd_0. Then, from the first site (ξ_1, η_1) of St_1 it goes to the right at speed 1 for sd_0 time steps before becoming stationary. It remains stationary for time st_1, which will be determined by means of signals at speed 1 starting from the sites of St_1. Signal S_1 will memorize st_1 until it meets the speed 1 signal D_f, emitted from (x_f, t_f). There, it will emit a stationary signal which will install the part Tst_1 of T.

  Moreover, a signal, at speed 1, will count the number sd_2 of sites of Sd_2, and bring it to S_1. This information will arrive at the end of the former stationary phase, and makes S_1 go to the right for sd_2 moves, and so on.
- S_2 : This will build the part Tst_2 of T. It will start from the site of S_1 where S_1 becomes stationary for the second time, and will receive from S_1 the number of previous part of right moves of S_1.
- A finite family of signals S_i is so obtained.
- When D_f does not get any more information from signals S_i, it becomes stationary and the process stops when the last DT diagonal meets it.

S being finite, the number of described signals is finite, but several of them are superimposed over one another. Actually, if S has at most h successive moves to the right and at most k successive stationary moves, then the states, outside S of a cellular automaton $\mathcal{A}$, which realizes the above algorithm, belong to $Q_{\mathcal{A}^*} = Q_\mathcal{B} \times (Q_h \times Q_h \times Q_k \times Q_s)^k$, where s represents the number of superimpositions. The space effectively used by all signals essentially lies between S, D_f, and a signal $\mathcal{P}$ which appears below all of them, and has vertical segments of the same length as vertical segments of S and diagonal segments of length $2sd_i$, $i \in \{1, \ldots, k\}$.

☐ **Fig. 3**
(a) Translation of $S : S_0$. (b) Translation of $S : S_0$ and S_1.

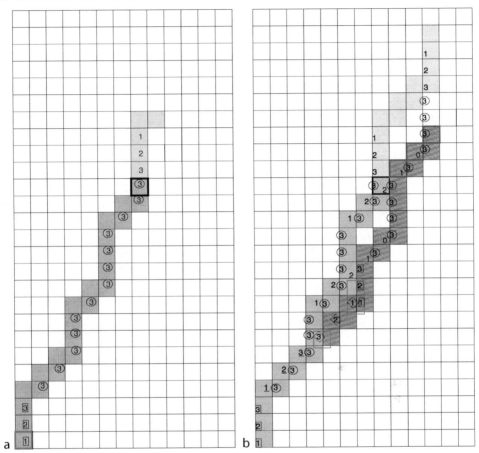

2.4   A Notion of Transformation by Cellular Automaton

In the construction of ❱ Sect. 2.3, the automaton $\mathcal{A}$ actually acts on the orbit $c_{\mathcal{B}}$ of $\mathcal{B}$, identifying all the states of $\mathcal{B}$ which are not in S, otherwise being uniform in $\mathcal{B}$. This leads one to generalize the situation and to consider a cellular automaton that *works* on a colored half plane satisfying some property (presenting a finite right signal, e.g., as in ❱ Sect. 2.3) and transforms it into another colored half plane satisfying another property (e.g., containing a finite right signal the upper part being a translation of the lower one as in ❱ Sect. 2.3). The following definitions are proposed and will be used in the next section.

Definition 4   Let $C \cup \{\lambda\}$ be some finite set.

1. A cellular automaton $\mathcal{A}$ is said to *work on* C if :
   - $Q_{\mathcal{A}} = \Theta_{\mathcal{A}} \times (C \cup \{\lambda\})$,

- $\delta_{Q_A} : Q_A^3 \to \Theta_A$,
- $\{L\} \subseteq \Theta_A$ where L is quiescent

2. To each application μ from $\mathbb{Z} \times \mathbb{N}$ into $C \cup \{\lambda\}$ and to each configuration $c_A \in \Theta_A^{\mathbb{Z}}$ one can associate a sequence $(c_A^t)_{t \geq 0}$ of A-configurations defined by $c_A^0 = c_A$

$$c_A^{t+1}(x) = \delta_{Q_A}((c_A^t(x-1), \mu(x-1, t)), (c_A^t(x), \mu(x, t)),$$
$$(c_A^t(x+1), \mu(x+1, t)))$$

which is called the *orbit of c_A on μ*.

If $c_A^0 = {}^\infty L^\infty$, the *orbit of A on μ* will be considered.

Moreover, one can associate to each orbit of a cellular automaton some coloring of the half plane.

Definition 5 Let A be some cellular automaton such that $Q_A = Q_A^\sharp \times Q_A^\flat$ (with $Q_A^\sharp \cap Q_A^\flat = \emptyset$) and let c_A be some A-configuration. To each pair of sets $C^\sharp \subseteq Q_A^\sharp$ and $C^\flat \subseteq Q_A^\flat$, one associates an application $\mu_{c_A, C^\sharp, C^\flat}$ from $\mathbb{Z} \times \mathbb{N}$ into $C^\sharp \cup C^\flat \cup \{\lambda\}$ as follows:

- If $\langle x, t \rangle \in Q_A^\sharp \times C^\flat$, then $\mu_{c_A, C^\sharp, C^\flat}(\langle x, t \rangle)$ is the component of $\langle x, t \rangle$ in $C^\flat$.
- If $\langle x, t \rangle \in C^\sharp \times Q_A^\flat$, then $\mu_{c_A, C^\sharp, C^\flat}(\langle x, t \rangle)$ is the component of $\langle x, t \rangle$ in $C^\sharp$.
- $\mu_{c_A, C^\sharp, C^\flat}(\langle x, t \rangle)$ is λ if both or neither of the two conditions above hold.

Then, one can see a cellular automaton that works on μ as an automaton which takes, as input, some picture $\mu : \mathbb{Z} \times \mathbb{N} \to C \cup \{\lambda\}$ defined by a finite part of sites in $\mathbb{Z} \times \mathbb{N}$ colored by C. As a result, one wants some finite picture colored by a new set $C^\star$, and one wants the process to be uniform. This leads first to a notion of transformation (❯ Definition 6) and the definition of a *uniformly working* cellular automaton (❯ Definition 7).

Definition 6 Let C and $C \cup C^\star$ be two finite sets, let M be the set of functions from $\mathbb{Z} \times \mathbb{N}$ into $C \cup \{\lambda\}$, and let $M^\star$ be the set of functions from $\mathbb{Z} \times \mathbb{N}$ into $C \cup C^\star \cup \{\lambda\}$. A transformation $\mathbb{T}_{C, C^\star}$ means a partial function from M into $M^\star$.

Definition 7 Let C and $C \cup C^\star$ be two finite sets, $(C \cap C^\star = \emptyset)$. Let $\mathbb{T}_{C, C^\star}$ be some transformation. A cellular automaton A which works on C *uniformly realizes* $\mathbb{T}_{C, C^\star}$ if:

1. Θ_A contains $C^\star$
2. For each μ belonging to the $\mathbb{T}_{C, C^\star}$-domain, $\mathbb{T}_{C, C^\star}(\mu) = \mu_{\infty L^\infty, C, C^\star}$.

Returning to the translation of finite right signals in this formal framework allows one to prove a first negative result.

3   Cellular Translations of Finite Right Signals

3.1   Setting Down the Problem

One considers two transformations:

$\mathbb{T}_{S, S \cup S^\star}$ Its domain is made up of finite colorings μ of $\mathbb{Z} \times \mathbb{N}$ presenting finite right signals S, while their images $\mathbb{T}_{S, S \cup S^\star}(\mu)$ present finite right signals $S^\star$ obtained by the translation $(x_f - x_0, t_f - t_0)$ applied to S as studied in ❯ Sect. 2.3, without any other condition on S.

Then $\mathbb{T}_{S,S \cup S^*}$ transforms some finite right signal S, starting at (x_0, t_0), ending at (x_f, t_f), into the finite right signal starting at (x_0, t_0), ending at $(2x_f - x_0, 2t_f - t_0 + 1)$ starting with S extended, via the above translation, by an instance of itself.

$\mathcal{U}_{S,S^*}$ Its domain is made up of finite colorings μ of $\mathbb{Z} \times \mathbb{N}$ presenting finite right signals S (with origin (x_0, t_0), end (x_f, t_f)). Their images $\mathbb{U}_{S,S^*}(\mu)$ present finite right signals S^* with origin (x_0^*, t_0^*), end $(x_0^* + 2(x_f - x_0), t_0^* + 2(t_f - t_0) + 1)$ with the following properties:

- $\tau_1(S) \subseteq S^*$, where τ_1 denotes the translation defined by $(x_0^* - x_0, t_0^* - t_0)$.
- $S^* = \tau_1(S) \cup \tau_2(\tau_1(S))$, where τ_2 denotes the translation given by $(x_f - x_0, t_f - t_0 + 1)$.

Then, $\mathbb{U}_{S,S^*}$ transforms some signal S into a finite right signal of the same shape as the signal S^* produced by $\mathbb{T}_{S,S \cup S^*}$ in the previous point, but with a starting site (x_0^*, t_0^*) elsewhere in $\mathbb{Z} \times \mathbb{N}$.

Proposition 2   *No cellular automaton can uniformly realize the transformation* $\mathbb{T}_{S,S \cup S^*}$.

Proof ❯ *Figure 4* can be seen as the significant part of an orbit μ of a cellular automaton $\mathcal{B}$, with six states including a quiescent one (the "white" one), all the transitions of which are represented on the diagram. The initial configuration is given on the lowest line. The orbit μ shows a finite right signal S in gray and black (made up of right moves followed by only one stationary move, beginning at (x_0, t_0), ending at (x_f, t_f)), the length, ℓ, of which depends on the initial configuration.

Suppose there is some automaton $\mathcal{A}$ that uniformly realizes the wanted transformation. As S is essentially diagonal, information that can start from its sites only goes above or on the diagonal D_S which carries it, and there is no way to get back information necessary for

◻ **Fig. 4**
Example of a signal S (gray and black) the length of which depends on the initial line : it springs from the meeting of the two striped signals and finishes a step after the meeting with the signal in stipple, arriving at speed 1 from the right.

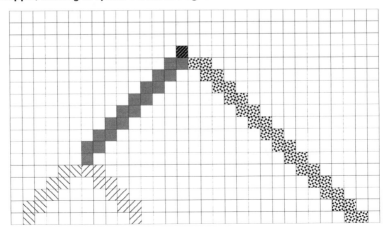

delimiting the translated signal $S^\star$ on $D_{S^\star}$. The states of the sites $(x_f + \zeta, t_f - 1 + \zeta)$ $(\zeta \in \mathbb{N}^\star)$ define an ultimately periodic sequence of period at most $|\Theta_A|$, after at most $|\Theta_A|$ steps. As a consequence (see ❷ Sect. 5.2), the states of the sites $(x_f + \zeta, t_f + \zeta)$ are ultimately periodic of period at most $|\Theta_A|^2$ and the sites $(x_f + \zeta, t_f + \zeta + 1)$ are ultimately periodic of period at most $|\Theta_A|^3$. As a consequence, the signal image of S in $\mathbb{T}_{S,S^\star}$ has to be infinite or of length bounded by $2|\Theta_A|^3$, a contradiction.

Then there is no cellular automaton which realizes $\mathbb{T}_{S,S^\star}$. This shows that, in the construction of ❷ Sect. 2.3, the automaton A needs the knowledge of automaton B.

3.2   Uniform Translation of Finite Right Signals

Proposition 3   *There is a cellular automaton which uniformly realizes the transformation $\mathbb{U}_{S,S^\star}$.*

Proof   The proof rests on a method of signal analysis and the possibility of rebuilding the signal starting with the analysis of its stationary and rightward moves. It is split into several parts dedicated to different algorithms.

▶   Point a : Analysis of a finite right signal.

The algorithm is illustrated with ❷ Fig. 5. It consists in describing the succession of stationary and rightward moves of the original signal S on a stationary signal that starts at the end of S.

A family $(S_i)_i$ of signals, parallel to S (which is S_0), is built. Signal S_i contributes to the construction of S_{i+1} only starting from its third move when it sends out to the left, at speed 1, a signal ς_3 that carries the nature (stationary or rightward) of this third move. Then, it indicates its moves to the right by means of a signal ς at speed 1 to the left. The signals S_i end at $(x_{f_i}, t_{f_i}) = (x_f - i, t_f + i)$ on a signal $\mathcal{E}_\searrow$ sent, at speed 1 to the left, from the end site (x_f, t_f) of S.

The states of a signal S_i are to be taken in a set $T \times \{0, 1\} \times \{a, b, \emptyset\}$. The value of the second component says whether the first move of S_i is stationary (0) or rightward (1), while the value of the third component indicates whether the second move is rightward (a), stationary (b) or is unknown ($\emptyset$). These pieces of data are collected on $\mathcal{E}_\searrow$, and sent to the right and collected again on a stationary signal $\mathcal{E}_\uparrow$ created at the end of S. Let it be stressed that each signal S_i sends pieces of information to be marked on signal $\mathcal{E}_\uparrow$.

▶   Point b : Analysis of the stationary moves (rightward moves) of a finite right signal.

What one has to do now is to extract the stationary moves of S, and to keep their number and the information of whether they follow a stationary or a rightward move, as shown in ❷ Fig. 6a. The idea is the same as the one in point a.

Signal S_{i+1} starts at the end of the second (from origin) stationary move on signal S_i, $i \geq 0$, and $S_0 = S$. It carries pieces of information relative to these first stationary moves : a if the successor move on S_i is stationary, b if it is rightward, $\emptyset$ if there is no successor move. All this data is collected on the signal $\mathcal{V}$ which goes to the left, at speed 1, from the end of S. Then, they are redirected to the line $x = x_f$ at speed 1 and symbolized in ❷ Fig. 6a by circles. A dark gray circle means that the indicated move (which was stationary) is followed by a rightward

◘ Fig. 5

Extraction of the moves of a finite right signal ($0a1a1b1a0a1a0\emptyset$). The original signal $\mathcal{S}$ is represented by an essentially thick black line. For $i \geq 1$, $\mathcal{S}_i$ is represented by an essentially medium-gray line. The signals ς_3 are represented by thick medium-gray lines, with a fine darker gray line inside if the move is to the right, or a fine lighter gray line inside if the move is stationary. The signals ς are represented by fine dark gray lines. Inside the $\mathcal{S}_i$-lines appear two lines: the first one, starting from the starting point of $\mathcal{S}_i$, indicates the first move of $\mathcal{S}_i$ stationary (0, light gray) or to the right (1, dark gray); the second one, starting one time unit later, indicates its second move stationary (a, light gray) or to the right (b, right gray), information which comes from $\mathcal{S}_{i-1}(i \geq 1)$. Component $\emptyset$ is indicated by the lack of a line. The signal $\mathcal{E}_\searrow$ and $\mathcal{E}_\uparrow$ are easily recognizable. Data sent from $\mathcal{E}_\searrow$ to $\mathcal{E}_\uparrow$ are in light gray for stationary moves and in dark gray for moves to the right. They are symbolized on $\mathcal{E}_\uparrow$ by light or dark circles. The end of signal $\mathcal{S}$, denoted by $\emptyset$, is indicated by a dark striped line and symbolized on $\mathcal{E}_\uparrow$ by a circle with stripes inside.

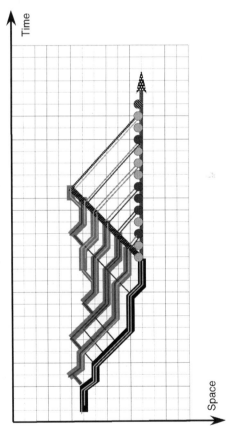

one (if any), while a white circle means that the indicated move (stationary) is followed by a stationary one (if any). As for the last circle with stripes, it also marks the end of $\mathcal{V}$. On $\mathcal{V}$, the information on successors of stationary moves is obtained, but not on their number, which appears on the vertical segment stemming from (x_f, t_f).

■ **Fig. 6**

(a) Analysis of the stationary moves of a finite right signal (*abbaabaa*). The star marks the end of the initial signal S, the same as in ❷ *Fig. 5*. (b) Analysis of the rightward moves of a finite right signal (*abbaa*), the same as in ❷ *Fig. 5*.

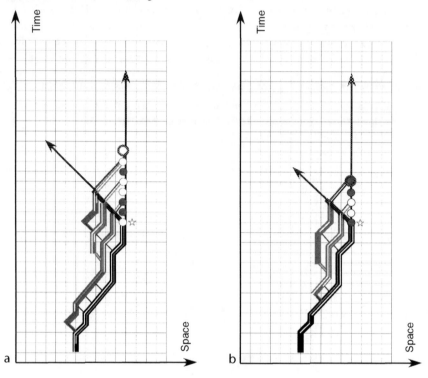

The moves to the right of S are similarly analyzed (❷ *Fig. 6b*). Finally, both algorithms acting in parallel bring the number of stationary and rightward moves onto the line $y - t_f = -(x - x_f)$ starting at (x_f, t_f).

▶ Point c : Reconstruction of a finite right signal.

Suppose that one is able to gather, at some site (x_I, t_I) data obtained in point b in the following way : data about stationary moves on the line $D_\uparrow$ ($x = x_I$), data about rightward moves on the line $D_\rightarrow$ ($y - t_I = x - x_I$), both lines starting from (x_I, t_I). Then, it is possible to rebuild the analyzed signal starting from (x_I, t_I), as shown in ❷ *Fig. 7*. From each site on $D_\uparrow$ a signal parallel to $D_\rightarrow$ is sent, which says whether the data indicates a change in the direction of the signal (*b*), a stationary move (*a*) or end of data. From each site on $D_\rightarrow$ is sent a signal parallel to $D_\uparrow$ which says whether the data indicates a change of direction of the signal (case *a*) or not (case *b*) or end of the data. Now, it is easy to rebuild a signal S' which is a translated copy of the initial segment S.

- From point b, it is known whether the first sequence of moves is stationary or not.
- A sequence of stationary moves goes on as long as it does not get a signal that says that it has to change to a right move, which is the beginning of a sequence of right moves.

□ Fig. 7

Reconstruction of a finite right signal. Analysis of the stationary (rightward) moves of the studied signal are coded on $D_\uparrow$ ($D_\rightarrow$, respectively). On $D_\uparrow$, a gray circle means "I was stationary and I become rightward," a white circle means "I was stationary and I remains stationary." On $D_\uparrow$, a gray circle means "I was rightward and I become stationary," a white circle means "I was rightward and I remain rightward." The star marks the origin of the new signal. In fact, in the figure, the construction of the wanted signal S is anticipated, in starting with data *abbaabaaabbaabaa* represented on $D_\uparrow$ and *abbaaabbaa* represented on $D_\rightarrow$, which correspond to the duplicated results of the analyses represented in **❷ Fig. 6a, b.**

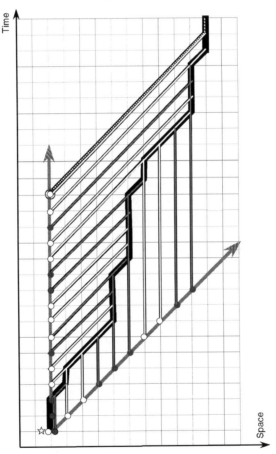

- A sequence of moves to the right goes on as long as it does not get a signal that says it has to change direction. Then, a stationary move is executed, which is the beginning of a sequence of stationary moves.
- The construction is achieved when both signals of end arrive.

It is clear that the new signal S' has the same sequences of moves as the original signal S. But it is not yet the wanted signal. In order to get the results, one has to duplicate the data concerning stationary and rightward moves obtained in point b, as well as to find a suitable point (x_I, t_I).

▶ Point d : The end of the proof, duplication, and transfer of sets of data.

The number of stationary (rightward) moves is denoted $n_\uparrow$ ($n_\rightarrow$, respectively), so the length of signals $\mathcal{V}$ in b) are $\left\lfloor \frac{n_\uparrow}{2} \right\rfloor + 1$ and $\left\lfloor \frac{n_\rightarrow}{2} \right\rfloor + 1$. Then several cases would have to be considered according to the parity of these numbers. As all the data resulting from the analyses of moves are on $D_\uparrow$, one wants to find (x_I, t_I) on this very line. Two cases have to be distinguished depending on whether $2n_\rightarrow \geq n_\uparrow$ or not. In the first case, the algorithm is the following one:

1. Data concerning the moves to the right, obtained, at the end of point b, on $D_\uparrow$ starting at site (x_f, t_f) are sent to the left at speed 1. They rebound on the vertical line drawn from the end of the signal $\mathcal{V}$ corresponding to rightward moves (actually, two cases are to be considered according to the parity of $\left\lfloor \frac{n_\rightarrow}{2} \right\rfloor$, see ❷ Sect. 6.1) corresponding to the rightward moves and come back to the right, at speed 1, up to $D_\uparrow$, the vertical line starting from site (x_f, t_f).

■ **Fig. 8**

(a) Duplication and setting up of the rightward moves. The sequence *abbaa* has become *abbaaabbaa*. The star represents the suitable origin of the duplication. (b) Setting up of the stationary moves. The sequence *abbaabaa* has become *abbaabaaabbaabaa*.

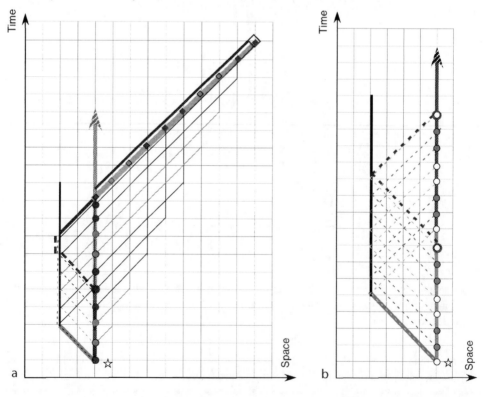

2. Then all the data corresponding to the rightward moves (on sites (x_f, t_f) up to $(x_f, t_f + n_{\rightarrow})$) are sent on the diagonal $\Delta_{\rightarrow}$, parallel to $D_{\rightarrow}$, starting from site $(x_f, t_f + n_{\rightarrow})$ (see ❷ *Fig. 8a* and ❷ *Sect. 6.1*).

3. Data corresponding to stationary moves are collected on $D_{\uparrow}$ starting from (x_f, t_f). They are sent at speed 1 to the left, stay on the vertical line $V_{\uparrow}$ created at the end of the signal $\mathcal{V}$ corresponding to the stationary moves for a time d and come back at speed 1 on $D_{\uparrow}$ where they rebound at speed 1 up to $V_{\uparrow}$, where they again rebound at speed 1 up to $D_{\uparrow}$ and stop. The delay d is determined, by $2n_{\rightarrow} - n_{\uparrow}$ and its parity, in such a way that the duplication of the data starts at site $(x_f, t_f + n_{\rightarrow})$. See ❷ *Fig. 8b*. Here two cases are to be considered, according to the parity of $\lfloor \frac{n_{\rightarrow}}{2} \rfloor$ (❷ *Fig. 8b* shows the even case).

4. Finally, it is possible to build the wanted signal $\mathcal{S}^*$ with origin $(x_f, t_f + n_{\rightarrow})$, as in point c.

The algorithm for the second case is analogous. Actually, the data concerning rightward signal are treated the same way except that they are dispatched on the right diagonal stemming from site $(x_f, t_f + n_{\uparrow})$. Data on stationary moves are duplicated from this very site by the usual means of rebounding between $V_{\uparrow}$ and $D_{\uparrow}$.

In the proof above, the origin of $\mathcal{S}^*$ is not the best possible one because the data could be sent on $\mathcal{V}$ not 2 by 2 but 2^k by 2^k for any $k \in \mathbb{N}$. Then the origin of $\mathcal{S}^*$ would be $\left(x_f, t_f + \frac{n_{\rightarrow}}{k}\right)$ or $\left(x_f, t_f + \frac{n_{\uparrow}}{k}\right)$. But the origin can never be (x_f, t_f) as proved in ❷ Proposition 2. One observes that computations are done in a part of the space–time diagram which is above the given signal, and never below. This is necessary according to ❷ Proposition 2.

Of course the same process can be applied to left signals (diagonals become $D_{\leftarrow}$ parallel to the straight line $y = -x$).

For any signal, the process can be easily adapted: it suffices to cut the signal into left and right signals and to shift the signals $D_{\uparrow}$ parallel to $\mathcal{S}$.

◻ **Fig. 9**
(a) Translation of a finite right signal according to a left diagonal. (b) The impossibility of translating a finite right signal according to some signal.

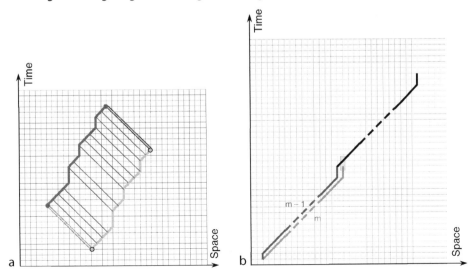

4 Various Constructions of Signals Starting from Signals

4.1 Translations and Other Modifications

❯ *Figure 9a* shows how to shift a finite right signal according to a diagonal to the left. Let S be a finite right signal (light gray in ❯ *Fig. 9a*) and let P be an arriving point on the diagonal $\{x_0 - \ell, t_0 + \ell | \ell \in \mathbb{N}\}$ stemming from the origin (x_0, t_0) of S (light stripes in ❯ *Fig. 9a*). Each time S "advances to the right", a signal $\mathcal{D}$ is emitted at speed 1 to the left (fine lines in ❯ *Fig. 9a*). From P is emitted a new signal $\mathcal{T}$ (dark gray in ❯ *Fig. 9a*) that stays stationary except when it gets a signal $\mathcal{D}$, then it goes to the right. A particular signal $\mathcal{D}^*$ (in dark stripes in ❯ *Fig. 9a*) is created on the end site of S; it goes at speed 1 to the left and kills $\mathcal{T}$ when they meet.

However, it is not trivial to generalize the previous construction. In the framework of ❯ Sect. 3.1, consider, for example, the signal S (in light gray in ❯ *Fig. 9b*) defined by m moves at speed 1 to the right followed by two stationary moves, and the signal $\mathcal{T}$ defined by one stationary move followed by $m-1$ moves at speed 1 to the right, followed by two stationary moves (dark gray in ❯ *Fig. 9b*). If one wants to translate S according to $\mathcal{T}$, the three sites $(x_f, t_f + \ell)$, $\ell \in \{0, \ldots, 3\}$ (where (x_f, t_f) is the end site of $\mathcal{T}$) have to know the encoding of m, which is impossible.

But it is always possible to translate any finite right signal according to a *periodic* left signal (for a number of times equal to an integer multiple of the period). ❯ *Figure 10a* shows an example of a corresponding geometric diagram. A left periodic signal $\mathcal{T}$ is defined by its period π, which can be seen as a sequence $(c_0, \ldots, c_{\pi-1})$ of elements of $\{-1, 0, 1\}$ (0 coding a stationary move, -1 a left one and 1 a right one). In this case, the signals are no more coded by one state, but by π states as it is shown in ❯ *Fig. 10b*.

◼ **Fig. 10**
(a) Translation of a finite right signal according to a periodic left signal, geometric diagram.
(b) Translation of a finite right signal according to a periodic left signal, representation of the left signal by states.

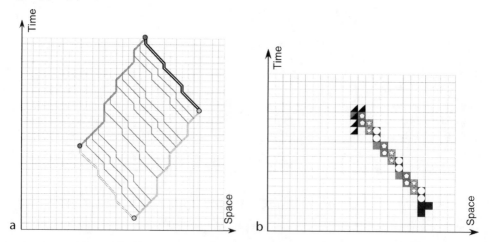

Numerous techniques exist in order to build signals by means of other signals. Some examples have already been seen. The most straightforward method is illustrated in ❷ *Fig. 11a*. Some signal S (fine lines in ❷ *Fig. 11a*) sends out, each time it goes to the right, a signal D which goes to the left at speed 1. When these signals D meet another signal T, they modify it.

The set of states on T is partitioned into two subsets (black and dark gray in ❷ *Fig. 11a*). It is stationary except when in the dark gray state it receives a signal D. Then it goes to the right. The color is swapped each time the signal receives D. In this way, T goes to the right "more or less" with half speed compared to S. Instead of partitioning T, it is also possible to distinguish

❑ **Fig. 11**
(a) Modification of a finite right signal using its own moves to the right. Parity is indicated on T.
(b) Modification of a right signal by means of its rightward moves, parity marked on S.
(c) Modification of a right signal by means of its stationary moves.

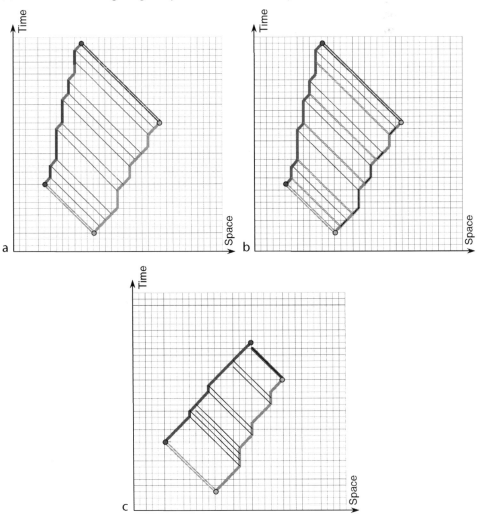

two types of states for S (light gray and light gray with a black line in ❯ *Fig. 11b*) and to consider two types of signals $\mathcal{D}$, say $\mathcal{D}_1$ and $\mathcal{D}_2$ (denoted by fine dark gray lines and light gray doubled lines in ❯ *Fig. 11b*). Then $\mathcal{T}$ only advances when it gets a signal $\mathcal{D}_1$.

One can finally think of using the stationary moves of S. It is possible as shown on ❯ *Fig. 11c*. Only one stationary move of S out of two induces a stationary move of $\mathcal{T}$. In this case, one has to partition T into six parts because a signal to the right and a signal to the left can meet in two different ways (see ❯ *Sect. 2.1*). These cases are highlighted in very light gray on ❯ *Fig. 11c*. One then gets a signal $\mathcal{T}$ which goes to the right "more or less" twice as fast as S.

4.2 Infinite Families of Signals

One can iterate the process and build slower and slower signals as described in ❯ Sect. 4.1 and in ❯ *Fig. 11b*. (This idea appears for the first time in Waksman (1996)). The only difficulty is to define the origins of the signals $\mathcal{T}$ (here denoted $\mathcal{S}_k$ with $\mathcal{S}_1 = \mathcal{S}$). Actually, it is possible to put these origins on any infinite left signal. In order to make the process more regular, one has to set these origins on a stationary signal (in black in ❯ *Fig. 12a, b*). One gets, for example, the geometric diagram in ❯ *Fig. 12a*, where the initial signal is the diagonal signal at speed 1 to the right. (The partitions of S are omitted in the figure.) ❯ *Figure 12b* shows the necessary states. The slope of S is 1 and the slope of $\mathcal{S}_k$ is $2^k - 1$, $k \geq 1$. If (x_0, t_0) denotes the origin of S, then the origin of $\mathcal{S}_k$ is $\left(x_0, t_0 + \sum_{i=1}^{i=k} 2^i\right)$.

◻ **Fig. 12**
(a) Infinite family of signals of slope $2^k - 1$: geometric diagram. **(b)** Infinite family of signals of slope $2^k - 1$: geometric diagram and states.

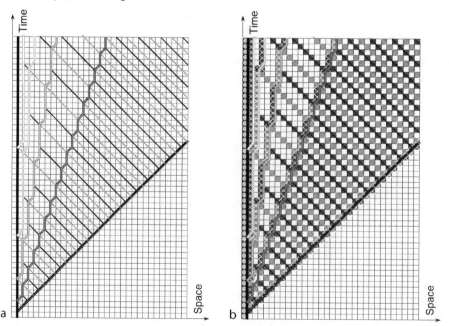

□ Fig. 13
(a) Minimization of the states number. (b) Another example.

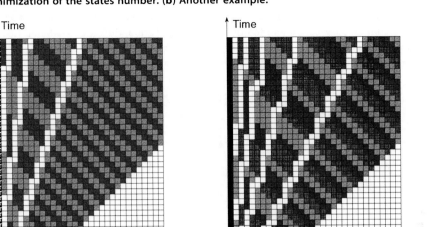

It is possible to minimize the number of necessary states, by means of clever encodings, as shown in ❷ *Fig. 13a*. A detailed study can be found in Delorme and Mazoyer (2002b).

Of course, it is possible to distinguish more than two types of diagonal signals to the left and then obtain other infinite families of signals as in ❷ *Fig. 13b*. In fact, for any integers p and q, with $p > q$, it is possible to build an infinite family of signals of slopes $\left(\frac{p}{q}\right)^k$ (see Mazoyer (1989b)).

What is noticeable is that segments of exponential length can be built by a set of finite automata, with computation divided up among all the cells.

4.3 Parabolas and Exponentials

It appears, for the first time in Fisher (1965), that it is possible to build signals of nonlinear slopes on cellular automata. A discrete parabolic branch is constructed in Fisher (1965), and other examples are also exhibited in the chapter ❷ Computations on Cellular Automata of this volume.

4.3.1 Parabolas

How to build a signal $\mathcal{S}$ of slope $\Pi_{\mathcal{S}}(t) = \lfloor x^2 \rfloor$ is shown here. The basic idea is to use the formula

$$1 + 2 + \ldots + k = \frac{k(k + 1)}{2}$$

◻ **Fig. 14**

(a) Construction of $y = \frac{x(x+1)}{2}$. The signal S is in fine black lines. The signal $\mathcal{Z}$ is in striped lines, it zigzags between the signals S and T. The signal T is in thick lines and presents three sorts of states : medium gray corresponding to rightward moves, light and dark gray when it is stationary. After a right move, it is in light gray before receiving $\mathcal{Z}$ and in dark gray afterward. **(b)** Construction of $y = x(x + 1)$. **(c)** Construction of $y = x^2$.

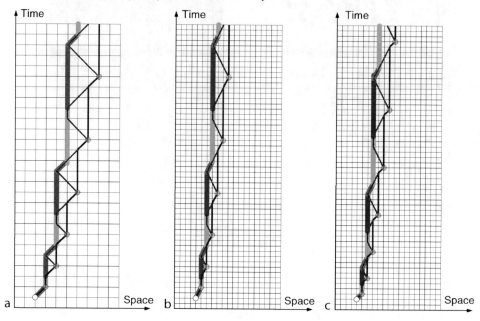

In a first step, one builds a signal S of slope $\Pi_S(x) = \left\lfloor \frac{x(x+1)}{2} \right\rfloor$ (see ❷ *Fig. 14a*). Signal S (black in ❷ *Fig. 14a*) moves one time to the right, then waits k units of time. That requires an implementation of a counter for k. If k is even, $k = 2\ell$, a signal $\mathcal{Z}$, zigzagging as fast as possible between two stationary signals T and S on cells x and $x + \ell$, counts 2ℓ. If k is odd, $k = 2\ell + 1$, in order to count $2\ell + 1$, it suffices that signal $\mathcal{Z}$ "waits" one time unit on T. Then, from sites $P_S = \left(x_0 + 2\ell + 1, t_0 + \frac{(2\ell+1)(2\ell+2)}{2} \right)$ (on S) a signal $\mathcal{Z}_1$ is sent along a diagonal to the left (black oblique stripes in ❷ *Fig. 14a*), it sends a signal $\mathcal{Z}_3$ which stays one unit of time on T (black horizontal stripes in ❷ *Fig. 14a*) and comes back along a diagonal to the right (black oblique stripes in ❷ *Fig. 14a*) as $\mathcal{Z}_2$, and meets S at time $t_0 + \frac{(2\ell+1)(2\ell+2)}{2} + 2\ell + 1$ as wanted. Moreover, when $\mathcal{Z}_2$ meets S, signal S moves once to the right, then sends a new signal $\mathcal{Z}_1$ that immediately sends a signal $\mathcal{Z}_2$ when arriving on T. The distance between S and T is then $2\ell + 2$ and $\mathcal{Z}_2$ arrives on S after k times as wanted. In order to know whether $\mathcal{Z}_1$ waits one time or not, it is sufficient to distinguish two types of states in T (light or dark gray in ❷ *Fig. 14a*). Then, in order to build S, it is sufficient to move the counter "wall" T of one cell each time S has moved twice.

In a second step, one builds a signal $S^\star$ of slope $\Pi_{S^\star}(x) = x(x + 1)$ (see ❷ *Fig. 14b*). The former signal S has then to become another signal $S^\star$. When S goes to the right, $S^\star$ goes to the right and waits for one unit of time and when S remains one unit of time on a cell, $S^\star$ remains two units of time on it. In order to establish this, it is sufficient to replace the speeds 1 and -1 of the counters in the former construction by speeds $\frac{1}{2}$ and $\frac{-1}{2}$ and to ask the signal $\mathcal{Z}_1$ to remain stationary two units of time instead of one. Then one gets the diagram shown in ❯ *Fig. 14b*.

Finally, in order to get the signal $\mathcal{S}^\sharp$ of slope $\Pi_{\mathcal{S}^\sharp}(x) = x^2$, it suffices to transform the moves at speed $\frac{1}{2}$ of $\mathcal{S}^*$ into moves at speed 1 as shown in ❯ *Fig. 14c*.

4.3.2 Signals of Exponential Slopes

Suppose one wants to build signals of exponential slopes. The idea is given by the formula

$$1 + 2 + 2^2 + \ldots + 2^{k-1} = 2^k - 1$$

One first builds a signal $\mathcal{S}$ of slope $\Pi_{\mathcal{S}}(x) = 2^x - 1$ by following step-by-step the process of the previous construction (see ❯ *Fig. 15a*). Between site $\left(x_0 + k, t_0 + \sum_{j=0}^{j=k-1} 2^j\right)$ and site $\left(x_0 + k + 1, t_0 + \sum_{j=1}^{j=k} 2^j\right)$, $\mathcal{S}$ has to remain stationary $2^k - 1$ units of time, before advancing by one step. This can be done with three signals $\mathcal{Z}_1$, $\mathcal{Z}_3$, and $\mathcal{Z}_2$ as follows. Signal $\mathcal{Z}_1$ has to go 2^{k-1} times to the left before giving birth to $\mathcal{Z}_3$ on the site

■ **Fig. 15**
(a) Construction of $y = 2^x - 1$ (first method). (b) Construction of $y = 2^x - 1$ (second method).

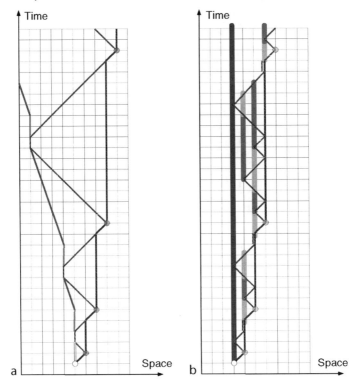

$M_k = (x_0 + k - 2^{k-1}, \ t_0 + \sum_{j=1}^{j=k-1} 2^j + 2^{k-1})$. With $\zeta = 2^{k-1}$ one has $M_k = (x_0 + k - \zeta, t_0 + 3\zeta - 1)$ and also $M_{k+1} = (x_0 + k + 1 - 2\zeta, t_0 + 6\zeta - 1)$. That leads to $M_{k+1} - M_k = (1 - \zeta, 3\zeta)$, that is also $M_{k+1} - M_k = (1 - \zeta, 3 + 3(\zeta - 1))$. Then, with $\eta = \zeta - 1$, $M_{k+1} - M_k = (-\eta, 3 + \eta)$. Between the sites M_k and M_{k+1}, the straight line $\{(-\eta, 3 + \eta) \mid \eta \in \mathbb{N}\}$ is built by a signal which stays three units of time on M_k followed by a signal of slope -3 (dark gray in ❷ *Fig. 15a*). One observes that the computing area arbitrarily stretches out to the right and to the left of cell x_0.

There are of course other methods to count 2^k. In particular, ❷ *Fig. 15b* shows one in which the computing area does not stretch to the left of x_0 and is contained between S and the straight line $x = x_0$. In using the results of ❷ Sect. 5.2, the computing area can be moved between S and the diagonal stemming from (x_0, t_0).

To conclude, it must be noted that similar methods allow one to build signals of slopes k^p and p^k for $p \geq 2$.

5 Functions Constructions

5.1 Fisher's Constructibility

Fisher (1965) showed that increasing functions could be computed on cellular automata by marking times on a cell (prime numbers in Fisher's paper, see the chapter ❷ Computations on Cellular Automata). Later, it was proved that DOL systems could be generated on one-dimensional cellular automata as in Čulik and Karhumäki (1983) and Čulik and Dube (1993). This work has been resumed and completed in Terrier (1991). Finally, "marking times" were studied in more detail in Mazoyer and Terrier (1999).

Definition 8 (to mark times) A cellular automaton $\mathcal{A} = (Q, \delta)$ (with a quiescent state L and a distinguished state G) is said to *mark times* according to some increasing function, f, if there is some subset M of Q such that, in the evolution of $\mathcal{A}$ on the basic initial configuration $(^\infty LGL^\infty)$, the state of cell 0 is in M at time t if and only if $t = f(n)$ for some $n \in \mathbb{N}$.

Many increasing functions can be "marked" by such cellular automata. ❷ *Figures 16a, b,* ❷ *17a, b,* and ❷ *18a, b* show how functions $n \to 2^n$, $n \to 3^n$, $n \to n^2$, and $n \to n^3$ can be marked. The algorithms are briefly described.

5.1.1 Marking Times k^n

In order to mark the times 2^n, one uses the formula

$$2^n - 1 = 1 + 2 + 2^2 + \ldots + 2^{n-1}$$

Two signals (with two types of stripes in ❷ *Fig. 16a*) oscillate at speed one between a stationary signal that marks cell 0 (dark gray in ❷ *Fig. 16a*) and a signal of slope 3 (white with black edges in ❷ *Fig. 16a*) that progresses by i cells in $3i$ time steps.

Obviously, using the formula of the geometric series, all the functions $n \to k^n$ allow us to mark times k^n, by means of "black and white signals" of slope $\frac{k+1}{k-1}$. ❷ *Figure 16b* illustrates the case $k = 3$.

◘ Fig. 16
(a) Marking times 2^n. (b) Marking times 3^n.

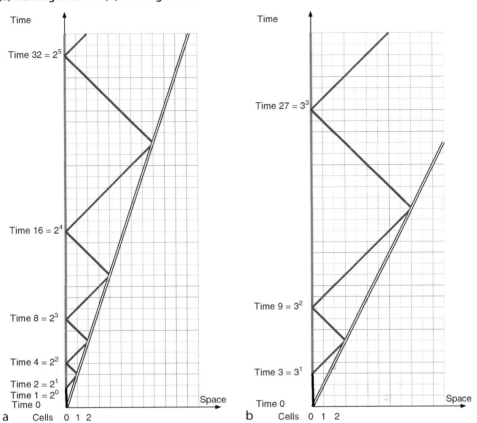

5.1.2 Marking Times n^k

In order to mark sites $(0, n^2)$, one uses the formula

$$(n+1)^2 = n^2 + 2n + 1$$

An idea of the method in the case where one wants to distinguish the sites $P_n = (n, n^2)$ is given and, consequently, manages to mark times $n^2 + n$. One observes that vector $P_n P_{n+1}$ is $(1, 2n + 1)$. It is possible to go from site P_n to site P_{n+1} by means of a zigzag between the current cell and cell 0 (striped lines in ❷ *Fig. 17a*) followed by a shift of one cell to the right (dark gray lines edged in black in ❷ *Fig. 17a*).

In a similar way, one can distinguish the sites (n, n^3) in using the formula

$$(n+1)^3 = n^3 + 3n^2 + 3n + 1$$

❷ *Figure 17b* illustrates the process, which is a little more complex. In order to build the vector $(n + 1, (n + 1)^3) - (n, n^3) = (1, 3n^2 + 3n + 1)$, one identifies the one step moves of the signal to be built. In ❷ *Fig. 17b*, the signal joining the sites (n, n^3) is a white black edged line, except

☐ **Fig. 17**

(a) Distinguishing times (n, n^2) and marking $n \to n^2 + n$. (b) Distinguishing times (n, n^3) and marking $n \to n^3 + n$.

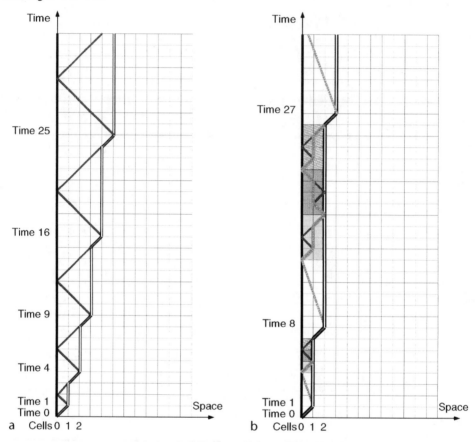

when it moves to the right for one step where it becomes a gray black edged line. Cell 0 is marked by a stationary black signal. By means of a signal of slope 3 stemming from site (n, n^3) (light and medium striped lines in ❷ *Fig. 17b*), times $3n$ are marked. In order to mark $3n^2$, one uses three times the method to distinguish (n, n^2) (see ❷ *Fig. 17a*). The signal marking the sites (n, n^2), starting from cell 0 (or from cell n) will reach cell n (or 0, respectively) in n^2 units of time. In ❷ *Fig. 17b*, n^2 is marked starting from the cell 0 (with a light gray background) then from the cell n (with a medium gray background) and finally, starting again from cell 0 (with a background striped in light and medium gray).

It is clear that by iterating the method, and using the binomial formula, one is able to mark (n, n^k) for all k.

Now, consider times n^2. One observes that sending a signal to the left at speed 1 from site $(n - 1, n^2 - (n - 1)) = (n - 1, n^2 - n + 1)$ ends in cell 0 where n^2 is then marked. So, one has to build signals that allow one to distinguish these sites, and this can be done as above: As $(n, (n + 1)^2 - (n + 1) + 1) - (n, n^2 - n + 1) = (1, 2n)$, from $(n - 1, n^2 - (n - 1))$ stems a signal $\mathcal{S}$ up to site $(n, n^2 - (n - 1) + 1)$, where it sends a signal $\mathcal{C}$ to the left, which has to

☑ **Fig. 18**

(a) **Marking times n^2. (b) Marking times n^3.**

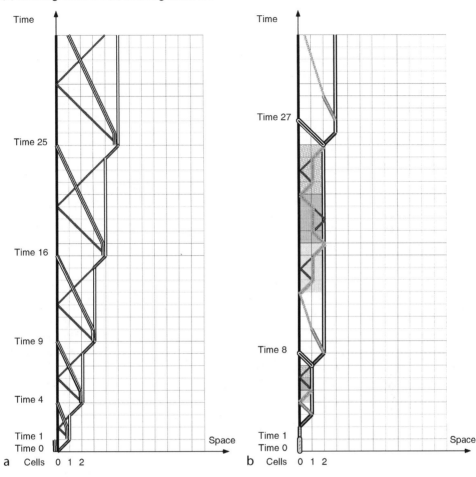

count $2n - 1$, and becomes itself stationary. In order to count $2n - 1$, $\mathcal{C}$ goes at speed 1 to the left up to cell 1, stays there for one time, starts again at speed one to the right and meets $\mathcal{S}$ on $(n, n^2 - n + 1)$. Of course the process has to be initiated (for $n = 1, 2$). This is illustrated in ❷ *Fig. 18a*. In order to mark times n^3 (and more generally n^k), one uses a different method because times $3n$ have to be marked. One loses one time step in replacing the signal of slope 3 in ❷ *Fig. 17a*) by a signal that moves from one cell to the left in two time steps (light and medium gray stripes, edged or not in dark gray in ❷ *Fig. 18b*). The process is initiated (for $n = 1$) as indicated in ❷ *Fig. 18b*.

5.1.3 Some Properties of Functions That Can Be Marked

More generally, it has been proved (Mazoyer and Terrier 1999) that the class of functions that can be marked is closed under addition, subtraction (with some additional conditions of growth), multiplication by a rational, multiplication, composition, and iteration. There is also an interesting characterization of these functions which links them to a class of complexity. (For the zoology of cellular classes, see Delorme and Mazoyer (1996).)

Proposition 4 (see Mazoyer and Terrier (1999)) *A function can be marked on the first cell if and only if the language* $\{1^{f(n)}|n \in \mathbb{N}\}$ *is recognizable in real time by some cellular automaton.*

Another interesting property is expressed in the following proposition, the proof of which is developed below.

Proposition 5 (see Mazoyer and Terrier (1999)) *If a function f can be marked on the first cell and if there is some integer k such that*

$$\forall n, \ (k-1)f(n) \geq kn$$

then there is a signal which characterizes the sites (n,f(n)), and is hence of slope f.

The condition that f has to satisfy is necessary – it is possible to mark 2^{2^n} and $n + \log\log n$ on the first cell, but impossible to mark $(n, n + \log\log n)$, because there is no signal with a slope between n and $n + \log n$ (see ❯ Proposition 6).

Proof A signal $\mathcal{E}$ is sent at speed 1 to the right from each site $(0, f(n))$ (see ❯ Fig. 19). Let k be assumed to be even. One begins by marking the sites $(kn, kf(n))$.

- From site $(0, 0)$ starts a signal $\mathcal{T}$ of slope $\frac{k+1}{k-1}$, a signal $\mathcal{D}_{\frac{k}{2}}$ to the right that progresses $\frac{k}{2}$ cells in $\frac{k}{2}$ steps and that becomes stationary on the cell $\frac{k}{2}$, and a signal $\mathcal{D}_k$ that progresses to the right, of k cells in k times, then remains on cell k.
- When a signal $\mathcal{E}$ and a signal $\mathcal{T}$ meet, the signal $\mathcal{E}$ disappears, a signal $\mathcal{E}_{-1}$ of slope -1 is created and $\mathcal{T}$ goes on with the same slope.
- When a signal $\mathcal{E}_{-1}$ and a signal $\mathcal{D}_{\frac{k}{2}}$ meet, the signal $\mathcal{E}_{-1}$ disappears, a signal $\mathcal{E}_1$ of slope 1 is created, the signal $\mathcal{D}_{\frac{k}{2}}$ goes on to the right for $\frac{k}{2}$ cells in $\frac{k}{2}$ times and then becomes stationary.
- When a signal $\mathcal{E}_1$ and a signal $\mathcal{D}_k$ meet, the signal $\mathcal{E}_1$ disappears, the signal $\mathcal{D}_k$ goes on to the right for k cells in k times and becomes again stationary.

It is not difficult to verify that the nth signal $\mathcal{E}$ meets signal $\mathcal{T}$ on the site $\left(\frac{(k-1)f(n)}{2}, \frac{(k+1)f(n)}{2}\right)$. Starting from this site, $\mathcal{E}_{-1}$ meets signal $\mathcal{D}_{\frac{k}{2}}$ on site $\left(\frac{kn}{2}, kf(n) - \frac{kn}{2}\right)$. Then signal $\mathcal{E}_1$, which stems from this site, meets $\mathcal{D}_k$ on site $(kn, kf(n))$.

Now, in order to mark the sites $(n, f(n))$, one groups the cells k by k. That means that one considers a new cellular automaton such that the state of site (i, j) represents the states of the sites

$$\{(ki + u, kj + v) \ ; \ 0 \leq u < k \ ; \ 0 \leq v < k\}$$

It is known that the converse of ❯ Proposition 5 is true; it comes from the stability under subtraction. From the sites $(n, f(n))$, the sites $(0, n + f(n))$ can be marked. According to the ❯ Proposition 5 of Mazoyer and Terrier (1999), with $a = b = 1$, $m = 1$, $g(n) = n$, and $\hat{f}(n) = f(n) + n$, the required condition $\hat{f} \setminus g \geq (mb + 1) \setminus ma$ can be obtained to say that $\hat{f}(n) - g(n) = f(n)$ is Fisher's constructible. ❯ Proposition 5 can also be derived from the ❯ Proposition 5 of Mazoyer and Terrier (1999).

☐ **Fig. 19**
Proof of ❷ Proposition 5. Signals $\mathcal{E}$ are drawn in black stripes on a medium gray background as well as signals $\mathcal{E}_{-1}$ (stripes in the other direction). Signal $\mathcal{T}$ is black. Signal $\mathcal{D}_{\frac{k}{2}}$ is light gray, while $\mathcal{D}_k$ is black edged white. Signals $\mathcal{E}_1$ dark striped.

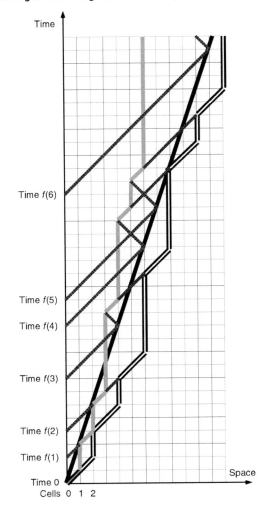

It should be noticed that although in the proof of ❷ Proposition 5 the space between the sites $(n, f(n))$ is used, it can be avoided in some cases. For example, ❷ *Fig. 20a, b* shows how to transform the algorithms of ❷ *Fig. 16a, b*, which marks times 2^n and 3^n into algorithms marking the sites $(n, 2^n)$ and $(n, 3^n)$. The idea is the same in both cases. The signal that was stationary on cell 0 progresses for one cell to the right each time it is marked (thick black edged medium or dark gray lines in ❷ *Fig. 20a, b*). This move is compensated by means of the zigzagging signal. The signal of slope 3 (or 2) is accelerated by one cell to the right; it progresses by two cells for three times (or by two cells for two times) (edged black medium gray in

■ **Fig. 20**

(a) Marking sites $(n, 2^n)$. **(b)** Marking sites $(n, 3^n)$.

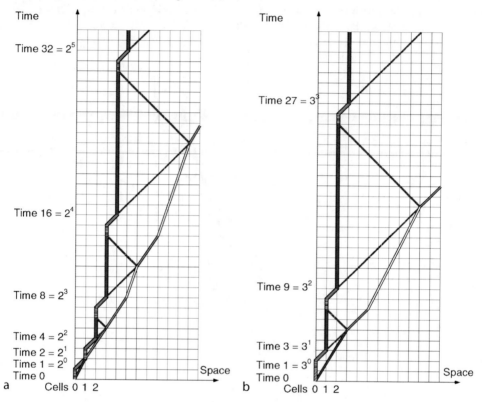

❯ *Fig. 20a, b)*, on time before coming back to its *normal* advance. The zigzagging signal has to lose one unit of time, so it waits one time before going on.

5.2 Limitation of Slopes

Until now, signals have been constructed starting from a basic initial configuration $^\infty LGL^\infty$ where L is a quiescent state and G a distinguished one. It can be proved here that, in this framework, it is not possible to get signals of all slopes.

Proposition 6 (Mazoyer and Terrier 1999) *There does not exist any cellular automaton that, starting from a basic initial configuration $(^\infty LGL^\infty)$, builds a signal of slope between 1 and $1 + \mathcal{O}(\log n)$.*

Proof First, prove a preliminary result.

Let D_k denote the diagonal $\{(j, j + k) \mid j \in \mathbb{N}\}$. *In the evolution of a cellular automaton $\mathcal{A} = (Q, \delta)$ on a basic configuration $^\infty LGL^\infty$ the sequence of the states on D_k is ultimately periodic of period $\pi_k \leq |Q|^{k+1}$ and its non-periodic part is of length $\iota_k \leq |Q|^{k+1}$. Moreover $\mathrm{lcm}(\pi_k, \pi_{k-1})$ is π_k (or π_{k-1} divides π_k).*

The proof is by induction on k and relies on the fact that the behavior of a cellular automaton on a diagonal D_k is the behavior of a transducer, the Fisher automaton (see Sutner (1997)).

- The sequence $(q_j)_j$ of D_0-states is given by $q_0 = G$ and $q_{j+1} = \delta(q_j, L, L)$. One has then a dynamical system on a finite set with $|Q|$ elements. It becomes ultimately periodic with $\pi_0 \leq |Q|$, and $\iota_0 \leq |Q|$.
- The sequence $(q_j^{k+1})_j$ of D_{k+1}-states is given by $q_0^{k+1} = q$ and $q_{j+1}^{k+1} = \delta(q_j^{k+1}, q_{j+1}^k, q_{j+2}^{k-1})$. Let π be $\mathrm{lcm}(\pi_k, \pi_{k-1})$. Then the dynamical system $q_j^{k+1} \to q_{j+\pi}^{k+1}$ on a finite set $|Q|$ becomes ultimately periodic of period less than $|Q|$. As a consequence, $\pi_{k+1} \leq |Q| \times |Q|^{k+1}$ and lcm (π_{k+1}, π_k) is either π_{k+1} or π_k. The length of the transitory is, of course, at most $|Q|^{k+2}$.

Now, assume that the evolution of some cellular automaton shows a signal S of slope $1 + o(\log n)$ but not 1. Then, for k large enough, segments of consecutive states in S of length more than $|Q|^{k+1}$ appear on D_k. If not, the asymptotic slope of S would be greater than, or equal to, $1 + \log_{|S|} n$, which contradicts the above property.

Proposition 7 (see Mazoyer and Terrier (1999)) *There is a cellular automaton $\mathcal{A} = (Q, \delta)$ which, starting from a basic initial configuration ($^\infty LGL^\infty$), builds a signal of slope $1 + \log_B n$ (with $B \geq 2$).*

Proof Let B be some base. Let $\bar{n}_B = a_0^n |B|^0 + a_1^n |B|^1 + \ldots + a_{\lfloor \log_B n \rfloor + 1}^n |B|^{\lfloor \log_B n \rfloor + 1}$ be the representation of n in base B. If one writes in an array the digits of the numbers in such a way that a_j^n is in square (n, j) (see ➋ *Fig. 21a* for $|B| = 2$), one gets an array that is locally constructible – the propagation of what to carry is local: the $(n + 1)$st column is obtained by adding 1 to the nth column from bottom to top. In square (n, j), this induces a carry over β_j^n with $\beta_{-1}^n = 1$. More precisely, a_{j+1}^n (and β_{j+1}^n) is the remainder (the quotient, respectively) of the division of $a_j^{n-1} + \beta_j^n$ by B. Moreover, this array is such that the kth line is periodic of period 2^{k+1} starting with $\sum_{i=0}^{i=k-1} 2^i = 2^k - 1$, and that the height of the nth column representing n is $\lfloor \log_B n \rfloor + 1$.

The local nature of the above array allows one to easily design a cellular automaton $\mathcal{A} = (Q, \delta)$ that builds it. The digits of n are on a diagonal to the right (see ➋ *Fig. 21b*) and the state $\langle j, n + j \rangle$ represents a_j^n and β_j^n (the latter is not shown in ➋ *Fig. 21b*). The state transition is given by $a_{j+1}^n = a_j^{n-1} + \beta_j^n \pmod{B}$.

The idea of the proof is to use the former process. In order to get the wanted result, if n is written in base B as

$$\bar{n}_B = a_0^n B^0 + a_1^n B^1 + \ldots + a_{\lfloor \log_B n \rfloor + 1}^n B^{\lfloor \log_B n \rfloor + 1},$$

it is sufficient to set a_j^n and β_j^n on site $(n, n + j)$. The former computations show that $\langle n, n + j + 1 \rangle$ is the remainder and the quotient of the addition of the first part of $\langle n - 1, n - 1 + j \rangle$ and the second part of $\langle n, n + j \rangle$ by $|B|$. But the arguments of δ are $\langle n - 1, n + j \rangle$, $\langle n, n + j \rangle$ and $\langle n + 1, n + j \rangle$, which means that the information contained in $\langle n - 1, n - 1 + j \rangle$ must be kept for one unit of time, in order to find it again in $\langle n - 1, n + j \rangle$. A way to do it is to choose $Q = B \cup \{0^{\text{with carry}}\} \cup \{L, S\}$ where L is quiescent and S represent the signal S of slope $1 + \log_B n$. The process is illustrated in ➋ *Fig. 21c*. In this case, $B = 2$, values of a_j^n are in dark gray (for 0) or black (for 1) while those of β_j^n are in light gray (for 0) or medium gray (for 1). Gray stripes represent S.

◘ Fig. 21

(a) The sequence of natural numbers written in base 2. (b) Generation of all the representations of natural numbers in base 2. (c) Signal of slope $1 + \log_2 n$.

															1	1	1	1	1	1	1	1	1	1	
							1	1	1	1	1	1	1	1	0	0	0	0	0	0	0	0	1	1	
			1	1	1	1	0	0	0	0	1	1	1	1	0	0	0	0	1	1	1	1	0	0	
	1	1	0	0	1	1	0	0	1	1→0	0	1	1	0	0	1	1	0	0	1	1	0	0		
a 0	1	0	1	0	1	0	1	0	1	0	1	0	1	0	1	0	1	0	1	0	1	0	1	0	1

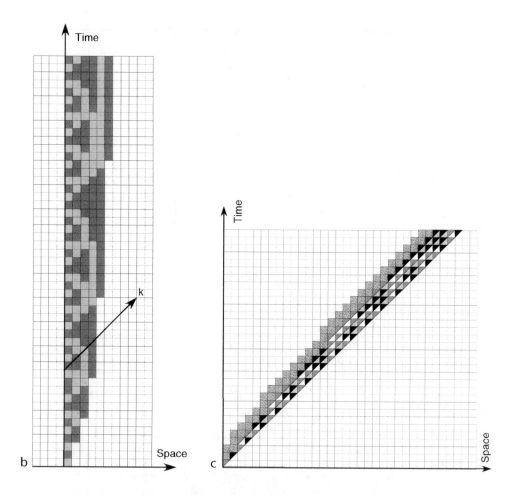

b   c

6   Basic Tools and Methods for Moving Information

6.1   Information Moves

In a space–time diagram, one identifies some states or some patterns of states as *information*. It is actually impossible to give a formal general definition of information. It is all the more difficult because one can imagine that information is given to the network from outside (see the chapter ❯ Computations on Cellular Automata, and also ❯ Definition 4). The following definition is used.

Definition 9 A signal S (in some orbit of a cellular automaton $\mathcal{A} = (Q, \delta)$) is said to carry pieces of information (or simply information) from Γ if $S = S^\star \times (\Gamma \cup \{\Lambda\})$.

Starting from such a definition, every geometric transformation on signals (as in ❯ Sect. 4.1) amounts to moving information. In particular, one can see signal translation in ❯ Sect. 4.1 (forgetting the stop point of the transformation, and considering signals as infinite "guides") as a way to create an infinite stream of information that moves along a rational direction. This stream can be "collected" on a signal parallel to the one which carries it, as in ❯ Sect. 4.1, or, a priori, on any other signal.

In certain cases, the process is simple, as in the examples of ❯ Fig. 22. But it can be more complicated and needs the data to be compressed. Consider a signal S of slope $\frac{p}{q}$ that contains k elementary pieces of information on k cells. The stream of information from these cells (meaning the information contained in a strip determined by these k cells of S and left diagonals of some rational slope $\frac{p'}{q'} \geq 1$) can mix pieces of information originally contained in these k successive cells. It is clear enough that if one wants to collect the original information (from Γ) carried by S and sent via this stream, onto a signal $S^\star$ of slope $\frac{p^\star}{q^\star}$, some pieces of information coming from different cells will arrive on $S^\star$ in the same cell. (The "arrival" cell is assumed to be the greater cell that has knowledge of the information). This is the case when the ratio $\frac{p^\star}{q^\star} / \frac{p}{q}$ is strictly more than 1. Information is said to be compressed. Conversely there is a phenomenon of decompression if the same ratio is strictly less than 1. In this case, pieces of information coming from a same cell of S possibly arrive on different cells of $S^\star$. Examples of rational compressions and decompression are illustrated in ❯ Figs. 23a–c.

❯ Figure 23a shows the case where the slope of S is 1 while the slope of $S^\star$ is 2. The data is compressed in a periodic but nonconstant way. In ❯ Fig. 23b, signal S is of slope 0 and $S^\star$ of slope 1. In this case the data is compressed in a constant way. ❯ Figure 23c shows the case of decompression.

■ **Fig. 22**
Information moves collected in a simple way. The original signal is in gray. It sends information to the black signal, along lines that determine a "stream" of information.

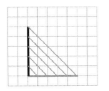

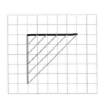
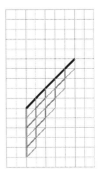

⬛ **Fig. 23**

(a) Rational compression. In ❷ *Figs. 23a–c*, pieces of information are represented by circles or stars in a different level of gray. The signal S from which information is sent out is light gray while the signal which collects them is dark gray. (b) Constant compression. (c) Decompression.

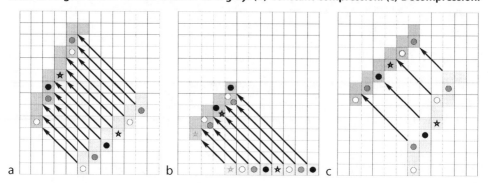

The phenomena of compression and decompression influence the general notions of "grouping" and "de-grouping", which are developed in the chapter ❷ Universalities in Cellular Automata.

Sometimes, one has to keep pieces of information on the same cell for some time (which means moving it along a line of slope ∞), or to move it from some cell to another cell at some finite distance in some finite number of time units. This can be done by the means of "counters." Two main types of counters appear in the literature: linear counters, as in ❷ Sect. 4.3, or exponential counters (see ❷ *Fig. 27b*). It is worth noting that signals of polynomial slopes, briefly discussed in ❷ Sect. 4.3, produce polynomial counters. Until now, these counters have not depended on the data size (that is the number of cells of S which carry information in Γ). However, this will happen, for example, when arithmetical operations have to be performed on this size or when synchronization is needed. Next, such arithmetical operations will be developed, but first briefly recall the definition and one main result concerning synchronization.

Definition 10   A cellular automaton $\mathcal{A} = (Q, \delta)$ is a solution to the firing squad synchronization problem (FSSP) if

- Q is the disjoint union of Q_L, Q_X, $\{G\}$, and $\{F\}$.
- (Q_L, δ) and (Q_X, δ) are sub-automata of $\mathcal{A}$. (In other words, δ is stable on Q_L and Q_X.)
- For each initial configuration $^{\infty}qGq^{\star^{n-1}}q^{\infty}$, where states q are in Q_L and states $q^{\star}$ in Q_X, there exists some integer θ such that:
  - For all times $t \in \{0, \ldots, \theta - 1\}$, configuration at time t is of the form $^{\infty}qq_1^{\star}\ldots q_n^{\star}q^{\infty}$ where states q are in Q_L and states $q_i^{\star}$ in $Q_X \cup \{G\}$.
  - The configuration at time θ is of the form $^{\infty}qF^nq^{\infty}$ where the states q are in Q_L. $\mathcal{A}$ is in optimal time if $\theta = 2n - 2$.

Synchronization is a well-known arithmetical problem (see Balzer (1967) and Mazoyer (1987), and a paper on the state of the art in 1989 (Mazoyer 1989a)).

Proposition 8

1. *There are solutions to the FSSP in optimal time* (Balzer 1967; Mazoyer 1987).
2. *There are solutions in a wide variety of times* (Gruska et al. 2006; Mazoyer 1989b).

It has to be noted that it is also possible to start from configurations $^\infty q(q^\star)^m \, G(q^\star)^{n-m-1} q^\infty$ (Vaskhavsky et al. 1969) or from configurations $^\infty qG(q^\star)^{n-2} Gq^\infty$ (Mazoyer and Reimen 1992), and that it is possible to synchronize according to a given slope (this has been introduced in Čulik (1989)).

6.2 Arithmetical Operations on Segments

In order to organize a space–time diagram, it is sometimes necessary to extract information from some signal. If the signal is horizontal, it can be said that the information carried by the signal is not pieces of information on its states, but that the information is its length ℓ (as in ❷ Sect. 4.3). Then, knowing two segments (that means, horizontal segments with marked beginning and end of "lengths" p and q), one can be asked to compute (under the form of segments) things such as $p \div q$ or $p \bmod q$. Two methods are described. The first one is very general and uses the techniques encountered in ❷ Sect. 6.1. However, it is not very fast. The second one, geometrical, comes down to cutting a segment in ratio $\frac{p}{q}$ where $\frac{p}{q} < 1$. In any case, the first step is to compute an integer multiple of q.

▶ The first operation : to put kq on a diagonal ($k, q \in \mathbb{N}$).

This task consists of duplicating k times the initial segment of length q. One has the choice to code k either by states or by a counter, and the result can be obtained with or without synchronization.

The general process is as follows:

- In the first step, the data contained between two vertical signals (G_0 and G_1 in ❷ *Figs. 24a–d*) is carried on a diagonal between sites A and B. The sites A and α are marked by means of two signals at speeds 1 and -1, stemming from G_0 and G_1. The data is also duplicated between sites B and C by means of a signal of slope 2 sent out from site α (dark striped line in ❷ *Fig. 24a*).
- The process is then repeated: the abscissa of A', the "new" site A, being the one of the former site B, and α', the "new" site α, being on the cell of site C and found at the intersection of a signal of slope $\frac{3}{2}$ sent out from site A and of a stationary signal stemming from site C. The start D on G_0 is obtained in sending a signal of slope -1 from the site B on G_1.

The algorithm is stopped after k repetitions of the process (see ❷ *Fig. 24a*) due to the encoding of k by states. If one wants to synchronize the process, one uses a signal that zigzags k times between G_0 and G_1, counting the beginning on G_0 if k is even, and on G_1 if k is odd, as shown in ❷ *Fig. 24b*. Of course, a general method of synchronizing in the sense of ❷ Definition 10 can also be used.

When k is given by some counter (there are different possibilities: as in ❷ Sect. 4.3, or in ❷ *Fig. 27b*), or in unary (as in ❷ *Figs. 24c, d*) a "freezing" technique is used. One assumes that the counter is situated at the left of G_0, between H and G_0 (see ❷ *Figs. 24c, d*). Each time the initial length is added, cell G_0 launches the process of decreasing the counter by 1, taking 0

■ **Fig. 24**

(a) Duplication of a horizontal segment *k* times, *k* being coded by states, without synchronization. (b) Duplication of a horizontal segment *k* times, *k* being coded by states, with synchronization. (c) Duplication of a horizontal segment *k* times by means of a counter, without synchronization. (d) Duplication of a horizontal segment *k* times by means of a counter, with synchronization.

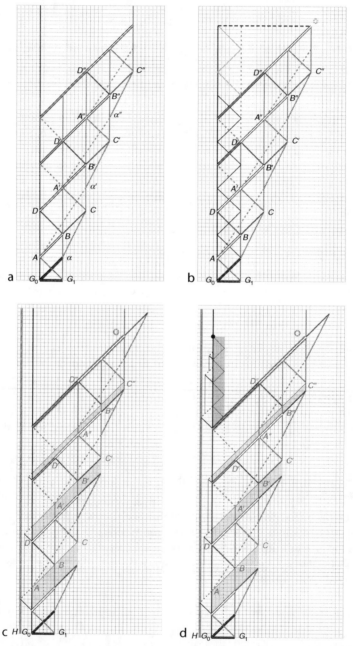

time units. During these θ time steps, the general algorithm is frozen (light areas in ❱ *Figs. 24c, d*).

Freezing an evolution of a sub-automaton $\mathcal{A}^* = (Q^*, \delta)$ of some automaton $\mathcal{A} = (Q_A, \delta)$ consists in duplicating Q^* in $Q^* \cup Q^\sharp$. At the time of freezing, a state q^* of Q^* becomes the corresponding state $q^\sharp$ of $Q^\sharp$. A state of $Q^\sharp$ remains stationary during the freezing. At the time of unfreezing, it becomes again the state q^* of Q^* and interacts with the states of $Q^\sharp$ as well as with the ones of Q^*. This makes the general algorithm to evolve with some delays (❱ *Fig. 24c*).

If one wants to synchronize the result, one uses general synchronization in the sense of ❱ Definition 10, starting from the site O in ❱ *Fig. 24d*. Also other methods are possible, but they depend on the counter. For example, if the response time of the counter is less than the length to duplicate, then one can use a signal zigzagging between G_0 and $G_0 + \frac{G_1 - G_0}{2}$ as in ❱ *Fig. 24d* (see ❱ Sect. 4.3).

Now, see how to find the quotient and the remainder of the division of ℓ by q when $\ell \geq q$.

▶ The division.

The data ℓ is given, as before, as a horizontal segment delimited by two stationary signals (in medium gray in ❱ *Figs. 25a–c*). A signal $\mathcal{D}$ is sent at speed 1. If q is coded by states (❱ *Figs. 25a, c*), D contains q counter using states $d_0, \ldots, d_{q-1}$. Otherwise, $D = \{d\}$, and q is given by a segment. In this case q is duplicated, as indicated in the first point, k times until $kq \geq \ell$. The duplication process is stopped when the diagonal that carries kq meets the stationary signal marking the end of the initial segment ℓ. In either case, the iq^{th} cells of $\mathcal{D}$ become marked (by d_{q-1} or by the a signal marking q).

A more elegant method is illustrated in ❱ *Fig. 25c*. It consists of using the states $d_0, \ldots, d_{q-1}$ to create a signal $\mathcal{S}$ of slope q, by the means of the process in ❱ Sect. 2.1. Then, the distance between the signal marking the beginning of segment ℓ and the signal $\mathcal{S}$ at the time when $\mathcal{S}$ meets the left diagonal indicating the end of ℓ, gives the quotient. A state of $\mathcal{S}$ (or a state of $\mathcal{D}$ if all is coded as in ❱ Sect. 4.2) gives the remainder of the division of ℓ by q.

We conclude with two remarks. First, a particular relative positioning of the segments coding ℓ, p, and q is assumed. But the data places can be modified by the techniques in ❱ Sect. 6.1. Secondly, only unary counters have been considered. This also is general because the computation of the base change is realized in an optimal time on a cellular automaton with neighborhood $\{-1, 0, +1\}$ (see the chapter ❱ Computations on Cellular Automata of this volume).

6.3 A Riddle

Starting from the possibility to mark times 2^k on cell 0, a riddle method is next developed, which allows one to get the binary expansion of each integer. This method can also be used with other functions.

❱ *Figure 26a* reproduces ❱ *Fig. 16a*. The initial configuration of the automaton that builds this geometric diagram is $^\infty LGL^{n-2}EL^\infty$, where L is quiescent, G is the state of cell 0 and a special state E is the initial state of cell $n - 1$. Cell $n - 1$ sends a signal $\mathcal{E}$, to the left, at speed 1. This signal meets one and only one signal of speed one stemming to the right from cell 0. If $\mathcal{E}$ meets the kth signal ($k \geq 1$), then $2^k \leq n < 2^{k+1}$. That allows one to know $\lfloor \log_2 n \rfloor$.

The process can be iterated (see ❱ *Fig. 26b*). Each time a signal at speed 1 to the right, coming from a site $(0, 2^k)$, meets the signal $\mathcal{D}$ of slope 3 (white on black in ❱ *Fig. 26a* and

■ **Fig. 25**

(a) General method for the division of ℓ by q, where q is coded by states. In ❯ *Fig. 25a–c*, data is represented by circles or stars. In ❯ *Fig. 25a, b*, stationary signals are indicated by triangles, and the signals of slope -1 are striped. The counting of the quotient is realized in the very dark gray. The remainder is the segment delimited by the two stationary signals at the right (dark and medium gray). (b) General method, q coded by segments. (c) Method using a signal, q coded by states.

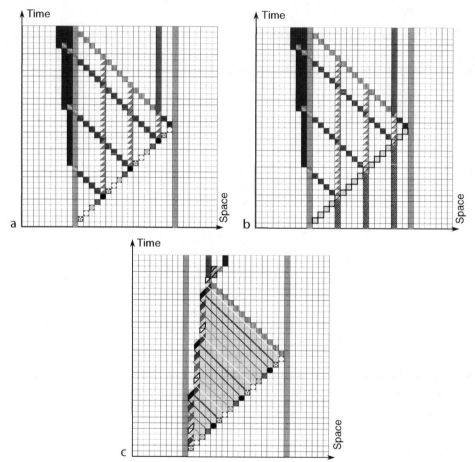

black in ❯ *Fig. 26b*) on site $(2^{k-1}, 3.2^{k-1})$, a new exponential construction is started: a stationary signal is launched (in white on black and medium gray in ❯ *Fig. 26b*) and the process is iterated. It stops on the signal at speed 1 coming from the site $(0, 2^{k+1})$. Successive iterations are represented in ❯ *Fig. 26b*. Let us call signals of type $\mathcal{L}$ the signals at speed 1 to the right.

The process above easily allows one to know the number of bits 1 in the binary expansion of n (see Terrier (1991)). Signal $\mathcal{E}$, sent from cell $n-1$, reaches cell 0 at time n and intersects several traces of right moving signals of type $\mathcal{L}$, in white and black vertical signals in ❯ *Fig. 26c*. The last intersected signal is emitted from site $(0, \lfloor \log_2 n \rfloor)$. If $(i_0 = 0, i_1, \ldots, i_v)$ is the sequence of the iteration levels of the intersected signals of type $\mathcal{L}$, one has $n = \sum_{j=0}^{j=v} \frac{2^{\lfloor \log_2 n \rfloor}}{2^{i_j}} + \epsilon$

■ **Fig. 26**

(a) $\lfloor \log_2 n \rfloor$. On ❷ *Fig. 26a–c*, signals to the right (left) are striped to the right (to the left). Signal $\mathcal{E}$, representing the integer n, comes at speed one from the left. (b) Iterations. (c) Number of 1's in $\overline{n}_2$.

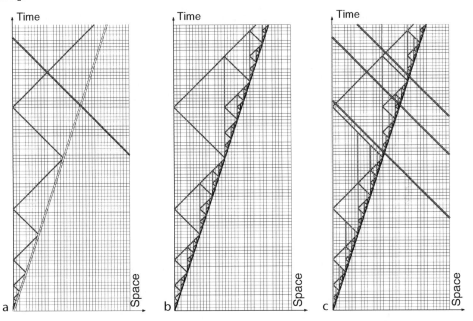

a   b   c

with $\epsilon \in \{0, 1\}$. One can determine ϵ by observing the intersection point of signals $\mathcal{D}$ and $\mathcal{E}$. In ❷ *Fig. 26c*, the binary expansion of $n = 66$, $n = 86$ and $n = 98$ has 2, 4, and 3 digits 1.

All the signals in ❷ *Fig. 26b* building the iterated exponential appear as "particular grid patterns." This allows one to uniformly determine some given properties, and can be used as a riddle.

The diagram in ❷ *Fig. 26b* can be complemented, as illustrated in ❷ *Fig. 27a*, in order to get the whole binary expansion of any integer. One proceeds as follows:

1. At each iteration $j \geq 2$ of the exponential computation, one considers the triangles $(0, 2^j)$, $(2^{j-1}, 3.2^{j-1})$ and $(0, 2^{j+1})$ (T_j, light gray in ❷ *Fig. 27a*) and $(2^{j-1}, 3.2^{j-1})$, $(0, 2^{j+1})$ and $(2^j, 3.2^j)$ (T'_j, medium gray in ❷ *Fig. 27a*).
2. At each iteration j, a stationary signal is sent from each site marking the end of a former iteration on T_j (white on gray in ❷ *Fig. 27a*), which stops when it enters T'_j and emits a new signal of slope 2 (light stripes on medium gray in ❷ *Fig. 27a*). It disappears at the end of the iteration (that means on the other side of the triangle where it meets a gray stationary signal, coming from the construction of the iterated exponential).
3. The process is then iterated.

Finally, a signal of slope 1 to the left, starting from cell $n - 1$, with $n = 2^k + 2^{k-\ell} + \dots$ intersects $\ell - 1$ signals of slope 2 (light stripes on medium gray in ❷ *Fig. 27a*) after having crossed the signal indicating that $n - 2^k \geq 2^{k-\ell}$. So, one gets $\overline{n}_2$. For example, if one considers a line corresponding to $n = 58$ (white on black in ❷ *Fig. 27a*), starting from cell 0, it meets two

■ **Fig. 27**

(a) Getting $\overline{n_2}$ by means of the riddle. (b) Getting $\overline{n_2}$ as in ❷ Sect. 5.2.

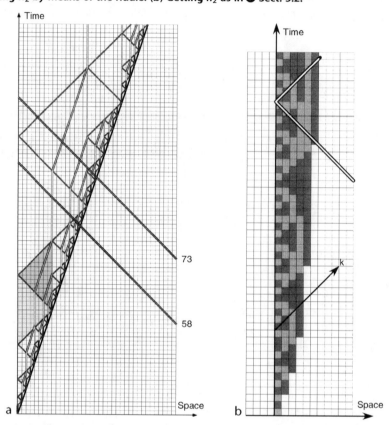

successive stationary (white on gray) signals (so one has 111), then one signal of slope 2 (gray stripes on black), then another stationary (gray) signal (so one gets 01). Finally, as the line arrives on $\mathcal{D}$ under the middle of the "little segment," one has 0 as the last bit. Altogether, we have read 111010, which actually represents 58!

Of course, in order to find $\overline{n_2}$, one also can use the method of generating binary words as in ❷ Sect. 5.2, illustrated in ❷ *Fig. 27b*. The difference between both methods is that the "riddle" allows one to *anticipate* the computation and to use, in what follows, the obtained information.

6.4   Folding Space–Time

The interest here is in the evolution of an initial line of type c_{L^*} of some automaton $\mathcal{A}$ where L^* is some quiescent state but possibly different from the usual quiescent state L. And the intention is to build an automaton $\mathcal{C}$ that simulates $\mathcal{A}$ and computes only on positive cells.

More precisely, if $\mathcal{A} = (Q_{\mathcal{A}}, \delta_{\mathcal{A}})$, one builds $\mathcal{C} = (\Theta, \Delta)$ as follows (see ❷ *Fig. 28a*).

1. $\Theta = Q_{\mathcal{A}} \cup (Q_{\mathcal{A}} \times Q_{\mathcal{A}})$, denoted as $\Theta = Q_{\mathcal{A}} \cup (Q_{\mathcal{A}, \rightarrow} \times Q_{\mathcal{A}, \leftarrow})$.

☐ **Fig. 28**

(a) Zones in an evolution of $\mathcal{A}$. State $L^\star$ is in white. The influence of states different from $L^\star$ on cells to the left of cell 0 are marked by stripes. (b) Zones of computation in the evolution of $\mathcal{C}$ simulating $\mathcal{A}$: states $(X, \leftarrow)$ are represented in the lower triangle while the upper triangle represents $(X, \leftarrow)$ and $(X, \rightarrow)$.

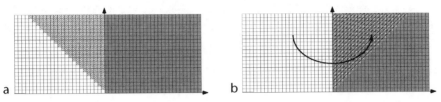

a   b

2. Initial configuration of $\mathcal{C}$ is the "same" as the one of $\mathcal{A}$, that is $^\infty L^\star q_0 q_1 \ldots$, where the q_i are in the subset Q_A of Θ.

3. • If q_ℓ, q_c, q_r are in the subset Q_A of Θ and all different from $L^\star$, $\Delta(q_\ell, q_c, q_r) = ((\delta_A(q_\ell, q_c, q_r), \rightarrow), (L^\star, \leftarrow))$.
   • If q_c, q_r are in the subset Q_A of Θ and all different from $L^\star$, $\Delta(L^\star, q_c, q_r) = ((\delta_A(L^\star, q_c, q_r), \rightarrow), (\delta_A(L^\star, L^\star, q_c), \leftarrow))$.
   • If q_c, q_r are in the subset Q_A of Θ and if q_c or q_r is equal to $L^\star$, $\Delta(L^\star, q_c, q_r) = L^\star$ (arbitrary choice).

   Then, if at time 1, the configuration of $\mathcal{A}$ has become $^\infty L^\star q^1_{-1}\, q^1_0\, q^1_1\, q^1_2 \ldots$, the configuration of $\mathcal{C}$ is
   $^\infty L^\star (q^1_0, \rightarrow), (q^1_{-1}, \leftarrow), (q^1_0, \rightarrow), (L^\star, \leftarrow), (q^1_2, \rightarrow), (L^\star, \leftarrow), \ldots$.

4. • $\forall (q_\ell, \rightarrow), (q_c, \rightarrow), (q_r, \rightarrow) \in Q_{A, \rightarrow}, \forall (q^\star_\ell, \leftarrow), (q^\star_c, \leftarrow), (q^\star_r, \leftarrow) \in Q_{A, \leftarrow},$
   $\Delta(((q\ell, \rightarrow), (q^\star_\ell, \leftarrow)), ((q_c, \rightarrow), (q^\star_c, \leftarrow)), ((q_r, \rightarrow), (q^\star_r, \leftarrow))) =$
   $((\delta_A(q_\ell, q_c, q_r), \rightarrow), (\delta_A(q_r, q_c, q_\ell), \leftarrow))$.
   • $\forall (q_c, \rightarrow), (q_r, \rightarrow) \in Q_{A, \rightarrow}, \forall (q^\star_c, \leftarrow), (q^\star_r, \leftarrow) \in Q_{A, \leftarrow},$
   $\Delta(L^\star, ((q_c, \rightarrow), (q^\star_c, \leftarrow)), ((q_r, \rightarrow), (q^\star_r, \leftarrow))) =$
   $((\delta_A(q^\star_c, q_c, q_r), \rightarrow), (\delta_A(q^\star_r, q^\star_c, q_c), \leftarrow))$.

   Then, if at time t the configuration of $\mathcal{A}$ has become $^\infty L^\star q^t_{-t}, \ldots, q^t_{-1} q^t_0 q^t_1 q^t_2 \ldots$, the configuration of $\mathcal{C}$ is
   $^\infty L^\star ((q^t_0, \rightarrow), (q^t_{-1}, \leftarrow)), \ldots, ((q^t_{-1}, \rightarrow), (q^t_{-t}, \leftarrow)), ((q^t_t, \rightarrow), (L^\star, \leftarrow)), ((q^t_{t+1}, \rightarrow), (L^\star, \leftarrow)), \ldots$.

5. Nonspecified transitions are arbitrarily fixed to give $L^\star$.

This technique of folding is used, for example, to prove that the class of languages recognized in linear time is exactly the class of languages recognized in linear time and space bounded by the input length (see Delorme and Mazoyer (1996) and Mazoyer (1999)).

6.5 Freezing and Clipping

The technique of "freezing" has already been evoked in ❷ Sect. 6.2. Here, we recall and generalize it.

1. ❷ *Figure 29a* symbolized the freezing technique. Given some cellular automaton $\mathcal{A}$ with a special state G and some integer k, there exists an automaton $\mathcal{B}_k$ such that $Q_A \subseteq Q_{B_k}$ and

■ Fig. 29

(a) Freezing. (b) A generalization of freezing. (c) An orbit of $\mathcal{A}$. Cells are represented by geometric forms and times by patterns. This orbit is periodic of temporal period 3 and spatial period 7. Gray zones are the "clipping zones" of ❷ Fig. 29d. They appear here only to make the comparison between the figures easier. (d) Action of the vector (1, 1) on the $\mathcal{A}$-orbit of ❷ Fig. 29c. States of $\mathcal{B}$, which are in $(Q_\mathcal{A} \cup \{\lambda\})^3$ (λ represents the absence of $\mathcal{A}$-states) are represented in the following way : state at the bottom, right, is the state of $\mathcal{A}$, if there was no clipping ; at the bottom, left, is a state of $\mathcal{A}$ moved by (1, 1), that is the state of the left neighbor, one time before in the $\mathcal{A}$-orbit ; the state at the top, left, is a state of $\mathcal{A}$ moved by (2, 2), that is the state of the second left neighbor two times before in the $\mathcal{A}$-orbit. (e) A stationary signal $\mathcal{S}_\mathcal{A}$, marked by circled states, in some $\mathcal{A}$-orbit. (f) Action of the clipping vector (1, 1) on the $\mathcal{A}$-orbit of ❷ Fig. 29e. (g) A signal with a right move $\mathcal{S}_\mathcal{A}$, marked by circled states, in some $\mathcal{A}$-orbit. (h) Action of the clipping vector (−1) on the $\mathcal{A}$-orbit of ❷ Fig. 29g. Cells of $\mathcal{S}_\mathcal{A}$ are marked by a circle. One or several components of $\mathcal{B}$ may carry states of $\mathcal{S}$.

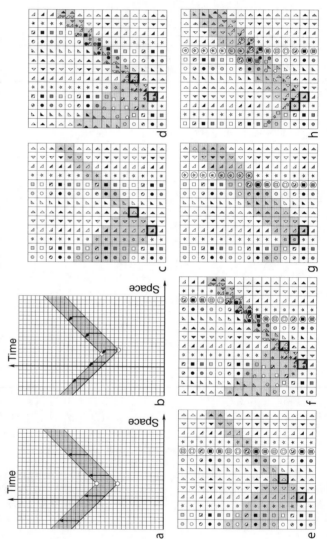

- On an initial configuration c_A, the orbit of which contains one – and only one – site (x_G, t_G) in state G, the evolution of $\mathcal{B}_k$ is described as follows:
  - Let D_G and $D_G^\star$ denote the diagonal and the anti-diagonal stemming from site (x_G, t_G), and let $D_{G,k}$ and $D_{G,k}^\star$ denote the diagonal and the anti-diagonal stemming from $(x_G, t_G + k)$.
  - If site (ξ, η) is under or on $D_G \cup D_{G,\,k}^\star$, then $\langle \xi, \eta \rangle_{\mathcal{B}_k} = \langle \xi, \eta \rangle_A$.
  - If site (ξ, η) is strictly above $D_{G,k} \cup D_{G,\,k}^\star$ then $\langle \xi, \eta \rangle_{\mathcal{B}_k} = \langle \xi, \eta - k \rangle_A$.
  - The other sites are used to carry, vertically and during k times, the states of the sites on $D_G \cup D_{G,\,k}^\star$, as well as on D_G' and $D_{G,\,k}^{\star'}$ (translations of D_G and $D_{G,\,k}^\star$ by $(0, 1)$). This is done (inside the gray stripes (see ❷ Fig. 29a)) by means of new states added to Q_A, in such a way that the evolution of A can normally resume. (One has to consider the diagonals D_G' and $D_G^{\star'}$ because of the dependencies implied by the neighborhood vector.)

2. The former method can be generalized by replacing integer k by some vector $\vec{V}$ of $\mathbb{Z} \times \mathbb{N}$, called the "clipping" vector, which is represented, as in ❷ Sect. 4.1, by a finite sequence $(c_0, \ldots, c_k)$ with $c_i \in \{-1, 0, +1\}$. Then, for each automaton A such that Q_A contains a special state G and for each vector $\vec{V} = (c_0, \ldots, c_k)$ there exists an automaton $\mathcal{B}_{\vec{V}}$ with $Q_A \subseteq Q_{\mathcal{B}_{\vec{V}}}$, such that,

   - On an initial configuration c_A, the orbit of which contains one – and only one – site (x_G, t_G) in state G, the behavior of $\mathcal{B}_{\vec{V}}$ is as follows:
     - Let D_G and $D_G^\star$ denote the diagonal and the anti-diagonal stemming from site (x_G, t_G), and let $D_{G,\vec{V}}$ and $D_{G,\vec{V}}^\star$ be the diagonal and the anti-diagonal stemming from $(x_G + \#_1(\vec{V}) - \#_{-1}(\vec{V}), t_G + \#_0(\vec{V}) + \#_1(\vec{V}) + \#_{-1}(\vec{V}))$, where $\#_\ell(\vec{V})$ is the number of ℓ's in the sequence $(c_0, \ldots, c_k)$.
     - If site (ξ, η) is under or on $D_G \cup D_G^\star$, then $\langle \xi, \eta \rangle_{\mathcal{B}_{\vec{V}}} = \langle \xi, \eta \rangle_A$.
     - If site (ξ, η) is strictly above $D_{G,\vec{V}} \cup D_{G,\vec{V}}^\star$ then $\langle \xi, \eta \rangle_{\mathcal{B}_{\vec{V}}} = \langle \xi - \#_1(\vec{V}) + \#_{-1}(\vec{V}), \eta - \#_0(\vec{V}) - \#_1(\vec{V}) - \#_{-1}(\vec{V}) \rangle_A$.
     - Otherwise (inside the gray stripes in ❷ Fig. 29b), the process consists in carrying, according to $\vec{V}$, the states of $D_G \cup D_G^\star$ and $D_G' \cup D_G^{\star'}$, by means of states added to the original states of A, in such a way that the evolution of A can normally resume.

   ❷ Figure 29d shows the clipping determined by vector $(1, 1)$. In the left part of the diagram (cells of abscissa $x < x_G$), the states of sites on the diagonal stemming from (x_G, t_G) (the lowest gray site), as well as of the diagonal immediately below are moved from $(2, 2)$. In the right part of the diagram, states of sites on the diagonal to the right stemming from (x_G, t_G) as well as of sites on the diagonal immediately above, are moved by $(2, 2)$. One also observes that the whole part of the A-orbit, (sites (x, t), $x > x_G, t \geq t_G + 4$, above the gray zone in ❷ Fig. 29c lays above the gray zone in the $\mathcal{B}$-orbit shifted by $(2, 2)$ in ❷ Fig. 29d.

3. An important particular case is the one in which, for each site (ξ, η) of $D_G \cup D_G'$ (resp. $D_G^\star \cup D_G^{\star'}$),

$$\langle \xi + \#_1(\vec{V}) - \#_{-1}(\vec{V}), \eta + \#_{-1}(\vec{V}) + \#_0(\vec{V}) + \#_1(\vec{V}) \rangle_A = \langle \xi, \eta \rangle_A.$$

Then, $\mathcal{B}_{\vec{V}}$ does not need to carry states of A between D_G and $D_{G,\vec{V}}$ (resp. D_G' and $D_{G,\vec{V}}^{\star'}$). So, the evolution of $\mathcal{B}_{\vec{V}}$ on c_A only differs from the evolution of A on c_A "on the other side."

4. In case of points 1 and 2, one may allow state G to appear on several sites (even infinitely many times) in the evolution of an initial configuration c_A provided it does not appear in the zone where $\mathcal{B}_{\vec{V}}$ has to carry states of $\mathcal{A}$ (gray parts in ❷ *Fig. 29a, b*).

   In case of point 3, sites $(x_{G_j}, t_{G_j}), j \in \mathbb{N}$, and $t_{G_j} < t_{G_{j+1}}$, where G appears, can be in the zone of the evolution which is not modified. A special case is the one where there is some integer k such that for all $j \in \mathbb{N}^*$, $(x_{G_j}, t_{G_j}) = (x_{G_0} + k, t_{G_0} + k)$, which means that the points (x_{G_j}, t_{G_j}) are regularly placed on D_{G_0} (resp. $(x_{G_j}, t_{G_j}) = (x_{G_0} - k, t_{G_0} + k)$ on $D^\star_{G_0}$).

5. In the case of an initial line c_A, the evolution shown state G on one – and only one – site (x_G, t_G). Assume now that the $\mathcal{A}$-orbit of c_A contains a signal $\mathcal{S}_A$, defined by a sequence of moves $\eta_0, \ldots$, which meets D_G and $D_{G,\vec{V}}$ but does not enter the zone between $D^\star_G$ and $D^\star_{G,\vec{V}}$ (resp. meets $D^\star_G$ and $D^\star_{G,\vec{V}}$, but does not enter the zone between D_G and $D_{G,\vec{V}}$). Processes of points 1 and 2 applied to c_A provides on the corresponding orbit of $\mathcal{B}_{\vec{V}}$ a signal that coincides with $\mathcal{S}_A$ outside the band determined by D_G and $D_{G,\vec{V}}$ (resp. $D^\star_G$ and $D^\star_{G,\vec{V}}$) and that follows $\vec{V}$ inside the corresponding diagonals. Of course, carrying the necessary data onto the diagonal on which the action of $\mathcal{A}$ resumes requires new states. To give two examples:

   (a) Action of a clipping vector $(1, 1, \ldots, 1)$ (sequence of k moves by 1) on a stationary original signal.

   ❷ *Figure 29f* shows the result of the clipping vector $(1, 1)$ on an $\mathcal{A}$-orbit carrying a signal $\mathcal{S}_A$ (see ❷ *Fig. 29e*), which is transformed into a signal $\mathcal{S}^\star$ on the corresponding $\mathcal{B}_{(1,1)}$-orbit. Actually, $\mathcal{S}^\star$ is $\mathcal{S}_A$ in which two moves to the right have been inserted in the clipping zone. More generally, one observes that, if the sequence of moves of $\mathcal{S}_A$ is $\eta_0, \ldots, \eta_{j_0}, \eta_{j_0+1}, \ldots$ and if η_{j_0} is the last move of $\mathcal{S}_A$ before reaching D_G, then the sequence of moves of $\mathcal{S}^\star$ is $\eta_0, \ldots, \eta_{j_0}, 1, 1, \ldots, 1, \eta_{j_0+1}, \ldots$, where the number of inserted 1's is k.

   (b) Action of a clipping vector (-1) on a move to the right followed by a stationary move. Consider the situation where the signal $\mathcal{S}_A$ arrives on the diagonal $D^\star_G$ on site (x, t), coming from site $(x - 1, t)$ and going to site $(x, t + 1)$. Then the crippling effect of vector -1 will be to transform site (x, t) of $\mathcal{A}$ into site $(x - 1, t + 1)$ of $\mathcal{B}_{-1}$ on the diagonal $D^\star_G$ and site $(x, t + 1)$ of $\mathcal{A}$ into $(x - 1, t + 2)$ of $\mathcal{B}_{-1}$. The result is that a move to the right has been replaced by two stationary moves. ❷ *Figures 29g, h* illustrate this point. In this case, the clipping allows the choice between two signals: an indentation to the left or a straight vertical line (the choice in what follows).

6. Let it be assumed now that the signal $\mathcal{S}_A$ is periodic of period $(\eta_0, \ldots, \eta_\ell)$. Then, by suitably implementing points 4 and 5, one is able to modify the period, and one gets the following result.

Proposition 9 *Let $\mathcal{A}$ be a cellular automaton whose set of states Q_A contains a special state G and a subset S. Let p and q be in $\mathbb{N}^*$, and let Π be a sequence $(\epsilon_0, \ldots, \epsilon_{\Pi-1})$, ϵ_i in $\{0,1\}$. Then, a cellular automaton $\mathcal{B}_{G,S,\Pi,p,q}$ exists such that $Q_A \subseteq Q_{\mathcal{B}_{G,S,\Pi,p,q}}$ and*

- *On each $\mathcal{A}$-configuration c_A the evolution of which has exactly one site (x_G, t_G) in state G and a signal $\mathcal{S}$ with states in S, ultimately periodic with period Π and ultimately above the diagonal D_G stemming from (x_G, t_G).*
- *The evolution of $\mathcal{B}_{G,S,\Pi,p,q}$ contains a signal $\mathcal{S}^\star$ with $S \subseteq S^\star$, of slope $\frac{p+q}{q}$.*

Proof If the period of S is (0), on the anti-diagonal D_G^* stemming from (x_G, t_G), starting from a suitable site, a clipping by vector $\mathbf{1}$ is launched each two sites. Then one gets a new signal of period $(1, 0)$, see point 5a. If S has period (1), on the diagonal D_G stemming from (x_G, t_G), starting from a suitable site, a clipping by vector $-\mathbf{1}$ is launched each two sites. Then one gets a new signal of period $(0, 0)$, see point 5b. By changing the starting point of the period, one may assume that $c_0 = 1$ and $c_1 = 0$. By regularly launching clippings by vector $-\mathbf{1}$ from the diagonal D_G, one may "transform" all c_i of value 1 into 00. One then gets a signal of period $(1, \underbrace{0, \ldots, 0}_{\alpha \text{ times}})$. One also may assume that α is even using a new clipping by vector $\mathbf{1}$ launched from D_G^*.

Now, in launching from D_G^* a clipping by vector $(\underbrace{1, \ldots, 1}_{\ell \text{ times}})$, one gets a signal of period $(\underbrace{1, \ldots, 1}_{\ell+1 \text{ times}}, \underbrace{0, \ldots, 0}_{\alpha \text{ times}})$. Finally one regularly launches k clippings by vector $-\mathbf{1}$ from D_G in order to get a period $(\underbrace{1, \ldots, 1}_{\ell+1-k \text{ times}}, \underbrace{0, \ldots, 0}_{\alpha+2k \text{ times}})$.

One can choose ℓ, k and β such that $\ell + 1 - k = \beta p$ and $\alpha + 2k = \beta q$. As α is even, let us say $\alpha = 2\alpha^\star$, one can use $\beta = 2\alpha^\star$, $k = \alpha^\star(q - 1)$ and $\ell = \alpha^\star(2p + q - 1) - 1$. In this way one gets a new signal, the slope of which is ultimately $\frac{p+q}{q}$.

It has to be noted that, at each main step of transformation, a new automaton is built.

7 Conclusion

Some methods in algorithmics on one-dimensional cellular automata have been shown. Other methods also exist, in particular, methods that are based on different possibilities of compressing states of some cellular automaton (see Mazoyer (1992) and Heen (1996)). Different methods have arisen in the course of studies of specific questions such as, for example, the synchronization of a firing squad, the recognition or computing power of cellular automata (see the chapter ❷ Language Recognition by Cellular Automata) or the computation of functions as in Fisher (1965) (see also the chapter ❷ Computations on Cellular Automata). But few general studies exist except for that by Delorme and Mazoyer (2002b). And if one chooses the geometric point of view, as in ❷ Sect. 3, a lot remains to be done.

In the case of two-dimensional cellular automata, things are still more complicated because it is very difficult to draw space–time diagrams. The notion of a signal may be generalized in two ways. Either signals can be viewed as surfaces (see, e.g., Delorme and Mazoyer (2002a, 2004)), or they can be thought of as lines (see, e.g., Delorme and Mazoyer (1999)). Some methods of dimension 1 can be used. The method that can be found in Terrier (2006) or the synchronization method of Balzer (1967) is mentioned, which is extended to the plane in Szwerinski (1982). Of course other specific methods exist (very close to discrete geometry), such as "thinness" Romani (1976).

The notion of a signal is a core piece of the tools presented in this chapter. But algorithms without signals also exist, and these are minimally investigated. A characteristic example is the proof that XOR synchronizes each exponential line (Kari 1994; Yunès 2008).

References

Balzer R (1966) Studies concerning minimal time solutions to the firing squad synchronization problem. Ph.D. thesis, Carnegie Institute of Technology

Balzer R (1967) An 8-states minimal time solution to the firing squad synchronization problem. Inform Control 10:22–42

Čulik K (1989) Variations of the firing squad synchronization problem. Inform Process Lett 30:153–157

Čulik K, Dube S (1993) L-systems and mutually recursive function systems. Acta Inform 30:279–302

Čulik K, Karhumäki J (1983) On the Ehrenfeucht conjecture for DOL languages. ITA 17(3):205–230

Delorme M, Mazoyer J (1996) Languages recognition and cellular automata. In: Almeida J, Gomez GMS, Silva PV (eds) World Scientific, Singapore, pp 85–100

Delorme M, Mazoyer J (2002a) Reconnaissance parallèle des langages rationnels sur automates cellulaires plans. Theor Comput Sci 281:251–289

Delorme M, Mazoyer J (2002b) Signals on cellular automata. In: Adamatzky A (ed) Springer, London, pp 231–274

Delorme M, Mazoyer J (2004) Real-time recognition of languages on a two-dimensional archimedean thread. Theor Comput Sci 322(2):335–354

Delorme M, Mazoyer J, Tougne L (1999) Discrete parabolas and circles on 2D cellular automata. Theor Comput Sci 218(2):347–417

Fisher PC (1965) Generation of primes by a one dimensional real time iterative array. J ACM 12:388–394

Gruska J, Salvatore L, Torre M, Parente N (2006) Different time solutions for the firing squad synchronization problem. Theor Inform Appl 40(2):177–206

Heen O (1996) Economie de ressources sur automates cellulaires. Ph.D. thesis, Université Paris Diderot. In French

Kari J (1994) Rice's theorem for limit sets of cellular automata. Theor Comp Sci 127(2):227–254

Mazoyer J (1987) A six states minimal time solution to the firing squad synchronization problem. Theor Comput Sci 50:183–238

Mazoyer J (1989a) An overview on the firing squad synchronization problem. In: Choffrut C (ed) Automata networks, Lecture notes in computer science. Springer, Heidelberg

Mazoyer J (1989b) Solutions au problème de la synchronisation d'une ligne de fusiliers. Habilitation à diriger des recherches (1989). In French

Mazoyer J (1992) Entrées et sorties sur lignes d'automates. In: Cosnard MNM, Robert Y (eds) Algorithmique parallèle, Masson, Paris, pp 47–64. In French

Mazoyer J (1999) Computations on cellular automata. In: Delorme M, Mazoyer J (eds) Springer-Verlag, London, pp 77–118

Mazoyer J, Reimen N (1992) A linear speed-up theorem for cellular automata. Theor Comput Sci 101:59–98

Mazoyer J, Terrier V (1999) Signals in one-dimensional cellular automata. Theor Comput Sci 217(1):53–80

Ollinger N (2002) Automates cellulaires: structures. Ph.D. thesis, Ecole Normale Supérieure de Lyon. In French

Rapaport I (1998) Ordre induit sur les automates cellulaires par l'opération de regroupement. Ph.D. thesis, Ecole Normale Supérieure de Lyon

Richard G (2008) Systèmes de particules et collisions discrètes dans les automates cellulaires. Ph.D. thesis, Aix-Marseille Université. In French

Romani F (1976) Cellular automata synchronization. Inform Sci 10:299–318

Sutner K (1997) Linear cellular automata and Fisher automaton. Parallel Comput 23(11):1613–1634

Szwerinski H (1982) Time optimal solution of the firing squad synchronization problem for n-dimensional rectangles with the general at any position. Theor Comput Sci 19:305–320

Terrier V (1991) Temps réel sur automates cellulaires. Ph.D. thesis, Université Lyon 1. In French

Terrier V (2006) Closure properties of cellular automata. Theor Comput Sci 352(1):97–107

Theyssier G (2005) Automates cellulaires: un modèle de complexité. Ph.D. thesis, Ecole Normale Supérieure de Lyon. In French

Ulam S (1957) The Scottish book: a collection of problems. Los Alamos

Vaskhavsky VI, Marakhovsky VB, Peschansky VA (1969) Synchronization of interacting automata. Math Syst Theory 14:212–230

von Neumann J (1966) Theory of self-reproducing. University of Illinois Press, Urbana, IL

Waksman A (1996) An optimum solution to the firing squad synchronization problem. Inform Control 9:66–78

Yunès JB (2008) Goto's construction and Pascal's triangle: new insights in cellular automata. In: Proceeding of JAC08. Uzès, France, pp 195–202

4 Language Recognition by Cellular Automata

Véronique Terrier
GREYC, UMR CNRS 6072, Université de Caen, France
veroniqu@info.unicaen.fr

G. Rozenberg et al. (eds.), *Handbook of Natural Computing*, DOI 10.1007/978-3-540-92910-9_4,
© Springer-Verlag Berlin Heidelberg 2012

Abstract

Cellular automata (CA) comprise a simple and well-formalized model of massively parallel computation, which is known to be capable of universal computation. Because of their parallel behavior, CA have rich abilities of information processing; however, it is not easy to define their power limitations. A convenient approach to characterizing the computation capacity of CA is to investigate their complexity classes. This chapter discusses the CA complexity classes and their relationships with other models of computations.

1   Preliminaries

Cellular automata (CA) provide an ideal framework for studying massively parallel computation. Their description is very simple and homogeneous; however, as emphasized by various examples, they allow one to distribute and synchronize the information in a very efficient way. Their ability to do fast computation raised complexity issues. Early works drew much attention to CA as language recognizers.

Obviously, CA are of the same computational power as Turing machines. Hence the major motivation for CA complexity study is to gain knowledge on the way parallelism acts and to explicitly explain the gain that may be achieved with CA. In particular, much interest is devoted to their low-time complexity classes because they may provide significant complexity benefits. But, as for the other models of computation, it is not a simple task to deeply evaluate their power and their limitations.

In this chapter, the essential developments on CA in the field of language recognition are reviewed. The purpose is to give insight into the performance of different types of CA, with a major focus on the low-time complexity classes. Of course, it will not be possible to report every outcome in detail and some choices have been made. First, only space-bounded computations are discussed. Specifically, because of the interest in fast computation, one will just consider space bound defined as the space consumed by low time computation. And, despite the interest in general complexity issues, we decided not to include non-deterministic or alternating variants of CA. As a matter of fact, they are beyond the scope of realistic devices, because even their minimal time classes contain *NP*-complete problems. For the same reasons, CA with tree-like array structures shall be ignored.

The rest of this chapter is organized as follows. ❷ Section 2 introduces the different variants of CA recognizers and their complexity classes. ❷ Section 3 reviews the known inclusions and equalities among these complexity classes. ❷ Section 4 recalls the limitations established on the recognition power of CA. ❷ Section 5 relates to the comparison with other models of computation. As a conclusion, ❷ Sect. 6 discusses some of the old open questions that remain up to now unsolved.

2   Definitions and Examples

Basically, a CA is a regular array of cells. These cells range over a finite set of states and evolve synchronously at discrete time steps. At each step, the new state of each cell is produced by a local transition rule according to the current states in its neighborhood. So a CA is completely specified by a tuple $(d, S, \mathcal{N}, \delta)$ where d is the dimension of the array of cells indexed by $\mathbb{Z}^d$, S is the finite set of states, the neighborhood $\mathcal{N}$ is a finite ordered subset of $\mathbb{Z}^d$, and δ is the transition function from $S^{|\mathcal{N}|}$ into S.

A cell is denoted by $\mathbf{c}$, $\mathbf{c} \in \mathbb{Z}^k$; a site $(\mathbf{c}, t)$ denotes the cell $\mathbf{c}$ at time t and $\langle \mathbf{c}, t \rangle$ denotes its state. And at each step $t \geq 0$, the state is updated in this way: $\langle \mathbf{c}, t+1 \rangle = \delta(\langle \mathbf{c} + \mathbf{v_1}, t \rangle, \dots, \langle \mathbf{c} + \mathbf{v_r}, t \rangle : \mathcal{N} = \{\mathbf{v_1}, \dots, \mathbf{v_r}\})$. Depending on the neighborhood, the flow of information may go in all directions of the cellular array as with the von Neumann neighborhood $\{\mathbf{v} \in \mathbb{Z}^k : \Sigma |v_i| \leq 1\}$ or with the Moore one $\{\mathbf{v} \in \mathbb{Z}^k : \max |v_i| \leq 1\}$ or it may be restricted to one-way in some directions as with the one-way von Neumann neighborhood $\{-\mathbf{v} : \mathbf{v} \in \mathbb{N}^k \text{ and } \Sigma v_i \leq 1\}$ or with the one-way Moore one $\{-\mathbf{v} : \mathbf{v} \in \mathbb{N}^k \text{ and } \max v_i \leq 1\}$. In what follows, one-way communication will refer to such restrictions whereas two-way communication will refer to the ability to transmit information everywhere in the cellular array.

In order for a CA to act as a language recognizer, one has to specify how the input is given to the cellular array and how the result of the computation is obtained.

The Input Mode

For the sake of simplicity, it is assumed that Σ the finite alphabet of input letters is a subset of the states set S of the CA. A quiescent state λ such that $\delta(\lambda, \dots, \lambda) = \lambda$ is also considered. Two modes are distinguished to give the input to the array: the parallel one and the sequential one. In the *parallel mode*, the whole input is supplied at initial time to the array. It implies an implicit synchronization at the beginning of the computation. All input symbols are fed into a distinct cell and are arranged in such a way as to keep the input structure. In the *sequential mode*, a specific cell is chosen to receive the input. All cells are initially quiescent and the input symbols are fed serially to the specific cell. When the whole input has been read by this input cell, an end-marker symbol $ is infinitely fed to it. Because the input cell takes into account not only its neighborhood but also its received input symbol, it behaves in a particular way.

The Output Mode

The advantage of recognition problems is that the output is of yes/no type. Hence, on a CA, a specific cell is enough to indicate acceptance of rejection of the input. The choice of this output cell is arbitrary in the case of two-way communication. But, in the case of one-way communication, it is subject to constraint: the output cell must be able to get all information from the input.

Language Recognition

For the purpose of recognition, two subsets of the states set S are specified: the set S_{acc} of accepting states and the set S_{rej} of rejecting states. A language L is said to be *recognized* by a CA, if on input w the output cell enters at some time t_e an accepting state if $w \in L$ or a rejecting state if $w \notin L$; and for all time $t < t_e$, the output cell is neither in an accepting state nor in a rejecting state. A time complexity T refers to a function relating the input size to an amount of time steps. A CA recognizer works in time T, if it accepts or rejects each word w of size s within $T(s)$ steps.

Usually, the focus is on acceptance. In this case, only the set S_{acc} of accepting states is specified. A language L is said to be *accepted* by a CA, if on input w the output cell enters an accepting state if and only if $w \in L$. A CA acceptor works in time T, if it accepts each word $w \in L$ of size s within $T(s)$ steps.

In fact, the notion of acceptance and recognition turns out to be equivalent for current time complexities. Indeed, a CA, on an input of size s, is able to distinguish the output cell at time $T(s)$ when T is any standard time complexity. Hence, the CA can reject at time $T(s)$ all non-accepted words of size s. So, except just below, one will not differentiate between CA recognizer and CA acceptor.

An example of a CA recognizer is now given.

Example 1 *The majority language that consists of the words over the alphabet* $\{a, b\}$ *in which there are strictly more a's than b's is recognized by a one-dimensional CA with parallel input mode and one-way neighborhood* $\{-1, 0\}$.

The CA is defined in this way. $\Sigma = \{a, b\}$ is the input alphabet, $S = \{a, b, A, B, 1, 2, n, x\}$ is the states set, $S_{acc} = \{A\}$ is the set of accepting states, $S_{rej} = \{B, x\}$ is the set of rejecting states, the cell -1 is assumed to remain in a persistent state $\#$, $\delta : S \cup \{\#\} \times S \to S$ is the transition function displayed below.

δ	a	b	A	B	1	2	n	x
a	a	b	A		a	n	A	
b	a	b		B	n	b	B	
A	a		A		a	n	A	
B		b		B	n	b	B	
1	1	2			1	2		
2	1	2			1	2		
n					1	2	n	
x	A	B	x	x	A	B	x	x
$\#$	1	2	x	x	A	B		x

Since the neighborhood is one-way, the output cell is chosen to be the rightmost one. The computations on inputs $w_1 = aaaaabbaaabbabbbbba$ and $w_2 = babbaaabbbba$ are depicted in ❷ *Fig. 1*. As the input mode is parallel, the words are supplied at initial time. On each cell, above the diagonal, the length of the sequence of a's and A's or, b's and B's, records the difference between the numbers of a's and b's of the input. The input word w_1 is accepted since the output cell $n - 1$ enters the accepting state A, whereas the input word w_2 is rejected since it enters the rejecting state B. The CA recognizer works within time complexity $T(n) = 2n$. As a matter of fact, the cell 0 knows at time 1 that it is the leftmost one; so by mean of a signal, which moves one cell to the right every two steps, it is simple to mark the output cell $n - 1$ at time $2n$. Thus the status of the words not accepted may be decided at this time even though the set of rejecting states is not specified. More generally, notice that, on an input of length n, the output cell is able to know the whole input at time $n - 1$, whereas it is able to know that the input is completed at time n.

Time Complexities

Among the time complexities, two functions are of major interest: the real time and the linear time. The *real-time* complexity, denoted by rt, means that each word is accepted or rejected "as soon as possible." Here, only a slight difference between acceptance and recognition is

□ Fig. 1
Recognition of the majority language.

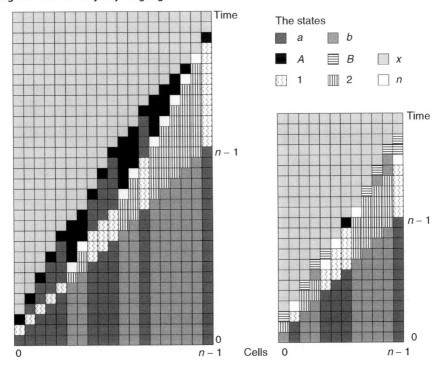

encountered. For acceptance, the real-time complexity corresponds to the minimal time for the output cell to read the whole input, whereas for recognition, one additional unit of time may be required in order that the output cell knows that the input is completed. In the sequel, we only deal with real-time acceptance. Practically, the real-time complexity is specified by the way the input is supplied, the choices of the output site and the neighborhood. The *linear-time* complexity, denoted by *lt*, is just defined as the real-time function multiplied by any constant strictly greater than 1.

Space-Bounded Computation

In this chapter, one will restrict oneself to space-bounded computation. During all the computation, only a fixed number of cells, depending on the size of the input, are active. All other cells remain in a persistent state ♯ from the start to the end. One may imagine many space bounds. However, in practice, the bounded space is *uniquely* defined as the space required to perform real-time computation. It means that the active part of the array is identical whatever the effective time complexity may be. Then, in the following, all classes defined in terms of time complexity classes will be actually both time and space bounded. As a matter of fact, space-bounded and unbounded computations become the same for real-time and linear-time complexities. On the other hand, because the space is bounded, the evolution becomes periodical after a number of steps that is exponential in the size of the space. That establishes an upper bound on the relevant time complexities.

The Dependency Graph

In order to reflect the neighborhood constraints on the sites involved in the computation, one will consider the graph induced by the dependencies between them. Precisely, a dependency graph is defined according to a given type of CA, a fixed time complexity, and the input size. It is a directed graph. Its set of vertices consists of the sites, which both are influenced by the input and can have an effect on the output site—in other words, all relevant sites regarding the computation. Its edges are the couples of sites $((\mathbf{c} + \mathbf{v}, t), (\mathbf{c}, t + 1))$, for all $\mathbf{v}$ belonging to the neighborhood $\mathcal{N}$.

The different types of CA recognizers are described in the following subsections. Their features are stated by the array dimension, the input dimension and, in the parallel mode, the way to place the input into the array.

2.1   One-Dimensional CA Language Recognizer

A one-dimensional CA recognizer is structured as a linear array and operates on words. The space is assumed to be linearly bounded: the number of active cells equals the length of the input. In practice, the active cells are numbered $0, \ldots, n - 1$, for an input of length n. Because the other cells always remain in the persistent state $\sharp$, one-dimensional CA with two-way communication have the same computational power as Turing machines working in $O(n)$ space.

Different variants of one-dimensional CA are currently defined. On the one hand, it depends on the neighborhood, which allows either two-way or one-way communication. Actually, from Poupet (2005), it is known that there are somewhat only two kinds of neighborhoods, the two-way one $\{-1, 0, 1\}$ and the one-way one $\{-1, 0\}$. On the other hand, it depends on the input mode being either parallel or sequential. Hence four variants of one-dimensional CA are characterized according to the neighborhood and input mode choices. They are named PCA, SCA, POCA, and SOCA. The first letter "P" or "S" stands for parallel or sequential input mode and the "O" occurrence makes distinctions between one-way or two-way communication (❷ *Table 1*).

❑ **Table 1**
The four variants of CA in dimension one

Automata	Neighborhood	Input mode	Output cell
PCA	$\{-1, 0, 1\}$	Parallel	0
SCA	$\{-1, 0, 1\}$	Sequential	0
POCA	$\{-1, 0\}$	Parallel	$n - 1$
SOCA	$\{-1, 0\}$	Sequential	$n - 1$

Different denominations have been used by other authors: SCA and SOCA are more often called iterative array (IA) and one-way iterative array (OIA); POCA are also called one-way cellular automata (OCA) or mentioned as trellis automata.

Let one describe the way to carry out the input in the two modes. In the parallel input mode, at initial time 0, the ith symbol of the input $w = x_0 \ldots x_{n-1}$ of size n, is fed to the cell i: $\langle i, 0 \rangle = x_i$. In the sequential input mode, all active cells are initially quiescent (i.e., set to λ). The cell indexed by 0 is chosen to read the input: it gets the ith symbol x_i of the input $w = x_0 \ldots x_{n-1}$

■ Fig. 2
The space–time diagram of the four one-dimensional CA variants.

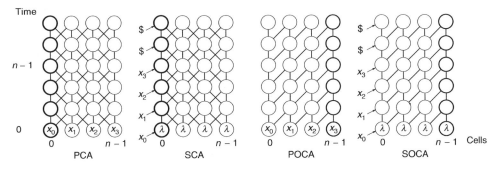

at time i; then it gets an end-marker $ at all time $t \geq n$. Note that the input cell 0 evolves according to a particular transition function $\delta_{init}: (\Sigma \cup \{\$\}) \times S^{|\mathcal{N}|} \to S$, so $\langle 0, t \rangle = \delta_{init}(x_t, \langle v_1, t-1 \rangle, \ldots, \langle v_r, t-1 \rangle)$ where $\mathcal{N} = \{v_1, \ldots, v_r\}$ refers to the neighborhood and $x_t = \$$ when $t \geq n$.

For the output cell that yields the result, the initial cell 0 in case of PCA and SCA that use two-way communication is chosen. The unique possibility for POCA and SOCA is the rightmost cell indexed $n - 1$ since their neighborhood $\{-1, 0\}$ is one-way.

❷ *Figure 2* above shows the customary representation of the four models.

Some notation is useful for dealing with the various complexity classes of one-dimensional CA recognizers. For a time complexity function T from $\mathbb{N}$ into $\mathbb{N}$, PCA(T) (resp., SCA(T), POCA(T), and SOCA(T)) will denote the class of languages recognizable in time T by PCA (resp., SCA, POCA, and SOCA). With some liberty, PCA will refer to the class of languages recognized in unbounded time by PCA; and in the same way, SCA, POCA, and SOCA will indicate the device type as well their corresponding unbounded time complexity classes.

Particular attention is devoted to the low-time complexity classes, namely the real time and the linear time. The real-time complexity $rt(n)$ is defined as the earliest time for the output cell to read the whole input of length n. More precisely, it corresponds to $n - 1$ in case of PCA, SCA, and POCA and to $2n - 2$ in case of SOCA. And $lt(n) = \tau \, rt(n)$, where τ is any constant strictly greater than 1, gives rise to linear-time complexity. In the following, the classes of language recognized in real time by PCA, SCA, POCA, and SOCA will be denoted by RPCA, RSCA, RPOCA, and RSOCA. For linear-time complexity, the corresponding classes will be designated by LPCA, LSCA, LPOCA, and LSOCA.

Some examples are now given to illustrate the computation ability of the real-time complexity classes.

Example 2 (Cole 1969) *The palindrome language* $\{w \in \Sigma^*: w = w^R\}$ *is a real-time SCA language* (w^R *denotes w read backward*).

The algorithm is described in a geometrical way. Its discretization to obtain the corresponding RSCA is straightforward. First of all, the input word received sequentially on the input cell 0 is sent at maximal speed to the right. Precisely the symbol x_i, which is fed on cell $c = 0$ at time $t = i$ follows the line A_i of equation $t = c + i$ with $c \geq 0$. Simultaneously, a signal F of speed $1/3$ starts from the input cell 0 at initial time and draws the line $t = 3c$. So the signals A_i and F

■ **Fig. 3**

Real-time recognition of the palindrome language by a SCA.

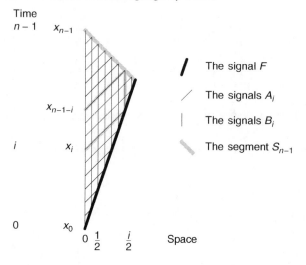

intersect on the point $(i/2, 3i/2)$. From this intersection, the symbol x_i carried by A_i goes further along the vertical line B_i: $c = i/2$ with $t \geq 3i/2$. In this way, the symbols x_j and x_i meet on the point $(i/2, j + i/2)$ where the signals A_j and B_i intersect. In particular, for any integer k, on the segment S_k: $c + t = k$ with $0 \leq c \leq k/4$, which runs at maximal speed to the left from the signal F to the output cell 0, one may compare all pairs of symbols $\{x_i, x_{k-i}\}$ with i such that $0 \leq i \leq k/2$. Now, on an input $x_0 \ldots x_{n-1}$ of length n, the sequence of comparisons between x_i and x_{n-1-i} with $0 \leq i \leq (n-1)/2$ determines whether the input is a palindrome or not. This sequence of comparisons is exactly the one that occurs on the segment S_{n-1}: $c + t = n - 1$. Finally, observe that the segment S_{n-1} reaches the cell $c = 0$ at time $t = n - 1$. In other words, the result is obtained in real time (❯ *Fig. 3*).

Example 3   (Cole 1969) *The square language* $\{ww: w \in \Sigma^*\}$ *is a real-time SCA language.*

On the one hand, each input symbol x_i is carried along the line A_i of equation $t = c + i$ with $c \geq 0$. On the other hand, each input symbol x_i is first sent at maximal speed to the left, following the signal B_i of equation $t = -c + i$ with $c \geq 0$. Simultaneously, initialized by the input cell c at time 0, a signal G of equation $t = -3c - 1$ starts from the point $(-1/2, 1/2)$ and moves with speed $1/3$ to the left. So the signals B_i and G intersect on the point $(-(i+1)/2, (3i+1)/2)$. From this intersection, the symbol x_i carried by B_i goes further along the signal C_i: $t = c + 1 + 2i$ with $t \geq (3i+1)/2$. Then the signal C_i intersects the initial cell $c = 0$ on the point $(0, 1 + 2i)$. From this intersection, the symbol x_i follows the signal D_i: $t = 3c + 1 + 2i$. In this way, for any i, j with $i < j$, the symbols x_i and x_j meet at the point $((j-1)/2 - i, (3j-1)/2 - i)$ where the signals D_i and A_j intersect. Now, on an input $x_0 \ldots x_{2n-1}$ of even length $2n$, the sequence of comparisons between x_i and x_{n+i} with $0 \leq i < n$ determines whether the input is a square word or not. These comparisons are performed on the points $((n-1-i)/2, (3n-1+i)/2)$. Hence they occur on the segment S_{2n-1}: $t + c = 2n - 1$ with $0 \leq c \leq (2n-1)/4$, which starts from the signal F: $t = 3c$, moves at maximal speed to the left and reaches the cell 0 at time $2n - 1$. Therefore the result is obtained in

☐ **Fig. 4**

Real-time recognition of the square language by a SCA.

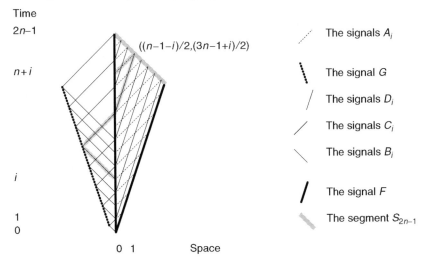

minimal time. With regard to space, note that negative cells are not necessarily active, since the negative side can be folded on the positive side, that is, all activity on the negative cells can be as well performed on the positive side (❍ *Fig. 4*).

Example 4   (Smith 1972) *The Dyck language is a real time POCA language.*

A RPOCA that recognizes the Dyck language over the alphabet $\{a, b\}$ can be defined in this way.

$\Sigma = \{a, b\}$ is the input alphabet, $S = \{a, b, d, o\}$ is the set of states, $S_{acc} = \{d\}$ is the set of accepting states and $\delta : S^2 \to S$ is the transition function displayed below.

δ	a	b	d	o
a	a	d	a	a
b	o	b		o
d		b		a
o	o	b	b	o

A significant feature of POCA is that the real-time computation on an input w contains the real-time computations of all its factors. An illustration is depicted in ❍ *Fig. 5*.

Example 5   (Culik 1989) *The language $L = \{a^i b^{i+j} a^j : i, j \in \mathbb{N}\}$ is a real-time POCA language.*

First, the RPOCA will reject all the words outside the regular language $\{a^i b^j a^k : i, j, k \in \mathbb{N}\}$. Second, notice that the factors of a word of shape $a^* b^n a^*$ that belong to L are on the one hand

◘ Fig. 5

Recognition of the Dyck language by a RPOCA.

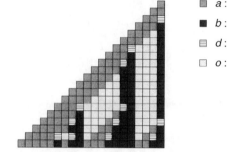

▦ *a* : Proper prefixes of Dyck words

■ *b* : Proper suffixes of Dyck words

▤ *d* : Dyck words

☐ *o* : Other words

Real time computation on
the word *aaaaababbaaabbbbaabb*

Real time computation on
the factor *aababbaaabbb*

◘ Fig. 6

Real-time recognition of the Culik language by POCA.

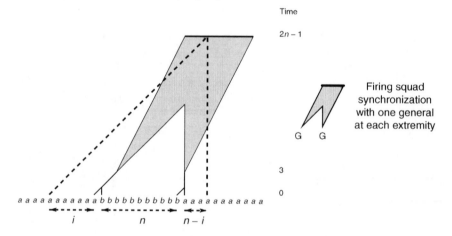

Firing squad
synchronization
with one general
at each extremity

Time

$2n-1$

3

0

$G \quad G$

$a\,a\,a\,a\,a\,a\,a\,a\,a\,a\,a\,a\,b\,b\,b\,b\,b\,b\,b\,b\,b\,b\,a\,a\,a\,a\,a\,a\,a\,a\,a\,a$

$i \qquad\qquad n \qquad\qquad n-i$

the words $a^i b^i$ and $b^i a^i$, which will be simply accepted by the RPOCA, and on the other hand the words $a^i b^n a^{n-i}$. Observe that to accept, in real time, all the factors $a^i b^n a^{n-i}$ with $i = 1, \ldots,$ $n-1$ of an input of shape $a^* b^n a^*$, means that $n-1$ consecutive cells simultaneously enter at time $2n-1$ and for the first time in an accepting state. In other words, a synchronization process is required to recognize this language. As sketched in ❷ *Fig. 6*, the synchronization can be set up using a firing squad synchronization process with two generals located respectively according to the boundary between the *a*'s and *b*'s and the boundary between the *b*'s and *a*'s.

2.2 Multidimensional CA Language Recognizer

Natural extensions to higher dimensional arrays have been early investigated in (Cole 1969; Chang et al. 1987; Ibarra and Palis 1988). In this framework, the space increases with the dimension, while the minimal time complexity remains linearly bounded by the length of the input word. Precisely, on an input of length n, the space of a d-dimensional CA recognizer is defined as a d-dimensional array of size n in each dimension. Hence, a d-CA with two-way neighborhoods has the same computation ability as Turing machines working in $O(n^d)$ space.

The diversity of neighborhoods also rises with the dimension. In the case of two-way neighborhoods and sequential input mode, it is known from Cole (1969) that the computation ability is preserved even though the neighborhood is restricted to the von Neumann one: $\{\mathbf{v} \in \mathbb{Z}^d : \Sigma |v_i| \leq 1\}$. Whether the same is true or not for the other variants is unclear. However, the impact of the neighborhood choice appears less crucial for language recognizers where almost all cells are initially quiescent than for picture language recognizers that will be defined below. Hence, as neighborhood issues are currently studied in this last context, we shall only regard the von Neumann neighborhood and its one-way counterpart $\{-\mathbf{v} : \mathbf{v} \in \mathbb{N}^d$ and $\Sigma v_i \leq 1\}$ in the case of a multidimensional CA language recognizer.

For an input of length n, the space area consists of the cells $\{\mathbf{c} : 0 \leq c_1, \dots, c_d < n\}$. The two-dimensional case is depicted in ❷ *Fig. 7*. The cells outside this area remain in a persistent

◻ **Fig. 7**
The space array of the four 2-CA variants.

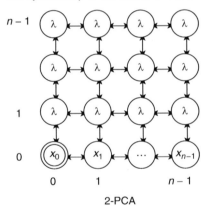

2-PCA

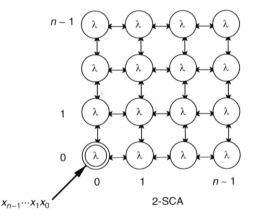

2-SCA

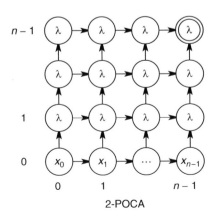

2-POCA

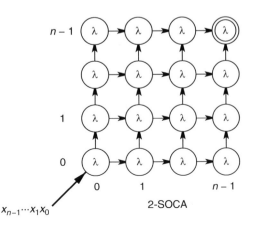

2-SOCA

state ♯ during all the computation and each cell inside the area is assumed to remain quiescent until the step when it may be affected by the input. In the parallel mode, the input is supplied on the first dimension of the array: the ith input symbol of the input $w = x_0 \ldots x_{n-1}$ is fed to the cell $(i, 0, \ldots, 0)$ at time 0. In the sequential mode, the specific cell that gets the input serially from time 0 is chosen to be the cell indexed by $\mathbf{0} = (0, \ldots, 0)$. Of course, this input cell evolves according to a particular transition function δ_{init}, which takes into account its neighborhood and the received input symbol. The choice of the output cell depends on the neighborhood. One chooses the cell $\mathbf{0} = (0, \ldots, 0)$ with the von Neumann neighborhood and the opposite corner $\mathbf{n} - \mathbf{1} = (n-1, \ldots, n-1)$ with the one-way von Neumann neighborhood.

Analogously to one-dimensional CA recognizer, $d - $ PCA, $d - $ SCA, $d - $POCA, and $d - $ SOCA denote the four d-dimensional variants according to whether the input mode is parallel or sequential and whether the neighborhood is the two-way von Neumann one or its one-way counterpart. They are denominated variously in different papers: $k - $ SCA is named k-dimensional iterative array in Cole (1969); $2 - $ SOCA is named one-way two-dimensional iterative array in Chang et al. (1987) and OIA in Ibarra and Palis (1988); $k - $ POCA is named k-dimensional one-way mesh connected array in Chang et al. (1989).

The real-time function $rt(n)$, which is defined as the minimal time for the output cell to read the whole input of size n, corresponds to $n - 1$ in the case of $d - $ PCA and $d - $ SCA, $d(n - 1)$ in the case of $d - $ POCA and $(d + 1)(n - 1)$ in the case of $d - $ SOCA.

2.3 Two-Dimensional CA Language Recognizer Equipped with a Thread

With the parallel mode in order to save time, the input word may be supplied in a more compact way than the linear one. The difficulty is that there are many ways to set up the one-dimensional input words into multidimensional arrays. All ways are arbitrary and lead to distinct devices with their own complexity classes. A proper approach has been proposed by Delorme and Mazoyer (2002). To set up the inputs in a uniform way, they equip the CA with a thread along which the inputs are written. Thus the thread is a given parameter of the CA, which is independent of the inputs and their length. It is defined as an infinite sequence of adjacent positions in the two-dimensional array without repetition. ❷ *Figure 8* depicts the

◻ **Fig. 8**
Archimedean and Hilbert threads.

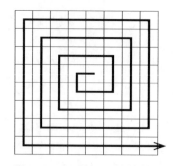 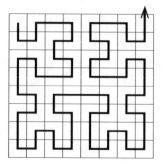

Archimedean and Hilbert threads. In practice, the device array is provided with an additional layer that codes the thread. In this layer, each cell records how the thread enters and exits this cell. At initial time, the input symbols are placed consecutively along the thread. Then the evolution of each cell depends on both the states and the thread components of its neighborhood. The output cell is chosen to be the first cell of the thread. Now, the real-time complexity depends on both the neighborhood and the thread (precisely, its significant part according to the input length). No further details on these device types will be given, but the interested reader can refer to Delorme and Mazoyer (2002, 2004) for complete definitions and examples.

2.4 Two-Dimensional CA Picture Recognizer

For higher dimensions, another point of view is to process multidimensional data instead of simple strings. Motivated by image processing issues, much interest has been devoted to picture language recognized by two-dimensional CA. Another stimulation comes from the developments of picture language theory (Giammarresi and Restivo 1997). Hence, in the following, attention is limited to this context, although investigations of arbitrary dimensional CA with arbitrary dimensional inputs would be fairly instructive.

Let us recall some definitions related to picture languages. A picture p over an alphabet Σ is defined as a rectangular $m \times n$ array of symbols of Σ. The couple (m, n) refers to the size of the picture and $p(\mathbf{c})$ denotes the symbol at position $\mathbf{c}$. The set Σ^{**} denotes the set of all pictures over Σ. A picture language over Σ is any subset of Σ^{**}.

A two-dimensional CA picture recognizer (in short a PictCA) designates a CA recognizer that operates on pictures. On PictCA, the input mode is parallel: at initial time 0 the symbol $p(\mathbf{c})$ of the input picture p is fed to the cell $\mathbf{c}$. For an input of size (m, n), the bounded space consists of the $m \times n$ cells $\mathbf{c} = (x, y)$ with $0 \le x < m$ and $0 \le y < n$. Outside, the cells remain in a persistent state $\sharp$ during all the computation.

A priori, the output cell is the cell indexed by $(0, 0)$. The time complexities T are functions defined from $\mathbb{N}^2$ to $\mathbb{N}$. And a PictCA is said to accept a picture language L in time T if it accepts the pictures $p \in L$ of size (m, n) in at most $T(m, n)$ steps. As usual, real time means "as soon as possible" and is conditional on the neighborhood. Precisely, the real-time function $rt_{\mathcal{N}}(m, n)$ is, for a PictCA with neighborhood $\mathcal{N}$, the minimal time needed by the output cell $(0, 0)$ to receive any particular part of a picture input of size (m, n). That means $rt_{\mathcal{N}}(m, n) = m + n - 2$ when $\mathcal{N}$ is the von Neumann neighborhood and $rt_{\mathcal{N}}(m, n) = \max(m, n) - 1$ when $\mathcal{N}$ is the Moore neighborhood. The linear-time complexities for PictCA with the neighborhood $\mathcal{N}$ are functions $lt_{\mathcal{N}}$ where $lt_{\mathcal{N}}(m, n) = \tau\, rt_{\mathcal{N}}(m, n)$ and τ is any constant strictly greater than 1. In what follows, the class of all pictures languages recognized by a PictCA with the neighborhood $\mathcal{N}$ in real time (or linear time) will just be named as the real-time (or linear-time) PictCA with the neighborhood $\mathcal{N}$.

Various algorithms for pictures have been proposed in the general context of mesh-connected arrays of processors. But, their processing elements are not necessarily finite-state contrary to CA. And specific examples, which illustrate the possibilities of processing the data on PictCA, are scarce. Anyway, Beyer (1969) and Levialdi (1972) have independently exhibited two real-time PictCA with the Moore neighborhood that recognize the set of connected pictures. The majority language that consists of the pictures over the alphabet $\{0, 1\}$ in which there are more 1's than 0's has also been examined. Savage (1988) has shown that it is recognized in linear time by PictCA with one-way neighborhoods.

3 Positive Results and Simulation

In this section, the main known equalities and inclusions among CA complexity classes will be examined. These positive results are essentially based on the geometrical characteristics inherited from the regularity of the network structure. The proofs are established by simulations that widely use the tools presented in the chapter ❷ Algorithmic Tools on Cellular Automata by Delorme and Mazoyer and exploit the malleability of the dependency graphs.

3.1 Basic Equivalences Among the Low-Complexity Classes

As an introduction, the figure below gives a general overview of the main relationships among the complexity classes in dimension one.

$$PCA = SCA = DSpace(n)$$
$$\cup|$$
$$POCA = SOCA$$
$$\cup|$$
$$LPCA = LSOCA = LSCA$$
$$\cup|$$
$$RPCA = RSOCA = LPOCA$$
$$\subsetneq \qquad \supsetneq$$
$$RPOCA \neq RSCA$$

In this section, we only focus on the various equivalences between the low-complexity classes. We shall return to the equality of POCA and SOCA in ❷ Sect. 3.4, to the incomparability of RPOCA and RSCA and their proper inclusions in RPCA in ❷ Sect. 4.1. Further discussions will also follow about the famous questions whether the inclusions RPCA ⊆ PCA and RPCA ⊆ LPCA are strict in ❷ Sect. 6.1 and in ❷ Sects. 3.5 and ❷ 6.2.

The original proofs of the positive relationships can be found in the following papers. The equality LPCA = LSOCA comes from Ibarra et al. (1985) where it was observed that every PCA working in time $T(n)$ can be simulated by SCA in time $n + T(n)$. The equality LSOCA = LSCA has been noticed in Ibarra and Jiang (1988). The equality LPOCA = RPCA is from Choffrut and Culik (1984) and the equality RSOCA = LPOCA from Ibarra and Jiang (1987). Further relationships between SOCA and POCA in higher dimension have been reported in Terrier (2006b), in particular the equality $d - RSOCA = d - LPOCA$ and the inclusions $d - RSOCA \subseteq (d+1) - RPOCA$ and $d - LSOCA \subseteq (d+1) - RSCA$. One can also notice that restricted to unary languages (i.e., languages over a one-letter alphabet), RSCA is as powerful as RPCA.

All these equalities are easily obtained by basic simulations. To construct such simulations between one device $\mathcal{A}$ and another device $\mathcal{B}$, a simple method consists in exhibiting a transformation, which maps the dependency graph of the initial device $\mathcal{A}$ into another directed graph and to verify that this mapped graph, modulo slight modifications, fits the dependency graph of the device $\mathcal{B}$.

To illustrate this, one example can be considered: the inclusion of RPCA into RSOCA. The question is how to simulate the real-time computations of a given PCA $\mathcal{A}$ by the real-time computations of a SOCA $\mathcal{B}$. The simulation is essentially based on the following transformation $g(c, t) = (t, 2t + c)$. Hence we should verify that g allows one, with slight modifications, to convert a real-time computation of $\mathcal{A}$ into a real-time computation of $\mathcal{B}$. For that, we must check that the conditions imposed by the features of the device $\mathcal{B}$ are respected, namely the conditions relating to the structure of the array, the finite memory nature of each cell, the input and output modes, and the neighborhood. First, $g(i, 0) = (0, i)$ guarantees the conversion from the parallel input mode to the sequential input mode and $g(0, n - 1) = (n - 1, 2(n - 1))$ ensures the correspondence between the output sites. Second g maps all sites of $\mathcal{A}$ into sites of $\mathcal{B}$, precisely a site of $\mathcal{B}$ is mapped to by at most one site of $\mathcal{A}$. So a finite memory capacity is enough on each cell of $\mathcal{B}$. Finally, on $\mathcal{A}$ governed by the neighborhood $\{-1, 0, 1\}$, the elementary data movements are $(-1, 1)$, $(0, 1)$, $(1, 1)$. They are converted into the movements $(1, 1)$, $(1, 2)$, $(1, 3)$, which satisfy the dependencies constraints on $\mathcal{B}$. Effectively, the data are transmitted through intermediate sites according to the elementary moves $(1, 1)$ and $(0, 1)$ and that without exceeding the finite memory capacity of each cell (❷ *Fig. 9*).

An interesting link between one-dimensional devices and two-dimensional devices equipped with a thread has been observed in Delorme and Mazoyer (2004). It states that the class of languages recognized in real time by a two-dimensional CA with the Archimedean thread and the Moore neighborhood is strictly contained in the class of languages recognized in real time by one-dimensional PCA. Although the minimal time in $O(\sqrt{n})$ for two-dimensional CA with the Archimedean thread is lower than n the minimal time for PCA, this result is far from being straightforward. Notably, during the simulation, each cell must recover its particular position from the output cell and its neighbors in the initial Archimedean spiral. In fact, it enlightens one on the impact of the way input data are space-distributed on the recognition power.

Finally, notice that for picture languages recognition, PictCA with the von Neumann and Moore neighborhoods can simulate each other with a linear time overhead and thus are linear time equivalent. Unfortunately, finer inclusions concerning PictCA are ignored.

❏ **Fig. 9**
Simulation of a RPCA $\mathcal{A}$ by a RSOCA $\mathcal{B}$.

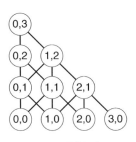
The RPCA $\mathcal{A}$

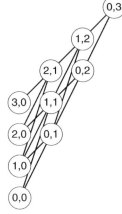
The result of
the transformation g

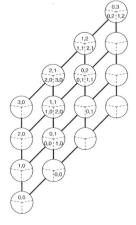
The RSOCA $\mathcal{B}$

3.2 Linear Speedup

When investigating complexity classes, an immediate question concerns linear acceleration of the running time. For CA, because the recognition time never goes below real time, a linear speedup corresponds to an acceleration by a constant factor of the running time beyond real time.

Theorem 1 *Let f be a function from $\mathbb{N}$ to $\mathbb{N}$ and rt be the real time function for PCA (resp. SCA, POCA or SOCA). For any constant R, a language recognized in time $rt(n) + f(n)$ by a PCA (resp. SCA, POCA or SOCA) is also recognized in time $rt(n) + \lceil f(n)/R \rceil$ by another PCA (resp. SCA, POCA or SOCA).*

In an early work (Beyer 1969), Beyer demonstrated the speed up result for PCA (included in the case of dimension two). Because Beyer (1969) was not widely circulated, the same result can be also found in Ibarra et al. (1985a) and in more general settings in Mazoyer and Reimen (1992). A proof for POCA is in Bucher and Culik (1984) and for SCA and SOCA in Ibarra et al. (1985b). A generalization to SCA in dimension two is given in Ibarra and Palis (1988) and to SOCA and POCA in arbitrary dimension in Terrier (2006b).

One can now recall the usual methods applied to speed up the computation in a linear way. The simplest case is when communication is one-way. Because any two cells are not mutually interdependent, the communication graph is a directed graph that is acyclic. This characteristic allows one to speed up computation easily. ❷ *Figure 10* depicts the scheme for dimension

❏ **Fig. 10**
Linear speedup in case of one-way communication.

one: once the cell gets the whole input part situated on its left, it can operate R times faster for any integer constant R. This principle can be generalized to higher dimension as long as communication is one-way.

When communication is two-way and all cells are interdependent, to reduce the running time requires compacting the space. In grouping cells into fewer ones, the time to exchange information between cells is reduced and so the computation can be achieved at a higher rate. The grouping operation differs according to the input mode.

It is immediate when the input mode is sequential, as all cells have the same initial quiescent state. Initially there is no difficulty in grouping information into fewer cells. Then the accelerated computation can take place immediately when the distinguished cell has obtained the whole input. ❯ *Figure 11* illustrates the situation in dimension one. The grouping operation is initially accomplished within each diagonal: for any integer constant R, the sites are grouped together by R. Once the input is read and the end marker is constantly fed, the information initially processed by R diagonals can be processed by a simple one and hence accelerate the computation by a factor R. The fact remains that the time space diagram is distorted. It involves keeping some redundant information in each cell but it preserves the data accessibility requirement while respecting the dependency constraints.

◻ **Fig. 11**
Linear speedup in case of two-way communication and sequential input mode.

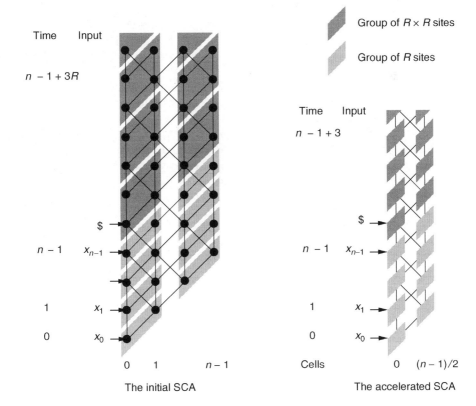

◧ Fig. 12

Linear speedup in case of two-way communication and parallel input mode.

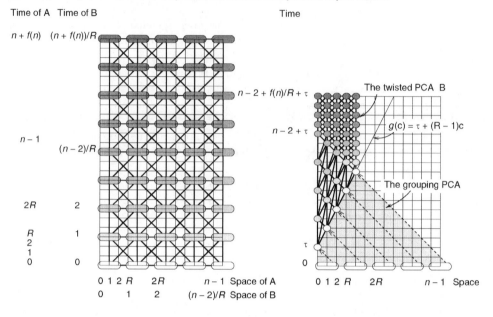

The case of parallel input mode (and two-way communication) is more tricky. The input data are initially fed into the array and some time must be spent grouping them together. In order to avoid losing more time, the accelerated computation must start as soon as possible. ❯ *Figure 12* illustrates the method for a PCA $\mathcal{A}$. On the one hand, observe that one can construct a PCA $\mathcal{B}$, which simulates R times faster the PCA $\mathcal{A}$ provided the input given to $\mathcal{A}$ is fed compacted with a factor R to $\mathcal{B}$. On the other hand, there is a grouping PCA that turns the initial ungrouped input into a grouped one. In this way, one can stick the grouping PCA with a twisted variant of the accelerated PCA $\mathcal{B}$ to obtain the desired PCA. The accelerated computation starts on each cell as soon as its neighbor cells are grouped.

In developing such techniques for picture language recognizer, Beyer (1969) has shown linear speedup result for CA with the von Neumann neighborhood. These techniques work for the Moore neighborhood and similar kinds of neighborhoods as well (see Terrier 2004).

Theorem 2 *Let f be a function from* $\mathbb{N} \times \mathbb{N}$ *to* $\mathbb{N}$, $\mathcal{N}$ *be the von Neumann or Moore neighborhood and* $rt_{\mathcal{N}}$ *be the corresponding real time function. For any constant R, a picture language recognized in time* $rt_{\mathcal{N}}(m, n) + f(m, n)$ *by a PCA with the neighborhood* $\mathcal{N}$ *is also recognized in time* $rt_{\mathcal{N}}(m, n) + \lceil f(m, n)/R \rceil$ *by another PCA with the same neighborhood.*

What is disturbing about picture language recognition is that some neighborhoods seem not to admit linear speedup. In the common method, each cell implicitly knows in which direction to send its content to achieve the grouping process. This direction corresponds to a shorter path toward the output cell. But for some neighborhoods, this direction differs according to the position of the cell into the array and no alternative efficient grouping process is currently known.

3.3   Constant Speedup

In addition to the speedup results of ❯ Sect. 3.2, real-time complexity could also be defined modulo a constant. In dimension one, this property was first observed in Choffrut and Culik (1984) and a whole complete generalization has be done in Poupet (2005). Excluding pathological neighborhoods for which the output cell is not able to read the whole input, we have constant time acceleration for PCA.

Proposition 1   *For any neighborhood $\mathcal{N}$, any function satisfying $f(n) \geq rt_{\mathcal{N}}(n)$ and any constant τ:*

$$\mathrm{PCA}_{\mathcal{N}}(f(n) + \tau) = \mathrm{PCA}_{\mathcal{N}}(f(n))$$

An interesting consequence, which emphasizes the robustness of one-dimensional CA model, follows from this speedup (see Poupet 2005). The computation ability in dimension one is somewhat independent of the underlying communication graph.

Theorem 3   *In dimension one, all neighborhoods are real-time equivalent to either the one-way neighborhood $\{-1, 0\}$ or the standard one $\{-1, 0, 1\}$.*

The situation turns out to be less satisfactory in higher dimensions. For the two classical von Neumann and Moore neighborhoods, the real-time complexity can also be defined modulo a constant. But a uniform approach does not yet exist to deal with various neighborhoods. And in spite of investigations about arbitrary neighborhoods reported in Delacourt and Poupet (2007), nothing much is known.

3.4   Equivalence Between the Parallel and Sequential Input Modes

A result that cannot be ignored says that, if one does not bother with the time comparison, the parallel and sequential input modes are equivalent. This result is fairly astonishing in the case of one-way communication. Indeed the parallel input mode combined with one-way communication induces strong limitations on the access to the data. As illustrated in ❯ *Fig. 13*, a cell of a POCA has no access on the input letters applied on its right whereas every cell of a

❏ **Fig. 13**
Influence of the ith input symbol on a POCA and on a SOCA.

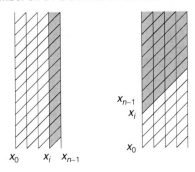

SOCA has access to the whole input. Moreover, for SOCA, the end marker $ supplied after the whole input letters provides information on the length of the input. On the other hand, such knowledge is impossible for POCA. Despite these outstanding differences, Ibarra and Jiang (1987) have shown that SOCA and POCA accept the same class of languages.

Theorem 4 POCA = SOCA.

Because of the advantages of SOCA over POCA, the simulation of a POCA by a SOCA is straightforward. In contrast, the reverse simulation is considerably more involved. Let one just present a rough idea of the construction and its difficulties. To simulate a SOCA by a POCA, the basic idea is to systematically generate each possible input, to simulate the SOCA on it, and check if it is actually the real input. For that, the working area is divided in two parts. The enumeration process takes place in the left part. First the inputs of length 1, then those of length 2, then those of length 3, etc., are generated in such a way that they can be obtained serially by the leftmost cell of the right part. Simultaneously, in the right part, the SOCA is simulated on the successive generated inputs. Moreover, the successive inputs are compared with the real input.

The delicate point is that the POCA does not have hints about the length of the real input and has to deal with inputs of all possible lengths. In particular, the time and the space required to simulate the SOCA on the current input depends on its length. Also this simulation time induces a delay between the generation of two inputs. And the demarcation of the two parts of the working area fluctuates according to the current length of the generated inputs. Then to distribute and to synchronize the different subcomputations into the working area entails many of the techniques described in the chapter ❷ Algorithmic Tools on Cellular Automata. A complete description can be found in the original paper (Ibarra and Jiang 1987) and with a direct construction in Delorme and Mazoyer (1994).

The drawback of this construction is its exponential cost. Indeed, whatever the time complexity of the initial SOCA, the POCA simulates the behavior of the SOCA on an exponential number of inputs. Additional questions related to the simulation cost of SOCA by POCA will be discussed in ❷ Sect. 6.2.

3.5 Closure Properties

Closure properties are naturally investigated in order to evaluate the computation ability of the different CA classes and to possibly achieve separation results. Obviously, all deterministic CA classes satisfy the closure under Boolean operations. More attention has been paid to the closure properties under other language operations as concatenation, reverse, or cycle. As you may recall, the reverse of a language L is $L^R = \{w^R : w \in L\}$ where w^R is the word w written backward and its cycle is $L^{Cy} = \{vu : uv \in L\}$. ❷ Table 2 sums up the known results in dimension one (Y stands for *yes*, N for *no*, and ? for *open*).

Let us give further details about these results and the open questions. Because PCA has the same computation power as linear-space Turing machine, it satisfies various closure properties: reverse, concatenation, Kleene star, ε-free homomorphism, inverse homomorphism, and cycle. From Chang et al. (1988) it is known that the one-way counterparts POCA and SOCA also share the above closure properties. The proofs make essential use of similar constructions as the one outlined in ❷ Sect. 3.4 to show that POCA = SOCA. In ❷ Sect. 4.1, we will account for the negative closure properties of the RSCA class and its higher

❑ Table 2
Closure properties in dimension one

Class	Reverse closure	Concatenation closure	Cycle closure
PCA	Y	Y	Y
POCA	Y	Y	Y
LPCA	Y	?	?
RPCA	?	?	?
RSCA	N	N	N
RPOCA	Y	N	N

dimension counterparts and then for the negative closure under concatenation of RPOCA. Curiously enough, this last result is the opposite in higher dimensions: for any dimension $d >$ 1, $d -$ RPOCA satisfies closure under concatenation as well as closure under Kleene star (Terrier 2006b). As observed in Choffrut and Culik (1984), the closure under reverse of RPOCA follows from the symmetry of its dependency graph. For the same structural reason, the result extends to the higher dimensions (Terrier 2006b). The closure under reverse of LPCA is an immediate consequence of the linear speedup results (Ibarra and Jiang 1988).

It is not known whether RPCA is closed under reverse or under concatenation. But a striking result due to Ibarra and Jiang (1988) relates the property of RPCA to be closed under reverse and its ability to be as powerful as LPCA. In a similar way, the cycle closure property of RPCA can be linked to the equality between RPCA and LPCA (Terrier 2006a). The following theorem gathers these relationships.

Theorem 5 *The following three statements are equivalent:*

- RPCA = LPCA.
- RPCA *is closed under reverse.*
- RPCA *is closed under cycle.*

This theorem is very interesting and deserves some explanations. Let one focus on the equivalence between the ability of RPCA to be as powerful as LPCA and its reverse closure property. The proof relating to the third statement makes essential use of similar arguments. As LPCA is closed under reverse, the equality RPCA = LPCA directly implies the closure under reverse of RPCA. To show the converse implication is more tricky. How may a closure property involve speedup of linear-time computation? First the assertion LPCA = RPCA can be restated in the following equivalent statement "Let c be any constant. If $\{w \sharp^{c2^{\lceil \log |w| \rceil}} : w \in L\}$ is a RPCA language then L is a RPCA language." The key argument developed in Ibarra and Jiang (1988) is that the reverse form of this statement is true.

Lemma 1 *Let c be any constant. If $\{\sharp^{c2^{\lceil \log |w| \rceil}} w : w \in L\}$ is a RPCA language then L is a RPCA language.*

The proof of this meaningful lemma is fairly involved. A whole description can be found in the original paper (Ibarra and Jiang 1988) and with a direct construction in Delorme and

Mazoyer (1994). Now according to this ❷ Lemma 1 and providing RPCA is closed under reverse, the following chain of implications is obtained:

$L \in$ LPCA

$\Rightarrow \widetilde{L} = \{w \sharp^{c2^{\lceil \log |w| \rceil}} : w \in L\} \in$ RPCA   (straightforward property)

$\Rightarrow \widetilde{L}^R = \{\sharp^{c2^{\lceil \log |w| \rceil}} w : w \in L^R\} \in$ RPCA   (reverse closure)

$\Rightarrow L^R \in$ RPCA   (Lemma 1)

$\Rightarrow L \in$ RPCA   (reverse closure)

An immediate consequence of ❷ Theorem 5 is that if RPCA is closed under reverse or under cycle then it is closed under concatenation (Ibarra and Jiang 1988). Whether the converse is true remains an open question. We just know a weaker implication also based on ❷ Lemma 1 (Terrier 2006a).

Proposition 2   *If* RPCA *is closed under concatenation then*

- LPCA *unary languages are* RPCA *unary languages.*
- LPCA *is closed under concatenation.*

Here let us have a look at closure properties of picture language recognizers. Up to now the closure properties under concatenation have not been studied. In the matter of the reverse operation, its counterpart for picture languages is the 180° rotation operation. As in the one-dimensional case, the linear speedup results entail the closure under rotation of linear-time PictCA with the von Neumann or Moore neighborhoods. In contrast, it can be seen in ❷ Sect. 4.1 that real-time PictCA with the Moore neighborhood, real-time PictCA and linear-time PictCA with the one-way von Neumann or one-way Moore neighborhood do not satisfy closure under rotation. Interestingly, the arguments developed by Ibarra and Jiang to relate, in dimension one, closure properties to computation ability can be extended to PictCA recognizer (Terrier 2003b).

Proposition 3   *Real time* PictCA *with the von Neumann neighborhood is closed under rotation if and only if real time and linear time* PictCA *with the von Neumann neighborhood are equivalent.*

At last a nice property pointed out by Szwerinski (1985) can be noticed. In dimension one, a PCA may confuse right and left without time loss. Formally, a local transition function $\delta : S^3 \to S$ is said symmetrical if for all $r, c, l \in S$ holds: $\delta(l, c, r) = \delta(r, c, l)$. So the following proposition, as shown in Szwerinski (1985) and in a more general setting in Kobuchi (1987), emphasizes that the orientation does not matter for one-dimensional PCA.

Proposition 4   *If a language L is recognized in time T by some* PCA, *then L is recognized in the same time T by another* PCA *whose transition function is symmetrical.*

4   Limitations

The purpose of this section is to present the known limitations on the recognition power of CA. It makes use of either algebraic arguments or diagonalization ones.

4.1 Algebraic Arguments

Let one start with a basic negative result. As observed in Culik et al. (1984), POCA operating in real time on languages over a one-letter alphabet are no more powerful than finite automata.

Proposition 5 *The class of unary languages recognized by* RPOCA *is the class of regular unary languages.*

This result has been strengthened in Buchholz and Kutrib (1998): it requires at least an amount of $n + \log n$ time to recognize non-regular unary languages on POCA.

An interesting consequence of ❷ Proposition 5, observed in Choffrut and Culik (1984), derives from the existence of non-regular unary languages that are recognized by RSCA and therefore by RPCA. For instance, the languages $\{1^{2^n}: n \in \mathbb{N}\}$ and $\{1^p : p \text{ is a prime}\}$ belong to RSCA (see Choffrut and Culik (1984) and Fischer (1965) and the chapter ❷ Computations on Cellular Automata by Mazoyer and Yunès). It yields the following relationships.

Corollary 1
- RPOCA $\subsetneq$ RPCA
- RSCA $\nsubseteq$ RPOCA

One can now concentrate on elaborate techniques to obtain lower bounds. These techniques were introduced by Hartmanis and Stearns (1965) for real-time Turing machines and adapted to real-time SCA by Cole (1969). Using counting arguments, they exploit limits on interaction between data.

4.1.1 The Method for Real-Time SCA

Let one recall the method used in Cole (1969) to get negative results on RSCA. As illustrated in ❷ *Fig. 14*, a characteristic of the real-time SCA computation on a given input w is the following one. The suffix of arbitrary length k of w only has an impact on the computation during the k last steps. In particular, the first part z and the suffix u of length k of the input $w = zu$ interact on a number of cells independent of the length of z. Hence significant information on z may be lost before the k last steps of the computation. Precisely, consider the configuration of the SCA at time $|z| - 1$ when the last symbol of z is fed. At this time, the useful part consists of the $k + 1$ first sites, which may influence the result of the computation obtained on the output cell, k steps after. Let $h(z)$ denote the sequence of states of length $k + 1$ of this useful part. Thus the result of the computation of the RSCA on the input zu is completely specified by $h(z)$ and u. That sets an upper bound on the number of distinct behaviors and thus implies the following condition on the structure of RSCA languages.

Proposition 6 *Let L be a language over some alphabet Σ. Let X be a finite subset of Σ^* and let k be the maximal length of the words in this set X. Consider, for each word $z \in \Sigma^*$, the indicator function p_z from X into $\{0, 1\}$ defined as:*

$$p_z(u) = \begin{cases} 1 & \text{if } zu \in L \\ 0 & \text{otherwise} \end{cases}$$

◘ **Fig. 14**

Real-time computation of a SCA on some input $w = zu$.

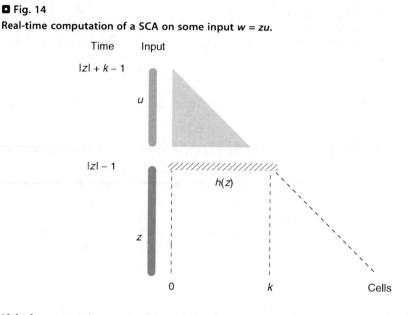

If the language L is recognized in real time by some SCA, then the number of distinct functions $|\{p_z: z \in \Sigma^*\}|$ is of order at most $2^{O(k)}$.

Proof   Suppose that L is recognized by some RSCA $\mathcal{A}$. For any words $z, z' \in \Sigma^*$, observe that if $h(z) = h(z')$ then for all $u \in X$ either zu and $z'u$ are both accepted or both rejected by $\mathcal{A}$. Indeed the words u in X have length at most k and $h(z)$, as defined above, consists of the information accessible to the k last steps. Hence if $h(z) = h(z')$, then $p_z = p_{z'}$. Now notice that every $h(z)$ is a word of length $k + 1$ on the finite set of states of $\mathcal{A}$. Therefore, the number of distinct functions p_z does not exceed the number of distinct sequences $h(z)$, which is of order $2^{O(k)}$.

A couple of examples are now given.

Example 6 $L = \{x_1 \sharp x_2 \sharp \ldots x_t \sharp x: x_1, x_2, \ldots, x_t, x \in \{0, 1\}^* \text{ and } x = x_j \text{ for some } j \text{ with } 1 \leq j \leq t\}$ is not a real-time SCA language.

Proof   Set $X = \{0, 1\}^k$. Associate to each subset A of X, the word $z_A = x_1 \sharp x_2 \sharp \ldots x_t \sharp$ where $x_1, x_2, \ldots, x_t$ is some enumeration of the words in A. Namely, z_A is defined in such a way that $z_A u \in L$ if and only if $u \in A$. Thus, if A and B are two distinct subsets of X then $p_{z_A} \neq p_{z_B}$. So the number of functions $|\{p_z: z \in \Sigma^*\}|$ is at least 2^{2^k}, the number of subsets of X. Therefore, according to ❷ Proposition 6, L is not a RSCA language.

The following example differs from the previous one in the number of functions p_z.

Example 7 $L = \{01^{a_1}01^{b_1}01^{a_2}01^{b_2} \cdots 01^{a_t}01^{b_t}01^a01^b : a_1, b_1, \ldots, a_t, b_t \geq 0 \text{ and } a = a_j, \ b = b_j$ for some j with $1 \leq j \leq t\}$ is not a real-time SCA language.

Proof   Set $X = \{01^a01^b: a, b \geq 0 \text{ and } a + b + 2 \leq k\}$. Associate to each subset A of X, the word $z_A = x_1 \ldots x_t$ where $x_1, \ldots, x_t$ is some enumeration of the words in A. As z_A is defined,

$z_A u \in L$ if and only if $u \in A$. Thus, if A and B are two distinct subsets of X, then $p_{z_A} \neq p_{z_B}$. So the number of functions $|\{p_z \colon z \in \Sigma^*\}|$ is at least $2^{k(k-1)}$, the number of subsets of X. Hence L is not a RSCA language.

Several other languages are known not to belong to RSCA. Among them, the languages that have been shown not real time recognizable by multitape Turing machines in Hartmanis and Stearns (1965) and Rosenberg (1967) are also not real time recognizable by SCA. Indeed both devices share the same features on data accessibility, which are exploited to get limitations. Actually, all these languages are RPCA languages. As a consequence, RSCA is less powerful than RPCA. Furthermore, specific examples exhibited in Hartmanis and Stearns (1965), Rosenberg (1967), Cole (1969), Dyer (1980), and Kutrib (2001) yield several negative properties. A representative one is the language of words, which end with a palindrome of length at least three: $L = \{w \in \Sigma^* \colon w = uv, v = v^R, |v| > 2\}$. Yet L is linear context free and belongs to RPOCA; furthermore, the reverse of L and the palindrome language belong to RSCA. Therefore RSCA contains neither all linear context-free languages nor RPOCA languages and is not closed under reversal and under concatenation. The following corollary summarizes the various results obtained.

Corollary 2
- RSCA *is strictly contained in* RPCA.
- RSCA *and* RPOCA *are incomparable.*
- RSCA *is not closed under reversal, concatenation, Kleene closure, sequential mapping, or the operations of taking derivatives and quotients.*
- RSCA *does not contain all deterministic linear context free languages.*

One further noteworthy result of Cole (1969) is that the power of real-time SCA increases with the dimension of the space.

Proposition 7   *For any dimension d, $d -$ RSCA $\subsetneq (d + 1) -$ RSCA.*

Proof   The number of cells that influence the k last steps of any $d -$ RSCA computation increases polynomially with the dimension d. Precisely, the counterpart of $h(z)$ in dimension d, is a d-dimensional word of diameter $O(k)$ and volume $O(k^d)$. Hence the number of distinct words $h(z)$ is in $2^{O(k^d)}$. Furthermore, the condition stated in ❥ Proposition 6 can be rewritten in this way: "If the language L is recognized in real time by some $d -$ SCA then the number of distinct functions $|\{p_z \colon z \in \Sigma^*\}|$ is of order at most $2^{O(k^d)}$." As a consequence the language presented in ❥ Example 6 is not a $d -$ RSCA language whatever the dimension d may be. On the other hand, the language given in ❥ Example 7 is not a $1 -$ RSCA language but it may be, and in fact is, a $2 -$ RSCA language. Of course, languages with the same kind of structure as the one of ❥ Example 7 can be built to separate $d -$ RSCA and $(d + 1) -$ RSCA.

4.1.2   The Method for Real-Time POCA

One can now look at the case of RPOCA. A first characteristic noticed in Culik (1989) is that the real-time computation on an input w contains the real-time computations of all its factors. Recall the ❥ Example 4 in ❥ Sect. 2: the RPOCA, which tests whether an input is a Dyck word, processes together all its factors. So the evolution on an input of size n decides the memberships of $n(n + 1)/2$ words. This constraint has been exploited in Terrier (1995) to get a non-RPOCA language.

◻ **Fig. 15**

Real-time computation of a POCA on some input *uzv*.

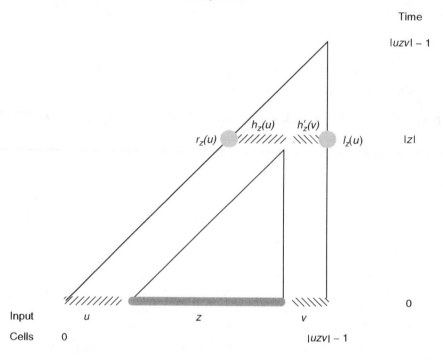

A second characteristic of RPOCA computation noticed in Terrier (1996) is the following one. As depicted in ❯ *Fig. 15*, on an input *w*, its prefixes *u* and suffixes *v* only interact during the $|uv| - 1$ last steps. More precisely, on input $w = uzv$, consider at time $|z|$ the useful part that consists of the $|uv|$ sites, which may have an impact on the result obtained $|uv| - 1$ steps after. This part subdivides into the first $|u|$ sites named $h_z(u)$ and the last $|v|$ sites named $h'_z(v)$. Note that the result of the computation on *uzv* is completely specified by $h_z(u)$ and $h'_z(v)$. Furthermore, according to the first characteristic, the results of all the factors, which contained *z* are determined by $h_z(u)$ and $h'_z(v)$. On the other hand, $h_z(u)$ is not influenced by the suffix *v* as well as $h'_z(v)$ by the prefix *u*. Hence, intuitively the information exchange between *u* and *v* takes place when significant information about *z* may be lost. This situation gives rise to the following condition on the structure of RPOCA languages.

Proposition 8 *Let L be a language over some alphabet* Σ. *Let X,Y be two sets of words on* Σ. *Denote the set of all suffixes of words in X by* Suff(X) *and the set of all prefixes of words in Y by* Pref(Y). *Consider, for each word* $z \in \Sigma^*$, *the indicator function* p_z *defined as:*

$$p_z \colon \text{Suff}(X) \times \text{Pref}(Y) \to \{0, 1\}$$

$$(u, v) \quad \mapsto \begin{cases} 1 \text{ if } uzv \in L. \\ 0 \text{ otherwise} \end{cases}$$

If L is recognized in real time by some POCA then the number of distinct functions $|\{p_z : z \in \Sigma^*\}|$ *is of order at most* $2^{O(|\text{Suff}(X)| + |\text{Pref}(Y)|)}$.

Proof Suppose that L is recognized by some RPOCA $\mathcal{A}$. Denote S the set of states of $\mathcal{A}$. Consider the two functions l_z from $\text{Suff}(X)$ into S and r_z from $\text{Pref}(Y)$ into S defined in this following way. To every u in $\text{Suff}(X)$, l_z associates the state entered at time $|z|$ by the leftmost cell involved in the real-time computation of $\mathcal{A}$ on input uz. Symmetrically, to every v in $\text{Pref}(Y)$, r_z associates the state entered at time $|z|$ by the rightmost cell involved in the real-time computation of $\mathcal{A}$ on input zv. Observe that $h_z(u)$, as defined above, is the sequence of $r_z(x)$ where x ranges over the set of all suffixes of u: $h_z(u_1 u_2 \ldots u_k) = r_z(u_1 u_2 \ldots u_k) r_z(u_2 \ldots u_k) \ldots r_z(u_k)$. In a symmetric way, $h'_z(v)$ is the sequence of $l_z(y)$ where y ranges over the set of all prefixes of v: $h'_z(v_1 \ldots v_{k-1} v_k) = l_z(v_1) \ldots l_z(v_1 \ldots v_{k-1}) l_z(v_1 v_2 \ldots v_k)$. Therefore, if $r_z = r_{z'}$ and $l_z = l_{z'}$, then for all $u \in \text{Suff}(X)$ and $v \in \text{Pref}(Y)$ either uzv and $uz'v$ are both accepted or both rejected by $\mathcal{A}$. In other words, if $r_z = r_{z'}$ and $l_z = l_{z'}$ then $p_z = p_{z'}$. Therefore, the number of distinct functions p_z does not exceed the product of the number of distinct functions r_z with the number of distinct functions l_z, which is in $2^{O(|\text{Suff}(X)| + |\text{Pref}(Y)|)}$.

Example 8 $L = \{x \sharp x_1 \$ y_1 \sharp x_2 \$ y_2 \sharp \ldots \sharp x_t \$ y_t \sharp y : x_1, y_1, x_2, y_2, \ldots, x_t, y_t, x, y \in \{0, 1\}^* \text{ and } x = x_j, y = y_j \text{ for some } j \text{ with } 1 \leq j \leq t\}$ is not a real time POCA *language*.

Proof Set $X = Y = \{0, 1\}^k$. Associate to each subset A of $\text{Suff}(X) \times \text{Pref}(Y)$, the word $z_A = \sharp x_1 \$ y_1 \sharp x_2 \$ y_2 \sharp \ldots \sharp x_t \$ y_t \sharp$ where $(x_1, y_1), \ldots, (x_t, y_t)$ is some enumeration of the words in A. By construction, $x z_A y \in L$ if and only if $(x, y) \in A$. Thus, if A, B are two distinct subsets of $\text{Suff}(X) \times \text{Pref}(Y)$ then $p_{z_A} \neq p_{z_B}$. Therefore, the number of distinct functions $|\{p_z : z \in \Sigma^*\}|$ is at least $2^{2^{2k}}$ the number of subsets of $\text{Suff}(X) \times \text{Pref}(Y)$. It is of greater order than $2^{O(|\text{Suff}(X)| + |\text{Pref}(Y)|)} = 2^{O(2^k)}$.

Example 9 Let us consider the linear context free language $L_1 = \{w: w = 1^u 0^u \text{ or } w = 1^u 0 y 10^u \text{ with } y \in \{0, 1\}^* \text{ and } u > 0\}$. The context free language $L = L_1 \cdot L_1$ is not a real time POCA *language*.

Proof Fix some integer k. Set $X = \{1^k\}$ and $Y = \{0^k\}$. Associate to each subset A of $\text{Suff}(X) \times \text{Pref}(Y)$, the word $z_A = 0^{i_1} 1^{j_1} \ldots 0^{i_t} 1^{j_t}$ where $(1^{i_1}, 0^{j_1}), \ldots, (1^{i_t}, 0^{j_t})$ is some enumeration of the words in A. As z_A is defined, $1^u z_A 0^v \in L$ if and only if $(1^u, 0^v) \in A$. Thus, if A, B are two distinct subsets of $\text{Suff}(X) \times \text{Pref}(Y)$ then $p_{z_A} \neq p_{z_B}$. Therefore, the number of distinct functions $|\{p_z: z \in \Sigma^*\}|$ is at least $2^{(k+1)^2}$ the number of subsets of $\text{Suff}(X) \times \text{Pref}(Y)$. It is of greater order than $2^{O(|\text{Suff}(X)| + |\text{Pref}(Y)|)} = 2^{O(k)}$. According to ❷ Proposition 8 we conclude that L is not a RPOCA language.

Since linear context-free languages are RPOCA languages, an immediate consequence of ❷ Example 9 is that RPOCA is not closed under concatenation and does not contain all context-free languages. With further results obtained in Klein and Kutrib (2003), the following corollary gives the main negative properties of RPOCA.

Corollary 3

- *The class of real-time POCA languages is not closed under concatenation, Kleene closure, ε-free homomorphisms.*
- *The class of real-time POCA languages does not contain all context-free languages.*

Furthermore, as shown in Buchholz et al. (2000) and Klein and Kutrib (2003), the previous arguments combined with padding techniques lead to an infinite hierarchy of separated classes between real-time SCA and linear-time SCA as well as another one between real-time POCA and linear-time POCA. For the sake of ❷ Proposition 9 below, the definition of Fischer's constructibility can be found in the chapter ❷ Algorithmic Tools on Cellular Automata by Delorme and Mazoyer. At least, the next function T can be viewed as a "reasonable" function of asymptotic order at most n.

Proposition 9

- Let T, T' be two functions from $\mathbb{N}$ to $\mathbb{N}$ such that the inverse of T is Fischer constructible and $T' \in o(T)$. We have the strict inclusion $\text{SCA}(n + T'(n)) \subsetneq \text{SCA}(n + T(n))$.
- Let T, T' be two functions from $\mathbb{N}$ to $\mathbb{N}$ such that the inverse of T is Fischer constructible and $T' \log(T') \in o(T)$. We have the strict inclusion $\text{POCA}(n + T'(n)) \subsetneq \text{POCA}(n + T(n))$.

Similarly, such algebraic techniques may be applied to exhibit limits on the computation ability of picture language recognizers. In this way, weakness of real-time computation with the Moore neighborhood has been observed in Terrier (1999). Precisely, there exists a picture language, which is real-time recognizable by no PictCA with the Moore neighborhood. Furthermore, this picture language is real-time recognizable by a PictCA with the von Neumann neighborhood and its corresponding language obtained by a rotation of 180° is accepted in real time with both the von Neumann and Moore neighborhoods.

In the same vein, restricted communication has been shown to reduce the computational power of low-complexity picture classes (Terrier 2006a). Precisely, there exists a language recognized in real time by PictCA with both the von Neumann and Moore neighborhoods but not recognized in linear time with any one-way neighborhoods $\mathcal{N}$ such that $a + b \geq 0$ for every (a, b) in $\mathcal{N}$. Furthermore, the corresponding language obtained by a rotation of 180° is recognized by PictCA with one-way neighborhoods $\mathcal{N}_1 = \{(0,0), (0,1), (1,0)\}$, $\mathcal{N}_2 = \{(0,0), (0,1), (1,0), (1,1)\}$ or $\mathcal{N}_3 = \{(0,0), (0,1), (1,0), (1,1), (-1,1), (1,-1)\}$.

The various consequences are summarized in the following proposition.

Proposition 10

- Real-time PictCA with the Moore neighborhood does not contain real-time PictCA with the von Neumann neighborhood.
- Real-time PictCA with the Moore neighborhood is not closed under rotation.
- Real-time PictCA with the Moore and von Neumann neighborhoods are not contained in linear time PictCA with a one-way neighborhood $\mathcal{N}$ where $a + b \geq 0$ for all $(a, b) \in \mathcal{N}$.
- Real-time and linear time PictCA with the one-way neighborhoods $\mathcal{N}_1$, $\mathcal{N}_2$ and $\mathcal{N}_3$ are not closed under rotation.

As emphasized by Cervelle and Formenti (2009), one can as well make use of Kolmogorov complexity to derive these results. This alternative approach expresses in a more direct manner the fact that significant information is lost. Here, we will not recall the formal definition of Kolmogorov complexity, but just give an example to illustrate the method. The reader is referred to Cervelle and Formenti (2009) for comprehensive definitions and for an example involving POCA device.

A basic example is the language not belonging to RSCA, which was presented in ❷ Example 6: $L = \{x_1 \sharp x_2 \sharp \ldots \sharp x_t \sharp x \colon x_1, x_2, \ldots, x_t, x \in \{0, 1\}^* \text{ and } x = x_j \text{ for some } j \text{ with } 1 \leq j \leq t\}$. Contrary to the algebraic method, we will not take into account all possible evolutions, but just focus on a "complex" one. Let one fix ω some Kolmogorov random number of length 2^k. Consider some bijection val between $X = \{0, 1\}^k$ and $\{0, \ldots, 2^k - 1\}$, which codes each word of X by a distinct integer below 2^k, for instance $val(a_0 \ldots a_{k-1}) = \sum a_i 2^i$. Consider the set $A_\omega = \{x \in X \colon \text{the}$

symbol of rank $val(x)$ in ω is a 1} and $z_\omega = x_1 \sharp \ldots x_t \sharp$ where $x_1, \ldots, x_t$ is some enumeration of the words in A_ω. Now if L is recognized in real time by some SCA $\mathcal{A}$ then ω can be reconstructed from the description of $\mathcal{A}$, the description of the bijection val and the sequence of states $h(z_\omega)$. Indeed, from $\mathcal{A}$ and $h(z_\omega)$, we can decide, for each $x \in X$, whether $z_\omega x$ is in L or not, and so whether the symbol of rank $val(x)$ in ω is a 1 or a 0. The respective lengths of these descriptions are in $O(1)$ for $\mathcal{A}$, in $O(\log k)$ for val and in $O(k)$ for $h(z_\omega)$ and their sum is less than the 2^k bits of the word ω. Hence it leads to a contradiction on the incompressibility of ω.

Undoubtedly, algebraic arguments and also Kolmogorov complexity are powerful tools to establish limits on the computation ability of restricted devices with low complexity. But it should be admitted that many questions on the power of these devices are left open. For instance, the same arguments are exploited to prove that some languages do not belong to real-time Turing machines nor RSCA. Does it mean that real-time SCA are no more powerful than real-time Turing machines? Moreover, most of the witness languages, which enable one to derive negative properties, are usually built in an *ad hoc* manner. Then we fail to determine the status of more "natural" languages like, for RPOCA, the majority language $\{w \in \{0, 1\}^* : w \text{ has more 1's than 0's}\}$ or the square language $\{ww : w \in \{0, 1\}^*\}$.

4.2   Diagonalization Arguments

In computational complexity, many separation results use diagonalization techniques. Diagonalization also works to separate CA classes, but essentially in an indirect way via the Turing machines. It consists in exhibiting efficient simulations of CA by Turing machines, which allow one to translate Turing machine results into CA results. In this regard, Goldschlager (1982) has shown that whatever its dimension may be, a CA that works in time T can be simulated by a Turing machine in space T.

Fact 1   *For any dimension d and any complexity function T, $d - \text{PCA}(T) \subseteq \text{DSpace}(T)$.*

Because the PCA dependency graphs are regular, the simulation which consists of a depth first search in these graphs with height T, can be performed in space T by a Turing machine. In addition, this result holds as well as for CA as a picture recognizer. As a consequence, for CA in dimension two and higher, separation follows from the Turing machine space hierarchy. In particular, for language and picture recognition, in dimension greater than one, CA working in linear time are strictly less powerful than CA working in unrestricted time and that within the same bounded space.

In the case of restricted communication, a better simulation has been obtained for POCA as language recognizer (Chang et al. 1989).

Fact 2   *For any dimension d, $d - \text{POCA} \subseteq \text{DSpace}(n^{2 - 1/d})$.*

It allows one to show that, in dimension two and higher, restricted communication reduces the language recognition ability. Precisely, there is a language accepted by a $2 - \text{PCA}$ that cannot be accepted by any $d - \text{POCA}$ whatever the dimension d. Yet the question is still open in the case of PictCA.

5 Comparison with Other Models

In this section, CA is compared with other computational models. Such investigations may help to identify significant features of CA.

5.1 Sequential Models

In order to determine to what extent the use of parallelism provides significant advantages, one of the major concerns is the relationship between parallel computation and sequential computation.

First, efficient simulations of CA by sequential devices permit one to render explicit limitations on the CA computation ability. Notably, as seen in ❷ Sect. 4.2, the simulations, which connect CA time complexities with Turing space complexities, entail separation results for CA. But to relate the CA time with the Turing time, there seems to be no better simulation than the trivial one, which states that the work of a CA performed within time T and space S can be done by a Turing machine within time $T \times S$ and space S. For the other representative sequential model, the random access machine, a lower overhead has been obtained in Ibarra and Palis (1987). By the way of a precomputation phase, it has been shown that any $d - \text{POCA}$ working in time $T(n)$ can be simulated by a unit cost random access machine in time $n^d T(n) / \log^{1+1/d} T(n)$.

Conversely, the simulation of sequential devices by CA is a great challenge. What gain may be achieved with CA? For specific sequential problems, the CA computation power is manifest. For instance, it has been observed in Ibarra and Kim (1984) that real-time POCA contains a language that is P-complete and in Chang et al. (1988) that the language QBF of true quantified Boolean formulas that is PSpace-complete is in POCA.

But when we try to get general simulations, we come up against the delicate question of whether parallel algorithms are always faster than sequential ones. Indeed, there is no guarantee that efficient parallelization is always possible. Or there might exist a faster CA for each singular sequential solution whereas no general simulation exists. Besides, one can recall Fischer's algorithm to recognize the set $\{1^p : p \text{ is a prime}\}$ (Fischer 1965) and Cole's and Culik's ones seen in ❷ Sect. 2. They suggest that clever strategies in parallel are inadequate in sequential and *vice versa*. Then when the conception of efficient parallel algorithms makes use of radically different techniques from the sequential ones, automatic parallelization appears highly improbable.

Hence without surprise, the known simulation of Turing machines by CA provides no parallel speedup. As viewed in the chapter ❷ Universalities in Cellular Automata by Ollinger, the early construction of Smith (1971) mimics one step of computation on a Turing machine (with an arbitrary number of tapes) in one step on a CA (with unbounded space). Furthermore, no effective simulations have been proposed, even for restricted variants. For instance, we do not know whether any finite automata with k heads can be simulated on CA in less than $O(n^k)$ steps, which is the sequential time complexity. And one wonders whether one-way multihead finite automata whose sequential time complexity is linear may be simulated in real time on CA.

In contrast to this great ignorance, the result of Kosaraju is noteworthy (Kosaraju 1979).

Proposition 11 *Any picture language recognized by a four-way finite automata can be recognized in linear time by a PictCA.*

Unfortunately, a complete proof has never been published. Let one just outline the basic idea. The key point is to code the behavior of the finite automaton on a block of size $n \times n$ by a directed graph. A vertex of such a graph is a couple (q, n) where q is an automaton state and n a boundary node of the block. Then the directed graph records for each couple of vertices $((q_1, n_1), (q_2, n_2))$, if, when the automaton enters in state q_1 at boundary node n_1, it exits in state q_2 at boundary node n_2. The trick is that the space to record the adjacency matrix of this graph is of the same order as the corresponding block. Furthermore, the adjacency matrix of a block of size $2^i \times 2^i$ can be effectively computed from the four adjacency matrices of the four sub-blocks of size $2^{i-1} \times 2^{i-1}$. To this end, the four adjacency matrices are reorganized in one, the transitive closure is computed and then the new non-boundary nodes are eliminated. Now using this procedure recursively, the adjacency matrix of the whole initial pattern can be computed in linear time.

5.2 Alternating Automata and Alternating Grammars

The correspondence between CA and alternating finite automata was first pointed out by Ito et al. (1989), who showed the equivalence between a particular variant of two-dimensional CA and a restricted type of two-dimensional alternating finite automata. In Terrier (2003a, 2006b), alternating analogues of real-time CA with sequential input mode were given as follows.

Proposition 12
- *Real time d — SCA are equivalent through reverse to real time one-way alternating finite automata with d counters.*
- *Real time d — SOCA are equivalent through reverse to one-way alternating finite automata with $d + 1$ heads.*

Similarly, for CA with parallel input mode, which implicitly induces a synchronization at initial time, one might search a characterization in terms of one-way synchronized finite automata. Yet, these equivalences are somewhat unsatisfying in the sense that the corresponding types of alternating finite automata provide no further information about the computation power of CA. Moreover, writing algorithms is more intuitive for CA than for alternating devices.

Introducing the notion of alternating grammar, Okhotin (2002) exhibited a characterization of RPOCA, which gives some insight on the relationship between CA and the Chomsky hierarchy. First let one briefly present the alternating grammars. An alternating grammar is a grammar enhanced with a conjunctive operation denoted by &. Each production is of the form $\alpha \rightarrow \alpha_1 \& \ldots \& \alpha_k$ where $\alpha, \alpha_1, \ldots, \alpha_k$ are strings over a set of variables and terminals. Such a production denotes that the language generated by α is the intersection of the languages generated by $\alpha_1, \ldots, \alpha_k$. Analogously to linear context-free grammar, a linear conjunctive grammar is defined as an alternating grammar with the restrictions that for every production $\alpha \rightarrow \alpha_1 \& \ldots \& \alpha_k$, α is a symbol and no α_i has more than one instance of variable.

From an algorithm given in Smith (1972), it was already known that POCA are able to recognize in real time every linear context-free languages. Finally, as shown in Okhotin (2002), extending linear context-free grammar with the conjunctive operation & leads to a complete characterization of RPOCA.

Proposition 13 *The languages recognized in real time by* POCA *are precisely the languages generated by linear conjunctive grammar.*

5.3 Other Massively Parallel Models

There exist other massively parallel computational models than the CA model. Among them, Boolean circuits, parallel random access machines (PRAM), and alternating Turing machines attract great attention. Curiously enough, CA and these parallel models seem not to recognize the existence of each other. Actually the way in which they modelize parallelism differs on several essential points. On CA, the network structure is homogeneous and the interactions are uniform and local. Furthermore, the d-dimensional array structures usually considered have a constant expansion rate: from a given cell, the number of cells accessible in $2t$ steps is linearly related to the number of cells accessible in t steps. In contrast, constraints on the network structures of uniform Boolean circuits are rather weak. Besides, most of the PRAM variants neglect the communication issue that is the bottleneck in physical machines. Another point of discord is the parallel computation thesis, which states that parallel time is polynomially equivalent to sequential space. This relationship satisfied by Boolean circuits, PRAM, and alternating Turing machines seems not to apply to CA whose network is structured as a d-dimensional array.

Hence, despite their common concern about massively parallel computation, very little is known about their links. At least, it has been proved in Chang et al. (1988) that POCA can simulate linear time bounded alternating Turing machine. As a consequence, remark also that POCA contains $\mathrm{NSpace}(\sqrt{n})$.

6 Questions

Central questions about CA as language recognizer have emerged from the very beginning and up to now remain without answer. To end this chapter, we will go back over some emblematic ones.

6.1 The Linear Time Versus Linear Space Question

The first important issue concerning the recognition power of CA is whether, for one-dimensional space-bounded CA, minimal time is less powerful than unrestricted time (Smith 1972). According to the intuition that more time gives more power, the equality RPCA = PCA between minimal time and unrestricted time CA seems very unlikely. But we fail to separate these classes. Actually, this flaw in knowledge is not specific to parallel computation when we think on similar questions in the computational complexity theory such that L $\overset{?}{=}$ P or P $\overset{?}{=}$ PSpace and more generally when we wonder how do time and space relate.

As a matter of fact, the equality RPCA = PCA would imply the equality P = PSpace since RPCA is included in P and some PSpace-complete language belongs to PCA. More precisely, using padding techniques, it is known from Poupet (2007) that if RPCA = PCA (in other words, if $\mathrm{PCA}(f) = \mathrm{DSpace}(f)$ for $f(n) = n$) then for every space constructible function f: PCA $(f) = \mathrm{DTime}(f^2) = \mathrm{DSpace}(f)$.

The common difficulty in establishing strict hierarchies lies in the fact that the amount of only one resource (here time) is varying while the amount of a second resource (here space) remains fixed. In that case, classical diagonalization arguments are of no help. And algebraic techniques only work for low-level complexity classes.

6.2   The Influence of the Input Mode

When the communication is two-way, the input mode, either parallel or sequential, does not have a great impact on the recognition time. Indeed, whatever the input mode may be, one is free to rearrange the input in various ways into the space–time diagram within linear time. So PCA and SCA are time-wise equivalent up to linear-time complexity.

The situation is not so simple when the communication is one-way. In this case, there exists a strong restriction for parallel input mode on the access to the input. The key point is that the ith cell of a POCA can only access to the first i symbols of the input, whereas every cell of a SOCA has access to the whole input within linear time. Despite this restriction, POCA and SOCA are equivalent. But taking time into account would make the difference. On the one hand, SOCA simulates POCA without time overhead. On the other hand, the only known algorithm to simulate a SOCA by a POCA is based on a brute-force strategy with an exponential cost even when the SOCA works in linear time (cf. ❷ Sect. 3.4).

Now we may wonder whether there exist less costly simulations by POCA for linear-time SOCA or polynomial-time SOCA. This question echoes another famous one: does RPCA equal LPCA? Indeed RPCA, RSOCA, and LPOCA are equivalent and also LPCA and LSOCA are equivalent. Thus the question whether RPCA is as powerful as LPCA is the same one as whether LPOCA can simulate LSOCA or not. Further, recall the result of Ibarra and Jiang, which relates this question to the closure properties under reverse and under cycle of RPCA (cf. ❷ Sect. 3.5). This result can be restated in this way: LPOCA is as powerful as LSOCA if and only if LPOCA is closed under reverse. Currently, the only idea to recognize on a POCA the reverse or the cycle of a language that is recognized by a LPOCA is the same one that for the simulation of a SOCA by a POCA: the exhaustive strategy, which consists in systematically generating all possible inputs. In other words, the same obstacle is encountered: an exponential cost, which is far from the expected linear cost.

Of course it reinforces the belief that LPCA is strictly more powerful than RPCA. Moreover, a close look at the method used by Ibarra and Jiang to link recognition ability and closure properties suggests that the result could be generalized in the following similar way. For POCA, the amount of time sufficient to simulate an arbitrary SOCA of complexity f equals the amount of time sufficient to recognize the reverse (or the cycle) of an arbitrary language accepted in time f by a POCA. The common difficulty is to make explicit the impact of the manner in which the input is supplied in case of one-way communication.

6.3   The Neighborhood Influence

Another major issue is the impact of the underlying communication graph on the computation ability. First of all, the difference between one-way communication and two-way communication is not well understood. We do not know whether POCA is as powerful as PCA. In fact, a strict inclusion between POCA and PCA would separate linear-time CA and linear-space CA. But also, as it was stressed in Ibarra and Jiang (1987), it would improve Savitch's theorem. Indeed, $\text{NSpace}(\sqrt{n}) \subseteq \text{POCA} \subseteq \text{PCA} = \text{DSpace}(n)$. Hence to distinguish POCA and PCA would distinguish $\text{NSpace}(\sqrt{n})$ and $\text{DSpace}(n)$. On the other hand, we are far from claiming the equality of POCA and PCA since we do not even know whether POCA is able to simulate PCA working in quasi-linear time.

However, in dimension higher than one, one-way communication sets limits on the language recognition ability. There exists a language accepted by some $2 - \text{PCA}$, which is accepted by no $d - \text{POCA}$, whatever the dimension d may be (cf. ❷ Sect. 4.2). Curiously enough, we fail to get such a result in the case of picture language recognition. The question of whether the inclusion between PictCA with one-way neighborhoods and PictCA with two-way neighborhoods is proper or not is still open, such as in dimension one.

In many aspects, the situation becomes much more complicated for computation on picture languages. Contrary to dimension one where all neighborhoods are real-time equivalent to either the one-way neighborhood $\{-1, 0\}$ or the two-way neighborhood $\{-1, 0, 1\}$, the recognition ability of PictCA appears more widely influenced by the choice of the neighborhood. For instance, there exists a picture language recognized in real time with the von Neumann neighborhood, which is not recognized in real time with the Moore neighborhood. On the other hand, it is an open question whether all picture languages recognized in real time with the Moore neighborhood are also recognized in real time with the von Neumann neighborhood. More generally, the precise relationships between the various neighborhoods are ignored. And worse, some neighborhoods seem not to admit linear speedup. Actually, the rules that lie behind the communication properties are not simple to grasp. One difficulty is the question of orientation in the two-dimensional underlying communication graph. Notably, Szwerinski's property, which states that a one-dimensional PCA may confuse right and left without time loss and emphasizes that the orientation does not matter in dimension one, seems not to apply to PictCA. And one factor, which might discriminate between the various neighborhoods, would be whether each cell of the array somehow knows the direction toward the output cell.

References

Beyer WT (1969) Recognition of topological invariants by iterative arrays. Technical Report AITR-229, MIT Artificial Intelligence Laboratory, October 1, 1969

Bucher W, Čulik K II (1984) On real time and linear time cellular automata. *RAIRO* Theor Inf Appl 81:307–325

Buchholz T, Kutrib M (1998) On time computability of functions in one-way cellular automata. Acta Inf 35(4):329–352

Buchholz T, Klein A, Kutrib M (2000) Iterative arrays with small time bounds. In: Nielsen M, Rovan B (eds) MFCS, Lecture notes in computer science, vol 1893. Springer, Berlin, Heidelberg, pp 243–252

Cervelle J, Formenti E (2009) Algorithmic complexity and cellular automata. In: Meyers RA (ed) Encyclopedia of complexity and system science. Springer, New York

Chang JH, Ibarra OH, Palis MA (1987) Parallel parsing on a one-way array of finite state machines. IEEE Trans Comput C-36(1):64–75

Chang JH, Ibarra OH, Vergis A (1988) On the power of one-way communication. J ACM 35(3):697–726

Chang JH, Ibarra OH, Palis MA (1989) Efficient simulations of simple models of parallel computation by time-bounded ATMs and space-bounded TMs. Theor Comput Sci 68(1):19–36

Choffrut C, Culik K II (1984) On real-time cellular automata and trellis automata. Acta Inf 21(4):393–407

Cole SN (1969) Real-time computation by n-dimensional iterative arrays of finite-state machine. IEEE Trans Comput 18:349–365

Culik K II (1989) Variations of the firing squad problem and applications. Inf Process Lett 30(3):153–157

Culik K II, Gruska J, Salomaa A (1984) Systolic trellis automata. I. Int J Comput Math 15(3–4):195–212

Delacourt M, Poupet V (2007) Real time language recognition on 2D cellular automata: Dealing with non-convex neighborhoods. In: Kucera L, Kucera A (eds) Mathematical foundations of computer science 2007, vol 4708 of Lecture Notes in Computer Science, pp 298–309

Delorme M, Mazoyer J (1994) Reconnaisance de langages sur automates cellulaires. Research Report 94–46, LIP, ENS Lyon, France

Delorme M, Mazoyer J (2002) Reconnaissance parallèle des langages rationnels sur automates cellulaires plans. [Parallel recognition of rational languages on plane cellular automata] Selected papers in honour of Maurice Nivat. Theor Comput Sci 281(1–2):251–289

Delorme M, Mazoyer J (2004) Real-time recognition of languages on an two-dimensional Archimedean thread. Theor Comput Sci 322(2):335–354

Dyer CR (1980) One-way bounded cellular automata. Inf Control 44(3):261–281

Fischer PC (1965) Generation of primes by one-dimensional real-time iterative array. J ACM 12:388–394

Giammarresi D, Restivo A (1997) Two-dimensional languages. In: Rozenberg G, Salomaa A (eds) Handbook of Formal Languages, vol 3. Springer, New York, pp 215–267

Goldschlager LM (1982) A universal interconnection pattern for parallel computers. J ACM 29(4):1073–1086

Hartmanis J, Stearns RE (1965) On the computational complexity of algorithms. Trans Am Math Soc (AMS) 117:285–306

Ibarra OH, Jiang T (1987) On one-way cellular arrays. SIAM J Comput 16(6):1135–1154

Ibarra OH, Jiang T (1988) Relating the power of cellular arrays to their closure properties. Theor Comput Sci 57(2–3):225–238

Ibarra OH, Kim SM (1984) Characterizations and computational complexity of systolic trellis automata. Theor Comput Sci 29(1–2):123–153

Ibarra OH, Palis MA (1987) On efficient simulations of systolic arrays of random-access machines. SIAM J Comput 16(2):367–377

Ibarra OH, Palis MA (1988) Two-dimensional iterative arrays: characterizations and applications. Theor Comput Sci 57(1):47–86

Ibarra OH, Kim SM, Moran S (1985a) Sequential machine characterizations of trellis and cellular automata and applications. SIAM J Comput 14(2):426–447

Ibarra OH, Palis MA, Kim SM (1985b) Some results concerning linear iterative (systolic) arrays. J Parallel Distrib Comput 2(2):182–218

Ito A, Inoue K, Takanami I (1989) Deterministic two-dimensional on-line tessellation acceptors are equivalent to two-way two-dimensional alternating finite automata through 180° rotation. Theor Comput Sci 66(3):273–287

Kobuchi Y (1987) A note on symmetrical cellular spaces. Inf Process Lett 25(6):413–415

Kosaraju SR (1979) Fast parallel processing array algorithms for some graph problems (preliminary version). In ACM conference record of the eleventh annual ACM symposium on theory of computing: papers presented at the symposium, Atlanta, Georgia, ACM Press, New York, 30 April–2 May 1979, pp 231–236

Klein A, Kutrib M (2003) Fast one-way cellular automata. Theor Comput Sci 295(1–3):233–250

Kutrib M (2001) Automata arrays and context-free languages. In Where mathematics, computer science, linguistics and biology meet, Kluwer, Dordrecht, The Netherlands, pp 139–148

Levialdi S (1972) On shrinking binary picture patterns. Commun ACM 15(1):7–10

Mazoyer J, Reimen N (1992) A linear speed-up theorem for cellular automata. Theor Comput Sci 101(1):59–98

Okhotin A (2002) Automaton representation of linear conjunctive languages. In International conference on developments in language theory (DLT), LNCS, vol 6. Kyoto, Japan

Poupet V (2005) Cellular automata: real-time equivalence between one-dimensional neighborhoods. In Diekert V, Durand B (eds) STACS 2005, 22nd annual symposium on theoretical aspects of computer science, Stuttgart, Germany, February 24–26, 2005, Proceedings, vol 3404 of Lecture Notes in Computer Science. Springer, pp 133–144

Poupet V (2007) A padding technique on cellular automata to transfer inclusions of complexity classes. In Diekert V, Volkov MV, Voronkov A (eds) Second international symposium on computer science in Russia, vol 4649 of Lecture Notes in Computer Science. Springer, pp 337–348

Rosenberg AL (1967) Real-time definable languages. J ACM 14(4):645–662

Savage C (1988) Recognizing majority on a one-way mesh. Inf Process Lett 27(5):221–225

Smith AR III (1971) Simple computation-universal cellular spaces. J ACM 18(3):339–353

Smith AR III (1972) Real-time language recognition by one-dimensional cellular automata. J Comput Syst Sci 6:233–253

Szwerinski H (1985) Symmetrical one-dimensional cellular spaces. Inf Control 67(1–3):163–172

Terrier V (1995) On real time one-way cellular array. Theor Comput Sci 141(1–2):331–335

Terrier V (1996) Language not recognizable in real time by one-way cellular automata. Theor Comput Sci 156(1–2):281–285

Terrier V (1999) Two-dimensional cellular automata recognizer. Theor Comput Sci 218(2):325–346

Terrier V (2003a) Characterization of real time iterative array by alternating device. Theor Comput Sci 290 (3):2075–2084

Terrier V (2003b) Two-dimensional cellular automata and deterministic on-line tessalation automata. Theor Comput Sci 301(1–3):167–186

Terrier V (2004) Two-dimensional cellular automata and their neighborhoods. Theor Comput Sci 312(2–3): 203–222

Terrier V (2006a) Closure properties of cellular automata. Theor Comput Sci 352(1–3):97–107

Terrier V (2006b) Low complexity classes of multi-dimensional cellular automata. Theor Comput Sci 369(1–3):142–156

5 Computations on Cellular Automata

Jacques Mazoyer[1] · *Jean-Baptiste Yunès*[2]
[1]Laboratoire d'Informatique Fondamentale de Marseille (LIF),
Aix-Marseille Université and CNRS, Marseille, France
mazoyerj2@orange.fr
[2]Laboratoire LIAFA, Université Paris 7 (Diderot), France
jean-baptiste.yunes@liafa.jussieu.fr

G. Rozenberg et al. (eds.), *Handbook of Natural Computing*, DOI 10.1007/978-3-540-92910-9_5,
© Springer-Verlag Berlin Heidelberg 2012

Abstract

This chapter shows how simple, common algorithms (multiplication and prime number sieve) lead to very *natural* cellular automata implementations. All these implementations are built with some *natural* basic tools: signals and grids. Attention is first focussed on the concept of signals and how simple and rich they are to realize computations. Looking closely at the space–time diagrams and the dependencies induced by the computations reveals the concept of grids, and shows how powerful they are in the sense of computability theory.

1 Introduction

This chapter shows how well-known, frequently used algorithms are fundamentally parallel and can be easily implemented on cellular automata (CAs). In fact, programming such simple algorithms with CAs reveals major issues in computability theory. How do inputs enter and outputs exit the machine? How does this affect the programming? What is the concept of time? It also raises questions about what exactly a CA is. Two main examples are considered.

First, the multiplication of two numbers is considered. Though one can consider this as a trivial task, it can be seen that implementing it with CAs is of great interest. It illustrates how CAs can catch the behavior at different scales of different kinds of machines: from circuits made of elementary gates to grids of processors.

Second, the CA implementation of the Eratosthenes prime number sieve is considered. This, unlike the preceding problem, is not a function mapping input values to a corresponding output, but is a "pure" computation, namely, a never ending loop, that produces beeps at times, which are prime numbers.

These considerations stress the importance of the dependency graph in cellular automata. It is sufficient to have a very simple graph to be able to compute any computable function (computable in the sense of Turing's definition). We also show that any distortion applied to a grid can be useful to compute a functional composition or a function call (in the sense of traditional computer programming).

2 Usual Multiplication Algorithm and CAs

There exist many different algorithms to multiply two numbers. A very common algorithm, which is taught in elementary classrooms of many countries, is considered. It has a very straightforward implementation in cellular automata.

❷ *Figure 1* is a sample of that classical algorithm used to compute 148 × 274. Each intermediate result is produced digit by digit from the right to left. The propagation of the necessary carries introduces "natural" dependencies between those digits, the horizontal arrows of ❷ *Fig. 2a*. One observes that in basis 2, no such carry is needed as the multiplicand is either multiplied by 1 or by 0, so that no delay is necessary for the computation of the horizontal lines.

The final result is also produced digit by digit from the right to left, propagating a carry (whatever the basis). A final digit is the last digit of the sum of all digits in its column and the carry of the preceding one – the vertical arrows of ❷ *Fig. 2a*. This induces a "natural" order

◻ Fig. 1

The usual multiplication.

```
        1 4 8
      × 2 7 4
      ─────────
        5 9 2      = 148 × 4
      1 0 3 6 .    = 148 × 70
      2 9 6 . .    = 148 × 200
      ─────────
      4 0 5 5 2
```

◻ Fig. 2

(a) Graph of dependencies. (b) Extended graph.

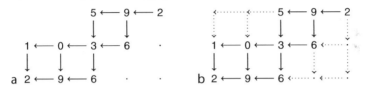

◻ Fig. 3

(a) Example as a trellis CA. (b) Multiplication as a trellis.

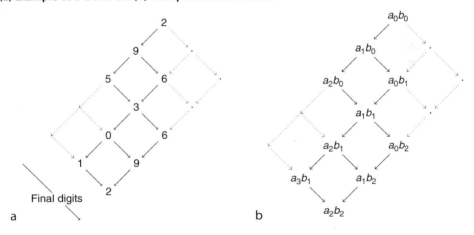

on the production of the digits of the result: they are produced from the least to the most significant one.

All those dependencies make the trellis shown in ❷ *Figs. 2* and ❷ *3.* ❷ *Figure 2b* shows the full dependency graph of which the only part in ❷ *Fig. 2a* is used for the particular multiplication. ❷ *Figure 2b* represents the hand execution of the algorithm. A rotation leads to ❷ *Fig. 3*, which is a classical view of a space–time diagram of a trellis CA. Replacing sample digits by generic ones gives a good hint of what really happens during the process as seen in ❷ *Fig. 3b*.

Of course, there still remains the problem of carry accumulation in the summing process to get a final digit. This leads to unbounded carries that contradict the finite nature of cells in cellular automata. To solve the problem, one can move elementary carries as soon as they are produced.

The details are left to the reader who can refer to ❷ Sect. 3 to get an idea of how this can be done. The stress now is rather on what the above analysis tells us about more general concerns in CAs.

2.1 Input/Output Mode

First, one can remark that ❷ *Fig. 2b* with "vertical" time is not the space–time diagram of a CA, since it would require that the ith digit of the multiplicator be repeated all along the line at time i.

❷ *Figure 3a* shows the convenient trellis CA, which is obtained by changing the arrow of time: the vertical of ❷ *Fig. 3b* is the diagonal of ❷ *Fig. 2b*.

As shown in ❷ *Fig. 3b*, in this trellis CA, the input mode is parallel for the multiplicand (a_i is located on column i of the trellis) and sequential for the multiplicator (the b_i's enter cell 0, one after the other). However, this is just a point of view as illustrated by ❷ *Fig. 4*: the two factors can be given as parallel inputs distributed left and right of the "central" cell. Moreover, from that trellis automaton, one can construct a usual linear (non-trellis) CA implementing the same computation as shown in ❷ *Fig. 4* (retain one line of cells out of two). This construction is in fact quite a general result, as illustrated by ❷ *Fig. 5a, b*.

The output is sequential in the trellis automata: all bits are successively obtained on the "central" cell.

◼ **Fig. 4**
Parallel inputs on multiplier trellis CA. Dashed arrows illustrate the flow of the bits of the multiplicator, and full arrows the flow of the bits of the multiplicand.

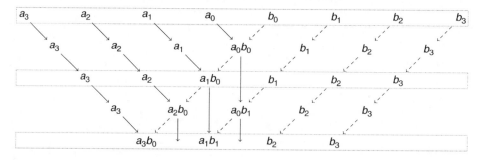

◼ **Fig. 5**
(a) Trellis CA. (b) Corresponding non-trellis CA.

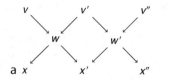 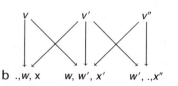

Thus, with this algorithm, there is a dissymmetry between input and output modes. This prevents any obvious composition of the algorithm with itself in order to iterate multiplications. Therefore, it cannot be considered as an elementary programming block.

2.2 About Time

The pertinent definition of a time step in a trellis CA is illustrated in ❷ *Fig. 5*. What is important is not the definition of time one can superimpose on the graph of dependencies, but the dependencies themselves, so that a time step is the shortest path from a cell to itself. For a more general discussion on this topic, cf. Wolfram's book (see Wolfram 2002, Chap. 9).

This reflects two different views on the topic. First, one is only interested in causalities intrinsic to computation and tries to synthesize a machine. Second, one is interested in programming machines. As seen in the following, these two views are both interesting to consider.

3 Real-Time Multiplication

Atrubin (1965) designed a cellular automaton that was able to multiply two binary numbers in real time, answering McNaughton's problem about the existence of such a machine. Recall that real-time computation was defined by McNaughton (see McNaughton 1961) as the following:

▶ If a machine has input sequence $i_0, i_1, i_2, \ldots$ and output sequence $o_0, o_1, o_2, \ldots$, where o_k is a function of inputs up to i_k, and if the machine computes o_k within a fixed time limit independent of n following the receipt of i_k, then the computation is said to be accomplished in real time.

Goyal (1976) modified Atrubin's machine so that all the cells were identical (the first cell of Atrubin's machine was slightly different than the other ones). And later on, in 1991, Even (1991) used that construction to compute modular multiplication in linear time.

Knuth (1997) also designed an iterative machine able to multiply in real time.

Of course, designing a fast multiplier circuit is not a new problem as it is one of the elementary building blocks of any computer. Previous references are mentioned just because they focus on models of machines not on circuit design: concerns are not exactly the same.

All those solutions use the same key idea: store two pairs of consecutive bits of the factors on each cell. Another way to understand it is to remark that to produce the n least significant bits of the product in real time, the number of elementary products, $a_i \cdot b_j$, that need to be computed is of the order of $n^2 / 2$. However, from ❷ *Fig. 6*, it is seen that there are only about $n^2 / 4$ locations in the space–time diagram, so that some grouping is needed.

Here, such constructions are exhibited, which differ in many details of the former ones but fully respect the scheme. The machine has two layers on which different flows of information take place. The first layer is essentially devoted to the space–time layout of the digits of the factors, and the second is used to produce the bits of the product.

In the following, the two factors of the multiplication are denoted by A and B, P, the product $A \times B$, and by extension, a_i (resp. b_i, p_i) denotes the ith bit of the binary writing of A (resp. B, P).

◘ **Fig. 6**
How data flow through the graph. Inputs propagate from bottom-left to top-right. Bits of the product propagate from bottom-right to top-left.

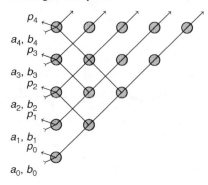

3.1 Input/Output Mode

The machine has a sequential input mode, that is, pairs of corresponding bits of the two factors successively enter the machine. The computation generates two main flows of information as illustrated in ❷ *Fig. 6* and described below.

Output mode of the machine is also sequential, that is, bits of the results are successively produced on the first cell in real time.

Thus, this machine can be composed with itself.

3.2 Information Flow

There are two main flows of information during the process as illustrated in ❷ *Fig. 6*. First, the input (a_i, b_i) at time i travels from the left to right (bottom-left to top-right arrows in the space–time diagram). Second, the output p_i is generated at time $i + 1$, and its computation is done bit by bit all along a right to left path in the space (bottom-right to top-left arrows in the space–time diagram).

3.3 Propagating Bits of Factors

As illustrated in ❷ *Fig. 7*, there are two internal write-once storages in each cell, say S^0 (at the bottom) and S^1 (at the top). When a value, (a_i, b_i), enters from the left to a cell, there are three cases:

- If S^0 is empty, then the input is pushed into, storing the value in S^0.
- Else if S^1 is empty, then the input is pushed into, storing the value in S^1.
- Else the input is pushed to the right, giving it as input to the next right cell at the next time step (this transient storage will be denoted by T).

Thus, (a_i, b_i) is stored in $S^{i \bmod 2}_{\lfloor i/2 \rfloor}$.

◻ Fig. 7

Layer 1: how factors' bits are stored. One can easily observe that a given input (a_i, b_i) propagates bottom-left to top-right until it finds its right location.

$t = 6$	(a_1, b_1) (a_0, b_0)	(a_3, b_3) (a_2, b_2)	(a_4, b_4)

(a_6, b_6) (a_5, b_5) (a_4, b_4)

$t = 5$	(a_1, b_1) (a_0, b_0)	(a_3, b_3) (a_2, b_2)	

(a_5, b_5) (a_4, b_4)

$t = 4$	(a_1, b_1) (a_0, b_0)	(a_3, b_3) (a_2, b_2)	

(a_4, b_4) (a_3, b_3)

$t = 3$	(a_1, b_1) (a_0, b_0)	(a_2, b_2)	

(a_3, b_3) (a_2, b_2)

$t = 2$	(a_1, b_1) (a_0, b_0)		

(a_2, b_2)

$t = 1$	(a_1, b_1) (a_0, b_0)		

(a_1, b_1)

$t = 0$	(a_0, b_0)		

(a_0, b_0)

3.4   Accumulating Product Bits

The second layer is used to carry the effective computation of the multiplication. In that layer, the information moves backward to the first cells, accumulating at each time step some new partial computation of the product. In each cell, there are two transient storages, one used to compute the partial product and another to manage the carries.

At a given time, stored values in the first layer are used to compute products accumulated in the second layer. For example, if there is (a, b) in T_i, (a', b') in S_i^1, and (a'', b'') in S_i^0 then it is possible to compute $a \cdot b' + a' \cdot b$ and $a \cdot b'' + a'' \cdot b$. As time goes, values (a, b) represent all digits with ranks greater than those of (a', b'), of (a'', b''). Thus, every possible product of digits $a_i \cdot b_j + a_j \cdot b_i$ can be computed as shown below.

◘ **Fig. 8**

Some cells at time $t + 2$ and $t + 3$ ($t > 2i + 5$). One can remark that all $a_i b_j$ such that $i + j = k$ which contribute to p_k are located on two successive diagonals. Here, bits contributing to p_{t+2i+4} have been boxed. The way they are cumulated is illustrated by arrows: in each cell necessary elementary products are computed and their values are propagated in order to be added at next time step.

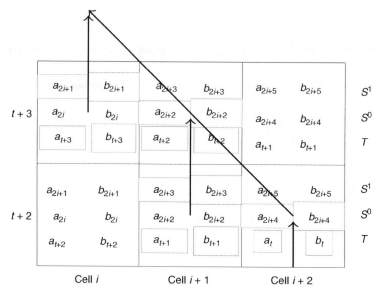

▶ *Figure 8* shows the contents of the three storages in the layer 1 of the machine at two successive time steps. Boxed contents exhibit the contribution of input bits to the computation of the bit $t + 2i + 4$ of the final product.

At some time $t + 2$ ($t > 2i + 5$), one can observe that one can compute:

$$\begin{cases} a_{t+1} \cdot b_{2i+3} + a_{2i+3} \cdot b_{t+1} & \text{on cell } i + 1 \\ a_t \cdot b_{2i+4} + a_{2i+4} \cdot b_t & \text{on cell } i + 2 \end{cases}$$

and, at time $t + 1$:

$$\begin{cases} a_{t+3} \cdot b_{2i+1} + a_{2i+1} \cdot b_{t+3} & \text{on cell } i \\ a_{t+2} \cdot b_{2i+2} + a_{2i+2} \cdot b_{t+2} & \text{on cell } i + 1 \end{cases}$$

From this, one can infer that there is a "natural" dependency between cells as illustrated in ▶ *Fig. 9*, where:

- P is a partial computation of some result bit p_n (with $n = k + 2i$) coming from the right (namely, $\sum_{j=0}^{k-1} a_j \cdot b_{n-j} + a_{n-j} \cdot b_j \bmod 2$).
- P′ is a contribution to bit p_n computed at the previous time step (namely, $a_{k-1} \cdot b_{2i+1} + a_{2i+1} \cdot b_{k-1}$).
- P″ is a partial computation of bit p_n (namely, $\sum_{j=0}^{k} a_j \cdot b_{n-j} + a_{n-j} \cdot b_j \bmod 2$).
- P* is a contribution to bit p_{n+1} (namely, $a_k \cdot b_{2i+1} + a_{2i+1} \cdot b_k$).

◻ Fig. 9
The multiplier.

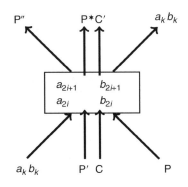

- C is the carry of the partial computation of bit p_{n-1} computed at a previous time step.
- C' is the carry of the partial computation of bit p_n which is sent to the computation of bit p_{n+1}.

This computation follows the relations (which more generally are written in base b)

$$\left(\sum_{i=0}^{n} d_i\right) \bmod b = \left(\left(\sum_{i=0}^{n-1} d_i\right) \bmod b + d_n\right) \bmod b \tag{1}$$

and

$$\left(\sum_{i=0}^{n} d_i\right) \Big/ b = \left(\sum_{i=0}^{n-1} d_i\right) \Big/ b + \left(\left(\sum_{i=0}^{n-1} d_i\right) \bmod b + d_n\right) \Big/ b \tag{2}$$

(where / denotes integer division). Thus, bits of the product are produced in real time.

To be compliant with the definition of cellular automata and their intrinsic finiteness, one must verify that P, P', P'', P*, C, and C' always carry a finite amount of information. The corresponding equations are

$$P'' = (a_k b_{2i} + a_{2i} b_k + C + P + P') \bmod 2 \tag{3}$$

$$P^\star = (a_k b_{2i+1} + a_{2i+1} b_k) \tag{4}$$

$$C' = (a_k b_{2i} + a_{2i} b_k + C + P + P')/2 \tag{5}$$

Observe that P'' is a single bit, and that P''' is two bits wide (three values). From ❷ Eqs. 4 and ❷ 5 it is seen that there are at most four different possible values since C = 4 is the smallest solution of $(C + 5)/2 \le C$. Thus, C and C' consist of 3 bits.

3.5 Boundary Conditions

Two special cases have to be considered when the storages of a_i, b_i occur in layer 1: this corresponds to the initiation of computations of the partial products.

First, when a cell is empty and a value (a_{2k}, b_{2k}) enters in (remember that if a cell number i is empty, the first value to store in is a pair of even indexed bits), the value $a_{2k} \cdot b_{2k}$ is produced as P′, initiating the production of bit $4k$ of the output.

Second, when a cell is half-empty and a value (a_{2k+1}, b_{2k+1}) enters in, the value $a_{2k+1} \cdot b_{2k+1}$ is produced as C′, initiating the production of bit $4k + 2$ of the output.

3.6 Analysis

It is shown that even if multiplying this way seems to be very tricky at first look, one can decompose the machine as the "superposition" of three different very simple machines. This is similar to modularization in programming. The three simpler machines are:

- A storing machine which places every bits of the factors at the right place
- A bit producing machine which computes every single elementary product
- A parallel additioner which sweeps the space–time to gather every bit and adds them appropriately.

4 Prime Numbers Sieve and CAs

Searching or enumerating prime numbers is a mathematical activity, at least as old as Eratosthene's algorithm which is a very elementary method to get, by exclusion, all prime numbers up to n:

▶ Write the sequence of every integer from 2 to n. Check every proper multiple of 2, then every proper multiple of 3, etc. The unchecked integers are the prime ones.

One sees that this very simple algorithm can be implemented with CAs. Such work was first done by Fischer (1965) and intensively studied later on, around 1990, by Korec (1997).

All these implementations heavily use the concept of signal. Signals are very important when programming CAs, it is commonly considered as an elementary brick to build a CA program. A signal can be roughly defined here as a *moving information* whose trace in the space–time diagram is generally a connected path. This is a very "natural" concept, and very simple signals are used in implementation of prime number sieves.

These implementations use clocks. If a clock has a period of length p steps, it is obvious to remark that it marks every multiple of p, if it has been set at time p. Thus combining multiple clocks, one is able to mark every multiple of every integer, as in Eratosthene's algorithm. Clocks are implemented with signals.

It is important to note that this computation is not a function, as commonly defined in mathematics. There is no input and no output, that is, no value is introduced nor produced. The result is a behavior, as one can observe that something happens during the computation. In the example, the computation will leave some cell in a given remarkable state at all prime times. Nevertheless, one can look at this behavior as the characteristic function of the set of prime numbers: identify all states associated with the production of a prime number to 1 and all other ones to 0. Otherwise said, the language $\{1^p | p \text{ prime}\}$. And this is very different of many computations, where more traditional mathematical functions are realized (see, e.g., ❷ Sect. 2).

□ Fig. 10

A naive prime number sieve. Dotted lines represent the newly launched clock. At time step *i* the new clock has a period of length *i*.

$$\times 2 \quad \times 2, \times 3 \quad \times 2, \times 3, \quad \times 2, \times 3, \quad \times 2, \times 3,$$
$$\times 4 \qquad \times 4, \times 5 \quad \times 4, \times 5,$$
$$\times 6$$

❯ *Figure 10* illustrates how Eratosthene's idea leads to a naive implementation that could not be implemented in CA's. This construction is a simple superposition of all appropriate clocks.

First, at time 2, a clock of period 2 is launched such that every even tick is marked. Such a clock is implemented *via* a signal (an information) which goes to the right one unit of time and then goes back one unit of time, thus bouncing on the initial cell one tick over two.

At time 3, a clock of period 3 is started. At the same time, it can be observed that no signal has entered the first cell showing that this tick is a prime integer (indeed 3 is prime). The clock of period 3 is implemented with a signal going to the right one half more than the previous zigzag signal (of period 2), since the difference of half the periods of the two signals is 0.5. It may seem strange to move an information one half in the two directions since space–time is discrete in both dimensions, but that can be simply simulated going one time in the chosen direction, staying one time at the same place. Such a pause may be viewed as an internal simulation of two half opposite moves.

At time 4, a clock of period 4 is started, and so on.

But this cannot be computed by CA, as it is easy to remark that unbounded many signals need to be started as time grows (at times, in cases such as $\Pi_{i=0}^{k}i$, k signals have to be differentiated), which is not possible with a finite automaton.

One would also like to simplify the process such that only necessary signals have to be launched (signals of periods p for any p prime). But at times such like $\Pi_{i=0}^{k}p_i$, k signals should also be launched. Another construction should be necessary to construct right borders (as this will not correct the problem, how these walls can be constructed is not shown and this is left as an exercise to the curious reader).

To solve the problem, one simply needs to create disjoint corridors in which individual clocks are built. As one can remark, the problem is to synchronize all those separated clocks in such a way that ticks can be reported at the right time on the first cell. ❯ *Figure 11a* shows how a clock running into a corridor located far away from the first cell generates delayed ticks.

❯ *Figure 12* illustrates the whole process involved in the "computation" of prime numbers. As in more traditional programming and mathematical reasoning, the whole problem is decomposed into a few simpler problems and then reassembled.

First, the construction of corridors is set up. Suppose height n has been built, then ❯ *Fig. 11b* shows how corridor with width n is built and how height $n + 1$ is built on the appropriate cell.

Second, in each corridor of width n, a clock of period $2n + 1$ will be set up to mark multiples of $2n + 1$ on the first cell. As distance from the left side of the corridor of width n to the first cell is $\sum_{i=1}^{n-1}i$ one needs to set up the clock of period $2n + 1$ such that it beats its corridor's left side at times $k \times (2n + 1) - \sum_{i=1}^{n-1}i$.

■ **Fig. 11**
(a) Corridors and delays. (b) Next corridor.

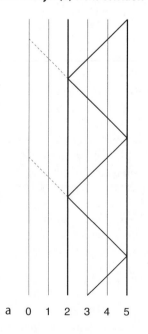

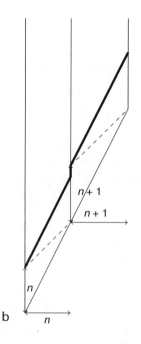

a 0 1 2 3 4 5   b n

◘ **Fig. 12**

Prime number sieve construction. Black disks represent "even ticks." Gray disks represent "composite non-even ticks." White disks correspond to "prime ticks."

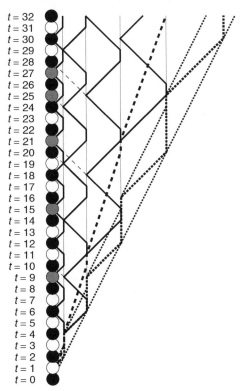

$t = 32$
$t = 31$
$t = 30$
$t = 29$
$t = 28$
$t = 27$
$t = 26$
$t = 25$
$t = 24$
$t = 23$
$t = 22$
$t = 21$
$t = 20$
$t = 19$
$t = 18$
$t = 17$
$t = 16$
$t = 15$
$t = 14$
$t = 13$
$t = 12$
$t = 11$
$t = 10$
$t = 9$
$t = 8$
$t = 7$
$t = 6$
$t = 5$
$t = 4$
$t = 3$
$t = 2$
$t = 1$
$t = 0$

5   CA Computations with Grids

5.1   Dependencies

As seen in ❷ Sects. 2 and ❷ 3 about multiplication on CA, to compute on CA is to construct and manage different flows of information. To multiply, two main flows are built so that, at their meeting, elementary computations are made. The computations are chained together following the "natural" dependency graph of the space–time diagram of a linear cellular automaton with neighborhood of nearest cells, $v = \{-1, 0, +1\}$. Obviously, the number of information flows is only bounded by imagination.

In this construction, the key idea is to manage the meeting of the first two flows so that a new flow is generated which, in its turn, meets the third flow. One can remind that these constructions are very similar to techniques used in systolic algorithms, up to one more constraint: the graph of data moves must be injected into the dependency graph of the cellular automaton.

Determining this dependency graph is a very important question to be solved. Historically, two approaches emerged. The first one is due to Cole (1969) who showed that every one-dimensional CA with any arbitrary neighborhood can be simulated by a 1D-CA with neighborhood $v = \{-1, 0\}$ or $v = \{0, +1\}$. The second one has been introduced by Čulik (1982) and is now known as trellis automaton with neighborhood $v = \{-1, +1\}$. In fact, these two approaches are very similar and tightly related (see Choffrut and Čulik 1984).

The basic idea is as follows:

▶ It is sufficient to have two independent communication channels to be able to simulate any one-dimensional CA whatever be its neighborhood.

To understand how these approaches are intrinsically the same, the following monoid is introduced:

$$\langle e, a, a^{-1}, \alpha_{-\ell}, \ldots, \alpha_0, \ldots, \alpha_r | aa^{-1} = a^{-1}a = e; \ \alpha_i = \alpha_0 a^i = a^i \alpha_0 \rangle \qquad (6)$$

where e is the neutral element, a and a^{-1} generate the initial line, and the influence of cell $c - i$ onto cell c is represented by generator α_i with $-\ell \leq i \leq r$. The dependency graph of a CA with a neighborhood $v = \{-\ell, \ldots, -1, 0, +1, \ldots, r\}$ is obtained from the Cayley graph of the monoid (❷ Eq. 6) by forgetting all arcs labeled a or a^{-1}.

In case of the usual neighborhood $v = \{-1, 0, +1\}$, the above monoid is

$$\langle e, a, a^{-1}, \alpha_{-1}, \alpha_0, \alpha_1 | aa^{-1} = a^{-1}a = e; \alpha_1 = \alpha_0 a = a\alpha_0; \alpha_{-1} = \alpha_0 a^{-1} = a^{-1}\alpha_0 \rangle \qquad (7)$$

In the case of neighborhood $\{0, +1\}$ (resp. $\{-1, 0\}$, resp. $\{-1, +1\}$), the associated monoid is

$$\langle e, b, b^{-1}, \beta, \gamma | bb^{-1} = b^{-1}b = e; \gamma = b\beta = \beta b \rangle \qquad (8)$$

where $b = a$, $\beta = \alpha_0$, and $\gamma = \alpha_1$ (resp. $b = a$, $\beta = \alpha_{-1}$, and $\gamma = \alpha_0$, resp. $b = a^2$, $\beta = \alpha_{-1}$, and $\gamma = \alpha_1$).

Now, the dependency graph associated with ❷ Eq. 7 can be embedded into the dependency graph associated with ❷ Eq. 8 using the relations $\alpha_0 = \beta\gamma = \gamma\beta$, $\alpha_1 = \beta^2$, and $\alpha_{-1} = \gamma^2$.

Consider that the inputs of the computation are not located on the initial line (cells $(x, 0)$), but on the two main diagonals issued from the initial cell $(0, 0)$ (i.e. cells $(x, |x|)$), then the dependency graph associated by ❷ Eq. 8, illustrated by ❷ Fig. 13a, is the one associated to the monoid:

$$\langle e, \alpha, \beta | \alpha\beta = \beta\alpha \rangle \qquad (9)$$

The embedding, in the general case, is left to the reader.

In the following, we will sometimes say that α is *stepping to the right*, β is *stepping to the left*, and their product ($\alpha\beta$ or $\beta\alpha$) is *staying in place* (cf. ❷ Fig. 13b).

The proof of the main Theorem 1 shows how the simulation of an automaton with neighborhood $\{-1, 0, +1\}$ can be done on such a grid, with an appropriate distribution of the initial configuration.

5.2 First Approach to the Notion of Grids

From an initial finite line of cellular automaton with neighborhood $\{-1, 0, +1\}$, one can construct many graphs associated to monoid of ❷ Eq. 9. Here, "construct" informally means drawing with signals in the space–time diagram (see the chapter ❷ Algorithmic Tools on Cellular Automata); where signals correspond to the generators of the semigroup (i.e., α, β) and their meeting corresponds to the elements of the semigroup.

◘ Fig. 13

(a) Graph induced by the monoid {e, α, β|αβ = βα} (point e not shown). (b) Relation α·β = β·α, otherwise said "stay in place is as the same as moving one step to the right then one step to the left (or the converse)."

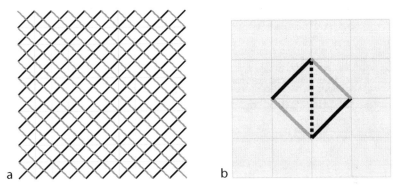

a   b

◘ Fig. 14

(a) A rational grid. (b) A nonrational grid (with point e).

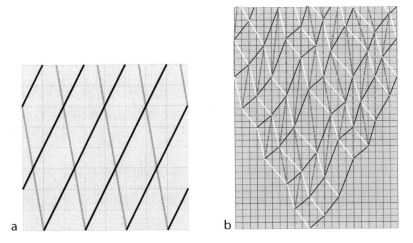

a   b

Such a graph is called a **grid**. It will be more formally defined in ❷ Sect. 5.3.

If the site corresponding to the element of the semigroup $\alpha^p \beta^t$ is denoted by (p, t), then, one observes that:

- Given p_e, t_e, λ, μ, one can construct a **rational grid** so that $(p, t) = (p_e + p \cdot \lambda, t_e + t \cdot \mu)$, see ❷ *Fig. 14a*.
- Even **nonrational grids** are constructible (see ❷ *Fig. 14b*).
- Not every recursive grid is constructible as the value of $(0, t)$ only depends on the value of sites $(-t, 0), \cdots, (0, 0) \cdots, (t, 0)$, and is then constructible in time by a Turing machine computing with a semi-infinite ribbon and simulating the cellular automaton from the following initial configuration:

$$(\langle 0,0 \rangle, \star), (\langle 1,0 \rangle, \langle -1, 0 \rangle) \dots (\langle t, 0 \rangle, \langle -t, 0 \rangle)$$

5.3   Definitions

Now, the definition of what is really meant by constructing or drawing a grid in the orbit of a particular configuration of a cellular automaton with neighborhood $\{-1, 0, +1\}$ is given. In the following, the concept of **signals**, introduced in the chapter ❷ Algorithmic Tools on Cellular Automata, is used intensively (see Adamatzky 2002; Delorme and Mazoyer 1999; Poupet 2005; Mazoyer and Terrier 1999; ❷ Definitions 6 and ❷ 8), but, the main goal to compute, these definitions will be restricted a little bit.

Definition 1   (consequences) The *immediate consequence* $\mathbb{C}(\mathscr{X})$ of $\mathscr{X} \subset \mathbb{Z} \times \mathbb{N}$ is defined as the set of points $(x, y) \in \mathbb{Z} \times \mathbb{N}$, such that $(x, y) \notin \mathscr{X}$ but such that all three sites $(x - 1, y - 1)$, $(x, y - 1)$, and $(x + 1, y - 1)$ are in $\mathscr{X}$.

The *consequence* of $\mathscr{X}$ is defined as $\mathbb{C}^\star(\mathscr{X}) = \lim_{n \to \infty} U_n$, where $U_0 = \mathscr{X}$ and for all $n > 0$, $U_{n+1} = U_n \bigcup \mathbb{C}(U_n)$.

Definition 2   (influences) The *immediate influence* $\mathbb{I}(\mathscr{X})$ of $\mathscr{X} \subset \mathbb{Z} \times \mathbb{N}$ is defined as the set of points $(x, t) \in \mathbb{Z} \times \mathbb{N}$, such that at least one of $(x - 1, t - 1)$, $(x, t - 1)$, or $(x + 1, t - 1)$ is in $\mathscr{X}$.

The *influence* of $\mathscr{X}$ is defined as $\mathbb{I}^\star(\mathscr{X}) = \lim_{n \to \infty} U_n$, where $U_0 = \mathscr{X}$ and for all $n > 0$, $U_{n+1} = U_n \bigcup \mathbb{I}(U_n)$.

Definition 3   (pluggable set) $\mathscr{X} \subset \mathbb{Z} \times \mathbb{N}$ is an *input/output pluggable set* if, letting t_0 be the least t such that $\exists x \, (x, t) \in \mathscr{X}$,

- There exists an **input point** $(x_0, t_0) \in \mathscr{X}$ such that $\mathscr{X} \cap \{(x, t) | x \in \mathbb{Z}, t > t_0\} \subset \mathbb{I}^\star(\{(x_0, t_0)\})$
- There exists an **output point** $(x_f, t_f) \in \mathscr{X}$ such that $\mathscr{X} \cap \{(x, t_f) | x \in \mathbb{Z}\} = \{(x_f, t_f)\}$
- $\mathscr{X} \cap \{(x, t) | x \in \mathbb{Z}\} \subset \mathbb{I}(\mathscr{X} \cap \{(x, t - 1) | x \in \mathbb{Z}\})$ for all t such that $t_0 < t \leq t_f$.

❷ *Figure 15* shows what a pluggable set looks like. One observes that, by definition, either there is a unique input point, or there are two input points (two direct neighbors or at distance 1) or there are three consecutive inputs points.

■ **Fig. 15**
A pluggable set. The two black points are the input and the output points of the set.

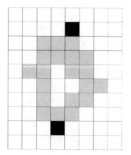

Definition 4 (signals) Given a cellular automaton $\mathcal{A} = (Q, \delta, \{-1, 0, +1\})$, $\mathcal{S} \subset \mathbb{Z} \times \mathbb{N}$ is a *signal* over $S \subset Q$, if

- $\mathcal{S}$ is indexed by an initial segment of $\mathbb{N}$. $\mathcal{S} = (x_k, t_k)_{k \in \{0, ..., \ell_{\mathcal{S}}\}}$ where $\ell_{\mathcal{S}}$ is either an integer or $+\infty$ and $t_{k+1} = t_k + 1$.
- $\langle x_k, t_k \rangle$, the state of the cell (x_k, t_k), verifies $\langle x_k, t_k \rangle \in S$.
- $\mathcal{S}$ is connected so that if $(x_k, t_k) \in \mathcal{S}$ either $x_{k+1} = x_k$, $x_{k+1} = x_k - 1$, or $x_{k+1} = x_k + 1$.
- If $(x, t) \notin \mathcal{S}$ and at least one of its eight neighbors $(x - 1, t)$, $(x + 1, t)$, $(x, t - 1)$, $(x, t + 1)$, $(x - 1, t - 1)$, $(x - 1, t + 1)$, $(x + 1, t - 1)$, or $(x + 1, t + 1)$ is in $\mathcal{S}$, then its state $\langle x, t \rangle$ is not in S.

The point (x_0, y_0) is the **initial point** of signal $\mathcal{S}$.

If $\ell_{\mathcal{S}}$ is finite, $(x_{\ell_{\mathcal{S}}}, y_{\ell_{\mathcal{S}}})$ is the **final point** of signal $\mathcal{S}$, which will be conveniently denoted by (x_f, y_f).

If $\ell_{\mathcal{S}}$ is finite, a signal is called a **left signal** (resp. **right signal**) if $x_f - x_0 < 0$ (resp. $x_f - x_0 > 0$).

It is worth noticing that from a unique configuration, it is possible to construct many signals defined over the same set of states.

Definition 5 (interacting signals) Signals $\mathcal{S}$ and $\mathcal{T}$ *interact* if

- $(x_{f_{\mathcal{T}}}, y_{f_{\mathcal{T}}}) = (x_{f_{\mathcal{S}}} + 2, y_{f_{\mathcal{S}}})$ and the interacting site is $(x_{f_{\mathcal{S}}} + 1, y_{f_{\mathcal{S}}} + 1)$.
- Or $(x_{f_{\mathcal{T}}}, y_{f_{\mathcal{T}}}) = (x_{f_{\mathcal{S}}} + 1, y_{f_{\mathcal{S}}})$ and the interacting sites are $(x_{f_{\mathcal{S}}}, y_{f_{\mathcal{S}}} + 1)$ and $(x_{f_{\mathcal{T}}}, y_{f_{\mathcal{T}}} + 1)$.
- Or $(x_{f_{\mathcal{T}}}, y_{f_{\mathcal{T}}}) = (x_{f_{\mathcal{S}}}, y_{f_{\mathcal{S}}})$ and the interacting site is $(x_{f_{\mathcal{S}}}, y_{f_{\mathcal{S}}} + 1)$.

The three cases are illustrated by ❷ *Fig. 16*.

Definition 6 (signals creation) Given a cellular automaton $\mathcal{A} = (Q, \delta, \{-1, 0, +1\})$, a signal $\mathcal{S}$ over S is created by a pluggable set $\mathcal{X} \subset \mathbb{Z} \times \mathbb{N}$ with respect to an output point $(x_{f_{\mathcal{X}}}, y_{f_{\mathcal{X}}})$ of $\mathcal{X}$, if $(x_{0_{\mathcal{S}}}, y_{0_{\mathcal{S}}})$ is in the immediate consequences of $(x_{f_{\mathcal{X}}}, y_{f_{\mathcal{X}}})$, that is, the input point of $\mathcal{S}$ is in the immediate consequence of the output point of $\mathcal{X}$.

Definition 7 (symbol encoding) A cellular automaton $\mathcal{A} = (Q, \delta, \{-1, 0, +1\})$ can encode symbols of alphabet A with μ channels if $Q = Q' \times (A \cap \{\varepsilon\})^{\mu}$ (with ε being a state used to denote the lack of encoding on the corresponding channel).

Definition 8 (grids) Given a cellular automaton $\mathcal{A} = (Q, \delta, \{-1, 0, +1\})$ and three disjoints subsets L, R, and W of Q, it can be said that $\mathcal{A}$ *constructs a grid* $\mathcal{G}_{L,R,W}$ if there are three integers x_{max}, h_{max} and t_{max} such that:

❐ **Fig. 16**
Possible signal interactions. The three different cases correspond to the three possible relative positions of the final points of the interacting signals.

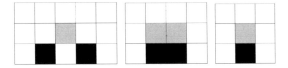

1. There exist pluggable sets $D_{(n,m)}$ of $\mathbb{Z} \times \mathbb{N}$ over W (with $n, m \in \mathbb{N}$) so that:
   (a) The input point $(x_{0_{(n+1,m)}}, t_{0_{(n+1,m)}})$ of $D_{(n+1,m)}$ is such that $x_{0_{(n+1,m)}} - x_{0_{(n,m)}} \geq x_{max}$ and $t_{0_{(n+1,m)}} - t_{0_{(n,m)}} \geq t_{max}$.
   (b) The input point $(x_{0_{(n,m+1)}}, t_{0_{(n,m+1)}})$ of $D_{(n,m+1)}$ is such that $x_{0_{(n,m)}} - x_{0_{(n,m+1)}} \geq x_{max}$ and $t_{0_{(n,m+1)}} - t_{0_{(n,m)}} \geq t_{max}$.
   (c) The output point $(x_{f_{(n,m)}}, t_{f_{(n,m)}})$ of $D_{(n,m)}$ is such that $|x_{f_{(n,m)}} - x_{0_{(n,m)}}| \leq x_{max}$ and $t_{f_{(n,m)}} - t_{0_{(n,m)}} \leq t_{max}$.
2. There exists an infinite family of *left* signals $\mathscr{L}_{(n,m)}$ over L, for n, m in $\mathbb{N}$, so that $\mathscr{L}_{(n,m)}$ is created by $D_{(n,m)}$. These signals are called *left moves*.
3. There exists an infinite family of *right* signals $\mathscr{R}_{(n,m)}$ over R, for n, m in $\mathbb{N}$, so that $\mathscr{R}_{(n,m)}$ is created by $D_{(n,m)}$. These signals are called *right moves*.
4. For n, m in $\mathbb{N}^\star$, $\mathscr{L}_{(n,m-1)}$ and $\mathscr{R}_{(n-1,m)}$ interact, and $(x_{0_{(n,m)}}, t_{0_{(n,m)}})$ the entry point of $D_{(n,m)}$ is one of the interacting sites (recall that there are one or two interacting sites).
5. *(Case of sites on the first right diagonal of the grid)* for $n \in \mathbb{N}^\star$, $(x_{0_{(n,0)}}, t_{0_{(n,0)}})$ is in the immediate consequences of $(x_{f_{\mathscr{R}_{(n,0)}}}, t_{f_{\mathscr{R}_{(n,0)}}})$.
6. *(Case of sites on the first left diagonal of the grid)* for $m \in \mathbb{N}^\star$, $(x_{0_{(0,m)}}, t_{0_{(0,m)}})$ is in the immediate consequences of $(x_{f_{\mathscr{L}_{(0,m)}}}, t_{f_{\mathscr{L}_{(0,m)}}})$.
7. Any left (resp. right) diagonal segment of a left signal $\mathscr{L}_{(0,i)}$ (resp. right signal $\mathscr{R}_{(i,0)}$) has length bounded by h_{max}.
8. The ending point of signal $\mathscr{L}_{(0,i)}$ (resp. $\mathscr{R}_{(i,0)}$) is on the left (resp. right) border of the left hull of $\mathscr{L}_{(0,i)}$ (resp. right hull of $\mathscr{R}_{(i,0)}$).

The initial point of $\mathscr{G}_{L,R,W}$ is $(x_{0_{(0,0)}}, t_{0_{(0,0)}})$ the initial point of $D_{(0,0)}$.

■ **Fig. 17**

A continuous geometric representation of a grid. Subsets $D_{(n,m)}$ are represented by polygons filled in gray, right moves are segments drawn in black and left ones in gray. $2x_{max}$ and t_{max} are, respectively, the width and height of enclosing rectangles in which $D_{(n,m)}$ are embedded.

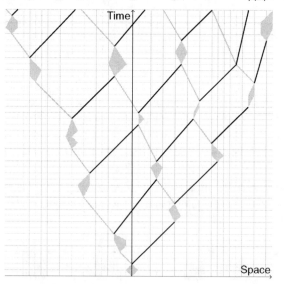

❯ *Figure 17* illustrates the geometrical representation of a grid. In the space–time diagram, $D_{(n,m)}$'s (indexed by $\mathbb{N} \times \mathbb{N}$) can be seen as *rough points*. Note that the size of these points is bounded due to conditions ❯ 1c of ❯ Definition 8 (this can be relaxed in a more general case).

One can imagine that a grid is a grillage, turned clockwise by $90°$ and then deformed. Signals $\mathscr{L}$ and $\mathscr{R}$ are wires that connect two neighboring cells in the grid. $\mathscr{L}$ connects cells with the same first coordinate, while $\mathscr{R}$ connects cells with the same second coordinate. ❯ *Figures 13a* and ❯ *14a, b* illustrate different grids with $x_{\max} = t_{\max} = 0$ (domains are reduced to their minimum: a simple interacting site).

Some intuition about the $D_{(n,m)}$'s is given. If such grids are considered as the support of some computation, the $D_{(n,m)}$'s are where local transition function computations occur. They are created by the interaction (as in ❯ Definition 5) of two signals $\mathscr{L}$ and $\mathscr{R}$ (item 4 of ❯ Definition 8), except those with a null coordinate that are created *ex-nihilo* from the initial configuration. If signals $\mathscr{L}_{(n,m-1)}$ and $\mathscr{R}_{(n-1,m)}$ carry values (see ❯ Definition 7) to the entry point of some $D_{(n,m)}$, then a computation can take place inside $D_{(n,m)}$, which produces a result that is finally encoded into its output point and transmitted to the neighbors through $\mathscr{L}_{(n,m+1)}$ and $\mathscr{R}_{(n+1,m)}$ that are created from the output point. The full computation is initiated on sites $D_{(n,0)}$ ($n \in \mathbb{N}$) and $D_{(0,m)}$ ($m \in \mathbb{N}$) and this initialization is not related to grids.

A grid can also be interpreted as a trellis cellular automaton where values $\cdots a_{-m} \cdots a_{-1} a_0 a_1 \cdots a_n \cdots$ are encoded (see ❯ Definition 7 below) as:

- a_i is encoded into states of signal $\mathscr{R}_{(i,0)}$, for $i \geq 0$.
- a_i is encoded into states of $\mathscr{L}_{(0,i-1)}$, for $i < 0$.

Thus, every $D_{(n,0)}$ (resp. $D_{(0,m)}$) gets a value encoded in $\mathscr{R}_{(n-1,0)}$ (resp. $\mathscr{L}_{(0,m-1)}$), and sends and encodes it in $\mathscr{L}_{(n,1)}$ (resp. in $\mathscr{R}_{(1,m)}$) but not in $\mathscr{R}_{(n,0)}$ (resp. $\mathscr{L}_{(0,n)}$).

Each $D_{(n,m)}$'s ($n, m \in \mathbb{N}^{\star}$) receives two values sent by $\mathscr{L}_{(n,m-1)}$ and $\mathscr{R}_{(n-1,m)}$, computes a new value in $D_{(n,m)}$ using the states of W and sends the result through $\mathscr{L}_{(n,m)}$ and $\mathscr{R}_{(n,m)}$.

Now, given the preceding considerations, the main proposition about grids is introduced. In the following, without loss of generality, cellular automata is considered with a fixed neighborhood $\{-1, 0, +1\}$.

Theorem 1 (CA simulation on grids) *Suppose given $\mathscr{B}$ and an initial configuration $c_{\mathscr{B}}$ which constructs a grid $\mathscr{G}$ in the sense of ❯ Definition 8. For every cellular automaton $\mathscr{A}$ there exists a CA $\mathscr{C}$ and a delay $d_{\mathscr{A}} \in \mathbb{N}$ such that for every initial configuration $c_{\mathscr{A}}$, there exists a configuration $c_{\mathscr{C}}$ so that automaton $\mathscr{C}$ on $c_{\mathscr{C}}$ constructs the grid $\mathscr{G}$ vertically translated by $d_{\mathscr{A}}$ on which output points of $D_{(n,m)}$ encode state $\langle n - m, \min\{n, m\} \rangle_{\mathscr{A}}$ of the space–time diagram of $\mathscr{A}$ on $c_{\mathscr{A}}$.*

Moreover, $\mathscr{C}$ depends on $\mathscr{A}$ and $\mathscr{B}$ (not on $c_{\mathscr{B}}$), and the maps $\mathscr{A}, \mathscr{B} \mapsto \mathscr{C}$, and $c_{\mathscr{A}}, c_{\mathscr{B}} \mapsto c_{\mathscr{C}}$ are computable.

Proof The key ideas are as follows:

1. The initial configuration of $\mathscr{A}$ is distributed on the initial configuration of the grid constructed by $\mathscr{B}$, so that the state of the output point of $D_{n,0}$ encodes state $\langle n, 0 \rangle_{\mathscr{A}}$, whereas that of $D_{0,n}$ encodes state $\langle -n, 0 \rangle_{\mathscr{A}}$ (see ❯ Fig. 18a). This necessitates to wedge $c_{\mathscr{A}}(0)$ with the entry point of $D_{0,0}$ (hence the delay $d_{\mathscr{A}}$), and then to appropriately distribute the remaining parts of $c_{\mathscr{A}}$ all along the borders of the grid.

◘ Fig. 18

(a) Distribution of configuration $c_{\mathcal{A}}$ on the grid generated by automaton $\mathcal{B}$. (b) How sites of $\mathcal{A}$ are located on domains D.

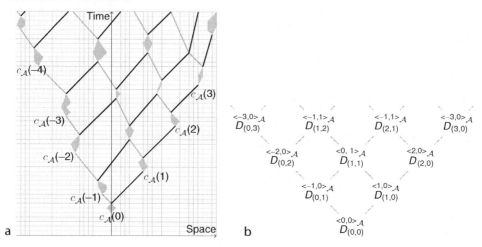

a

b

2. The simulation of automaton $\mathcal{A}$ with neighborhood $\{-1, 0, +1\}$ is done using an automaton computing on a grid generated by $\mathcal{B}$ on configuration $c_{\mathcal{B}}$, which is exactly the grid generated by the monoid of ❯ Eq. 9. In that simulation, the state $\langle n, t \rangle_{\mathcal{A}}$ (with $n \in \mathbb{N}$) is encoded into the state of the output point of $D_{(n+t,t)}$ whereas the state $\langle -n, t \rangle_{\mathcal{A}}$ is encoded into the state of the output point of $D_{(t,n+t)}$, see ❯ Fig. 18b.

These ideas will now be developed.

5.3.1 Wedging the Initial Configuration of $\mathcal{A}$

First, one must remark that the entry point $(x_{0_{(0,0)}}, y_{0_{(0,0)}})$ of $D_{0,0}$ can be easily characterized as the first point in the orbit of $\mathcal{B}$ that is in a state of W and that is not in the influence of any point in state in W, L, or R.

The value of $x_{0_{(0,0)}}$ is not known: the sole way to determine is to let automaton $\mathcal{B}$ run. This leads one to shift $c_{\mathcal{A}}$ in both left and right directions simultaneously. The speed of the shifts cannot be the maximum speed 1 as it would then be impossible to stop them when needed. Though any other speed <1 can be chosen, for the sake of simplicity, speed $\frac{1}{2}$ is used to shift the initial configuration of $\mathcal{A}$. Only the right shift construction is treated and assumed that it is the right shift of $c_{\mathcal{A}}(0)$ which eventually meets the vertical with abscissa $x_{0_{(0,0)}}$ at site $(x_{0_{(0,0)}}, y)$.

Consider the three possible cases:

- *Case $y = y_{0_{(0,0)}}$.* ❯ *Figure 19* illustrates this simple case and shows how the flow of data of the shift is stopped. When $c_{\mathcal{A}}(0)$ meets $(x_{0_{(0,0)}}, y_{0_{(0,0)}})$ two maximum speed opposite signals are launched. These signals then meet the flow of data and then stop their moves (this is why the shift cannot be done at maximum speed, otherwise it would be impossible to stop the moves). One can remark that this construction compresses the left part and sparses the right part of $c_{\mathcal{A}}$ (see comments of ❯ *Fig. 19*).

■ **Fig. 19**

Data shift and how this flow is stopped. For $i \in \mathbb{N}$ the $c_{\mathscr{A}}(i)$'s are distributed on one cell out of two: those with coordinates $x_{0_{(0,0)}} + 2i$. Also, $c_{\mathscr{A}}(-3i)$ is distributed on $x_{0_{(0,0)}} - 2i$ whereas $c_{\mathscr{A}}(-3i-1)$ and $c_{\mathscr{A}}(-3i-2)$ are both distributed on $x_{0_{(0,0)}} - 1 - 2i$. The width of vertical gray lines illustrates this property (thick lines represent two data on the same cell).

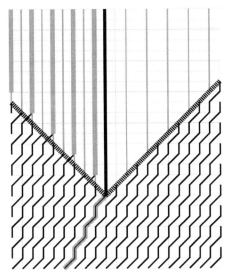

- *Case $y > y_{0_{(0,0)}}$.* In that case (see ❯ *Fig. 20a*), as cell $x_{0_{(0,0)}}$ has not met $c_{\mathscr{A}}(0)$, automaton $\mathscr{C}$ freezes the computation of $\mathscr{B}$ from time $y_{0_{(0,0)}}$ until $c_{\mathscr{A}}(0)$ is met. Then automaton $\mathscr{C}$ thaws the computation of $\mathscr{B}$.
- *Case $y < y_{0_{(0,0)}}$.* In that case (see ❯ *Fig. 20b*), cell $x_{0_{(0,0)}}$ has already met $c_{\mathscr{A}}(0)$. This means that the shift has gone too far on the right so that one has to shift it to the left. Then $\mathscr{C}$ freezes the computation of $\mathscr{B}$ from time $y_{0_{(0,0)}}$ until $c_{\mathscr{A}}(0)$ has gone back to cell $x_{0_{(0,0)}}$ at which time $\mathscr{C}$ thaws the computation of $\mathscr{B}$. In the additional left shift of $c_{\mathscr{A}}$, two cases occur:
  - if $i > 0$, the additional left shift of $c_{\mathscr{A}}(i)$ hits the right thawing signal, and the flow is simply straightened up.
  - if $i < 0$ there are two subcases. For $|i|$ big enough, when $c_{\mathscr{A}}(i)$ meets the left freezing signal, data are compressed as in case $y = y_{0_{(0,0)}}$. For $|i|$ small enough, the additional left shift of $c_{\mathscr{A}}(i)$ hits the vertical $x_{0_{(0,0)}}$. Then a new vertical is inserted which necessitates to push one step to the left all the already-built verticals.

The delay $d_{\mathscr{A}}$ mentioned in the proposition is exactly the time during which $\mathscr{C}$ freezes the computation of $\mathscr{B}$.

5.3.2 Distributing the Initial Configuration of $\mathscr{A}$

Now, the initial configuration, $c_{\mathscr{A}}$, has been lifted. And so one has to position the $c_{\mathscr{A}}(i)$'s and $c_{\mathscr{A}}(-i)$'s ($i \in \mathbb{N}$) in the appropriate input points of the $D_{(i,0)}$'s and $D_{(0,i)}$'s of the grid as illustrated in ❯ *Fig. 18a*.

5 Computations on Cellular Automata

■ Fig. 20

(a) $c_{\mathscr{A}}(0)$ passes after $(x_{0_{(0,0)}}, y_{0_{(0,0)}})$. In the gray part, the construction of the grid is frozen.

(b) $c_{\mathscr{A}}(0)$ passes before $(x_{0_{(0,0)}}, y_{0_{(0,0)}})$. In the gray part, the construction of the grid is frozen. $c_{\mathscr{A}}$'s are sent back appropriately.

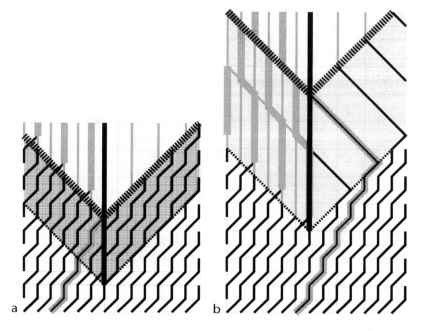

To do so, the idea is to make data follow the "hull" of the grid (cf. ❷ Definition 9 below). Left (resp. right) data must follow the leftmost (resp. rightmost) border of signals $\mathscr{L}_{(0,i)}$ (resp. $\mathscr{R}_{(i,0)}$) and of domains $D_{(0,i)}$ (resp. $D_{(i,0)}$). Thus the data for the left (resp. right) border never moves to the right (resp. left).

For clarity, the focus is only on the right side.

The distribution is as illustrated in ❷ Fig. 22. The distribution of the $c_{\mathscr{A}}(i+k)$'s ($k \geq 1$) on the right diagonal issued from the entry point of $D_{(i+1,0)}$ is the same as that on the diagonal issued from the entry point of $D_{(i,0)}$.

The idea is to push every $c_{\mathscr{A}}(i)$ to the right each time the right border of the right hull goes to the right. Thus, on a diagonal segment of length ℓ on the right border, up to 2ℓ distinct successive $c_{\mathscr{A}}(i)$'s have to be coded (recall that up to two values can be encoded into a single cell, cf. ❷ Fig. 19). To do so, a stack is necessary and the size of the stack is bounded by the longest diagonal segment of the border. This is a reason for the constraint in item 7 of ❷ Definition 8.

Definition 9 (hull of a set of sites) The hull of a set of sites:$\mathscr{X}$ is defined as the set of sites:

$$\{(x, t) | \exists t' \leq t, (x, t') \in \mathscr{X}\}$$

The hull can be characterized by a simple construction.

Now, for each diagonal segment of the right hull, one has to compute its length. This value is used all along the diagonal of the space–time diagram issued from the last point of the

□ **Fig. 21**

(a) Constructing the right hull of a set of sites and propagating the lengths of right diagonal segments. (b) Pushing $c_{\mathscr{A}}(i)$ to the right. For the sake of visibility, right 0 counters have been omitted.

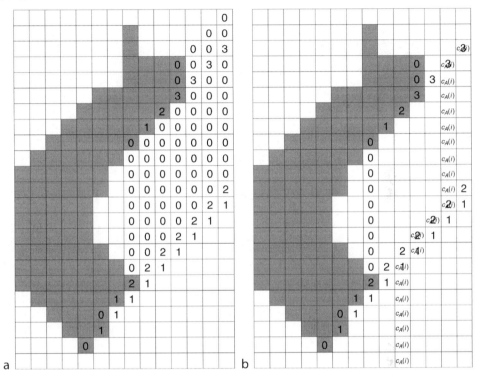

a   b

considered segment to shift every value of $c_{\mathscr{A}}$ (see ❯ *Fig. 21a*). Starting from the initial point of the grid a counter is incremented and moved right each time the grid expands to the right. When the grid does not expand to the right, the counter stops its incrementation and its value is broadcasted to the right; the counter is then reset and moved straight up until it meets the grid again (see ❯ *Fig. 21a*).

❯ *Figure 21b* illustrates how a value of $c_{\mathscr{A}}$ is shifted to follow the right border, the right hull of the grid.

❯ *Figure 22* shows how the full process takes place.

5.3.3 Simulation of $\mathscr{A}$ on the Grid

Recall that one wants the state $\langle i, t \rangle_{\mathscr{A}}$ to be encoded in the output point of $D_{(i+t,t)}$ and the state $\langle -i, t \rangle_{\mathscr{A}}$ on the output point of $D_{(t,i+t)}$, for $i \in \mathbb{N}$. Since the value $\langle j, t \rangle_{\mathscr{A}}$ is necessary to compute the values $\langle j-1, t+1 \rangle_{\mathscr{A}}, \langle j, t+1 \rangle_{\mathscr{A}}$, and $\langle j+1, t+1 \rangle_{\mathscr{A}}$ ($j \in \mathbb{Z}$), one has to show that one can carry this value $\langle j, t \rangle_{\mathscr{A}}$ from the output point of the domain which encodes it

◘ **Fig. 22**

How the right hull of a grid is built and how values of the initial configuration of $\mathscr{A}$ are shifted to follow the shape of the hull as closely as possible and how they are injected into appropriate domains. Only the right part of the process is illustrated for convenient reasons. $c_{\mathscr{A}}(i)$ is injected into $D_{(0,i)}$. $c_{\mathscr{A}}(i+1)$ is used to build the right hull and is then injected into $D_{(0,i+1)}$. $c_{\mathscr{A}}(i+k)$ (for $k \geq 2$) are shifted to the right according to the path constructed by $c_{\mathscr{A}}(i+1)$. One can easily show that if the longest diagonal segment followed by $c_{\mathscr{A}}(i+1)$ is of length ℓ, at most ℓ values are stacked on the diagonals issued from the hull.

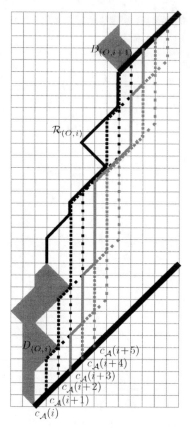

(namely, $D_{(j+t,t)}$ if $j > 0$, $D_{(t,|j|+t)}$ if $j < 0$, or $D_{(t,t)}$ if $j = 0$) to the input points of the appropriate domains. This leads to the following three cases:

- *Case* $j > 0$. Let $\langle j, t \rangle_{\mathscr{A}}$ follow signal $\mathscr{L}_{(j+t,t)}$, travel through $D_{(j+t,t+1)}$, then follow signal $\mathscr{R}_{(j+t,t+1)}$, travel through $D_{(j+t+1,t+1)}$, then follow $\mathscr{R}_{(j+t+1,t+1)}$. This allows us to carry $\langle j, t \rangle_{\mathscr{A}}$ onto the input points of $D_{(j+t,t+1)}$, $D_{(j+t+1,t+1)}$, and $D_{(j+t+2,t+1)}$. With the notations in the end of ❷ Sect. 5.1, this is done following the three paths α, $\alpha\beta$, and $\alpha\beta\beta$ of the dependencies Cayley graph.
- *Case* $j < 0$. This case is illustrated by ❷ *Fig. 23*. Let $\langle j, t \rangle_{\mathscr{A}}$ follow signal $\mathscr{R}_{(t,|j|+t)}$, travel through $D_{(t+1,|j|+t)}$, then follow signal $\mathscr{L}_{(t+1,|j|+t)}$, travel through $D_{(t+1,|j|+t+1)}$, then follow

◘ Fig. 23
How the computation of the simulated CA is embedded into a grid. The figure only shows case
$j < 0$. **Except for the central cell, the state of a simulated cell is broadcasted through the right**
output wire, then to the two following left wires (see the black dashed line for example). This
carries the state to the three corresponding "neighbors" at next time step.

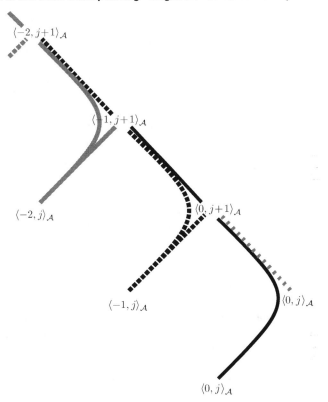

$\mathscr{L}_{(t+1,|j|+t+1)}$. This allows us to carry $\langle j, t \rangle_{\mathscr{A}}$ onto the input points of $D_{(t+1,|j|+t)}$, $D_{(t+1,|j|+t+1)}$, and $D_{(t+1,|j|+t+2)}$. With the notations in the end of ❯ Sect. 5.1, this is done following the three paths β, $\beta\alpha$, and $\beta\alpha\alpha$ of the dependencies Cayley graph.

- *Case* $j = 0$. This is a particular case since the left and right neighbors are on two different diagonals. Let $\langle 0, t \rangle_{\mathscr{A}}$ follow signal $\mathscr{R}_{(t,t)}$, travel through $D_{(t+1,t)}$, then follow signal $\mathscr{L}_{(t+1,t)}$, travel through $D_{(t+1,t+1)}$, then at this point, the value is distributed on the two opposite directions, namely: $\mathscr{L}_{(t+1,t+1)}$ and $\mathscr{R}_{(t+1,t+1)}$. This allows us to carry $\langle 0, t \rangle_{\mathscr{A}}$ onto the input points of $D_{(t+1,t+1)}$, $D_{(t+2,t+1)}$, and $D_{(t+1,t+2)}$. With the notations in the end of ❯ Sect. 5.1, this is done following the three paths $\beta\alpha$, $\beta\alpha\alpha$, and $\beta\alpha\beta$ of the dependencies Cayley graph.

Now, every domain receives the three needed states at its input point and can compute the transition function of automaton $\mathscr{A}$ and encode the resulting value at its output point.

6   Beyond Grids

As there exists many kinds of intrinsic simulations of CAs by CAs, one can ask what makes such simulation via grids especially interesting. The answer is that this construction is very flexible. The authors think that it is closer to an hypothetical, massively parallel programming tool. The theorem in itself does not give much more power to compute, but can be used as a basic tool to understand how recursive schemas can be implemented. Some arguments in favor of this thesis is given.

6.1   Going Back to the Multiplication: Composition

Go back and start again with a simple algorithm: the multiplication. Recall that at the beginning of this chapter, it was observed that a multiplication algorithm naturally leads to a cellular automaton synthesis, but that such a construction is not always (or at least not easily) composable with itself. Note that inputs and outputs are a recurrent problem in computer science.

Now, in the light of grid computation, a multiplication is simply a computation taking place on a grid. No matter what the real shape of that grid is, provided that the grid has some necessary properties, the computation can take place on it. In other words, it can be said that if the grid realizes the dependency graph, the computation is doable within it.

From ❷ *Fig. 24a, b*, one can show that it could be possible to view the multiplication as a process taking place on a grid, with inputs given onto the first diagonal and the

■ **Fig. 24**
(a) Simple multiplication algorithm of 14 by 23. This figure illustrates the main flows of bits of the multiplicand (gray flows) and the multiplier (black flows). Details are omitted. (b) The same algorithm implemented as a trellis. It is clear that every single vertical move has been replaced by a right move followed by a left move.

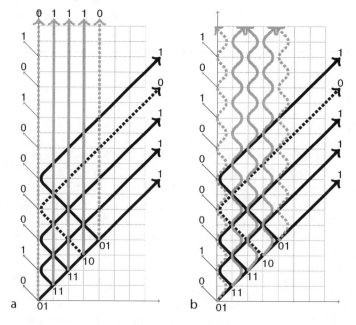

result produced onto the main vertical line of the grid (the left part of the grid is of no usage).

Now, one can bend the grid as wanted, provided that the bending leads to a constructible grid. ❷ *Figure 25a, b* show two different possible bendings.

It is easy to show that the two bendings are compatible, that is, that the two functions can be combined to provide a multiplication of three numbers. The two first are multiplied together, and the third is injected to the output location of the result to make them the input of the second multiplication. This implements the fact that to compute $n_1 \times n_2 \times n_3$, it is sufficient to have a binary multiplier and to combine it to compute $(n_1 \times n_2) \times n_3$.

Remark that the former construction is such that two grids linked together were constructed so that the right move of the second grid, $\beta_{\mathcal{G}_2}$, is exactly the vertical move of the first grid, $\beta_{\mathcal{G}_2} = \beta_{\mathcal{G}_1} \alpha_{\mathcal{G}_1}$. The exact values of these parameters relatively to the underlying space–time diagram are left to the implementation. In the provided example, they have been chosen so that the computation time is minimized, and so the result is produced on the first cell of the underlying space–time diagram.

This can be extended as far as required. ❷ *Figure 26a* illustrates the case where three grids are fitted together. The first, $\mathcal{G}_1$, has $\alpha_{\mathcal{G}_1} = \alpha$ and $\beta_{\mathcal{G}_1} = \beta^3$ (so that the vertical move is $v_{\mathcal{G}_1} = \beta^3 \alpha$). The second, $\mathcal{G}_2$, has $\alpha_{\mathcal{G}_2} = \alpha$ and $\beta_{\mathcal{G}_2} = v_{\mathcal{G}_1} = \beta^3 \alpha$ (so that the vertical move is $v_{\mathcal{G}_2} = \beta^3 \alpha^2$). The third, $\mathcal{G}_3$, has $\alpha_{\mathcal{G}_3} = \alpha$ and $\beta_{\mathcal{G}_3} = v_{\mathcal{G}_2} = \beta^3 \alpha^2$ (so that the vertical move

◻ **Fig. 25**
(a) Bended grid of ❷ Fig. 24b. A right move is obtained by two right moves ($\beta_{\text{bended}} = \beta^2$).
A left move is a simple left move ($\alpha_{\text{bended}} = \alpha$). (b) Bended grid of ❷ Fig. 24b. A right move is
obtained by two right moves and a left move ($\beta_{\text{bended}} = \beta^2 \alpha$). A left move is a simple left move
($\alpha_{\text{bended}} = \alpha$).

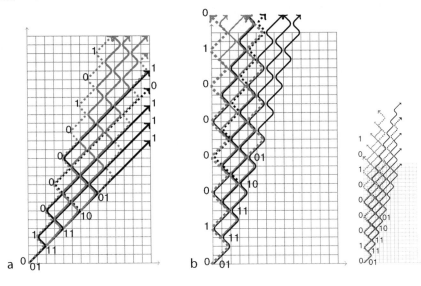

◘ Fig. 26

(a) Three fitted grids to compute the composition of three functions. For example, one can compute the product of four numbers as $(((n_1 \cdot n_2) \cdot n_3) \cdot n_4)$. (b) Three grids to compute a tree of three functions: two interleaved grids are fitted to the third. For example, one can compute the product of four numbers as $((n_1 \cdot n_2) \cdot (n_3 \cdot n_4))$.

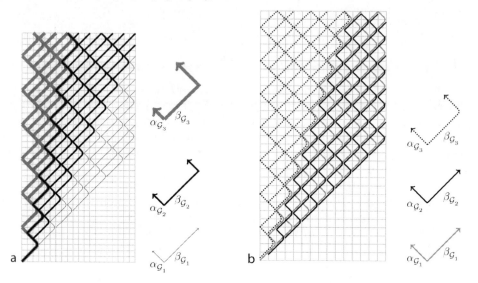

is $v_{\mathcal{G}_3} = \beta^3 \alpha^3$). Details such as input locations are left to the reader. Observe that to multiply n numbers, it is sufficient to have $n - 1$ grids, and that the bit p of the result is produced at time $2(n - 1)p$.

❷ *Figure 26b* illustrates how parallelism can be really achieved to compute the product of 2^n numbers more efficiently than the previous method. The idea is to decompose the product as a tree of computations. Each level of the tree is implemented by interleaving the necessary grids (2^m numbers leads to 2^{m-1} products computed by 2^{m-1} grids).

The reader is referred to Mazoyer (1987) to see how an infinite family of signals can be built and used to split the space–time diagram in a regular manner.

6.2 Achieving Recursion

Now one can ask how recursion can be achieved using such constructions. The answer is (at least theoretically) very simple: it is sufficient to view a domain as a grid. Since domains are defined as sets of sites, there can be room to do computation within them. For example, one can construct a grid inside a domain. Now to implement such a construction, one needs to be very careful, as there are many difficulties involved. ❷ *Figure 27* illustrates how a function call can be implemented.

▫ Fig. 27
Implementing a function call. A domain is used to compute a function. As no one can know how long that *internal* computation will take, the calling process is frozen and then thawed appropriately. Black domains are *internal* call domains. The gray hull is here to represent the *external* view of the domain.

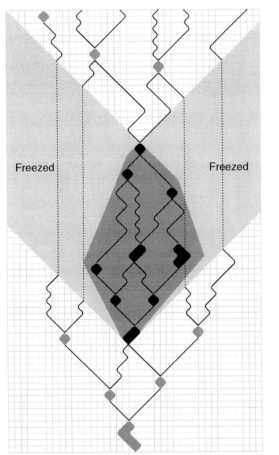

References

Adamatzky A (2002) Collision-based computing. Springer, London. ISBN 978-1852335403

Atrubin AJ (1965) A one-dimensional real-time iterative multiplier. IEEE Trans Electron Comput 14:394–399

Choffrut C, Čulik K II (1984) On real-time cellular automata and trellis automata. Acta Inform 21:393–407

Cole SN (1969) Real-time computation by n-dimensional iterative arrays of finite-states machines. IEEE Trans Comput C-18(4):349–365

Čulik K II, Gruska J, Salomaa A (1982) Systolic trellis automata for VSLI on balanced trees. Acta Inform 18:335–344

Delorme M, Mazoyer J (1999) Cellular automata: A parallel model. Mathematics and its applications, vol 460. Kluwer, Dordrecht. ISBN 0-7923-5493-1

Even S (1991) Systolic modular multiplication. In: Advances in cryptology: CRYPT0'90. Lecture notes on computer science, vol 537. Springer, New York, pp 620–624

Fischer PC (1965) Generation of primes by a one-dimensional real-time iterative array. J ACM 12(3):388–394

Goyal LN (1976) A note on Atrubin's real-time iterative multiplier. IEEE Trans Electron Comput C25(5):546–548

Knuth DE (1997) Seminumerical algorithms, the art of computer programming. vol 2, 2nd edn. Addison-Wesley, Reading, MA. ISBN 0-201-03822-6

Korec I (1997) Real-time generation of primes by a one-dimensional cellular automaton with 9 States. Preprint Series of Mathematical Institute of Slovak Academy of Sciences, Bratislava, Preprint 13/1997

Mazoyer J (1987) A six states minimal time solution to the firing squad synchronization problem. Theor Comput Sci 50:183–328

Mazoyer J, Terrier V (1999) Signals in one-dimensional cellular automata. Theor Comput Sci 217(1):53–80 DOI 10.1016/S0304-3975(98)00150-9

McNaughton R (1961) The theory of automata, a survey. Adv Comput 2:379–421

Poupet V (2005) Cellular automata: Real-time equivalence between one-dimensional neighborhoods. In: STACS 2005. Proceedings of the 22nd annual symposium on theoretical aspects of computer science, Stuttgart, Germany, February 24–26, 2005 pp 133–144

Wolfram S (2002) A new kind of science. Wolfram Media, Champaign, IL

6 Universalities in Cellular Automata*

Nicolas Ollinger
Laboratoire d'informatique fondamentale de Marseille (LIF),
Aix-Marseille Université, CNRS, Marseille, France
nicolas.ollinger@lif.univ-mrs.fr

*A preliminary version of this work appeared under the title *Universalities in Cellular Automata: A (Short) Survey* in the proceedings of the first edition of the JAC conference (Ollinger 2008).

G. Rozenberg et al. (eds.), *Handbook of Natural Computing*, DOI 10.1007/978-3-540-92910-9_6,
© Springer-Verlag Berlin Heidelberg 2012

Abstract

This chapter is dedicated to computational universalities in cellular automata, essentially Turing universality, the ability to compute any recursive function, and intrinsic universality, the ability to simulate any other cellular automaton. Constructions of Boolean circuits simulation in the two-dimensional case are explained in detail to achieve both kinds of universality. A detailed chronology of seminal papers is given, followed by a brief discussion of the formalization of universalities. The more difficult one-dimensional case is then discussed. Seminal universal cellular automata and encoding techniques are presented in both dimensions.

1   Introduction

Universality is a key ingredient of the theory of computing, grounding its roots in the work of Gödel (1931), that describes the ability for some formal systems to manipulate themselves through proper encodings. By applying such an encoding to machines, Turing (1936) introduces a universal machine with the capacity to simulate every other machine of its class. Stated in an axiomatic way, an acceptable enumeration (Odifreddi 1989) of partial recursive functions, $\varphi_n : \mathbb{N} \to \mathbb{N}$, is endowed with a bijective and recursive bracket function, $\langle\rangle : \mathbb{N}^2 \to \mathbb{N}$, for which there exists u such that, for all pairs of integers $m, n \in \mathbb{N}$:

$$\varphi_u(\langle m, n \rangle) \simeq \varphi_m(n)$$

where $x \simeq y$ if either both x and y are undefined, or $x = y$. Notice that this does not define a property of being universal but rather provides one universal function. Given a machine in a well-defined family admitting a universal machine, a natural question is to identify the computational power of the machine: what can it compute, up to some admissible encoding? Is it universal, in the sense that it can compute every computable function? As such, universality is a first criterion to discriminate among machines. For Turing machines, a proper formal definition of universality that captures the intuition of what is universal and what is not and encompasses previous commonly admitted constructions remains to be found. That does not stop people from searching and designing very clever small universal machines.

As Turing machines, cellular automata are machines, in the sense that they are described as finite state machines interacting locally with an environment, receiving finite local information from the environment as inputs and interacting locally with the environment through finite outputs. Whereas a Turing machine consists of just one finite state machine behaving sequentially, a cellular automaton consists of an infinite regular grid of copies of a same finite state machine interacting synchronously, uniformly, and locally. Investigating the respective computational power of cellular automata leads to consideration of universality.

The idea and construction of a universal cellular automaton is as old as the formal study of the object itself, starting with the work of von Neumann (1966) on self-reproduction in the 1940s, using cellular automata under suggestions by Ulam. Following the work of Turing, a Turing universal cellular automaton is an automaton encompassing the whole computational power of the class of Turing machines, or by the so-called Church–Turing thesis the class of recursive functions. To encode complex behaviors in a cellular automaton's dynamics, one can describe how to encode any computing device of a universal class of machines (Turing machine, tag systems, etc.) and use classical tools of computability theory to shape wanted behaviors of the object. This is basically what von Neumann did. He designed a cellular automaton able to encode any Turing machine, the machine being moreover equipped with a construction arm controlled by the machine's head.

But Turing universality is not the only reasonable kind of universality one might expect from cellular automata. It is quite unnatural to consider a universality of highly parallel potentially infinite devices as cellular automata by simulation of the dynamics of sequential finite machines – indeed, as we will discuss, to give a both widely acceptable yet precise definition of Turing universality is a very difficult and unfulfilled challenge. As the study of cellular automata shifted both to dimension 1 and to the study of its dynamics, another kind of universality emerged. An intrinsically universal cellular automaton is an automaton able to properly simulate the behavior of any other cellular automaton on any type of configuration (might it be infinite). It turns out that most of the historical constructions in dimension 2 and more, designed as Turing universality, are intrinsically universal by the simple fact that they are designed to encode any Boolean circuit.

A formal definition of universality might not seem so important. In fact, when building a precise cellular automaton from scratch, to be universal, a definition is often implicit: the obtained behavior is the one engineered by the designer. The definition turns out to be required more when proceeding by analysis: given a cellular automaton rule, is it universal?

The present chapter is constructed as follows. ❷ Section 2 gently introduces universalities through the study of an original two-dimensional simple universal cellular automaton using Boolean circuits encoding. ❷ Section 3 provides the formal definitions and notations used for cellular automata, configurations, and dynamics. ❷ Section 4 is an annotated chronology of seminal papers concerning universality and universal cellular automata. ❷ Section 5 discusses the right definition of universalities in cellular automata. ❷ Section 6 discusses the construction and analysis of universal cellular automata in dimensions 2 and more, mostly using Boolean circuits simulation. ❷ Section 7 discusses Turing universality, its links with universal Turing machines, and the main techniques of construction. ❷ Section 8 discusses intrinsic universality and the main techniques of construction. ❷ Section 9 discusses universality in the special restricted case of reversible cellular automata.

Remark Cellular automata are dynamical objects; the figures in this chapter are static. For a better experience, the reader is strongly encouraged to complete his reading by experimenting with some reasonable cellular automata simulator software. While writing this chapter, the author enjoyed using Golly (2008), both to verify ideas and produce pictures for the two-dimensional automata.

2 Copper: A Simple Universal Cellular Automaton

Before discussing formal definitions of different kinds of universalities, an example of a simple universal cellular automaton is considered. The automaton that is considered is called *Copper*, abbreviated as **Cu**. It is modeled after classical universal 2D rules, in the spirit of (albeit different to) the automata of Banks (1970, 1971).

2.1 Rule of the Game

Configurations of **Cu** consist of an infinite two-dimensional grid of cells. Each cell is in one of three possible states: 0, 1, or 2. Thus, a configuration can be described as a matrix of states, as

◻ **Fig. 1**

Cu configuration drawing. (a) Configuration. (b) Symbolic representation.

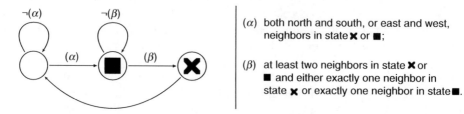

1	1	1	1	1	1	1	0	0
0	2	0	0	0	1	0	0	0
0	0	0	0	0	1	0	0	0
0	1	1	1	1	1	1	1	1
0	1	0	0	2	0	0	1	0
0	1	0	0	0	0	0	1	0
1	1	1	1	1	1	1	1	1

a   b

◻ **Fig. 2**

Cu local rule: evolution of a cell state depending on its neighbors.

(α) both north and south, or east and west, neighbors in state ✖ or ■;

(β) at least two neighbors in state ✖ or ■ and either exactly one neighbor in state ✖ or exactly one neighbor in state ■.

depicted in ❯ *Fig. 1a*. For the sake of readability, a symbolic representation inspired by the usage of states is preferred, depicting state 0 as *void*, state 1 as *wire*, and state 2 as *particle*, as depicted in ❯ *Fig. 1b*.

The world of **Cu** is equipped with a global clock controlling the evolution of configurations. At each time step, all cells change their states synchronously according to a same *local rule*. The local rule consists of looking at the states of the four cardinal neighbors (north, east, south, and west) to choose the new state of a cell. The local rule of **Cu**, invariant by rotation, is explained in ❯ *Fig. 2* and reads as follows. Cells in void state stay in a void state but enter a wire state if two opposite neighbors (north and south or west and east) are both not in void state. Cells in a wire state stay in the wire state but enter a particle state if the number of neighbors not in the void state is greater than or equal to two, and there is exactly one neighbor in the wire state or in the particle state. Cells in the particle state always enter the void state.

2.2   Particles on Wires

The rule of **Cu** is designed so that properly constructed configuration patches behave as networks of wires on which particles move, interact, and vanish. A *wire* consists of a straight line of wire cells of width 1 bordered by void states. For a particle to move on a wire, it should be oriented. Orientation is obtained by encoding *moving particles* on a wire as a pair of a particle state and a void state. The direction of the movement is given by the particle state as depicted in ❯ *Fig. 3*.

When a particle arrives at the end of a wire, a *dead end*, the particle vanishes, as depicted in ❯ *Fig. 4*. If two particles arrive from opposite directions on a same wire and *collide*, both particles vanish, as depicted in ❯ *Fig. 5*. Notice that the collisions should happen with odd spacing, otherwise the wire would be damaged.

■ Fig. 3

Cu particle moving on a wire from left to right.

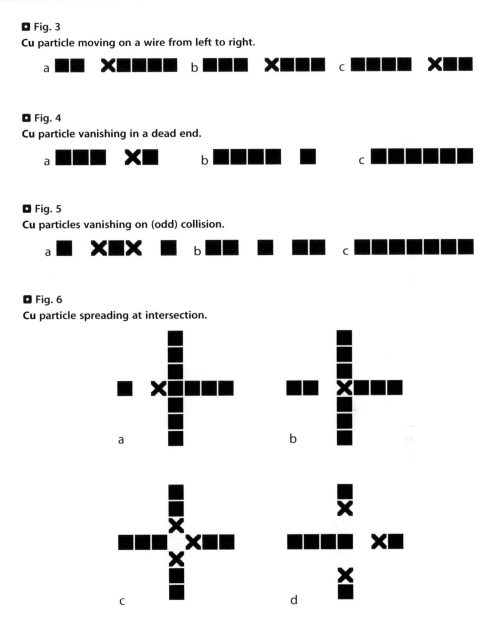

■ Fig. 4

Cu particle vanishing in a dead end.

■ Fig. 5

Cu particles vanishing on (odd) collision.

■ Fig. 6

Cu particle spreading at intersection.

Wires can be connected at *intersection points* where three or four wires meet, The connection of two wires is possible but does not behave in an interesting way from the particle point of view (a particle destroys a two-wire connection when it moves through it), it can be replaced by a three-wire intersection plus a dead end. At an intersection, a single particle entering the intersection generates a leaving particle on every other end, as depicted in ❷ *Fig. 6*. Symmetrically, when particles arrive synchronously from every other end at an intersection, they join as a single leaving particle on the empty end, as depicted in ❷ *Fig. 7*. Every other scenarios of particles entering an intersection leads to no leaving particle, as depicted in ❷ *Fig. 8*.

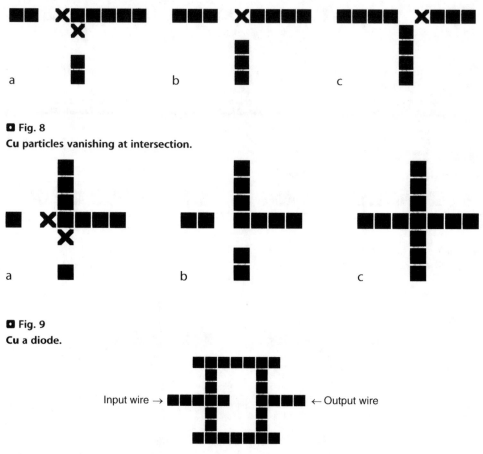

◘ Fig. 7
Cu particles joining at intersection.

a   b   c

◘ Fig. 8
Cu particles vanishing at intersection.

a   b   c

◘ Fig. 9
Cu a diode.

Input wire →   ← Output wire

2.3   Computing Boolean Functions

Using particles and networks of wires makes it possible to encode bits and gates to encode every Boolean function. This is done by constructing and combining widgets, exploiting properties of intersections. Each widget connects input wires to output wires. By entering synchronized particles on the input wires, after some *traversal delay*, synchronized leaving particles appear on output wires. Moreover, each widget has a proper *safety delay* ensuring that if inputs are spaced by at least this delay, the widget behaves according to its specification (the safety delay can be smaller or bigger than the traversal delay). The synchronization of particles is obtained by adding delays on the path by introducing turns on wires using intersections.

A first widget is the *diode*. A diode is a widget computing the identity function from an input wire to an output wire, with the special property that if a particle improperly enters the output wire (respecting the safety delay), it is guaranteed that no particle will leave the input wire. Combining a four-wire intersection and a three-wire intersection, one obtains the diode depicted in ❷ *Fig. 9.*

Combining two diodes with an intersection provides two elementary widgets computing an OR gate or a XOR gate, as depicted in ➋ *Fig. 10*. The diodes ensure that no particles leave the input wires and the intersection combines the input particles into an output particle.

Planar acyclic circuits consisting of fan-outs, OR, and XOR gates are known to be able to compute any Boolean function, using a proper encoding of Boolean values. For the sake of completeness, a description is given on how this can be used in **Cu** to construct wire crossing, NOT, AND, and OR gated. Planar crossing is obtained using two fan-outs and three XOR gates, as depicted in ➋ *Fig. 11* and implemented more compactly in **Cu** as depicted in ➋ *Fig. 12*. All monotone Boolean functions are computable using OR and XOR gates because $A \wedge B = (A \vee B) \oplus (A \oplus B)$. To encode all Boolean functions with monotone functions, a classical trick is to encode each Boolean value A as the pair of values $(A, \bar{A})$. Using this encoding, Boolean values are represented on two wires: no particle means no value, values are encoded by exactly one particle on the corresponding wire. A Boolean NOT gate is encoded as a wire crossing, implemented in **Cu** as depicted in ➋ *Fig. 12*. A Boolean AND gate with inputs $(A, \bar{A})$ and $(B, \bar{B})$ outputs $(A \wedge B, \bar{A} \vee \bar{B})$ is implemented more compactly in **Cu** as depicted in ➋ *Fig. 13*. A Boolean OR gate is encoded symmetrically, outputing $(A \vee B, \bar{A} \wedge \bar{B})$. Notice that, using this construction scheme, one can encode every Boolean function into a widget with a traversal delay that depends on the size of the associated circuit but with a safety delay that can be uniformly bounded by the maximal safety delay of the elementary widgets composing the circuit. The safety delay does not depend on the computed function.

Remark Different encodings are possible. For example, a more compact encoding of a Boolean function $f: \{0, 1\}^k \rightarrow \{0, 1\}^l$ adds one input and one output with constant values 1. Using this encoding, NOT maps $(1, A)$ into $(1, 1 \oplus A)$, AND maps $(1, A, B)$ into $(1, A \wedge B)$, and OR maps $(1, A, B)$ into $(1, A \vee B)$. The constant value 1 acts like a global constant, an asynchronous clock signal indicating that a computation is occurring. This solution provides simpler gates at

◻ Fig. 10
Cu elementary particle gates. (a) OR gate. (b) XOR gate.

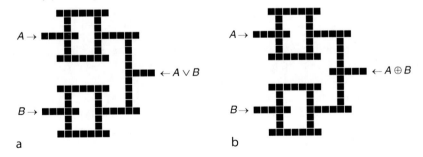

a   b

◻ Fig. 11
Planar crossing with fan-outs and XOR gates.

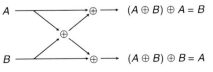

▪ Fig. 12
Cu particles crossing.

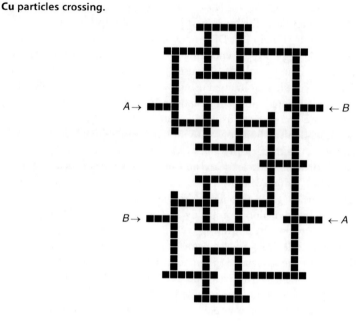

the cost of more complicated routing problems to route and synchronize the constant to each NOT gate.

 Remark Different gates are possible. For example, a more compact logic can be derived from the compact diode depicted in ❷ *Fig. 14*. Sample OR and XOR gates are depicted in ❷ *Fig. 15a, b*. In this case, better delays are obtained at the cost of more complicated routing and synchronization issues.

2.4 Encoding Finite State Machines

Using Boolean functions make it possible to encode finite state machines. A finite state machine can be seen as a map $f : S \times I \to S \times O$ where S is a finite set of states, I is a finite set of input values, and O is a finite set of output values. At each time step, reading input i, the machine evolves from states s to state s' and outputs o where $f(s, i) = (s', o)$. Once a binary encoding is chosen for S, I, and O, f can be seen as a Boolean function. The finite state machine is encoded by looping the S bits from the output to the input of the corresponding encoding of the Boolean function, as depicted in ❷ *Fig. 16*.

 Some details should be considered carefully. Each finite state machine constructed this way has a proper clocking: it is the traversal time of the Boolean function plus the time for the state to go from the output back to the input of the circuit. By construction, the circuit is synchronized on this clocking: inputs and outputs are considered synchronized on the circuits and no garbage signal is produced in between two heartbeats on any outside wire (state, input, and output). For this to work, the clocking has to be larger than the safety delay of the Boolean function (this can be fixed by adding extra delays if required). Notice that the input and output signals are synchronized with the circuit, so a direct binary encoding on wires is possible.

Fig. 13
Cu Boolean AND gate.

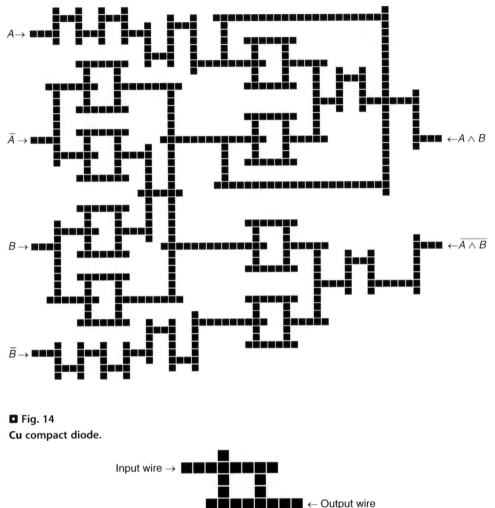

Fig. 14
Cu compact diode.

To handle complex computations, finite state machines have to be connected via input and output wires to their environment, encoded as other dedicated finite state machines. The easiest way to connect two finite state machines is to have them share the same clocking. When considering infinite sequences of finite state machines, this might lead to unbounded clocking. To avoid such pitfalls, one might insert in between the two computing finite state machines a simple memory finite state machine with a small fixed clocking, τ, independent of the number of wires. By ensuring that the memory does not send signals on output wires without being ordered to do so on input wires, and by fixing the clocking of the two computing finite state machine to be a multiple of τ, as depicted in ● *Fig. 17*, one obtains better clockings.

Remark As the interest here is only in the computational potentiality of cellular automata, such a naive encoding is sufficient. For efficient implementations, one might consider efficient **Cu**

◘ Fig. 15

Cu compact elementary gates. (a) Compact OR gate. (b) Compact XOR gate.

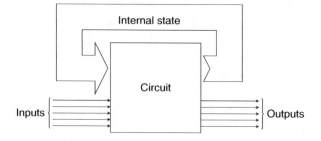

a B b B

◘ Fig. 16

Encoding a finite state machine with a Boolean circuit.

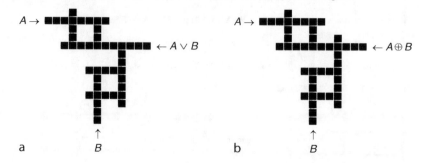

◘ Fig. 17

Synchronization using memory.

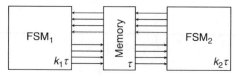

encodings of multiplexers, demultiplexers, memory lattices, busses, etc. – mimicking classical circuit design.

2.5 Turing Universality

Using finite state machines makes it possible to achieve Turing universality by encoding classical computational models. One possibility is to consider Turing machines (a finite state machine plus a tape, represented as a biinfinite sequence of finite state machines encoding tape cells). Another possibility is to encode a Minsky machine: a finite state machine equipped with two counters. On each counter, the machine can increment the counter, decrement the counter if it is not zero, and test if the counter value is zero or positive. Such a counter can be encoded as an infinite sequence of identical finite state machines. The complete encoding is depicted symbolically in ❯ *Fig. 18.*

Given a Minsky machine, one can recursively construct an ultimately periodic configuration of **Cu** that simulates the behavior of the machine. Thus, **Cu** is Turing universal, it can compute every recursive function.

Remark In this chapter, the use of infinite configurations is allowed, provided that the configurations are ultimately periodic. After all, what is void? Is it really empty or is it somehow regular? It is possible to consider universality in the restricted case of finite configurations, the universal cellular automata have a few more states. The notion of intrinsic universality requires ultimately periodic configurations.

2.6 Intrinsic Universality

Using finite state machines makes it possible to achieve a simpler yet stronger form of universality: intrinsic universality, that is the ability for **Cu** to simulate every two-dimensional cellular automaton. The idea is simple: encode the local rule of a given cellular automaton as a finite state machine and put infinitely many copies of that same machine on a regular grid, using wires to synchronously connect outputs to the proper neighbor inputs, as depicted in ❯ *Fig. 19.*

Given a cellular automaton, one can recursively find a delay T (the clocking of the finite state machine) and associate to each state s of the automaton a square patch of configuration $\tau(s)$ of **Cu** (the finite state machine with s encoded on its wires) so that for each configuration x of the automaton, the configuration $\tau(x)$ of **Cu** evolves in T steps into $\tau(y)$ where y is the image configuration of x by the encoded automaton.

3 Definitions

It is now time to associate formal definitions to the objects.

A *cellular automaton* $\mathscr{A}$ is a tuple (d, S, N, f) where d is the *dimension* of space; S is a finite set of *states*; N, a finite subset of $\mathbb{Z}^d$, is the *neighborhood*; and $f: S^N \to S$ is the *local rule*, or *transition function*, of the automaton. A *configuration* of a cellular automaton is a coloring of

◻ **Fig. 18**
Two-counters machine encoding.

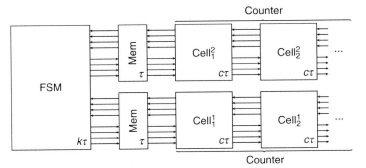

■ Fig. 19

Intrinsic universality encoding.

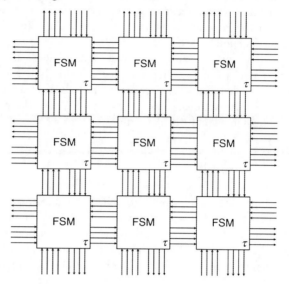

■ Fig. 20

Partial space–time diagram of the $(\mathbb{Z}_2, +)$ rule.

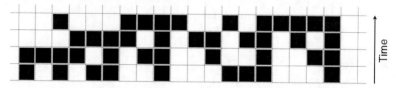

the space by S, an element of $S^{\mathbb{Z}^d}$. The *global rule* $G : S^{\mathbb{Z}^d} \to S^{\mathbb{Z}^d}$ of a cellular automaton maps a configuration $c \in S^{\mathbb{Z}^d}$ to the configuration $G(c)$ obtained by applying f uniformly in each cell: for all position $z \in \mathbb{Z}^d$, $G(c)(z) = f(c(z + v_1), \ldots, c(z + v_k))$ where $N = \{v_1, \ldots, v_k\}$. A *space–time diagram* of a given cellular automaton is a mapping $\Delta \in S^{\mathbb{N} \times \mathbb{Z}^d}$ such that for all time step $t \in \mathbb{N}$, $\Delta(t + 1) = G(\Delta(t))$.

Example 1 ❯ *Figure 20* is a partial representation of a space–time diagram of the cellular automaton $(1, \mathbb{Z}_2, \{0, 1\}, f)$, where $f(x, y) = x + y$. State 0 is represented by white, state 1 by black. Time goes from bottom to top.

This chapter, considers, for the most part, cellular automata of dimensions 1 and 2 with the typical neighborhoods depicted in ❯ *Fig. 21*: von Neumann $\{(-1, 0), (1, 0), (0, -1), (0, 1)\}$ and Moore $\{-1, 0, 1\}^2$ in dimension 2, first neighbors $\{-1, 0, 1\}$ and one way $\{-1, 0\}$ in dimension 1.

Several subsets of the space of configurations are considered. Given a *quiescent state* q satisfying $f(q, \ldots, q) = q$, a *q-finite configuration* c is a configuration equal to q in all but finitely

☐ **Fig. 21**

Typical neighborhoods. (a) von Neumann. (b) Moore. (c) First neighbors. (d) One way.

many cells: there exists α such that for all position $z \in \mathbb{Z}^d$, $\|z\|_\infty > \alpha \rightarrow c(z) = q$. A configuration c admits p as a *periodicity vector* if for all position $z \in \mathbb{Z}^d$, $c(z + p) = c(z)$. A configuration c in dimension d is *periodic* if it admits a family of d non-colinear periodicity vectors: there exists $p \in \mathbb{N}^d$ such that $(p_1, 0, \ldots, 0)$, $(0, p_2, 0, \ldots, 0)$, $\ldots$, and $(0, \ldots, 0, p_d)$ are periodicity vectors of c. A configuration c in dimension d is *ultimately periodic* if there exists α and d non-colinear vectors v_i such that for all position $z \in \mathbb{Z}^d$ and all vector v_i, $\|z\|_\infty > \alpha \rightarrow c(z + v_i) = c(z)$. Notice that in dimension 1, an ultimately periodic configuration can have two different ultimately periodic patterns on each side.

Constraints can also be added to the local rule. Symmetries are usually considered to obtain more natural rules mimicking physical systems. A symmetry rule can be seen as a one-to-one mapping $\rho : \mathbb{Z}^d \rightarrow \mathbb{Z}^d$: the image of a configuration $c \in S^{\mathbb{Z}^d}$ by the symmetry rule ρ is the configuration $\rho(c)$ satisfying for all position $z \in \mathbb{Z}^d$, $\rho(c)(z) = c(\rho(z))$. A cellular automaton $\mathscr{A}$ respects a symmetry rule ρ if ρ and G commute, that is, $\rho(G(c)) = G(\rho(c))$. Typical symmetries include reflections around point $(\rho_0(x, y) = (-x, -y))$, around axes $(\rho_x(x, y) = (-x, y))$ and rotations $(\theta(x, y) = (-y, x))$. A cellular automaton is *totalistic* if its set of states is a subset of $\mathbb{N}$ and the local rule f can be written as $f(s_1, \ldots, s_k) = g(\sum_{i=1}^k s_i)$. Totalistic rules respect all symmetries that preserve the neighborhood (*i.e.*, such that the image of the neighborhood by the symmetry rule is equal to the neighborhood): totalistic cellular automata with the von Neumann or Moore neighborhood are reflection and rotation invariants.

A cellular automaton is injective (resp. surjective, one-to-one) if its global rule is injective (resp. surjective, one-to-one). A cellular automaton $\mathscr{A}$ is reversible if there exists a cellular automaton $\mathscr{B}$ that reverts it, that is such that $G_{\mathscr{B}} \circ G_{\mathscr{A}}$ is the identity map.

Theorem 1 (Hedlund 1969; Richardson 1972) *A cellular automaton is reversible if and only if it is injective.*

Theorem 2 (Amoroso and Patt 1972) *It is decidable given a one-dimensional cellular automaton to decide whether it is reversible.*

Theorem 3 (Kari 1990, 1994) *It is undecidable given a two-dimensional cellular automaton to decide whether it is reversible.*

Whereas reversibility is an undecidable question, the construction of reversible cellular automata is possible, provided that the backward rule is constructed at the same time as the forward rule. Partitioned cellular automata provide a convenient way to construct reversible cellular automata. A *partitioned cellular automaton* is a cellular automaton with state set $S_1 \times S_2 \times \ldots \times S_k$, whose local rule can be rewritten as $f((s_1^1, \ldots, s_1^k), \ldots, (s_k^1, \ldots, s_k^k)) = \varphi(s_1^1, s_2^2, \ldots, s_k^k)$ where $\varphi : \prod S_i \rightarrow \prod S_i$ is the partitioned rule. As it is straightforward to verify, a partitioned cellular automaton is reversible iff partitioned rule is one-to-one. As the partitioned rule is a mapping from a finite set to itself, any partially defined injective rule can be completed to a reversible cellular automaton.

For a better and more complete introduction to the theory of cellular automata, see the chapter ❸ Basic Concepts of Cellular Automata, and/or Delorme (1999), and/or Kari (2005).

4 Chronology

It is a difficult task to give a fair and complete chronology of a research topic. This section proposes an exploration of the history of the field in three main eras:

1. The *computation and machines* era describes seminal papers outside the realm of cellular automata that lead to the main tools necessary to consider computation in the context of abstract machines.
2. The *universality and cellular automata* era is the core part of the chronology: it describes seminal papers along the path of universality study in the realm of cellular automata, from the early work of von Neumann in the 1950s to the end of the 1990s.
3. The *recent trends* era is a more subjective choice of some papers in the field in the twenty-first century.

4.1 Computation and Machines

Gödel (1931) in his now classical paper describing incompleteness theorems, introduces the so-called Gödel numberings: the ability to encode and manipulate a formal system inside itself if the system is complex enough. The concept of universality directly depends on such an encoding: a universal machine simulates a machine through its encoding. For a precise analysis from a logic and computer science point of view of Gödel's paper, see Lafitte (2008).
Turing (1936), while introducing Turing machines and proving the undecidability of the halting problem by a diagonal argument, also introduced his universal machine. Fixing an enumeration of Turing machines and a recursive bijective pairing function $\langle .,. \rangle : \mathbb{N}^2 \rightarrow \mathbb{N}$, he describes a machine U that, on input $\langle m, n \rangle$, computes the same value as the machine encoded m on input n. Universality as a property is not discussed: a unique universal Turing machine U is given. For a discussion of the development of ideas from Leibniz to Turing results, see Davis (2000).
Post (1943) At that time, many different models of computation were proposed and proved equivalent, leading to the so-called Church–Turing thesis. Post introduces tag systems, a combinatorial word-based system successfully used since to construct size-efficient universal Turing machines. For a modern definition and discussion of tag systems, see Minsky (1967).
Kleene (1956) Finite state machines are at the hearth of many models of computation. Kleene's paper proves the equivalence between three different families of objects: regular languages, finite automata, and Boolean circuits. Boolean circuits are modeled after the formal study of abstract neurons by McCulloch and Pitts (1943). This equivalence is fundamental both to concrete computer design and discrete models of computation like cellular automata. For a modern discussion on this equivalence and its consequences from the point of view of

computation and machines, see Minsky (1967). Perrin (1995) gives a history of this period and important achievements with respect to the field of finite automata and formal languages.

Minsky (1967) In the spirit of the question from Shannon (1956) about the size of a smallest Turing machine, Minsky explains how to efficiently encode tag systems computations into Turing machines and describe a universal Turing machine with four symbols and seven states. This marks the real start of a (still running) competition.

Lecerf (1963), Bennett (1973) Reversible computation is concerned with computing devices that can unroll their computation, going back in time. In their independent papers, Lecerf and Bennett prove that reversible Turing machines are able to simulate just any Turing machine. Thus, there exists reversible universal Turing machines.

Fredkin and Toffoli (1982) To encode classical computations into discrete models, Kleene's (1956) theorem permits to go freely from circuits to finite state machine, which is an essential ingredient for computation. Fredkin and Toffoli discuss an analogous for reversible computation: elementary building blocks to encode any reversible finite state machine as a reversible circuit. This paper also introduces the so-called billiard ball model of computation: a discrete cellular automata model to encode reversible computations. The encoding of reversible finite state machines into circuits was later improved by Morita (1990).

4.2 Universality and Cellular Automata

von Neumann (1966) Introducing cellular automata in order to construct a self-reproducing machine, reflecting the nature of life, von Neumann takes a fixed-point approach. His two-dimensional, 29 states, von Neumann neighborhood cellular automaton is able to simulate a particular kind of Turing machine that can also control a construction arm. The power of the construction arm is rich enough to construct with finitely many instructions a copy of the Turing machine itself. Whereas the machine is constructed with a form of Turing universality in mind, the simulation of the Turing machine is done with very simple components wiring down a particular family of Boolean circuits. As a consequence, the original cellular automaton is also, *a posteriori*, intrinsically universal. The construction of von Neumann leads to various improvements and discussions on the encoding of Boolean circuits, the different organs that compose the machine and the transmission of signals. A non-exhaustive list of interesting papers might be Arbib (1966), Burks (1970a, b, c), Moore (1962, 1970), and Thatcher (1970a, b).

Codd (1968) Following the principle of von Neumann's idea on self-reproduction, Codd drastically reduces the complexity of the automaton. Codd's two-dimensional rule uses eight states with the von Neumann neighborhood. Signals are conveyed by pairs of states (an oriented particle) moving between walls and reacting upon collision. This cellular automaton is also universal for Boolean circuits and so intrinsically universal. A later construction by Langton (1984), based on Codd ideas, has fewer states and a very simple family of self-reproducing loops but loses its computation universal capabilities.

Banks (1970) The work of Banks is noticeable with respect to several aspects and also because of its relatively small diffusion in the cellular automata community. Banks constructs a family of very small cellular automata (two-dimensional, von Neumann neighborhood, very symmetric, four to two states) simulating Boolean circuits in a very simple and modern way (signals moving in wires, Boolean gates on collisions), he identified and used explicitly the property of intrinsic universality and gave a transformation to construct relatively small universal one-dimensional cellular automata with large neighborhoods starting from two-dimensional ones (re-encoding it into a one-dimensional first-neighbors automaton with 18 states). The construction of a two-dimensional four state universal cellular automaton in the spirit of Banks is provided by Noural and Kashef (1975).

Conway (1970) (Gardner, 1970; Berlekamp 1982) The Game of Life introduced by Conway is certainly among the most famous cellular automata and the first rule to be proven universal by analysis of a given rule rather than on purpose construction. A modern exposition of the Game of Life universality and a proof of its intrinsic universality was later proposed by Durand and Róka (1999).

Smith III (1971) The simulation of Turing machine by cellular automata to construct one-dimensional Turing universal cellular automata is studied by Smith III. Among several results, he explains how to construct a one-dimensional Turing universal cellular automaton with first neighbors and 18 states.

Toffoli (1977) Any cellular automaton of dimension d can be simulated, in a certain sense, by a cellular automaton of dimension $d + 1$. Using this assertion, Toffoli shows that two-dimensional reversible cellular automata can be Turing universal. The result was later improved by Hertling (1998).

Margolus (1984) While Toffoli transforms any Turing machine into a two-dimensional cellular automaton by using a new spatial dimension to store computational choices, Margolus constructs a Turing universal two-dimensional reversible cellular automaton by simulation of a bouncing billiard ball, complex enough to compute any reversible Boolean function of conservative logic. The billiard ball model cellular automaton has 16 states defined as two-by-two blocks of binary cells and von Neumann neighborhood.

Albert and Čulik (1987) Each cellular automaton can be simulated by a totalistic cellular automaton with one-way neigborhood. With the help of the last proposition, Albert and Čulik construct the first universal cellular automaton obtained by simulation of any cellular automaton of the same dimension. The automaton works along the following principle: each macro-cell copies the state of its left neighbor and adds it to its state obtaining some n, then by copying the nth element of a reference table, it selects its new state. Whereas the spirit of intrinsic universality is definitely there, the technical implementation is less clear. The one-dimensional first-neighbors automaton obtained has 14 states. The construction was later improved by Martin (1993, 1994) with better transition time complexity and using the smn theorem.

Morita and Harao (1989) Introducing partitioned cellular automata, Morita and Harao explicitly simulate any reversible Turing machine on a one-dimensional reversible cellular automaton, proving that one-dimensional reversible cellular automata can be Turing universal. The construction was later improved by Dubacq (1995), simulating any Turing machine in real time (without loss of time).

Lindgren and Nordahl (1990) The direct simulation of Turing machine on one-dimensional cellular automata proposed by Smith III can be improved and any m states n symbols machine can be simulated by a $(m + n + 2)$-states cellular automaton following Lindgren and Nordhal. Applying this to Minsky's seven states and four symbols machine and then transforming the simple simulation into a macro-state signal-based simulation, Lindgren and Nordhal obtain a one-dimensional first neighbors seven-state Turing universal cellular automaton. The intrinsic universality status of this automaton is unknown.

Durand and Róka (1996) Revisiting the Game of Life and filling holes in the universality proof, Durand and Róka publish the first discussion on the different kinds of universality for cellular automata and the problem of formal definition.

Durand-Lose (1997) Using a modern definition of intrinsic universality, Durand-Lose goes one step further than Morita and Harao by constructing a reversible one-dimensional cellular automata intrinsically simulating any reversible cellular automaton.

4.3 Recent Trends

Imai and Morita (2000) The improvement in the construction of small and simple two-dimensional reversible cellular automata continues. Imai and Morita use partitioned cellular automata to define an eight-state universal automaton.

Ollinger (2002b) Using simulation techniques between cellular automata, strong intrinsically universal cellular automata with few states can be constructed, here six states.

Cook (2004) Very small universal cellular automata cannot be constructed, they have to be obtained by analysis. Realizing a real tour de force, Cook was able to prove the Turing universality of the two-states first-neighbors so-called rule 110 by analyzing signals generated by the rule and their collisions. The intrinsic universality of this automaton remains open. The original construction, simulation of a variant of tag system, was exponentially slow. For a proof of Cook's result using signals, see Richard (2008).

Neary and Woods (2006) Recently, the prediction problem of rule 110 was proven P-complete by Neary and Woods by a careful analysis and modification of Turing machines simulation techniques by tag systems. As P-completeness is required for intrinsic universality, this is another hint of the potential strong universality of rule 110.

Richard (2008) The limits of constructed small intrinsically universal cellular automata are converging toward analyzed cellular automata. Using particles and collisions, Richard was recently able to construct a four-state intrinsically universal first-neighbors one-dimensional cellular automaton.

5 Formalizing Universalities

What is a universal cellular automaton? At first, the question might seem simple and superficial: a universal cellular automaton is an automaton able to compute anything recursive. In fact, a formal definition is both required and difficult to obtain. The requirement for such a definition is needed to define a frontier between cellular automata of maximal complexity and the others: in particular when considering the simplest cellular automata, to be able to identify the most complex cellular automata. The difficulty arises from the fact that a definition broad enough to encapsulate all constructions of the literature and all *fair enough* future constructions is required. For more details concerning this philosophical question, see Durand and Róka's (1999) attempt to give formal definitions.

5.1 Turing Universality

Turing universality is the easiest form of universality one might think about, that is, with a computability culture: let the cellular automaton *simulate* a well-known universal model of computation, either simulating one universal object of the family or any object of the family.

The first approach pushes back the problem to the following one: what is a universal Turing machine? What is a universal tag system? In its original work, Turing did not define universal machines but *a* unique universal machine. The definition of universality for Turing machines was later discussed by Davis (1956) who proposed to rely on recursive degrees, defining universal machines as machines with maximal recursive degree. This definition, while formal, lacks precise practical view of encoding problems: the issue continues to be discussed in the world of Turing machines becoming more important, as smaller and smaller universal machines are proposed. For a view on the universality of Turing machines and pointers to literature related to the topic, see Woods (2007).

The second approach leads to the problem of heterogeneous simulation: classical models of computation have inputs, step function, halting condition, and output. Cellular automata have no halting condition and no output. As pointed out by Durand and Róka (1999), this leads to very tricky encoding problems: their own attempt at a Turing universality based on this criterion as encoding flaw permitting us counterintuitively to consider very simple cellular automata as universal.

Turing universality of dynamical systems in general and cellular automata in particular has been further discussed by Delvenne et al. (2006) and Sutner (2004). None of the proposed definitions is completely convincing so far, so it has been chosen on purpose not to provide the reader with yet another weak formal definition.

5.2 Intrinsic Universality

Intrinsic universality, on the other hand, is easier to formalize, yet a more robust notion (in the sense that variations along the lines of the definition lead to the same set of universal automata). Consider a homogenous type of simulation: cellular automata simulated by cellular automata in a shift invariant, time invariant way. A natural type of universal object exists in this context: cellular automata that are able to simulate each cellular automaton. Following the ideas of grouping and bulking (Rapaport 1998; Mazoyer and Rapaport 1999; Ollinger 2002a), a general notion of simulation broad enough to scope all reasonable constructions of the literature is introduced.

Direct simulation between two cellular automata can be formalized as follows. A cellular automaton $\mathcal{B}$ *directly simulates* a cellular automaton $\mathcal{A}$, denoted $G_{\mathcal{A}} \prec G_{\mathcal{B}}$, of the same dimension according to a mapping $\varphi : S_{\mathcal{A}} \to 2^{S_{\mathcal{B}}}$ if for any pair of states $a, b \in S_{\mathcal{A}}$, $\varphi(a) \cap \varphi(b) = \emptyset$ and for any configuration $c \in S_{\mathcal{A}}^{\mathbb{Z}^d}$, $G_{\mathcal{B}}(\varphi(c)) \subseteq \varphi(G_{\mathcal{A}}(c))$.

For any state set S, let $(m_1, \ldots, m_d)$ be a tuple positive integers, the *unpacking bijective map* $o_{(m_1,\ldots,m_d)} : \left(S^{\prod_i m_i} \right)^{\mathbb{Z}^d} \to S^{\mathbb{Z}^d}$ is defined for any configuration $c \in \left(S^{\prod m_i} \right)^{\mathbb{Z}^d}$ and any position $z \in \mathbb{Z}^d$ and $r \in \prod_i \mathbb{Z}_{m_i}$ as $o_{(m_1,\ldots,m_d)}(c)(m_1 z_1 + r_1, \ldots, m_d z_d + r_d) = c(z)(r)$. The *translation* of vector $v \in \mathbb{Z}^d$ is defined for any configuration $c \in S^{\mathbb{Z}^d}$ and position $z \in \mathbb{Z}^d$ as $\sigma_v(c)(z) = c(z - v)$.

Simulation between two cellular automata is extended by considering packing, cutting, and shifting of the two cellular automata such that direct simulation occur between both transformed objects. Universal objects are then maximum in the induced preorder. In fact, it can be proved that simulation on one side is sufficient for universal objects.

Definition 1 (intrinsic universality) A cellular automaton $\mathcal{U}$ is intrinsically universal if for each cellular automaton $\mathcal{A}$ of the same dimension there exists an unpacking map o_m, a positive integer $n \in \mathbb{N}$, and a translation vector $v \in \mathbb{Z}^d$ such that $G_{\mathcal{A}} \prec o_m^{-1} \circ G_{\mathcal{U}}^n \circ o_m \circ \sigma_v$.

Theorem 4   (Rapaport 1998; Mazoyer and Rapaport 1999) *No cellular automaton is intrinsically universal in real time (i.e., when constraining cutting constant n to be equal to* $\max(m)$*): simulation cannot perform both information displacement and transition computation at the same time.*

Theorem 5   (Ollinger 2003) *Given a cellular automaton, it is undecidable to determine whether it is intrinsically universal.*

Turing universality and intrinsic universality notions are really different notions. Some erroneous claims by Wolfram (1984, 2002) affirm, for example, that rule 110 is intrinsically universal. In fact, the question is yet open, Turing universality is the only proven thing.

Theorem 6   (Ollinger 2002a; Theyssier 2005a) *There exists Turing universal cellular automata which are not intrinsically universal. Moreover, some of them are at the bottom of an infinite increasing chain of equivalences classes of the preorder.*

Universality can also be discussed when considering language recognition or computation on grids. This topic is out of scope of the present paper. For more on this topic, see Mazoyer (1996, 1999).

6   High Dimensions: 2D and More

In two and more dimensions, an easy way to construct both intrinsically and Turing universal cellular automata is to go through Boolean circuit simulation. Boolean circuits can encode any finite state machine, and a cell of a cellular automaton or the control and tape of a Turing machine can be described as finite state machines.

6.1   Boolean Circuits Simulation

The topic of Boolean circuits simulation with cellular automata is quite popular and a lot has been written on it, see for example, recreations around the wireworld cellular automaton designed by Silverman and discussed in Dewdney (1990). This was already discussed with **Cu** in ❷ Sect. 2. Technical hints and possible exotic extensions without entering details are given here.

To simulate Boolean circuits, one typically needs to mix the following ingredients.

Wires Boolean signals travel in piecewise straight line in space, their paths are the wires. Several encoding of Boolean signals with or without explicit wires are possible: moving particles encoded as states with a direction vector, bouncing on walls to turn as in game of life (Gardner 1970); wire path encoded as wire cells with explicit direction vector on each wire cell as in von Neumann (1966); undirected wire cells on which directed signals travel; undirected wire cells on which pairs of two different signal cells travel, the direction being given by the orientation of the pair as in wireworld (Dewdney 1990); pairs of undirected wire paths in between which pairs of two different signal cells travel as in Codd (1968) or Banks (1971).

Turn and delay Boolean signals should be able to turn in space and delay their arrival to permit signal synchronization.

Signal crossing In order to encode all Boolean circuits, crossing of signals has to be encoded either explicitly (adding crossing states) or implicitly using delaying techniques (as in von Neumann 1966) or Boolean logic tricks.

Gates Signals must be combined using Boolean gates at least taken in a Boolean universal family of gates. AND, OR, NOT are the classical ones but NAND or NOR is sufficient alone.

Fan-out Signals must be duplicated in some way either with an explicit fan-out state or using specific wire split rules.

Remarks and encoding tricks regarding Boolean circuit simulation:

- Universal Boolean function families and their expressive power are described in Post (1941). But, in cellular automata encoding, it is easy to use constants and multiple wires encoding, thus the number of Boolean classes depending on the implemented gates is finite and small.

- Clocks are only needed when dealing with some form of synchronized logic simulation. It is often used because Boolean signals are encoded with two values: empty wire or signal on wire. With such an encoding, NOT gate has to generate new signal on wire, and clock signal is used to determine at which time steps to do so. However, a classical coding trick to avoid the use of clocks and diodes is to only implement OR and AND gates and use the two wires trick to gain Boolean universality: a signal is encoded as one signal on one wire, the second being empty (thus no signal is encoded as no signal on both wires), then the NOT gate is just the wire crossing $(x, y) \mapsto (y, x)$, the AND gate can be encoded as $(x, y) \mapsto (x \wedge y, x \vee y)$ and the OR gate as $(x, y) \mapsto (x \vee y, x \wedge y)$. As both OR and AND produce signal only if there is at least one signal in input, the need for clock vanishes.

- Wire crossing can be gained for free by using the XOR gate as planar crossing can be implemented with XORs.

- Delays come for free if the wires can turn in all directions.

- In dimension 3, wire crossing is not needed, use the third dimension to route wires.

- Signal encoding can be done using signal constructions, in the spirit of Mazoyer and Terrier (1999), in order to reduce the number of states, see the chapter ❷ Algorithmic Tools on Cellular Automata.

Remark Boolean circuit simulation is not restricted to square grids. As an example of a more exotic lattice, Gajardo and Goles (2001) encoded a Boolean circuit simulator on a hexagonal lattice (with proper cellular automata definition).

6.2 The Game of Life

❷ Section 2 provides an original example of a universal 2D cellular automaton. One of the most famous and celebrated cellular automata, the Game of Life, introduced by Conway (Gardner 1970; Berlekamp et al. 1982) is now discussed. For a hint of the richness of the game, the reader is referred to the Life lexicon (Silver S, http://www.argentum.freeserve.co.uk/lex_home.htm).

The Game of life is a two-dimensional cellular automaton with two states (dead or alive) and a Moore neighborhood. Each cell evolves according to the state of its eight neighbors. When a dead cell has exactly three alive neighbors, it becomes alive. Survival occurs when an alive cell has exactly two or three alive neighbors. Death occurs either by overcrowding when an alive cell has five or more alive neighbors, or by exposure when an alive cell has one or no alive neighbor. The rule is completely rotation and mirror image invariant.

The Turing universality of the Game of Life was sketched by Conway et al. (1982). A more detailed proof including a simplification considering intrinsic universality was provided by Durand and Róka (1999). An explanation on how to use almost the same techniques as for **Cu** to improve their construction and avoid the clocking problems is given here.

6.2.1 Objects

To encode Boolean circuit's components, one has to choose proper elementary gate encoding elements.

Glider In the Game of Life, a particle on a wire is represented by a glider moving in straight line, as depicted in ❷ *Fig. 22*. There is no physical representation of the wire itself. The glider slides diagonally one cell every four steps.

Crash When two orthogonal gliders meet, under proper synchronization conditions, a crash occurs, destroying both gliders without leaving garbage, as depicted in ❷ *Fig. 23*.

Eater A special static object called an eater is used to destroy, under proper synchronization conditions, a colliding glider, leaving the eater in place, as depicted in ❷ *Fig. 24*.

Duplicator A more elaborated object is the duplicator. Its static part consists only of a two-by-two block. When SE and NE gliders meet, under proper synchronization conditions, near the block, both gliders are destroyed, the block remains in place, and a new NE glider, with a different synchronization, is created, as depicted in ❷ *Fig. 25*. If the input SE glider is not present, then the NE glider continues its way without being perturbed by the block. If the SE glider is considered an input x, one might see the duplicator as a widget with two outputs, an NE output with value $\neg x$ and an SE output with value x.

Gosper p46 gun A gun is a device emitting gliders at a given rate. The construction here uses the Gosper p46 gun, one of the first discovered guns, emitting gliders with period 46, as depicted in ❷ *Fig. 26*. The construction would work with any gun with a period large enough to permit glider streams crossing.

◧ Fig. 22
GoL glider.

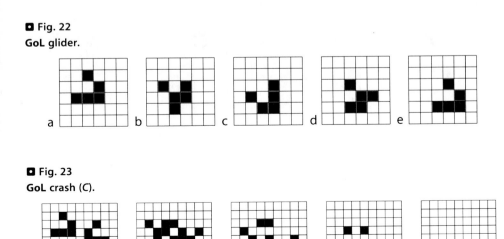

a   b   c   d   e

◧ Fig. 23
GoL crash (C).

a   b   c   d   e

◨ Fig. 24
GoL eater (E).

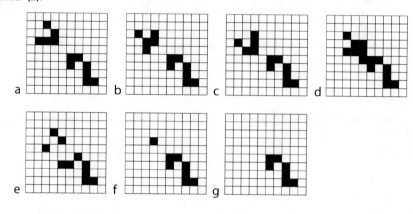

◨ Fig. 25
GoL duplicator (D).

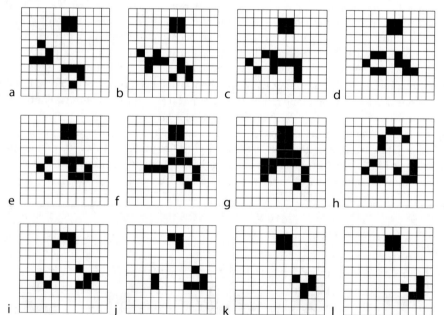

Crossing The use of guns will force one to synchronize gliders: on a wire, particles are encoded every period as a glider or no glider. The chosen period 46 permits that two properly synchronized streams cross, as depicted in ❷ *Fig. 27*.

6.2.2 Gates

Using these objects, one can construct elementary gates. Wires are straight empty paths along which glider streams synchronized with period 46 can move straight forward.

◼ Fig. 26
GoL Gosper p46 gun (G).

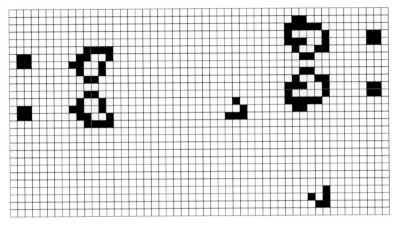

◼ Fig. 27
GoL p46 crossing (X).

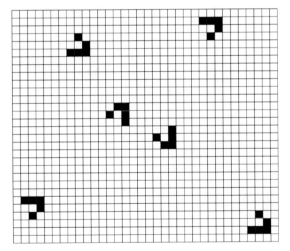

A first widget is the NOT turn widget, the principle of which is depicted in ❷ *Fig. 28a.* Using a gun and crashes, it transforms an incoming stream of gliders into its negation, turning the stream by 90°. Several such widgets can be combined to achieve delays, side translation, and synchronization.

Using the objects and proper synchronization, eventually adding delays and translations, one can encode crossing, fan-out, AND gate, and OR gate as symbolically depicted in ❷ *Fig. 28.*

Once these elementary gates are obtained, using the same encoding tricks as in ❷ Sect. 2, every Boolean circuit can be encoded (e.g., a turn of a Boolean value encoded as a pair of glider streams is obtained by combining NOT turns and crossing).

■ **Fig. 28**
GoL (symbolic) elementary gates. **(a)** NOT turn. **(b)** Crossing. **(c)** Fan-out. **(d)** AND gate. **(e)** OR gate.

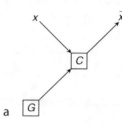

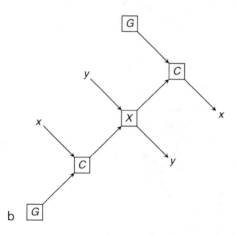

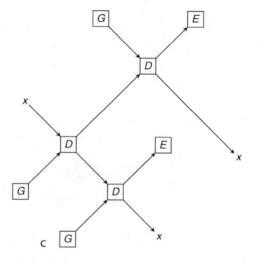

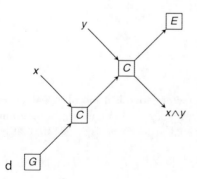

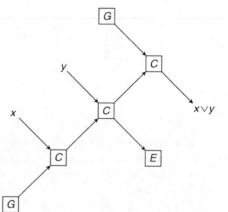

6.2.3 Universalities

Using exactly the same techniques as in ❯ Sect. 2, encoding finite state machines, one concludes that the Game of Life is intrinsically universal. To obtain the Turing universality of the Game of Life on finite configurations, in the style of Conway et al. (Berlekamp et al. 1982), one further needs to encode a counter on a finite configuration. As the main interest here is in intrinsic universality, universality on finite configurations is not discussed.

6.3 Banks' Two-State Universal CA

Small intrinsically universal cellular automata are quite simple to construct in dimension 2 with few states: Banks (1970, 1971) does it with two states and von Neumann neighborhood with reflection and rotation symmetry. Universality is yet again obtained using Boolean circuits simulation. The construction is briefly sketched.

The local rule is rotation and mirror image invariant. An alive cell with exactly two alive neighbors dies if the two alive neighbors are not aligned. A dead cell becomes alive if exactly three or four of its neighbors are alive. In all the remaining cases, the cell keeps its current state.
Particles on wires Wires are encoded by thick bars of alive cells of thickness three, bordered by dead cells. Dead ends are encoded by extending the extremity column of the wire to be of thickness five. A particle on one side of a wire is encoded as two diagonal holes, as depicted in ❯ Fig. 29.
Widgets To achieve all possible paths and fan-out, two widgets are provided, as depicted in ❯ Fig. 30a, b. Moreover, one can construct an AND-NOT gate, as depicted in ❯ Fig. 30c. Such a gate is not universal by itself. Fortunately, it can be combined by a clock to construct NOT gate and then a universal NOR gate. A clock is depicted in ❯ Fig. 30d. Using NOR gates, one can construct an almost-crossing gate (the almost-crossing requires one of the inputs to have value 0). Combining almost-crossing and delays, one obtains a crossing.
Gates Combining fan-out, crossing and NOR gates, one can encode every Boolean circuit, as per ❯ Sect. 2. This cellular automaton is intrinsically universal.

7 Turing Universality in 1D

In dimension one, Boolean circuit encoding is more puzzling as wire-crossing capabilities are bounded by the local rule. Thus, historically, computation universality is achieved by direct simulation of universal models of computations.

◻ Fig. 29
Banks particle moving on a wire from left to right.

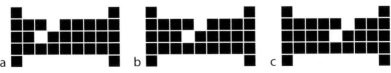

a   b   c

◘ Fig. 30
Banks elementary widgets. (a) Fan-out. (b) Turn. (c) AND-NOT gate. (d) 16-Clock.

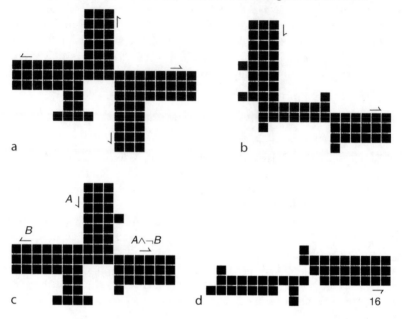

a

b

c

d

16

7.1 Universal Models of Computation

Turing machines Turing machines are easy to encode on cellular automata (see below) as an infinite tape really looks like a configuration of a cellular automaton. In fact, several variants of Turing machines exist and an important literature on universality in the Turing world provide useful objects to build small universal automaton based on this model. The question of existence of small universal Turing machines was first raised by Shannon (1956), different variants of Turing machines are discussed by Fischer (1965). For a survey on small Turing machine construction, see Woods and Neary (2007).

Tag systems Tag systems provide a better model to design very small universal objects. In fact, very small universal Turing machines are constructed by simulation of tag systems and their variants as originally proposed by Minsky (1967) and Cocke and Minsky (1964). The original drawback of tag system was its exponential slowdown when simulating Turing machines. This drawback was removed recently by Woods and Neary (2006, 2007) by achieving polynomial time simulation. The Turing universality of rule 110 is obtained by Cook (2004) by direct simulation of a proper variant of tag systems.

The variant of Turing machine used is the following. A *Turing machine* is a tuple (S, Σ, B, s_0, T) where S is a finite set of states, Σ is a finite alphabet with a special blank symbol $B \in \Sigma$, $s_0 \in S$ is the initial state, and $T : S \times \Sigma \rightarrow S \times \Sigma \times \{\leftarrow, \rightarrow\}$ is a partial transition map. A transition rule $T(s, a) = (s', b, d)$ reads as follows: when reading a from state s, write b on the tape, move in direction d, and enter state s'. A configuration of the machine is a triple (s, z, c) where $s \in S$ is the current state of the machine, $z \in \mathbb{Z}$ is the position of the head, and $c \in S^{\mathbb{Z}}$ is the content of the tape. The machine goes in one step, from a configuration (s, z, c) to a configuration (s', z', c'), if the transition rule $T(s, c(z)) = (s'', d, b)$ is defined and verifies $s' = s''$, $z' - z = d$, $c'(z) = b$ and for all position $z'' \neq z$, $c'(z) = c(z)$. Starting from a configuration c, a halting computation of the machine in time t consists of a sequence of configurations $(c_i)_{i=0}^{t}$ such that $c_0 = c$, the machine cannot reach any configuration from c_t and for all i, the machine goes in one step from c_i to c_{i+1}. The configuration c_t is the output of the computation.

7.2 *à la* Smith III

Following Smith III (1971), a given Turing machine (S, Σ, B, s_0, T) can be simulated by a cellular automaton $(1, S', \{-1, 0, 1\}, f)$ as follows, as depicted in ❷ *Fig. 31*. Let $S' = \Sigma \cup S \times \Sigma$. A configuration (s, z, c) of the Turing machine is encoded as a configuration $c' = \tau(s, z, c)$ of the cellular automaton in the following way: $c'(z) = (s, c(z))$ and for all positions $z' \neq z$, $c'(z') = c(z)$. The local rule encodes the transition function of the Turing machine. For each transition $T(s, a) = (s', b, \leftarrow)$, for all states $x, y \in S$, $f(x, y, (s, a)) = (s', y)$, and $f(x, (s, a), y) = b$. Symmetrically, for each transition $T(s, a) = (s', b, \rightarrow)$, for all states $x, y \in S$, $f((s, a), y, x) = (s', y)$, and $f(x, (s, a), y) = b$. All undefined transitions apply identity: $f(x, y, z) = y$. With this encoding, starting from an encoded configuration $\tau(c)$, the configuration evolves in one step to a configuration $\tau(c')$ where c' is the next computation step of the Turing machine if it exists, $c' = c$ otherwise. Using this simulation, a Turing machine with m states and n symbols is simulated by a one-dimensional cellular automaton with first-neighbors and $(m + 1)n$ states.

7.3 *à la* Lindgren and Nordahl

To lower the number of states, Lindgren and Nordahl (1990) introduce a simulation scheme where each step of the Turing machine computation is emulated by two time steps in the cellular automaton. A given Turing machine (S, Σ, B, s_0, T) can be simulated by a cellular automaton $(1, S', \{-1, 0, 1\}, f)$ as follows, as depicted in ❷ *Fig. 32*. Let $S' = \Sigma \cup S \cup \{\bullet, \leftrightarrow\}$. A configuration (s, z, c) of the Turing machine is encoded as a configuration $c' = \tau(s, z, c)$ of the cellular automaton in the following way: for all $z' < z$, $c'(2z') = \bullet$, $c'(2z' + 1) = c(z)$; for all $z' > z$, $c'(2z' + 1) = \bullet$, $c'(2z' + 2) = c(z)$; $c'(2z) = \bullet$ and either $c'(2z + 1) = s$, $c'(2z + 2) = c(z)$ or $c'(2z + 1) = c(z)$, $c'(2z + 2) = s$ (two possible encodings). The local rule encodes the transition function of the Turing machine. Applying the rule: for each

■ **Fig. 31**
Turing machine simulation *à la* Smith III.

B	B	(q_0, a)	a	B
B	B	a	(q_1, B)	B
B	B	(q_1, b)	B	B
B	(q_0, B)	b	B	B
B	B	(q_0, a)	B	B

◘ Fig. 32

Turing machine simulation à la Lindgren and Nordahl.

B	•	B	•	a	q_0	•	a	•	B
B	↔	B	↔	a	↔	q_0	a	↔	B
B	•	B	•	a	•	q_1	B	•	B
B	↔	B	↔	a	q_1	↔	B	↔	B
B	•	B	•	q_1	b	•	B	•	B
B	↔	B	q_1	↔	b	↔	B	↔	B
B	•	B	q_0	•	b	•	B	•	B
B	↔	B	↔	q_0	b	↔	B	↔	B
B	•	B	•	q_0	a	•	B	•	B

transition $T(s, a) = (s', b, \leftarrow)$, $f(\bullet, s, a) = s'$, $f(s, a, \bullet) = b$, $f(\bullet, a, s) = s'$, $f(a, s, \bullet) = b$, for each transition $T(s, a) = (s', b, \rightarrow)$, $f(\bullet, s, a) = b$, $f(s, a, \bullet) = s'$, $f(\bullet, a, s) = b$, $f(a, s, \bullet) = s'$. Moving: for all $s \in S$, $a, b \in \Sigma$, $f(\leftrightarrow, s, a) = \bullet$, $f(a, \leftrightarrow, s) = sf(a, s, \leftrightarrow) = \bullet$, $f(s, \leftrightarrow, a) = s$. All undefined transitions apply the identity rule but for $\bullet$ and $\leftrightarrow$ that alternates: for all states $x, y \in S$, $f(x, \bullet, y) = \leftrightarrow$ and $f(x, \leftrightarrow, y) = \bullet$ With this encoding, starting from an encoded configuration $\tau(c)$, the configuration evolves in two steps to a configuration $\tau(c')$ where c' is the next computation step of the Turing machine if it exists, $c' = c$ otherwise. Using this simulation, a Turing machine with m states and n symbols is simulated by a one-dimensional cellular automaton with first-neighbors and $m + n + 2$ states.

7.4 à la Cook

Simulation of tag systems is more tricky due to the non-locality of one computation step of the system. Following Cook (2004), one can consider cyclic tag systems. A cyclic tag system is given as a finite set of words $(w_0, \ldots, w_{N-1})$ on the alphabet $\{\circ, \bullet\}$. A configuration of the system is a word $u \in \{\circ, \bullet\}^*$. At time step t, the configuration u evolves to a configuration v according to either $u_0 = \circ$ and $u_0 v = u$, or $u_0 = \bullet$ and $u_0 v = u w_{t \bmod N}$. Cyclic tag systems can encode any recursive function. To encode all cyclic tag systems in a same cellular automaton $(1, S, \{-1, 0, 1\}, f)$, one can follow the following principle. Encode each configuration u of a cyclic tag system $(w_0, \ldots, w_{N-1})$ as a configuration of the kind $^\omega(T_-^k) \cdot u \cdot \blacksquare (w_0 \blacktriangle \ldots \blacktriangle w_{N-1})^\omega$ where intuitively T is a clock signal, $\blacksquare$ is the frontier between u and the rule and the rule is repeated on the right, each word separated by a $\blacktriangle$. Giving the complete local rule is tedious but its principle can be sketched: each time a clock signal hits the first letter of u, it erases it and sends a signal to the right transporting the value of that letter; when the signal meets the $\blacksquare$ it removes it and begins to treat the w word on the right, either erasing it or just crossing it; when the signal meets a $\blacktriangle$, it changes it into a $\blacksquare$ and the signal disappears. This principle is used by Cook to simulate cyclic tag systems with rule 110 particles and collisions.

7.5 Rule 110

The universality of rule 110 is certainly one of the most complicated construction and a real tour de force. It is yet the smallest known Turing universal cellular automaton. Due to its size, universality was obtained by analysis. There is not enough room here to completely describe the construction, so the reader will only be given an idea of its taste. For a complete exposure of the proof, read the original paper by Cook (2004) or the modern particle-based exposure by Richard (2008).

Rule 110 is a two state one-dimensional cellular automaton with a first-neighbors neighborhood. Its local rule can be given as follows:

$$f(x, y, z) = \lfloor 110/2^{4x+2y+z} \rfloor \bmod 2$$

To encode every cyclic tag system (to be precise, a subfamily of cyclic tag systems rich enough to encode any recursive function) into space–time diagrams of rule 110 following the technique described above, Cook selected, by careful analysis, 18 particles, depicted in ❷ *Fig. 33*, and 23 collisions, depicted in ❷ *Figs. 34–36*. These particles are arranged into groups of parallel particles to encode bits and other signals. Collisions are grouped together to obtain the proper interaction in between particle groups. The tedious part of the proof consists in proving that, with proper synchronization, particles combine to follow the simulation scheme, as depicted in ❷ *Fig. 37*.

Cook's construction proved the Turing universality of rule 110. Intrinsic universality is yet still open.

Open Problem *Is rule 110 intrinsically universal?*

Another step was obtained by Neary and Woods, showing that rule 110 is indeed complicated. P-completeness of the prediction problem is a necessary condition for intrinsic universality. The prediction problem consists, given a segment of states, to decide the state at the top of the computation triangle obtained by iterating the local rule of a fixed cellular automaton.

Theorem 7 (Neary and Woods 2006) *The prediction problem of rule 110 is P-complete.*

7.6 Encoding Troubles

The main point of discussion here is to decide which kind of configurations are acceptable for encoding Turing universal computation. Finite configurations are certainly not a problem and using any potentially non-recursive configuration would permit trivial cellular automata to be misleadingly called universal. The previous constructions involving Turing machines use finite or ultimately periodic configurations with a same period on both sides, the same one for all simulated machines, whereas the tag system encoding uses ultimately periodic configurations with different periods, moreover these periodic parts depend on the simulated tag system. The tag system really needs this ultimate information, as transforming the simulating cellular automaton into one working on finite configurations would have a constant but large impact on the number of states. As pointed in Durand and Róka (1999), this configuration encoding problem adds difficulties to the formal definition of Turing universality.

8 Intrinsic Universality in 1D

Even if the concept of intrinsically universal cellular automata took some time to emerge, intrinsic universality does not require more complex constructions to be achieved.

☐ **Fig. 33**

110 particles used in the construction (with highlighted finite coloring).

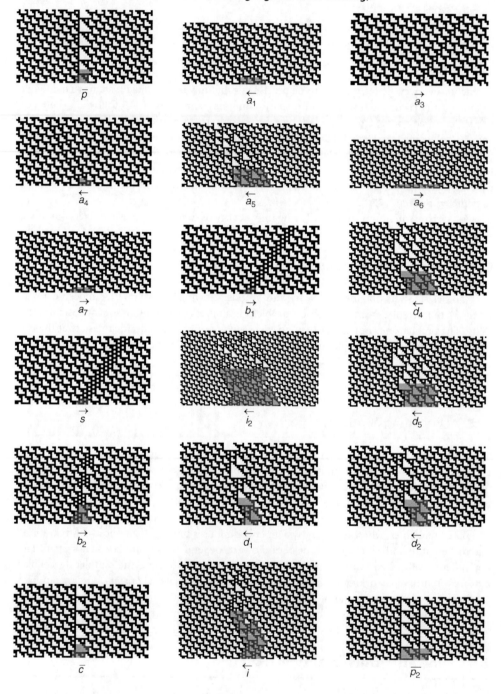

◘ **Fig. 34**

110 collisions used in the construction.

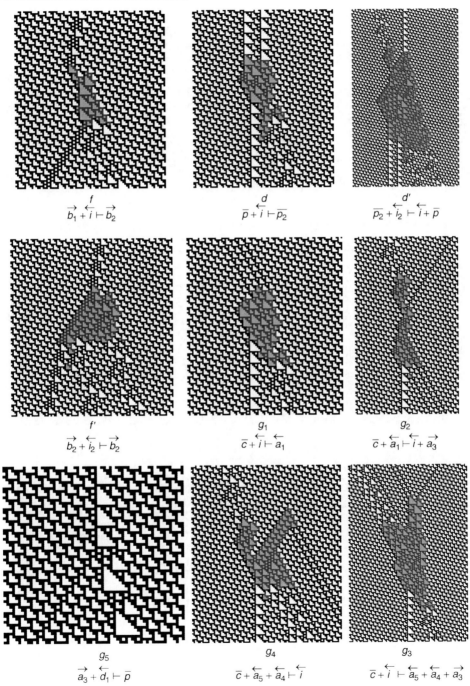

$$\overset{f}{\underset{}{}}$$
$$\overrightarrow{b_1} + \overleftarrow{i} \vdash \overrightarrow{b_2}$$

$$\overset{d}{\underset{}{}}$$
$$\overline{p} + i \vdash \overline{p_2}$$

$$\overset{d'}{\underset{}{}}$$
$$\overline{p_2} + \overleftarrow{i_2} \vdash \overleftarrow{i} + \overline{p}$$

$$\overset{f'}{\underset{}{}}$$
$$\overrightarrow{b_2} + \overleftarrow{i_2} \vdash \overrightarrow{b_2}$$

$$\overset{g_1}{\underset{}{}}$$
$$\overline{c} + \overleftarrow{i} \vdash \overleftarrow{a_1}$$

$$\overset{g_2}{\underset{}{}}$$
$$\overline{c} + \overleftarrow{a_1} \vdash \overleftarrow{i} + \overrightarrow{a_3}$$

$$\overset{g_5}{\underset{}{}}$$
$$\overrightarrow{a_3} + \overleftarrow{d_1} \vdash \overline{p}$$

$$\overset{g_4}{\underset{}{}}$$
$$\overline{c} + \overleftarrow{a_5} + \overleftarrow{a_4} \vdash \overleftarrow{i}$$

$$\overset{g_3}{\underset{}{}}$$
$$\overline{c} + \overleftarrow{i} \vdash \overleftarrow{a_5} + \overleftarrow{a_4} + \overrightarrow{a_3}$$

☐ **Fig. 35**

110 collisions used in the construction (cont.).

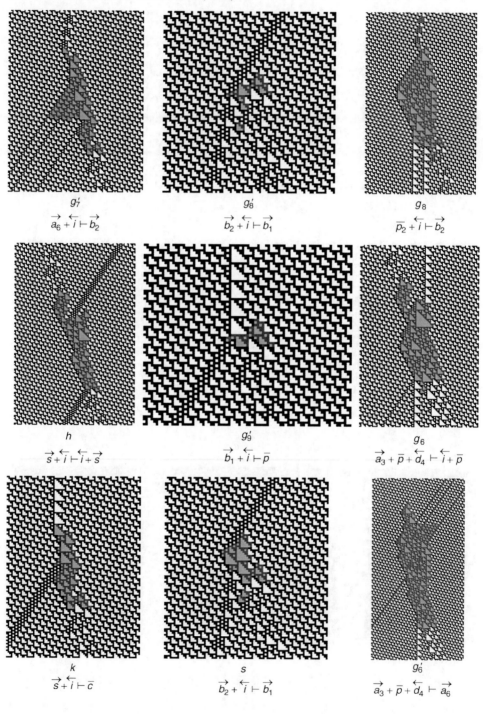

g_7'

$\overrightarrow{a_6} + \overleftarrow{i} \vdash \overrightarrow{b_2}$

g_8'

$\overrightarrow{b_2} + \overleftarrow{i} \vdash \overrightarrow{b_1}$

g_8

$\overline{p_2} + \overleftarrow{i} \vdash \overrightarrow{b_2}$

h

$\overrightarrow{s} + \overleftarrow{i} \vdash \overleftarrow{i} + \overrightarrow{s}$

g_9'

$\overrightarrow{b_1} + \overleftarrow{i} \vdash \overline{p}$

g_6

$\overrightarrow{a_3} + \overline{p} + \overrightarrow{d_4} \vdash \overleftarrow{i} + \overline{p}$

k

$\overrightarrow{s} + \overleftarrow{i} \vdash \overline{c}$

s

$\overrightarrow{b_2} + \overleftarrow{i} \vdash \overrightarrow{b_1}$

g_6'

$\overrightarrow{a_3} + \overline{p} + \overrightarrow{d_4} \vdash \overrightarrow{a_6}$

☐ **Fig. 36**
110 collisions used in the construction (end).

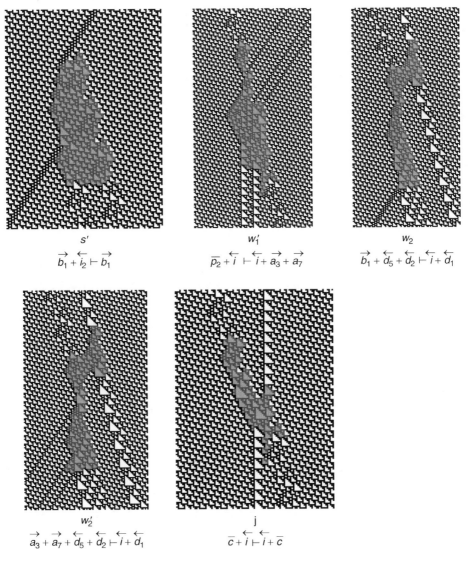

$$s'$$
$$\overrightarrow{b_1} + \overleftarrow{i_2} \vdash \overrightarrow{b_1}$$

$$w_1'$$
$$\overline{p_2} + \overrightarrow{i} \vdash \overleftarrow{i} + \overrightarrow{a_3} + \overrightarrow{a_7}$$

$$w_2$$
$$\overrightarrow{b_1} + \overleftarrow{d_5} + \overleftarrow{d_2} \vdash \overleftarrow{i} + \overleftarrow{d_1}$$

$$w_2'$$
$$\overrightarrow{a_3} + \overrightarrow{a_7} + \overleftarrow{d_5} + \overleftarrow{d_2} \vdash \overleftarrow{i} + \overleftarrow{d_1}$$

$$j$$
$$\overline{c} + \overrightarrow{i} \vdash \overleftarrow{i} + \overline{c}$$

8.1   Techniques

Several techniques are used to construct them:

Parallel Turing machines table lookup A simple way to achieve intrinsic universality is to use synchronized parallel Turing heads (one copy of the same Turing machine per encoded cell) to lookup in the transition table (one copy in each encoded cell) of the encoded cellular automaton. Notice that the Turing machines used for this are not the same ones that are

⬛ Fig. 37

110 symbolic description of the simulation.

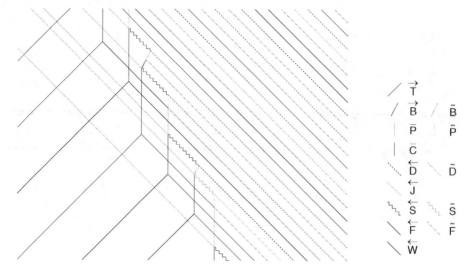

Turing universal. In fact, their computational power is very small but they can carry precise information movement tasks.

One-way totalistic lookup Another more cellular automata-centric way to achieve intrinsic universality is, following Albert and Čulik (1987), to simplify the task of the previous machine by simulating only one-way totalistic cellular automata that are sufficient to simulate all cellular automata.

Signals The previous models are complex because the information displacement involved is still complex due to the sequential behavior of head-based machines. Following Ollinger (2002b) and Ollinger and Richard (2008), particles and collisions, that is, signals in the style of Mazoyer and Terrier (1999), can be used to encode the information and perform the lookup task with parallel information displacement.

8.2 Parallel Turing Machines Table Lookup

The parallel Turing machines table lookup technique is explained here, the other ones being refinements based on it. The one-dimensional first-neighbors universal cellular automaton $\mathcal{U}$ simulates a cellular automaton $(1, S, \{-1, 0, 1\}, f)$ the following way, as depicted in ❷ *Fig. 38.* Each configuration c is encoded as the concatenation of $\psi_{\mathcal{A}}(c(z))$ for all z. For each state $s \in S$, $\psi_{\mathcal{A}}(s)$ is a word of the kind $\blacksquare \tau(f(1, 1, 1)) \bullet \tau(f(1, 1, 2)) \bullet \ldots \bullet \tau(f(N, N, N)) \blacktriangle 0^k \tau(s) 0^k 0^k$ where N is the size of S, k is the number of bits needed to encode numbers from 1 to N, and $\tau(s)$ is a binary encoding of the state s. The simulation proceeds as follows so that the movement of each head is the same, up to translation, on each well encoded configuration. First the $\blacksquare$ letter is activated as a Turing head in initial state. The Turing head then moves to the left and copies the state $\tau(s_L)$ of the left neighbor in place of the first 0^k block. Then it symmetrically copies the state $\tau(s_R)$ of the right neighbor in place of the second 0^k block.

■ **Fig. 38**

Principle of parallel Turing machine table lookup.

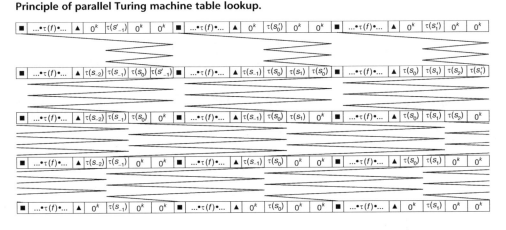

This being done, the head scans the entire transition table, incrementing a counter (e.g., stored on top of the encoded states) at each step: if at some point the counter is equal to the triple of states, the result is copied from the transition table to the third 0^k block. At the end of the scan, the counter information is cleared, the result of the transition is copied in the $\tau(s)$ place and all three 0^k blocks are restored. The head then goes back to the ■. Using this simulation, one step of the simulated cellular automaton is simulated in a constant number of steps for each cell by each Turing head. The universal automaton does not depend on the simulated automaton and is so intrinsically universal. A careful design can lead to less than 20 states. If the simulation uses the one-way totalistic technique, encoding states in unary, then it is easy to go under 10 states.

Notice that general Turing universal cellular automata construction schemes from previous section concerning Turing machines can be adapted to produce small intrinsically universal cellular automata: apply the encoding schemes on machines performing the cell simulation task. However, the tag system simulations do not provide direct way to obtain intrinsic universality. Moreover, it is possible to design cellular automata Turing universal by tag system simulation and not intrinsically universal.

8.3   Four States

The smallest known intrinsically universal one-dimensional cellular automaton with first neighbors has four states. Due to its size, universality was obtained by a combination of design and analysis. There is not enough room here to completely describe the construction, so the reader will only be given an idea of its taste. For a complete description of the proof, read Ollinger and Richard (2008).

The local rule is given in ❷ *Fig. 39* with five particles and several families of signals (groups of particles encoding integers and integer sequences). The principle of the construction (well, in fact it is a bit more complicated due to encoding problems) is to simulate a one-way cellular automaton encoding states as an integer encoded into a signal. The simulation sums the signals from the cell and its neighbors and reads the result of the transition in a transition table

⬛ **Fig. 39**

4st local rule, background, particles, and signals.

Local rule

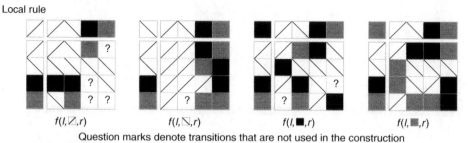

$f(l,\boxslash,r)$ $f(l,\boxbslash,r)$ $f(l,\blacksquare,r)$ $f(l,\blacksquare,r)$

Question marks denote transitions that are not used in the construction

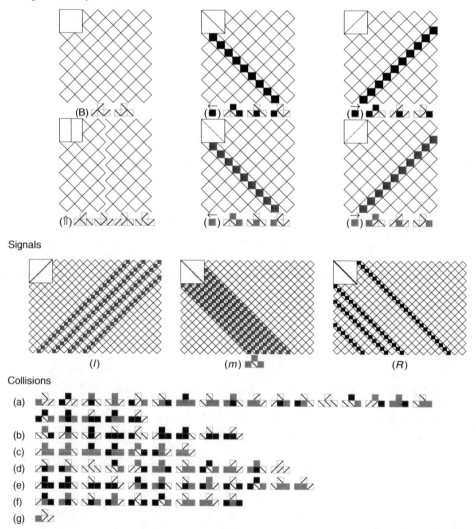

Background and particles

Signals

Collisions

■ Fig. 40
4st simulation scheme.

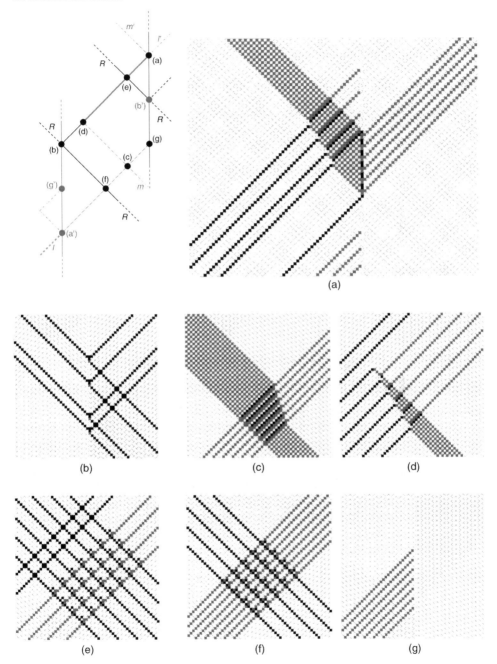

(a)

(b)   (c)   (d)

(e)   (f)   (g)

encoded in a signal as a sequence of integers. Finally, it duplicates the result. The simulation is depicted in ❷ *Fig. 40* including the different collisions.

8.4 Computing Under Constraints

More exotic intrinsically universal cellular automata have been studied on constrained rules. As an example, Moreira (2003) constructed number-conserving intrinsically universal automata, Bartlett and Garzon (1993, 1995) and Ollinger (2001) did the same for bilinear cellular automata, and Theyssier (2005b) for captive cellular automata for which he proves that almost all captive automata are intrinsically universal.

9 Sub-universalities: The Reversible Case

Universality might also be considered a proper subfamily of cellular automata, giving birth to some notions of sub-universality. Reversible cellular automata are special in the sense that they can achieve Turing universality as any Turing machine can be simulated by a reversible Turing machine but they cannot achieve intrinsic universality: reversible cellular automata only simulate reversible cellular automata. However, there exists reversible cellular automata which are universal with respect to the class of reversible cellular automata.

The chapter ❷ Reversible Cellular Automata of the present volume, authored by Morita, is dedicated to reversible cellular automata, the reader is referred to that chapter for a proper discussion on reversible computing and reversible Boolean circuits.

Definition 2   (reversible intrinsic universality) A reversible cellular automaton $\mathscr{U}$ is intrinsically universal for reversible cellular automata if for each reversible cellular automaton $\mathscr{A}$ of the same dimension there exists an unpacking map o_{m}, a positive integer $n \in \mathbb{N}$, and a translation vector $v \in \mathbb{Z}^d$ such that $G_{\mathscr{A}} \prec o_{\mathrm{m}}^{-1} \circ G_{\mathscr{U}}^n \circ o_{\mathrm{m}} \circ \sigma_v$.

Turing universality and a weak form of intrinsic universality have been proposed by Morita (1995), Morita and Imai (1996, 2001), Durand-Lose (1995, 1996), and Miller and Fredkin (2005). As for classical cellular automata in higher dimensions, the simulation of reversible Boolean circuits automatically gives reversible intrinsic universality.

For one-dimensional cellular automata, reversible intrinsic universality can be achieved by simulating any one-way-reversible partitioned-reversible cellular automaton with a first-neigbor-reversible partitioned-reversible cellular automaton. How to use a scheme similar to parallel Turing machine table lookup with reversible Turing machines to achieve this goal is briefly sketched. The first adaptation is to remark that the local partition rule of a reversible automaton is a permutation, thus it can be encoded as a finite sequence of permutation pairs. So, the table of transition is encoded as a finite sequence of pairs of states. The reversible Turing-machine task is to scan the transition table and for each pair it contains to replace the actual state by the second element of the pair if the state appears in the current pair. It is technical but straightforward to see that a reversible Turing machine can achieve this. The information movement is reversible, as in partitioned automata each cell gives half its state to a neighbor and takes half a state from the other without erasing any information.

Developing this simulation scheme, one constructs a reversible intrinsically universal cellular automaton.

10 Conclusion

This chapter has given the reader keys both to further explore the literature and to construct by himself conceptually simple Turing universal and/or intrinsically universal cellular automata in one and two dimensions. The following broad bibliography can be explored with the help of the main text, all references being cited.

Acknowledgment

The author would like to thank G. Richard for providing and preparing figures to illustrate both rule 110 and the 4-state intrinsically universal cellular automaton.

References

Albert J, Čulik II K (1987) A simple universal cellular automaton and its one-way and totalistic version. Complex Syst 1(1):1–16

Amoroso S, Patt YN (1972) Decision procedures for surjectivity and injectivity of parallel maps for tessellation structures. J Comput Syst Sci 6:448–464

Arbib MA (1966) Simple self-reproducing universal automata. Info Control 9(2):177–189

Banks ER (1970) Universality in cellular automata. In: Symposium on switching and automata theory, Santa Monica, 1970. IEEE, New York, pp 194–215

Banks ER (1971) Information processing and transmission in cellular automata. Ph.D. thesis, Massachusetts Institute of Technology

Bartlett R, Garzon M (1993) Monomial cellular automata. Complex Syst 7(5):367–388

Bartlett R, Garzon M (1995) Bilinear cellular automata. Complex Syst 9(6):455–476

Bennett CH (1973) Logical reversibility of computation. IBM J Res Dev 17(6):525–532

Berlekamp ER, Conway JH, Guy RK (1982) Winning ways for your mathematical plays, vol 2, Games in particular. Academic Press [Harcourt Brace Jovanovich Publishers], London

Burks AW (1970a) Von Neumann's self-reproducing automata. In: Burks AW (ed) Essays on cellular automata. University of Illinois Press, Urbana, IL, pp 3–64 (Essay one)

Burks AW (1970b) Programming and the theory of automata. In: Burks AW (ed) Essays on cellular automata. University of Illinois Press, Urbana, IL, pp 65–83 (Essay two)

Burks AW (1970c) Towards a theory of automata based on more realistic primitive elements. In: Burks AW (ed) Essays on cellular automata. University of Illinois Press, Urbana, IL, pp 84–102 (Essay three)

Cocke J, Minsky M (1964) Universality of tag systems with p=2. J ACM 11(1):15–20

Codd EF (1968) Cellular automata. Academic Press, New York

Cook M (2004) Universality in elementary cellular automata. Complex Syst 15:1–40

Davis MD (1956) A note on universal turing machines. In: Shannon CE, McCarthy J (eds) Automata studies. Princeton University Press, Princeton, NJ, pp 167–175

Davis MD (2000) The universal computer: The road from Leibniz to Turing. W.W. Norton, New York

Delorme M (1999) An introduction to cellular automata: some basic definitions and concepts. In: Delorme M, Mazoyer J (eds) Cellular automata (Saissac, 1996). Kluwer, Dordrecht, pp 5–49

Delvenne J, Kurka P, Blondel VD (2006) Decidability and universality in symbolic dynamical systems. Fundam Info 74(4):463–490

Dewdney AK (1990) The cellular automata programs that create Wireworld, Rugworld and other diversions. Sci Am 262:146–149

Dubacq JC (1995) How to simulate Turing machines by invertible one-dimensional cellular automata. Int J Found Comput Sci 6(4):395–402

Durand B, Róka Z (1999) The Game of Life: universality revisited. In: Delorme M, Mazoyer J (eds) Cellular automata (Saissac, 1996). Kluwer, Dordrecht, pp 51–74

Durand-Lose J (1995) Reversible cellular automaton able to simulate any other reversible one using partitioning automata. In: Baeza-Yates RA, Goles E, Poblete PB (eds) LATIN '95, Valparaiso, 1995. Lecture notes in computer science, vol 911. Springer, Berlin/New York, pp 230–244

Durand-Lose J (1996) Automates cellulaires, automates partitions et tas de sable. Ph.D. thesis, Université Bordeaux I

Durand-Lose J (1997) Intrinsic universality of a 1-dimensional reversible cellular automaton. In: STACS 97 (Lübeck), Lecture notes in computer science, vol 1200. Springer, Berlin, pp 439–450

Fischer PC (1965) On formalisms for Turing machines. J ACM 12(4):570–580

Fredkin E, Toffoli T (1982) Conservative logic. Int J Theor Phys 21:219–253

Gajardo A, Ch EG (2001) Universal cellular automaton over a hexagonal tiling with 3 states. IJAC 11(3):335–354

Gardner M (1970) Mathematical games: The fantastic combinations of John Conway's new solitaire game 'Life'. Sci Am 223(4):120–123

Gödel K (1931) Über formal unentscheidbare Sätze der Principia Mathematica und verwandter Systeme I. Monatshefte für Mathematik und Physik 38:173–198. Reprinted in Feferman S, Dawson W, Kleene SC, Moore G, Solovay RM, van Heijendort J (eds) (1986) Kurt Gödel: Collected works, vol 1. Oxford University Press, Oxford, pp 144–195

Golly Gang: Golly version 2.0 (2008) An open-source, cross-platform application for exploring the Game of Life and other cellular automata. http://golly.sf.net/

Hedlund GA (1969) Endormorphisms and automorphisms of the shift dynamical system. Math Syst Theory 3:320–375

Hertling P (1998) Embedding cellular automata into reversible ones. In: Calude CS, Casti J, Dinneen MJ (eds) Unconventional models of computation. Springer

Imai K, Morita K (2000) A computation-universal two-dimensional 8-state triangular reversible cellular automaton. Theor Comput Sci 231(2):181–191

Kari J (1990) Reversibility of 2D cellular automata is undecidable. Phys D. Nonlinear Phenomena 45(1–3):379–385. Cellular automata: theory and experiment, Los Alamos, NM, 1989

Kari J (1994) Reversibility and surjectivity problems of cellular automata. J Comput Syst Sci 48(1):149–182

Kari J (2005) Theory of cellular automata: a survey. Theor Comput Sci 334:3–33

Kleene SC (1956) Representation of events in nerve nets and finite automata. In: Shannon CE, McCarthy J (eds) Automata studies. Princeton University Press, Princeton, NJ, pp 3–41

Lafitte G (2008) Gödel incompleteness revisited. In: Durand B (ed) Symposium on cellular automata journes automates cellulaires, JAC'2008, Uzès, 2008. MCCME, Moscow, pp 74–89

Langton C (1984) Self-reproduction in cellular automata. Phys D 10(1–2):135–144

Lecerf Y (1963) Machines de Turing réversibles. C R Acad Sci Paris 257:2597–2600

Lindgren K, Nordahl MG (1990) Universal computation in simple one-dimensional cellular automata. Complex Syst 4(3):299–318

Margolus N (1984) Physics-like models of computation. Phys D 10:81–95

Martin B (1993) Construction modulaire d'automates cellulaires. Ph.D. thesis, École Normale Supérieure de Lyon

Martin B (1994) A universal cellular automaton in quasi-linear time and its S-m-n form. Theor Comput Sci 123(2):199–237

Mazoyer J (1996) Computations on one dimensional cellular automata. Ann Math Artif Intell 16:285–309

Mazoyer J (1999) Computations on grids. In: Cellular automata (Saissac, 1996). Kluwer, Dordrecht, pp 119–149

Mazoyer J, Rapaport I (1999) Inducing an order on cellular automata by a grouping operation. Discrete Appl Math 91(1–3):177–196

Mazoyer J, Terrier V (1999) Signals in one-dimensional cellular automata. Theor Comput Sci 217(1):53–80. Cellular automata, Milan, 1996

McCulloch WS, Pitts W (1943) A logical calculus of the ideas immanent in nervous activity. Bull Math Biophys 5:115–133

Miller DB, Fredkin E (2005) Two-state, reversible, universal cellular automata in three dimensions. In: Bagherzadeh N, Valero M, Ramírez A (eds) Proceedings of the conference on computing frontiers, Ischia, pp 45–51. ACM

Minsky M (1967) Computation: finite and infinite machines. Prentice Hall, Englewoods Cliffs, NJ

Moore EF (1962) Machine models of self-reproduction. In: Proceedings of symposia in applied mathematics, vol 14, 1962. American Mathematical Society, Providence, pp 17–33

Moore EF (1970) Machine models of self-reproduction. In: Burks AW (ed) Essays on cellular automata. University of Illinois Press, Urbana, IL, pp 187–203 (Essay six)

Moreira A (2003) Universality and decidability of number-conserving cellular automata. Theor Comput Sci 292(3):711–721

Morita K (1990) A simple construction method of a reversible finite automaton out of Fredkin gates, and its related problem. IEICE Trans Info Syst E73(6):978–984

Morita K (1995) Reversible simulation of one-dimensional irreversible cellular automata. Theor Comput Sci 148(1):157–163

Morita K, Harao M (1989) Computation universality of one-dimensional reversible (injective) cellular automata. IEICE Trans Info Syst E72:758–762

Morita K, Imai K (1996) Self-reproduction in a reversible cellular space. Theor Comput Sci 168(2):337–366

Morita K, Imai K (2001) Number-conserving reversible cellular automata and their computation-universality. Theor Info Appl 35(3):239–258

Neary T, Woods D (2006) P-completeness of cellular automaton rule 110. In: Proceedings of ICALP 2006, Venice. Lecture notes in Computer science, vol 4051. Springer, Berlin, pp 132–143

von Neumann J (1966) In: Burks AW (ed) Theory of self-reproducing automata. University of Illinois Press, Urbana, IL

Noural F, Kashef R (1975) A universal four-state cellular computer. IEEE Trans Comput 24(8):766–776

Odifreddi PG (1989) Classical recursion theory. Elsevier, Amsterdam

Ollinger N (2001) Two-states bilinear intrinsically universal cellular automata. In: Freivalds R (ed) Fundamentals of computation theory, Riga, 2001. Lecture notes in computer science, vol 2138. Springer, Berlin, pp 396–399

Ollinger N (2002a) Automates cellulaires: structures. Ph.D. thesis, École Normale Supérieure de Lyon

Ollinger N (2002b) The quest for small universal cellular automata. In: Widmayer P, Triguero F, Morales R, Hennessy M, Eidenbenz S, Conejo R (eds) International colloquium on automata, languages and programming, (Málaga, 2002). Lecture notes in computer science, vol 2380. Springer, Berlin, pp 318–329

Ollinger N (2003) The intrinsic universality problem of one-dimensional cellular automata. In: Symposium on theoretical aspects of computer science, Berlin, 2003. Lecture notes in computer science. Springer, Berlin, doi:10.1007/3-540-36494-3_55

Ollinger N (2008) Universalities in cellular automata: a (short) survey. In: Durand B (ed) Symposium on cellular automata journes automates cellulaires, JAC'2008, Użes. MCCME, Moscow, pp 102–118

Ollinger N, Richard G (2008) A particular universal cellular automaton. In: Neary T, Woods D, Seda AK, Murphy N (eds) CSP. Cork University Press, Cork, Ireland

Perrin D (1995) Les débuts de la théorie des automates. Tech Sci Info 14:409–433

Post E (1941) The two-valued iterative systems of mathematical logic. Princeton University Press, Princeton, NJ

Post E (1943) Formal reductions of the general combinatorial decision problem. Am J Math 65(2):197–215

Rapaport I (1998) Inducing an order on cellular automata by a grouping operation. Ph.D. thesis, École Normale Supérieure de Lyon

Richard G (2008) Rule 110: Universality and catenations. In: Durand B (ed) Symposium on cellular automata journes automates cellulaires, JAC'2008, Użes. MCCME, Moscow, pp 141–160

Richardson D (1972) Tessellations with local transformations. J Comput Syst Sci 6:373–388

Shannon CE (1956) A universal Turing machine with two internal states. In: Shannon CE, McCarthy J (eds) Automata studies. Princeton University Press, Princeton, NJ, pp 157–165

Smith AR III (1971) Simple computation-universal cellular spaces. J ACM 18:339–353

Sutner K (2004) Universality and cellular automata. In: Margenstern M (ed) MCU 2004, Saint-Petersburg, 2004. Lecture notes in computer science, vol 3354. Springer, Berlin/Heidelberg, pp 50–59

Thatcher JW (1970a) Self-describing Turing machines and self-reproducing cellular automata. In: Burks AW (ed) Essays on cellular automata. University of Illinois Press, Urbana, IL, pp 103–131 (Essay four)

Thatcher JW (1970b) Universality in the von Neumann cellular model. In: Burks AW (ed) Essays on cellular automata. University of Illinois Press, Urbana, IL, pp 132–186 (Essay five)

Theyssier G (2005a) Automates cellulaires: un modèle de complexités. Ph.D. thesis, École Normale Supérieure de Lyon

Theyssier G (2005b) How common can be universality for cellular automata? In: STACS 2005, Stuttgart, 2005. Lecture notes in computer science, vol 3404. Springer, Berlin/Heidelberg, pp 121–132

Toffoli T (1977) Computation and construction universality of reversible cellular automata. J Comput Syst Sci 15(2):213–231

Turing AM (1936) On computable numbers with an application to the Entscheidungsproblem. Proc Lond Math Soc 42(2):230–265

Wolfram S (1984) Universality and complexity in cellular automata. Phys D Nonlinear Phenomena 10(1–2):1–35. Cellular automata: Los Alamos, NM, 1983

Wolfram S (2002) A new kind of science. Wolfram Media, Champaign, IL

Woods D, Neary T (2006) On the time complexity of 2-tag systems and small universal Turing machines. In: FOCS 2006, Berkeley, 2006. IEEE Computer Society, Washington, DC, pp 439–448

Woods D, Neary T (2007) The complexity of small universal Turing machines. In: Cooper SB, Löwe B, Sorbi A (eds) Computability in Europe, CiE 2007, Siena, 2007. Lecture notes in computer science, vol 4497. Springer, Berlin/Heidelberg, pp 791–799

7 Reversible Cellular Automata

Kenichi Morita
Department of Information Engineering, Graduate School of
Engineering, Hiroshima University, Japan
morita@iec.hiroshima-u.ac.jp

|---|---|---|
2	Definitions and Basic Properties	233
3	Finding Reversible Cellular Automata	236
4	Computing Ability of Reversible Cellular Automata	240
5	Concluding Remarks	254

G. Rozenberg et al. (eds.), *Handbook of Natural Computing*, DOI 10.1007/978-3-540-92910-9_7,
© Springer-Verlag Berlin Heidelberg 2012

Abstract

A reversible cellular automaton (RCA) is a special type of cellular automaton (CA) such that every configuration of it has only one previous configuration, and hence its evolution process can be traced backward uniquely. Here, we discuss how RCAs are defined, their properties, how one can find and design them, and their computing abilities. After describing definitions on RCAs, a survey is given on basic properties on injectivity and surjectivity of their global functions. Three design methods of RCAs are then given: using CAs with block rules, partitioned CAs, and second-order CAs. Next, the computing ability of RCAs is discussed. In particular, we present simulation methods for irreversible CAs, reversible Turing machines, and some other universal systems by RCAs, in order to clarify the universality of RCAs. In spite of the strong constraint of reversibility, it can be seen that RCAs have rich abilities in computing and information processing, and even very simple RCAs have universal computing ability.

1   Introduction

An RCA is a CA whose global transition function (i.e., a mapping from configurations to configurations) is injective. The study of RCAs has a relatively long history. In the early 1960s, Moore (1962) and Myhill (1963) showed the so-called Garden-of-Eden theorem, which gives a relation between injectivity and surjectivity of a global function. Note that a Garden-of-Eden configuration is one that can exist only at time 0, and hence if there is such a configuration in a CA, then its global function is not surjective. Since then, injectivity and surjectivity of global functions were studied extensively and more generally. But the term "reversible" CA was not used in the early stage of the history.

The notion of reversibility comes from physics. It is one of the fundamental microscopic physical laws of nature. Landauer (1961) argued how physical reversibility is related to "logical reversibility" (roughly speaking, logical reversibility is a property of "backward determinism" in a computing system). He posed Landauer's principle that states that an irreversible logical operation inevitably causes heat generation in a physical computing system. After that, various models of reversible computing, such as reversible Turing machines (Bennett 1973; Lecerf 1963) and reversible logic circuits (Fredkin and Toffoli 1982; Toffoli 1980), appeared, and the research field of reversible computing was formed.

Toffoli (1977) first studied RCAs from such a viewpoint. He showed that every k-dimensional (irreversible) CA can be embedded in a $k + 1$-dimensional RCA, and hence a two-dimensional RCA is computation-universal. An RCA can be thought of as an abstract model of physically reversible space. Therefore, it is a useful framework to formalize and study the problem of how computation and information processing are performed efficiently in physically reversible spaces. Since then, much research on RCAs was performed, and it gradually turned out that RCAs have rich abilities in computing, though an RCA is a very special subclass of CA.

An outline of this chapter is as follows. In ❷ Sect. 2, it is first shown how RCAs are defined, and then there is a discussion on their basic properties of injectivity and surjectivity. In ❷ Sect. 3, useful frameworks for designing RCAs are discussed. This is because it is in general hard to find RCAs if the standard framework of CAs is used. Here, three frameworks are explained: CAs with block rules, partitioned CAs, and second-order CAs. In ❷ Sect. 4, simulation methods of other (irreversible) CAs, reversible Turing machines, cyclic tag systems, and reversible logic circuits, by one- and two-dimensional RCAs, are given. By this, one can see computation universality of such RCAs, and some of them can be very simple.

2 Definitions and Basic Properties

In this section, formal definitions on cellular automata (CAs), and RCAs are given. To define reversibility of a CA, the notions of injectivity and invertibility are needed, which later turn out to be equivalent. Basic properties related to RCAs are also discussed.

2.1 Definitions

A CA is a system consisting of infinitely many finite automata (called cells) placed and interconnected uniformly in a space. Each cell changes its state depending on the current states of the neighbor cells, according to a given local transition function. It is assumed that all the cells change their states synchronously at every discrete time step. Hence, the whole state (called configuration) of the cellular space also changes at every discrete time, and this gives a global transition function. Until now many interesting variants of CAs, such as asynchronous CAs, CAs in a non-Euclidian space, and so on, have been proposed and studied, but here we restrict them only to standard ones, that is, deterministic CAs in the k-dimensional Euclidian space. A formal definition of them is given as follows.

Definition 1 A *k-dimensional m-neighbor cellular automaton (CA)* is a system defined by

$$A = (\mathbb{Z}^k, Q, (n_1, \ldots, n_m), f)$$

Here, $\mathbb{Z}$ is the set of all integers, and thus $\mathbb{Z}^k$ is the set of all k-dimensional points with integer coordinates at which cells are placed. Q is a nonempty finite set of states of each cell, and $(n_1, \ldots, n_m)$ is an element of $(\mathbb{Z}^k)^m$ called a *neighborhood* $(m = 1, 2, \ldots)$. $f: Q^m \to Q$ is a *local function*, which determines how each cell changes its state depending on the states of m neighboring cells.

A *k-dimensional configuration* over Q is a mapping $\alpha : \mathbb{Z}^k \to Q$. Let $\mathrm{Conf}(Q)$ denote the set of all k-dimensional configurations over Q, that is, $\mathrm{Conf}(Q) = \{\alpha \mid \alpha : \mathbb{Z}^k \to Q\}$. If some state $\# \in Q$ that satisfy $f(\#, \ldots, \#) = \#$ is specified as a *quiescent state*, then the notion of finiteness of a configuration can be defined as follows. A configuration α is called *finite* iff the set $\{x \mid x \in \mathbb{Z}^k \wedge \alpha(x) \neq \#\}$ is finite. Otherwise it is *infinite*. Let $\mathrm{Conf_{fin}}(Q)$ denote the set of all k-dimensional finite configurations over Q, that is, $\mathrm{Conf_{fin}}(Q) = \{\alpha \mid \alpha \in \mathrm{Conf}(Q) \wedge \alpha$ is finite$\}$.

The *global function* $F: \mathrm{Conf}(Q) \to \mathrm{Conf}(Q)$ of A determines how a configuration changes to other configuration. It is defined by the following formula.

$$\forall \alpha \in \mathrm{Conf}(Q), \ x \in \mathbb{Z}^k : \ F(\alpha)(x) = f(\alpha(x + n_1), \ldots, \alpha(x + n_m))$$

Let F_{fin} denote the restriction of F to $\mathrm{Conf_{fin}}(Q)$ provided that a quiescent state is specified. One can see every finite configuration is mapped to a finite configuration by F. Therefore, F_{fin} defines a mapping $\mathrm{Conf_{fin}}(Q) \to \mathrm{Conf_{fin}}(Q)$.

A reversible CA will be defined as a special subclass of a (standard) CA. To do so, two notions related to reversibility of CAs are first introduced. They are injectivity and invertibility.

Definition 2 Let $A = (\mathbb{Z}^k, Q, (n_1, \ldots, n_m), f)$ be a CA.

1. A is called an *injective CA* iff its global function F is one-to-one.
2. A is called an *invertible CA* iff there exists a CA $A' = (\mathbb{Z}^k, Q, N', f')$ that satisfies the following condition.

$$\forall \alpha, \beta \in \text{Conf}(Q) : \ F(\alpha) = \beta \text{ iff } F'(\beta) = \alpha$$

where F and F' are the global functions of A and A', respectively. Here, A' is called the *inverse CA* of A.

It is clear from the above definition that if a CA is invertible then it is injective. One can see that its converse also holds from the results independently proved by Hedlund (1969) and Richardson (1972). This is stated by the following theorem.

Theorem 1 (Hedlund 1969 and Richardson 1972) *Let A be a CA. A is injective iff it is invertible.*

Since the notions of injectivity and invertibility in CAs are thus proved to be equivalent, the terminology *reversible CA* (RCA) for such a CA will be used hereafter. Note that, in the case of reversible Turing machines (Bennett 1973), or reversible logic circuits (Fredkin and Toffoli 1982), the situation is somewhat simpler. In these models, injectivity is defined on their "local" operations. For example, a reversible logic circuit is a one consisting of logic elements with injective logical functions. Hence, the operation (or the state transition) of the whole circuit is also injective. Furthermore, by replacing each logic element by the one with the inverse logical function an inverse logic circuit that "undoes" the original one can be easily obtained. Thus, injectivity of a circuit is trivially equivalent to its invertibility. Therefore, in these models, reversibility can be directly defined without introducing the two notions of injectivity and invertibility.

2.2   Injectivity and Surjectivity of CAs

Historically, injectivity and surjectivity of a global function of CAs were first studied in the Garden-of-Eden problem (Moore 1962; Myhill 1963). A configuration α is called a *Garden-of-Eden configuration* of a CA A iff it has no preimage under the global function F, that is, $\forall \alpha' \in \text{Conf}(Q) : F(\alpha') \neq \alpha$. Hence, existence of a Garden-of-Eden configuration is equivalent to non-surjectivity of the global function F. Moore (1962) first showed the following proposition: if there are "erasable configurations" (which are mapped to the same configuration by the global function) then there is a Garden-of-Eden configuration (it is sometimes called Moore's Garden-of-Eden theorem). Later, Myhill (1963) proved its converse. Their results are now combined and stated in the following theorem called the Garden-of-Eden theorem.

Theorem 2 (Moore 1962 and Myhill 1963) *Let A be a CA, and F be its global function. F is surjective iff F_{fin} is injective.*

Since it is clear if F is injective then F_{fin} is also so, the following corollary is derived.

Corollary 1 *If F is injective, then it is surjective.*

Note that there is a CA that is surjective but not injective (Amoroso and Cooper 1970). Hence the converse of this corollary does not hold. After the works of Moore and Myhill, relations among injectivity and surjectivity in unrestricted, finite, and periodic configurations have been extensively studied (Amoroso and Cooper 1970; Maruoka and Kimura 1976, 1979; Richardson 1972).

Next, decision problems on injectivity and surjectivity on CAs are considered. For the case of one-dimensional CAs, injectivity and surjectivity of the global function are both decidable.

Theorem 3 (Amoroso and Patt 1972) *There is an algorithm to determine whether the global function of a given one-dimensional CA is injective (surjective) or not.*

Later, a quadratic time algorithm to determine the injectivity (surjectivity) of a global function using de Bruijn graphs was given by Sutner (1991).

For two-dimensional CAs, injectivity (i.e., equivalently reversibility) and surjectivity are both undecidable. Thus, they are also undecidable for the higher dimensional case. The following theorem was proved by a reduction from the tiling problem, which is a well-known undecidable problem.

Theorem 4 (Kari 1994) *The problem whether the global function of a given two-dimensional CA is injective (surjective) is undecidable.*

2.3 Inverse CAs and Their Neighborhood Size

For each reversible CA, there is an inverse CA, which undoes the operations of the former. In the two-dimensional case, the following fact on the inverse CAs (Kari 1994) can be observed. Let A be an arbitrary two-dimensional reversible CA, and A' be the inverse CA of A. One cannot, in general, compute an upper-bound of the neighborhood size $|N'|$ from A. If one can assume, on the contrary, such an upper-bound is computable, then the total number of candidates of the inverse CA of A is also bounded. Since it is easy to test if a CA $\tilde{A}$ is the inverse of the CA A for any given $\tilde{A}$, the inverse of A can be found among the above candidates in finite steps. From this, one can design an algorithm to decide whether a given CA is reversible or not, which contradicts ❷ Theorem 4. It is stated in the following theorem.

Theorem 5 (Kari 1994) *There is no computable function g that satisfies $|N'| \leq g(|Q|, |N|)$ for any two-dimensional reversible CA $A = (\mathbb{Z}^2, Q, N, f)$ and its inverse $A' = (\mathbb{Z}^2, Q, N', f')$.*

On the other hand, for the case of one-dimensional CAs, the following result has been shown.

Theorem 6 (Czeizler and Kari 2007) *Let $A = (\mathbb{Z}, Q, N, f)$ be an arbitrary one-dimensional reversible CA, and $A' = (\mathbb{Z}, Q, N', f')$ be its inverse, where $|Q| = n$ and $|N| = m$. If N consists of m consecutive cells, then $|N'| \leq n^{m-1} - (m-1)$. Furthermore, when $m = 2$, this upper-bound is tight.*

3 Finding Reversible Cellular Automata

As seen in ❷ Sect. 2.2, it is in general hard to find RCAs. From ❷ Theorem 4, it is not possible to test if a given two-dimensional CA is reversible or not. It causes a difficulty to design RCAs with specific properties like computation-universality. In the case of one-dimensional CAs, there is a decision algorithm for reversibility (❷ Theorem 3), and there are also several studies on enumerating all one-dimensional RCAs (e.g., Boykett 2004 and Mora et al. 2005). But, since the decision algorithm is not so simple, it is still difficult to design RCAs with specific properties by using the standard framework of CAs even in the one-dimensional case.

So far several methods have been proposed to make it feasible to design RCAs. They are CAs with block rules (Margolus 1984; Toffoli and Margolus 1990), partitioned CAs (Morita and Harao 1989), CAs with second-order rules (Margolus 1984; Toffoli and Margolus 1990; Toffoli et al. 2004), and others.

3.1 CAs with Block Rules

A local function of a standard CA is a mapping $f: Q^m \rightarrow Q$, and therefore it is impossible to make f itself be one-to-one except the trivial case where $|Q| = 1$ or $m = 1$. Instead, if a local function is of the form $f: Q^m \rightarrow Q^m$, then it can be one-to-one. Assume all the cells are grouped into blocks each of which consists of m cells. Such a local function maps a state of each block to a new state of the same block. If f is one-to-one, and is applied to all the blocks in a given configuration in parallel, then the induced global function will be also one-to-one. Of course, at the next time step, the grouping into blocks should be changed. Otherwise, interaction between blocks is impossible. Such a CA is often called a block CA.

Margolus (1984) first proposed this type of CA, by which he composed a specific model of a computation-universal two-dimensional two-state RCA. In his model, all the cells are grouped into blocks of size 2×2 as in ❷ Fig. 1. Its local function is specified by "block rules," and is shown in ❷ Fig. 2. This CA evolves as follows: At time 0, the local function is applied to every solid line block, then at time 1 to every dotted line block, and so on, alternately. This kind of neighborhood is called Margolus neighborhood. One can see that the local function defined by the block rules in ❷ Fig. 2 is one-to-one (such a one-to-one local

◘ **Fig. 1**
Grouping cells into blocks of size 2 × 2 by solid or dotted lines, which form the Margolus neighborhood.

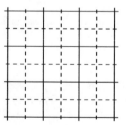

☐ Fig. 2

An example of block rules for the Margolus CA (Margolus 1984). (Since rotation-symmetry is assumed here, rules obtained by rotating both sides of a rule are also included.)

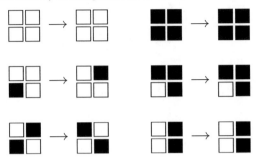

function is sometimes called a block permutation). Therefore the Margolus CA is reversible. Note that CAs with Margolus neighborhood are not conventional CAs, because each cell should know the relative position in a block and the parity of time besides its own state.

A k-dimensional block RCA can be also defined for any k in a similar manner. ❷ *Figure 3* gives a one-dimensional case of block size 2. It shows how a one-to-one local function f is applied to blocks at each time step.

3.2 Partitioned CAs

A partitioned cellular automaton (PCA) (Morita and Harao 1989) is also a framework where the notion of "local reversibility" can be defined, and hence has some similarity to a CA with block rules. But, it is noted that PCAs are in the framework of conventional CAs (i.e., a PCA is a special case of a CA). In addition, flexibility of neighborhood is rather high. So, it is convenient to use PCAs, for example, for showing results on computation-universality of RCAs. Therefore, in the following sections, this framework is mainly used to give specific examples of RCAs. A shortcoming of PCA is, in general, the number of states per cell becomes large.

Definition 3 A *deterministic k-dimensional m-neighbor partitioned cellular automaton (PCA)* is a system defined by

$$P = (\mathbb{Z}^k, (Q_1, \ldots, Q_m), (n_1, \ldots, n_m), f)$$

Here, Q_i ($i = 1, \ldots, m$) is a nonempty finite set of states of the ith part of each cell (thus the state set of each cell is $Q = Q_1 \times \ldots \times Q_m$), the m-tuple $(n_1, \ldots, n_m) \in (\mathbb{Z}^k)^m$ is a neighborhood, and $f : Q \rightarrow Q$ is a local function.

Let $p_i : Q \rightarrow Q_i$ be the projection function such that $p_i(q_1, \ldots, q_m) = q_i$ for all $(q_1, \ldots, q_m) \in Q$. The global function $F : \mathrm{Conf}(Q) \rightarrow \mathrm{Conf}(Q)$ of P is defined as the one that satisfies the following formula.

$$\forall \alpha \in \mathrm{Conf}(Q), x \in \mathbb{Z}^k : F(\alpha)(x) = f(p_1(\alpha(x + n_1)), \ldots, p_m(\alpha(x + n_m)))$$

7

◾ **Fig. 3**

A one-dimensional RCA with block rules defined by a one-to-one local function *f*. Here the block size is 2.

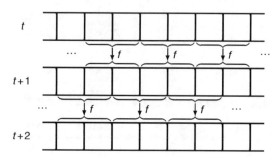

If some state in Q is specified as a quiescent state, a finite (and infinite) configuration can be defined in a similar manner as in CAs. A quiescent state $(\#_1, \ldots, \#_m) \in Q$ is a one such that $f(\#_1, \ldots, \#_m) = (\#_1, \ldots, \#_m)$. (In general, the states $\#_1, \ldots, \#_m$ may be different from each other. However, by renaming the states in each part appropriately, the states $\#_1, \ldots, \#_m$ can be identified as representing the same state $\#$. In what follows, it is often assumed so, and the quiescent state is written by $(\#, \ldots, \#)$.)

A one-dimensional three-neighbor (i.e., radius 1) PCA P_{1D} is therefore defined as follows.

$$P_{1D} = (\mathbb{Z}, (L, C, R), (1, 0, -1), f)$$

Each cell has three parts, that is, left, center, and right, and their state sets are L, C, and R. The next state of a cell is determined by the present states of the left part of the right-neighboring cell, the center part of this cell, and the right part of the left-neighboring cell (not depending on all three parts of the three cells). ❷ *Figure 4* shows its cellular space, and how the local function *f* is applied.

Let $(l, c, r), (l', c', r') \in L \times C \times R$. If $f(l, c, r) = (l', c', r')$, then this equation is called a local rule (or simply a rule) of the PCA P_{1D}. It can be written in a pictorial form as shown in ❷ *Fig. 5*. Note that, in the pictorial representation, the arguments of the left-hand side of $f(l, c, r) = (l', c', r')$ appear in a reverse order.

Similarly, a two-dimensional PCA P_{2D} with Neumann-like neighborhood is defined as follows.

$$P_{2D} = (\mathbb{Z}^2, (C, U, R, D, L), ((0,0), (0,-1), (-1,0), (0,1), (1,0)), f)$$

❷ *Figure 6* shows the cellular space of P_{2D}, and a pictorial representation of a rule $f(c, u, r, d, l) = (c', u', r', d', l')$.

It is easy to show the following proposition stating the equivalence of local and global injectivity (a proof for the one-dimensional case given in Morita and Harao (1989) can be extended to higher dimensions).

Proposition 1 *Let $P = (\mathbb{Z}^k, (Q_1, \ldots, Q_m), (n_1, \ldots, n_m), f)$ be a k-dimensional PCA, and F be its global function. Then the local function f is one-to-one, iff the global function F is one-to-one.*

◻ Fig. 4
Cellular space of a one-dimensional 3-neighbor PCA P_{1D}, and its local function f.

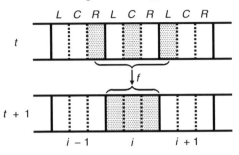

◻ Fig. 5
A pictorial representation of a local rule $f(l, c, r) = (l', c', r')$ of a one-dimensional 3-neighbor PCA P_{1D}.

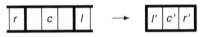

◻ Fig. 6
Cellular space of a two-dimensional 5-neighbor PCA P_{2D} and its local rule. Each cell is divided into five parts. The next state of each cell is determined by the present states of the center part of the cell, the downward part of the upper neighbor, the left part of the right neighbor, the upper part of the downward neighbor, and the right part of the left neighbor.

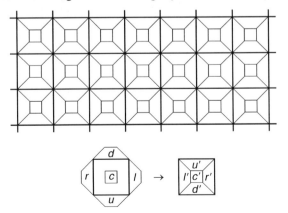

It is also easy to see that the class of PCAs is regarded as a subclass of CAs (Morita and Harao 1989).

Proposition 2   *For any k-dimensional m-neighbor PCA P, a k-dimensional m-neighbor CA A whose global function is identical with that of P can be obtained.*

From the above, if one wants to construct an RCA it is sufficient to give a PCA whose local function f is one-to-one. This makes a design of an RCA feasible.

3.3   Second-Order CAs

A second-order CA is one where the state of a cell at time $t+1$ is determined not only by the states of neighbor cells at time t but also by the state of the cell at $t-1$ (Margolus 1984; Toffoli and Margolus 1990; Toffoli et al. 2004). Margolus first proposed a method of using second-order CAs to design a reversible one, and gave a specific model of a computation-universal two-dimensional three-state reversible second-order CA (Margolus 1984).

It is assumed that a binary operator $\ominus$ is defined on the state set Q, which satisfies $a \ominus b = c$ iff $a \ominus c = b$ for all $a, b, c \in Q$. A typical example is $Q = \{0, 1, \ldots, r-1\}$ and the operation $\ominus$ is the mod-r subtraction. Let $N = (n_1, \ldots, n_m) \in (\mathbb{Z}^k)^m$ be a neighbor in the k-dimensional cellular space, and let $f: Q^m \to Q$ be an arbitrary mapping. Consider the local transition of the state of each cell defined by the following equation, where $\alpha^t \in \mathrm{Conf}(Q)$ denotes a configuration at time t, and $x \in \mathbb{Z}^k$.

$$\alpha^{t+1}(x) = f(\alpha^t(x+n_1), \ldots, \alpha^t(x+n_m)) \ominus \alpha^{t-1}(x)$$

It naturally induces a global function that determines the next configuration from the present and the previous configurations. From the above equation, the following one is easily derived, which elucidates that the state of each cell at time $t-1$ is determined by the states of neighbor cells at t and the state of the cell at $t+1$ by the *same* formula as above.

$$\alpha^{t-1}(x) = f(\alpha^t(x+n_1), \ldots, \alpha^t(x+n_m)) \ominus \alpha^{t+1}(x)$$

Hence, a second-order CA defined in such a method is reversible in this sense for *any* mapping $f: Q^m \to Q$. Though second-order CAs are not in the standard framework of CAs, they are also useful for giving RCAs. The relation between second-order CAs and the lattice gas model is discussed in Toffoli et al. (2004).

4   Computing Ability of Reversible Cellular Automata

A CA is called *computation-universal* (or Turing-universal) if it can simulate any Turing machine. From the beginning of the history of CA studies, it has been one of the main research topics. In the 1950s, von Neumann (1966) designed a 29-state two-dimensional CA, and showed that any Turing machine can be embedded in its cellular space, and it can reproduce itself. In this section, the problem of how reversible CAs can have universality is investigated. In what follows, there is a description on how various universal systems can be simulated by RCAs.

Though a precise definition of "simulation" is not given here, it is explained as follows. When we consider the case that a system A simulates another system B, we should first give an encoding method of computational configurations of B into those of A. Assume A starts from a configuration α that contains an encoding of an initial configuration β of B. We say A simulates B, if for every configuration β' that appears in the computing process of B starting from β, a configuration α' containing an encoding of β' appears at some time step in the computing process of A.

See also the chapter ❯ Universalities in Cellular Automata on computing ability and universality of CAs.

4.1 Simulating Irreversible CAs by RCAs

Toffoli (1977) first showed the following theorem stating that every irreversible CA can be simulated by an RCA by increasing the dimension by one.

Theorem 7 (Toffoli 1977) *For any k-dimensional (irreversible) CA A, a k + 1-dimensional RCA A' that simulates A in real time can be constructed.*

The intuitive idea for proving it is as follows. A' simulates A by using a k-dimensional subspace, say $\{(x_1,\ldots,x_k,0) \mid x_i \in \mathbb{Z}, i = 1,\ldots,k\}$, in real time. Since this subspace always keeps the present configuration of A, A' can compute the new configuration of A using only this subspace. At the same time, A' sends the information of the old configuration to the subspace $\{(x_1,\ldots,x_k,1) \mid x_i \in \mathbb{Z}, i = 1,\ldots,k\}$. The previous information stored in $\{(x_1,\ldots,x_k,1) \mid x_i \in \mathbb{Z}, i = 1,\ldots,k\}$ is also sent to the subspace $\{(x_1,\ldots,x_k,2) \mid x_i \in \mathbb{Z}, i = 1,\ldots,k\}$, and so on. In other words, the additional dimension is used to record *all* the past history of the evolution of A from $t = 0$ up to now, and thus A' can be an RCA. (If the framework of PCA is used, then to construct an RCA A' is rather simple (Morita 2001a).)

From this result, it can be seen that a computation-universal two-dimensional RCA exists, since there is a (irreversible) one-dimensional CA that can simulate a universal Turing machine.

As for one-dimensional CA with finite configurations, reversible simulation is possible without increasing the dimension. Hence, a computation-universal one-dimensional RCA also exists.

Theorem 8 (Morita 1995) *For any one-dimensional (irreversible) CA A with finite configurations, a one-dimensional reversible CA A' that simulates A (but not in real time) can be constructed.*

4.2 Simulating Reversible Turing Machines by One-Dimensional RCAs

It is also possible to design an RCA that can simulate a reversible Turing machine directly. Bennett (1973) proved the universality of a reversible Turing machine by showing a conversion method of a given irreversible Turing machine to an equivalent reversible one, which never leaves garbage symbols except an input and an output on its tape at the end of computation. Bennett (1973) used a quadruple formalism to define a reversible Turing machine, because it is useful to define an *inverse* Turing machine for a given reversible Turing machine. But here, a quintuple formalism is employed, because it is commonly used in the theory of irreversible Turing machines, and these two formalisms are convertible to each other (Morita and Yamaguchi 2007).

Definition 4   A *1-tape Turing machine in the quintuple form* (TM) is defined by

$$T = (Q, S, q_0, q_f, s_0, \delta)$$

where Q is a nonempty finite set of states, S is a nonempty finite set of symbols, q_0 is an initial state ($q_0 \in Q$), q_f is a final (halting) state ($q_f \in Q$), s_0 is a special blank symbol ($s_0 \in S$). δ is a move relation, which is a subset of ($Q \times S \times S \times \{-, 0, +\} \times Q$). Each element of δ is a quintuple of the form $[p, s, s', d, q]$. It means if T reads the symbol s in the state p, then write s', shift the head to the direction d ($-$, 0, and $+$ means left-, zero-, right-shift, respectively), and go to the state q.

T is called a *deterministic TM* iff the following condition holds for any pair of distinct quintuples $[p_1, s_1, s'_1, d_1, q_1]$ and $[p_2, s_2, s'_2, d_2, q_2]$ in δ.

$$\text{If } p_1 = p_2, \text{ then } s_1 \neq s_2$$

T is called a *reversible TM* iff the following condition holds for any pair of distinct quintuples $[p_1, s_1, s'_1, d_1, q_1]$ and $[p_2, s_2, s'_2, d_2, q_2]$ in δ.

$$\text{If } q_1 = q_2, \text{ then } s'_1 \neq s'_2 \wedge d_1 = d_2$$

The next theorem shows computation-universality of a reversible three-tape Turing machine.

Theorem 9   (Bennett 1973) *For any deterministic (irreversible) one-tape TM, there is a deterministic reversible three-tape TM which simulates the former.*

It is also shown in Morita et al. (1989) that for any irreversible one-tape TM, there is a reversible one-tape two-symbol TM, which simulates the former. In fact, to prove computation-universality of a one-dimensional reversible PCA, it is convenient to simulate a reversible one-tape TM.

Theorem 10   (Morita and Harao 1989) *For any deterministic reversible one-tape TM T, there is a one-dimensional reversible PCA P that simulates the former.*

It is now explained how the reversible PCA P is constructed (here, the method in Morita and Harao (1989) is modified to adopt the quintuple formalism, and to reduce the number of states of P). Let $T = (Q, S, q_0, q_f, s_0, \delta)$. It is assumed that q_0 does not appear as the fifth element in any quintuple in δ (such a reversible TM can always be constructed from an irreversible one (Morita et al. 1989)). It can also be assumed that for any quintuple $[p, s, t, d, q]$ in δ,

$d \in \{-,+\}$ iff q is a non-halting state, and $d \in \{0\}$ iff q is a halting state. A reversible PCA $P = (\mathbb{Z}, (L, C, R), (1, 0, -1), f)$ that simulates T is as follows. The state sets L, C, and R are

$$L = R = Q \cup \{\#\}, C = S$$

Let Q_+, Q_-, and Q_H be as follows.

$$Q_+ = \{q \mid \exists p \in Q \; \exists s, t \in S \, ([p, s, t, +, q] \in \delta)\}$$
$$Q_- = \{q \mid \exists p \in Q \; \exists s, t \in S \, ([p, s, t, -, q] \in \delta)\}$$
$$Q_H = \{p \mid \neg (\exists q \in Q \; \exists s, t \in S \; \exists d \in \{-, 0, +\} \, ([p, s, t, d, q] \in \delta))\}$$

Since T is a reversible TM, Q_+, Q_-, and Q_H are mutually disjoint. The local function f is as below.

1. For every $s, t \in S$, and $q \in Q - (Q_H \cup \{q_0\})$

$$f(\#, s, \#) = (\#, s, \#)$$
$$f(\#, s, q_0) = (\#, s, q_0)$$
$$f(q_0, s, \#) = (q_0, s, \#)$$
$$f(q_0, s, q_0) = (\#, t, q) \text{ if } [q_0, s, t, +, q] \in \delta$$
$$f(q_0, s, q_0) = (q, t, \#) \text{ if } [q_0, s, t, -, q] \in \delta$$

2. For every $p, q \in Q - (Q_H \cup \{q_0\})$, and $s, t \in S$

$$f(\#, s, p) = (\#, t, q) \text{ if } p \in Q_+ \text{ and } [p, s, t, +, q] \in \delta$$
$$f(p, s, \#) = (\#, t, q) \text{ if } p \in Q_- \text{ and } [p, s, t, +, q] \in \delta$$
$$f(\#, s, p) = (q, t, \#) \text{ if } p \in Q_+ \text{ and } [p, s, t, -, q] \in \delta$$
$$f(p, s, \#) = (q, t, \#) \text{ if } p \in Q_- \text{ and } [p, s, t, -, q] \in \delta$$

3. For every $p \in Q - (Q_H \cup \{q_0\})$, $q \in Q_H$, and $s, t \in S$

$$f(\#, s, p) = (q, t, q) \text{ if } p \in Q_+ \text{ and } [p, s, t, 0, q] \in \delta$$
$$f(p, s, \#) = (q, t, q) \text{ if } p \in Q_- \text{ and } [p, s, t, 0, q] \in \delta$$

4. For every $q \in Q_H$ and $s \in S$

$$f(\#, s, q) = (\#, s, q)$$
$$f(q, s, \#) = (q, s, \#)$$

One can verify that the right-hand side of each rule differs from that of any other rule, because T is deterministic and reversible. If the initial computational configuration of T is

$$\cdots s_0 t_1 \cdots q_0 t_i \cdots t_n s_0 \cdots$$

then set P to the following configuration.

$$\ldots, (\#, s_0, \#), (\#, t_1, \#), \ldots, (\#, t_{i-1}, q_0), (\#, t_i, \#), (q_0, t_{i+1}, \#), \ldots, (\#, t_n, \#), (\#, s_0, \#), \ldots$$

The simulation process starts when left- and right-moving signals q_0's meet at the cell containing t_i. It is easily seen that, from this configuration, P can correctly simulate T by the rules in (2) step by step. If T becomes a halting state $q \, (\in Q_H)$, then the two signals q's are created, and travel leftward and rightward indefinitely. Note that P itself cannot halt, because P is reversible, but the final result is kept unchanged.

■ **Fig. 7**

The initial and the final computational configuration of T_{2n} for a given unary input 1.

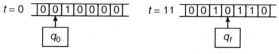

Example 1 Consider a reversible TM $T_{2n} = (Q, \{0, 1\}, q_0, q_f, 0, \delta)$, where $Q = \{q_0, \dots, q_7, q_f\}$, and δ is as below.

$$\delta = \{[q_0, 0, 0, +, q_1],$$
$$[q_1, 0, 0, 0, q_f], [q_1, 1, 0, +, q_2],$$
$$[q_2, 0, 0, +, q_3], [q_2, 1, 1, +, q_2],$$
$$[q_3, 0, 1, +, q_4], [q_3, 1, 1, +, q_3],$$
$$[q_4, 0, 1, +, q_5],$$
$$[q_5, 0, 0, -, q_6],$$
$$[q_6, 0, 0, -, q_7], [q_6, 1, 1, -, q_6],$$
$$[q_7, 0, 1, +, q_1], [q_7, 1, 1, -, q_7]\}.$$

For a given unary number n on the tape, T_{2n} computes the function $2n$ and writes it on the tape as in ❷ *Fig. 7*. A simulation process of T_{2n} by a reversible PCA P_{2n}, which is constructed by the method described above, is in ❷ *Fig. 8*.

Each cell of the reversible PCA P constructed in ❷ Theorem 10 has $(m + 1)^2 n$ states, if T has m states and n symbols. Since there is a 15-state 6-symbol universal reversible TM (Morita 2008a), a 1536-state universal RCA is obtained by this method.

4.3 Simulating Cyclic Tag Systems by One-Dimensional RCAs

A cyclic tag system (CTS) is a kind of string rewriting system proposed by Cook (2004) to show universality of the elementary CA of rule 110. This system is also useful for constructing computation-universal one-dimensional RCAs with a small number of states.

Definition 5 A CTS is defined by $C = (k, \{Y, N\}, (p_0, \dots, p_{k-1}))$, where k ($k = 1, 2, \dots$) is the length of a cycle (i.e., period), $\{Y, N\}$ is the alphabet used in this system, and $(p_0, \dots, p_{k-1}) \in (\{Y, N\}^*)^k$ is a k-tuple of production rules. A pair (v, m) is called an *instantaneous description* (ID) of C, where $v \in \{Y, N\}^*$ and $m \in \{0, \dots, k-1\}$. m is called a *phase* of the ID. A transition relation $\Rightarrow$ on the set of IDs is defined as follows. For any $(v, m), (v', m') \in \{Y, N\}^* \times \{0, \dots, k-1\}$

$$(Yv, m) \Rightarrow (v', m') \text{ iff } [m' = m + 1 \bmod k] \wedge [v' = vp_m]$$
$$(Nv, m) \Rightarrow (v', m') \text{ iff } [m' = m + 1 \bmod k] \wedge [v' = v]$$

A sequence of IDs $(v_0, m_0), (v_1, m_1), \dots$ is called a *computation starting from* $v \in \{Y, N\}^*$ iff $(v_0, m_0) = (v, 0)$ and $(v_i, m_i) \Rightarrow (v_{i+1}, m_{i+1})$ $(i = 0, 1, \dots)$. We also write a computation by $(v_0, m_0) \Rightarrow (v_1, m_1) \Rightarrow \dots$.

◻ **Fig. 8**

Simulating T_{2n} by a one-dimensional reversible PCA P_{2n}. The state # is indicated by a blank.

$t=0$	0 q_0	0	q_0 1	0	0	0	0
$t=1$	0	0 q_1	1	0	0	0	0
$t=2$	0	0	0 q_2	0	0	0	0
$t=3$	0	0	0	0 q_3	0	0	0
$t=4$	0	0	0	0	1 q_4	0	0
$t=5$	0	0	0	0	1	1 q_5	0
$t=6$	0	0	0	0	1	1	q_6 0
$t=7$	0	0	0	0	1	q_6 1	0
$t=8$	0	0	0	0	q_6 1	1	0
$t=9$	0	0	0	q_7 0	1	1	0
$t=10$	0	0	1 q_1	0	1	1	0
$t=11$	0	0	1	q_f 0 q_f	1	1	0
$t=12$	0	0	q_f 1	0	1 q_f	1	0
$t=13$	0	q_f 0	1	0	1	1 q_f	0
$t=14$	q_f 0	0	1	0	1	1	0 q_f

In a CTS, production rules are applied cyclically to rewrite a given string. If the first symbol of the host string is Y, then it is removed and a specified string at that phase is attached to the end of the host string. If it is N, then it is simply removed and no string is attached.

Example 2 Consider the following CTS.

$$C_0 = (3, \{Y, N\}, (Y, NN, YN))$$

If NYY is given as an initial string, then

$$(NYY, 0) \Rightarrow (YY, 1) \Rightarrow (YNN, 2) \Rightarrow (NNYN, 0) \Rightarrow (NYN, 1) \Rightarrow (YN, 2)$$

is an initial segment of a computation starting from NYY.

A 2-tag system is a special class of classical tag systems, and Minsky (1967) proved that for any TM, there is a 2-tag system that simulates any computing process of the TM. Hence a 2-tag system is computation-universal. Cook (2004) proved the following theorem, and thus it can be seen that a CTS is also computation-universal.

Theorem 11 (Cook 2004) *For any 2-tag system, there is a CTS that simulates the former.*

In Morita (2007), two models of one-dimensional reversible PCAs that can simulate any CTS are given. The first one is a 36-state model that works on infinite configurations, and the second one is a 98-state model that works on finite configurations. As shown in the following theorem, the former result was improved in Morita (2008b).

Theorem 12 (Morita 2008b) *There is a 24-state one-dimensional two-neighbor reversible PCA* P_{24} *that can simulate any CTS on infinite (but ultimately periodic) configurations.*

The reversible PCA P_{24} in ❷ Theorem 12 is given below.

$$P_{24} = (\mathbb{Z}, (\{Y, N, +, -\}, \{y, n, +, -, *, /\}), (0, -1), f)$$

where f is defined as follows.

$$
\begin{aligned}
f(\mathbf{c}, \mathbf{r}) &= (\mathbf{c}, \mathbf{r}) \quad \text{for } \mathbf{c} \in \{Y, N\}, \text{ and } \mathbf{r} \in \{y, n, +, -, /\} \\
f(Y, *) &= (+, /) \\
f(N, *) &= (-, /) \\
f(-, \mathbf{r}) &= (-, \mathbf{r}) \quad \text{for } \mathbf{r} \in \{y, n, *\} \\
f(\mathbf{c}, \mathbf{r}) &= (\mathbf{r}, \mathbf{c}) \quad \text{for } \mathbf{c}, \mathbf{r} \in \{+, -\} \\
f(+, y) &= (Y, *) \\
f(+, n) &= (N, *) \\
f(+, /) &= (+, y) \\
f(-, /) &= (+, n) \\
f(+, *) &= (+, *)
\end{aligned}
$$

Note that a cell of P_{24} has only the center and the right parts: $C = \{Y, N, +, -\}$, and $R = \{y, n, +, -, *, /\}$. It is easily seen that $f \colon C \times R \to C \times R$ is one-to-one.

In the following, there is an explanation as to how P_{24} can simulate a CTS by using the previous example C_0 with an initial string *NYY*. The initial configuration of P_{24} is set as shown in the first row of ❷ *Fig. 9*. The string *NYY* is given in the center parts of three consecutive cells. The right-part states of these three cells are set to $-$. The states of the cells right to the three cells are set to $(-, -), (Y, -), (Y, -), \ldots$. The production rules (Y, NN, YN) is given by a sequence of the right-part states $y, n, *$, and $-$ in a reverse order, where the sequence $- *$ is used as a delimiter indicating the beginning of a rule. Thus, one cycle of the rules (Y, NN, YN) is represented by the sequence $ny - * nn - * y - *$. Since these rules are applied cyclically, infinite copies of the sequence $ny - * nn - * y - *$ must be given. These right-part states $y, n, *$, and $-$ act as right-moving signals until they reach the first symbol of a rewritten string.

P_{24} can simulate a rewriting process of C_0 as shown below. If the signal $*$ meets the state Y (or N, respectively), which is the head symbol of a rewritten string, then the signal changes Y (N) to the state $+$ ($-$), and the signal itself becomes $/$ that is sent rightward as a used (garbage) signal. At the next time step, the center-part state $+$ ($-$) meets the signal $-$, and the former becomes a right-moving signal $+$ ($-$), and the latter (i.e., $-$) is fixed as a center-part state at this position. The right-moving signal $+$ ($-$) travels through the rewritten string consisting of Y's and N's, and when it meets the center-part state $+$ or $-$, then it is fixed as a new center-part state, which indicates the last head symbol is Y (N). Note that the old center-part state $+$ ($-$) is sent rightward as a used signal.

◻ Fig. 9

Simulating the computing process $(NYY,0) \Rightarrow^* (YN,2)$ of the CTS C_0 in ❷ Example 2 by the reversible PCA P_{24} (Morita 2008b).

Signals y and n go rightward through the rewritten string consisting Y's and N's until it meets $+$ or $-$. If y (n, respectively) meets $+$, then the signal becomes Y (N) and is fixed at this position (since the last head symbol is Y), and $+$ is shifted to the right by one cell. If y or n meets $-$, then the signal simply continues to travel rightward without being fixed (since the last head symbol is N). Note that all the used and unused information are sent rightward as garbage signals, because they cannot be deleted by the constraint of reversibility. ❷ *Figure 9* shows how a computation of C_0 is performed in P_{24}. In this way, any CTS can be simulated.

4.4 Simulating Reversible Logic Elements by Two-Dimensional RCAs

The universality of a CA can also be proved by showing any logic circuit is simulated in its cellular space. In the case of two-dimensional CAs, such a method is often used. A set of logic elements is called *logically universal*, if any logic function can be realized by using only these elements. For example, {AND, NOT} is a well-known logically universal set. Furthermore, if delay elements can be used (or memory elements) besides these elements, then any sequential machine (i.e., finite automaton with outputs) can be composed from them. Since a finite-state control and tape cells of a Turing machine are regarded as sequential machines, any Turing machine can be constructed by using these elements.

In the following, several computation-universal two-dimensional RCAs are shown in which universal reversible logic elements are embedded. In particular, the Fredkin gate, and the rotary element (RE) are taken into consideration.

4.4.1   Simulating the Fredkin Gate

A typical example of a universal reversible logic gate is the Fredkin gate (Fredkin and Toffoli 1982). It is a reversible (i.e., its logical function is one-to-one) and bit-conserving (i.e., the number of 1's is conserved between inputs and outputs) logic gate shown in ❯ *Fig. 10*. It is known that any combinational logic circuit is composed only of Fredkin gates. ❯ *Figure 11* shows that AND, NOT, and Fan-out elements can be implemented only with Fredkin gates (Fredkin and Toffoli 1982) by allowing constant inputs and garbage outputs (e.g., in the implementation of AND, constant 0 should be supplied to the third input line, and the first and the third output lines produce garbage signals x and $\bar{x}y$). Hence, {Fredkin gate} is logically universal, and any sequential machine can be constructed with Fredkin gate and delay elements.

Though the Fredkin gate is a simple element, it can be decomposed into a much simpler gate called a switch gate (❯ *Fig. 12*) and its inverse gate (Fredkin and Toffoli 1982). A switch gate is also a reversible and bit-conserving logic gate. In addition, it is known to be realized by the billiard ball model (BBM) of computation (Fredkin and Toffoli 1982). The BBM is a kind

◪ Fig. 10
A Fredkin gate.

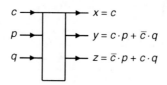

◪ Fig. 11
AND, NOT, and Fan-out made of Fredkin gates (Fredkin and Toffoli 1982).

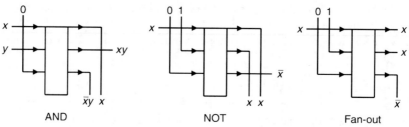

◪ Fig. 12
A switch gate (*left*), and a Fredkin gate implemented by two switch gates and two inverse switch gates (*right*) (Fredkin and Toffoli 1982).

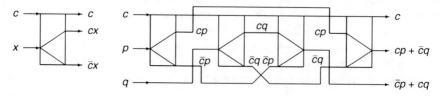

◘ Fig. 13
A switch gate realized in the Billiard Ball Model (Fredkin and Toffoli 1982).

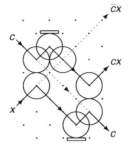

of physical model of computation where a signal "1" is represented by an ideal ball, and logical operations and routing can be performed by their elastic collisions and reflections by reflectors. ❷ *Figure 13* shows a BBM realization of a switch gate.

Now, consider the two-dimensional 2-state RCA with Margolus neighborhood, which has block rules shown in ❷ *Fig. 2* (Margolus 1984). Margolus showed that the BBM can be realized in this cellular space. ❷ *Figure 14* shows a reflection of a ball by a mirror in the Margolus CA. Hence, the following theorem is arrived at.

Theorem 13 (Margolus 1984) *There is a computation-universal two-dimensional 2-state RCA with Margolus neighborhood.*

Margolus (1984) also constructed a two-dimensional 3-state second-order RCA in which BBM can be realized.

Theorem 14 (Margolus 1984) *There is a computation-universal two-dimensional 3-state second-order RCA.*

There are also other RCAs that can simulate the Fredkin gate in their cellular spaces. Here, an 8-state reversible PCA model T_1 on a triangular grid (Imai and Morita 2000) is considered. Its local function is extremely simple as shown in ❷ *Fig. 15*. In this PCA, each of the switch gate and the inverse switch gate is simply simulated by using only one cell (hence, it does not simulate the BBM). On the other hand, signal routing, crossing, and delay are very complex to realize, because a kind of "wall" is necessary to make a signal go straight ahead. ❷ *Figure 16* shows the configuration that simulates the Fredkin gate. Thus, we have the following theorem.

Theorem 15 (Imai and Morita 2000) *There is a computation-universal two-dimensional 8-state reversible triangular PCA.*

4.4.2 Simulating a Reversible Logic Element with Memory

Besides reversible logic gates, there are also universal reversible logic elements with memory, with which reversible TMs are constructed rather concisely. A RE (Morita 2001b) is one such element. Conceptually, it has four input lines $\{n, e, s, w\}$ and four output lines $\{n', e', s', w'\}$,

◘ Fig. 14

Reflection of a ball by a reflector in the Margolus RCA (Margolus 1984).

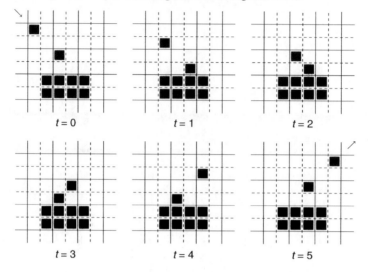

◘ Fig. 15

The local function of the 2^3-state rotation-symmetric reversible triangular PCA T_1.

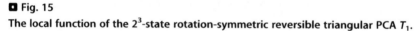
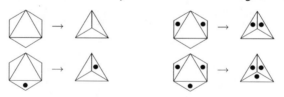

and two states called H-state and V-state as in ❯ *Fig. 17* (hence it has a 1-bit memory). The values of inputs and outputs are either 0 or 1, but the input (and the output) are restricted as follows: at most one "1" appears as an input (output) at a time. The operation of an RE is undefined for the cases where signal 1's are given to two or more input lines. Signals 1 and 0 are interpreted as existence and nonexistence of a particle. An RE has a "rotating bar" to control the moving direction of a particle. When no particle exists, nothing happens. If a particle comes from a direction parallel to the rotating bar, then it goes out from the output line of the opposite side without affecting the direction of the bar (❯ *Fig. 18a*). If a particle comes from a direction orthogonal to the bar, then it makes a right turn, and rotates the bar by 90° (❯ *Fig. 18b*). It is clear its operation is reversible.

It is known that a Fredkin gate can be simulated by a circuit composed only of REs (Morita 2001b). Hence, {RE} is logically universal. It is also shown that any reversible TM can be constructed by using only REs (Morita 2001b).

Consider the following two-dimensional 4-neighbor reversible PCA P_3 (Morita et al. 2002).

$$P_3 = (\mathbb{Z}^2, (\{0, 1, 2\}^4), ((0, -1), (-1, 0), (0, 1), (1, 0)), f)$$

◻ Fig. 16

A Fredkin gate realized in the 2^3-state reversible triangular PCA T_1 (Imai and Morita 2000).

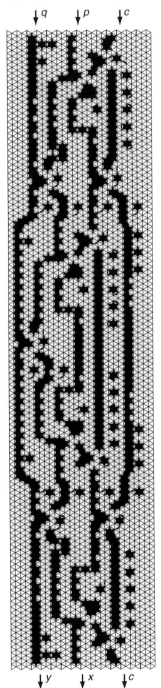

◘ **Fig. 17**

Two states of a rotary element (RE).

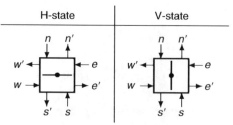

◘ **Fig. 18**

Operations of an RE: (a) the parallel case, and (b) the orthogonal case.

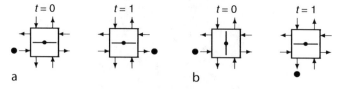

where f is given in ❷ *Fig. 19*. In the cellular space of P_3, an RE is realized as shown in ❷ *Fig. 20*. It is also possible to design subconfigurations for signal routing. By using them, any circuit composed of REs can be simulated in the P_3 space, and thus it is computation-universal.

Furthermore, in P_3, any reversible counter machine, which is also known to be computation-universal (Morita 1996), can be embedded as stated in the following theorem. Here, we only show an example of a configuration realizing a counter machine in ❷ *Fig. 21* (its detail is in Morita et al. (2002)).

Theorem 16 (Morita et al. 2002) *There is a computation-universal two-dimensional 81-state reversible PCA in which any reversible counter machine is simulated in finite configurations.*

4.5 Intrinsically Universal RCAs

There is another notion of universality for CAs called intrinsic universality. A CA is *intrinsically universal* if it can simulate any CA. Von Neumann's 29-state CA (von Neumann 1966) is an example of an intrinsically universal irreversible CA, because any logic circuit composed of AND, OR, NOT, and memory elements is simulated in its cellular space, and every CA can be implemented by such a logic circuit. Studies on intrinsically universal one-dimensional CAs are found in Ollinger (2002) and Ollinger and Richard (2006).

In the case of RCAs, intrinsic universality can be defined as follows: An RCA is *intrinsically universal* if it can simulate any RCA. One such two-dimensional RCA is given in Durand-Lose (1995). Related to this topic, there is a problem whether every RCA is structurally reversible, that is, whether it can be expressed by locally reversible transformations or not (Toffoli and

■ Fig. 19

The local function of the 3^4-state rotation-symmetric reversible PCA P_3 (Morita et al. 2002). The states 0, 1, and 2 are represented by blank, white dot, and black dot, respectively. The rule scheme (m) represents 33 rules not specified by (a)–(l), where $w,x,y,z \in$ {blank, ∘, ·}.

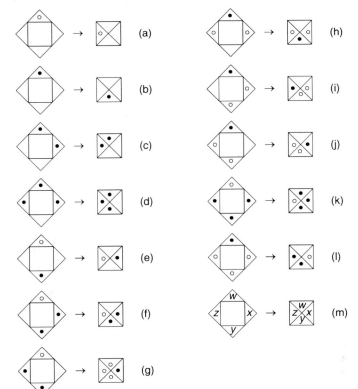

■ Fig. 20

Operations of an RE in the reversible PCA P_3: (a) the parallel case, and (b) the orthogonal case.

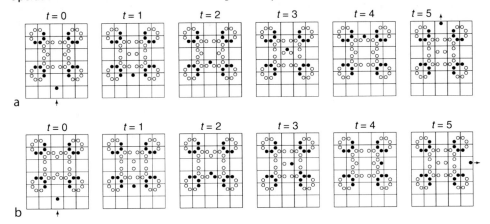

◘ **Fig. 21**

An example of a reversible counter machine, which computes the function 2x+2, embedded in the reversible PCA P_3 (Morita et al. 2002).

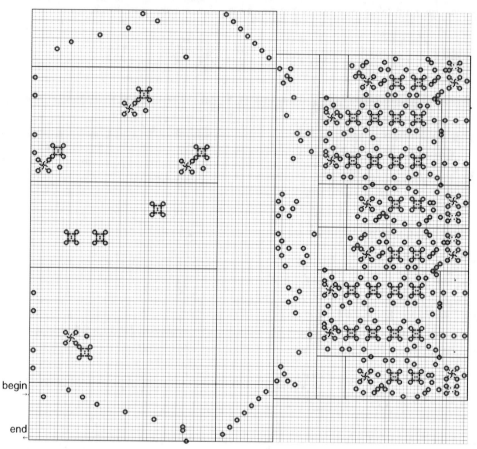

Margolus 1990). Kari (1996) solved this problem affirmatively by showing that every one- and two-dimensional RCA can be represented by block permutations and translations. Hence, an RCA in which any operation of block permutation and translation can be realized (e.g., by implementing a reversible logic circuit made of Fredkin gates) is intrinsically universal.

5   Concluding Remarks

In this chapter, several aspects and properties of RCAs, especially their computing ability, are discussed. In spite of the strong constraint of reversibility, RCAs have a rich ability in computing and information processing, and even very simple RCAs have computation-universality.

There are many open problems left for future studies. For example, consider computation-universality of RCAs. All the universal RCAs obtained so far use the framework of

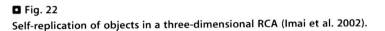

◻ Fig. 22
Self-replication of objects in a three-dimensional RCA (Imai et al. 2002).

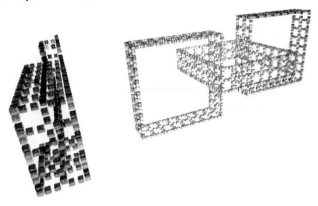

either CAs with block rules, partitioned CAs (PCAs), or second-order CAs, as seen in the previous sections. In the one-dimensional case, a 24-state computation-universal RCA exists (❯ Theorem 12). But the number of states seems to be reduced much more, and the conjecture is that there is a universal one-dimensional RCA with fewer than ten states. To obtain such a result and develop the theory of RCAs, new techniques of designing RCAs will be necessary.

There are many topics on RCAs that are not stated in the previous sections. Some of them are briefly discussed. An early work on quantum CAs, which is also a kind of RCA, is found in the paper by Watrous (1995). Complexity of RCAs is studied by Sutner (2004). Kutrib and Malcher (2008) deal with RCAs as a model of formal language recognition systems. Imai and Morita (1996) showed a 3n-step solution of the firing squad synchronization problem on an RCA. Self-replication of a pattern is also possible in two- and three-dimensional RCAs (Imai et al. 2002; Morita and Imai 1996) (❯ *Fig. 22*).

There are several survey papers and books related to RCAs that are written from different viewpoints. Toffoli and Margolus (1990), and Kari (2005b) give good survey papers on RCAs. The paper by Kari (2005a), the books by Toffoli and Margolus (1987), and by Wolfram (2001) are references on (general) CAs, which also describe RCAs to some extent. On universality of general CAs, see the chapter ❯ Universalities in Cellular Automata by Ollinger.

References

Amoroso S, Cooper G (1970) The Garden of Eden theorem for finite configurations. Proc Am Math Soc 26:158–164

Amoroso S, Patt YN (1972) Decision procedures for surjectivity and injectivity of parallel maps for tessellation structures. J Comput Syst Sci 6:448–464

Bennett CH (1973) Logical reversibility of computation. IBM J Res Dev 17:525–532

Boykett T (2004) Efficient exhaustive listings of reversible one dimensional cellular automata. Theor Comput Sci 325:215–247

Cook M (2004) Universality in elementary cellular automata. Complex Syst 15:1–40

Czeizler E, Kari J (2007) A tight linear bound for the synchronization delay of bijective automata. Theor Comput Sci 380:23–36

Durand-Lose J (1995) Reversible cellular automaton able to simulate any other reversible one using partitioning automata. In: Proceedings of LATIN 95, Valparaiso, Chile, LNCS 911. Springer, pp 230–244

Fredkin E, Toffoli T (1982) Conservative logic. Int J Theor Phys 21:219–253

Hedlund GA (1969) Endomorphisms and automorphisms of the shift dynamical system. Math Syst Theory 3:320–375

Imai K, Hori T, Morita K (2002) Self-reproduction in three-dimensional reversible cellular space. Artif Life 8:155–174

Imai K, Morita K (1996) Firing squad synchronization problem in reversible cellular automata. Theor Comput Sci 165:475–482

Imai K, Morita K (2000) A computation-universal two-dimensional 8-state triangular reversible cellular automaton. Theor Comput Sci 231:181–191

Kari J (1994) Reversibility and surjectivity problems of cellular automata. J Comput Syst Sci 48:149–182

Kari J (1996) Representation of reversible cellular automata with block permutations. Math Syst Theory 29:47–61

Kari J (2005a) Theory of cellular automata: a survey. Theor Comput Sci 334:3–33

Kari J (2005b) Reversible cellular automata. In: Proceedings of the DLT 2005, Palermo, Italy, LNCS 3572. Springer, pp 57–68

Kutrib M, Malcher A (2008) Fast reversible language recognition using cellular automata. Inf Comput 206:1142–1151

Landauer R (1961) Irreversibility and heat generation in the computing process. IBM J Res Dev 5:183–191

Lecerf Y (1963) Machines de Turing réversibles — Reursive insolubilité en $n \in \mathbf{N}$ de l'équation $u = \theta^n u$, où θ est un isomorphisme de codes. Comptes Rendus Hebdomadaires des Séances de l'Académie des Sciences 257:2597–2600

Margolus N (1984) Physics-like model of computation. Physica 10D:81–95

Maruoka A, Kimura M (1976) Condition for injectivity of global maps for tessellation automata. Inf Control 32:158–162

Maruoka A, Kimura M (1979) Injectivity and surjectivity of parallel maps for cellular automata. J Comput Syst Sci 18:47–64

Minsky ML (1967) Computation: finite and infinite machines. Prentice-Hall, Englewood Cliffs, NJ

Moore EF (1962) Machine models of self-reproduction. In: Proceedings of the symposia in applied mathematics, vol 14. American Mathematical Society, New York, pp 17–33

Mora JCST, Vergara SVC, Martinez GJ, McIntosh HV (2005) Procedures for calculating reversible one-dimensional cellular automata. Physica D 202:134–141

Morita K (1995) Reversible simulation of one-dimensional irreversible cellular automata. Theor Comput Sci 148:157–163

Morita K (1996) Universality of a reversible two-counter machine. Theor Comput Sci 168:303–320

Morita K (2001a) Cellular automata and artificial life — computation and life in reversible cellular automata. In: Goles E, Martinez S (eds) Complex systems. Kluwer, Dordrecht, pp 151–200

Morita K (2001b) A simple reversible logic element and cellular automata for reversible computing. In: Proceedings of the 3rd international conference on machines, computations, and universality, LNCS 2055. Springer, Heidelberg, Germany, pp 102–113

Morita K (2007) Simple universal one-dimensional reversible cellular automata. J Cell Autom 2:159–165

Morita K (2008a) Reversible computing and cellular automata — a survey. Theor Comput Sci 395:101–131

Morita K (2008b) A 24-state universal one-dimensional reversible cellular automaton. In: Adamatzky A et al. (eds) Proceedings of AUTOMATA-2008. Luniver Press, Bristol, UK, pp 106–112

Morita K, Harao M (1989) Computation universality of one-dimensional reversible (injective) cellular automata. Trans IEICE Jpn E-72:758–762

Morita K, Imai K (1996) Self-reproduction in a reversible cellular space. Theor Comput Sci 168:337–366

Morita K, Yamaguchi Y (2007) A universal reversible Turing machine. In: Proceedings of the 5th international conference on machines, computations, and universality, LNCS 4664. Springer, pp 90–98

Morita K, Shirasaki A, Gono Y (1989) A 1-tape 2-symbol reversible Turing machine. Trans IEICE Jpn E-72:223–228

Morita K, Tojima Y, Imai K, Ogiro T (2002) Universal computing in reversible and number-conserving two-dimensional cellular spaces. In: Adamatzky A (ed) Collision-based computing. Springer, London, UK, pp 161–199

Myhill J (1963) The converse of Moore's Garden-of-Eden theorem. Proc Am Math Soc 14:658–686

von Neumann J (1966) In: Burks AW (ed) Theory of self-reproducing automata. The University of Illinois Press, Urbana, IL

Ollinger N (2002) The quest for small universal cellular automata. In: Proceedings of ICALP, LNCS 2380. Lyon, France, pp 318–329

Ollinger N, Richard G (2006) A particular universal cellular automaton, oai:hal.archives-ouvertes.fr:hal-00095821_v2

Richardson D (1972) Tessellations with local transformations. J Comput Syst Sci 6:373–388

Sutner K (1991) De Bruijn graphs and linear cellular automata. Complex Syst 5:19–31

Sutner K (2004) The complexity of reversible cellular automata. Theor Comput Sci 325:317–328

Toffoli T (1977) Computation and construction universality of reversible cellular automata. J Comput Syst Sci 15:213–231

Toffoli T (1980) Reversible computing. In: de Bakker JW, van Leeuwen J (eds) Automata, languages and programming, LNCS 85. Springer, Berlin, Germany, pp 632–644

Toffoli T, Margolus N (1987) Cellular automata machines. The MIT Press, Cambridge, MA

Toffoli T, Margolus N (1990) Invertible cellular automata: a review. Physica D 45:229–253

Toffoli T, Capobianco S, Mentrasti P (2004) How to turn a second-order cellular automaton into lattice gas: a new inversion scheme. Theor Comput Sci 325:329–344

Watrous J (1995) On one-dimensional quantum cellular automata. In: Proceedings of the 36th symposium on foundations of computer science. Las Vegas, NV. IEEE, pp 528–537

Wolfram S (2001) A new kind of science. Wolfram Media, Champaign, IL

8 Conservation Laws in Cellular Automata

Siamak Taati
Department of Mathematics, University of Turku, Finland
siamak.taati@gmail.com

G. Rozenberg et al. (eds.), *Handbook of Natural Computing*, DOI 10.1007/978-3-540-92910-9_8,
© Springer-Verlag Berlin Heidelberg 2012

Abstract

A conservation law in a cellular automaton (CA) is the statement of the invariance of a local and additive energy-like quantity. This chapter reviews the basic theory of conservation laws in cellular automata. A general mathematical framework for formulating conservation laws in cellular automata is presented and several characterizations of them are summarized. Computational problems regarding conservation laws (verification and existence problems) are discussed. Microscopic explanations of the dynamics of the conserved quantities in terms of flows and particle flows are explored. The related concept of dissipating energy-like quantities is also addressed.

1   Introduction

A cellular automaton (CA) is an abstract structure, consisting of a d-dimensional checkerboard ($d = 1, 2, 3, \ldots$). Each cell of the board has a state chosen from a finite set of states. The state of each cell changes with time, according to a uniform, deterministic rule, which takes into account the previous state of the cell itself and those in its neighborhood. The changes, however, happen synchronously, and in discrete time steps.

In mathematics and computer science, cellular automata are studied as abstract models of computation, in the same way that Turing machines are (see the chapters ❯ Algorithmic Tools on Cellular Automata, Language Recognition by Cellular Automata, Computations on Cellular Automata, and Universalities in Cellular Automata). They are also treated as paradigms of symmetric dynamical systems on the Cantor space (see the chapter ❯ Cellular Automata Dynamical Systems). However, the chief reason for the interest in cellular automata comes from their characteristic similarities with nature: they are spatially extended dynamical systems; they are uniform (the same laws are applied everywhere in the space); the interactions are local (no action-at-a-distance); the amount of information in a finite region of space is finite (cf. Fredkin et al. 1982). Further characteristics of nature, such as the microscopic reversibility and conservation laws also arise in CA in a natural way (see the chapter ❯ Reversible Cellular Automata). This makes cellular automata exceptionally suitable for modeling physical and biological phenomena on the one hand (see the chapter ❯ Cellular Automata and Lattice Boltzmann Modeling of Physical Systems), and as a design framework for natural computing, on the other hand (see, e.g., Margolus 1984).

This chapter is a survey of the basic known results about local and additive conservation laws in cellular automata. Studying conservation laws in cellular automata may be beneficial from different aspects. When modeling a physical system, conservation laws of the system serve as design constraints that one would like to program in the model (see the chapter ❯ Cellular Automata and Lattice Boltzmann Modeling of Physical Systems). In physically realistic models of computation, conservation laws should naturally be taken into account (see Fredkin and Toffoli 1982 and Margolus 1984). Moreover, conservation laws in a CA may provide the same kind of "physical" insight about its dynamics as the insight that conservation laws in physics provide about the physical world.

In the rest of this section, the concept of conservation laws in cellular automata is illustrated by a number of examples. ❯ Section 2 provides a precise mathematical formulation of conservation laws in cellular automata. ❯ Section 3 is dedicated to the algorithmic problems that arise from studying conservation laws: how to discover them, and how to verify their validity. ❯ Section 4 discusses local explanations of conservation laws in terms of the flow of the conserved quantity. Finally, ❷ Sect. 5 comments on some related issues. Through-out the exposition, the closely related concept of nonincreasing energy-like quantities is also discussed.

1.1   Few Examples

Conservation laws in cellular automata are obtained in a similar fashion as in physics. A real value is associated to each local pattern of cell states, interpreted as the "energy" (or "electric charge", or . . .) of that particular arrangement of states. The total energy of a configuration is the sum of the energy values of the patterns seen in different places on it. Intuitively, a conservation law asserts that the total energy of each configuration remains unchanged under the iterations of the CA.

As an example, consider the well-known *Traffic* CA, which resembles cars moving on a highway. The Traffic CA is a one-dimensional CA, consisting of an infinite number of cells arranged next to each other on a line. Each cell has two possible states: ■ (interpreted as a "car") or □ ("empty space"). At each step, a car moves one cell forward if and only if its front cell is empty. ❷ *Figure 1* shows a typical space–time diagram of the evolution of the Traffic CA. Not surprisingly, the number of cars on the highway remains constant along the evolution of the CA. To state this more precisely, the various positions on the line are indexed with integers $i \in \mathbb{Z}$. A configuration of the model is an assignment of values ■ or □ to every position on the line. For each configuration $i \mapsto x[i]$, Fx is written for the configuration after one step. The car conservation law can now be stated, by saying that, for any configuration x, the following equality holds:

$$\sum_{i=-\infty}^{+\infty} \theta(x[i]) = \sum_{i=-\infty}^{+\infty} \theta((Fx)[i]) \tag{1}$$

where $\theta(■) = 1$ and $\theta(□) = 0$. Note that if the number of cars on a configuration x is infinite, the sum $\sum_i \theta(x[i])$ becomes $+\infty$. However, in this case, the configuration Fx has also infinitely many cars on it, and the equality still holds. (To learn more about cellular automata models of car traffic, see for example Nagel and Schreckenberg (1992) and Nagel and Herrmann (1993), and the relevant discussion in the chapter ❷ Cellular Automata and Lattice Boltzmann Modeling of Physical Systems.)

◻ **Fig. 1**
A typical space–time diagram of the Traffic CA. Time evolves downward. The highway is directed toward the left.

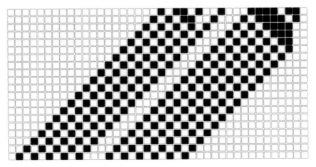

☐ **Fig. 2**

A typical space–time diagram of the Just Gliders CA. The moving objects annihilate on encounter.

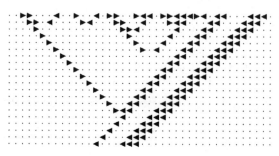

Another example is the *Just Gliders* CA. This is also one-dimensional. Each cell can be in either state ◄ (a "particle" moving to the left), or ► (a "particle" moving to the right), or · ("empty space"). At each step, each particle moves one cell ahead. Particles moving in opposite directions annihilate when they meet. See ❷ *Fig. 2* for a typical space–time diagram. If the momentum of a right-moving particle ► is defined to be 1, and the momentum of a left-moving particle ◄ to be −1, one can easily see that the total momentum of a configuration remains constant with time. More precisely, setting

$$\theta(a) \overset{\triangle}{=} \begin{cases} 1 & \text{if } a = \blacktriangleright \\ -1 & \text{if } a = \blacktriangleleft \\ 0 & \text{if } a = \cdot \end{cases} \tag{2}$$

❷ Equation 1 is valid for any configuration x with a finite number of particles. For an infinite configuration (i.e., a configuration with infinitely many particles on it), the sum $\sum_i \theta(x[i])$ is not necessarily meaningful anymore. Therefore, the conservation law is expressed in terms of finite configurations only.

Alternatively, the above conservation law can be formulated in terms of the average momentum per cell of spatially periodic configurations. Namely, if a configuration x has period $p > 0$ (i.e., if $x[i + p] = x[i]$ for every i), one can say that

$$\frac{\sum_{i=1}^{p} \theta(x[i])}{p} = \frac{\sum_{i=1}^{p} \theta((Fx)[i])}{p} \tag{3}$$

As discussed later, these two formulations are equivalent in general; a CA conserves the energy of every finite configuration if and only if it preserves the average energy per cell of each spatially periodic configuration.

Next, a few physically interesting examples will be discussed. The *Ising* model was introduced by Wilhelm Lenz (1888–1957) and Ernst Ising (1900–1998) to explain the phenomenon of phase transition in ferromagnetic materials. It is a stochastic model and is extensively studied in statistical mechanics (see, e.g., Kindermann and Snell (1980); Sinai (1982); Georgii et al. (2000)). Gérard Vichniac has introduced a deterministic CA-like dynamics on it (Vichniac 1984) (see also Toffoli and Margolus 1987; Chopard and Droz 1998).

In the Ising model, each cell represents a tiny piece of ferromagnetic material having a spin (i.e., a magnetic moment resulting from the angular momentum of the electrons). For simplicity, each spin is approximated by either of two values: ↑ (spin-up) or ↓ (spin-down).

Adjacent spins tend to align. This tendency is depicted by assigning an energy 1 to each pair of adjacent spins that are anti-aligned, and energy -1 to those that are aligned.

Vichniac's dynamics is specially designed in such a way as to conserve this energy, hence emulating the regime where there is no heat transfer in and out of the material. The states of the cells are updated in two stages. The cells are colored black and white as on the chess board. At the first stage, all the black cells are updated in the following way: a spin on a black cell is flipped (from $\uparrow$ to $\downarrow$, or from $\downarrow$ to $\uparrow$) if and only if the change does not affect the total energy of the bonds with its adjacent spins. At the second stage, the white cells are updated in a similar fashion. ❯ *Figure 3* shows few snapshots from a simulation of this CA-like dynamics.

To put things formally, the set of possible states for each cell is $S \triangleq \{\uparrow, \downarrow\}$. In the d-dimensional model, the cells are indexed by the elements $i = (i_1, i_2, \ldots, i_d)$ in $\mathbb{Z}^d$, the d-dimensional square lattice. Hence, a configuration of the model is an assignment $x : \mathbb{Z}^d \to S$. The dynamics is defined in terms of two mappings $x \mapsto F_0 x$ (for updating the black cells) and $x \mapsto F_1 x$ (for updating the white cells). For every cell $i \in \mathbb{Z}^d$,

$$(F_0 x)[i] \triangleq \begin{cases} \neg x[i] & \text{if } i \text{ black and } \sum_{j \in N(i)} \varsigma(x[j]) = 0 \\ x[i] & \text{otherwise} \end{cases} \tag{4}$$

$$(F_1 x)[i] \triangleq \begin{cases} \neg x[i] & \text{if } i \text{ white and } \sum_{j \in N(i)} \varsigma(x[j]) = 0 \\ x[i] & \text{otherwise} \end{cases} \tag{5}$$

Here, a few shorthand notations have been used. First, $\neg a$ denotes the reverse of a; that is, $\neg\uparrow \triangleq \downarrow$ and $\neg\downarrow \triangleq \uparrow$. Next, $\varsigma(a)$ is used to denote the sign of a spin; that is, $\varsigma(\uparrow) \triangleq +1$ and $\varsigma(\downarrow) \triangleq -1$. Finally, $N(i)$ represents the set of immediate neighbors of cell i. In summary, the dynamics is obtained by alternately applying F_0 and F_1 on a configuration:

$$x_0 \xrightarrow{F_0} x_0' \xrightarrow{F_1} x_1 \xrightarrow{F_0} x_1' \xrightarrow{F_1} x_2 \xrightarrow{F_0} \cdots \tag{6}$$

The conservation of energy can now be formulated by saying that each of F_0 and F_1 preserves the sum

$$\Theta(x) \triangleq \sum_{\substack{i,j \\ \text{adjacent}}} [1 - \varsigma(x[i])\varsigma(x[j])] \tag{7}$$

◘ **Fig. 3**
Simulation of Vichniac's dynamics on a spatially periodic configuration of the two-dimensional Ising model. Blue represents $\uparrow$. Green represents $\downarrow$. (a) The initial configuration. (b) The configuration at time $t = 10$. (c) The configuration at time $t = 60$.

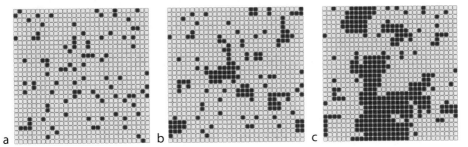

a   b   c

for any configuration x with only finitely many downward spins ↓, or finitely many upward spins ↑. A similar statement can be formulated for spatially periodic configurations.

Notice that Vichniac's dynamics is reversible; one can completely regenerate the configuration at time t, knowing the configuration at time $t + 1$. In fact, all one needs to do is to apply F_0 and F_1 in the reverse order. Reversibility is apparently a fundamental feature of nature. Even though the macroscopic world as one perceives it looks irreversible, every known physical process behaves reversibly in the ultimate microscopic scale (cf. Feynman 1965).

The next example, due to Pomeau (1984) and Margolus (1987), identifies an interesting energy-like invariant in a large class of reversible models. Each cell of the lattice has a state from the finite ring $\mathbb{Z}_m = \{0, 1, 2, \ldots, m - 1\}$. The dynamics is of second order; that is, the configuration at time $t + 1$ depends not only on the configuration at time t, but also on the configuration at time $t - 1$; that is, $c_{t+1} = F(c_t, c_{t-1})$. The state of a cell $i \in \mathbb{Z}^d$ at time $t + 1$ is obtained by a rule of the form

$$c_{t+1}[i] = c_{t-1}[i] + f(c_t[N(i)]) \tag{8}$$

Here $N(i)$ is a finite set of cells that is called the neighborhood of cell i. The neighborhood is assumed to be uniform; that is, there is a finite set $N \subseteq \mathbb{Z}^d$ such that $N(i) \triangleq \{i + k : k \in N\}$ for every cell i. $c_t[N(i)]$ means the pattern of the states seen on the neighborhood of cell i in configuration c_t. Mathematically, this can be seen as an element of $\mathbb{Z}_m^N$, the set of all possible assignments $p : N \to \mathbb{Z}_m$. The function $f : \mathbb{Z}_m^N \to \mathbb{Z}_m$ assigns a value $f(p) \in \mathbb{Z}_m$ to each neighborhood pattern $p : N \to \mathbb{Z}_m$. Intuitively, in order to update its state, a cell i applies a function f on the current state of its neighbors, and depending on the result, permutes the state it used to have one step before.

Notice that any automaton that is defined this way is automatically reversible; one can retrace an orbit $\ldots, c_{t-1}, c_t, c_{t+1}, \ldots$ backward using the rule

$$c_{t-1}[i] = c_{t+1}[i] - f(c_t[N(i)]) \tag{9}$$

Now, suppose that the neighborhood N is symmetric, meaning that $k \in N$ if and only if $-k \in N$. Suppose further that one finds a function $g : \mathbb{Z}_m^N \to \mathbb{Z}_m$ of the form

$$g(p) \triangleq \sum_{k \in N} \beta_k p[k] \tag{10}$$

($\beta_k \in \mathbb{Z}_m$) that has the following two properties:

(i) It is symmetric; that is, $\beta_{-k} = \beta_k$ for every $k \in N$.
(ii) It is orthogonal to f, in the sense that $f(p)g(p) = 0$ for every $p \in \mathbb{Z}_m^N$.

Denote the configuration in which every cell is in state 0 by **0**. For simplicity, assume that $\ldots, \mathbf{0}, \mathbf{0}, \mathbf{0}, \ldots$ is a valid orbit of the automaton. Equivalently, this means that f maps the uniformly-0 pattern into 0. (One could avoid this requirement by formulating the conservation law in terms of spatially periodic configurations.) A configuration is called finite if only a finite number of cells have nonzero states in it.

Let $\ldots, c_{t-1}, c_t, c_{t+1}, \ldots$ be an arbitrary orbit consisting of finite configurations. The value of the sum

$$\Theta(c_{t-1}, c_t) \triangleq \sum_{i \in \mathbb{Z}^d} c_{t-1}[i] g\left(c_t[N(i)]\right) \tag{11}$$

is claimed to be independent of the time t.

From ❷ Eq. 8 and property (ii) one can write

$$(c_{t+1}[i] - c_{t-1}[i])g\ (c_t[N(i)]) = 0 \tag{12}$$

Summing over all cells i, one obtains

$$\sum_{i\in\mathbb{Z}^d} c_{t-1}[i]g\ (c_t[N(i)]) = \sum_{i\in\mathbb{Z}^d} c_{t+1}[i]g\ (c_t[N(i)]) \tag{13}$$

By the symmetry of g (property (i)) one can rewrite the right-hand side as follows:

$$\sum_{i\in\mathbb{Z}^d} c_{t+1}[i]g\ (c_t[N(i)]) = \sum_{i\in\mathbb{Z}^d}\sum_{k\in N} \beta_k c_{t+1}[i]c_t[i+k] \tag{14}$$

$$= \sum_{i'\in\mathbb{Z}^d}\sum_{k'\in N} \beta_{-k'} c_{t+1}[i'+k']c_t[i'] \tag{15}$$

$$= \sum_{i'\in\mathbb{Z}^d}\sum_{k'\in N} \beta_{k'} c_{t+1}[i'+k']c_t[i'] \tag{16}$$

$$= \sum_{i'\in\mathbb{Z}^d} c_t[i']g\ (c_{t+1}[N(i')]) \tag{17}$$

Therefore, one obtains

$$\Theta(c_{t-1}, c_t) = \sum_{i\in\mathbb{Z}^d} c_{t-1}[i]g\ (c_t[N(i)]) \tag{18}$$

$$= \sum_{i\in\mathbb{Z}^d} c_t[i]g\ (c_{t+1}[N(i)]) \tag{19}$$

$$= \Theta(c_t, c_{t+1}) \tag{20}$$

proving the claim.

Yet another beautiful example is the following discrete model of an excitable medium, due to Greenberg and Hastings (1978b) (see the relevant part in the chapter ❷ Cellular Automata and Lattice Boltzmann Modeling of Physical Systems, and also Greenberg et al. (1978a, 1980). The CA runs on a two-dimensional board. Each cell is either "at rest" (state ☐), "excited" (state ■), or in a "refractory phase" (state ▦). A cell that is at rest remains so unless it is "stimulated" by one or more of its four neighbors (i.e., if at least one of its neighbors is excited). An excited cell undergoes a 1-step refractory phase before going back to rest and starting to respond to stimulations again. Typically, a configuration of the infinite board contains a number of "singularities" with waves continuously swirling around them (❷ Fig. 4). The singularities are never created or destroyed. Therefore, the number of such singularities remain constant throughout time. To be more precise, the singularities are the 2×2 blocks of cells with states ▦■/☐▷, ■■/▦▷, or ■■/▦☐, or the rotations or mirror images of these blocks. One can easily verify that a singular 2×2 block remains singular after one step, and a non-singular block remains non-singular.

Sometimes, one can find an energy-like function that is not perfectly conserved by the evolution of a CA, yet whose total value is never increased (or never decreased) with time. Physically, such a situation is comparable with a system that is isolated from its environment, except that it may dissipate heat, resulting in a decrease of its total energy. Mathematically, a nonincreasing energy function may be helpful in studying stability properties of a CA.

◘ **Fig. 4**

Simulation of Greenberg–Hastings model on a spatially periodic configuration. **(a)** The initial configuration. **(b)** The configuration at time $t = 10$. **(c)** The configuration at time $t = 60$.

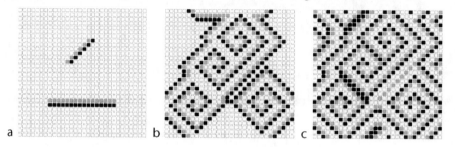

a   b   c

As an example, in the Just Gliders CA that was discussed before, the number of left-moving particles is never increased, though it may decrease. Formally, one has

$$\sum_{i=-\infty}^{+\infty} \theta_L(x[i]) \geq \sum_{i=-\infty}^{+\infty} \theta_L((Fx)[i]) \tag{21}$$

where

$$\theta_L(a) \triangleq \begin{cases} 1 & \text{if } a = \blacktriangleleft \\ 0 & \text{otherwise} \end{cases} \tag{22}$$

for any configuration x with a finite number of particles on it. Equivalently, one can write

$$\frac{\sum_{i=1}^{P} \theta_L(x[i])}{p} \geq \frac{\sum_{i=1}^{P} \theta_L((Fx)[i])}{p} \tag{23}$$

for every spatially periodic configuration x with period $p > 0$. In the Traffic CA, it is easy to verify that the number of blocks ■■ of two consecutive cars is never increased; that is,

$$\sum_{i=-\infty}^{+\infty} \theta'(x[i, i+1]) \geq \sum_{i=-\infty}^{+\infty} \theta'((Fx)[i, i+1]) \tag{24}$$

where

$$\theta'(ab) \triangleq \begin{cases} 1 & \text{if } ab = \blacksquare\blacksquare \\ 0 & \text{otherwise} \end{cases} \tag{25}$$

A more interesting example is the *Sand Pile* model due to Bak et al. (1987, 1988). On each cell of the board, there is a stack of sand grains. One can consider the height of this stack as the state of the cell. So, each cell $i \in \mathbb{Z}^d$ has a state $h[i] \in \{0, 1, \ldots, N\}$. (To keep the state set finite, it has been assumed that each cell may contain no more than $N > 0$ grains.) If the difference between the height of the stacks in two adjacent cells is more than a threshold $K \geq 4d$, one grain of sand from the higher cell tumbles down onto the lower cell. More precisely, if $h : \mathbb{Z}^d \to \{0, 1, \ldots, N\}$ is a configuration of the automaton, its configuration after one step would be $Fh : \mathbb{Z}^d \to \{0, 1, \ldots, N\}$, where

$$(Fh)[i] = h[i] - |\{j \in N(i) : h[i] - h[j] \geq K\}| \\ + |\{j \in N(i) : h[j] - h[i] \geq K\}| \tag{26}$$

◻ Fig. 5

Two consecutive configurations of the one-dimensional Sand Pile model with parameter $K = 5$.
(a) The initial configuration. (b) The configuration after one step.

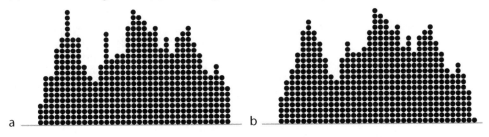

a   b

Here $N(i)$ denotes the set of immediate neighbors of cell i (❂ Fig. 5). Clearly, the total number of sands on a finite configuration is never changed; that is,

$$\sum_{i\in\mathbb{Z}^d}(Fh)[i] = \sum_{i\in\mathbb{Z}^d} h[i] \tag{27}$$

A bit less trivial is the fact that the sum of the squares of the number of sands on the cells is nonincreasing:

$$\sum_{i\in\mathbb{Z}^d}((Fh)[i])^2 \leq \sum_{i\in\mathbb{Z}^d}(h[i])^2 \tag{28}$$

To see this, note that the height of a single grain of sand does never increase with time. The sum of the heights of all grains on configuration h is

$$\sum_{i\in\mathbb{Z}^d}\sum_{k=1}^{h[i]} k = \sum_{i\in\mathbb{Z}^d}\frac{1}{2}h[i]\,(h[i]+1) \tag{29}$$

$$= \frac{1}{2}\sum_{i\in\mathbb{Z}^d}(h[i])^2 + \frac{1}{2}\sum_{i\in\mathbb{Z}^d}h[i] \tag{30}$$

and does never increase with time. From ❂ Eq. 27 one knows that the second term in the above sum remains constant with time. Therefore, the first term cannot increase with time. (The original Sand Pile automaton of Bak et al. (1987, 1988) is more elegant in that its cell states store not the height $h[i]$, but the difference $h[i] - h[i+1]$ between the heights of two consecutive cells. This way, one can represent sand piles of arbitrary height using a bounded number of cell states. As remarked by Goles (1992), a similar nonincreasing energy can be found for such CA, provided the energy is allowed to depend on the previous configurations, too. Nonuniform nonincreasing energies are used in Anderson et al. (1989) and Goles and Kiwi (1993) to analyze Sand Pile and similar automata.)

2   Mathematical Formulation

In this section, what is exactly meant by a conservation law in a CA is formulated in a more precise language.

2.1 Cellular Automata

The discussion is restricted to cellular automata on the infinite d-dimensional square lattice. Thus the cells of the *lattice* are indexed by the elements of $\mathbb{Z}^d$. The *states* of the cells are chosen from a finite set S that contains at least two elements. A *configuration* of the lattice is a mapping $x : \mathbb{Z}^d \to S$ which assigns a state to each cell on the lattice. A *pattern* means an assignment $p : D \to S$ of states to a subset $D \subseteq \mathbb{Z}^d$ of cells. A *finite* pattern is a pattern that has a finite domain. If $p : D \to S$ is a pattern and $E \subseteq D$, $p[E]$ is written to denote the restriction of p to E; that is, $p[E]$ stands for the pattern $q : E \to S$ seen over the set E under p.

For each $k \in \mathbb{Z}^d$, σ^k denotes the *translation* by k. That is, for each pattern $p : D \to S$, $\sigma^k p$ is the pattern with $(\sigma^k p)[i] = p[k + i]$ whenever $k + i \in D$.

To specify a *cellular automaton* (CA), one further needs to specify a *neighborhood* and a *local rule*. The neighborhood is presented by a finite set $0 \in N \subseteq \mathbb{Z}^d$. The neighborhood of a cell i is the set $N(i) \triangleq \{i + k : k \in N\}$. The local rule is a function $f : S^N \to S$ that provides a new state for each cell i by looking at its neighborhood pattern. Hence, for every configuration $x : \mathbb{Z}^d \to S$, one obtains a new configuration $Fx : \mathbb{Z}^d \to S$ where

$$(Fx)[i] = f((\sigma^i x)[N]) \tag{31}$$

$$= f(\sigma^i(x[N(i)])) \tag{32}$$

for each cell i. The dynamics of the CA is realized by iterating the *global mapping* F on an initial configuration x (This definition does not cover some of the examples ❷ Sect. 1 (namely, Vichniac's dynamics on the Ising model, and the second order model). However, those models can be easily transformed into a standard CA as defined here (cf. Toffoli and Margolus 1987).)

$$x \xrightarrow{F} Fx \xrightarrow{F} F^2 x \xrightarrow{F} F^3 x \xrightarrow{F} \cdots . \tag{33}$$

One often identifies a CA with its global mapping.

It is useful to see the configuration space $S^{\mathbb{Z}^d}$ as a topological space. The product topology on $S^{\mathbb{Z}^d}$ is the smallest topology with respect to which all the projections $x \mapsto x[i]$ are continuous. In this topology, a sequence $\{x_t\}_{t=1}^{\infty}$ converges to a configuration x if and only if for each cell i, $x_t[i] = x[i]$ for all sufficiently large t. The space $S^{\mathbb{Z}^d}$ is compact and metrizable, and is homeomorphic to the Cantor set. A *cylinder* is a set of the form

$$[p] \triangleq \{x \in S^{\mathbb{Z}^d} : x[D] = p\} \tag{34}$$

where $p : D \to S$ is a finite pattern. Cylinders are both open and closed, and form a basis for the product topology. The global mapping of a cellular automaton is continuous with respect to the product topology. Moreover, every translation-invariant continuous mapping on $S^{\mathbb{Z}^d}$ is the global mapping of a CA (Hedlund 1969). The topological aspects of cellular automata are discussed in the chapter ❷ Cellular Automata Dynamical Systems.

The Borel σ-algebra on the space $S^{\mathbb{Z}^d}$ of configurations is the σ-algebra generated by the cylinders. A Borel probability measure π is completely determined by assigning a probability $0 \leq \pi([p]) \leq 1$ to each cylinder $[p]$ in a consistent way (cf. Parthasarathy 1967). The space $\mathcal{M}$ of all Borel probability measures on $S^{\mathbb{Z}^d}$ can be topologized by the vague topology (aka the weak* topology) (see, e.g., Walters 1982). This space is also compact and metrizable. A sequence $\{\pi_t\}_{t=0}^{\infty}$ converges to a measure π if and only if for each cylinder $[p]$, the sequence

$\{\pi_t([p])\}_{t=0}^{\infty}$ of real numbers converges to $\pi([p])$. A CA F induces a continuous mapping on $\mathcal{M}$ via $F\pi \triangleq \pi \circ F^{-1}$. A translation-invariant measure is a measure $\pi \in \mathcal{M}$ such that $\sigma^k \pi = \pi$ for every translation σ^k. The set of all translation-invariant Borel probability measures on $S^{\mathbb{Z}^d}$ is a closed and convex subspace of $\mathcal{M}$ and is denoted by $\mathcal{M}_{\sigma}$.

2.2 Energy

Let $S^{\#}$ denote the set of all finite patterns *modulo* translations (i.e., forgetting their exact positions). To be strict, an element of $S^{\#}$ is a class $\langle p \rangle$ of patterns that can be obtained by translating p. However, unless there is a risk of confusion, one often abuses the notations and identifies a class $\langle p \rangle$ with any of its elements. The elements of $S^{\#}$ are seen as generalized words. In particular, when $d = 1$ (i.e., on the one-dimensional lattice $\mathbb{Z}$), one has $S^{*} \subseteq S^{\#}$, where S^{*} stands for the set of finite words on the alphabet S. Let $\emptyset$ be the (unique) pattern with an empty domain.

A (local and additive) *energy* is specified by assigning an *interaction potential* $\theta(p) \in \mathbb{R}$ to each finite pattern $p \in S^{\#}$. One requires that $\theta(\emptyset) = 0$, and that the set $\{p : \theta(p) \neq 0\} \subseteq S^{\#} \backslash \{\emptyset\}$ is finite. The latter set is called the *support* of θ and is denoted by $\operatorname{supp}(\theta)$. For a configuration $x : \mathbb{Z}^d \rightarrow S$ and a finite collection A of cells, the value $\theta(x[A])$ is interpreted as the energy resulting from the interaction of the cells in A. The *total energy* of x (whenever meaningful) is simply the sum

$$\Theta(x) \triangleq \sum_{\substack{A \subset \mathbb{Z}^d \\ \text{finite}}} \theta(x[A]) \tag{35}$$

However, the sum (❯ Eq. 35) typically does not have a well-defined value or is infinite.

There are essentially three ways around this problem. The first approach is to consider only the difference between the energy of two configurations that are only slightly perturbed from each other (more precisely, differ on only a finite number of cells). The second approach is to work with the average or expected energy per cell in a configuration. The third approach is to avoid any global notion of energy and instead describe a conservation law in terms of the local redistribution of energy at each step. Fortunately, all these lead to equivalent concepts of a conservation law. Now, the first two approaches and their equivalence are discussed. The local approach is discussed later in ❯ Sect. 4.

It is said that two configurations $x, y : \mathbb{Z}^d \rightarrow S$ are *asymptotic* if they disagree on no more than a finite number of cells. The *difference* between the energy of two asymptotic configurations x and y is defined to be

$$\delta\Theta(x, y) \triangleq \sum_{\substack{A \subset \mathbb{Z}^d \\ \text{finite}}} [\theta(y[A]) - \theta(x[A])] \tag{36}$$

It is worth noting how the *locality* and *additivity* of energy translate in this framework. The *interaction range* of an energy θ can be identified by a minimal neighborhood $0 \in M \subseteq \mathbb{Z}^d$ such that, for every pattern $p : D \rightarrow S$ in the support of θ and every $i \in D$, one has $D \subseteq M(i)$. Since $\operatorname{supp}(\theta)$ is finite, the interaction range of θ is also finite.

If two patterns $p : D \rightarrow S$ and $q : E \rightarrow S$ agree on the intersection $D \cap E$ of their domains (in particular, if $D \cap E = \emptyset$), one can merge them together and obtain a pattern $p \vee q : D \cup E \rightarrow S$ that agrees with p and q on their domains. The *boundary* of a set A (with respect to the neighborhood M) is the set $\partial M(A) \triangleq M(A) \backslash A$.

Proposition 1 (locality) *Let $p, q : D \to S$, $r : \partial M(D) \to S$, and $u, v : \mathbb{Z}^d \backslash M(D) \to S$ be arbitrary patterns. Then,*

$$\delta\Theta(p \vee r \vee u, q \vee r \vee u) = \delta\Theta(p \vee r \vee v, q \vee r \vee v) \qquad (37)$$

Proposition 2 (additivity) *Let $D, E \subseteq \mathbb{Z}^d$ be two finite nonempty sets such that $M(D) \cap E = D \cap M(E) = \varnothing$. Let $p_0, p_1 : D \to S$, $q_0, q_1 : E \to S$ and $w : \mathbb{Z}^d \backslash D \backslash E \to S$ be arbitrary patterns. Then,*

$$\begin{aligned}
\delta\Theta(p_0 \vee q_0 \vee w, p_1 \vee q_1 \vee w) &= \delta\Theta(p_0 \vee q_0 \vee w, p_1 \vee q_0 \vee w) \\
&\quad + \delta\Theta(p_0 \vee q_0 \vee w, p_0 \vee q_1 \vee w)
\end{aligned} \qquad (38)$$

A *local observable* (or a locally observable property) is a mapping $\mu : S^{\mathbb{Z}^d} \to \Gamma$ (Γ being an arbitrary set) that depends only on the states of a finite number of cells. That is, μ is a local observable if there is a finite neighborhood $W \subseteq \mathbb{Z}^d$ (the observation *window*) and a local rule $g : S^W \to \Gamma$ such that $\mu(x) = g(x[W])$. Observing a configuration x *around* a cell i one gets the value $\mu(\sigma^i x) = g(x[W(i)])$.

Every real-valued local observable μ with local rule $g : S^W \to \mathbb{R}$ defines an interaction potential θ via

$$\theta(p) \triangleq \begin{cases} g(p) & \text{if } p \in S^W \\ 0 & \text{otherwise} \end{cases} \qquad (39)$$

The energy difference $\delta\Theta(x, y)$ can then be calculated using

$$\delta\Theta(x, y) = \sum_{i \in \mathbb{Z}^d} \left[\mu(\sigma^i y) - \mu(\sigma^i x)\right] \qquad (40)$$

Conversely, given an interaction potential θ, one can construct a real-valued local observable μ that generates $\delta\Theta$ via ❷ Eq. 40. For example, one can choose the interaction range M of θ as the observation window and define

$$\mu_\theta(x) \triangleq \sum_{0 \in A \subseteq M} \frac{1}{|A|} \theta(x[A]) \qquad (41)$$

Hence, one can equivalently specify an energy using a real-valued local observable.

Consider an energy which is formalized using a local observable μ. For every $n \geq 0$, let $I_n \triangleq [-n, n]^d$ be the centered hypercube of size $(2n + 1)^d$ on the lattice. The *average* energy per cell in a configuration x is obtained by taking the limit of the finite averages

$$\frac{\sum_{i \in I_n} \mu(\sigma^i x)}{|I_n|} \qquad (42)$$

when $n \to \infty$. Since the limit does not always exist, one uses the *upper* or *lower* average energy per cell

$$\overline{\mu}(x) \triangleq \limsup_{n \to \infty} \frac{\sum_{i \in I_n} \mu(\sigma^i x)}{|I_n|} \qquad (43)$$

$$\underline{\mu}(x) \triangleq \liminf_{n \to \infty} \frac{\sum_{i \in I_n} \mu(\sigma^i x)}{|I_n|} \qquad (44)$$

For every translation-invariant Borel probability measure $\pi \in \mathcal{M}_\sigma$ on $S^{\mathbb{Z}^d}$, one can also define the *expected* energy per cell

$$\pi(\mu) \triangleq \int \mu d\pi \triangleq \sum_{p:W \to S} g(p)\pi([p]) \tag{45}$$

The mapping $\pi \mapsto \pi(\mu)$ is uniformly continuous. According to the pointwise ergodic theorem (see, e.g., Walters 1982), for every ergodic measure $\pi \in \mathcal{M}_\sigma$ (i.e., ergodic with respect to σ) and π-almost every configuration $x \in S^{\mathbb{Z}^d}$, one has $\overline{\mu}(x) = \underline{\mu}(x) = \pi(\mu)$. Ergodic measures are the extremal points of the convex set $\mathcal{M}_\sigma$.

2.3   Conservation of Energy

One can now formulate conservation laws in several different, but equivalent ways. One says that a CA F *conserves* an energy defined in terms of an interaction potential θ or an observable μ if any of the following equivalent conditions hold.

Theorem 1   (Hattori and Takesue 1991; Boykett and Moore 1998; Boccara and Fuks 1998; 2002; Pivato 2002; Kůrka 2003a; Durand et al. 2003; Moreira et al. 2004) *Let* $F : S^{\mathbb{Z}^d} \to S^{\mathbb{Z}^d}$ *be a cellular automaton and* $\theta : S^{\#} \to \mathbb{R}$ *an interaction potential. Let* $\mu : S^{\mathbb{Z}^d} \to \mathbb{R}$ *be a local observable that generates the same energy as* θ. *Let* $\diamond : \mathbb{Z}^d \to S$ *be an arbitrary configuration. The following conditions are equivalent.*

C-1 $\delta\Theta(Fx, Fy) = \delta\Theta(x, y)$ *for every two asymptotic configurations x and y.*
C-2 $\delta\Theta(F\diamond, Fx) = \delta\Theta(\diamond, x)$ *for every configuration x asymptotic to* $\diamond$.
C-3 $\overline{\mu}(Fx) = \overline{\mu}(x)$ *for every configuration x.*
C-4 $\overline{\mu}(Fx) = \overline{\mu}(x)$ *for every periodic configuration x.*
C-5 $(F\pi)(\mu) = \pi(\mu)$ *for every translation-invariant probability measure* π.
C-6 $(F\pi)(\mu) = \pi(\mu)$ *for every ergodic translation-invariant probability measure* π.

Condition (C-1) states that the CA preserves the difference between the energy of every two asymptotic configurations, while condition (C-2) requires that the energy of a configuration relative to a fixed orbit remains constant. Condition (C-3) means that the (upper) average energy per cell of each configuration remains unchanged with time. Condition (C-3) only requires the CA to preserve the average energy per cell of the periodic configurations. Conditions (C-5) and (C-6) express a conservation law in terms of the expected energy per cell.

The equivalence of the conditions (C-1) and (C-2) is an immediate consequence of the locality of energy. To prove that (C-1) implies (C-3), one first shows that $\overline{\mu}(Fz) = \overline{\mu}(z)$ in the special case that z is a uniform configuration. An arbitrary configuration x can then be approximated by configurations x_n that agree with x on larger and larger centered hypercubes I_n and have a fixed state s on all the other cells. Similarly, the implication (C-4) $\Rightarrow$ (C-1) follows from the locality of energy, by considering periodic configurations $\hat{x}$ and $\hat{y}$ that agree with x and y on a sufficiently large hypercube I_n. That (C-3) implies (C-6) follows from the ergodic theorem. The implication (C-6) $\Rightarrow$ (C-5) is a consequence of the fact that every translation-invariant measure π is a limit of convex combinations of ergodic measures (cf. Walters 1982). To see that (C-5) implies (C-4), for every periodic configuration x, consider a measure π_x whose probability mass is concentrated on the σ-orbit of x. More details can be found, for example, in Taati (2009). Yet other characterizations of conservation laws can be found in Pivato (2002) and Durand et al. (2003).

A uniform configuration $\Diamond$ for which $F\Diamond = \Diamond$ is called a *quiescent* configuration. For example, in the Traffic CA, the configuration with no car is quiescent; so is the configuration with cars on every cell. There is often a distinguished quiescent configuration $\Diamond$, which is seen as the "background" or the "vacuum," and one is interested in the evolution of the configurations that are asymptotic to $\Diamond$. A configuration that is asymptotic to $\Diamond$ is then called a *finite* configuration. The image of a finite configuration under F is obviously finite.

Choosing the interaction potential θ properly, one can ensure that this distinguished quiescent configuration $\Diamond$ has zero total energy. Then every finite configuration x would have a finite well-defined total energy $\Theta(x) = \delta\Theta(\Diamond, x)$. Hence, one can express the conservation law in terms of the total energy of the finite configurations: the CA F conserves θ if and only if

C-7 $\Theta(Fx) = \Theta(x)$ for every finite configuration x.

2.4 Dissipation of Energy

Recall, from the previous section, the Sand Pile example, in which an energy was presented that was not conserved, but its value was never increased. Such *dissipation laws* are now formalized. An energy θ is said to be *dissipative* (or nonincreasing) under a CA F if any of the following equivalent conditions hold.

Theorem 2 (Kůrka 2003a; Moreira et al. 2004; Bernardi et al. 2005) *Let* $F : S^{\mathbb{Z}^d} \to S^{\mathbb{Z}^d}$ *be a cellular automaton and* $\theta : S^\# \to \mathbb{R}$ *an interaction potential. Let* $\mu : S^{\mathbb{Z}^d} \to \mathbb{R}$ *be a local observable that generates the same energy as* θ. *The following conditions are equivalent.*

D-1 $\overline{\mu}(Fx) \leq \overline{\mu}(x)$ *for every configuration* x.
D-2 $\overline{\mu}(Fx) \leq \overline{\mu}(x)$ *for every periodic configuration* x.
D-3 $(F\pi)(\mu) \leq \pi(\mu)$ *for every translation-invariant probability measure* π.
D-4 $(F\pi)(\mu) \leq \pi(\mu)$ *for every ergodic translation-invariant probability measure* π.

Furthermore, if $\Diamond : \mathbb{Z}^d \to S$ *is a quiescent configuration, the following condition is also equivalent to the above.*

D-5 $\delta\Theta(\Diamond, Fx) \leq \delta\Theta(\Diamond, x)$ *for every finite configuration* x.

The proofs are similar to those for conservation laws. In case the quiescent configuration $\Diamond$ has total energy zero, the condition (C-5) takes the following concise form:

D-6 $\Theta(Fx) \leq \Theta(x)$ for every finite configuration x.

3 Algorithmics

In this section, two algorithmic questions related to conservation laws in cellular automata are discussed. It is shown how one can verify the validity of a conservation law algorithmically. Next, the more difficult question of finding conservation laws in a CA is discussed. Though one can enumerate all the possible candidates and verify their validity one by one, there is no algorithm to tell beforehand whether a CA has any nontrivial conservation law or not, even when restricted to one-dimensional CA.

Dissipative energies are more complicated to recognize. Fortunately, in one dimension, one can still decide whether a given energy is dissipative or not. The question is, however, undecidable in the higher-dimensional case.

3.1 Conservation Laws

Suppose that one is given a CA F and an interaction potential θ, and asked whether F conserves the energy defined by θ. Verifying either of the characterizations (C-1)–(C-6) requires establishing infinitely many equalities. However, as was first noticed by Hatori and Takesue (1991), it suffices to verify the equality

$$\delta\Theta(Fx, Fy) = \delta\Theta(x, y) \tag{46}$$

for all x and y that differ on exactly one cell.

Indeed, suppose that x and y are two configurations that disagree only on a finite set D of cells. One can then find a sequence

$$x = x_0, \; x_1, \; x_2, \; \ldots, \; x_n = y \tag{47}$$

of configurations, where $n = |D|$, such that x_i and x_{i+1} disagree on exactly one cell. If ❽ Eq. 46 holds whenever two configurations differ on a single cell, one can write

$$\delta\Theta(x, y) = \sum_{i=0}^{n-1} \delta\Theta(x_i, x_{i+1}) \tag{48}$$

$$= \sum_{i=0}^{n-1} \delta\Theta(Fx_i, Fx_{i+1}) \tag{49}$$

$$= \delta\Theta(Fx, Fy) \tag{50}$$

(Note that $\delta\Theta(a, c) = \delta\Theta(a, b) + \delta\Theta(b, c)$ for every a, b, c.)

Proposition 3 (Hattori and Takesue 1991) *Let $F : S^{\mathbb{Z}^d} \to S^{\mathbb{Z}^d}$ be a cellular automaton and $\theta : S^{\#} \to \mathbb{R}$ an interaction potential. The following conditions are equivalent.*

C-1 $\delta\Theta(Fx, Fy) = \delta\Theta(x, y)$ *for every two asymptotic configurations x and y.*
C-8 $\delta\Theta(Fx, Fy) = \delta\Theta(x, y)$ *for every two configurations x and y that disagree on exactly one cell.*

Since θ (and hence $\delta\Theta$) is translation-invariant, one can, in fact, consider only configurations x and y that disagree on cell 0.

Now let $0 \in M \subseteq \mathbb{Z}^d$ be the interaction range of θ, and let $0 \in N \subseteq \mathbb{Z}^d$ be the neighborhood of F. For every two configurations x and y that disagree only on cell 0, the value $\delta\Theta(x, y)$ depends only on the state of the cells in x and y that are in the neighborhood $M(0)$. Similarly, since Fx and Fy may disagree only on the set $N^{-1}(0)$ (i.e., the set of cells that have cell 0 as neighbor) the value $\delta\Theta(Fx, Fy)$ depends on the state of the cells in Fx and Fy that are in the set $M(N^{-1}(0))$. Therefore, condition (C-8) reduces to a finite number of equalities, which can each be verified in a finite time.

The following algorithm exploits the above discussion to answer whether F conserves θ. Let $f: S^N \to S$ be the local rule of F. There is a natural way to extend the application of f to any finite pattern $p: N(A) \to S$. Namely, the image of p under f is a pattern $f(p): A \to S$, where $f(p)$ $[i] \triangleq f((\sigma^i p)[N])$ for every $i \in A$. For every finite pattern $p: D \to S$, let one use the shorthand

$$\Theta(p) \triangleq \sum_{A \subseteq D} \theta(p[A]) \tag{51}$$

```
let W = M(N^{-1});
for every pattern p : N(W)  → S and every state s ∈ S,
    let q : N(W) → S and
        set q[0] = s and q[i] = p[i] for i ≠ 0;
    let p', q' : W → S and
        set p' = f(p) and q' = f(q);
    set δΘ = Θ(q) − Θ(p);
    set δΘ' = Θ(q') − Θ(p');
    if δΘ' ≠ δΘ,
        return "no" and halt;
return "yes";
```

There are few ways to improve the efficiency of the above algorithm. One can order the cells on the lattice lexicographically. Namely, in this ordering, a cell $i = (i_1, i_2, \ldots, i_d)$ precedes a cell $j = (j_1, j_2, \ldots, j_d)$, written $i \prec j$, if there is a $1 \leq k \leq d$ such that $i_k < j_k$ and $i_l = j_l$ for all $1 \leq l < k$. Let $\diamond \in S$ be a fixed state. Using a more clever argument, one can restrict the main loop of the algorithm to all patterns $p: N(W) \to S$ for which $p[i] = \diamond$ for all cells $i \succeq 0$.

The set of all interaction potentials θ that are conserved by F is a linear space $\mathscr{W}_F$. It is interesting to note that, given a finite set $P \subseteq S^\#$, one can use a variant of the above algorithm to construct the space $\mathscr{W}_F[P]$ of all those interaction potentials conserved by F that have support P. To do so, one considers $\theta: P \to \mathbb{R}$ as a vector of unknowns, and collects all the equations $\delta\Theta' = \delta\Theta$ given by the algorithm. The space $\mathscr{W}_F[P]$ is simply the solution space of this system of equations.

In Durand et al. (2003), a different approach has led to another efficient algorithm for verifying the validity of conservation laws.

As already mentioned at the beginning of this section, given a CA F, one can enumerate all the candidate energies θ and verify whether they are conserved by F. (Strictly speaking, one cannot enumerate all the energies, since the set of real-valued energies is uncountable. However, note that for each energy θ, the set of possible values that $\delta\Theta$ can take is a finitely generated subgroup of $\mathbb{R}$, and hence isomorphic to $\mathbb{Z}^m$ for some $m > 0$. The discussion merely uses the group structure of the energy values. Therefore, one can equivalently work with energies whose values are in $\mathbb{Z}^m$. Such energies can indeed be enumerated.) Though every conservation law for F is eventually discovered by this algorithm, the algorithm never ends, and there is no way to predict whether, continuing to run the algorithm, one is going to find any new conservation law. In fact, it is undecidable whether a given CA has any nontrivial conservation law or not.

Proposition 4 *There is no algorithm that, given a cellular automaton, decides whether it has any nontrivial conservation law or not.*

This is an immediate consequence of the undecidability of the finite tiling problem. The finite tiling problem is a variant of the tiling problem, which asks whether one can tile the entire plane using decorated tiles chosen from a finite number of given types (see the chapter ❷ Basic Concepts of Cellular Automata).

A *Wang tile*, called after the Chinese–American mathematician Hao Wang (1921–1995), is a unit square with colored edges. Two Wang tiles can be placed next to each other if their abutting edges have the same color. A set of Wang tiles is given by a finite set T, and four mappings $n, w, s, e : T \rightarrow C$ that identify the colors of the northern, western, southern, and eastern edges of the tiles. A *valid tiling* of the plane is a configuration $c : \mathbb{Z}^2 \rightarrow T$ that respects the tiling rule; that is, $n(c[i,j]) = s(c[i,j+1])$ and $w(c[i,j]) = e(c[i-1,j])$ for every $i, j \in \mathbb{Z}$.

The finite tiling problem asks, given a set T of Wang tiles and a designated *blank* tile $\diamond \in T$, whether there is a nontrivial *finite* valid tiling of the plane; that is, a valid tiling in which all but a finite number of tiles are blank. Here a *nontrivial* tiling means a tiling that uses at least one non-blank tile. Reducing the halting problem of Turing machines to this problem, one can show that the finite tiling problem is undecidable.

Given a tile set T and a designated blank tile $\diamond \in T$, a two-dimensional CA F is constructed as follows: the state set of F is T. On every configuration $c : \mathbb{Z}^2 \rightarrow T$, the CA looks at each cell i and its four immediate neighbors. If there is a tiling error on cell i (i.e., if the tile on position i does not match with at least one of its adjacent tiles), the CA changes the state of i to blank. Otherwise, the state of cell i is kept unchanged.

Let $\diamond$ be the uniformly blank configuration. If the tile set T admits no nontrivial finite valid tiling, every finite configuration of F is eventually turned into $\diamond$. Therefore, following the characterization (C-2), in this case F has no nontrivial conservation law. On the other hand, suppose that T admits a nontrivial finite valid tiling c. Let p be the finite pattern consisting of all the non-blank cells of c, as well as a margin of blank cells around them. Clearly, F preserves the number of the occurrences of p on any configuration c. In summary, T admits a nontrivial finite valid tiling if and only if F has a nontrivial conservation law. Since no algorithm can decide whether T admits a nontrivial finite valid tiling, no algorithm can either tell whether F has a nontrivial conservation law.

Note that the above argument does not rule out the existence of an algorithm that solves the problem only for one-dimensional cellular automata. Unfortunately, even when restricted to one-dimensional CA, the question remains undecidable.

Theorem 3 (Formenti et al. 2008) *There is no algorithm that, given a one-dimensional cellular automaton, decides whether it has any nontrivial conservation law or not.*

The proof of this fact relies on the undecidability of the existence of periodic orbits in 2-counter machines (Blondel et al. 2002). A 2-counter machine is a finite automaton equipped with two unbounded counters. At each step, the automaton can increase or decrease either of the counters and check whether its value is zero. It is well-known that 2-counter machines are equivalent, in power, with Turing machines (see, e.g., Minsky 1967); any algorithm can be implemented by a suitable 2-counter machine. As it is proven in Blondel (2002), there is no algorithm that decides whether a given 2-counter machine has a configuration whose orbit is periodic.

To prove ❷ Theorem 3, one (algorithmically) transforms the problem of deciding whether a given 2-counter machine has a periodic configuration to the problem of deciding whether a given CA has a nontrivial conservation law. Since any algorithm solving the latter problem would lead to an algorithm solving the former, one can conclude that no algorithm can solve the latter problem.

The idea of the transformation is simple: given a 2-counter machine M, one constructs a one-dimensional CA F with a distinguished quiescent configuration $\diamond$ with the property that:

(i) If M has no periodic orbit, F eventually transforms each finite configuration to the quiescent configuration $\diamond$.

(ii) If M has a periodic orbit, F has a nontrivial conservation law.

Let $\diamond$ be the state of the cells in $\diamond$. Every finite configuration in F is uniquely partitioned into disjoint segments. Each segment is either syntactically valid, in which case it simulates M on some configuration, or it is not syntactically valid, in which case the CA gradually turns all its cells into $\diamond$. The size of a valid segment, however, remains unchanged. So if at some point the simulation requires more cells than available in the segment, the segment overflows and becomes syntactically invalid.

Now, if the machine M does not have any periodic orbit, every valid segment eventually overflows. Hence, every finite configuration eventually changes into $\diamond$. On the other hand, if M does have a periodic configuration, a segment simulating M on such a configuration does never overflow. Therefore, one can construct a nontrivial energy that counts the number of such simulation segments, and that is conserved by F. See Formenti et al. (2008) and Taati (2009) for the details.

3.2 Dissipation Laws

Next, the problem of verifying whether a given energy is non-increasing under the iteration of a given CA have discussed. While there is an algorithm that solves the problem for one-dimensional CA (Kůrka 2003a) (see also Moreira et al. 2004, Bernardi et al. 2005, and Bernardi 2007), the problem is undecidable in higher dimensions (Bernardi et al. 2005).

One-dimensional CA have a convenient representation (up to composition with translations) using edge-labeled De Bruijn graphs. The *De Bruijn* graph of order k ($k > 0$) over an alphabet S is a graph $B_k[S]$ with vertex set $V \stackrel{\triangle}{=} S^k$ and edge set $E \stackrel{\triangle}{=} S^{k+1}$, where for every $a, b \in S$ and $u \in S^{k-1}$, there is an edge aub from au to ub.

Let $F: S^{\mathbb{Z}} \to S^{\mathbb{Z}}$ be a one-dimensional CA with neighborhood $[-l, r] = \{-l, -l+1, \ldots, r\}$, and local rule $f: S^{[-l, r]} \to S$. For every $k \geq l + r$, the CA can be represented on the De Bruijn graph $B_k[S]$ with labeling $\lambda : E \to S^{k-(l+r)}$ that is defined as follows: for every edge $u_0 u_1 \ldots u_k \in S^{k+1}$, let $\lambda(u_0 u_1 \cdots u_k) = v_l v_{l+1} \cdots v_{k-r}$, where $v_i = f(u_{i-l} u_{i-l+1} \cdots u_{i+r})$.

The edge sequence $p = \{p[i]\}_{i \in \mathbb{Z}}$ of a bi-infinite walk on $B_k[S]$ represents a unique (up to translation) configuration $c : \mathbb{Z} \to S$, while its label sequence $\lambda(p) \stackrel{\triangle}{=} \{\lambda(p[i])\}_{i \in \mathbb{Z}}$ represents Fc. Conversely, every configuration $c : \mathbb{Z} \to S$ corresponds to a unique bi-infinite walk on $B_k[S]$.

Consider now an energy defined in terms of a local observable $\mu : S^{\mathbb{Z}} \to \mathbb{R}$. Without loss of generality, one can assume that μ has observation window $[0, m] = \{0, 1, \ldots, m\}$, for some $m \geq 0$, and local rule $g : S^{[0,m]} \to \mathbb{R}$. If $k \geq l + m + r$, one can also represent the energy μ on $B_k[S]$ in a suitable way. To each edge $u_0 u_1 \ldots u_k \in S^{k+1}$, one can assign two real numbers

$$\alpha(u_0 u_1 \cdots u_k) \stackrel{\triangle}{=} g(u_0 u_1 \cdots u_m) \tag{52}$$

and

$$\beta(u_0 u_1 \cdots u_k) \overset{\triangle}{=} g(v_l v_{l+1} \ldots v_{l+m}) \tag{53}$$

where $v_l v_{l+1} \cdots v_{k-r} = \lambda(u_0 u_1 \cdots u_k)$ is the label of $u_0 u_1 \ldots u_k$.

The average energy per cell of a periodic configuration c can be calculated by averaging the value of α over the edges of the closed walk p corresponding to c on the graph. The average of β over the same closed walk is the average energy per cell of the configuration Fc. Therefore, the energy μ is nonincreasing under F if and only if

$$\frac{\beta(p_1) + \beta(p_2) + \cdots + \beta(p_n)}{n} \le \frac{\alpha(p_1) + \alpha(p_2) + \cdots + \alpha(p_n)}{n} \tag{54}$$

for every closed walk $p_1 p_2 \ldots p_n$. It is not difficult to verify that in order to test the above property, one only needs to test ❷ Eq. 54 for every p that is a cycle (i.e., a closed walk with no repeating vertices). Since the number of cycles on $B_k[S]$ is finite, this gives rise to the following simple algorithm for testing whether the energy μ is nonincreasing or not.

> for every cycle $p_1 p_2 \ldots p_n$ on $B_k[S]$,
>   if $\beta(p_1) + \beta(p_2) + \cdots + \beta(p_n) > \alpha(p_1) + \alpha(p_2) + \cdots + \alpha(p_n)$,
>   return "no" and halt;
> return "yes";

For higher-dimensional CA, there is no algorithmic way to verify whether a given energy is dissipative. Intuitively, this means that, for an energy to be non-increasing, cells that are very far from each other may need to collaborate. Any local increase in energy is compensated by a local decrease somewhere on the lattice, but there is no computable upper bound on the distance one may need to look to see such a decrease.

Theorem 4 (Bernardi et al. 2005) *There is no algorithm that, given a (two-dimensional) cellular automaton F and an energy θ, decides whether θ is nonincreasing under F.*

The proof is via a reduction from the finite tiling problem, which was introduced after ❷ Proposition 4. Let $(T, \diamond)$ be an instance of the finite tiling problem, where T is a set of Wang tiles and $\diamond \in T$ is the distinguished blank tile. First, the tile set P in ❷ Fig. 6 is used to

◻ **Fig. 6**
(a) Tile set *P*. (b) A typical finite valid tiling using tile set *P*.

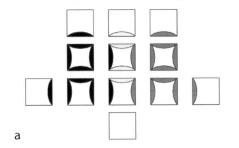

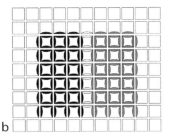

a   b

construct a tile set $S \subseteq P \times T$. The tile set S consists of all tiles (p, t) satisfying the following two conditions:

- If $p \in \{\square, \square, \square, \square, \square, \square\}$, then $t = \diamond$.
- If $p = \square$, then $t \neq \diamond$.

The color of an edge of tile (p, t) is simply the combination of the colors of that edge in the components p and t; that is, $\alpha(p, t) \triangleq (\alpha(p), \alpha(t))$ for $\alpha = n, e, w, s$. The tile $(\square, \diamond)$ is designated as the blank tile for S.

It is easy to see that S admits a nontrivial finite valid tiling if and only if T admits a nontrivial finite valid tiling. Also, note that every nontrivial finite valid tiling using the tiles in S contains at least one occurrence of $\square$. On the other hand, one can verify that if T (and hence S) does not admit a nontrivial finite valid tiling, on every finite configuration $c : \mathbb{Z}^2 \rightarrow S$, the number of occurrences of $\square$ cannot exceed the number of tiling errors (i.e., those positions $i \in \mathbb{Z}^2$ where the tile on i does not match with at least one of its adjacent tiles).

Next, let a two-dimensional CA F and an interaction potential θ be constructed. The CA F has S as state set. On each configuration c, the CA looks at each cell i and its four immediate neighbors. If there is a tiling error on cell i, the CA changes the state of i to blank. On the other hand, if there is no tiling error on cell i, the CA keeps the state of i unchanged unless i has a state of the form $(\square, t)$ with $t \in T$. In the latter case, the state of i is turned into an arbitrary value (p', t') with $p' \notin \{\square, \square\}$. The energy $\theta : S^{\#} \rightarrow \mathbb{R}$ is, in fact, context free; that is, it assigns nonzero potentials only to the single-cell patterns. For every state $s = (p, t)$ one defines

$$\theta(s) \triangleq \begin{cases} 0 & \text{if } (p, t) = (\square, \diamond) \\ 1 & \text{if } p = \square \\ 2 & \text{otherwise} \end{cases} \tag{55}$$

For an arbitrary finite configuration c, let $\gamma(c) \triangleq \Theta(Fc) - \Theta(c)$ denote the change in the total energy of c after one step. By construction, every tiling error contributes either -1 or -2 to γ, while every correctly tiled occurrence of $\square$ contributes $+1$ to γ. If T does not admit any nontrivial finite valid tiling, the number of tiling errors on any configuration c is greater than or equal to the number of occurrences of $\square$ on c. Therefore, for every configuration c, $\gamma(c) \leq 0$. On the other hand, suppose that T admits a nontrivial finite valid tiling. Then, S also admits a nontrivial finite valid tiling c. While there is no tiling error on c, there is at least one occurrence of $\square$. Hence, $\gamma(c) > 0$.

Since there is no general algorithm for deciding whether T admits a nontrivial finite valid tiling, it is concluded that no algorithm can decide whether θ is nonincreasing under F. See Bernardi et al. (2005) and Bernardi (2007).

4   Flows and Particles

A conservation law, as discussed in the previous sections, is a global property of a CA. It asserts that certain local additive quantity, which is called energy, is globally preserved. It does not, however, provide any microscopic mechanism behind this. Namely, it does not elaborate how the energy is manipulated locally so that its global quantity remains intact. Microscopic explanations of conservation laws can be given in terms of "flow" of energy from one cell to another. In fact, every conservation law in cellular automata has such flow explanations, as it was first noted in Hattori and Takesue (1991). However, these flows are not unique.

4.1   Local Conservation Laws

In physics, every local and additive conserved quantity (such as energy, momentum, electric charge, etc.) is *locally* conserved. In fact, any conservation law in physics must inevitably be expressible in a local form, in order to be compatible with the principle of relativity (cf. Feynman 1965).

In cellular automata, local conservation laws are formalized in terms of the concept of flow. The amount of *flow* from cell i to cell j on a configuration x is specified by a real number $\Phi_{i\to j}(x)$. The mapping $x, i, j \mapsto \Phi_{i\to j}(x) \in \mathbb{R}$ is required to satisfy the following natural conditions:

(i) For every $i, j \in \mathbb{Z}^d$, the mapping $x \mapsto \Phi_{i\to j}(x)$ is a local observable.
(ii) For every configuration x, all cells $i, j \in \mathbb{Z}^d$, and any displacement $a \in \mathbb{Z}^d$,

$$\Phi_{a+i\to a+j}(x) = \Phi_{i\to j}(\sigma^a x) \tag{56}$$

(iii) There is a finite set $I \subseteq \mathbb{Z}^d$ such that $\Phi_{i\to j} = 0$ unless $i - j \in I$.

Equivalently, a mapping $x, i, j \mapsto \Phi_{i\to j}(x)$ is called a flow if there exist finite sets $K, I \subseteq \mathbb{Z}^d$ and a rule $\varphi : S^K \times I \to \mathbb{R}$ such that

$$\Phi_{i\to j}(x) = \begin{cases} \varphi(x[K(j)], i - j) & \text{if } i - j \in I \\ 0 & \text{otherwise} \end{cases} \tag{57}$$

for every $x : \mathbb{Z}^d \to S$ and $i, j \in \mathbb{Z}^d$. In summary, the amount of flow to each cell is decided locally, by looking at a finite neighborhood K of that cell. The set I is the set of directions from which energy flows to a cell.

Let $F : S^{\mathbb{Z}^d} \to S^{\mathbb{Z}^d}$ be a CA and $\mu : S^{\mathbb{Z}^d} \to \mathbb{R}$ a local observable defining an energy. A flow Φ is said to be *compatible* with μ and F (or, Φ is a flow for μ under the dynamics of F) if the following *continuity equations* hold (see ❷ *Fig. 7a*):

(a) For every configuration x and every cell a,

$$\mu(\sigma^a x) = \sum_{j\in\mathbb{Z}^d} \Phi_{a\to j}(x) \tag{58}$$

(b) For every configuration x and every cell a,

$$\sum_{i\in\mathbb{Z}^d} \Phi_{i\to a}(x) = \mu(\sigma^a F x) \tag{59}$$

An energy μ is *locally conserved* by F if it has a flow under F. Conservation laws and local conservation laws are equivalent concepts in cellular automata (Hattori and Takesue 1991):

Theorem 5   *Let $F : S^{\mathbb{Z}^d} \to S^{\mathbb{Z}^d}$ be a cellular automaton and $\mu : S^{\mathbb{Z}^d} \to \mathbb{R}$ a local observable. There is a flow Φ compatible with μ and F if and only if F conserves the energy generated by μ.*

That the existence of a flow compatible with μ and F implies that F conserves μ is easy to see. To prove the converse, one needs to construct, for any CA F and any energy μ conserved by F, a flow Φ compatible with μ and F. One such construction can be found, for example, in Taati (2009). The idea is now illustrated by an example, namely, the car conservation law in the Traffic CA.

◻ **Fig. 7**

(a) Continuity of the flow: $\sum_i \varphi_i = \mu = \sum_j \psi_j$. **(b, c)** Two different flows for the car conservation law in the Traffic CA.

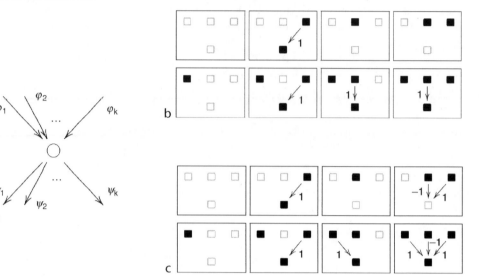

◻ **Fig. 8**

The identified effect of each car on the configuration after one step. **(a)** Effects identified from left to right. **(b)** Effects identified from right to left.

Let x be an arbitrary configuration consisting of a finite number of cars on the highway. Start from the empty highway and place the cars, one by one from left to right, on the highway to obtain x. At each step, identify the effect of placing one new car on position i on the configuration of the highway after one time step. This effect is expressed by flows from cell i to the neighboring cells (see ❷ *Fig. 8a*). Note that placing a new car on the cell i may only change the state of the cells i and $i-1$ in the following configuration. Furthermore, these changes depend only on whether there has already been a car on the cell $i-1$ or not. Therefore, the effect of each car can be identified locally as depicted in ❷ *Fig. 7b*. However, note that this is not the only way to identify the effect of each car. For example, if the cars are placed from right to left, a different flow is obtained as illustrated in ❷ *Figs. 8b* and ❷ *7c*.

Even though flows provide intuitive ways to think about conservation laws, their nonuniqueness is not plausible. Hence, one may want to identify a flow that is the most natural in some sense.

4.2   Particle Flows

Consider the following game played with pebbles on configurations of a CA. Let $x : \mathbb{Z}^d \to S$ be an arbitrary configuration. On each cell of the lattice, a number of pebbles have been placed. The number of pebbles on cell i depends only on the state of cell i and is denoted by $\eta(x[i])$. At each iteration of the CA, the state of each cell i changes to its new value $(Fx)[i]$. The goal is to redistribute the pebbles on the lattice so that the number of pebbles on each cell matches with its new state; that is, each cell i obtains $\eta((Fx)[i])$ pebbles. Is there a local and uniform strategy to do this? "Local" means that each pebble can be moved within a bounded distance of its original position, and one is only allowed to look at the states of a bounded number of cells around it in order to decide where a pebble should be moved. "Uniform" means that the same strategy should be used for redistributing the pebbles on every cell and over every configuration.

The assignment $\eta : S \to \mathbb{N}$ ($\mathbb{N} \triangleq \{0, 1, 2, \ldots\}$) defines an energy in its usual sense. A pebble can be interpreted as the "quantum" of energy; that is, the tiniest bit of energy, which is indecomposable. The desired strategy for the game is simply a flow Φ that takes its values in the set $\mathbb{N}$ of nonnegative integers, and that is compatible with η and F. For example, the flow depicted in ❷ *Figs. 7b* and ❷ *8a* defines a valid strategy, while the flow in ❷ *Figs. 7c* and ❷ *8b* does not.

Let a flow Φ be called a *particle flow* if its values are from the set of nonnegative integers. Any function $\eta : S \to \mathbb{N}$ is called a *particle assignment*. A necessary condition for the existence of particle flow compatible with a particle assignment η and a CA F is, of course, that F conserves the energy generated by η. For one-dimensional CA, this condition is also known to be sufficient; Fukś (2000) and Pivato (2002) have shown that any particle assignment η conserved by a one-dimensional CA F has a particle flow (see also Moreira et al. 2004). For higher-dimensional CA, however, the question is open. See Kari and Taati (2008) for a partial solution in two dimensions.

Theorem 6   (Fukś 2000, Pivato 2002) *Let* $F : S^{\mathbb{Z}} \to S^{\mathbb{Z}}$ *be a one-dimensional cellular automaton and* $\eta : S \to \mathbb{N}$ *a particle assignment. There is a particle flow* Φ *compatible with* η *and* F *if and only if* F *conserves* η.

Theorem 7   (Kari and Taati 2008) *Let* $F : S^{\mathbb{Z}^2} \to S^{\mathbb{Z}^2}$ *be a two-dimensional cellular automaton with radius-$\frac{1}{2}$ neighborhood, and let* $\eta : S \to \mathbb{N}$ *be a particle assignment. There is a particle flow* Φ *compatible with* η *and* F *if and only if* F *conserves* η.

The radius-$\frac{1}{2}$ neighborhood refers to the four-element neighborhood

$$\{(0,0), (0,1), (1,0), (1,1)\} \tag{60}$$

Particle flows, if exist, are not unique; a particle flow compatible with a particle assignment and a CA can be modified in infinitely many ways to obtain new particle flows compatible with the same assignment and CA. In one dimension, however, there are natural criteria that ensure the uniqueness. For example, it is shown in Moreira et al. (2004) that for each particle assignment conserved by a one-dimensional CA, there is a unique particle flow that preserves the order of the particles.

An argument, due to Pivato (2002), is now mentioned that strongly suggests that ❯ Theorem 6 should be valid also for higher-dimensional CA.

Theorem 8 (Pivato 2002) *Let* $F : S^{\mathbb{Z}^d} \to S^{\mathbb{Z}^d}$ *be a cellular automaton with neighborhood* $0 \in N \subseteq \mathbb{Z}^d$. *A particle assignment* $\eta : S \to \mathbb{N}$ *is conserved by* F *if and only if*

$$\sum_{i \in A} \eta\,(x[i]) \leq \sum_{i \in N^{-1}(A)} \eta((Fx)[i]) \tag{61}$$

and

$$\sum_{i \in A} \eta\,((Fx)[i]) \leq \sum_{i \in N(A)} \eta(x[i]) \tag{62}$$

for every configuration x *and every finite set* $A \subseteq \mathbb{Z}^d$.

For every two consecutive configurations, x and $y = Fx$, a bipartite graph $G_N[\eta, x, y] = (U, V, E)$ is constructed as follows. For every particle on x, the graph has a vertex in U, and for every particle on y, there is a vertex in V. There is an edge between a particle $u \in U$ coming from a cell i and a particle $v \in V$ coming from a cell j if and only if i is a neighbor of j; that is, if and only if $i - j \in N$.

A perfect matching in graph $G_N[\eta, x, y]$ is a way of identifying particles on x with particles on y in such a way that the position of each particle on x is a neighbor of its position on y. A necessary and sufficient condition for a (possibly infinite, but locally finite) bipartite graph to have a perfect matching is given by Hall's Marriage Theorem (see, e.g., van Lint and Wilson 1992): a bipartite graph $G = (U, V, E)$ has a matching that covers U if and only if for every finite set $A \subseteq U$, the number of vertices in V that are adjacent to A is at least $|A|$. If G has a matching that covers U and a matching that covers V, G also has a perfect matching.

It follows from ❯ Theorem 8 that if a CA F with neighborhood N conserves a particle assignment η, then for every configuration x, the graph $G_N[\eta, x, Fx]$ must have a perfect matching. Therefore, if F conserves η, the particles on any configuration x can be identified with those on Fx in such a way that the two positions of each particle are within bounded distance from each other. It remains open whether, in general, such identification can be done in a local and uniform fashion.

4.3   Flows for Dissipative Energies

In the light of ❯ Theorem 5, it is natural to ask whether nonincreasing energies can also be explained locally. A flow explanation of a nonincreasing energy may be possible by relaxing the continuity equations (❷ Eqs. 58 and ❷ 59). In one dimension, this is always possible. In higher dimensions, however, the undecidability result of ❷ Theorem 4 imposes an obstacle to the general existence of such flows.

Let $F : S^{\mathbb{Z}^d} \to S^{\mathbb{Z}^d}$ be a CA and $\mu : S^{\mathbb{Z}^d} \to \mathbb{R}$ a local observable defining an energy. A flow Φ is said to be *semi-compatible* with μ and F if

(a) For every configuration x and every cell a,

$$\mu(\sigma^a x) \geq \sum_{j \in \mathbb{Z}^d} \Phi_{a \to j}(x) \tag{63}$$

(b) For every configuration x and every cell a,

$$\sum_{i \in \mathbb{Z}^d} \Phi_{i \to a}(x) \geq \mu(\sigma^a F x) \tag{64}$$

Proposition 5   *There exists a two-dimensional cellular automaton $F : S^{\mathbb{Z}^2} \to S^{\mathbb{Z}^2}$ and a local observable $\mu : S^{\mathbb{Z}^2} \to \mathbb{R}$ such that the energy generated by μ is non-increasing under F, but there is no flow semi-compatible with μ and F.*

As mentioned above, this is a consequence of ❷ Theorem 4. First, note that there is a semi-algorithm (a semi-algorithm means an algorithmic process that does not necessarily halt on every input) that recognizes those energies that are not nonincreasing: given an energy μ and a CA F, such a semi-algorithm tests the inequality $\bar{\mu}(Fx) \leq \bar{\mu}(x)$ for periodic configurations x with larger and larger periods. According to the characterization (D-2), if μ is not nonincreasing, the inequality fails on some periodic configuration, which will eventually be found. Now, suppose that every nonincreasing energy μ has a semi-compatible flow. Then, one can construct a semi-algorithm that recognizes those energies that are nonincreasing: given an energy μ and a CA F, one simply enumerates all the possible flows and verifies ❷ Eqs. 63 and ❷ 64 for them one by one. If μ is nonincreasing, the process will eventually find a semi-compatible flow. Running these two semi-algorithms in parallel one obtains an algorithm for deciding whether a given energy is nonincreasing; hence contradicting ❷ Theorem 4.

The above argument does not provide any explicit example of a nonincreasing energy with no semi-compatible flow. An explicit construction has been obtained by A. Rumyantsev (see Bernardi 2007).

In one-dimensional case, every nonincreasing energy has a flow explanation.

Theorem 9   *Let $F : S^{\mathbb{Z}} \to S^{\mathbb{Z}}$ be a one-dimensional cellular automaton and $\mu : S^{\mathbb{Z}} \to \mathbb{R}$ a local observable. There is a flow Φ semi-compatible with μ and F if and only if the energy generated by μ is nonincreasing under F.*

A proof in a rather different setting is given in the chapter ❷ Cellular Automata Dynamical Systems. In one dimension, in addition, any nonincreasing energy defined using a particle assignment has a semi-compatible particle flow.

Theorem 10   (Moreira et al. 2004) *Let $F : S^{\mathbb{Z}} \to S^{\mathbb{Z}}$ be a one-dimensional cellular automaton and $\eta : S \to \mathbb{N}$ a particle assignment. If η is nonincreasing under F, there is a particle flow Φ semi-compatible with η and F.*

5   Further Topics

This chapter concludes with some miscellaneous remarks.

A large body of research done on conservation laws is concentrated on number-conserving automata (see, e.g., Boccara and Fukś 1998, 2002, 2006; Fukś 2000; Durand et al. 2003; Formenti and Grange 2003; Moreira et al. 2004; Bernardi et al. 2005; Fukś and Sullivan 2007). In a *number-conserving* CA, the state of a cell represents the number of particles in that cell, and the dynamics is such that the total number of particles is preserved. Most results on number-conserving CA have counterparts in the general setting of conservation laws, which can be proven in more or less similar manners.

There are, however, results that are specific to the framework of number-conserving CA. For example, Moreira (2003) has shown that every one-dimensional CA can be simulated by a number-conserving one. In particular, this implies that there are number-conserving one-dimensional CA that are intrinsically universal (cf. the chapter ❷ Reversible Cellular Automata). Morita and Imai have constructed computationally universal reversible CA that conserve a natural quantity associated to each cell (Morita and Imai 2001).

An interesting issue, which needs further investigation, is the possible connections between the conservation laws of a CA and its dynamical properties. Specifically, it is interesting to know what kind of restrictions a conservation law may impose on the dynamics of a CA? Dynamical properties of number-conserving CA are studied in Formenti and Grange (2003). There, among other things, it has been proved that in every surjective number-conserving CA, the set of temporally periodic configurations is dense. The same holds for any surjective one-dimensional CA that has a conserved energy with a unique ground configuration (Formenti et al.; Taati 2009). A ground configuration for an energy Θ is a configuration x such that $\delta\Theta(x, y) \geq 0$ for any configuration y asymptotic to x. There is a long-standing open question, asking whether every surjective CA has a dense set of temporally periodic configurations (see, e.g., Kari 2005 and Boyle 2008). It is also shown that positively expansive CA cannot have any nontrivial conservation laws (Formenti et al.; Taati 2009). See the chapter ❷ Cellular Automata Dynamical Systems or Kůrka (2001, 2003b) for information on cellular automata as dynamical systems.

Nonincreasing energies have a more established connection with the dynamics, as they define Lyapunov functions for the measure dynamical system $(\mathcal{M}_\sigma, F)$. See the chapter ❷ Cellular Automata Dynamical Systems for more on this point of view.

Another trend of research has been to find counterparts of Noether's theorem for cellular automata (e.g., Margolus 1984; Baez and Gilliam 1994; Boykett 2005; Boykett et al. 2008; personal communication with N. Ollinger). Noether's theorem – after the German mathematician Emmy Noether (1882–1935) – establishes a correspondence between the conservation laws of a dynamical system defined in terms of a Hamiltonian or a Lagrangian and its symmetries (see, e.g., Landau and Lifshitz 1976 and Arnold 1989).

Finally, the interesting connection between conservation laws and invariant Gibbs measures in reversible cellular automata is worth mentioning. Gibbs measures are especially important in statistical mechanics, as they characterize the equilibrium state of lattice Hamiltonian systems (see, e.g., Georgii 1988 and Ruelle 2004). In fact, early study of conservation laws in cellular automata was motivated by the issue of ergodicity in statistical mechanical systems (Takesue 1987).

In reversible cellular automata, there is a one-to-one correspondence between conservation laws and invariant Gibbsian specifications (Kari and Taati; Taati 2009). As a special case, a reversible CA conserves a context-free energy $\eta : S \rightarrow \mathbb{R}$ if and only if it preserves the Bernoulli measure with probabilities proportional to $2^{-\eta}$. This is at least partially valid also for surjective CA. Similar results, though in different settings, are obtained in Helvik et al. (2007) (see also Toffoli 1988). It is an interesting open issue to investigate the questions of statistical mechanics in the framework of (deterministic) reversible cellular automata.

References

Anderson R, Lovász L, Shor P, Spencer J, Tardos E, Winograd S (1989) Disks, balls, and walls: analysis of a combinatorial game. Am Math Mon 96(6): 481–493

Arnold VI (1989) Mathematical methods of classical mechanics, 2nd edn. Springer, New York. English Translation

Baez, JC, Gilliam J (1994) An algebraic approach to discrete mechanics. Lett Math Phys 31:205–212

Bak P, Tang C, Wiesenfeld K (1987) Self-organized criticality: an explanation of $1/f$ noise. Phys Rev Lett 59(4):381–384

Bak P, Tang C, Wiesenfeld K (1998) Self-organized criticality. Phys Rev A 38(1):364–374

Bernardi V (2007) Lois de conservation sur automates cellulaires. Ph.D. thesis, Université de Provence

Bernardi V, Durand B, Formenti E, Kari J (2005) A new dimension sensitive property for cellular automata. Theor Comp Sci 345:235–247

Blondel VD, Cassaigne J, Nichitiu C (2002) On the presence of periodic configurations in Turing machines and in counter machines. Theor Comput Sci 289:573–590

Boccara N, Fukś H (1998) Cellular automaton rules conserving the number of active sites. J Phys A Math Gen 31(28):6007–6018

Boccara N, Fukś H (2002) Number-conserving cellular automaton rules. Fundam Info 52:1–13

Boccara N, Fukś H (2006) Motion representation of one-dimensional cellular automata rules. Int J Mod Phys C 17:1605–1611

Boykett T (2005) Towards a Noether-like conservation law theorem for one dimensional reversible cellular automata. arXiv:nlin.CG/0312003

Boykett T, Moore C (1998) Conserved quantities in one-dimensional cellular automata. Unpublished manuscript

Boykett T, Kari J, Taati S (2008) Conservation laws in rectangular CA. J Cell Automata 3(2):115–122

Boyle M (2008) Open problems in symbolic dynamics. In: Burns K, Dolgopyat D, Pesin Y (eds) Geometric and probabilistic structures in dynamics, contemporary mathematics. American Mathematical Society, Providence, RI

Chopard B, Droz M (1998) Cellular automata modeling of physical systems. Cambridge University Press, Cambridge

Durand B, Formenti E, Róka Z (2003) Number conserving cellular automata I: decidability. Theor Comput Sci 299:523–535

Feynman R (1965) The character of physical law. MIT Press, Cambridge, MA

Formenti E, Grange A (2003) Number conserving cellular automata II: dynamics. Theor Comput Sci 304:269–290

Formenti E, Kari J, Taati S (2008) The most general conservation law for a cellular automaton. In: Hirsch EA, Razborov AA, Semenov A, Slissenko A (eds) Proceedings of CSR 2008, Moscow, June 7–12, 2008. LNCS, vol 5010. Springer, Berlin/Heidelberg, pp 194–203, doi: 10.1007/978-3-540-79709-8

Formenti E, Kari J, Taati S Preprint

Fredkin E, Toffoli T (1982) Conservative logic. Int J Theor Phys 21:219–253

Fukś H (2000) A class of cellular automata equivalent to deterministic particle systems. In: Feng S, Lawniczak AT, Varadhan SRS (eds) Hydrodynamic limits and related topics. Fields Institute communications, vol 27. American Mathematical Society, Providence, RI, pp 57–69

Fukś H, Sullivan K (2007) Enumeration of number-conserving cellular automata rules with two inputs. J Cell Automata 2:141–148

Georgii HO (1988) Gibbs measures and phase transitions. Walter de Gruyter, Berlin

Georgii HO, Häggström O, Maes C (2000) The random geometry of equilibrium phases. In: Domb C, Lebowitz J (eds) Phase transitions and critical phenomena, vol 18. Academic Press, London, pp 1–42

Goles E (1992) Sand pile automata. Ann Institut Henri Poincaré (A) Phys théor 59(1):75–90

Goles E, Kiwi MA (1993) Games on line graphs and sand piles. Theor Comput Sci 115:321–349

Greenberg JM, Hassard BD, Hastings SP (1978a) Pattern formation and periodic structures in systems modeled by reaction-diffusion equations. Bull Am Math Soc 84(6):1296–1327

Greenberg JM, Hastings SP (1978b) Spatial patterns for discrete models of diffusion in excitable media. SIAM J Appl Math 34(3):515–523

Greenberg J, Greene C, Hastings S (1980) A combinatorial problem arising in the study of reaction-diffusion equations. SIAM J Algebraic Discrete Meth 1(1):34–42

Hattori T, Takesue S (1991) Additive conserved quantities in discrete-time lattice dynamical systems. Phys D 49:295–322

Hedlund GA (1969) Endomorphisms and automorphisms of the shift dynamical system. Math Syst Theory 3:320–375

Helvik T, Lindgren K, Nordahl MG (2007) Continuity of information transport in surjective cellular automata. Commun Math Phys 272:53–74

Kari J (2005) Theory of cellular automata: a survey. Theor Comput Sci 334:3–33

Kari J, Taati S In preparation

Kari J, Taati S (2008) A particle displacement representation for conservation laws in two-dimensional cellular automata. In: Durand B (ed) Proceedings of JAC 2008. MCCME Publishing House, Moscow, pp 65–73

Kindermann R, Snell JL (1980) Markov random fields and their applications. American Mathematical Society, Providence, RI

Kůrka P (2001) Topological dynamics of cellular automata. In: Codes, systems and graphical models. Springer, Berlin

Kůrka P (2003a) Cellular automata with vanishing particles. Fundam Info 58:1–19

Kůrka P (2003b) Topological and symbolic dynamics. Cours spécialisés, vol. 11. Société Mathématique de France, Paris

Landau LD, Lifshitz EM (1976) Course of theoretical physics: mechanics, 3rd edn. Butterworth-Heinemann, Oxford. English Translation

van Lint JH, Wilson RM (1992) A course in combinatorics. Cambridge University Press, Cambridge

Margolus N (1984) Physics-like models of computation. Phys D 10:81–95

Margolus N (1987) Physics and computation. Ph.D. thesis, Massachusetts Institute of Technology

Minsky M (1967) Computation: finite and infinite machines. Prentice-Hall, Englewood Cliffs, NJ

Moreira A (2003) Universality and decidability of number-conserving cellular automata. Theor Comput Sci 292:711–723

Moreira A, Boccara N, Goles E (2004) On conservative and monotone one-dimensional cellular automata and their particle representation. Theor Comput Sci 325:285–316

Morita K, Imai K (2001) Number-conserving reversible cellular automata and their computation-universality. Theor Info Appl 35:239–258

Nagel K, Herrmann HJ (1993) Deterministic models for traffic jams. Phys A 199:254–269

Nagel K, Schreckenberg M (1992) A cellular automaton model for freeway traffic. J Phys I 2(12):2221–2229

Parthasarathy KR (1967) Probability measures on metric spaces. Academic Press, New York

Pivato M (2002) Conservation laws in cellular automata. Nonlinearity 15:1781–1793

Pomeau Y (1984) Invariant in cellular automata. J Phys A Math Gen 17(8):L415–L418

Ruelle D (2004) Thermodynamic formalism, 2nd edn. Cambridge University Press, Cambridge

Sinai YG (1982) Theory of phase transitions: rigorous results. Pergamon Press, Oxford

Taati S (2009) Conservation laws in cellular automata. Ph.D. thesis, University of Turku

Takesue S (1987) Reversible cellular automata and statistical mechanics. Phys Rev Lett 59(22):2499–2502

Toffoli T (1988) Information transport obeying the continuity equation. IBM J Res Dev 32(1):29–36

Toffoli T, Margolus N (1987) Cellular automata machines: a new environment for modeling. MIT Press, Cambridge

Toffoli T, Margolus N (1990) Invertible cellular automata: a review. Phys D 45:229–253

Vichniac GY (1984) Simulating physics with cellular automata. Phys D 10:96–116

Walters P (1982) An introduction to ergodic theory. Springer, New York

9 Cellular Automata and Lattice Boltzmann Modeling of Physical Systems

Bastien Chopard
Scientific and Parallel Computing Group, University of Geneva, Switzerland
bastien.chopard@unige.ch

G. Rozenberg et al. (eds.), *Handbook of Natural Computing*, DOI 10.1007/978-3-540-92910-9_9,
© Springer-Verlag Berlin Heidelberg 2012

Abstract

Cellular automata (CA) and lattice Boltzmann (LB) methods provide a natural modeling framework to describe and study many physical systems composed of interacting components. The reason for this success is the close relation between these methods and a mesoscopic abstraction of many natural phenomena. The theoretical basis of the CA and LB approaches are introduced and their potential is illustrated for several applications in physics, biophysics, environmental sciences, traffic models, and multiscale modeling.

1   Introduction

As one can observe everyday, nature is made up of a large number of interacting parts, distributed over space and evolving in time. In many cases, one is interested in describing a natural system at a scale which is much larger than its elementary constituents. Then, often, the behavior of these parts can be reduced to a set of rather simple rules, without affecting the behavior of the whole. For instance, when modeling a fluid flow at a macroscopic scale, one does not have to account for the detailed microscopic interactions between the atoms of that fluid. Instead, one can assume the existence of abstract "fluid elements" interacting in such a way as to conserve mass and momentum.

Cellular automata (CA) can be thought of as a mathematical abstraction of the physical world, an abstraction in which time is discrete and space is made of little blocks, or cells. These cells are organized as a regular lattice, in such a way as to fully cover the spatial domain of interest. In such an approach, spatiotemporal physical quantities are introduced as numerical values associated with each cell. These quantities are called the *states* of the cells. Formally, the definition of a CA also assumes that these sates can only take a finite number of discrete values. On the other hand, lattice Boltzmann (LB) methods are more flexible and allow cells to have real-valued quantities.

CA were proposed in the late 1940s by von Neumann and Ulam (Burks 1970) as an abstraction of a biological system, in order to study the algorithmic mechanisms leading to the self-reproduction of living organisms. Since then, the CA approach has been applied to a wide range of problems and is an acknowledged modeling technique (Toffoli and Margolus 1987; Wolfram 1994; Rothman and Zaleski 1997; Chopard and Droz 1998; Deutsch and Dormann 2005; Ilachinski 2001; Gaylord and Nishidate 1996; Weimar 1998; Wolfram 2002). Moreover, it is still a quite active field of research: international conferences (e.g., ACRI) and dedicated journals (e.g., Journal of Cellular Automata) describe current developments.

LB methods developed later, in the 1990s, are an extension of the concept of lattice gas automata (LGA), a family of CA designed to describe hydrodynamic processes in terms of a discrete kinetic model. LB is now recognized as an alternative way to simulate on a computer complex fluid flows or other complex systems such as reaction–diffusion and advection–diffusion phenomena, as well as wave propagation in complicated geometries (Chopard and Droz 1998; Wolf-Gladrow 2000; Succi 2001; Sukop and Thorne 2005). LB methods are also an important research topic and two international conferences (DSFD and ICMMES) are places to disseminate new results in this field.

A key conceptual ingredient of CA is that they are not meant to be a space–time discretization of the partial differential equation (PDE) representing a given physical process. Instead, CA implement a mesoscopic model of this process, in terms of behavioral rules

◘ **Fig. 1**

The solution process, from problem to numerical results for either a PDE (*top*) or a CA (*bottom*) approach. In this figure, fluid dynamics is used to illustrate the point. In an appropriate limit, fluid dynamics can be either described by the Navier–Stokes equations or by a discrete model of interacting particles. This corresponds to two different languages to represent the same physical problem.

$$\partial_t \mathbf{u} + (\mathbf{u} \cdot \nabla)\mathbf{u} = -\frac{1}{\rho}\nabla p + \nu \nabla^2 \mathbf{u}$$

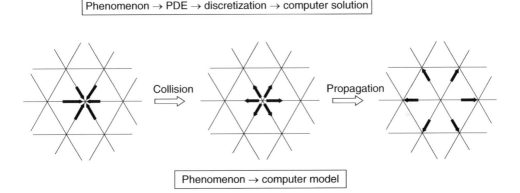

mimicking the physical interactions and translating them into a fully discrete universe. Concretely, CA evolution rules are transition functions, which synchronously change the state of each cell according to its value and those of the adjacent cells.

In other words, CA are based on an idealized, virtual microscopic version of the real world. Statistical physics teaches that the macroscopic behavior of many systems depends very little on the details of the microscopic interactions between the elementary constituents. This suggests that, in view of modeling efficiency, one can consider a new, fictitious microscopic universe whose numerical implementation is easy and fast, as long as this fictitious system has the same macroscopic behavior as the real one. The recipe to achieve this goal is to build a model with the right conservation laws and symmetries. These properties are indeed those that are preserved at all scales of description.

The above principles make the design of CA models quite intuitive and natural. Rule-based interactions are often easier to understand and discuss than a PDE, especially for researchers outside mathematics or physics. Since the level of description of a CA model is mesoscopic, the rules are close to the underlying physical interaction and it is rather easy to add new features to a model and to describe systems for which no PDE apply. On the other hand, CA and LB models heavily rely on computer simulations to derive results.

The diagram in ❷ *Fig. 1* sketches the solution process of a CA–LB approach compared to that of the more traditional PDE approach.

2 Definition of a Cellular Automata

In order to give a definition of a cellular automaton, a simple example is first presented. Although it is very basic, the rule discussed here exhibits a surprisingly rich behavior. It was proposed initially by Edward Fredkin in the 1970s (Banks 1971) and is defined on a two-dimensional square lattice.

Each site of the lattice is a cell, which is labeled by its position $\mathbf{r} = (i, j)$ where i and j are the row and column indices. A function $\psi(\mathbf{r}, t)$ is associated with the lattice to describe the state of each cell $\mathbf{r}$ at iteration t. This quantity ψ can be either 0 or 1.

The CA rule specifies how the states $\psi(\mathbf{r}, t + 1)$ are to be computed from the states at iteration t. We start from an initial condition at time $t = 0$ with a given configuration of the values $\psi(\mathbf{r}, t = 0)$ on the lattice. The state at time $t = 1$ is obtained as follows

1. Each site $\mathbf{r}$ computes the sum of the values $\psi(\mathbf{r}', 0)$ on the four nearest neighbor sites $\mathbf{r}'$ at north, west, south, and east. The system is supposed to be periodic in both i and j directions (like on a torus) so that this calculation is well defined for all sites.
2. If this sum is even, the new state $\psi(\mathbf{r}, t = 1)$ is 0 (white) and, else, it is 1 (black).

The same rule (steps 1 and 2) is repeated to find the states at time $t = 2, 3, 4, \ldots$.

From a mathematical point of view, this *parity rule* can be expressed by the following relation

$$\psi(i, j, t + 1) = \psi(i + 1, j, t) \oplus \psi(i - 1, j, t) \oplus \psi(i, j + 1, t) \oplus \psi(i, j - 1, t) \qquad (1)$$

where the symbol $\oplus$ stands for the exclusive OR logical operation. It is also the sum modulo 2: $1 \oplus 1 = 0 \oplus 0 = 0$ and $1 \oplus 0 = 0 \oplus 1 = 1$.

When this rule is iterated, very nice geometric patterns are observed, as shown in ❷ *Fig. 2*. This property of producing complex patterns starting from a simple rule is generic of many CA rules. Here, complexity results from some spatial organization, which builds up as the rule is iterated. The various contributions of successive iterations combine together in a specific way. The spatial patterns that are observed reflect how the terms are combined algebraically.

Based on this example a definition of a cellular automata is now given. Formally, a cellular automata is a tuple $(A, \Psi, R, \mathcal{N})$ where

i. A is a regular lattice of cells covering a portion of a d-dimensional space.
ii. $\Psi(\mathbf{r}, t) = \{\Psi_1(\mathbf{r}, t), \Psi_2(\mathbf{r}, t), \ldots, \Psi_m(\mathbf{r}, t)\}$ is a set of m Boolean variables attached to each site $\mathbf{r}$ of the lattice and giving the local state of the cells at time t.
iii. R is a set of rules, $R = \{R_1, R_2, \ldots, R_m\}$, which specifies the time evolution of the states $\Psi(\mathbf{r}, t)$ in the following way

$$\Psi_j(\mathbf{r}, t + \delta_t) = R_j(\Psi(\mathbf{r}, t), \Psi(\mathbf{r} + \mathbf{v}_1, t), \Psi(\mathbf{r} + \mathbf{v}_2, t), \ldots, \Psi(\mathbf{r} + \mathbf{v}_q, t)) \qquad (2)$$

where $\mathbf{r} + \mathbf{v}_k$ designate the cells belonging to the neighborhood $\mathcal{N}$ of cell $\mathbf{r}$ and δ_t is the duration of one time step.

In the above definition, the rule R is identical for all sites and is applied simultaneously to each of them, leading to a synchronous dynamics. As the number of configurations of the neighborhood is finite, it is common to pre-compute all the values of R in a lookup table. Otherwise, an algebraic expression can be used and evaluated at each iteration, for each cell, as in ❷ Eq. 1.

It is important to notice that the rule is *homogeneous*, that is, it cannot depend explicitly on the cell position $\mathbf{r}$. However, spatial (or even temporal) inhomogeneities can be introduced anyway by having some $\Psi_j(\mathbf{r})$ systematically 1 in some given locations of the lattice to mark particular cells on which a different rule applies. Boundary cells are a typical example of spatial inhomogeneities. Similarly, it is easy to alternate between two rules by having a bit, which is 1

■ Fig. 2

Several snapshots of the parity rule on a 196 ×196 periodic lattice. The upper left image correspond to an initial configuration make up of a square of 27 × 34 cells in state 1.

$t = 0$	$t = 31$	$t = 43$
$t = 75$	$t = 248$	$t = 292$
$t = 357$	$t = 358$	$t = 359$
$t = 360$	$t = 511$	$t = 571$

at even time steps and 0 at odd time steps. Finally, memory states can be included by simply copying the current state into a "past state" during the update.

The neighborhood $\mathcal{N}$ of each cell (i.e., the spatial region around each cell used to compute the next state) is usually made of its adjacent cells. It is often restricted to the nearest or next-to-nearest neighbors, to keep the complexity of the rule reasonable. For a two-dimensional cellular automaton, two neighborhoods are often considered, as illustrated in ❷ *Fig. 3*: the *von Neumann neighborhood*, which consists of a central cell (the one which is to be updated)

◻ Fig. 3

(a) Von Neumann and (b) Moore neighborhoods. The shaded region indicates the central cell, which is updated according to the state of the cells located within the domain marked with the bold line.

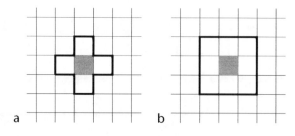

◻ Fig. 4

Various types of boundary conditions obtained by extending the neighborhood. The shaded block represents a virtual cell, which is added at the extremity of the lattice (left extremity, here) to complete the neighborhood.

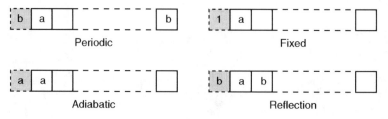

and its four geographical neighbors north, west, south, and east. The *Moore neighborhood* contains, in addition, the second nearest neighbor north-east, north-west, south-east, and south-west, which is a total of nine cells. See Chopard and Droz (1998) for more details. In practice, when simulating a given CA rule, one cannot deal with an infinite lattice. The system must be finite and have boundaries. Clearly, a site belonging to the lattice boundary does not have the same neighborhood as other internal sites. In order to define the behavior of these sites, a different evolution rule can be considered, which sees the appropriate neighborhood. This means that the information of being a boundary cell or not is coded at the site, using a particular value of Ψ_j, for a chosen j. Depending on this information, a different rule is selected. Following this approach, it is possible to define several types of boundaries, all with a different behavior.

Instead of having a different rule at the limits of the system, another possibility is to extend the neighborhood for the sites at the boundary. For instance, a very common solution is to assume *periodic* (or cyclic) boundary conditions, that is, one supposes that the lattice is embedded in a torus-like topology. In the case of a two-dimensional lattice, this means that the left and right sides are connected, and so are the upper and lower sides.

Other possible types of boundary conditions are illustrated in ❷ *Fig. 4*, for a one-dimensional lattice. It is assumed that the lattice is augmented by a set of virtual cells beyond its limits. *A fixed* boundary is defined so that the neighborhood is completed with cells having a preassigned value. An adiabatic boundary condition (or zero-gradient) is obtained by duplicating the value of the site to the extra virtual cells. A reflecting boundary amounts to copying the value of the other neighbor in the virtual cell.

According to the definition, a cellular automaton is deterministic. The rule is some well-defined function and a given initial configuration will always evolve identically. However, it may be very convenient for some applications to have a certain degree of randomness in the rule. For instance, it may be desirable that a rule selects one outcome among several possible states, with a probability p. CA whose updating rule is driven by some external probabilities are called *probabilistic* CA. On the other hand, those which strictly comply with the definition given above, are referred to as *deterministic* CA.

3 Cellular Automata and Complex Systems

3.1 Game of Life and Langton Ant

Complex systems are now an important domain in sciences. They are systems made of many interacting constituents which often exhibit spatiotemporal patterns and collective behaviors.

A standard and successful methodology in research has been to isolate phenomena from each other and to study them independently. This leads to a deep understanding of the phenomena themselves but also leads to a compartmentalized view of nature. The real world is made of interacting processes and these interactions bring new phenomena that are not present in the individual constituents. Therefore, the whole is more than the sum of its parts and new scientific tools and concepts may be required to analyze complex systems.

CAs offer such a possibility by being themselves simple, fully discrete complex systems. In this section, two CA, the so-called *Game of Life* and *Langton's Ant* model are introduced. Both illustrate interesting aspects of complex systems.

3.1.1 The Game of Life

In 1970, the mathematician John Conway proposed the now famous Game of Life (Gardner 1970) CA. The motivation was to find a simple rule leading complex behaviors in a system of fictitious one-cell organisms evolving in a fully discrete universe. The Game of Life rule is defined on a two-dimensional square lattice in which each spatial cell can be either occupied by a living organism (state one) or empty (state zero). The updating rule of the Game of Life is as follows: an empty cell surrounded by exactly three living cells gets alive; a living cell surrounded by less than two or more than three neighbors dies of isolation or overcrowdedness. Here, the surrounding cells correspond to a Moore neighborhood composed of the four nearest cells (north, south, east and west), plus the four second nearest neighbors, along the diagonals. It turns out that the Game of Life automaton has an unexpectedly rich behavior. Complex structures emerge out of a primitive "soup" and evolve so as to develop some new skills (see ❷ *Fig. 5*). For instance a particular spatial assembly of cells has the property to move across the lattice. Such an object, called *glider,* can be seen as higher level organism because it is composed of several simple elementary cells. Its detailed structure is shown in ❷ *Fig. 6*. Thus, by assembling in a clever way cells that are unable to move, it is possible to produce, at a larger scale, a new capability. This is a signature of complex systems. Of course, more complex objects can be built, such as for instance *glider guns*, which are arrangements of cell-producing gliders rhythmically.

■ **Fig. 5**
The Game of Life automaton. Black dots represent living cells whereas dead cells are white. The figure shows the evolution of a random initial configuration and the formation of spatial structures, with possibly some emerging functionalities.

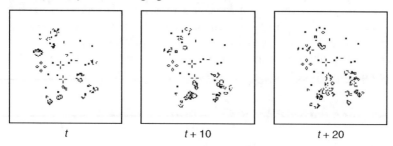

t ⠀⠀⠀⠀⠀⠀⠀⠀⠀⠀ $t + 10$ ⠀⠀⠀⠀⠀⠀⠀⠀⠀⠀ $t + 20$

■ **Fig. 6**
The detailed structure of a glider, over two consecutive iterations. A glider is an assembly of cells that has a higher functionality than its constituent, namely, the capability to move in space by changing its internal structure in a periodic way.

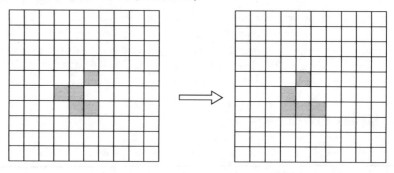

The Game of Life is a cellular automata capable of *universal computations*: it is always possible to find an initial configuration of the cellular space reproducing the behavior of any electronic gate and, thus, to mimic any computation process. Although this observation has little practical interest, it is very important from a theoretical point of view since it assesses the ability of CAs to be a nonrestrictive computational technique. As an illustration of this fact, the Game of Life has been used to compute prime numbers.

3.1.2 Langton's Ant

As just discussed, CAs exemplify the fact that a collective behavior can emerge out of the sum of many, simply interacting, components. Even if the basic and local interactions are perfectly known, it is possible that the global behavior obeys new laws that are not obviously extrapolated from the individual properties. The Langton's ant model further illustrates this aspect.

The ant rule is a cellular automata invented by Chris Langton (Stewart 1994) and Greg Turk, which models the behavior of a hypothetical animal (ant) having a very simple algorithm of motion. The ant moves on a square lattice whose sites are either white or gray. When the ant enters a white cell, it turns 90° to the left and paints the cell in gray. Similarly, if it enters a gray cell, it paints it in white and turns 90° to the right. This is illustrated in ❯ *Fig. 7.*

It turns out that the motion of this ant exhibits a very complex behavior. Suppose the ant starts in a completely white space, after a series of about 500 steps where it essentially keeps returning to its initial position, it enters a chaotic phase during which its motion is unpredictable. Then, after about 10,000 steps of this very irregular motion, the ant suddenly performs a very regular motion, which brings it far away from where it started.

❯ *Figure 8* illustrates the ant motion. The path the ant creates to escape the chaotic initial region has been called a highway (Propp 1994). Although this highway is oriented at 45° with respect to the lattice direction, it is traveled by the ant in a way, which makes very much think of a sewing machine: the pattern is a sequence of 104 steps that are repeated indefinitely.

The Langton ant is a good example of a cellular automata whose rule is very simple and yet generates a complex behavior, which seems beyond understanding. Somehow, this fact is typical of the cellular automata approach: although everything is known about the fundamental laws governing a system (because the rules are man-made!), it is difficult to explain its macroscopic behavior.

There is anyway a global property of the ant motion: The ant visits an unbounded region of space, *whatever* the initial space texture is (configuration of gray and white cells).

The proof (due to Bunimovitch and Troubetzkoy) goes as follows: suppose the region the ant visits is bounded. Then, it contains a finite number of cells. Since the number of iteration is infinite, there is a domain of cells that are visited infinitely often. Moreover, due to the rule of motion, a cell is either entered horizontally (we call it a H cell) or vertically (we call it a V cell). Since the ant turns by 90° after each step, a H cell is surrounded by four V cells and conversely. As a consequence, the H and V cells tile the lattice in a fixed checkerboard pattern. Now, the upper rightmost cell of the domain is considered, that is a cell whose right and upper neighbor

◻ **Fig. 7**
The Langton's ant rule.

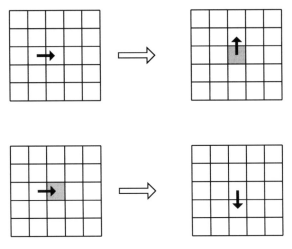

☐ **Fig. 8**

The Langton's ant rule. The motion of a single ant starts with a chaotic phase of about 10,000 time steps, followed by the formation of a highway. The figure shows the state of each lattice cell (gray or white) and the ant position (marked by the black dot). In the initial condition, all cells are white and the ant is located in the middle of the image.

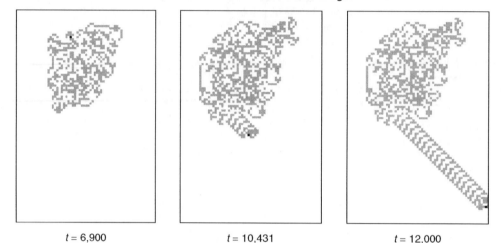

| $t = 6,900$ | $t = 10,431$ | $t = 12,000$ |

is not visited. This cell exists if the trajectory is bounded. If this cell is a H cell (and be so for ever), it has to be entered horizontally from left and exited vertically downward and, consequently, be gray. However, after the ant has left, the cell is white and there is a contradiction. The same contradiction appears if the cell is a V cell. Therefore, the ant trajectory is not bounded.

Beyond the technical aspect of this proof, it is interesting to realize that its conclusion is based on symmetry properties of the rule. Although one is not able to predict the detailed motion of the ant analytically (the only way is to perform the simulation), something can be learned about the global behavior of the system: the ant goes to infinity. This observation illustrates the fact that, often, global features are related to symmetries more than to details.

The case where several ants coexists on the lattice can be now considered. The rule defined in ❷ *Fig. 7* is only valid for a single ant. It can be generalized for situations where up to four ants enter the same site at the same time, from different sides.

To mathematically describe the ant motion, we introduce $n_i(\mathbf{r}, t)$, a Boolean variable representing the presence ($n_i = 1$) or the absence ($n_i = 0$) of an ant entering site $\mathbf{r}$ at time t along lattice direction c_i, where c_1, c_2, c_3, and c_4 stand for direction right, up, left, and down, respectively. If the color $\mu(\mathbf{r}, t)$ of the site is gray ($\mu = 0$), *all* entering ants turn 90° to the right. On the other hand, if the site is white ($\mu = 1$), then all ants turn 90° to the left. The color of each cell is modified after one or more ants have gone through. Here, we chose to switch $\mu \rightarrow 1 - \mu$ only when an odd number of ants are present.

When several ants travel simultaneously on the lattice, cooperative and destructive behaviors are observed. First, the erratic motion of several ants favors the formation of a local arrangement of colors allowing the creation of a highway. One has to wait much less time before the first highway appears. Second, once a highway is being created, other ants may use it to travel very fast (they do not have to follow the complicated pattern of the highway builder).

In this way, the term "highway" is very appropriate. Third, a destructive effect occurs as the second ant gets to the highway builder. It breaks the pattern and several situations may be observed. For instance, both ants may enter a new chaotic motion; or the highway is traveled in the other direction (note that the rule is time reversal invariant) and destroyed. ❷ *Figure 9* illustrates the multi-ant behavior.

The problem of an unbounded trajectory pauses again with this generalized motion. The assumption of Bunimovitch–Troubetzkoy's proof no longer holds in this case because a cell may be both a H cell or a V cell. Indeed, two different ants may enter the same cell, one vertically and the other horizontally. Actually, the theorem of an unbounded motion is wrong in several cases where two ants are present. Periodic motions may occur when the initial positions are well chosen.

For instance, when the relative location of the second ant with respect to the first one is $(\Delta x, \Delta y) = (2, 3)$, the two ants return to their initial position after 478 iterations of the rule (provided they started in a uniformly white substrate, with the same direction of motion). A very complicated periodic behavior is observed when $(\Delta x, \Delta y) = (1, 24)$: the two ants start a chaotic-like motion for several thousands of steps. Then, one ant builds a highway and escapes from the central region. After a while, the second ant finds the entrance of the highway and rapidly catches the first one. After the two ants meet, they start undoing their previous paths and return to their original position. This complete cycle takes about 30,000 iterations.

More generally, it is found empirically that, when $\Delta x + \Delta y$ is odd and the ants enter their site with the same initial direction, the two-ant motion is likely to be periodic. However, this is not a rule and the configuration $(\Delta x, \Delta y) = (1, 0)$ yields an unbounded motion, a diamond pattern of increasing diameter, which is traveled in the same direction by the two ants.

◻ **Fig. 9**
Motion of several Langton's ants. Gray and white indicate the colors of the cells at the current time. Ant locations are marked by the black dots. At the initial time, all cells are white and a few ants are randomly distributed in the central region, with random directions of motion. The first highway appears much earlier than when the ant is alone. In addition the highway can be used by other ants to travel much faster. However, the "highway builder" is usually prevented from continuing its construction as soon as it is reached by the following ants. For instance, the highway heading north-west after 4,900 steps gets destroyed. A new highway emerges later on from the rest, as seen from the snapshot at time $t = 8,564$.

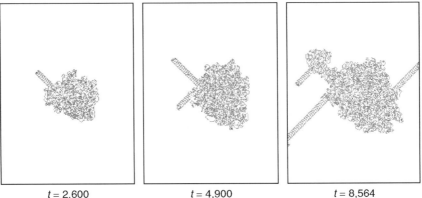

| $t = 2{,}600$ | $t = 4{,}900$ | $t = 8{,}564$ |

It turns out that the periodic behavior of a two-ant configuration is not so surprising. The rule defined is reversible in time, provided that there is never more than one ant at the same site. Time reversal symmetry means that if the direction of motion of all ants are reversed, they will move backward through their own sequence of steps, with an opposite direction of motion. Therefore, if at some point of their motion the two ants cross each other (on a lattice link, not on a site), the first ant will go through the past of the second one, and vice versa. They will return to the initial situation (the two ants being exchanged) and build a new pattern, symmetrical to the first one, due to the inversion of the directions of motion. The whole process then cycles forever. Periodic trajectories are therefore related to the probability that the two ants will, at some time, cross each other in a suitable way. The conditions for this to happen are fulfilled when the ants sit on a different sublattice (black or white sites on the checkerboard) and exit two adjacent sites against each other. This explains why a periodic motion is likely to occur when $\Delta x + \Delta y$ is odd.

An interesting conclusion is that, again, it is a symmetry of the rule (time reversal invariance) that allows one to draw conclusions about the global behavior. The details of periodic motions are not known but it is known that they are possible.

3.2   Cellular Automata as Simple Dynamical Systems

CA can also be seen as simple prototypes of dynamical systems. In physics, the time evolution of physical quantities is often governed by nonlinear equations. Due to the nonlinearities, solution of these dynamical systems can be very complex. In particular, the solution of these equations can be strongly sensitive to the initial conditions, leading to what is called a chaotic behavior. Similar complications can occur in discrete dynamical systems. Models based on cellular automata provide an alternative approach to study the behavior of dynamical systems. By virtue of their discrete nature, the numerical studies are free of rounding approximations and thus lead to exact results. Also, exhaustive searches in the space of possible rule can be considered for simple CA.

Crudely speaking, two classes of problems can be posed. First, given a cellular automaton rule, predict its properties. Second, find a cellular automaton rule that will have some prescribed properties. These two closely related problems are usually difficult to solve as seen on simple examples.

The simplest cellular automata rules are one-dimensional ones for which each site has only two possible states and the rule involves only the nearest-neighbors sites. They are easily programmable on a personal computer and offer a nice "toy model" to start the study of cellular automata.

A systematic study of these rules was undertaken by Wolfram in 1983 (Wolfram 1986, 1994). Each cell at location r has, at a given time, two possible states $s(r) = 0$ or $s(r) = 1$. The state s at time $t + 1$ depends only on the triplet $(s(r-1), s(r), s(r+1))$ at time t:

$$s(r, t+1) = \Phi(s(r-1, t), s(r, t), s(r+1, t)) \tag{3}$$

Thus to each possible values of the triplet $(s(r-1), s(r), s(r+1))$, one associates a value $\alpha_k = 0$ or 1 according to the following list:

$$\underbrace{111}_{\alpha_7} \quad \underbrace{110}_{\alpha_6} \quad \underbrace{101}_{\alpha_5} \quad \underbrace{100}_{\alpha_4} \quad \underbrace{011}_{\alpha_3} \quad \underbrace{010}_{\alpha_2} \quad \underbrace{001}_{\alpha_1} \quad \underbrace{000}_{\alpha_0} \tag{4}$$

Each possible CA rule $\mathscr{R}$ is characterized by the values $\alpha_0, \ldots, \alpha_7$. There are clearly 256 possible choices. Each rule can be identified by an index $\mathscr{N}_{\mathscr{R}}$ computed as follows

$$\mathscr{N}_{\mathscr{R}} = \sum_{i=0}^{7} 2^{(i)} \alpha_i \tag{5}$$

which corresponds to the binary representation $\alpha_7 \alpha_6 \alpha_5 \alpha_4 \alpha_3 \alpha_2 \alpha_1 \alpha_0$.

Giving a rule and an initial state, one can study the time evolution of the system. Some results can be deduced analytically using algebraic techniques, but most of the conclusions follow from numerical iterations of the rules. One can start from a simple initial state (i.e., only one cell in the state 1) or with a typical random initial state. According to their behavior, the different rules have been grouped in four different classes.

1. Class 1. These CA evolve after a finite number of time steps from almost all initial states to a unique homogeneous state (all the sites have the same value). The set of exceptional initial configurations, which behave differently is of measure zero when the number of cells N goes to infinity. An example is given by the rule 40 (see **❯** *Fig. 10a*). From the point of view of dynamical systems, these automata evolve toward a simple *limit point* in the phase space.

2. Class 2. A pattern consisting of separated periodic regions is produced from almost all the initial states. The simple structures generated are either stable or periodic with small periods. An example is given by rule 56 (see **❯** *Fig. 10b*) Here again, some particular initial states (set of measure zero) can lead to unbounded growth. The evolution of these automata is analogous to the evolution of some continuous dynamical systems to *limit cycles*.

3. Class 3. These CA evolve from almost all initial states to chaotic, aperiodic patterns. An example is given by rule 18 (see **❯** *Fig. 10c*). Small changes in the initial conditions almost always lead to increasingly large changes in the later stages. The evolution of these automata is analogous to the evolution of some continuous dynamical systems to *strange attractors*.

■ **Fig. 10**
Example of the four Wolfram rules with a random initial configuration. Horizontal lines correspond to consecutive iterations. The initial state is the uppermost line. (a) Rule 40 belonging to class 1 reaches a fixed point (stable configuration) very quickly. (b) Rule 56 of class 2 reaches a pattern composed of stripes that move from left to right. (c) Rule 18 is in class 3 and exhibits a self-similar pattern. (d) Rule 110 is an example of a class 4 cellular automaton. Its behavior is not predictable and, as a consequence, a rupture in the pattern on the left part is observed.

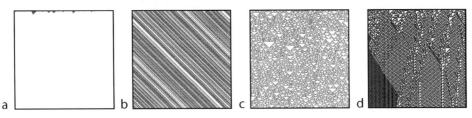

a b c d

4. Class 4. For these CA, persistent complex structures are formed for a large class of initial states. An example is given by the rule 110 (see ❷ *Fig. 10d*). The behavior of such CA can generally be determined only by an explicit simulation of their time evolution.

The "toy rules" considered by Wolfram, although very simple in construction, are capable of very complex behavior. The validity of this classification is not restricted to the simple rules described above but is somehow generic for more complicated rules. For example, one can consider rules for which each cell can have k different states and involve a neighborhood of radius ℓ (thus the rule depends on the values of $2\ell + 1$ cells). In this case, the number of possible rules is $k^{(k^{(2\ell+1)})}$. Several cases have been studied in the literature and the different rules can be classified in one of the four above classes. Many of the class 4 CA (starting with $k = 2, \ell = 2$) have the property of computational universality and initial configurations can specify arbitrary algorithmic procedures.

However the above "phenomenological" classification suffers drawbacks, the most serious of which is its non-decidability. See Culick and Yu (1988) for more details.

3.3 Competition, Cooperation, Contamination

In this section, a few CA rules that are very simple and mimic very natural interactions are briefly described: competition between adjacent cells, cooperation between them, or contamination of neighbors. All these ideas can be implemented within a discrete universe, with cells having only a few possible states. Other models of cooperation–competition are discussed in Galam et al. (1998).

3.3.1 Cooperation Models

For instance, in a simple cooperation model, a cell may want to evolve by copying the behavior of the majority of its neighbors. If the possible states are 0 or 1, then clearly an all 0's configuration or an all 1's configuration are both stable. But what happens if the initial configuration contains cells that are 1 with probability p and 0 with probability $1 - p$. It is likely that such a system will evolve to one of the two stable configuration. It would be very nice if the all 1's configuration is reached whenever $p > 1/2$ and the all 0's configuration would be the final stage of the case $p < 1/2$. Then, one would have built a system with only *local* calculation that is able to solve a global problem: deciding if the initial density of cells in state 1 is larger or smaller than 1/2. This problem is known as the *density task* and, in general, a simple majority rule is not able to give a reliable answer. See Sipper (1997) for more details.

From the point of view of modeling physical systems, a slight variant of the majority rule produces interesting patterns. The *twisted majority rule* proposed by Vichniac (1984) is defined on a two-dimensional lattice where each cell considers its Moore neighborhood. The evolution rule first computes the sum of the cells in state 1. This sum can be any value between 0 and 9. The new state $s(t + 1)$ of each cell is then determined from this local sum, according to the following table

$$
\begin{array}{ll}
\text{sum}(t) & 0\ 1\ 2\ 3\ 4\ 5\ 6\ 7\ 8\ 9 \\
s(t+1) & 0\ 0\ 0\ 0\ 1\ 0\ 1\ 1\ 1\ 1
\end{array}
\tag{6}
$$

☐ **Fig. 11**

Evolution of the twisted majority rule. The inherent "surface tension" present in the rule tends to separate the red phases $s = 1$ from the blue phase $s = 0$. The snapshots **(a)**, **(b)**, and **(c)** correspond to $t = 0$, $t = 72$, and $t = 270$ iterations, respectively. The other colors indicate how "capes" have been eroded and "bays" filled: light blue shows the blue regions that have been eroded during the last few iterations and yellow marks the red regions that have been filled.

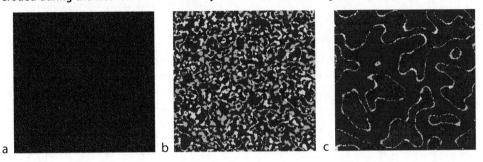

As opposed to the plain majority rule, here, the two middle entries of the table have been swapped. Therefore, when there is a slight majority of 1 around a cell, it turns to 0. Conversely, if there is a slight majority of 0, the cell becomes 1.

Surprisingly enough, this rule describes the interface motion between two phases, as illustrated in ❷ *Fig. 11*. It is observed that the normal velocity of the interface is proportional to its local curvature, as required by several physical theories. Of course, due to its discrete and local nature, the rule cannot detect the curvature of the interface directly. However, as the rule is iterated, local information is propagated to the nearest neighbors and the radius of curvature emerges as a collective effect.

3.3.2 Competition Models

In some sense, the twisted majority rule corresponds to a cooperative behavior between the cells. A quite different situation can be obtained if the cells obey a competitive dynamics. For instance, one may imagine that the cells compete for some resources at the expense of their nearest neighbors. A winner is a cell of state 1 and a loser a cell of state 0. No two winner cells can be neighbors and any loser cell must have at least one winner neighbor (otherwise nothing would have prevented it from winning).

This problem has a direct application in biology, to study cells differentiation. It has been observed in the development of the drosophila that about 25% of the cells forming the embryo are evolving to the state of neuroblast, while the remaining 75% does not. How can we explain this differentiation and the observed fraction since, at the beginning of the process all cells can be assumed equivalent? A possible mechanism (Luthi et al. 1998) is that some competition takes place between the adjacent biological cells. In other words, each cell produces some substance S but the production rate is inhibited by the amount of S already present in the neighboring cells. Differentiation occurs when a cell reaches a level of S above a given threshold.

Following this interpretation, the following CA model of competition can be considered. First, a hexagonal lattice, which is a reasonable approximation of the cell arrangement observed in the drosophila embryo, is considered. It is assumed that the values of S can be 0 (inhibited) or 1 (active) in each lattice cell.

- A $S = 0$ cell will grow (i.e., turn to $S = 1$) with probability p_{growth} provided that all its neighbors are 0. Otherwise, it stays inhibited.
- A cell in state $S = 1$ will decay (i.e., turn to $S = 0$) with probability p_{anihil} if it is surrounded by at least one active cell. If the active cell is isolated (all the neighbors are in state 0) it remains in state 1.

The evolution stops (stationary process) when no $S = 1$ cell feels any more inhibition from its neighbors and when all $S = 0$ cells are inhibited by their neighborhood. Then, with the biological interpretation, cells with $S = 1$ are those that will differentiate.

What is the expected fraction of these $S = 1$ cells in the final configuration? Clearly, from ❷ *Fig. 12*, the maximum value is 1/3. According to the inhibition condition imposed, this is the close-packed situation on the hexagonal lattice. On the other hand, the minimal value is 1/7, corresponding to a situation where the lattice is partitioned in blocks with one active cell surrounded by six inhibited cells. In practice, any of these two limits are not expected to occur spontaneously after the automaton evolution. On the contrary, one can observe clusters of close-packed active cells surrounded by defects, that is, regions of low density of active cells.

As illustrated in ❷ *Fig. 13*, CA simulations show indeed that the final fraction s of active cells is a mix of the two limiting situations of ❷ *Fig. 12*

$$.23 \leq s \leq .24$$

almost irrespectively of the values chosen for p_{anihil} and p_{growth}.

This is exactly the value expected from the biological observations made on the drosophila embryo. Thus, cell differentiation can be explained by a geometrical competition without having to specify the inhibitory couplings between adjacent cell and the production rate (i.e., the values of p_{anihil} and p_{growth}): the result is quite robust against any possible choices.

3.3.3 Contamination Models

Finally, after cooperation and competition dynamics, we also consider a contamination process. To make the model more interesting, we consider cells with at least three possible states. These are the resting state, the excited state, and the refractory state.

◼ **Fig. 12**
The hexagonal lattice used for the competition-inhibition CA rule. Black cells are cells of state 1 (winners) and white cells are cells of state 0 (losers). The two possible final states with a fully regular structure are illustrated with density 1/3 and 1/7 of winner, respectively.

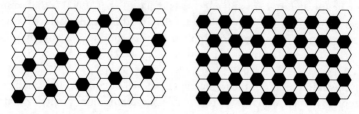

□ **Fig. 13**

Final configuration of the competition CA model. (a) A typical situation with about 23% of active cells, obtained with almost any value of p_{anihil} and p_{growth}. (b) Configuration obtained with $p_{anihil} = 1$ and $p_{growth} = 0.8$ and yielding a fraction of 28% of active cells; one clearly sees the close-packed regions and the defects.

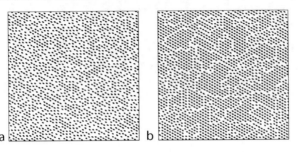

a   b

The resting state is a stable state of the system. But a resting state can respond to a local perturbation and become excited. Then, the excited state evolves to a refractory state where it no longer influences its neighbors and, finally, returns to the resting state.

The Greenberg–Hastings model is an example of a cellular automata model with a contamination mechanism. It is also called a model for an *excitable media* in the context of reactive systems and chemical waves.

The Greenberg–Hastings model can be defined as follows: the state $\psi(\mathbf{r}, t)$ of site $\mathbf{r}$ at time t takes its value in the set $\{0, 1, 2, \ldots, n - 1\}$. The state $\psi = 0$ is the resting state. The states $\psi = 1, \ldots, n/2$ (n is assumed to be even) correspond to excited states. The rest, $\psi = n/2 + 1, \ldots, n - 1$ are the refractory states. The CA evolution rule is the following:

1. If $\psi(\mathbf{r}, t)$ is excited or refractory, then $\psi(\mathbf{r}, t + 1) = \psi(\mathbf{r}, t) + 1 \mod n$.
2. If $\psi(\mathbf{r}, t) = 0$ (resting state) it remains so, unless there are at least k excited sites in the Moore neighborhood of site $\mathbf{r}$. In this case $\psi(\mathbf{r}, t + 1) = 1$.

The n states play the role of a clock: An excited state evolves through the sequence of all possible states until it returns to 0, which corresponds to a stable situation.

The behavior of this rule is quite sensitive to the value of n and the excitation threshold k. ❯ *Figure 14* shows the evolution of this CA for a given set of the parameters n and k. The simulation is started with a uniform configuration of resting states, perturbed by some excited sites randomly distributed over the system. Note that if the concentration of perturbation is low enough, excitation dies out rapidly and the system returns to the rest state. Increasing the number of perturbed states leads to the formation of traveling waves and self-sustained oscillations may appear in the form of ring or spiral waves.

The Greenberg–Hastings model has some similarity with the "tube-worms" rule proposed by Toffoli and Margolus (1987). This rule is intended to model the Belousov–Zhabotinsky reaction and is as follows. The state of each site is either 0 (refractory) or 1 (excited) and a local timer (whose value is 3, 2, 1, or 0) controls the refractory period. Each iteration of the rule can be expressed by the following sequence of operations: (1) where the timer is zero, the state is excited; (2) the timer is decreased by 1 unless it is 0; (3) a site becomes refractory whenever the timer is equal to 2; (4) the timer is reset to 3 for the excited sites, which have two, or more than four, excited sites in their Moore neighborhood.

◘ Fig. 14

Excitable medium: evolution of a configuration with 5% of excited states $\phi = 1$, and 95% of resting states (black), for $n = 8$ and $k = 3$.

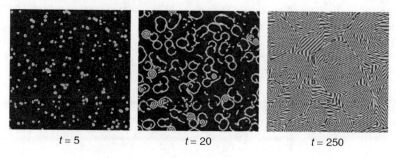

$t = 5$ $t = 20$ $t = 250$

◘ Fig. 15

The tube-worms rule for an excitable media.

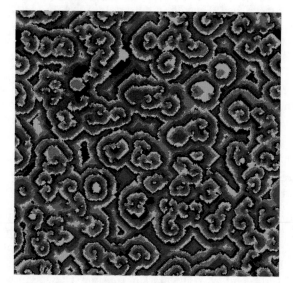

❯ *Figure 15* shows a simulation of this automaton, starting from a random initial configuration of the timers and the excited states. One can observe the formation of spiral pairs of excitations. Note that this rule is very sensitive to small modifications (in particular to the order of operations (1)–(4)).

Another rule which is also similar to Greenberg–Hastings and Margolus–Toffoli tubeworm models is the so-called forest-fire model. This rule describes the propagation of a fire or, in a different context, may also be used to mimic contagion in the case of an epidemic. Here, the case of a forest-fire rule is described. The forest-fire rule is a probabilistic CA defined on a two-dimensional square lattice. Initially, each site is occupied by either a tree, a burning tree, or it is empty. The state of the system is parallel updated according to the

◻ Fig. 16

The forest-fire rule: green sites correspond to a grown tree, black pixels represent burned sites, and the yellow color indicates a burning tree. The snapshots given here represents three situations after a few hundred iterations. The parameters of the rule are $p = 0.3$ and $f = 6 \times 10^{-5}$.

following rule: (1) a burning tree becomes an empty site; (2) a green tree becomes a burning tree if at least one of its nearest neighbors is burning; (3) at an empty site, a tree grows with probability p; (4) A tree without a burning neighbor becomes a burning tree with probability f (so as to mimic an effect of lightning). ❷ *Figure 16* illustrates the behavior of this rule, in a two-dimensional situation.

3.4   Traffic Models

CA models for road traffic have received a great deal of interest during the past few years (see Yukawa et al. 1994, Schreckenberg and Wolf 1998, Chopard et al. 1996, Schreckenberg et al. 1995, and Nagel and Herrmann 1993 for instance).

CA models for a single lane car motions are quite simple. The road is represented as a line of cells, each of them being occupied or not by a vehicle. All cars travel in the same direction (say to the right). Their positions are updated synchronously. During the motion, each car can be at rest or jump to the nearest neighbor site, along the direction of motion. The rule is simply that a car moves only if its destination cell is empty. This means that the drivers do not know whether the car in front will move or will be blocked by another car. Therefore, the state s_i of each cell at location i is entirely determined by the occupancy of the cell itself and that of its two nearest neighbors s_{i-1} and s_{i+1}. The motion rule can be summarized by the following table, where all eight possible configurations $(s_{i-1}s_is_{i+1})_t \rightarrow (s_i)_{t+1}$ are given

$$\underbrace{(111)}_{1} \underbrace{(110)}_{0} \underbrace{(101)}_{1} \underbrace{(100)}_{1} \underbrace{(011)}_{1} \underbrace{(010)}_{0} \underbrace{(001)}_{0} \underbrace{(000)}_{0} \qquad (7)$$

This cellular automaton rule turns out to be Wolfram rule 184 (Wolfram 1986; Yukawa et al. 1994). It is illustrated in ❷ *Fig. 17*.

This simple dynamics captures an interesting feature of real car motion: traffic congestion. Suppose there is a low car density ρ in the system, for instance something like

$$\dots 0010000010010000010 \dots \qquad (8)$$

◻ **Fig. 17**

Illustration of the basic traffic rule: car with a free cell in front can move. The other ones stay at rest.

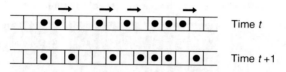

This is a free traffic regime in which all the cars are able to move. The average velocity $< v >$ defined as the number of motions divided by the number of cars is then

$$< v_{\text{free}} > = 1 \qquad (9)$$

On the other hand, in a high density configuration such as

$$\ldots 110101110101101110 \ldots \qquad (10)$$

only six cars over 12 will move and $< v > = 1/2$. This is a partially jammed regime.

In this case, since a car needs a hole to move to, one can expect that the number of moving cars simply equals the number of empty cells (Yukawa et al. 1994). Thus, the number of motions is $L(1 - \rho)$, where L is the number of cells. Since the total number of cars is ρL, the average velocity in the jammed phase is

$$< v_{\text{jam}} > = \frac{1 - \rho}{\rho} \qquad (11)$$

From the above relations the so-called fundamental flow diagram can be computed, that is, the relation between the flow of cars $\rho < v >$ as a function of the car density ρ: for $\rho \leq 1/2$, one can use the free regime expression and $\rho < v > = \rho$. For densities $\rho > 1/2$, one can use the jammed expression and $\rho < v > = 1 - \rho$. The resulting diagram is shown in ❷ *Fig. 18*. As in real traffic, we observe that the flow of cars reaches a maximum value before decreasing.

A richer version of the above CA traffic model is due to Nagel and Schreckenberg (1992). The cars may have several possible velocities $u = 0, 1, 2, \ldots, u_{\text{max}}$. Let u_i be the velocity of car i and d_i the distance, along the road, separating cars i and $i + 1$. The updating rule is

- The cars accelerate when possible: $u_i \rightarrow u_i' = u_i + 1$, if $u_i < u_{\text{max}}$.
- The cars slow down when required: $u_i' \rightarrow u_i'' = d_i - 1$, if $u_i' \geq d_i$.
- The cars have a random behavior: $u_i'' \rightarrow u_i''' = u_i'' - 1$, with probability p_i if $u_i'' > 0$.
- Finally the cars move u_i''' sites ahead.

This rule captures some important behaviors of real traffic on a highway: velocity fluctuations due to a nondeterministic behavior of the drivers, and "stop-and-go" waves observed in high density traffic regime.

Note that a street network can also be described using a CA. A possible approach is to couple several 1D CA model at each road intersection using a roundabout (Chopard et al. 1996; Chopard and Dupuis 2003). This is illustrated in ❷ *Fig. 19* for a Manhattan-like configuration of streets.

The reader can refer to recent literature for the new developments in this topic. See for instance Kanai et al. (2005, 2006).

⬛ **Fig. 18**

Traffic flow diagram. *Left:* **for the simple CA traffic rule.** *Right:* **for an urban CA traffic model, for a configuration of streets as shown in ❷** *Fig. 19.*

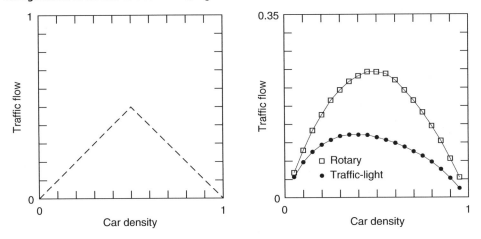

⬛ **Fig. 19**

Traffic configuration after 600 iterations, for a car density of 30%. Situation (a) corresponds to a situation where junctions are modeled as roundabouts, whereas image (b) mimics the presence of traffic lights. In the second case, car queues are more likely to form and the global mobility is less than in the first case, as shown in the right part of ❷ *Fig. 18.*

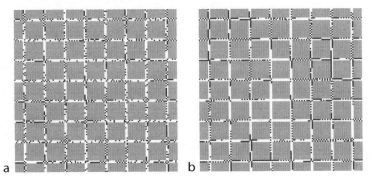

4   A Simple Model for a Gas of Particles

The Hardy, Pomeau, de Pazzis (HPP) rule is a simple example of an important class of cellular automata models: LGA. The basic ingredient of such models are point particles that move on a lattice, according to appropriate rules so as to mimic a fully discrete "molecular dynamics."

This model is mostly interesting for pedagogical reasons as it illustrates many important features of LGA and LB models in a simple way. However, HPP is of little practical interest because its physical behavior has many flaws (see for instance ❷ *Fig. 22*) that are cured in more sophisticated models, such as the famous FHP model (Frisch et al. 1986), which has been shown to reproduce the Navier–Stokes equations.

The HPP LGA is defined on a two-dimensional square lattice. Particles can move along the main directions of the lattice, as shown in ❯ *Fig. 20*. The model limits to 1 the number of particles entering a given site with a given direction of motion. This is the exclusion principle, which is common in most LGA (LGA models without exclusion principle are called multiparticle models (Chopard and Droz 1998)).

With at most one particle per site and direction, four bits of information at each site are enough to describe the system during its evolution. For instance, if at iteration t site $\mathbf{r}$ has the following state $s(\mathbf{r}, t) = (1011)$, it means that three particles are entering the site along direction 1, 3 and 4, respectively.

The CA rule describing the evolution of $s(\mathbf{r}, t)$ is often split in two steps: collision and propagation (or streaming). The collision phase specifies how the particles entering the same site will interact and change their trajectories. During the propagation phase, the particles actually move to the nearest neighbor site they are traveling to. This decomposition into two phases is a quite convenient way to partition the space so that the collision rule is purely local.

❯ *Figure 21* illustrates the HPP rules. According to the Boolean representation of the particles at each site, the collision rules for the two-particle head-on collisions are expressed as

$$(1010) \rightarrow (0101) \qquad (0101) \rightarrow (1010) \tag{12}$$

All other configurations are unchanged by the collision process.

After the collision, the propagation phase moves information to the nearest neighbors: The first bit of the state variable is shifted to the east neighbor cell, the second bit to the north, and so on. This gives the new state of the system, at time $t + 1$. Remember that both collision and propagation are applied simultaneously to all lattice sites.

The aim of the HPP rule is to reproduce some aspects of the real interactions between particles, namely, that momentum and particle number are conserved during a collision. From ❯ *Fig. 21*, it is easy to check that these properties are obeyed: A pair of zero momentum particles along a given direction is transformed into another pair of zero momentum along the perpendicular axis.

It is easy to express the HPP model in a mathematical form. For this purpose, the so-called occupation number $n_i(\mathbf{r}, t)$ is introduced for each lattice site $\mathbf{r}$ and each time step t. The index i

❏ **Fig. 20**
Example of a configuration of HPP particles.

■ **Fig. 21**
The HPP rule: (a) a single particle has a ballistic motion until it experiences a collision; (b) and (c) the two nontrivial collisions of the HPP model: two particles experiencing a head-on collision are deflected in the perpendicular direction. In the other situations, the motion is ballistic, that is, the particles are transparent to each other when they cross the same site.

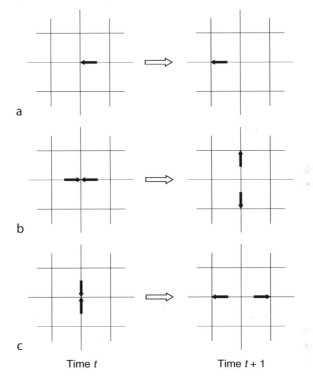

a

b

c

Time *t*   Time *t* + 1

labels the lattice directions (or the possible velocities of the particles). In the HPP model, the lattice has four directions (north, west, south, and east) and i runs from 1 to 4.

By definition and due to the exclusion principle, the n_i's are Boolean variables

$$n_i(\mathbf{r}, t) = \begin{cases} 1 & \text{if a particle is entering site } \mathbf{r} \text{ at time t along lattice direction } i \\ 0 & \text{otherwise} \end{cases}$$

From this definition it is clear that, for HPP, the n_i's are simply the components of the state s introduced above

$$s = (n_1, n_2, n_3, n_4)$$

In a LGA model, the microdynamics can be naturally expressed in terms of the occupation numbers n_i as

$$n_i(\mathbf{r} + \mathbf{v}_i \delta_t, t + \delta_t) = n_i(\mathbf{r}, t) + \Omega_i(n(\mathbf{r}, t)) \tag{13}$$

where $\mathbf{v}_i$ is a vector denoting the speed of the particle in the ith lattice direction and δ_t is the duration of the time step. The function Ω is called the collision term and it describes the interaction of the particles, which meet at the same time and same location.

Note that another way to express ❯ Eq. 13 is through the so-called collision and propagation operators C and P

$$n(t + \delta_t) = PCn(t) \tag{14}$$

where $n(t)$ describe the set of values $n_i(\mathbf{r}, t)$ for all i and $\mathbf{r}$. The quantities C and P act over the entire lattice. They are defined as

$$(Pn)_i(\mathbf{r}) = n_i(\mathbf{r} - \mathbf{v}_i\delta_t) \qquad (Cn)_i(\mathbf{r}) = n_i(\mathbf{r}) + \Omega_i$$

More specifically, for the HPP model, it can be shown (Chopard and Droz 1998) that the collision and propagation phase can be expressed as

$$n_i(\mathbf{r} + \mathbf{v}_i\delta_t, t + \delta_t) = n_i - n_i n_{i+2}(1 - n_{i+1})(1 - n_{i+3}) + n_{i+1}n_{i+3}(1 - n_i)(1 - n_{i+2}) \tag{15}$$

In this equation, the values $i + m$ are wrapped onto the values 1–4 and the right-hand term is computed at position $\mathbf{r}$ and time t. From this relation, it is easy to show that, for any values of n_i,

$$\sum_{i=1}^{4} n_i(\mathbf{r} + \mathbf{v}_i\delta_t, t + \delta_t) = \sum_{i=1}^{4} n_i(\mathbf{r}, t) \tag{16}$$

that expresses the conservation of the number of particle during the collision and the propagation. Similarly, it can be shown ($\mathbf{v}_1 = -\mathbf{v}_3$ and $\mathbf{v}_2 = -\mathbf{v}_4$) that

$$\sum_{i=1}^{4} n_i(\mathbf{r} + \mathbf{v}_i\delta_t, t + \delta_t)\mathbf{v}_i = \sum_{i=1}^{4} n_i(\mathbf{r}, t)\mathbf{v}_i \tag{17}$$

which reflects that momentum is conserved.

The behavior of the HPP model is illustrated in ❯ Fig. 22. From this simulation, it is clear that some spatially anisotropic behavior builds up during the time evolution of the rule. A square lattice is actually too poor to represent correctly a fluid system. The FHP model (Frisch et al. 1986; Chopard and Droz 1998), in essence similar to the HPP model, is based on a hexagonal lattice and also includes three-particle collision rules.

◰ Fig. 22

Time evolution of a HPP gas with a density wave. (a) The initial state is a homogeneous gas with a higher density of particles in the middle region (*dark area*) (b) After several iterations, the initial perturbation propagates as a wave across the system. As can be observed, there is a clear lack of isotropy in this propagation.

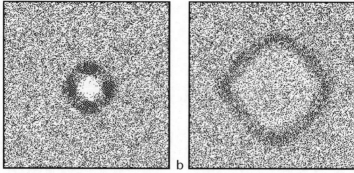

a   b

5   Lattice Boltzmann Models

Historically LB was developed as an extension of the CA-fluids described in ❿ Sect. 4. Another approach is to derive LB models from a discretization of the classical continuous Boltzmann equation (He and Luo 1997; Shan et al. 2006).

Here we stick to the historical approach as it better illustrates the close relation between the numerical scheme and the underlying discrete physical model of interacting particles. The main conceptual difference between LGA and LB models is that in the latter the cell state is no longer the Boolean variable n_i but a real-valued quantity f_i for each lattice directions i. Instead of describing the presence or absence of a particle, the interpretation of f_i is the density distribution function of particles traveling in lattice directions i.

From a practical point of view, the advantages of suppressing the Boolean constraint are several: less statistical noise, more numerical accuracy and, importantly, more flexibility to choose the lattice topology, the collision operator and boundary conditions. Thus, for many applications, the LB approach is preferred to the LGA one. On the other hand, LB models do not integrate local fluctuations that are naturally present in a LGA and that can have relevant physical effects (Chopard and Droz 1998).

Several textbooks (Chen and Doolen 1998; Succi 2001; Chopard and Droz 1998; Wolf-Gladrow 2000; Sukop and Thorne 2005) exist, which describe in great detail the LB approach. The method has been used extensively in the literature to simulate complex flows and other physical processes (Chopard and Droz 1998). For hydrodynamics, the LB method is now recognized as a serious competitor to the more traditional approaches based on the computer solution of the Navier–Stokes partial differential equations. Among the advantages of the LB method over more traditional numerical schemes, we mention its simplicity, its flexibility to describe complex flows, and its local nature (no need to solve a Poisson equation). Another feature of the LB method is its extended range of validity when the Knudsen number is not negligible (e.g., in microflows) (Ansumali et al. 2007).

5.1   General Principles

5.1.1   Definitions

The key quantities to define an LB model are the density distributions $f_i(\mathbf{r}, t)$ and the "molecular velocities" $\mathbf{v}_i$, for $i = 0 \ldots z$ where z is the lattice coordination number of the chosen lattice topology and $z + 1$ is the number of discrete velocities. The quantity f_i then denotes the number of particles entering lattice site $\mathbf{r}$ at time t with discrete velocity $\mathbf{v}_i$. Note that $\mathbf{v}_i$ is a vector so that molecular velocities have a norm and a direction. For instance, a common choice of velocities in 2D problem is

$$\begin{aligned}
\mathbf{v}_0 &= (0,0) \quad \mathbf{v}_1 = v(1,0) \quad \mathbf{v}_2 = v(1,1) \quad \mathbf{v}_3 = v(0,1) \quad \mathbf{v}_4 = v(-1,1) \\
\mathbf{v}_5 &= v(-1,0) \quad \mathbf{v}_6 = v(-1,-1) \quad \mathbf{v}_7 = v(0,-1) \quad \mathbf{v}_8 = v(1,-1)
\end{aligned} \tag{18}$$

In these expressions, v is a velocity norm defined as $v = \delta_r/\delta_t$, with δ_r being the lattice spacing and δ_t the duration of the time step. Both δ_r and δ_t can be expressed in any desired unit system.

From the f_i's and the $\mathbf{v}_i$'s the standard physical quantities such as particle density ρ, particle current $\rho\mathbf{u}$ can be defined, by taking various moments of the distribution

$$\rho(\mathbf{r}, t) = \sum_i f_i(\mathbf{r}, t) \qquad \rho(\mathbf{r}, t)\mathbf{u}(\mathbf{r}, t) = \sum_i f_i(\mathbf{r}, t)\mathbf{v}_i \tag{19}$$

The intuitive interpretation of these relations is obvious: The number of particles at point $\mathbf{r}$ and time t is the sum of all particles coming with all velocities; and the total momentum is the sum of momentum carried by each f_i.

In hydrodynamics, it is also important to define higher moments, such as the *momentum tensor*

$$\Pi_{\alpha\beta} = \sum_i f_i(\mathbf{r}, t)v_{i\alpha}v_{i\beta}$$

where Greek subscripts label spatial coordinates. The tensor $\Pi_{\alpha\beta}$ describes the amount of α-momentum transported along the β-axis.

Following the particle interpretation, one can say that in an LB model, all particles entering the same site at the same time from different directions (i.e., particles with different molecular velocities $\mathbf{v}_i$) collide. As a consequence, a new distribution of particles results. Then, during the next time step δ_t, the particles emerging from the collision move to a new lattice site, according to their new speed. Therefore, the dynamics of an LB model is the alternation of collision and propagation phases.

This is illustrated in ❯ *Figs. 23* and ❯ *24*, for two different lattice topologies in two dimensions (hexagonal and square lattices). In accordance with ❯ *Figs. 23* and ❯ *24*, the LB dynamics can be written as a collision phase

$$f_i^{\text{out}}(\mathbf{r}, t) = f_i^{\text{in}}(\mathbf{r}, t) + \Omega_i\big(f^{\text{in}}(\mathbf{r}, t)\big) \tag{20}$$

and a propagation (propagation is often termed *streaming* in the literature) phase

$$f_i^{\text{in}}(\mathbf{r} + \delta_t\mathbf{v}_i, t + \delta_t) = f_i^{\text{out}}(\mathbf{r}, t) \tag{21}$$

where Ω, the collision term is a model-specific function describing the outcome of the particle collision. When subscript i is omitted, Ω denotes the entire set of Ω_i. Its form will be discussed below. Here, an upper-script in or out to the distribution function f_i is added to distinguish the pre-collision distribution f_i^{in} from the post-collision ones f_i^{out}. When no upper-script is used,

◻ **Fig. 23**
Illustration of the collision and propagation phases in an LB model defined on a 2D hexagonal lattice, with six possible velocities. The arrows represent the particles, their directions correspond to the $\mathbf{v}_i$ and their length is proportional f_i.

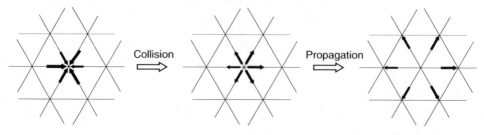

◻ Fig. 24
Illustration of the collision and propagation phases in an LB model defined on a 2D square lattice with eight velocities.

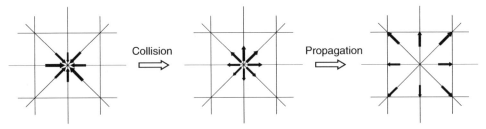

we define that $f = f^{in}$. Therefore, by combining ❷ Eqs. 20 and ❷ 21, an LB model can also be expressed as

$$f_i(\mathbf{r} + \delta_t \mathbf{v}_i, t + \delta_t) = f_i(\mathbf{r}, t) + \Omega_i(f(\mathbf{r}, t)) \tag{22}$$

Conservation laws play an important role in building an LB model. When some physical quantities are known to be conserved in a given phenomena, this conservation must be reflected exactly by the dynamics of the corresponding LB equation. For instance, if the number of particle ρ is conserved in the collision process, it is required to have

$$\sum_{i=0}^{z} f_i^{out}(\mathbf{r}, t) = \sum_{i=0}^{z} f_i^{in}(\mathbf{r}, t)$$

for all $\mathbf{r}$ and all t.

From ❷ Eq. 22, this means that the collision term must obey

$$\sum_{i=0}^{z} \Omega_i = 0 \tag{23}$$

Similarly, if momentum is also conserved (as in a fluid), it is required to have

$$\sum_{i=0}^{z} \mathbf{v}_i f_i^{out}(\mathbf{r}, t) = \sum_{i=0}^{z} \mathbf{v}_i f_i^{in}(\mathbf{r}, t)$$

and then

$$\sum_{i=0}^{z} \mathbf{v}_i \Omega_i = 0 \tag{24}$$

5.1.2 Lattice Properties

As seen previously, propagation moves particles with velocity $\mathbf{v}_i$ from one lattice site to a neighboring one. As a consequence, $\mathbf{r} + \delta_t \mathbf{v}_i$ must also be a point of the lattice. Thus, there is a tight connection between the spatial lattice and the discrete set of molecular velocities.

In the LB framework, the choice of velocity $\mathbf{v}_i$ (and consequently the corresponding spatial lattice) is commonly labeled as DdQq, where d is the spatial dimension ($d = 2$ for a two-dimensional problem) and q is the number of discrete velocities (or quantities).

When q is odd, it is assumed that the model includes a rest speed $\mathbf{v}_0 = 0$. Then velocities $\mathbf{v}_i$ are labeled from $i = 0$ to $q - 1$ and $q = z + 1$, z being the lattice coordination number. When q is even, the model contains no rest speed and the velocities $\mathbf{v}_i$ are labeled from $i = 1$ to $i = q$ and $q = z$. For instance, the D2Q9 lattice corresponds to the velocities given in ❷ Eq. 18 and illustrated in ❷ *Fig. 25*. A D2Q8 lattice is the same, but without $\mathbf{v}_0$. To build a proper LB model, the $\mathbf{v}_i$ should be carefully chosen. In addition to the fact that $\mathbf{r} + \delta_t \mathbf{v}_i$ must correspond to a lattice site, the molecular velocities must have enough symmetry and isotropy properties. In short, tensors built by summing the velocity components should have enough rotational invariance to represent the physical process under consideration. To achieve this goal, it is often necessary to add weights w_i to velocity vector $\mathbf{v}_i$, as suggested in the right panel of ❷ *Fig. 25*. In practice it is required that

$$\sum_i w_i = 1 \qquad \sum_i w_i \mathbf{v}_i = 0 \qquad \sum_i w_i v_{i\alpha} v_{i\beta} = c_s^2 \delta_{\alpha\beta} \tag{25}$$

where $\delta_{\alpha\beta}$ is the Kronecker symbol and c_s a coefficient to be determined. For a D2Q5 lattice $\mathbf{v}_0 = 0$, $\mathbf{v}_1 = (v, 0)$, $\mathbf{v}_2 = (0, v)$, $\mathbf{v}_3 = (-v, 0)$, and $\mathbf{v}_4 = (0, -v)$, the above set of conditions is easily fulfilled by choosing $w_1 = w_2 = w_3 = w_4 = (1 - w_0)/4 > 0$ and $c_s^2 = v^2(1 - w_0)/2$ because $\mathbf{v}_i$ and $\mathbf{v}_{i+1}$ are orthogonal 2D vectors.

Conditions (❷ 25) are sufficient to model diffusion processes or wave propagation (Chopard and Droz 1998). However, to model nonthermal hydrodynamic flows, they are not enough and must be supplemented by conditions on the third-, fourth-, and fifth-order tensors that can be built out of the $\mathbf{v}_i$'s. These conditions read

$$\sum_i w_i v_{i\alpha} v_{i\beta} v_{i\gamma} = 0$$

$$\sum_i w_i v_{i\alpha} v_{i\beta} v_{i\gamma} v_{i\delta} = c_s^4 \left(\delta_{\alpha\beta} \delta_{\gamma\delta} + \delta_{\alpha\gamma} \delta_{\beta\delta} + \delta_{\alpha\delta} \delta_{\beta\gamma} \right) \tag{26}$$

$$\sum_i w_i v_{i\alpha} v_{i\beta} v_{i\gamma} v_{i\delta} v_{i\epsilon} = 0$$

For thermo-hydrodynamic models, even higher order conditions must be considered, forcing one to add more discrete velocities to the system (Shan et al. 2006). The reader is directed to

◻ **Fig. 25**

The D2Q9 lattice with nine velocities, corresponding to a square lattice, including diagonals. Note that $\mathbf{v}_0 = (0, 0)$ is not shown on the figure. The right panel shows the ratio of the weights associated with every direction: the diagonal directions should have a weight four times smaller than the main directions in order to ensure the isotropy of the fourth-order tensor.

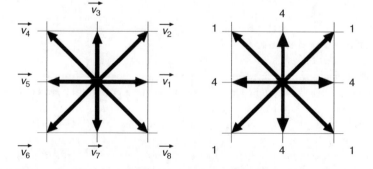

LB textbooks to better understand the origin and meaning of these isotropy conditions. Here, they shall be just accepted as a requirement on a proper choice of velocity sets. They can be satisfied for a D2Q9 model by taking

$$w_0 = 4/9 \qquad w_1 = w_3 = w_5 = w_7 = 1/9 \qquad w_2 = w_4 = w_6 = w_8 = 1/36$$

for which one gets $c_s^2/v^2 = 1/3$.

Three-dimensional models, such as the well-known D3Q19 model, can also be constructed to satisfy ❯ Eqs. 25 and ❯ 26. In the D3Q19 model, the velocity vectors are defined as

$$\mathbf{v}_0 = 0$$

$$\mathbf{v}_1 = v(-1,0,0) \quad \mathbf{v}_2 = v(0,-1,0) \quad \mathbf{v}_3 = v(0,0,-1)$$
$$\mathbf{v}_4 = v(-1,-1,0) \quad \mathbf{v}_5 = v(-1,1,0) \quad \mathbf{v}_6 = v(-1,0,-1)$$
$$\mathbf{v}_7 = v(-1,0,1) \quad \mathbf{v}_8 = v(0,-1,-1) \quad \mathbf{v}_9 = v(0,-1,1) \tag{27}$$
$$\mathbf{v}_{10} = v(1,0,0) \quad \mathbf{v}_{11} = v(0,1,0) \quad \mathbf{v}_{12} = v(0,0,1)$$
$$\mathbf{v}_{13} = v(1,1,0) \quad \mathbf{v}_{14} = v(1,-1,0) \quad \mathbf{v}_{15} = v(1,0,1)$$
$$\mathbf{v}_{16} = v(1,0,-1) \quad \mathbf{v}_{17} = v(0,1,1) \quad \mathbf{v}_{15} = v(0,1,-1)$$

and the lattice properties are

$$c_s^2/v^2 = 1/3 \qquad w_0 = 1/3 \qquad w_{\text{slow}} = 1/18 \qquad w_{\text{fast}} = 1/36$$

where w_{slow} concerns the $\mathbf{v}_i$ of norm v and w_{fast} the $\mathbf{v}_i$ of norm $\sqrt{2}v$.

5.2   Lattice BGK Models

We now return to the LB ❯ Eq. 22

$$f_i(\mathbf{r} + \delta_t \mathbf{v}_i, t + \delta_t) = f_i(\mathbf{r}, t) + \Omega_i(f(\mathbf{r}, t)) \tag{28}$$

and consider a special family of collision terms: the so-called *single relaxation time* models, also termed lattice BGK models (LBGK) for its correspondence with the BGK form of the continuous Boltzmann equation (Bhatnager et al. 1954).

Although more sophisticated models exist (d'Humières et al. 2002; Latt and Chopard 2006; Chikatamarla et al. 2006), the LBGK is still the most popular version of an LB model. It reads

$$f_i(\mathbf{r} + \delta_t \mathbf{v}_i, t + \delta_t) = f_i(\mathbf{r}, t) + \frac{1}{\tau}\left(f_i^{\text{eq}} - f_i\right) \tag{29}$$

where f^{eq} is called the *local equilibrium* distribution; it is a given function, which depends on the phenomena that one wants to model (note that when one refers to all f_i or all f_i^{eq}, the subscript i is dropped). The quantity τ is the so-called *relaxation time*. It is a parameter of the model, which is actually related to the transport coefficient of the model: viscosity for a fluid model, diffusion constant in case of a diffusion model.

In ❯ Eq. 29, it is important to note that the local equilibrium distribution f^{eq} depends on space and time only through the conserved quantities. This is a common assumption of statistical physics. In a hydrodynamic process, where both mass and momentum are conserved, f^{eq} will then be a function of ρ and $\mathbf{u}$.

Thus, in ❯ Eq. 29, to compute $f_i(\mathbf{r} + \delta_t \mathbf{v}_i, t + \delta_t)$ from the $f_i(\mathbf{r}, t)$ one first has to compute $\rho = \sum f_i$ and $\mathbf{u} = (1/\rho)\sum f_i \mathbf{v}_i$ before computing $f_i^{\text{eq}}(\rho, \mathbf{u})$. Only then can f_i be updated.

It is beyond the scope of this chapter to show the equivalence between the LB model and the differential equations representing the corresponding physical phenomena. This derivation requires rather heavy mathematical calculations and can be found in several textbooks. See for instance Chopard and Droz (1998), Chopard et al. (2002), and Lätt (2007) for a derivation based on the so-called multiscale *Chapman–Enskog* formalism. Or, see Junk et al. (2005) for a derivation based on the *asymptotic expansion*. Here the important results are simply given, without demonstration.

5.2.1   LBGK Fluid Models

A first central ingredient of LB models is to properly enforce the physical conservation laws in the collision term. Hydrodynamics is characterized by mass and momentum conservation, which, in the differential equation language, are expressed by the *continuity* and *Navier–Stokes* equations.

From ❯ Eqs. 23 and ❯ 24, conservation laws impose conditions on f_i^{eq} when an LBGK model is considered, namely

$$\sum_i f_i^{eq} = \sum_i f_i = \rho \qquad \sum_i \mathbf{v}_i f_i^{eq} = \sum_i \mathbf{v}_i f_i = \rho \mathbf{u} \tag{30}$$

In addition, in order to recover a hydrodynamic behavior, one imposes that $\Pi_{\alpha\beta}^{eq}$, the second moment of f^{eq}, which is the non-dissipative part of the momentum tensor, has the standard Euler form

$$\Pi_{\alpha\beta}^{eq} = \sum_i f_i^{eq} v_{i\alpha} v_{i\beta} = p\delta_{\alpha\beta} + \rho u_\alpha u_\beta \tag{31}$$

where p is the pressure.

Using ❯ Eqs. 25 and ❯ 26, it is easy to show that the following expression for f^{eq} satisfies the conservation laws (❯ 30)

$$f_i^{eq} = f_i^{eq}(\rho, \mathbf{u}) = \rho w_i \left(1 + \frac{v_{i\alpha} u_\alpha}{c_s^2} + \frac{1}{2c_s^4} Q_{i\alpha\beta} u_\alpha u_\beta\right) \tag{32}$$

where the Q_i's are tensors whose spatial components are

$$Q_{i\alpha\beta} = v_{i\alpha} v_{i\beta} - c_s^2 \delta_{\alpha\beta} \tag{33}$$

Note that in this equation and in what follows, the Einstein summation convention is used over repeated Greek indices

$$v_{i\alpha} u_\alpha = \sum_{\alpha=x,y,z} v_{i\alpha} u_\alpha$$

and

$$Q_{i\alpha\beta} u_\alpha u_\beta \equiv \sum_{\alpha,\beta \in \{x,y,z\}} Q_{i\alpha\beta} u_\alpha u_\beta$$

Note that f_i^{eq} can also be interpreted as a discretization of the Maxwell–Boltzmann distribution function of statistical physics.

One can also check that the second moment of ❷ Eq. 32 gives the correct expression for the Euler momentum tensor (❷ Eq. 31), provided that the pressure is related to the density ρ through an ideal gas relation

$$p = \rho c_s^2$$

From this expression, one can interpret the lattice parameter c_s as the speed of sound.

The fact that, in an LB model, the pressure is directly obtained from the density is an important observation. It means that in an LB fluid model, there is no need to solve a (nonlocal) Poisson equation for the pressure, as is the case when solving Navier–Stokes equations.

Using expression (❷ 32) for f^{eq}, the behavior of the LB model (❷ 29) can be analyzed mathematically with, for instance a Chapman–Enskog method. Several important results are obtained.

It is found that, to order δ_t^2 and δ_r^2, and for small Mach number ($\mathbf{u} \ll c_s$), the LB dynamics implies that ρ and $\mathbf{u}$ obey the continuity equation

$$\partial_t \rho + \partial_\alpha \rho u_\alpha = 0 \tag{34}$$

and the Navier–Stokes equation

$$\partial_t \mathbf{u} + (\mathbf{u} \cdot \nabla)\mathbf{u} = -\frac{1}{\rho}\nabla p + \nu \nabla^2 \mathbf{u} \tag{35}$$

with a kinematic viscosity ν depending on the relaxation time τ as

$$\nu = c_s^2 \delta_t (\tau - 1/2)$$

The last question that needs to be addressed here is how to obtain the expression of f in terms of the hydrodynamic quantities. It is easy to obtain ρ and $\mathbf{u}$ from the f_i, using ❷ Eq. 19. But the inverse relations, expressing f_i as a function of ρ and $\mathbf{u}$ is more difficult. There are $z + 1$ variables f_i and only $1 + d$ hydrodynamic quantities in d dimensions. However, in the hydrodynamic limit, it turns out that the derivatives of $\mathbf{u}$ are precisely the missing pieces of information.

The first step is to split the density distributions f_i as

$$f_i = f_i^{eq} + f_i^{neq} \qquad \text{assuming } f_i^{neq} \ll f_i^{eq}$$

where f_i^{eq}, by its definition (❷ 32) is already a function of ρ and $\mathbf{u}$

$$f_i^{eq} = f_i^{eq}(\rho, \mathbf{u}) = \rho w_i \left(1 + \frac{\mathbf{v}_i \cdot \mathbf{u}}{c_s^2} + \frac{1}{2c_s^4} Q_{i\alpha\beta} u_\alpha u_\beta\right) \tag{36}$$

The Chapman–Enskog expansion then gives

$$f_i^{neq} = -\delta_t \tau \frac{w_i}{c_s^2} \rho Q_{i\alpha\beta} \partial_\alpha u_\beta = -\delta_t \tau \frac{w_i}{c_s^2} \rho Q_{i\alpha\beta} S_{\alpha\beta} \tag{37}$$

where $S_{\alpha\beta} = (1/2)(\partial_\alpha u_\beta - \partial_\beta u_\alpha)$ is the so-called strain rate tensor. As one can see from this relation, the derivatives of $\mathbf{u}$ are part of the LB variables.

By taking the second moment of ❷ Eq. 37 one can obtain $\Pi_{\alpha\beta}^{neq}$, the nonequilibrium part of the momentum tensor. Due to the lattice properties (❷ 25) and (❷ 26),

$$\Pi_{\alpha\beta}^{neq} = \sum_i v_{i\alpha} v_{i\beta} f_i^{neq} = -2\delta_t \tau c_s^2 \rho S_{\alpha\beta} \tag{38}$$

Thus, the strain rate tensor can be directly obtained from $f^{\mathrm{neq}} = f - f^{\mathrm{eq}}$, without the need to compute finite differences.

Therefore, with ❷ Eqs. 19, ❷ 36, and ❷ 37, a relation that allows one to translate the hydrodynamic quantities to LB quantities and vice-versa is established

$$
\begin{pmatrix} \rho \\ \mathbf{u} \\ S_{\alpha\beta} \end{pmatrix} \leftrightarrow (f_i) \tag{39}
$$

They are valid in the hydrodynamic regime, with $u \ll c_s$ and for $\delta_t \to 0$ and $\delta_r \to 0$.

In order to illustrate the LBGK method, ❷ Fig. 26 shows an example of an LB fluid simulation, the flow past an obstacle.

5.3　Diffusion and Reaction–Diffusion LBGK Models

It is actually very easy to devise an LBGK model to describe other physical processes. The basic equation remains the same, namely

$$
f_i(\mathbf{r} + \delta_t \mathbf{v}_i, t + \delta_t) = f_i(\mathbf{r}, t) + \frac{1}{\tau}\left(f_i^{\mathrm{eq}} - f_i\right) \tag{40}
$$

What changes is the expression for f^{eq} and also the isotropy requirements of the lattice. For instance, for a diffusion model, ❷ Eq. 25 is sufficient because diffusion does not involve fourth-order isotropy constraints. In 2D, a D2Q4 lattice (having $w_i = 1/4$, $c_s^2/v^2 = 1/2$) is enough to model a diffusion process and, in 3D, a D3Q6 lattice ($w_i = 1/6$, $c_s^2/v^2 = 1/2$) has enough isotropy.

◻ **Fig. 26**
Several stages (from left to right and top to bottom) of the evolution of a flow past an obstacle, using a D2Q9 LBGK fluid model.

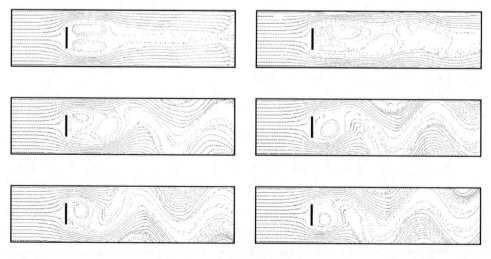

5.3.1 Diffusion

To build the local equilibrium distribution corresponding to a diffusion model, we notice that only the density $\rho = \sum f_i$ is conserved in the process. Then, it is found that

$$f_i^{eq} = w_i\rho \tag{41}$$

produces the diffusion equation

$$\partial_t\rho = D\nabla^2\rho$$

where

$$D = c_s^2\delta_t\left(\tau - \frac{1}{2}\right)$$

is the diffusion coefficient. Also, it is found that the particle current $\mathbf{j} = -D\nabla\rho$ can be computed from the non-equilibrium part of the distribution function, $f^{neq} = f - f^{eq} = -\tau\,\delta_t w_i v_{i\alpha}\,\partial_\alpha\rho$ as

$$\mathbf{j} = \left(1 - \frac{1}{2\tau}\right)\sum_i \mathbf{v}_i f_i^{neq} = \left(1 - \frac{1}{2\tau}\right)\sum_i \mathbf{v}_i f_i$$

Advection–diffusion processes

$$\partial_t\rho + \partial_\alpha\rho u_\alpha = D\nabla^2\rho$$

where $\mathbf{u}(\mathbf{r}, t)$ is a given velocity field, can be modeled by adding an additional term to the local equilibrium (❷ Eq. 41)

$$f_i^{eq} = w_i\rho\left(1 + \frac{1}{c_s^2}\mathbf{u}\cdot\mathbf{v}_i\right) \tag{42}$$

If $\mathbf{u}$ is the solution of a fluid flow, it is appropriate to add a term $w_i\rho(1/(2c_s^4))Q_{i\alpha\beta}u_\alpha u_\beta$ to ❷ Eq. 42. See Van der Sman and Ernst (2000), Suga (2001), Chopard et al. (2009), Ginzburg (2005), and Servan-Camas and Tsai (2008) for more details.

5.3.2 Reaction–Diffusion

In order to simulate a reaction–diffusion process

$$\partial_t\rho = D\nabla^2\rho + R(\rho)$$

where R is any reaction term, the LB diffusion model can be modified as

$$f_i(\mathbf{r} + \delta_t\mathbf{v}_i, t + \delta_t) = f_i(\mathbf{r}, t) + \frac{1}{\tau}(f_i^{eq} - f_i) + \delta_t w_i R(\rho) \tag{43}$$

For instance, the reaction $A + B \to C$ where C is some inert product can be simulated by two sets of equations (❷ Eq. 43), one for species A and one for species B. The reaction term is chosen as $R = -k\rho_A\rho_B$. ❷ *Figure 27* shows the evolution of the concentrations of both species when they are initially mixed.

Another example of a reaction–diffusion process is shown in ❷ *Fig. 28*. See Chopard et al. (1994) for more details.

■ Fig. 27

A reaction–diffusion model simulating the A + B → C reaction. The left panel shows an early stage and the right panel a later stage of the reaction process. Red denotes regions very rich in A, yellow indicates regions where A dominates over B. Similarly, green regions are those very rich in B and blue those where B is slightly more abundant than A.

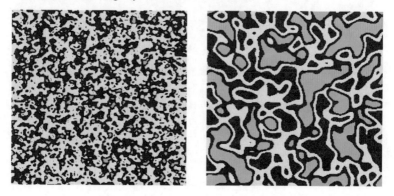

■ Fig. 28

A reaction–diffusion model simulating the formation of Liesegang bands in an A + B → C process, where C can precipitate and forms bands at positions that have interesting geometrical properties.

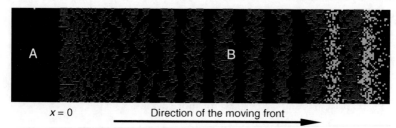

5.4   Wave Propagation

Finally, *wave propagation* can also be described by an LBGK model (Chopard and Droz 1998; Chopard et al. 2002). In this case, it is a must to choose $\tau = 1/2$, which ensures the time reversibility of the LB dynamics (Chopard and Droz 1998), a required symmetry of the wave equation.

Wave processes have two conserved quantities, $\rho = \sum f_i$, which can be any scalar quantity obeying a wave process, and its current $\mathbf{j} = \sum f_i \mathbf{v}_i$. However, as opposed to flow and diffusion models, the f_i are no longer positive quantities and they oscillate between a minimal negative value and a maximum positive value.

The appropriate form of the local equilibrium is found to be

$$f_i^{\text{eq}} = w_i \rho + w_i \frac{\mathbf{j} \cdot \mathbf{v}_i}{c_s^2}$$

$$f_0^{\text{eq}} = w_0 \rho$$

(44)

In the continuous limit, ➋ Eq. 40 with ➋ Eq. 44 yield

$$\partial_t \rho + \partial_\beta j_\beta = 0$$
$$\partial_t j_\alpha - c_s^2 \partial_\alpha \rho = 0 \qquad (45)$$

When combined, these two equations give the wave equation

$$\partial_t^2 \rho - c_s^2 \nabla^2 \rho = 0 \qquad (46)$$

The rest population f_0 allows us to adjust the speed of the wave from place to place by having the value of w_0 depend on spatial location $\mathbf{r}$ or time t.

As for the diffusion case, a second-order isotropy is enough for the wave equation. Therefore D2Q5 and D3Q7 lattices are appropriate. For other topologies, it is important to use the same weight $w_i = w$ for all nonzero velocities. It means that the second-order isotropy condition now reads

$$\sum_{i \geq 1} v_{i\alpha} v_{i\beta} = z c_{max}^2 \delta_{\alpha\beta} \qquad (47)$$

for some coefficient c_{max} and for z the lattice coordination number. For D2Q5 and D3Q7, it is easy to show that $c_{max}^2 / v^2 = 2/z$.

When $w_i = w$, one can conclude from the condition

$$1 = \sum_{i \geq 0} w_i = w_0 + zw$$

that

$$w = \frac{1 - w_0}{z} \qquad (48)$$

Therefore

$$\sum_{i \geq 1} w_i v_{i\alpha} v_{i\beta} = (1 - w_0) c_{max}^2 \delta_{\alpha\beta} \qquad (49)$$

and we obtain

$$c_s^2 = (1 - w_0) c_{max}^2 \qquad (50)$$

A consequence of this relation is that $w_0 < 1$ must be imposed. Note that this way of adjusting c_s is only possible for processes that do not require fourth-order isotropy. For a fluid model, there is however a way to adjust the speed of sound (Alexander et al. 1992; Yu and Zhao 2000; Chopard et al. 2002).

The numerical stability of the LBGK wave model is guaranteed because a quantity, termed *energy*,

$$E = \frac{w}{w_0} f_0^2 + \sum_{i \geq 1} f_i^2 \qquad (51)$$

is conserved during the collision step

$$E^{out} = \frac{w}{w_0} \left(f_0^{out} \right)^2 + \sum_{i \geq 1} \left(f_i^{out} \right)^2 = \frac{w}{w_0} \left(f_0^{in} \right)^2 + \sum_{i \geq 1} \left(f_i^{in} \right)^2 = E^{in} \qquad (52)$$

The proof of this condition also requires that $w_i = w$ for $i \neq 0$. When $(w/w_0) \geq 0$, the numerical stability of the model is guaranteed because, with E given, the f_i cannot diverge to $\pm\infty$.

With $0 \le w_0 < 1$ we obtain from (❯ Eq. 50) that the speed of the wave is such that

$$0 \le c_s \le c_{max}$$

the maximum speed being achieved with $w_0 = 0$ that is, without a rest population f_0. The refraction index is defined as

$$n = \frac{c_{max}}{c_s}$$

Using ❯ Eqs. 48 and ❯ 50, one then obtains

$$w = \frac{1}{z}\frac{c_s^2}{c_{max}^2} = \frac{1}{n^2 z} \qquad w_0 = 1 - \frac{c_s^2}{c_{max}^2} = \frac{n^2 - 1}{n^2} \tag{53}$$

With the above value of w_i and the fact that $\tau = 1/2$, the LBGK wave model can also be written as

$$f_i^{out} = \frac{2}{n^2 z}\rho + \frac{2}{z c_{max}^2}\mathbf{v}_i \cdot \mathbf{j} - f_i$$

$$f_0^{out} = 2\frac{n^2 - 1}{n^2}\rho - f_0 \tag{54}$$

❯ *Figure 29* illustrates this model with a D2Q5 lattice. A plane wave is produced on the left side of the domain and propagates to the right where it penetrates in a lens-shaped media with slower propagation speed.

Note that in Chopard and Droz (1998), Chopard et al. (2002), Marconi and Chopard (2003) the above model is investigated for its capability to model deformable elastic solids. ❯ *Figure 30* illustrates an application of the wave model to describe a fracture process.

Finally, note that the LBGK wave model is also known in the literature as the transmission line matrix model (Hoeffer 1985) and has been derived in several different contexts (Vanneste et al. 1992).

◻ Fig. 29

LB model simulating the propagation of a wave in a lens. The colors represent the energy *E*.

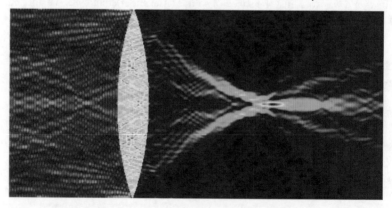

◘ Fig. 30
Fracture process in an LB model for an elastic solid body.

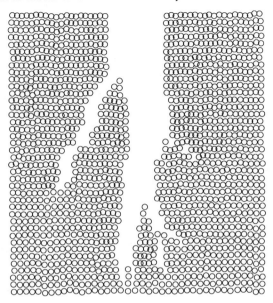

◘ Fig. 31
At the boundary of the domain, the density distributions coming from outside the system are not known. Specific boundary conditions must be used to define them.

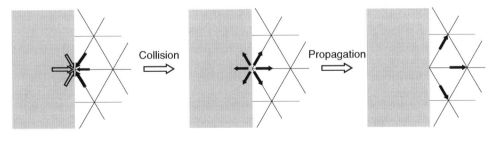

5.5 Boundary Conditions

Boundary conditions are an important aspect of an LB model. It is not an easy question to properly specify the values of the distributions f_i at the limit of the computational domain. Clearly, to apply the collision phase, all f_i must be defined. But at a boundary cell, the propagation phase does not provide any information from outside the domain. This is illustrated in ❷ *Fig. 31*. These unknown distributions must be specified according to the desired behavior of the system at the boundary.

Following this procedure, the time evolution of an LB model can then be represented by the following loop in a computer program:

```
for t=0 to tmax
  boundary
  observation
  collision
  propagation
endfor
```

Note that here a new operation, termed *observation* in the LB execution loop, has been introduced. This is where measurement can be done on the system. It is important to remember that f_i actually denotes f^{in} so that the theoretical results showing the correspondence between LB and physical quantities are only valid after the propagation step and before collision. Obviously, quantities such as ρ that are conserved during the collision can be measured after collision too. But, in hydrodynamics, this is not the case of the strain rate $S_{\alpha\beta}$.

In what follows, we focus on the discussion on simple hydrodynamical LB models. In practice, the way to impose a boundary condition is to use the correspondence expressed in ❯ Eq. 39 to build the missing f_i from the desired values of the fluid variables at the boundary.

A standard situation is shown in ❯ *Fig. 32*, for a D2Q9 lattice. The density distributions f_2, f_3, and f_4 must be determined before applying the collision operator. Assuming that the y-axis is vertical and pointing upward, we have the following relations (in lattice units where $v = 1$)

$$\rho = f_0 + f_1 + f_2 + f_3 + f_4 + f_5 + f_6 + f_7 + f_8$$
$$\rho u_y = f_2 + f_3 + f_4 - f_6 - f_7 - f_8 \tag{55}$$

From these equations, we get

$$\rho - \rho u_y = f_0 + f_1 + f_5 + 2(f_6 + f_7 + f_8) \tag{56}$$

The right-hand side of this equation is fully known.

◻ **Fig. 32**
For a flat wall at the lower boundary of a D2Q9 lattice, the distributions f_2, f_3, and f_4 are unknown and must be computed according to the desired boundary condition.

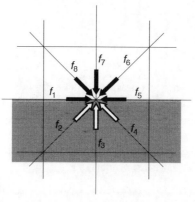

In case $\mathbf{u} = (u_x, u_y)$ is specified at the boundary (velocity boundary condition), we can compute ρ consistently (Inamuro et al. 1995)

$$\rho = \frac{f_0 + f_1 + f_5 + 2(f_6 + f_7 + f_8)}{1 - u_y}$$

Once ρ and $\mathbf{u}$ are known, $f_i^{eq}(\rho, \mathbf{u})$ is also known for all i, due to ❯ Eq. 36. Then, we can certainly compute $f_i^{neq} = f_i - f_i^{eq}$ for $i \in \{0, 1, 5, 6, 7, 8\}$.

From these quantities, we can now compute f_i^{neq} for the missing directions $i = 2, 3, 4$ by imposing that their nonequilibrium part obeys ❯ Eq. 37, which tells us that $f_i^{neq} = f_{opp(i)}^{neq}$, where $j = opp(i)$ is the index such that $\mathbf{v}_j = -\mathbf{v}_i$.

Unfortunately, there is no guarantee that this choice of the nonequilibrium parts of the distributions is globally consistent. For instance, one may well find that $\sum_i f_i^{neq} \neq 0$, which is inconsistent with ❯ Eq. 55. The solution to this problem is a *regularization* step in which the f_i^{neq} are redistributed over all directions. This step can be explained as follows: first we compute

$$\Pi_{\alpha\beta}^{neq} = \sum_i v_{i\alpha} v_{i\beta} f_i^{neq}$$

from the f_i^{neq} obtained above. Second, by combining ❯ Eqs. 37 and ❯ 38 we get

$$f_i^{neq} = -\delta_t \tau \frac{w_i}{c_s^2} \rho Q_{i\alpha\beta} S_{\alpha\beta}$$

$$= \frac{w_i}{2c_s^4} Q_{i\alpha\beta} \Pi_{\alpha\beta}^{neq} \qquad (57)$$

This equation allows one to recompute all $f_i^{neq}(i = 0, \ldots, z)$ from the previous ones. All the f_i are then redefined to their regularized value

$$f_i = f_i^{eq} + f_i^{neq} \qquad i = 0, \ldots, z$$

This terminates the calculation of the boundary condition for $\mathbf{u}$ imposed at the wall and guarantees the proper values of ρ and $\mathbf{u}$ at the wall since $\sum_i w_i Q_{i\alpha\beta} = \sum_i w_i \mathbf{v}_i Q_{i\alpha\beta} = 0$.

If, instead of $\mathbf{u}$, ρ is the prescribed quantity at the boundary (pressure boundary condition), the consistent flow speed u_y can be determined from ❯ Eq. 56

$$u_y = 1 - \frac{f_0 + f_1 + f_5 + 2(f_6 + f_7 + f_8)}{\rho}$$

By choosing for instance $u_x = 0$ we then obtain f_i^{eq}, for all i. Then, the same regularization procedure as explained above can be used to compute all the f_i at the wall.

The reader can refer to Zou and He (1997) and Latt et al. (2008) for a detailed discussion of the above on-site boundary condition and to Bouzidi et al. (2001), Guo et al. (2002b), Lallemand and Luo (2003), and Kao and Yang (2008) for a discussion of boundary conditions that are not located on lattice sites, or moving boundary conditions.

However, another simple solution to specify a boundary condition is to exploit the mesoscopic interpretation of the f_i as particles traveling with velocity $\mathbf{v}_i$. Following this idea, a very popular way to impose a boundary with zero velocity (no-slip condition) is to bounce back the particles from where they came. This is illustrated in ❯ *Fig. 33b*. Using such

◘ **Fig. 33**

Boundary conditions based on the particle interpretation of the LB method. The white arrows represent f^{in} and the black ones indicate f^{out}. (a) A free slip condition (specular reflection). (b) The bounce back rule to create a no-slip boundary condition on a wall.

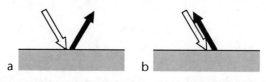

a   b

◘ **Fig. 34**

A lattice Boltzmann simulation of a Rayleigh–Benard convection.

a bounce-back condition actually means a redefinition of the collision operator of the LB model on the boundary cells

$$f_i^{out} = f_{opp(i)}^{in}$$

where $opp(i)$ is the direction such that $\mathbf{v}_i = -\mathbf{v}_{opp(i)}$.

In addition to the bounce-back rule, periodic boundary condition can be used when appropriate. For instance, when simulating the flow in a straight tube, one can sometimes say that the particles leaving the tube through the outlet are reinjected in the inlet. In such a situation it is also convenient to add a body force $\mathbf{F}$ to create and maintain the flow in the tube. The LBGK fluid model can then be modified as follows

$$f_i(\mathbf{r} + \delta_t\mathbf{v}_i, t + \delta_t) = f_i(\mathbf{r}, t) + \frac{1}{\tau}\left(f_i^{eq} - f_i\right) + w_i\frac{\delta_t}{c_s^2}\rho\mathbf{v}_i \cdot \mathbf{F} \qquad (58)$$

in order to reproduce a Navier–Stokes equation with external force

$$\partial_t\mathbf{u} + (\mathbf{u} \cdot \nabla)\mathbf{u} = -\frac{1}{\rho}\nabla p + \nu\nabla^2\mathbf{u} + \mathbf{F} \qquad (59)$$

Note that this way of adding a body force on an LBGK fluid model is only possible if $\mathbf{F}$ is constant in space and time. For a more general body force, refer to Guo et al. (2002a).

5.6   Examples of LB Modeling

The LB method has been used in many cases, from sediment transport to blood flow and clotting processes, to shallow water flow, just to mention a few applications. For the sake of illustration, the result of some simulations obtained with an LB model for complex flows is shown in this section without detailed explanations. More examples and on-line animations can be found on the Web (http://cui.unige.ch/~chopard/CA/Animations/root.html and http://www.lbmethod.org) and in a vast body of literature on LB models.

■ **Fig. 35**
Rayleigh–Taylor instability with two immiscible LB fluids. The blue fluid is heavier than the black one.

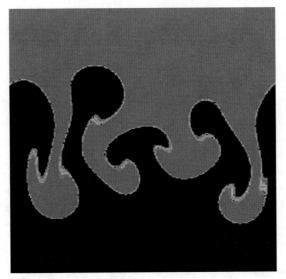

■ **Fig. 36**
Evolution of bubbles in a shear flow.

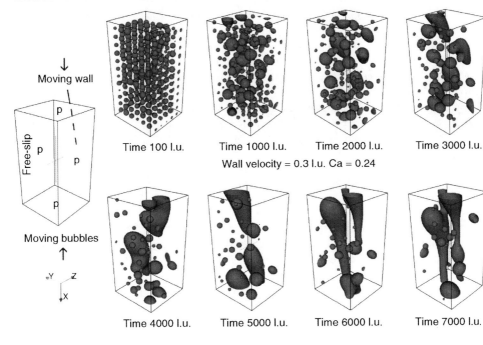

◘ Fig. 37

Bubbles rising in a fluid inside a porous media.

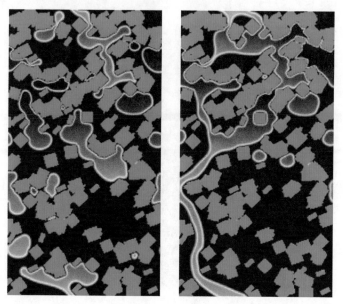

By coupling an advection–diffusion LB model for a temperature field with a fluid LB model with gravity force, it is possible to model a thermal fluid in the Boussinesq approximation and, in particular, the well-known Rayleigh–Benard convection (instability that appears when a fluid is heated from below). This is illustrated in ❷ *Fig. 34*.

An interesting capability of the LB fluid models is to consider multicomponent flows. In short, two-fluid models are implemented with, in addition, a repulsive coupling between them. Several ways exist to create the separation of the components (Martis and Chen 1996; Swift et al. 1996). ❷ *Figure 35* shows a simulation of the Rayleigh–Taylor instability, which occurs when a heavy fluid layer is on top of a lighter one. ❷ *Figure 36* represents the evolution of droplets in suspension in another fluid subject to a shear flow.

❷ *Figure 37* illustrates a two-component system (gas+liquid), in a porous media, with a buoyancy force field. Bubbles of gas are produced at the lower part of the system and they rise inside a porous media.

6 Conclusions

In this chapter, the concepts underlying the CA and LB approaches are presented. The ideas have been developed using a mesoscopic level of description. Physical systems are represented by idealized entities evolving on a discrete space–time universe. Implementing the correct conservation laws and symmetry in this virtual discrete mesoscopic universe ensures that the observed macroscopic behavior of the system is close enough to that of the corresponding physical system.

The clear relation that exists between the model and the real process makes CA and LB methods very intuitive. As such, they offer a new language to describe physical systems (as illustrated in ❷ *Fig. 1*), which has proven quite powerful in interdisciplinary research.

The simplicity and flexibility of the approach makes it very appropriate to describe many complex systems for which more traditional numerical methods are difficult to apply.

CA are by definition fully discrete numerical models. Usually, they have only a few possible states per cell and they can be implemented very efficiently on low cost dedicated hardware. Since CA models only require integer values, their numerical implementation is exact. No numerical instabilities or truncation error affect the simulation.

LB models offer a higher level of abstraction and consider real-valued variables. They offer much more flexibility than CA to implement a given interaction rule. However, they can be numerically unstable. LB methods are now acknowledged as a powerful way to simulate hydrodynamics and specifically complex fluids in time-dependent regimes. The reader interested to use the LB method can learn more in the several textbooks mentioned earlier and by exploring the Web site www.lbmethod.org.

A current challenging issue in computational science is multiscale, multiscience modeling. Many real-life problems cover a wide range of spatial and temporal scales and include different processes. Biomedical applications are a good example where physical processes (e.g., hemodynamics) interact with much slower biological ones at a microscopic scale (e.g., tissue modifications). Often, scales can be separated in the sense that the full system can be represented as several coupled submodels, each of them corresponding to a given scale and a given process.

Coupling different CA and LB models together is possible and has been investigated recently (Hoekstra et al. 2008a, b) using the concept of CxA (*Complex Automata*). A CxA is a graph whose nodes are CA or LB models and the edges implement the coupling strategies. A software framework (Hegewald et al. 2008) offers a way to realize such a graph of submodels and to simulate challenging biomedical applications (Evans et al. 2008).

Acknowledgments

I thank Jonas Latt, Orestis Malaspinas, Andrea Parmigiani, and Chris Huber for stimulating discussions and for providing many of the figures illustrating ❷ Sect. 5.6.

References

Alexander FJ, Chen H, Chen S, Doolen GD (Aug 1992) Lattice Boltzmann model for compressible fluids. Phys Rev A 46(4):1967–1970

Ansumali S, Karlin IV, Arcidiacono S, Abbas A, Prasianakis NI (2007) Hydrodynamics beyond Navier-Stokes: exact solution to the lattice Boltzmann hierarchy. Phys Rev Lett 98:124502

Banks E (1971) Information processing and transmission in cellular automata. Technical report, MIT. MAC TR-81

Bhatnagar P, Gross EP, Krook MK (1954) A model for collision process in gases. Phys Rev 94:511

Bouzidi M, Firdaouss M, Lallemand P (2001) Momentum transfer of a Boltzmann-lattice fluid with boundaries. Phys Fluids 13(11):3452–3459

Burks AW (1970) Von Neumann's self-reproducing automata. In: Burks AW (ed) Essays on cellular automata, University of Illinois Press, Urbana, IL, pp 3–64

Chen S, Doolen GD (1998) Lattice Boltzmann methods for fluid flows. Annu Rev Fluid Mech 30:329

Chikatamarla SS, Ansumali S, Karlin IV (2006) Entropic lattice Boltzmann models for hydrodynamics in three dimensions. Phys Rev Lett 97:010201

Chopard B, Droz M (1998) Cellular automata modeling of physical systems. Cambridge University Press, Cambridge, UK

Chopard B, Dupuis A (2003) Cellular automata simulations of traffic: a model for the city of Geneva. Netw Spatial Econ 3:9–21

Chopard B, Falcone J-L, Latt J (2009) The lattice Boltzmann advection-diffusion model revisited. Eur Phys J 171:245–249

Chopard B, Luthi P, Droz M (1994) Reaction-diffusion cellular automata model for the formation of Liesegang patterns. Phys Rev Lett 72(9): 1384–1387

Chopard B, Luthi PO, Queloz P-A (1996) Cellular automata model of car traffic in two-dimensional street networks. J Phys A 29:2325–2336

Chopard B, Luthi P, Masselot A, Dupuis A (2002) Cellular automata and lattice Boltzmann techniques: an approach to model and simulate complex systems. Adv Complex Syst 5(2):103–246. http://cui.unige.ch/~chopard/FTP/CA/acs.pdf

Culick K, Yu S (1988) Undecidability of CA classification scheme. Complex Syst 2:177–190

Deutsch A, Dormann S (2005) Cellular automaton modeling of biological pattern formation. Birkäuser, Boston, MA

d'Humières D, Ginzburg I, Krafczyk M, Lallemand P, Luo L-S (2002) Multiple-relaxation-time lattice Boltzmann models in three dimensions. Phil Trans R Soc A 360:437–451

Evans D, Lawford P-V, Gunn J, Walker D, Hose D-R, Smallwood RH, Chopard B, Krafczyk M, Bernsdorf J, Hoekstra A (2008) The application of multi-scale modelling to the process of development and prevention of stenosis in a stented coronary artery. Phil Trans R Soc 366(1879):3343–3360

Frisch U, Hasslacher B, Pomeau Y (1986) Lattice-gas automata for the Navier-Stokes equation. Phys Rev Lett 56:1505

Galam S, Chopard B, Masselot A, Droz M (1998) Competing species dynamics: qualitative advantage versus geography. Eur Phys J B 4:529–531

Gardner M (1970) The fantastic combinations of John Conway's new solitaire game "Life." Sci Am 220(4):120

Gaylord RJ, Nishidate K (1996) Modeling nature with cellular automata using Mathematica. Springer, New York

Ginzburg I (2005) Equilibrium-type and link-type lattice Boltzmann models for generic advection and anisotropic-dispersion equation. Adv Water Resour 28(11):1171–1195

Guo Z, Zheng C, Shi B (2002a) Discrete lattice effects on forcing terms in the lattice Boltzmann method. Phys Rev E 65:046308

Guo Z, Zheng C, Shi B (2002b) An extrapolation method for boundary conditions in lattice Boltzmann method. Phys Fluids 14:2007–2010

He X, Luo L-S (1997) A priori derivation of the lattice Boltzmann equation. Phys Rev E 55:R6333–R6336

Hegewald J, Krafczyk M, Tölke J, Hoekstra A, Chopard B (2008) An agent-based coupling platform for complex automata. In: Bubak M et al (ed) ICCS 2008, vol LNCS 5102, Springer, Berlin, Germany, pp 291–300

Hoeffer WJR (Oct 1985) The transmission-line matrix method. theory and applications. IEEE Trans Microw Theory Tech MTT-33(10):882–893

Hoekstra A, Falcone J-L, Caiazzo A, Chopard B (2008a) Multi-scale modeling with cellular automata: the complex automata approach. In: Umeo H et al (ed) ACRI 2008, vol LNCS 5191, Springer, Berlin, Germany, pp 192–199

Hoekstra A, Lorenz E, Falcone J-L, Chopard B (2008b) Towards a complex automata formalism for multi-scale modeling. Int J Multiscale Comput Eng 5(6):491–502

Ilachinski A (2001) Cellular automata: a discrete universe. World Scientific, River Edge, NJ

Inamuro T, Yoshino M, Ogino F (1995) A non-slip boundary condition for lattice Boltzmann simulations. Phys Fluids 7(12):2928–2930

Junk M, Klar A, Luo L-S (2005) Asymptotic analysis of the lattice Boltzmann equation. J Comput Phys 210(2)

Kanai M, Nishinari K, Tokihiro T (2005) Stochastic optimal velocity model and its long-lived metastability. Phys Rev E 72:035102(R)

Kanai M, Nishinari K, Tokihiro T (2006) Stochastic cellular automaton model for traffic flow. In: El Yacoubi S, Chopard B, Bandini S (eds) Cellular automata: 7th ACRI conference, Perpignan, France, vol 4173. LNCS, Springer, pp 538–547

Kao P-H, Yang R-J (2008) An investigation into curved and moving boundary treatments in the lattice Boltzmann method. J Comput Phy 227(11):5671–5690

Lallemand P, Luo L-S (2003) Lattice Boltzmann method for moving boundaries. J Comput Phys 184(2): 406–421

Lätt J (2007) Hydrodynamic limit of lattice Boltzmann equations. Ph.D. thesis, University of Geneva, Switzerland. http://www.unige.ch/cyberdocuments/theses2007/LattJ/meta.html

Latt J, Chopard B (2006) Lattice Boltzmann method with regularized non-equilibrium distribution functions. Math Comp Sim 72:165–168

Latt J, Chopard B, Malaspinas O, Deville M, Michler A (2008) Straight velocity boundaries in the lattice Boltzmann method. Phys Rev E 77:056703

Luthi PO, Preiss A, Ramsden J, Chopard B (1998) A cellular automaton model for neurogenesis in drosophila. Physica D 118:151–160

Marconi S, Chopard B (2003) A lattice Boltzmann model for a solid body. Int J Mod Phys B 17(1/2):153–156

Martis NS, Chen H (1996) Simulation of multi-component fluids in complex 3D geometries by the lattice Boltzmann method. Phys Rev E 53:743–749

Nagel K, Herrmann HJ (1993) Deterministic models for traffic jams. Physica A 199:254

Nagel K, Schreckenberg M (1992) Cellular automaton model for freeway traffic. J Phys I (Paris) 2:2221

Propp J (1994) Trajectory of generalized ants. Math Intell 16(1):37–42

Rothman D, Zaleski S (1997) Lattice-gas cellular automata: simple models of complex hydrodynamics. Collection Aléa. Cambridge University Press, Cambridge, UK

Schreckenberg M, Schadschneider A, Nagel K, Ito N (1995) Discrete stochastic models for traffic flow. Phys Rev E 51:2939

Schreckenberg M, Wolf DE (ed) (1998) Traffic and granular flow '97. Springer, Singapore

Servan-Camas B, Tsai FTC (2008) Lattice Boltzmann method for two relaxation times for advection-diffusion equation:third order analysis and stability analysis. Adv Water Resour 31:1113–1126

Shan X, Yuan X-F, Chen H (2006) Kinetic theory representation of hydrodynamics: a way beyond the Navier-Stokes equation. J Fluid Mech 550:413–441

Sipper M (1997) Evolution of parallel cellular machines: the cellular programming approach. Springer, Berlin, Germany. Lecture notes in computer science, vol 1194

Stewart I (July 1994) The ultimate in anty-particle. Sci Am 270:88–91

Succi S (2001) The lattice Boltzmann equation, for fluid dynamics and beyond. Oxford University Press, New York

Suga S (2006) Numerical schemes obtained from lattice Boltzmann equations for advection diffusion equations. Int J Mod Phys C 17(11):1563–1577

Sukop MC, Thorne DT (2005) Lattice Boltzmann modeling: an introduction for geoscientists and engineers. Springer, Berlin, Germany

Swift MR, Olrandini E, Osborn WR, Yeomans JM (1996) Lattice Boltzmann simulations of liquid-gas and binary fluid systems. Phys Rev E 54:5041–5052

Toffoli T, Margolus N (1987) Cellular automata machines: a new environment for modeling. The MIT Press, Cambridge, MA

Van der Sman RGM, Ernst MH (2000) Convection-diffusion lattice Boltzmann scheme for irregular lattices. J Comp Phys 160:766–782

Vanneste C, Sebbah P, Sornette D (1992) A wave automaton for time-dependent wave propagation in random media. Europhys Lett 17:715

Vichniac G (1984) Simulating physics with cellular automata. Physica D 10:96–115

Weimar JR (1998) Simulation with cellular automata. Logos, Berlin, Germany

Wolf-Gladrow DA (2000) Lattice-gas cellular automata and lattice Boltzmann models: an introduction. Lecture notes in mathematics, 1725. Springer, Berlin, Germany

Wolfram S (1986) Theory and application of cellular automata. World Scientific, Singapore

Wolfram S (1994) Cellular automata and complexity. Addison-Wesley, Reading MA

Wolfram S (2002) A new kind of science. Wolfram Sciences, New York

Yu H, Zhao K (Apr 2000) Lattice Boltzmann method for compressible flows with high Mach numbers. Phys Rev E 61(4):3867–3870

Yukawa S, Kikuchi M, Tadaki S (1994) Dynamical phase transition in one-dimensional traffic flow model with blockage. J Phys Soc Jpn 63(10):3609–3618

Zou Q, He X (1997) On pressure and velocity boundary conditions for the lattice Boltzmann BGK model. Phys Fluids 7:2998

Neural Computation

Tom Heskes and Joost N. Kok

10 Computing with Spiking Neuron Networks

Hélène Paugam-Moisy[1,2] · *Sander Bohte*[3]
[1]Laboratoire LIRIS – CNRS, Université Lumière Lyon 2, Lyon, France
[2]INRIA Saclay – lle-de-France, Université Paris-Sud, Orsay, France
helene.paugam-moisy@univ-lyon2.fr
hpaugam@lri.fr
[3]Research Group Life Sciences, CWI, Amsterdam, The Netherlands
s.m.bohte@cwi.nl

G. Rozenberg et al. (eds.), *Handbook of Natural Computing*, DOI 10.1007/978-3-540-92910-9_10,
© Springer-Verlag Berlin Heidelberg 2012

Abstract

Spiking Neuron Networks (SNNs) are often referred to as the third generation of neural networks. Highly inspired by natural computing in the brain and recent advances in neurosciences, they derive their strength and interest from an accurate modeling of synaptic interactions between neurons, taking into account the time of spike firing. SNNs overcome the computational power of neural networks made of threshold or sigmoidal units. Based on dynamic event-driven processing, they open up new horizons for developing models with an exponential capacity to memorize and a strong ability to do fast adaptation. Today, the main challenge is to discover efficient learning rules that might take advantage of the specific features of SNNs while keeping the nice properties (general-purpose, easy-to-use, available simulators, etc.) of traditional connectionist models. This chapter relates the history of the "spiking neuron" in ❷ Sect. 1 and summarizes the most currently-in-use models of neurons and synaptic plasticity in ❷ Sect. 2. The computational power of SNNs is addressed in ❷ Sect. 3 and the problem of learning in networks of spiking neurons is tackled in ❷ Sect. 4, with insights into the tracks currently explored for solving it. Finally, ❷ Sect. 5 discusses application domains, implementation issues and proposes several simulation frameworks.

1   From Natural Computing to Artificial Neural Networks

1.1   Traditional Neural Networks

Since the human brain is made up of a great many intricately connected neurons, its detailed workings are the subject of interest in fields as diverse as the study of neurophysiology, consciousness, and of course artificial intelligence. Less grand in scope, and more focused on the functional detail, artificial neural networks attempt to capture the essential computations that take place in these dense networks of interconnected neurons making up the central nervous systems in living creatures.

The original work of McCulloch and Pitts (1943) proposed a neural network model based on simplified "binary" neurons, where a single neuron implements a simple thresholding function: a neuron's state is either "active" or "not active," and at each neural computation step, this state is determined by calculating the weighted sum of the states of all the afferent neurons that connect to the neuron. For this purpose, connections between neurons are directed (*from* neuron N_i *to* neuron N_j), and have a weight (w_{ij}). If the weighted sum of the states of all the neurons N_i connected to a neuron N_j exceeds the characteristic threshold of N_j, the state of N_j is set to active, otherwise it is not (❷ *Fig. 1*, where index j has been omitted).

Subsequent neuronal models evolved where inputs and outputs were real-valued, and the nonlinear threshold function (Perceptron) was replaced by a linear input–output mapping (Adaline) or nonlinear functions like the sigmoid (Multilayer Perceptron). Alternatively, several connectionist models (e.g., RBF networks, Kohonen self-organizing maps (Kohonen 1982; Van Hulle 2000)) make use of "distance neurons" where the neuron output results from applying a transfer function to the (usually quadratic) distance $\| X - W \|$ between the weights W and inputs X, instead of the dot product, usually denoted by $< X, W >$ (❷ *Fig. 2*).

Remarkably, networks of such simple, connected computational elements can implement a wide range of mathematical functions relating input states to output states: With algorithms for setting the weights between neurons, these artificial neural networks can "learn" such relations.

□ Fig. 1

The first model of a neuron picked up the most significant features of a natural neuron: All-or-none output resulting from a nonlinear transfer function applied to a weighted sum of inputs.

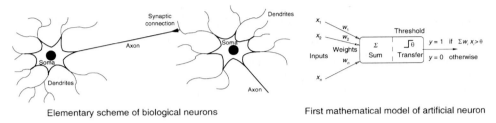

Elementary scheme of biological neurons   First mathematical model of artificial neuron

□ Fig. 2

Several variants of neuron models, based on a dot product or a distance computation, with different transfer functions.

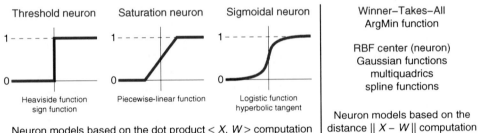

A large number of learning rules have been proposed, both for teaching a network explicitly to perform some task (supervised learning), and for learning interesting features "on its own" (unsupervised learning). Supervised learning algorithms include gradient descent algorithms (e.g., error backpropagation) (Rumelhart et al. 1986) that fit the neural network behavior to some target function. Many ideas on local unsupervised learning in neural networks can be traced back to the original work on synaptic plasticity by Hebb (1949), and his famous, oft-repeated quote:

▶ When an axon of cell A is near enough to excite cell B or repeatedly or persistently takes part in firing it, some growth process or metabolic change takes place in one or both cells such that A's efficiency, as one of the cells firing B, is increased.

Unsupervised learning rules inspired by this type of natural neural processing are referred to as Hebbian rules (e.g., in Hopfield's (1982) network model).

In general, artificial neural networks (NNs) have been proved to be very powerful as engineering tools in many domains (pattern recognition, control, bioinformatics, and robotics), and also in many theoretical cases:

• Calculability: NNs computational power outperforms a Turing machine (Siegelmann 1999).

• Complexity: The "loading problem" is NP-complete (Blum and Rivest 1989; Judd 1990).

- Capacity: Multilayer Perceptrons (MLP), Radial Basis Function (RBF) network, and Wavelet Neural Networks (WNN) are universal approximators (Cybenko 1988; Funahashi 1989; Hornik et al. 1989).
- Regularization theory (Poggio and Girosi 1989); Probably Approximately Correct learning (PAC-learning) (Valiant 1984); Statistical learning theory, Vapnik–Chervonenkis dimension (VC-dimension), and Support Vector Machines (SVM) (Vapnik 1998).

Nevertheless, traditional neural networks suffer from intrinsic limitations, mainly for processing large amounts of data or for fast adaptation to a changing environment. Several characteristics, such as iterative learning algorithms or artificially designed neuron models and network architectures, are strongly restrictive compared with biological processing in naturals neural networks.

1.2 The Biological Inspiration, Revisited

A new investigation in natural neuronal processing is motivated by the evolution of thinking regarding the basic principles of brain processing. When the first neural networks were modeled, the prevailing belief was that intelligence is based on reasoning, and that logic is the foundation of reasoning. In 1943, McCulloch and Pitts designed their model of the neuron in order to prove that the elementary components of the brain were able to compute elementary logic functions: Their first application of thresholded binary neurons was to build networks for computing Boolean functions. In the tradition of Turing's work (1939, 1950), they thought that complex, "intelligent" behavior could emerge from a very large network of neurons, combining huge numbers of elementary logic gates. History shows that such basic ideas have been very productive, even if effective learning rules for large networks (e.g., backpropagation for MLP) were discovered only at the end of the 1980s, and even if the idea of Boolean decomposition of tasks has been abandoned for a long time.

Separately, neurobiological research has greatly progressed. Notions such as associative memory, learning, adaptation, attention, and emotions have unseated the notion of logic and reasoning as being fundamental to understand how the brain processes information. *Time* has become a central feature in cognitive processing (Abeles 1991). Brain imaging and a host of new technologies (microelectrode, LFP (local field potential) or EEG (electroencephalogram) recordings, fMRI (functional magnetic resonance imaging)) can now record rapid changes in the internal activity of the brain, and help elucidate the relation between the brain activity and the perception of a given stimulus. The current consensus is that cognitive processes are most likely based on the activation of transient assemblies of neurons (see ❷ Sect. 3.2), although the underlying mechanisms are not yet well understood.

With these advances in mind, it is worth recalling some neurobiological detail: real neurons spike, at least most biological neurons rely on pulses as an important part of information transmission from one neuron to another neuron. In a rough and non-exhaustive outline, a neuron can generate an action potential – the *spike* – at the soma, the cell body of the neuron. This brief electric pulse (1 or 2 ms duration) then travels along the neuron's axon, which in turn is linked to the receiving end of other neurons, the dendrites (see ❷ *Fig. 1*, left view). At the end of the axon, synapses connect one neuron to another, and at the arrival of each individual spike the synapses may release neurotransmitters along the *synaptic* cleft. These neurotransmitters are taken up by the neuron at the receiving end, and modify the state of that *postsynaptic* neuron, in particular the *membrane potential*, typically making the neuron more or less likely to fire for some duration of time.

The transient impact a spike has on the neuron's membrane potential is generally referred to as the *postsynaptic potential*, or *PSP*, and the PSP can either inhibit the future firing – inhibitory postsynaptic potential, *IPSP* – or excite the neuron, making it more likely to fire – an excitatory postsynaptic potential, *EPSP*. Depending on the neuron, and the specific type of connection, a PSP may directly influence the membrane potential for anywhere between tens of microseconds and hundreds of milliseconds. A brief sketch of the typical way a *spiking neuron* processes is depicted in ❷ *Fig. 3*. It is important to note that the firing of a neuron may be a deterministic or stochastic function of its internal state.

Many biological details are omitted in this broad outline, and they may or may not be relevant for computing. Examples are the stochastic release of neurotransmitter at the synapses: depending on the firing history, a synaptic connection may be more or less reliable, and more or less effective. Inputs into different parts of the dendrite of a neuron may sum nonlinearly, or even multiply. More detailed accounts can be found in, for example, Maass and Bishop (1999).

Evidence from the field of neuroscience has made it increasingly clear that in many situations information is carried in the individual action potentials, rather than aggregate

◻ **Fig. 3**
A model of spiking neuron: N_j fires a spike whenever the weighted sum of incoming EPSPs generated by its presynaptic neurons reaches a given threshold. The graphic (*bottom*) shows how the membrane potential of N_j varies through time, under the action of the four incoming spikes (*top*).

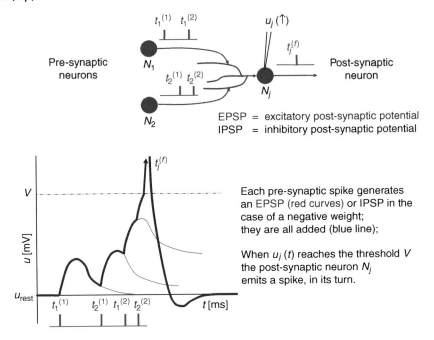

Each pre-synaptic spike generates an EPSP (red curves) or IPSP in the case of a negative weight; they are all added (blue line);

When $u_j(t)$ reaches the threshold V the post-synaptic neuron N_j emits a spike, in its turn.

measures such as "firing rate." Rather than the form of the action potential, it is the number and the timing of spikes that matter. In fact, it has been established that *the exact timing of spikes* can be a means for coding information, for instance in the electrosensory system of electric fish (Heiligenberg 1991), in the auditory system of echo-locating bats (Kuwabara and Suga 1993), and in the visual system of flies (Bialek et al. 1991).

1.3   Time as Basis of Information Coding

The relevance of the timing of individual spikes has been at the center of the debate about rate coding versus spike coding. Strong arguments against rate coding have been given by Thorpe et al. (1996) and van Rullen and Thorpe (2001) in the context of visual processing. Many physiologists subscribe to the idea of a Poisson-like rate code to describe the way neurons transmit information. However, as pointed out by Thorpe et al., Poisson rate codes seem hard to reconcile with the impressively efficient rapid information transmission required for sensory processing in human vision. Only 100–150 ms are sufficient for a human to respond selectively to complex visual stimuli (e.g., faces or food), but due to the feedforward architecture of the visual system, made up of multiple layers of neurons firing at an average rate of 10 ms, realistically only one spike or none could be fired by each neuron involved in the process during this time window. A pool of neurons firing spikes stochastically as a function of the stimulus could realize an instantaneous rate code: a *spike density code*. However, maintaining such a set of neurons is expensive, as is the energetic cost of firing so many spikes to encode a single variable (Olshausen 1996). It seems clear from this argument alone that the presence and possibly timing of individual spikes is likely to convey information, and not just the number, or rate, of spikes.

From a combinatorial point of view, precisely timed spikes have a far greater encoding capacity, given a small set of spiking neurons. The representational power of alternative coding schemes was pointed out by Recce (1999) and analyzed by Thorpe et al. (2001). For instance, consider that a stimulus has been presented to a set of n spiking neurons and that each of them fires at most one spike in the next T (ms) time window (❷ *Fig. 4*).

◻ **Fig. 4**

Comparing the representational power of spiking neurons, for different coding schemes. Count code: 6/7 spike per 7 ms, that is $\approx$ 122 spikes per s; Binary code: 1111101; Timing code: latency, here with a 1 ms precision; Rank order code: $E \geq G \geq A \geq D \geq B \geq C \geq F$.

	Count	Latency	Rank
(A)	1	5	3
(B)	1	6	5
(C)	1	7	6
(D)	1	5	4
(E)	1	1	1
(F)	0	–	–
(G)	1	3	2

Numeric examples:	Count code	Binary code	Timing code	Rank order
Left (opposite) figure $n = 7$, $T = 7$ ms	3	7	$\approx$ 19	12.3
Thorpe et al. (2001) $n = 10$, $T = 10$ ms	3.6	10	$\approx$ 33	21.8

Number of bits that can be transmitted by n neurons in a T time window.

Consider some different ways to decode the temporal information that can be transmitted by the n spiking neurons. If the code is to *count* the overall number of spikes fired by the set of neurons (population rate coding), the maximum amount of available information is $\log_2(n + 1)$, since only $n + 1$ different events can occur. In the case of a *binary code*, the output is an n-digits binary number, with obviously n as the information-coding capacity. A greater amount of information is transmitted with a *timing code*, provided that an efficient decoding mechanism is available for determining the precise times of each spike. In practical cases, the available code size depends on the decoding precision, for example for a 1 ms precision, an amount of information of $n \times \log_2(T)$ can be transmitted in the T time window. Finally, in *rank order coding*, information is encoded in the order of the sequence of spike emissions, that is one among the $n!$ orders that can be obtained from n neurons, thus $\log_2(n!)$ bits can be transmitted, meaning that the order of magnitude of the capacity is $n \log(n)$. However, this theoretical estimate must be reevaluated when considering the unavoidable bound on precision required for distinguishing two spike times, even in computer simulation (Cessac et al. 2010).

1.4 Spiking Neuron Networks

In *Spiking Neuron Networks* (SNNs), sometimes referred to as Pulsed-Coupled Neural Networks (PCNNs) in the literature, the presence and *timing* of individual spikes is considered as the means of communication and neural computation. This compares with traditional neuron models where analog values are considered, representing the *rate* at which spikes are fired.

In SNNs, new input–output notions have to be developed that assign meaning to the presence and timing of spikes. One example of such coding that easily compares to traditional neural coding is *temporal coding* (sometimes referred to as "latency coding" or "time-to-first-spike"). Temporal coding is a straightforward method for translating a vector of real numbers into a spike train, for example, for simulating traditional connectionist models using SNNs, as in Maass (1997a). The basic idea is biologically well founded: the more intensive the input, the earlier the spike transmission (e.g., in visual systems). Hence, a network of spiking neurons can be designed with n input neurons N_i whose firing times are determined through some external mechanism. The network is fed by successive n-dimensional input patterns $\mathbf{x} = (x_1, \ldots, x_n)$ – with all x_i inside a bounded interval of $\mathbb{R}$, for example $[0, 1]$ – that are translated into spike trains through successive temporal windows (comparable to successive steps of traditional NNs computation). In each time window, a pattern $\mathbf{x}$ is temporally coded relative to a fixed time T_{in} by one spike emission of neuron N_i at time $t_i = T_{in} - x_i$, for all i (❯ Fig. 5). It is straightforward to show that with such temporal coding, and some mild assumptions, any traditional neural network can be emulated by an SNN. However, temporal coding obviously does not apply readily to more continuous computing where neurons fire multiple spikes, in spike trains.

Many SNN approaches focus on the continuous computation that is carried out on such spike trains. Assigning meaning is then less straightforward, and depends on the approach. However, a way to visualize the temporal computation processed by an SNN is by displaying a complete representation of the network activity on a *spike raster plot* (❯ Fig. 6): With time on the abscissa, a small bar is plotted each time a neuron fires a spike (one line per neuron, numbered on the Y-axis). Variations and frequencies of neuronal activity can be observed in such diagrams, in the same way as natural neurons activities can be observed in spike raster

◧ Fig. 5

Illustration of the temporal coding principle for encoding and decoding real vectors in spike trains.

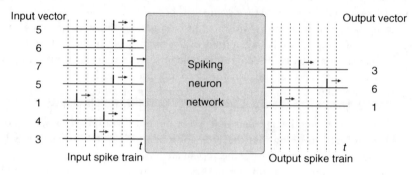

◧ Fig. 6

On a spike raster plot, a small bar is plotted each time (in abscissa) that a neuron (numbered in ordinates) fires a spike. For computational purposes, time is often discretized in temporal Δt units (*left*). The dynamic answer of an SNN, stimulated by an input pattern in temporal coding – diagonal patterns, bottom – can be observed on a spike raster plot (*right*). (From Paugam-Moisy et al. 2008.)

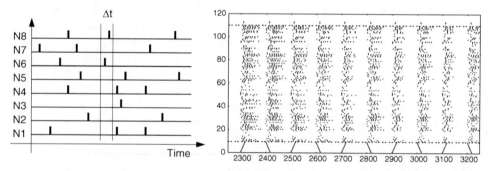

plots drawn from multielectrode recordings. Likewise, other representations (e.g., time-frequency diagrams) can be drawn from simulations of artificial networks of spiking neurons, as is done in neuroscience from experimental data.

Since the basic principle underlying SNNs is so radically different, it is not surprising that much of the work on traditional neural networks, such as learning rules and theoretical results, has to be adapted, or even has to be fundamentally rethought. The main purpose of this chapter is to give an exposition on important state-of-the-art aspects of computing with SNNs, from theory to practice and implementation.

The first difficult task is to define "the" model of neuron, as there exist numerous variants already. Models of spiking neurons and synaptic plasticity are the subject of ❷ Sect. 2. It is worth mentioning that the question of network architecture has become less important in SNNs than in traditional neural networks. ❷ Section 3 proposes a survey of theoretical results (capacity, complexity, and learnability) that argue for SNNs being a new generation of neural networks that are more powerful than the previous ones, and considers some of the ideas on how the increased complexity and dynamics could be exploited. ❥ Section 4 addresses different methods for learning in SNNs and presents the paradigm of *Reservoir Computing*. Finally, ❥ Sect. 5 focuses on practical issues concerning the implementation and use of SNNs for applications, in particular with respect to temporal pattern recognition.

2 Models of Spiking Neurons and Synaptic Plasticity

A spiking neuron model accounts for the impact of impinging action potentials – spikes – on the targeted neuron in terms of the internal state of the neuron, as well as how this state relates to the spikes the neuron fires. There are many models of spiking neurons, and this section only describes some of the models that have so far been most influential in Spiking Neuron Networks.

2.1 Hodgkin–Huxley Model

The fathers of the spiking neurons are the conductance-based neuron models, such as the well-known electrical model defined by Hodgkin and Huxley (1952) (❥ *Fig. 7*). Hodgkin and Huxley modeled the electrochemical information transmission of natural neurons with electrical circuits consisting of capacitors and resistors: C is the capacitance of the membrane, g_{Na}, g_K, and g_L denote the conductance parameters for the different ion channels (sodium Na, potassium K, etc.) and E_{Na}, E_K, and E_L are the corresponding equilibrium potentials. The variables m, h, and n describe the opening and closing of the voltage-dependent channels.

$$C\frac{du}{dt} = -g_{Na}m^3h(u - E_{Na}) - g_Kn^4(u - E_K) - g_L(u - E_L) + I(t) \qquad (1)$$

$$\tau_n\frac{dn}{dt} = -[n - n_0(u)], \quad \tau_m\frac{dm}{dt} = -[m - m_0(u)], \quad \tau_h\frac{dh}{dt} = -[h - h_0(u)]$$

Appropriately calibrated, the Hodgkin–Huxley model has been successfully compared to numerous data from biological experiments on the giant axon of the squid. More generally, it has been shown that the Hodgkin–Huxley neuron is able to model biophysically meaningful

■ Fig. 7
Electrical model of "spiking" neuron as defined by Hodgkin and Huxley. The model is able to produce realistic variations of the membrane potential and the dynamics of a spike firing, for example in response to an input current $I(t)$ sent during a small time, at $t < 0$.

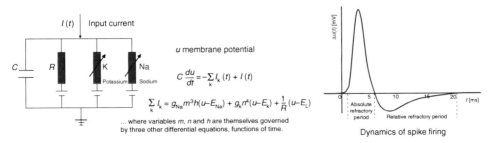

10

properties of the membrane potential, respecting the behavior recordable from natural neurons: an abrupt, large increase at firing time, followed by a short period where the neuron is unable to spike again, the *absolute refractoriness*, and a further time period where the membrane is depolarized, which makes renewed firing more difficult, that is, the *relative refractory period* (❷ *Fig. 7*).

The *Hodgkin–Huxley model* (HH) is realistic but far too complex for the simulation of SNNs. Although ordinary differential equations (ODE) solvers can be applied directly to the system of differential equations, it would be intractable to compute temporal interactions between neurons in a large network of Hodgkin–Huxley models.

2.2 Integrate-and-Fire Model and Variants

2.2.1 Integrate-and-Fire (I&F) and Leaky-Integrate-and-Fire (LIF)

Simpler than the Hodgkin–Huxley neuron model are Integrate-and-Fire (I&F) neuron models, which are much more computationally tractable (see ❷ *Fig. 8* for equation and electrical model).

An important I&F neuron type is the *Leaky-Integrate-and-Fire* (LIF) neuron (Abbott 1999; Stein 1965). Compared to the Hodgkin–Huxley model, the most important simplification in the LIF neuron implies that the shape of the action potentials is neglected and every spike is considered as a uniform event defined only by the time of its appearance. The electrical circuit equivalent for a LIF neuron consists of a capacitor C in parallel with a resistor R driven by an input current $I(t)$. In this model, the dynamics of the membrane potential in the LIF neuron are described by a single first-order linear differential equation:

$$\tau_m \frac{du}{dt} = u_{rest} - u(t) + RI(t) \tag{2}$$

where $\tau_m = RC$ is taken as the time constant of the neuron membrane, modeling the voltage leakage. Additionally, the firing time $t^{(f)}$ of the neuron is defined by a threshold crossing equation $u(t^{(f)}) = \vartheta$, under the condition $u'(t^{(f)}) > 0$. Immediately after $t^{(f)}$, the potential is reset to a given value u_{rest} (with $u_{rest} = 0$ as a common assumption). An absolute refractory period can be modeled by forcing the neuron to a value $u = -u_{abs}$ during a time d_{abs} after a spike emission, and then restarting the integration with initial value $u = u_{rest}$.

◻ **Fig. 8**

Electrical circuit and equation of the Integrate-and-Fire model (I&F).

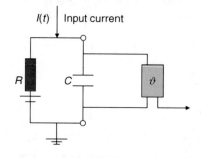

u being the membrane potential,

$$C\frac{du}{dt} = -\frac{1}{R}(u(t) - u_{rest}) + I(t)$$

Spike firing time $t^{(f)}$ is defined by

$$u(t^{(f)}) = \vartheta \text{ with } u'(t^{(f)}) > 0$$

2.2.2 Quadratic-Integrate-and-Fire (QIF) and Theta Neuron

Quadratic-Integrate-and-Fire (QIF) neurons, a variant where $\frac{du}{dt}$ depends on u^2, may be a somewhat better, and still computationally efficient, compromise. Compared to LIF neurons, QIF neurons exhibit many dynamic properties such as delayed spiking, bi-stable spiking modes, and activity-dependent thresholding. They further exhibit a frequency response that better matches biological observations (Brunel and Latham 2003). Via a simple transformation of the membrane potential u to a phase θ, the QIF neuron can be transformed to a Theta-neuron model (Ermentrout and Kopell 1986).

In the Theta-neuron model, the neuron's state is determined by a *phase*, θ. The Theta neuron produces a spike with the phase passing through π. Being one-dimensional, the Theta-neuron dynamics can be plotted simply on a phase circle (❿ *Fig. 9*).

The phase trajectory in a Theta neuron evolves according to:

$$\frac{d\theta}{dt} = (1 - \cos(\theta)) + \alpha I(t)(1 + \cos(\theta)) \tag{3}$$

where θ is the neuron phase, α is a scaling constant, and $I(t)$ is the input current.

The main advantage of the Theta-neuron model is that neuronal spiking is described in a continuous manner, allowing for more advanced gradient approaches, as illustrated in ❿ Sect. 4.1.

2.2.3 Izhikevich's Neuron Model

In the class of spiking neurons defined by differential equations, the two-dimensional *Izhikevich neuron model* (Izhikevich 2003) is a good compromise between biophysical plausibility and computational cost. It is defined by the coupled equations

$$\frac{du}{dt} = 0.04u(t)^2 + 5u(t) + 140 - w(t) + I(t) \qquad \frac{dw}{dt} = a(bu(t) - w(t)) \tag{4}$$

$$\text{with after-spike resetting}: \qquad \text{if } u \geq \vartheta \text{ then } u \leftarrow c \text{ and } w \leftarrow w + d$$

◻ **Fig. 9**
Phase circle of the Theta-neuron model, for the case where the baseline current $I(t) < 0$. When the phase goes through π, a spike is fired. The neuron has two fixed points: a saddle point θ_{FP}^+, and an attractor θ_{FP}^-. In the spiking region, the neuron will fire after some time, whereas in the quiescent region, the phase decays back to θ_{FP}^- unless the input pushes the phase into the spiking region. The refractory phase follows after spiking, and in this phase it is more difficult for the neuron to fire again.

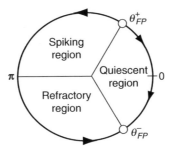

This neuron model is capable of reproducing many different firing behaviors that can occur in biological spiking neurons (❯ *Fig. 10*). (An electronic version of the original figure and reproduction permission are freely available at www.izhikevich.com.)

2.2.4 On Spiking Neuron Model Variants

Besides the models discussed here, there exist many different spiking neuron models that cover the complexity range between the Hodgkin–Huxley model and LIF models, with decreasing biophysical plausibility, but also with decreasing computational cost (see, e.g., Izhikevich (2004) for a comprehensive review, or Standage and Trappenberg (2005) for an in-depth comparison of Hodgkin–Huxley and LIF subthreshold dynamics).

Whereas the Hodgkin–Huxley models are the most biologically realistic, the LIF and – to a lesser extent – QIF models have been studied extensively due to their low complexity, making them relatively easy to understand. However, as argued by Izhikevich (2004), LIF neurons are a simplification that no longer exhibit many important spiking neuron properties. Where the full Hodgkin–Huxley model is able to reproduce many different neuro-computational properties and firing behaviors, the LIF model has been shown to be able to reproduce only three out of the 20 firing schemes displayed in ❯ *Fig. 10*: the "tonic spiking" (A), the "class 1 excitable" (G) and the "integrator" (L). Note that although some behaviors are mutually exclusive for a particular instantiation of a spiking neuron model – for example (K) "resonator" and (L) "integrator" – many such behaviors may be reachable with different parameter choices, for the same neuron model. The QIF model is already able to capture more realistic behavior, and the Izhikevich neuron model can reproduce all of the 20 firing schemes displayed in ❯ *Fig. 10*. Other intermediate models are currently being studied, such as the gIF model (Rudolph and Destexhe 2006).

The complexity range can also be expressed in terms of the computational requirements for simulation. Since it is defined by four differential equations, the Hodgkin–Huxley model requires about 1,200 floating point computations (FLOPS) per 1 ms simulation. Simplified to two differential equations, the Morris–Lecar or FitzHugh–Nagumo models still have a computational cost of one to several hundred FLOPS. Only five FLOPS are required by the LIF model, around 10 FLOPS for variants such as LIF-with-adaptation and quadratic or exponential Integrate-and-Fire neurons, and around 13 FLOPS for Izhikevich's model.

2.3 Spike Response Model

Compared to the neuron models governed by coupled differential equations, the *Spike Response Model* (SRM) as defined by Gerstner (1995) and Kistler et al. (1997) is more intuitive and more straightforward to implement. The SRM model expresses the membrane potential u at time t as an integral over the past, including a model of refractoriness. The SRM is a phenomenological model of neuron, based on the occurrence of spike emissions. Let $\mathcal{F}_j = \left\{ t_j^{(f)}; 1 \leq f \leq n \right\} = \left\{ t \mid u_j(t) = \vartheta \ \wedge \ u_j'(t) > 0 \right\}$ denote the set of all firing times of neuron N_j, and $\Gamma_j = \{i \mid N_i \text{ is presynaptic to } N_j\}$ define its set of presynaptic neurons. The state $u_j(t)$ of neuron N_j at time t is given by

$$u_j(t) = \sum_{t_j^{(f)} \in \mathcal{F}_j} \eta_j(t - t_j^{(f)}) + \sum_{i \in \Gamma_j} \sum_{t_i^{(f)} \in \mathcal{F}_i} w_{ij}\varepsilon_{ij}(t - t_i^{(f)}) + \underbrace{\int_0^\infty \kappa_j(r)I(t - r)dr}_{\text{if external input current}} \quad (5)$$

◘ Fig. 10
Many firing behaviors can occur in biological spiking neurons. Shown are simulations of the
Izhikevich neuron model, for different external input currents (displayed under each temporal
firing pattern). (From Izhikevich [2004].)

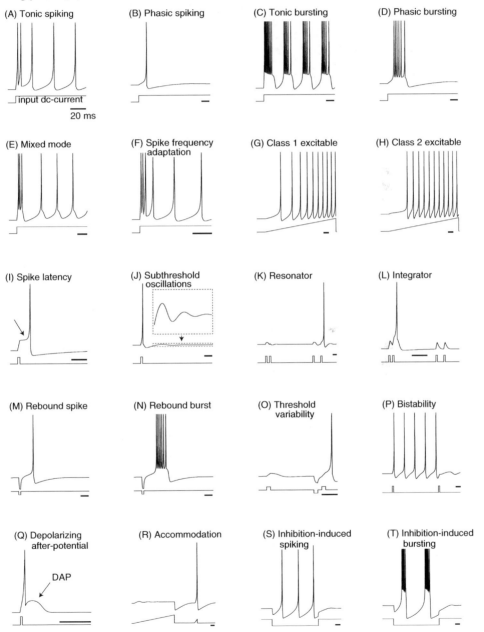

with the following kernel functions: η_j is non-positive for $s > 0$ and models the potential reset after a spike emission, ε_{ij} describes the membrane's potential response to presynaptic spikes, and κ_j describes the response of the membrane potential to an external input current (❷ *Fig. 11*). Some common choices for the kernel functions are:

$$\eta_j(s) = -\vartheta \exp\left(-\frac{s}{\tau}\right) \mathcal{H}(s)$$

or, somewhat more involved,

$$\eta_j(s) = -\eta_0 \exp\left(-\frac{s - \delta^{abs}}{\tau}\right) \mathcal{H}(s - \delta^{abs}) - K\mathcal{H}(s)\mathcal{H}(\delta^{abs} - s)$$

where $\mathcal{H}$ is the Heaviside function, ϑ is the threshold and τ a time constant, for neuron N_j. Setting $K \to \infty$ ensures an absolute refractory period δ^{abs} and η_0 scales the amplitude of relative refractoriness.

Kernel ε_{ij} describes the generic response of neuron N_j to spikes coming from presynaptic neurons N_i, and is generally taken as a variant of an α-function. (An α-function is like $\alpha(x) = x \exp^{-x}$.)

$$\varepsilon_{ij}(s) = \frac{s - d_{ij}^{ax}}{\tau_s} \exp\left(-\frac{s - d_{ij}^{ax}}{\tau_s}\right) \mathcal{H}(s - d_{ij}^{ax})$$

or, in a more general description,

$$\varepsilon_{ij}(s) = \left[\exp\left(-\frac{s - d_{ij}^{ax}}{\tau_m}\right) - \exp\left(-\frac{s - d_{ij}^{ax}}{\tau_s}\right)\right] \mathcal{H}(s - d_{ij}^{ax})$$

where τ_m and τ_s are time constants, and d_{ij}^{ax} describes the axonal transmission delay.

For the sake of simplicity, $\varepsilon_{ij}(s)$ can be assumed to have the same form $\varepsilon(s - d_{ij}^{ax})$ for any pair of neurons, only modulated in amplitude and sign by the weight w_{ij} (excitatory EPSP for $w_{ij} > 0$, inhibitory IPSP for $w_{ij} < 0$).

A short-term memory variant of SRM results from assuming that only the last firing $\hat{t}_j$ of N_j contributes to refractoriness, $\eta_j(t - \hat{t}_j)$ replacing the sum in formula (❷ *Eq. 5*) by a

☐ **Fig. 11**

The Spike Response Model (SRM) is a generic framework to describe the spike process. (Redrawn after Gerstner and Kistler 2002b.)

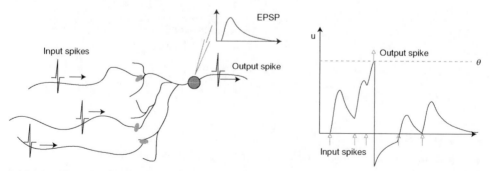

single contribution. Moreover, integrating the equation on a small time window of 1 ms and assuming that each presynaptic neuron fires at most once in the time window (reasonable because of the refractoriness of presynaptic neurons) reduces the SRM to the simplified SRM_0 model:

$$u_j(t) = \eta_j\left(t - \hat{t}_j\right) + \sum_{i \in \Gamma_j} w_{ij}\varepsilon(t - \hat{t}_i - d_{ij}^{ax}) \quad \text{next firing time} \quad t_j^{(f)} = t \Longleftrightarrow u_j(t) = \vartheta \quad (6)$$

Despite its simplicity, the SRM is more general than Integrate-and-Fire neuron models and is often able to compete with the Hodgkin–Huxley model for simulating complex neuro-computational properties.

2.4 Synaptic Plasticity and STDP

In all the models of neurons, most of the parameters are constant values, and specific to each neuron. The exception is the synaptic connections that are the basis of adaptation and learning, even in traditional neural network models where several synaptic weight updating rules are based on Hebb's law (Hebb 1949) (see ❷ Sect. 1). *Synaptic plasticity* refers to the adjustments and even formation or removal of synapses between neurons in the brain. In the biological context of natural neurons, the changes of synaptic weights with effects lasting several hours are referred to as Long-Term Potentiation (LTP) if the weight values (also called *efficacies*) are strengthened, and Long-Term Depression (LTD) if the weight values are decreased. On the timescale of seconds or minutes, the weight changes are denoted as Short-Term Potentiation (STP) and Short-Term Depression (STD). Abbott and Nelson (2000) give a good review of the main synaptic plasticity mechanisms for regulating levels of activity in conjunction with Hebbian synaptic modification, for example, redistribution of synaptic efficacy (Markram and Tsodyks 1996) or synaptic scaling. Neurobiological research has also increasingly demonstrated that synaptic plasticity in networks of spiking neurons is sensitive to the presence and precise timing of spikes (Markram et al. 1997; Bi and Poo 1998; Kempter et al. 1999).

One important finding that is receiving increasing attention is *Spike-Timing Dependent Plasticity*, STDP, as discovered in neuroscientific studies (Markram et al. 1997; Kempter et al. 1999), especially in detailed experiments performed by Bi and Poo (1998, 2001). Often referred to as a *temporal Hebbian rule*, STDP is a form of synaptic plasticity sensitive to the precise timing of spike firing relative to impinging presynaptic spike times. It relies on local information driven by backpropagation of action potential (BPAP) through the dendrites of the postsynaptic neuron. Although the type and amount of long-term synaptic modification induced by repeated pairing of pre- and postsynaptic action potential as a function of their relative timing vary from one neuroscience experiment to another, a basic computational principle has emerged: a maximal increase of synaptic weight occurs on a connection when the presynaptic neuron fires a short time before the postsynaptic neuron, whereas a late presynaptic spike (just after the postsynaptic firing) leads to a decrease in the weight. If the two spikes (pre- and post-) are too distant in time, the weight remains unchanged. This type of LTP/LTD timing dependency should reflect a form of causal relationship in information transmission through action potentials.

For computational purposes, STDP is most commonly modeled in SNNs using temporal windows for controlling the weight LTP and LTD that are derived from neurobiological experiments. Different shapes of STDP windows have been used in recent literature (Markram et al. 1997; Kempter et al. 1999; Song et al. 2000; Senn et al. 2001; Buchs and Senn 2002; Izhikevich et al. 2004; Kistler 2002; Gerstner and Kistler 2002a; Nowotny et al. 2003; Izhikerich and Desai 2003; Saudargiene et al. 2004; Meunier and Paugam-Moisy 2005; Mouraud and Paugam-Moisy 2006): They are smooth versions of the shapes schematized by polygons in ❷ *Fig. 12*. The spike timing (X-axis) is the difference $\Delta t = t_{post} - t_{pre}$ of firing times between the pre- and postsynaptic neurons. The synaptic change ΔW (Y-axis) operates on the weight update. For excitatory synapses, the weight w_{ij} is increased when the presynaptic spike is supposed to have a causal influence on the postsynaptic spike, that is when $\Delta t > 0$ and is close to zero (windows 1–3 in ❷ *Fig. 12*), and decreased otherwise. The main differences in shapes 1–3 concern the symmetry or asymmetry of the LTP and LTD subwindows, and the discontinuity or not of the ΔW function of Δt, near $\Delta t = 0$. For inhibitory synaptic connections, it is common to use a standard Hebbian rule, just strengthening the weight when the pre- and postsynaptic spikes occur close in time, regardless of the sign of the difference $t_{post} - t_{pre}$ (window 4 in ❷ *Fig. 12*).

There exist at least two ways to compute with STDP: The modification ΔW can be applied to a weight w according to either an additive update rule $w \leftarrow w + \Delta W$ or a multiplicative update rule $w \leftarrow w(1 + \Delta W)$.

The notion of temporal Hebbian learning in the form of STDP appears as a possible new direction for investigating innovative learning rules in SNNs. However, many questions arise and many problems remain unresolved. For example, weight modifications according to STDP windows cannot be applied repeatedly in the same direction (e.g., always potentiation) without fixing bounds for the weight values, for example an arbitrary fixed range [0, w_{max}] for excitatory synapses. Bounding both the weight increase and decrease is necessary to avoid either silencing the overall network (when all weights are down) or have "epileptic" network activity (all weights are up, causing disordered and frequent firing of almost all neurons). However, in many STDP driven SNN models, a saturation of the weight values to 0 or w_{max} has been observed, which strongly reduces further adaptation of the network to new events. Among other solutions, a regulatory mechanism, based on a triplet of spikes, has been described by Nowotny et al. (2003), for a smooth version of the temporal window 3 of ❷ *Fig. 12*, with an additive STDP learning rule. On the other hand, applying a multiplicative weight update also effectively applies a self-regulatory mechanism. For deeper insights into the influence of the nature of the update rule and the shape of STDP windows, the reader could refer to Song et al. (2000), Rubin et al. (2000), and Câteau and Fukai (2003).

◘ Fig. 12

Various shapes of STDP windows with LTP in blue and LTD in red for excitatory connections (1–3). More realistic and smooth ΔW function of Δt are mathematically described by sharp rising slope near $\Delta t = 0$ and fast exponential decrease (or increase) toward $\pm\infty$. Standard Hebbian rule (window 4) with brown LTP and green LTD are usually applied to inhibitory connections.

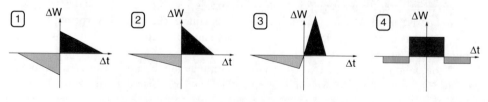

3   Computational Power of Neurons and Networks

Since information processing in spiking neuron networks is based on the precise timing of spike emissions (pulse coding) rather than the average numbers of spikes in a given time window (rate coding), there are two straightforward advantages of SNN processing. First, SNN processing allows for the very fast decoding of sensory information, as in the human visual system (Thorpe et al. 1996), where real-time signal processing is paramount. Second, it allows for the possibility of multiplexing information, for example, like the auditory system combines amplitude and frequency very efficiently over one channel. More abstractly, SNNs add a new dimension, the temporal axis, to the representation capacity and the processing abilities of neural networks. Here, different approaches to determining the computational power and complexity of SNNs are described, and current thinking on how to exploit these properties is outlined, in particular with regard to dynamic cell assemblies.

In 1997, Maass (1997b, 2001) proposed to classify neural networks as follows:

- *1st generation:* Networks based on McCulloch and Pitts' neurons as computational units, that is, threshold gates, with only digital outputs (e.g., Perceptron, Hopfield network, Boltzmann machine, multilayer networks with threshold units).
- *2nd generation:* Networks based on computational units that apply an activation function with a continuous set of possible output values, such as sigmoid or polynomial or exponential functions (e.g., MLP and RBF networks). The real-valued outputs of such networks can be interpreted as *firing rates* of natural neurons.
- *3rd generation of neural network models:* Networks which employ spiking neurons as computational units, taking into account the precise *firing times* of neurons for information coding. Related to SNNs are also pulse stream very large-scale integrated (VLSI) circuits, new types of electronic solutions for encoding analog variables using time differences between pulses.

Exploiting the full capacity of this new generation of neural network models raises many fascinating and challenging questions that will be discussed in the following sections.

3.1   Complexity and Learnability Results

3.1.1   Tractability

To facilitate the derivation of theoretical proofs on the complexity of computing with spiking neurons, Maass proposed a simplified spiking neuron model with a rectangular EPSP shape, the "type_A spiking neuron" (❷ *Fig. 13*). The type_A neuron model can for instance be justified as providing a link to silicon implementations of spiking neurons in analog VLSI neural microcircuits. Central to the complexity results is the notion of transmission delays: different transmission delays, d_{ij}, can be assigned to different presynaptic neurons, N_i, connected to a postsynaptic neuron N_j.

Let Boolean input vectors, $(x_1, \ldots, x_n)$, be presented to a spiking neuron by a set of input neurons $(N_1, \ldots, N_n)$ such that N_i fires at a specific time T_{in} if $x_i = 1$ and does not fire if $x_i = 0$. A type_A neuron is at least as powerful as a threshold gate (Maass 1997b; Schmitt 1998). Since spiking neurons can behave as coincidence detectors (for a proper choice of weights, a spiking neuron can only fire when two or more input spikes are effectively coincident in time) it is

◻ **Fig. 13**

Very simple versions of spiking neurons: "type_A spiking neuron" (rectangular-shaped pulse) and "type_B spiking neuron" (triangular-shaped pulse), with elementary representation of refractoriness (threshold goes to infinity), as defined in Maass (1997b).

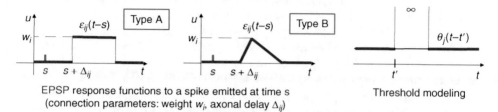

EPSP response functions to a spike emitted at time s   Threshold modeling
(connection parameters: weight w_i, axonal delay Δ_{ij})

straightforward to prove that the Boolean function CD_n *(Coincidence Detection function)* can be computed by a single spiking neuron of type_A (the proof relies on a suitable choice of the transmission delays d_{ij}):

$$CD_n(x_1, \ldots, x_n, y_1, \ldots, y_n) = \begin{cases} 1, \text{if } (\exists i) \; x_i = y_i \\ 0, \text{otherwise} \end{cases}$$

In previous neural network generations, the computation of the Boolean function CD_n required many more neurons: at least $\frac{n}{\log(n+1)}$ threshold gates and at least an order of magnitude of $\Omega(n^{1/4})$ sigmoidal units.

Of special interest is the *Element Distinctness function, ED_n*:

$$ED_n(x_1, \ldots, x_n) = \begin{cases} 1, \text{if } (\exists i \neq j) \; x_i = x_j \\ 0, \text{if } (\forall i \neq j)|x_i - x_j| \geq 1 \\ \text{arbitrary}, \text{otherwise} \end{cases}$$

Let real-valued inputs $(x_1, \ldots, x_n)$ be presented to a spiking neuron by a set of input neurons $(N_1, \ldots, N_n)$ such that N_i fires at time $T_{in} - cx_i$ (cf. temporal coding, defined in ❷ Sect. 1.4). With positive real-valued inputs and a binary output, the ED_n function can be computed by a single type_A neuron, whereas at least $\Omega(n \log (n))$ threshold gates and at least $\frac{n-4}{2} - 1$ sigmoidal hidden units are required.

However, for arbitrary real-valued inputs, type_A neurons are no longer able to compute threshold circuits. For such settings, the "type_B spiking neuron" (❷ *Fig. 13*) has been proposed, as its triangular EPSP can shift the firing time of a targeted postsynaptic neuron in a continuous manner. It is easy to see that any threshold gate can be computed by $O(1)$ type_B spiking neurons. Furthermore, at the network level, any threshold circuit with s gates, for real-valued inputs $x_i \in [0, 1]$, can be simulated by a network of $O(s)$ type_B spiking neurons.

From these results, Maass concludes that spiking neuron networks are computationally more powerful than both the 1st and the 2nd generations of neural networks.

Schmitt develops a deeper study of type_A neurons with programmable delays in Schmitt (1998) and Maass and Schmitt (1997). Some results are:

● Every Boolean function of n variables, computable by a single spiking neuron, can be computed by a disjunction of at most $2n-1$ threshold gates.

- There is no $\Sigma\Pi$-unit with fixed degree that can simulate a spiking neuron.
- The *threshold number* of a spiking neuron with n inputs is $\Theta(n)$.
- The following relation holds: $(\forall n \geq 2)\ \exists$ a Boolean function on n variables that has threshold number 2 and cannot be computed by a spiking neuron.
- The *threshold order* of a spiking neuron with n inputs is $\Omega(n^{1/3})$.
- The *threshold order* of a spiking neuron with $n \geq 2$ inputs is at most $n-1$.

3.1.2 Capacity

Maass (2001) considers *noisy spiking neurons*, a neuron model close to the SRM (cf. ❷ Sect. 2.3), with a probability of *spontaneous* firing (even under threshold) or not firing (even above threshold) governed by the difference:

$$\sum_{i\in\Gamma_j}\sum_{s\in\mathscr{F}_i, s<t} w_{ij}\varepsilon_{ij}(t-s) - \underbrace{\eta_j(t-t')}_{\text{threshold function}}$$

The main result from Maass (2001) is that for any given $\varepsilon, \delta > 0$ one can simulate any given feedforward sigmoidal neural network $\mathscr{N}$ of s units with linear saturated activation function by a network $\mathscr{N}_{\varepsilon,\delta}$ of $s + O(1)$ noisy spiking neurons, in temporal coding. An immediate consequence of this result is that SNNs are *universal approximators*, in the sense that any given continuous function $F : [0, 1]^n \rightarrow [0, 1]^k$ can be approximated within any given precision $\varepsilon > 0$ with arbitrarily high reliability, in temporal coding, by a network of noisy spiking neurons with a single hidden layer.

With regard to synaptic plasticity, Legenstein et al. (2005) studied STDP learnability. They define a *Spiking Neuron Convergence Conjecture* (SNCC) and compare the behavior of STDP learning by teacher-forcing with the Perceptron convergence theorem. They state that a spiking neuron can learn with STDP, basically, any map from input to output spike trains that it could possibly implement in a stable manner. They interpret the result as saying that STDP endows spiking neurons with *universal learning capabilities* for Poisson input spike trains.

Beyond these and other encouraging results, Maass (2001) points out that SNNs are able to encode time series in spike trains, but there are, in computational complexity theory, no standard reference models yet for analyzing computations on time series.

3.1.3 VC-Dimension

The first attempt to estimate the VC-dimension (see http://en.wikipedia.org/wiki/VC_dimension for a definition) of spiking neurons is probably the work of Zador and Pearlmutter (1996), where they studied a family of integrate-and-fire neurons (cf. ❷ Sect. 2.2) with threshold and time constants as parameters. Zador and Pearlmutter proved that for an Integrate-and-Fire (I&F) model, $VC_{\text{dim}}(\text{I\&F})$ grows as $\log(B)$ with the input signal bandwidth B, which means that the VC_{dim} of a signal with infinite bandwidth is unbounded, but the divergence to infinity is weak (logarithmic).

More conventional approaches (Maass and Schmitt 1997; Maass 2001) estimate bounds on the VC-dimension of neurons as functions of their programmable/learnable parameters, such as the synaptic weights, the transmission delays and the membrane threshold:

- With m variable positive delays, VC_{dim}(type_A neuron) is $\Omega(m \log (m))$ – even with fixed weights – whereas, with m variable weights, VC_{dim}(threshold gate) is $\Omega(m)$.
- With n real-valued inputs and a binary output, VC_{dim}(type_A neuron) is $O(n \log (n))$.
- With n real-valued inputs and a real-valued output, $pseudo_{\mathrm{dim}}$(type_A neuron) is $O(n \log (n))$.

The implication is that the learning complexity of a single spiking neuron is greater than the learning complexity of a single threshold gate. As Maass and Schmitt (1999) argue, this should not be interpreted as saying that supervised learning is impossible for a spiking neuron, but rather that it is likely quite difficult to formulate rigorously provable learning results for spiking neurons.

To summarize Maass and Schmitt's work: let the class of Boolean functions, with n inputs and 1 output, that can be computed by a spiking neuron be denoted by $\mathscr{S}_n^{xy}$, where x is b for Boolean values and a for analog (real) values and idem for y. Then the following holds:

- The classes $\mathscr{S}_n^{bb}$ and $\mathscr{S}_n^{ab}$ have VC-dimension $\Theta(n \log (n))$.
- The class $\mathscr{S}_n^{aa}$ has pseudo-dimension $\Theta(n \log (n))$.

At the network level, if the weights and thresholds are the only programmable parameters, then an SNN with temporal coding seems to be nearly equivalent to traditional Neural Networks (NNs) with the same architecture, for traditional computation. However, transmission delays are a new relevant component in spiking neural computation and SNNs with programmable delays appear to be more powerful than NNs.

Let $\mathscr{N}$ be an SNN of neurons with rectangular pulses (e.g., type_A), where all delays, weights and thresholds are programmable parameters, and let E be the number of edges of the $\mathscr{N}$ directed acyclic graph. (The directed acyclic graph is the network topology that underlies the spiking neuron network dynamics.) Then $VC_{\mathrm{dim}}(\mathscr{N})$ is $O(E^2)$, even for analog coding of the inputs (Maass and Schmitt 1999). Schmitt (2004) derived more precise results by considering a feedforward architecture of depth D, with nonlinear synaptic interactions between neurons.

It follows that the sample sizes required for the networks of fixed depth are not significantly larger than traditional neural networks. With regard to the generalization performance in pattern recognition applications, the models studied by Schmitt can be expected to be at least as good as traditional network models (Schmitt 2004).

3.1.4 Loading Problem

In the framework of PAC-learnability (Valiant 1984; Blumer et al. 1989), only hypotheses from $\mathscr{S}_n^{bb}$ may be used by the learner. Then, the computational complexity of training a spiking neuron can be analyzed within the formulation of the *consistency* or *loading problem* (cf. Judd 1990):

> ▶ *Given a training set T of labeled binary examples (X,b) with n inputs, do there exist parameters defining a neuron $\mathscr{N}$ in $\mathscr{S}_n^{bb}$ such that $(\forall (X, b) \in T)\ y_{\mathscr{N}} = b$?*

In this PAC-learnability setting, the following results are proved in Maass and Schmitt (1999):

- The *consistency problem* for a spiking neuron with binary delays is *NP*-complete.
- The *consistency problem* for a spiking neuron with binary delays and fixed weights is *NP*-complete.

Several extended results were developed by Šíma and Sgall (2005), such as:

- The *consistency problem* for a spiking neuron with nonnegative delays is *NP*-complete ($d_{ij} \in \mathbb{R}^+$). The result holds even with some restrictions (see Šíma and Sgall (2005) for precise conditions) on bounded delays, unit weights or fixed threshold.
- A single spiking neuron with programmable weights, delays and threshold does not allow robust learning unless $RP = NP$. The *approximation problem* is not better solved even if the same restrictions as above are applied.

3.1.5 Complexity Results Versus Real-World Performance

Non-learnability results, such as those outlined above, have of course been derived for classic NNs already, for example in Blum and Rivest (1989) and Judd (1990). Moreover, the results presented in this section apply only to a restricted set of SNN models and, apart from the programmability of transmission delays of synaptic connections, they do not cover all the capabilities of SNNs that could result from computational units based on firing times. Such restrictions on SNNs can rather be explained by a lack of practice for building proofs in such a context or, even more, by an incomplete and unadapted computational complexity theory or learning theory. Indeed, learning in biological neural systems may employ rather different mechanisms and algorithms than common computational learning systems. Therefore, several characteristics, especially the features related to computing in continuously changing time, will have to be fundamentally rethought to develop efficient learning algorithms and ad hoc theoretical models to understand and master the computational power of SNNs.

3.2 Cell Assemblies and Synchrony

One way to take a fresh look at SNNs complexity is to consider their dynamics, especially the spatial localization and the temporal variations of their activity. From this point of view, SNNs behave as *complex systems*, with emergent macroscopic-level properties resulting from the complex dynamic interactions between neurons, but hard to understand just looking at the microscopic level of each neuron processing. As biological studies highlight the presence of a specific organization in the brain (Sporns et al. 2005; Eguíluz 2005; Achard and Bullmore 2007), the complex networks research area appears to provide valuable tools ("Small-Word" connectivity (Watts and Strogatz 1998), presence of clusters (Newman and Girvan 2004; Meunier and Paugam-Moisy 2006), presence of hubs (Barabasi and Albert 1999), etc., see Newman (2003) for a survey) for studying topological and dynamic complexities of SNNs, both in natural and artificial networks of spiking neurons. Another promising direction for research takes its inspiration from the area of dynamic systems: Several methods and measures, based on the notions of phase transition, edge-of-chaos, Lyapunov exponents, or mean-field predictors, are currently proposed to estimate and control the computational performance of SNNs (Legenstein and Maass 2005; Verstraeten et al. 2007; Schrauwen et al. 2009). Although these directions of research are still in their infancy, an alternative is to revisit older and more biological notions that are already related to the network topology and dynamics.

The concept of the *cell assembly* was introduced by Hebb (1949), more than half a century ago. (The word "cell" was used at that time, instead of "neuron.") However, the idea was not further developed, neither by neurobiologists – since they could not record the activity of more than one or a few neurons at a time, until recently – nor by computer scientists. New techniques of brain imaging and recording have boosted this area of research in neuroscience only recently (cf. Wenneker et al. 2003). In computer science, a theoretical analysis of assembly formation in spiking neuron network dynamics (with SRM neurons) has been discussed by Gerstner and van Hemmen (1994), where they contrast ensemble code, rate code, and spike code as descriptions of neuronal activity.

A cell assembly can be defined as a group of neurons with strong mutual excitatory connections. Since a cell assembly, once a subset of its neurons are stimulated, tends to be activated as a whole, it can be considered as an operational unit in the brain. An association can be viewed as the activation of an assembly by a stimulus or another assembly. Then, short-term memory would be a persistent activity maintained by reverberations in assemblies, whereas long-term memory would correspond to the formation of new assemblies, for example, by a Hebb's rule mechanism. Inherited from Hebb, current thinking about cell assemblies is that they could play the role of "grandmother neural groups" as a basis for memory encoding, instead of the old controversial notion of "grandmother cell," and that material entities (e.g., a book, a cup, or a dog) and even more abstract entities such as concepts or ideas could be represented by cell assemblies.

Within this context, synchronization of firing times for subsets of neurons inside a network has received much attention. Abeles (1991) developed the notion of *synfire chains*, which describes activity in a pool of neurons as a succession of synchronized firing by specific subsets of these neurons. Hopfield and Brody demonstrated *transient synchrony* as a means for collective spatiotemporal integration in neuronal circuits (Hopfield and Brody 2000, 2001). The authors claim that the event of collective synchronization of specific pools of neurons in response to a given stimulus may constitute a basic computational building block, at the network level, for which there is no resemblance in traditional neural computing (❷ *Fig. 14*).

However, synchronization per se – even transient synchrony – appears to be too restrictive a notion for fully understanding the potential capabilities of information processing in cell assemblies. This has been comprehensively pointed out by Izhikevich (2006) who proposes the extended notion of *polychronization* within a group of neurons that are sparsely connected with various axonal delays. Based on the connectivity between neurons, a polychronous group is a possible stereotypical time-locked firing pattern. Since the neurons in a polychronous group have matching axonal conduction delays, the group can be activated in response to a specific temporal pattern triggering very few neurons in the group, other ones being activated in a chain reaction. Since any given neuron can be activated within several polychronous groups, the number of coexisting polychronous groups can be far greater than the number of neurons in the network. Izhikevich argues that networks with delays are "infinite-dimensional" from a purely mathematical point of view, thus resulting in much greater information capacity as compared to synchrony-based assembly coding. Polychronous groups represent good

□ Fig. 14

A spike raster plot showing the dynamics of an artificial SNN: Erratic background activity is disrupted by a stimulus presented between 1,000 and 2,000 ms. (By courtesy of D. Meunier, reprint from his PhD thesis (2007), Université Lyon 2, France)

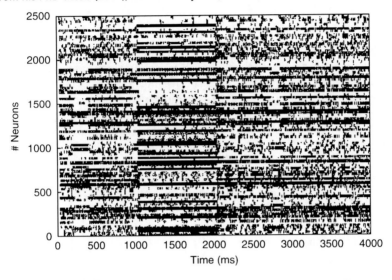

candidates for modeling multiple trace memory and they could be viewed as a computational implementation of cell assemblies.

Notions of cell assemblies and synchrony, derived from natural computing in the brain and biological observations, are inspiring and challenging computer scientists and theoretical researchers to search for and define new concepts and measures of complexity and learnability in dynamic systems. This will likely bring a much deeper understanding of neural computations that include the time dimension, and will likely benefit both computer science as well as neuroscience.

4 Learning in Spiking Neuron Networks

Traditionally, neural networks have been applied to pattern recognition, in various guises. For example, carefully crafted layers of neurons can perform highly accurate handwritten character recognition (LeCun et al. 1995). Similarly, traditional neural networks are preferred tools for function approximation, or regression. The best-known learning rules for achieving such networks are of course the class of error-backpropagation rules for supervised learning. There also exist learning rules for unsupervised learning, such as Hebbian learning, or distance-based variants like Kohonen self-organizing maps.

Within the class of computationally oriented spiking neuron networks, two main directions are distinguished. First, there is the development of learning methods equivalent to those developed for traditional neural networks. By substituting traditional neurons with spiking neuron models, augmenting weights with delay lines, and using temporal coding, algorithms for supervised and unsupervised learning have been developed. Second, there are networks and computational algorithms that are uniquely developed for networks of spiking neurons. These networks and algorithms use the temporal domain as well as the increased complexity of SNNs to arrive at novel methods for temporal pattern detection with spiking neuron networks.

4.1   Simulation of Traditional Models

Maass and Natschläger (1997) propose a theoretical model for emulating arbitrary Hopfield networks in temporal coding (see ❷ Sect. 1.4). Maass (1997a) studies a "relatively realistic" mathematical model for biological neurons that can simulate arbitrary feedforward sigmoidal neural networks. Emphasis is put on the fast computation time that depends only on the number of layers of the sigmoidal network, and no longer on the number of neurons or weights. Within this framework, SNNs are validated as universal approximators (see ❷ Sect. 3.1), and traditional supervised and unsupervised learning rules can be applied for training the synaptic weights.

It is worth remarking that, to enable theoretical results, Maass and Natschläger's model uses static reference times T_{in} and T_{out} and auxiliary neurons. Even if such artifacts can be removed in practical computation, the method rather appears to be an artificial attempt to make SNNs computing like traditional neural networks, without taking advantage of SNNs intrinsic abilities to compute with time.

4.1.1   Unsupervised Learning in Spiking Neuron Networks

Within this paradigm of computing in SNNs equivalently to traditional neural network computing, a number of approaches for unsupervised learning in spiking neuron networks have been developed, based mostly on variants of Hebbian learning. Extending on Hopfield's (1995) idea, Natschläger and Ruf (1998a) propose a learning algorithm that performs unsupervised clustering in spiking neuron networks, akin to RBF network, using spike times as input. Natschläger and Ruf's spiking neural network for unsupervised learning is a simple two-layer network of SRM neurons, with the addition of multiple delays between the neurons: An individual connection from a neuron i to a neuron j consists of a fixed number of m synaptic terminals, where each terminal serves as a sub-connection that is associated with a different delay d^k and weight w_{ij}^k (❷ Fig. 15). The delay d^k of a synaptic terminal k is defined by the difference between the firing time of the presynaptic neuron i, and the time the postsynaptic potential of neuron j starts rising.

A winner-takes-all learning rule modifies the weights between the source neurons and the neuron first to fire in the target layer using a time-variant of Hebbian learning: If the start of the PSP at a synapse slightly precedes a spike in the target neuron, the weight of this synapse is increased, as it exerted significant influence on the spike time via a relatively large contribution to the membrane potential. Earlier and later synapses are decreased in weight, reflecting their lesser impact on the target neuron's spike time. With such a learning rule, input patterns can be encoded in the synaptic weights such that, after learning, the firing time of an output neuron reflects the distance of the evaluated pattern to its learned input pattern thus realizing a kind of RBF neuron (Natschläger and Ruf 1998a).

Bohte et al. (2002b) extend on this approach to enhance the precision, capacity, and clustering capability of a network of spiking neurons by developing a temporal version of population coding. To extend the encoding precision and clustering capacity, input data is

■ Fig. 15

Unsupervised learning rule in SNNs: Any single connection can be considered as being multisynaptic, with random weights and a set of increasing delays, as defined in Natschläger and Ruf (1998b).

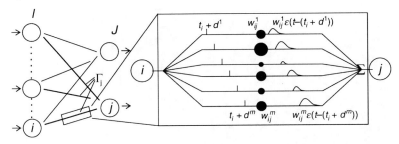

■ Fig. 16

Encoding with overlapping Gaussian receptive fields. An input value a is translated into firing times for the input-neurons encoding this input-variable. The highest stimulated neuron (neuron 5), fires at a time close to $T = 0$, whereas less-stimulated neurons, as for instance neuron 3, fire at increasingly later times.

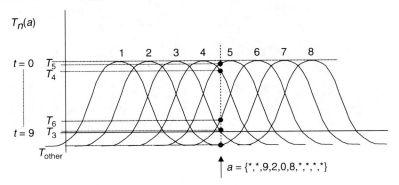

encoded into temporal spike-time patterns by population coding, using multiple local receptive fields like Radial Basis Functions. The translation of inputs into relative firing times is straightforward: An optimally stimulated neuron fires at $t = 0$, whereas a value up to say $t = 9$ is assigned to less optimally stimulated neurons (depicted in ❷ *Fig. 16*). With such encoding, spiking neural networks were shown to be effective for clustering tasks, for example ❷ *Fig. 17*.

4.1.2 Supervised Learning in Multilayer Networks

A number of approaches for supervised learning in standard multilayer feedforward networks have been developed based on gradient descent methods, the best known being error back-propagation. As developed in Bohte et al. (2002a), *SpikeProp* starts from error backpropagation to derive a supervised learning rule for networks of spiking neurons that transfer the information in the timing of a single spike. This learning rule is analogous to the derivation rule by Rumelhart et al. (1986), but SpikeProp applies to spiking neurons of the SRM type. To overcome the discontinuous nature of spiking neurons, the thresholding function is

■ **Fig. 17**
Unsupervised classification of remote sensing data. (a) The full image. Inset: image cutout that
is actually clustered. (b) Classification of the cutout as obtained by clustering with a
Self-Organizing Map (SOM). (c) Spiking Neuron Network RBF classification of the cutout image.

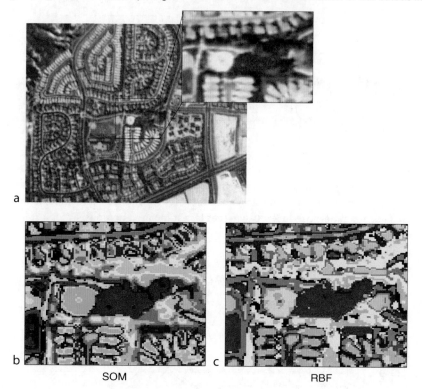

a

b   c
        SOM   RBF

approximated, thus linearizing the model at a neuron's output spike times. As in the unsupervised
SNN described above, each connection between neurons may have multiple delayed synapses with
varying weights (see ❷ *Fig. 15*). The SpikeProp algorithm has been shown to be capable of
learning complex nonlinear tasks in spiking neural networks with similar accuracy as traditional sigmoidal neural networks, including the archetypal XOR classification task (❷ *Fig. 18*).

The SpikProp method has been successfully extended to adapt the synaptic delays along
the error gradient, as well as the decay for the α-function and the threshold (Schrauwen and
Van Campenhout 2004a, b). Xin and Embrechts (2001) have further shown that the addition
of a simple momentum term significantly speeds up convergence of the SpikeProp algorithm.
Booij and Nguyen (2005) have, analogously to the method for BackPropagation-Through-
Time, extended SpikeProp to account for neurons in the input and hidden layer to fire
multiple spikes.

McKennoch et al. (2009) derived a supervised Theta-learning rule for multilayer networks
of Theta neurons. By mapping QIF neurons to the canonical Theta-neuron model (a nonlinear phase model, see ❷ Sect. 2.2), a more dynamic spiking neuron model is placed at the heart
of the spiking neuron network. The Theta-neuron phase model is cyclic and allows for
a continuous reset. Derivatives can then be computed without any local linearization
assumptions.

◘ **Fig. 18**
Interpolated XOR function $f(t_1, t_2) : [0, 6] \rightarrow [10, 16]$. **(a)** Target function. **(b)** Spiking Neuron Network output after training.

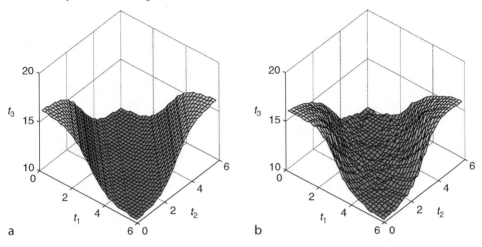

a   b

Some sample results showing the performance of both SpikeProp and the Theta Neuron learning rule as compared to error backpropagation in traditional neural networks is shown in ❯ *Table 1*. The more complex Theta-neuron learning allows for a smaller neural network to optimally perform classification.

As with SpikeProp, Theta learning requires some careful fine-tuning of the network. In particular, both algorithms are sensitive to *spike loss*, in that no error gradient is defined when the neuron does not fire for any pattern, and hence will never recover. McKennoch et al. (2009) heuristically deal with this issue by applying alternating periods of coarse learning, with a greater learning rate, and fine tuning, with a small learning rate.

As demonstrated in Belatreche et al. (2007), non-gradient-based methods like evolutionary strategies do not suffer from these tuning issues. For MLP networks based on various spiking neuron models, performance comparable to SpikeProp is shown. An evolutionary strategy is, however, very time consuming for large-scale networks.

4.2   Reservoir Computing

Clearly, the architecture and dynamics of an SNN can be matched, by temporal coding, to traditional connectionist models, such as multilayer feedforward networks or recurrent networks. However, since networks of spiking neurons behave decidedly different as compared to traditional neural networks, there is no pressing reason to design SNNs within such rigid schemes.

According to biological observations, the neurons of biological SNNs are sparsely and irregularly connected in space (network topology) and the variability of spike flows implies that they communicate irregularly in time (network dynamics) with a low average activity. It is important to note that the network topology becomes a simple underlying support to the neural dynamics, but that only active neurons contribute to information processing. At a given time t, the sub-topology defined by active neurons can be very sparse and different from the

☐ Table 1

Classification results for the SpikeProp and Theta neuron supervised learning methods on two benchmarks, the Fisher Iris dataset and the Wisconsin Breast Cancer dataset. The results are compared to standard error backpropagation, *BP A* and *BP B* denoting the standard Matlab backprop implementation with default parameters, where their respective network sizes are set to correspond to either the SpikeProp or the Theta-neuron networks (taken from McKennoch et al. (2009))

Learning method	Network size	Epochs	Train (%)	Test (%)
Fisher Iris Dataset				
SpikeProp	$50 \times 10 \times 3$	1,000	97.4	96.1
BP A	$50 \times 10 \times 3$	2.6e6	98.2	95.5
BP B	$4 \times 8 \times 1$	1e5	98.0	90.0
Theta-Neuron BP	$4 \times 8 \times 1$	1,080	100	98.0
Wisconsin Breast Cancer Dataset				
SpikeProp	$64 \times 15 \times 2$	1,500	97.6	97.0
BP A	$64 \times 15 \times 2$	9.2e6	98.1	96.3
BP B	$9 \times 8 \times 1$	1e5	97.2	99.0
Theta-Neuron BP	$9 \times 8 \times 1$	3,130	98.3	99.0

underlying network architecture (e.g., local clusters, short or long path loops, and synchronized cell assemblies), comparable to the active brain regions that appear colored in brain imaging scanners. Clearly, an SNN architecture has no need to be regular. A network of spiking neurons can even be defined randomly (Maass et al. 2002b; Jaeger 2002) or by a loosely specified architecture, such as a set of neuron groups that are linked by projections, with a given probability of connection from one group to the other (Meunier and Paugam-Moisy 2005). However, the nature of a connection has to be prior defined as an excitatory or inhibitory synaptic link, without subsequent change, except for the synaptic efficacy. That is, the weight value can be modified, but not the weight sign.

With this in mind, a new family of networks has been developed that is specifically suited to processing temporal input/output patterns with spiking neurons. The new paradigm is named *Reservoir Computing* as a unifying term for which the precursor models are Echo State Networks (ESNs) and Liquid State Machines (LSMs). Note that the term "reservoir computing" is not reserved to SNNs, since ESN was first designed with sigmoidal neurons, but this chapter mainly presents reservoir computing with SNNs.

4.2.1 Main Characteristics of Reservoir Computing Models

The topology of a reservoir computing model (❱ *Fig. 19*) can be defined as follows:

- A layer of K neurons with input connections toward the reservoir
- A recurrent network of M neurons, interconnected by a random and sparse set of weighted links: the so-called *reservoir*, which is usually left untrained
- A layer of L *readout neurons* with trained connections from the reservoir

◻ Fig. 19

Architecture of a reservoir computing network: the "reservoir" is a set of *M* internal neurons, with random and sparse connectivity.

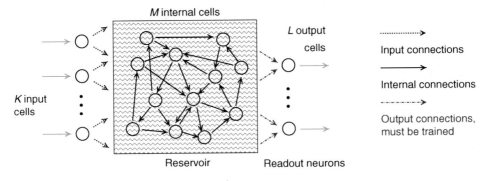

The early motivation of reservoir computing is the well-known difficulty to find efficient supervised learning rules to train recurrent neural networks, as attested by the limited success of methods like Backpropagation Through Time (BPTT), Real-Time Recurrent Learning (RTRL) or Extended Kalman Filtering (EKF). The difficulty stems from lack of knowledge of how to control the behavior of the complex dynamic system resulting from the presence of cyclic connections in the network architecture. The main idea of reservoir computing is to renounce training the internal recurrent network and only to pick out, by way of the readout neurons, the relevant part of the dynamic states induced in the reservoir by the network inputs. Only the reading out of this information is subject to training, usually by very simple learning rules, such as linear regression. The success of the method is based on the high power and accuracy of self-organization inherent to a random recurrent network.

In SNN versions of reservoir computing, a soft kind of unsupervised, local training is often added by applying a synaptic plasticity rule like STDP inside the reservoir. Since STDP was directly inspired by the observation of natural processing in the brain (see ❷ Sect. 2.4), its computation does not require supervised control or an understanding of the network dynamics.

The paradigm of "reservoir computing" is commonly referred to as such since approximately 2007, and it encompasses several seminal models in the literature that predate this generalized notion by a few years. The next section describes the two founding models that were designed concurrently in the early 2000s, by Jaeger (2001) for the ESN and by Maass et al. (2002b) for the LSM.

4.2.2 Echo State Network (ESN) and Liquid State Machine (LSM)

The original design of *Echo State Network*, proposed by Jaeger (2001), was intended to learn time series $\mathbf{u}(1)$, $\mathbf{d}(1)$, ..., $\mathbf{u}(T)$, and $\mathbf{d}(T)$ with recurrent neural networks. The internal states of the reservoir are supposed to reflect, as an "echo," the concurrent effect of a new teacher input $u(t + 1)$ and a teacher-forcing output $d(t)$, related to the previous time. Therefore, Jaeger's model includes backward connections from the output layer toward the reservoir (see ❷ *Fig. 20a*) and the network training dynamics is governed by the following equation:

$$\mathbf{x}(t + 1) = f\left(W^{\text{in}}\mathbf{u}(t + 1) + W\mathbf{x}(t) + W^{\text{back}}\mathbf{d}(t)\right) \tag{7}$$

10 Computing with Spiking Neuron Networks

◘ **Fig. 20**

Architecture of the two founding models of reservoir computing: ESN and LSM.

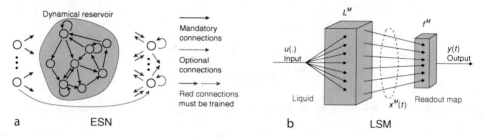

a   ESN   b   LSM

where $x(t+1)$ is the new state of the reservoir, W^{in} is the input weight matrix, W the matrix of weights in the reservoir, and W^{back} the matrix of feedback weights, from the output layer to the reservoir. The learning rule for output weights W^{out} (feedforward connections from reservoir to output) consists of a linear regression algorithm, for example least mean squares: At each step, the network states $x(t)$ are collected into a matrix M, after a washout time t_0, and the sigmoid-inverted teacher output $\tanh^{-1} \mathbf{d}(n)$ into a matrix T, in order to obtain $(W^{out})^t = M^\dagger T$ where $M^\dagger$ is the pseudo-inverse of M. In the exploitation phase, the network is driven by novel input sequences $u(t)$, with $t \geq T$ (desired output $d(t)$ is unknown), and it produces the computed output $y(t)$ with coupled equations like:

$$\mathbf{x}(t+1) = f\left(W^{in}\mathbf{u}(t+1) + W\mathbf{x}(t) + W^{back}\mathbf{y}(t)\right) \tag{8}$$

$$\mathbf{y}(t+1) = f^{out}(W^{out}[\mathbf{u}(t+1), \mathbf{x}(t+1), \mathbf{y}(t)]) \tag{9}$$

For the method to be efficient, the network must have the "Echo State Property," that is, the property of being state contracting, state forgetting, and input forgetting, which gives it a behavior of "fading memory." Choosing a reservoir weight matrix W with a spectral radius $|\lambda_{max}|$ slightly lower than 1 is neither a necessary nor a sufficient condition (as discussed in Lukoševičius and Jaeger 2009) but can be applied as a rule of thumb for most practical cases: the common practice is to rescale W after randomly initializing the network connections. An important remark must be made: the condition on the spectral radius is no longer clearly relevant when the reservoir is an SNN with fixed weights, and it totally vanishes when an STDP rule is applied to the reservoir. A comparative study of several measures for the reservoir dynamics, with different neuron models, can be found in Verstraeten et al. (2007).

ESNs have been successfully applied in many experimental settings, with networks no larger than 20–400 internal units, for example, in mastering the benchmark task of learning the Mackey–Glass chaotic attractor (Jaeger 2001). Although the first design of ESN was for networks of sigmoid units, Jaeger has also introduced spiking neurons (LIF model) in the ESNs (Jaeger 2002; Jaeger et al. 2007). Results improve substantially over standard ESNs, for example, in the task of generating a slow sinewave ($\mathbf{d}(n) = 1/5\sin(n/100)$), which becomes easy with a leaky integrator network (Jaeger 2002).

The basic motivation of the *Liquid State Machine*, defined by Maass et al. (2002b), was to explain how a continuous stream of inputs $u(\cdot)$ from a rapidly changing environment can be processed in real time by recurrent circuits of Integrate-and-Fire neurons (❷ *Fig. 20b*). The solution they propose is to build a "liquid filter" L^M – the reservoir – that operates similarly to water undertaking the transformation from the low-dimensional space of a set of motors stimulating its surface into a higher-dimensional space of waves in parallel. The liquid states, $x^M(t)$, are transformed by a readout map, f^M, to generate an output, $y(t)$, that can appear to be stable and appropriately scaled responses given by the network, even if the internal state never converges to a stable attractor. Simulating such a device on neural microcircuits, Maass et al. have shown that a readout neuron receiving inputs from hundreds or thousands of neurons can learn to extract salient information from the high-dimensional transient states of the circuit and can transform transient circuit states into stable outputs.

In mathematical terms, the liquid state is simply the current output of some operator L^M that maps input functions $u(.)$ onto functions $x^M(t)$. The L^M operator can be implemented by a randomly connected recurrent neural network. The second component of an LSM is a "memoryless readout map" f^M that transforms, at every time t, the current liquid state into the machine output, according to the equations:

$$x^M(t) = (L^M(u))(t) \tag{10}$$

$$y(t) = f^M(x^M(t)) \tag{11}$$

The readout is usually implemented by one or several Integrate-and-Fire neurons that can be trained to perform a specific task using very simple learning rules, such as a linear regression or the p-delta rule (Auer et al. 2008).

Often, in implementation, the neural network playing the role of the liquid filter is inspired by biological modeling cortical columns. Therefore, the reservoir has a 3D topology, with a probability of connection that decreases as a Gaussian function of the distance between the neurons.

The readout map is commonly task-specific. However, the hallmark of neural microcircuits is their ability to carry out several parallel real-time computations within the same circuitry. It appears that a readout neuron is able to build a sort of equivalence class among dynamical states, and then to well recognize similar (but not equal) states. Moreover, several readout neurons, trained to perform different tasks, may enable parallel real-time computing.

LSMs have been successfully applied to several nonlinear problems, such as the XOR and many others. LSMs and ESNs are very similar models of reservoir computing that promise to be convenient for both exploiting and capturing most temporal features of spiking neuron processing, especially for time series prediction and for temporal pattern recognition. Both models are good candidates for engineering applications that process temporally changing information.

4.2.3 Related Reservoir Computing Work

An additional work that was linked to the family of "reservoir computing" models after being published is the Backpropagation DeCorrelation rule (BPDC), proposed by Steil (2004). As an extension of the Atiya–Parlos learning rule in recurrent neural networks (Atiya and Parlos 2000), the BPDC model is based on a multilayer network with fixed weights until the last layer. Only this layer has learnable weights both from the reservoir (the multilayer network) to the readout (the last layer) and recurrently inside the readout layer. However, the BPDC model has not been proposed with spiking neurons so far, even if that appears to be readily feasible.

Another approach, by Paugam-Moisy et al. (2008), takes advantage of the theoretical results proving the importance of delays in computing with spiking neurons (see ❯ Sect. 3) for defining a supervised learning rule acting on the delays of connections (instead of weights) between the reservoir and the readout neurons. The reservoir is an SNN, with an STDP rule for adapting the weights to the task at hand, where it can be observed that polychronous groups (see ❯ Sect. 3.2) are activated more and more selectively as training goes on. The learning rule of the readout delays is based on a temporal margin criterion inspired by Vapnik's theory.

There exist reservoir computing networks that make use of evolutionary computation for training the weights of the reservoir, such as Evolino (Schmidhuber et al. 2007), and several other models that are currently proposed, with or without spiking neurons (Devert et al. 2007; Jiang et al. 2008a, b). Although the research area is in rapid expansion, several papers (Verstraeten et al. 2007; Schrauwen et al. 2007b; Lukoševičius and Jaeger 2009) offer valuable surveys.

4.3   Other SNN Research Tracks

Besides trying to apply traditional learning rules to SNNs and the development of reservoir computing, there are many research efforts that relate to learning with spiking neurons.

Much research is, for instance, being carried out on deriving theoretically principled learning rules for spiking neurons, for instance on Information Bottleneck learning rules that attempt to maximize measures of mutual information between input and output spike trains (Barber 2003; Chechik 2003; Pfister et al. 2003, 2006; Bell and Parra 2005; Toyoizumi et al. 2005a, b; Pfister and Gerstner 2006; Bohte and Mozer 2007; Büsing and Maass 2008). The aim of this work on theoretically principled learning is, typically, to arrive at easily understood methods that have spiking neurons carry out some form of Independent Component Analysis (ICA) (Toyoizumi et al. 2005a; Klampfl et al. 2009), or Principal Component Analysis (PCA) (Büsing and Maass 2008), or focus on sparse efficient coding (Olshausen 1996; Volkmer 2004; Lewicki 2002; Smith 2006).

These methods have variable applicability to real world problems, though some have demonstrated excellent performance: Smith and Lewicki (2006) develop an efficient encoding of auditory signals in spike-trains based on sparse over-complete dictionaries that outperforms many standard, filter-based approaches. Forms of reinforcement learning have been developed based on the combination of reward modulation and STDP (Xie and Seung 2001; Izhikevich 2007; Legenstein et al. 2008). Many of these algorithms are highly technical and much of this research is still converging can practical algorithms. We only mention these directions here, the reader can pursue the state of the art in these areas.

Just as advances in neurosciences have contributed to the reevaluation of the significance of the timing and presence of single spikes in neuronal activity, advances in neuropsychology suggest that brain-like systems are able to carry out at least some forms of Bayesian inference (Koerding and Wolpert 2004; Daw and Courvillie 2008). As a result, the implementation of Bayesian inference algorithms in neural networks has received much attention, with a particular emphasis on networks of spiking neurons.

In this line of research, the activity in neural networks is somehow related to representing probability distributions. Much of this research, however, relies on noisy, stochastic spiking neurons that are characterized by a spike density, and Bayesian inference is implicitly carried out by large populations of such neurons (Barber et al. 2005; Zemel et al. 1998; Sahani and Dayar 2003; Wu et al. 2003; Rao 2005; Gerwinn 2007; Hays et al. 2007; Ma et al. 2008). As noted by Deneve (2008a), coding probabilities with stochastic neurons "has two major drawbacks. First, [...], it adds uncertainty, and therefore noise, to an otherwise deterministic probability computation. Second, [...], the resulting model would not be self-consistent since the input and output firing rates have different meanings and different dynamics."

In Deneve (2008a, b), an alternative approach is developed for binary log-likelihood estimation in an SNN. Such binary log-likelihood estimation in an SNN has some known limitations: It can only perform exact inference in a limited family of generative models, and in a hierarchical model only the objects highest in the hierarchy truly have temporal dynamics. Interestingly, in this model neurons still exhibit a Poisson-like distribution of synaptic events. However, rather than reflecting stochasticity due to a noisy firing mechanism, it reflects the sensory input noise. Still, this type of SNN is at the forefront of current developments and many advances in this direction are to be expected.

5 Discussion

This chapter has given an outline of some of the most important ideas, models, and methods in the development of Spiking Neuron Networks, with a focus on pattern recognition and temporal data processing, such as time series. By necessity, many related subjects are not treated in detail here. Variants of the models and methods described in this chapter, and variants thereof, are increasingly being applied to real world pattern recognition. ❯ Section 4.1 listed some results on SNN algorithms applied to traditional pattern recognition, where a dataset of numeric vectors is mapped to a classification. However, as was emphasized in the section on reservoir computing, many interesting application domains have an additional temporal dimension: Not just the immediate data are important, but the *sequence* of data. An increasing amount of work deals with applying SNN concepts to various application domains with important temporal dynamics, such as speech processing, active vision for computers, and autonomous robotics.

5.1 Pattern Recognition with SNNs

Considerable work has focused on developing SNNs that are suitable for speech processing (Verstraeten et al. 2005; Holmberg et al. 2005; Wang and Parel 2005; Loiselle et al. 2005). Verstraeten et al. (2005) have developed a model based on Liquid State Machines that is trained to recognize isolated words. They compare several front-end signal encoding methods, and find that a nature-inspired front-end like a "Lyon Passive Ear" outperforms other methods when an LSM is applied. Similarly, Holmberg et al. (2005) developed a method for automatic speech recognition grounded in SNNs. As a "front end," they simulated a part of the inner ear, and then simulated octopus spiking neurons to encode the inner-ear signal in a spike train. They subsequently used a fairly simple classifier to recognize speech from both the inner-ear simulation and the spiking neuron spike trains. Wang and Pavel (2005) used an SNN to represent auditory signals based on using the properties of the spiking neuron refractory period. In their SNN, they converted amplitude to temporal code while maintaining the phase information of the carrier. They proposed that for auditory signals the narrow band envelope information could be encoded simply in the temporal inter-spike intervals. Rank order coding with spiking neural networks has been explored for speech recognition by Loiselle et al. (2005). They show it is an efficient method (fast response/adaptation ability) when having only small training sets.

One important fact that the speech-processing case studies highlight is that traditional preprocessing techniques do not provide optimal front ends and back ends for subsequent SNN processing. Still, many promising features have been pointed out, like robustness to noise, and combining SNN processing with other methods is proposed as a promising research area.

In parallel, a number of SNN-based systems have been developed for computer vision, for example, using spike asynchrony (Thorpe and Gautrais 1997); sparse image coding using an asynchronous spiking neural network (Perrinet and Samuelides 2002); a synchronization-based dynamic vision model for image segmentation (Azhar 2005); saliency extraction with a distributed spiking neural network (Chevallier et al. 2006; Masquelier et al. 2007; Chavallier and Tarroux 2008); and SNNs applied to character recognition (Wysoski et al. 2008).

SNN-based systems also develop increasingly in the area of robotics, where fast processing is a key issue (Maass et al. 2002a; Tenore 2004; Floreano et al. 2005, 2006; Panchev and Wermter 2006; Hartland and Bredeche 2007), from wheels to wings, or legged locomotion. The special abilities of SNNs for fast computing transient temporal patterns make them the first choice for designing efficient systems in the area of autonomous robotics. This perspective is often cited but not yet fully developed.

Other research domains mention the use of SNNs, such as Echo State Networks for motor control (e.g., Salmen and Plöger 2005), prediction in the context of wireless telecommunications (e.g., Jaeger and Haas 2004), or neuromorphic approaches to rehabilitation in medicine (e.g., Kutch 2004). In this context, the *Neuromorphic Engineer* newsletter often publishes articles on applications developed with SNNs (Institute of Neuromorphic Engineering newsletter: http://www.ine-web.org/).

It is worth remarking that SNNs are ideal candidates for designing *multimodal interfaces*, since they can represent and process very diverse information in a unifying manner based on time, from such different sources as visual, auditory, speech, or other sensory data. An application of SNNs to audiovisual speech recognition was proposed by Séguier and Mercier (2002). Crépet et al. (2000) developed, with traditional NNs, a modular connectionist model of multimodal associative memory including temporal aspects of visual and auditory data processing (Bouchut et al. 2003). Such a multimodal framework, applied to a virtual robotic prey–predator environment, with spiking neuron networks as functional modules, has proved capable of simulating high-level natural behavior such as cross-modal priming (Meunier and Paugam-Moisy 2004) or real-time perceptive adaptation to changing environments (Chevallier et al. 2005).

5.2   Implementing SNNs

Since SNNs perform computations in such a different way as compared to traditional NNs, the way to program an SNN model for application purposes has to be revised also. The main interest of SNN simulation is to take into account the precise timing of the spike firing, hence the width of the time window used for discrete computation of the successive network states must remain narrow (see ❷ Sect. 1.4), and consequently only a few spike events occur at each time step: In ❷ *Fig. 6*, only two spikes were fired inside the Δt time range, among the 64 potential connections linking the eight neurons. Hence, inspecting all the neurons and synapses of the network at each time step is exceedingly time consuming: In this example, a clock-based simulation (i.e., based on a time window) computes zero activity in 97% of the computations! An event-driven simulation is clearly more suitable for sequential simulations of spiking neural networks (Watts 1994; Mattia and Del Giudice 2000; Makino 2003; Rochel and Martinez 2003; Reutimann et al. 2003; McKennoch 2009), as long as the activity of an SNN can be fully described by a set of dated spikes. Nevertheless, event-driven programming that requires the next spike time can be explicitly computed in reasonable time, so that not all models of neurons can be used.

At the same time, SNN simulation can highly benefit from parallel computing, substantially more so than traditional NNs. Unlike a traditional neuron in rate coding, a spiking neuron does not need to receive weight values from each presynaptic neuron at each computation step. Since at each time step only a few neurons are active in an SNN, the classic bottleneck of message passing is removed. Moreover, computing the updated state of the membrane potential (e.g., for an SRM or LIF model neuron) is more complex than computing a weighted sum (e.g., for the threshold unit). Therefore, a communication time and computation cost are much better balanced in SNN parallel implementation as compared to traditional NNs, as proved by the parallel implementation of the *SpikeNET* software (Delorme 1999).

Well-known simulators of spiking neurons include *GENESIS* (Bower and Beeman 1998) and *NEURON* (Hines and Carnevale 1997), but they were designed principally for programming detailed biophysical models of isolated neurons rather than for fast simulation of very large-scale SNNs. However, NEURON has been updated with an event-driven mechanism on the one hand (Hines and Carnevale 2004) and a version for parallel machines on the other hand (Hines and Carnevale 2008). *BRIAN* (http://brian.di.ens.fr/) is a mainly clock-based simulator with an optional event-driven tool, whereas *MVASpike* (http://mvaspike.gforge. inria.fr/) is a purely event-driven simulator (Roche and Mortinez 2003). *DAMNED* is a parallel event-driven simulator (Mouraud and Puzenat 2009). A comparative and experimental study of several SNN simulators can be found in Brette et al. (2007). Most simulators are currently programmed in C or C++. Others are Matlab toolboxes, such as Jaeger's toolbox for ESNs available from the web page (http://www.faculty.jacobs-university.de/hjaeger/esn_research. html), or the *Reservoir Computing Toolbox,* available at http://snn.elis.ugent.be/node/59 and briefly presented in the last section of Verstraeten (2007). A valuable tool for developers could be *PyNN* (http://neuralensemble.org/trac/PyNN), a Python package for simulator-independent specification of neuronal network models.

Hardware implementations of SNNs are also being actively pursued. Several chapters in the book by Maass and Bishop (1999) are dedicated to this subject, and more recent work can be found in Upegui et al. (2004), Hellmich et al. (2005), Johnston et al. (2005), Chicca et al. (2003), Oster et al. (2005), Mitra et al. (2006), and Schrauwen et al. (2007a).

5.3   Conclusion

This chapter has given an overview of the state of the art in Spiking Neuron Networks: its biological inspiration, the models that underlie the networks, some theoretical results on computational complexity and learnability, learning rules, both traditional and novel, and some current application areas and results. The novelty of the concept of SNNs means that many lines of research are still open and are actively being pursued.

References

Abbott LF (1999) Brain Res Bull 50(5/6):303–304

Abbott LF, Nelson SB (2000) Synaptic plasticity: taming the beast. Nat Neurosci 3:1178–1183

Abeles M (1991) Corticonics: neural circuits of the cerebral cortex. Cambridge University Press, Cambridge

Achard S, Bullmore E (2007) Efficiency and cost of economical brain functional networks. PLoS Comput Biol 3(2):e17

Atiya A, Parlos AG (2000) New results on recurrent network training: unifying the algorithms and accelerating convergence. IEEE Trans Neural Netw 11(3): 697–709

Auer P, Burgsteiner H, Maass W (2008) A learning rule for very simple universal approximators consisting of a single layer of perceptrons. Neural Netw 21(5): 786–795

Azhar H, Iftekharuddin K, Kozma R (2005) A chaos synchronization-based dynamic vision model for image segmentation. In: IJCNN 2005, International joint conference on neural networks. IEEE–INNS, Montreal, pp 3075–3080

Barabasi AL, Albert R (1999) Emergence of scaling in random networks. Science 286(5439):509–512

Barber D (2003) Learning in spiking neural assemblies. In: Becker S, Thrun S, Obermayer K (eds) NIPS 2002, Advances in neural information processing systems, vol 15. MIT Press, Cambridge, MA, pp 165–172

Barber MJ, Clark JW, Anderson CH (2005) Neural representation of probabilistic information, vol 15. MIT Press, Cambridge, MA

Belatreche A, Maguire LP, McGinnity M (2007) Advances in design and application of spiking neural networks. Soft Comput-A Fusion Found Methodol Appl 11:239–248

Bell A, Parra L (2005) Maximising information yields spike timing dependent plasticity. In: Saul LK, Weiss Y, Bottou L (eds) NIPS 2004, Advances in neural information processing systems, vol 17. MIT Press, Cambridge, MA, pp 121–128

Bi G-q, Poo M-m (1998) Synaptic modification in cultured hippocampal neurons: dependence on spike timing, synaptic strength, and polysynaptic cell type. J Neurosci 18(24):10464–10472

Bi G-q, Poo M-m (2001) Synaptic modification of correlated activity: Hebb's postulate revisited. Annu Rev Neurosci 24:139–166

Bialek W, Rieke F, de Ruyter R, van Steveninck RR, Warland D (1991) Reading a neural code. Science 252:1854–1857

Blum A, Rivest R (1989) Training a 3-node neural net is NP-complete. In: Proceedings of NIPS 1988, advances in neural information processing systems. MIT Press, Cambridge, MA, pp 494–501

Blumer A, Ehrenfeucht A, Haussler D, Warmuth MK (1989) Learnability and the Vapnik-Chervonenkis dimension. J ACM 36(4):929–965

Bobrowski O, Meir R, Shoham S, Eldar YC (2007) A neural network implementing optimal state estimation based on dynamic spike train decoding. In: Schölkopf B, Platt JC, Hoffman T (eds) NIPS 2006, Advances in neural information processing systems, vol 20. MIT Press, Cambridge, MA, pp 145–152

Bohte SM, Mozer MC (2007) Reducing the variability of neural responses: a computational theory of spike-timing-dependent plasticity. Neural Comput 19:371–403

Bohte SM, Kok JN, La Poutre H (2002a) Spike-prop: error-backpropagation in multi-layer networks of spiking neurons. Neurocomputing 48:17–37

Bohte SM, La Poutre H, Kok JN (2002b) Unsupervised clustering with spiking neurons by sparse temporal coding and multilayer RBF networks. IEEE Trans Neural Netw 13:426–435

Booij O, tat Nguyen H (2005) A gradient descent rule for spiking neurons emitting multiple spikes. Info Process Lett 95:552–558

Bouchut Y, Paugam-Moisy H, Puzenat D (2003) Asynchrony in a distributed modular neural network for multimodal integration. In: PDCS 2003, International conference on parallel and distributed computing and systems. ACTA Press, Calgary, pp 588–593

Bower JM, Beeman D (1998) The book of GENESIS: exploring realistic neural models with the General Simulation system, 2nd edn. Springer, New York

Brette R, Rudolph M, Hines T, Beeman D, Bower JM et al. (2007) Simulation of networks of spiking neurons: a review of tools and strategies. J Comput Neurosci 23(3):349–398

Brunel N, Latham PE (2003) Firing rate of the noisy quadratic integrate-and-fire neuron, vol 15. MIT Press, Cambridge

Buchs NJ, Senn W (2002) Spike-based synaptic plasticity and the emergence of direction selective simple cells: simulation results. J Comput Neurosci 13:167–186

Büsing L, Maass W (2008) Simplified rules and theoretical analysis for information bottleneck optimization and PCA with spiking neurons. In: NIPS 2007, Advances in neural information processing systems, vol 20. MIT Press, Cambridge

Câteau H, Fukai T (2003) A stochastic method to predict the consequence of arbitrary forms of spike-timing-dependent plasticity. Neural Comput 15(3):597–620

Cessac B, Paugam-Moisy H, Viéville T (2010) Overview of facts and issues about neural coding by spikes. J Physiol (Paris) 104:5–18

Chechik G (2003) Spike-timing dependent plasticity and relevant mutual information maximization. Neural Comput 15(7):1481–1510

Chevallier S, Tarroux P (2008) Covert attention with a spiking neural network. In: ICVS'08, Computer Vision Systems, Santorini, 2008. Lecture notes in computer science, vol 5008. Springer, Heidelberg, pp 56–65

Chevallier S, Paugam-Moisy H, Lemaître F (2005) Distributed processing for modelling real-time multimodal perception in a virtual robot. In: PDCN 2005, International conference on parallel and distributed computing and networks. ACTA Press, Calgary, pp 393–398

Chevallier S, Tarroux P, Paugam-Moisy H (2006) Saliency extraction with a distributed spiking neuron network. In: Verleysen M (ed) ESANN'06, Advances in computational intelligence and learning. D-Side Publishing, Evere, Belgium, pp 209–214

Chicca E, Badoni D, Dante V, d'Andreagiovanni M, Salina G, Carota L, Fusi S, Del Giudice P (2003) A VLSI recurrent network of integrate-and-fire neurons connected by plastic synapses with long-term memory. IEEE Trans Neural Netw 14(5):1297–1307

Crépet A, Paugam-Moisy H, Reynaud E, Puzenat D (2000) A modular neural model for binding several modalities. In: Arabnia HR (ed) IC-AI 2000, International conference on artificial intelligence. CSREA Press, Las Vegas, pp 921–928

Cybenko G (1988) Approximation by superpositions of a sigmoidal function. Math Control Signal Syst 2:303–314

Daw ND, Courville AC (2008) The pigeon as particle filter. In: NIPS 2007, Advances in neural information processing systems, vol 20. MIT Press, Cambridge, MA

Delorme A, Gautrais J, Van Rullen R, Thorpe S (1999) SpikeNET: a simulator for modeling large networks of integrate and fire neurons. Neurocomputing 26–27:989–996

Deneve S (2008a) Bayesian spiking neurons. I. Inference. Neural Comput 20:91–117

Deneve S (2008b) Bayesian spiking neurons. II. Learning. Neural Comput 20:118–145

Devert A, Bredeche N, Schoenauer M (2007) Unsupervised learning of echo state networks: a case study in artificial embryogeny. In: Montmarché N et al. (eds) Artificial evolution, selected papers, Lecture Notes in Computer Science, vol 4926/2008, pp 278–290

Eguíluz VM, Chialvo GA, Cecchi DR, Baliki M, Apkarian AV (2005) Scale-free brain functional networks. Phys Rev Lett 94(1):018102

Ermentrout GB, Kopell N (1986) Parabolic bursting in an excitable system coupled with a slow oscillation. SIAM J Appl Math 46:233

Floreano D, Zufferey JC, Nicoud JD (2005) From wheels to wings with evolutionary spiking neurons. Artif Life 11(1–2):121–138

Floreano D, Epars Y, Zufferey J-C, Mattiussi C (2006) Evolution of spiking neural circuits in autonomous mobile robots. Int J Intell Syst 21(9):1005–1024

Funahashi K (1989) On the approximate realization of continuous mappings by neural networks. Neural Netw 2(3):183–192

Gerstner W (1995) Time structure of the activity in neural network models. Phys Rev E 51:738–758

Gerstner W, Kistler WM (2002a) Mathematical formulations of Hebbian learning. Biol Cybern 87(5–6):404–415

Gerstner W, Kistler W (2002b) Spiking neuron models: single neurons, populations, plasticity. Cambridge University Press, Cambridge

Gerstner W, van Hemmen JL (1994) How to describe neuronal activity: spikes, rates or assemblies? In: Cowan JD, Tesauro G, Alspector J (eds) NIPS 1993, Advances in neural information processing system, vol 6. MIT Press, Cambridge, MA, pp 463–470

Gerwinn S, Macke JH, Seeger M, Bethge M (2007) Bayesian inference for spiking neuron models with a sparsity prior. In: NIPS 2006, Advances in neural information processing systems, vol 19. MIT Press, Cambridge, MA

Hartland C, Bredeche N (2007) Using echo state networks for robot navigation behavior acquisition. In: ROBIO'07, Sanya, China

Hebb DO (1949) The organization of behavior. Wiley, New York

Heiligenberg W (1991) Neural nets in electric fish. MIT Press, Cambridge, MA

Hellmich HH, Geike M, Griep P, Rafanelli M, Klar H (2005) Emulation engine for spiking neurons and adaptive synaptic weights. In: IJCNN 2005, International joint conference on neural networks. IEEE-INNS, Montreal, pp 3261–3266

Hines ML, Carnevale NT (1997) The NEURON simulation environment. Neural Comput 9:1179–1209

Hines ML, Carnevale NT (2004) Discrete event simulation in the NEURON environment. Neurocomputing 58–60:1117–1122

Hines ML, Carnevale NT (2008) Translating network models to parallel hardware in NEURON. J Neurosci Methods 169:425–455

Hodgkin AL, Huxley AF (1952) A quantitative description of ion currents and its applications to conduction and excitation in nerve membranes. J Physiol 117:500–544

Holmberg M, Gelbart D, Ramacher U, Hemmert W (2005) Automatic speech recognition with neural spike trains. In: Interspeech 2005 – Eurospeech, 9th European conference on speech communication and technology, Lisbon, pp 1253–1256

Hopfield JJ (1982) Neural networks and physical systems with emergent collective computational abilities. Proc Natl Acad Sci 79(8):2554–2558

Hopfield JJ (1995) Pattern recognition computation using action potential timing for stimulus representation. Nature 376:33–36

Hopfield JJ, Brody CD (2000) What is a moment? "Cortical" sensory integration over a brief interval. Proc Natl Acad Sci 97(25):13919–13924

Hopfield JJ, Brody CD (2001) What is a moment? Transient synchrony as a collective mechanism for spatiotemporal integration. Proc Natl Acad Sci 98(3):1282–1287

Hornik K, Stinchcombe M, White H (1989) Multilayer feedforward networks are universal approximators. Neural Netw 2(5):359–366

Huys QJM, Zemel RS, Natarajan R, Dayan P (2007) Fast population coding. Neural Comput 19:404–441

Izhikevich EM (2003) Simple model of spiking neurons. IEEE Trans Neural Netw 14(6):1569–1572

Izhikevich EM (2004) Which model to use for cortical spiking neurons? IEEE Trans Neural Netw 15(5): 1063–1070

Izhikevich EM (2006) Polychronization: computation with spikes. Neural Comput 18(2):245–282

Izhikevich EM (2007) Solving the distal reward problem through linkage of STDP and dopamine signaling. Cereb Cortex 17(10):2443–2452

Izhikevich EM, Desai NS (2003) Relating STDP and BCM. Neural Comput 15(7):1511–1523

Izhikevich EM, Gally JA, Edelman GM (2004) Spike-timing dynamics of neuronal groups. Cereb Cortex 14:933–944

Jaeger H (2001) The "echo state" approach to analysing and training recurrent neural networks. Technical Report TR-GMD-148, German National Research Center for Information Technology

Jaeger H (2002) Tutorial on training recurrent neural networks, covering BPTT, RTRL, EKF and the "echo state network" approach. Technical Report TR-GMD-159, German National Research Center for Information Technology

Jaeger H, Haas H (2004) Harnessing nonlinearity: predicting chaotic systems and saving energy in wireless telecommunication. Science 304(5667):78–80

Jaeger H, Lukoševičius M, Popovici D, Siewert U (2007) Optimization and applications of echo state networks with leaky-integrator neurons. Neural Netw 20(3):335–352

Jiang F, Berry H, Schoenauer M (2008a) Supervised and evolutionary learning of echo state networks. In: Rudolph G et al. (eds) Proceedings of the 10th international conference on parallel problem solving from nature: PPSN X. Lecture notes in computer science, vol 5199. Springer, pp 215–224

Jiang F, Berry H, Schoenauer M (2008b) Unsupervised learning of echo state networks: balancing the double pole. In: GECCO'08: Proceedings of the 10th annual conference on genetic and evolutionary computation, ACM, New York, pp 869–870

Johnston S, Prasad G, Maguire L, McGinnity T (2005) Comparative investigation into classical and spiking neuron implementations on FPGAs. In: ICANN 2005, International conference on artificial neural networks. Lecture notes in computer science, vol 3696. Springer, New York, pp 269–274

Judd JS (1990) Neural network design and the complexity of learning. MIT Press, Cambridge, MA

Kempter R, Gerstner W, van Hemmen JL (1999) Hebbian learning and spiking neurons. Phys Rev E 59(4): 4498–4514

Kistler WM (2002) Spike-timing dependent synaptic plasticity: a phenomenological framework. Biol Cyber 87(5–6):416–427

Kistler WM, Gerstner W, van Hemmen JL (1997) Reduction of Hodgkin–Huxley equations to a single-variable threshold model. Neural Comput 9:1015–1045

Klampfl S, Legenstein R, Maass W (2009) Spiking neurons can learn to solve information bottleneck problems and extract independent components. Neural Comput 21(4):911–959

Koerding KP, Wolpert DM (2004) Bayesian integration in sensorimotor learning. Nature 427:244–247

Kohonen T (1982) Self-organized formation of topologically correct feature maps. Biol Cybern 43:59–69

Kutch JJ (2004) Neuromorphic approaches to rehabilitation. Neuromorphic Eng 1(2):1–2

Kuwabara N, Suga N (1993) Delay lines and amplitude selectivity are created in subthalamic auditory nuclei: the brachium of the inferior colliculus of the mustached bat. J Neurophysiol 69:1713–1724

LeCun Y, Jackel LD, Bottou L, Cortes C, Denker JS, Drucker H, Guyon I, Muller UA, Sackinger E, Simard P (1995) Learning algorithms for classification: a comparison on handwritten digit recognition, vol 276. World Scientific, Singapore

Legenstein R, Maass W (2005) What makes a dynamical system computationally powerful? In: Haykin S, Principe JC, Sejnowski TJ, McWhirter JG (eds) New directions in statistical signal processing: from systems to brain. MIT Press, Cambridge, MA

Legenstein R, Näger C, Maass W (2005) What can a neuron learn with spike-time-dependent plasticity? Neural Comput 17(11):2337–2382

Legenstein R, Pecevski D, Maass W (2008) Theoretical analysis of learning with reward-modulated spike-timing-dependent plasticity. In: NIPS 2007, Advances in neural information processing systems, vol 20. MIT Press, Cambridge, MA

Lewicki MS (2002) Efficient coding of natural sounds. Nat Neurosci 5:356–363

Loiselle S, Rouat J, Pressnitzer D, Thorpe S (2005) Exploration of rank order coding with spiking neural networks for speech recognition. In: IJCNN 2005, International joint conference on neural networks. IEEE–INNS Montreal, pp 2076–2080

Lukoševičius M, Jaeger H (July 2007) Overview of reservoir recipes. Technical Report 11, Jacobs University Bremen

Ma WJ, Beck JM, Pouget A (2008) Spiking networks for Bayesian inference and choice. Curr Opin Neurobiol 18(2):217–222

Maass W (1997a) Fast sigmoidal networks via spiking neurons. Neural Comput 10:1659–1671

Maass W (1997b) Networks of spiking neurons: the third generation of neural network models. Neural Netw 10:1659–1671

Maass W (2001) On the relevance of time in neural computation and learning. Theor Comput Sci 261:157–178 (extended version of ALT'97, in LNAI 1316:364–384)

Maass W, Bishop CM (eds) (1999) Pulsed neural networks. MIT Press, Cambridge, MA

Maass W, Natschläger T (1997) Networks of spiking neurons can emulate arbitrary Hopfield nets in temporal coding. Netw: Comput Neural Syst 8(4):355–372

Maass W, Schmitt M (1997) On the complexity of learning for a spiking neuron. In: COLT'97, Conference on computational learning theory. ACM Press, New York, pp 54–61

Maass W, Schmitt M (1999) On the complexity of learning for spiking neurons with temporal coding. Info Comput 153:26–46

Maass W, Steinbauer G, Koholka R (2002a) Autonomous fast learning in a mobile robot. In: Hager GD, Christensen HI, Bunke H, Klein R (eds) Sensor based intelligent robots, vol 2238. Springer, Berlin, pp 345–356

Maass W, Natschläger T, Markram H (2002b) Real-time computing without stable states: a new framework for neural computation based on perturbations. Neural Comput 14(11):2531–2560

Makino T (2003) A discrete event neural network simulator for general neuron model. Neural Comput Appl 11(2):210–223

Markram H, Tsodyks MV (1996) Redistribution of synaptic efficacy between neocortical pyramidal neurones. Nature 382:807–809

Markram H, Lübke J, Frotscher M, Sakmann B (1997) Regulation of synaptic efficacy by coincidence of postsynaptic APs and EPSPs. Science 275:213–215

Masquelier T, Thorpe SJ, Friston KJ (2007) Unsupervised learning of visual features through spike timing dependent plasticity. PLoS Comput Biol 3:e31

Mattia M, Del Giudice P (2000) Efficient event-driven simulation of large networks of spiking neurons and dynamical synapses. Neural Comput 12:2305–2329

McCulloch WS, Pitts W (1943) A logical calculus of the ideas immanent in nervous activity. Bull Math Biophys 5:115–133

McKennoch S, Voegtlin T, Bushnell L (2009) Spike-timing error backpropagation in theta neuron networks. Neural Comput 21(1):9–45

Meunier D (2007) Une modélisation évolutionniste du liage temporel (in French). PhD thesis, University Lyon 2, http://demeter.univ-lyon2.fr/sdx/theses/lyon2/2007/meunier_d, 2007

Meunier D, Paugam-Moisy H (2004) A "spiking" bidirectional associative memory for modeling intermodal priming. In: NCI 2004, International conference on neural networks and computational intelligence. ACTA Press, Calgary, pp 25–30

Meunier D, Paugam-Moisy H (2005) Evolutionary supervision of a dynamical neural network allows learning with on-going weights. In: IJCNN 2005, International joint conference on neural networks. IEEE–INNS, Montreal, pp 1493–1498

Meunier D, Paugam-Moisy H (2006) Cluster detection algorithm in neural networks. In: Verleysen M (ed) ESANN'06, Advances in computational intelligence and learning. D-Side Publishing, Evere, Belgium, pp 19–24

Mitra S, Fusi S, Indiveri G (2006) A VLSI spike-driven dynamic synapse which learns only when necessary. In: Proceedings of IEEE international symposium on circuits and systems (ISCAS) 2006. IEEE Press, New York, p 4

Mouraud A, Paugam-Moisy H (2006) Learning and discrimination through STDP in a top-down modulated associative memory. In: Verleysen M (ed) ESANN'06, Advances in computational intelligence and learning. D-Side Publishing, Evere, Belgium, pp 611–616

Mouraud A, Paugam-Moisy H, Puzenat D (2006) A distributed and multithreaded neural event driven simulation framework. In: PDCN 2006, International conference on parallel and distributed computing and networks, Innsbruck, Austria, February 2006. ACTA Press, Calgary, 2006

Mouraud A, Puzenat D (2009) Simulation of large spiking neuron networks on distributed architectures, the "DAMNED" simulator. In: Palmer-Brown D, Draganova C, Pimenidis E, Mouratidis H (eds) EANN 2009, Engineering applications of neural networks. Communications in computer and information science, vol 43. Springer, pp 359–370

Natschläger T, Ruf B (1998a) Online clustering with spiking neurons using radial basis functions. In: Hamilton A, Smith LS (eds) Neuromorphic systems: engineering silicon from neurobiology. World Scientific, Singapore, Chap 4

Natschläger T, Ruf B (1998b) Spatial and temporal pattern analysis via spiking neurons. Netw: Comp Neural Syst 9(3):319–332

Newman MEJ (2003) The structure and function of complex networks. SIAM Rev 45:167–256

Newman MEJ, Girvan M (2004) Finding and evaluating community structure in networks. Phys Rev E 69:026113

Nowotny T, Zhigulin VP, Selverston AI, Abardanel HDI, Rabinovich MI (2003) Enhancement of synchronization in a hybrid neural circuit by spike-time-dependent plasticity. J Neurosci 23(30):9776–9785

Olshausen BA, Fields DJ (1996) Emergence of simple-cell receptive field properties by learning a sparse code for natural images. Nature 381:607–609

Oster M, Whatley AM, Liu S-C, Douglas RJ (2005) A hardware/software framework for real-time spiking systems. In: ICANN 2005, International conference on artificial neural networks. Lecture notes in computer science, vol 3696. Springer, New York, pp 161–166

Panchev C, Wermter S (2006) Temporal sequence detection with spiking neurons: towards recognizing robot language instructions. Connect Sci 18:1–22

Paugam-Moisy H, Martinez R, Bengio S (2008) Delay learning and polychronization for reservoir computing. Neurocomputing 71(7–9):1143–1158

Perrinet L, Samuelides M (2002) Sparse image coding using an asynchronous spiking neural network. In: Verleysen M (ed) ESANN 2002, European symposium on artificial neural networks. D-Side Publishing, Evere, Belgium, pp 313–318

Pfister J-P, Gerstner W (2006) Beyond pair-based STDP: a phenomenological rule for spike triplet and frequency effects. In: NIPS 2005, Advances in neural information processing systems, vol 18. MIT Press, Cambridge, MA, pp 1083–1090

Pfister J-P, Barber D, Gerstner W (2003) Optimal Hebbian learning: a probabilistic point of view. In: Kaynak O, Alpaydin E, Oja E, Xu L (eds) ICANN/ICONIP 2003, International conference on artificial neural networks. Lecture notes in computer science, vol 2714. Springer, Heidelberg, pp 92–98

Pfister J-P, Toyoizumi T, Barber D, Gerstner W (2006) Optimal spike-timing-dependent plasticity for precise action potential firing in supervised learning. Neural Comput 18(6):1318–1348

Poggio T, Girosi F (1989) Networks for approximation and learning. Proc IEEE 78(9):1481–1497

Rao RPN (2005) Hierarchical Bayesian inference in networks of spiking neurons. In: Saul LK, Weiss Y, Bottou L (eds) NIPS 2004, Advances in neural information processing systems, vol 17. MIT Press, Cambridge, MA, pp 1113–1120

Recce M (1999) Encoding information in neuronal activity. In: Maass W, Bishop CM (eds) Pulsed neural networks. MIT Press, Cambridge

Reutimann J, Giugliano M, Fusi S (2003) Event-driven simulation of spiking neurons with stochastic dynamics. Neural Comput 15(4):811–830

Rochel O, Martinez D (2003) An event-driven framework for the simulation of networks of spiking neurons. In: Verleysen M (ed) ESANN'03, European symposium on artificial neural networks. D-Side Publishing, Evere, Belgium, pp 295–300

Rubin J, Lee DD, Sompolinsky H (2001) Equilibrium properties of temporal asymmetric Hebbian plasticity. Phys Rev Lett 86:364–366

Rudolph M, Destexhe A (2006) Event-based simulation strategy for conductance-based synaptic interactions and plasticity. Neurocomputing 69: 1130–1133

Rumelhart DE, Hinton GE, Williams RJ (1986) Learning internal representations by back-propagating errors. Nature 323:533–536

Sahani M, Dayan P (2003) Doubly distributional population codes: Simultaneous representation of uncertainty and multiplicity. Neural Comput 15:2255–2279

Salmen M, Plöger PG (2005) Echo state networks used for motor control. In: ICRA 2005, International joint conference on robotics and automation. IEEE, New York, pp 1953–1958

Saudargiene A, Porr B, Wörgötter F (2004) How the shape of pre- and postsynaptic signals can influence STDP: a biophysical model. Neural Comput 16(3):595–625

Schmidhuber J, Wiestra D, Gagliolo D, Gomez M (2007) Training recurrent networks by Evolino. Neural Comput 19(3):757–779

Schmitt M (1998) On computing Boolean functions by a spiking neuron. Ann Math Artif Intell 24:181–191

Schmitt M (2004) On the sample complexity of learning for networks of spiking neurons with nonlinear synaptic interactions. IEEE Trans Neural Netw 15(5):995–1001

Schrauwen B, Van Campenhout J (2004a) Extending SpikeProp. In: Proceedings of the international joint conference on neural networks, vol 1. IEEE Press, New York, pp 471–476

Schrauwen B, Van Campenhout J (2004b) Improving spikeprop: enhancements to an error-backpropagation rule for spiking neural networks. In: Proceedings of the 15th ProRISC workshop, vol 11

Schrauwen B, D'Haene M, Verstraeten D, Van Campenhout J (2007a) Compact hardware for real-time speech recognition using a liquid state machine. In: IJCNN 2007, International joint conference on neural networks, 2007, pp 1097–1102

Schrauwen B, Verstraeten D, Van Campenhout J (2007b) An overview of reservoir computing: theory, applications and implementations. In: Verleysen M (ed) ESANN'07, Advances in computational intelligence and learning. D-Side Publishing, Evere, Belgium, pp 471–482

Schrauwen B, Büsing L, Legenstein R (2009) On computational power and the order-chaos phase transition in reservoir computing. In: Koller D, Schuurmans D, Bengio Y, Bottou L (eds) NIPS'08, advances in neural information processing systems, vol 21. MIT Press, Cambridge, MA, pp 1425–1432

Séguie R, Mercier D (2002) Audio-visual speech recognition one pass learning with spiking neurons. In: ICANN'02, International conference on artificial neural networks. Springer, Berlin, pp 1207–1212

Senn W, Markram H, Tsodyks M (2001) An algorithm for modifying neurotransmitter release probability based on pre- and post-synaptic spike timing. Neural Comput 13(1):35–68

Siegelmann HT (1999) Neural networks and analog computation, beyond the Turing limit. Birkhauser, Boston, MA

Sima J, Sgall J (2005) On the nonlearnability of a single spiking neuron. Neural Comput 17(12):2635–2647

Smith EC, Lewicki MS (2006) Efficient auditory coding. Nature 439:978–982

Song S, Miller KD, Abbott LF (2000) Competitive Hebbian learning through spike-time dependent synaptic plasticity. Nat Neurosci 3(9):919–926

Sporns O, Tononi G, Kotter R (2005) The human connectome: a structural description of the human brain. PLoS Comp Biol 1(4):e42

Standage DI, Trappenberg TP (2005) Differences in the subthreshold dynamics of leaky integrate-and-fire and Hodgkin-Huxley neuron models. In: IJCNN 2005, International joint conference on neural networks. IEEE–INNS, Montreal, pp 396–399

Steil JJ (2004) Backpropagation-decorrelation: Online recurrent learning with O(n) complexity. In: IJCNN 2004, International joint conference on neural networks, vol 1. IEEE–INNS, Montreal, pp 843–848

Stein RB (1965) A theoretical analysis of neuronal variability. Biophys J 5:173–194

Tenore F (2004) Prototyping neural networks for legged locomotion using custom aVLSI chips. Neuromorphic Eng 1(2):4, 8

Thorpe SJ, Gautrais J (1997) Rapid visual processing using spike asynchrony. In: Mozer M, Jordan MI, Petsche T (eds) NIPS 1996, Advances in neural information processing systems, volume 9. MIT Press, Cambridge, MA, pp 901–907

Thorpe S, Fize D, Marlot C (1996) Speed of processing in the human visual system. Nature 381(6582):520–522

Thorpe S, Delorme A, Van Rullen R (2001) Spike-based strategies for rapid processing. Neural Netw 14:715–725

Toyoizumi T, Pfister J-P, Aihara K, Gerstner W (2005a) Generalized Bienenstock-Cooper-Munro rule for spiking neurons that maximizes information transmission. Proc Natl Acad Sci USA 102(14):5239–5244

Toyoizumi T, Pfister J-P, Aihara K, Gerstner W (2005b) Spike-timing dependent plasticity and mutual information maximization for a spiking neuron model. In: Saul LK, Weiss Y, Bottou L (eds) NIPS 2004, Advances in neural information processing systems, vol 17. MIT Press, Cambridge, MA, pp 1409–1416

Turing AM (1939) Systems of logic based on ordinals. Proc Lond Math Soc 45(2):161–228

Turing AM (1950) Computing machinery and intelligence. Mind 59:433–460

Upegui A, Peña Reyes CA, Sanchez E (2004) An FPGA platform for on-line topology exploration of spiking neural networks. Microprocess Microsyst 29:211–223

Valiant LG (1984) A theory of the learnable. Commun ACM 27(11):1134–1142

van Hulle M (2000) Faithful representations and topographic maps: from distortion- to information-based self-organization. Wiley, New York

Van Rullen R, Thorpe S (2001) Rate coding versus temporal order coding: what the retinal ganglion cells tell the visual cortex. Neural Comput 13:1255–1283

Vapnik VN (1998) Statistical learning theory. Wiley, New York

Verstraeten D, Schrauwen B, Stroobandt D (2005) Isolated word recognition using a liquid state machine. In: Verleysen M (ed) ESANN'05, European symposium on artificial neural networks. D-Side Publishing, Evere, Belgium, pp 435–440

Verstraeten D, Schrauwen B, D'Haene M, Stroobandt D (2007) An experimental unification of reservoir computing methods. Neural Netw 20(3):391–403

Viéville T, Crahay S (2004) Using an Hebbian learning rule for multi-class SVM classifiers. J Comput Neurosci 17(3):271–287

Volkmer M (2004) A pulsed neural network model of spectro-temporal receptive fields and population coding in auditory cortex. Nat Comput 3:177–193

Wang G, Pavel M (2005) A spiking neuron representation of auditory signals. In: IJCNN 2005, International joint conference on neural networks. IEEE–INNS, Montreal, pp 416–421

Watts L (1994) Event-driven simulation of networks of spiking neurons. In: Cowan JD, Tesauro G, Alspector J (eds) NIPS 1993, Advances in neural information processing systems, vol 6. MIT Press, Cambridge, MA, pp 927–934

Watts D, Strogatz S (1998) Collective dynamics of "small-world" networks. Nature 393:440–442

Wennekers T, Sommer F, Aertsen A (2003) Editorial: cell assemblies. Theory Biosci (special issue) 122:1–4

Wu S, Chen D, Niranjan M, Amari S (2003) Sequential Bayesian decoding with a population of neurons. Neural Comput 15:993–1012

Wysoski SG, Benuskova L, Kasabov N (2008) Fast and adaptive network of spiking neurons for multi-view visual pattern recognition. Neurocomputing 71(13–15):2563–2575

Xie X, Seung HS (2004) Learning in neural networks by reinforcement of irregular spiking. Phys Rev E 69 (041909)

Xin J, Embrechts MJ (2001) Supervised learning with spiking neuron networks. In: Proceedings of the IJCNN 2001 IEEE international joint conference on neural networks, Washington, DC, vol 3. IEEE Press, New York, pp 1772–1777

Zador AM, Pearlmutter BA (1996) VC dimension of an integrate-and-fire neuron model. Neural Comput 8(3):611–624

Zemel RS, Dayan P, Pouget A (1998) Probabilistic interpretation of population codes. Neural Comput 10:403–430

11 Image Quality Assessment — A Multiscale Geometric Analysis-Based Framework and Examples

Xinbo Gao[1] · *Wen Lu*[2] · *Dacheng Tao*[3] · *Xuelong Li*[4]

[1]Video and Image Processing System Lab, School of Electronic Engineering, Xidian University, China
xbgao@ieee.org

[2]Video and Image Processing System Lab, School of Electronic Engineering, Xidian University, China
luwen@mail.xidian.edu.cn

[3]School of Computer Engineering, Nanyang Technological University, Singapore
dacheng.tao@gmail.com

[4]Center for OPTical IMagery Analysis and Learning (OPTIMAL), State Key Laboratory of Transient Optics and Photonics, Xi'an Institute of Optics and Precision Mechanics, Chinese Academy of Sciences, Xi'an, Shaanxi, China
xuelong_li@opt.ac.cn

G. Rozenberg et al. (eds.), *Handbook of Natural Computing*, DOI 10.1007/978-3-540-92910-9_11,
© Springer-Verlag Berlin Heidelberg 2012

Abstract

This chapter is about objective *image quality assessment* (IQA), which has been recognized as an effective and efficient way to predict the visual quality of distorted images. Basically, IQA has three different dependent degrees on original images, namely, *full-reference* (FR), *no-reference* (NR), and *reduced-reference* (RR). To begin with, we introduce the fundamentals of IQA and give a broad treatment of the state-of-the-art. We focus on a novel framework for IQA to mimic the *human visual system* (HVS) by incorporating the merits from *multiscale geometric analysis* (MGA), *contrast sensitivity function* (CSF), and Weber's law of *just noticeable difference* (JND). Thorough empirical studies were carried out using the laboratory for image and video engineering (LIVE) database against subjective *mean opinion score* (MOS), and these demonstrate that the proposed framework has good consistency with subjective perception values and the objective assessment results well reflect the visual quality of the images.

1　Introduction

With the rapid development of information technology, digital image as a media for representing and communicating has witnessed tremendous growth. A huge number of processing methods have been proposed to treat images for different purposes. The performance of these methods highly depends on the quality of the images after processing. Therefore, how to evaluate image quality has become a burning question in recent years. Problems of *image quality assessment* (IQA) (Wang and Bovik 2006) occur in many applications, such as image compression, enhancement, communication, storage and watermarking, etc. In the process of image compression, lossy compression techniques may introduce artificial block, blurring and ringing effects, which can lead to image degradation. In poor transmission channels, transmission errors or data dropping may happen, which can lead to imperfect quality and distortion of the received video data. The past 5 years have witnessed tremendous demands for IQA methods in the following three ways: (1) they can be used to monitor image quality for quality control systems; (2) they can be employed to benchmark image processing systems and algorithms; (3) they can also be embedded into image processing systems to optimize algorithms and parameter settings.

Existing IQA metrics can be categorized into subjective (Rec. ITU-R) and objective methods (Wang and Bovik 2006). The former is based on quality, which is assessed by human observers; the latter depends on quantified parameters, which are obtained from the model to measure the image quality.

Because human observers are the ultimate receivers of the visual information contained in an image, a subjective method whose results are directly given by human observers is probably the best way to assess the quality of an image. The subjective method is one in which the observers are asked to evaluate the picture quality of sequences using a continuous grading scale and to give one score for each sequence. A number of different subjective methods are represented by ITU-R Recommendation BT.500 (Rec. ITU-R). The subjective quality measurement has been used for many years as the most reliable form of quality measurement. However, subjective experiment requires human viewers working over a long period and repeated experiments are needed for many image objects. Furthermore, it is expensive, time consuming, and cannot be easily and routinely performed for many scenarios, for example, real-time systems. Moreover, there currently is no precise mathematical model for subjective assessment. As a consequence, there is a need for an objective quality metric that accurately matches the subjective quality and can be easily implemented in various image systems, leading to the emergence of objective IQA.

Objective IQA makes use of the variation of several original or distorted image characteristics caused by degradation to represent the variation of image perceptual quality. Many objective quality metrics for predicting image distortions have been investigated. According to the availability of original image information, there is a general agreement (Wang and Bovik 2006) that objective quality metrics can be classified into three categories: *full-reference* (FR), *no-reference* (NR), and *reduced-reference* (RR).

To evaluate the quality of a distorted image, FR metrics, which have access to both whole original and reconstructed information, provide the most precise evaluation results compared with NR and RR. Generally, FR metrics can be divided into two steps: one tends to construct the errors between original and distorted images, then produce distortion maps; the other provides the global IQA by pooling the errors. Conventional FR IQA methods (Avcibas et al. 2002; Eskicioglu and Fisher 1995) calculate pixel-wise distances, for example, *peak signal-to-noise ratio* (PSNR) and *mean square error* (MSE), between the distorted image and the corresponding reference. These measurements are attractive because of their simplicity and good performance when images with the same type of degradation are compared. However, when images with different types of degradations are compared, the results may not be consistent with that of subjective IQA. Besides, images with various types of degradations but the same value could have very different subjective qualities. This is mainly because they are based on pixel-to-pixel difference calculations, which disregard image content and some characteristics of human visual perception.

Recently, a great deal of effort has been made to develop visual models (Mannos and Sakrison 1974) that take advantage of the known characteristics of the *human visual system* (HVS). The aim of HVS-based IQA is to evaluate how strong the error signal is perceived by the HVS according to the characteristics of human visual error sensitivity. A number of IQA methods have been proposed to evaluate the perceptual quality. Two famous models, the Daly *Visible Differences Predictor* (VDP) and Sarnoff *Visual Discrimination Model* (VDM), were proposed by Daly (1993) and Lubin (1995), respectively. The Daly VDP receives the distorted and corresponding reference as input and produces a difference map as output, which predicts the probability of detection for dissimilarities throughout the whole image. If two images vary substantially, the probability of prediction will be one, and as the differences aggravate, the probability does not increase further. The Sarnoff VDM was designed for physiological plausibility as well as for speed and simplicity. While the Daly VDP is an example of a frequency domain visual model, the Sarnoff VDM operates in the spatial domain. In Karunasekera and Kingsbury (1995), these two excellent models were comparatively evaluated. The Watson (1993) metric uses the DCT-based perceptual error measurement. The quantified errors for every coefficient in every block are scaled by the corresponding visual sensitivities of every DCT basis function in each block. Miyahara et al. (1998) reported a new methodology for the determination of an objective metric. This methodology is applied to obtain a *picture quality scale* (PQS) for the coding of achromatic images over the full range of the image quality presented by the *mean opinion score* (MOS). Damera-Venkata et al. (2000) modeled the degraded image as a reference image polluted by linear frequency distortion and additive noise contamination. Since the psycho-visual effects of frequency distortion and noise contamination are independent, they decouple these two sources of degradation and measure their effect on the HVS. Instead of computing a residual image, they compute a model restored image by applying the restoration algorithm on the original image, using the same parameters as those used while restoring the degraded image. A number of limitations of HVS-based IQA methods were discussed in Wang and Bovik (2006), because they must rely on several strong assumptions and generalizations.

Different from traditional HVS-based error-sensitivity IQA, structural similarity-based IQA (Wang and Bovik 2006) is based on the following philosophy: *The main function of the HVS is to extract structural information from the viewing field, and the HVS is highly adapted for this purpose.* Therefore, measurements of structural information loss can provide a good approximation to image perceived distortion. Wang and Bovik (2006) introduces a complementary framework for IQA based on the degradation of structural information and develops a structural similarity (SSIM) index to demonstrate its promise through a set of intuitive examples. Extensive experimental results have demonstrated that SSIM achieves better performance compared with the traditional methods. Sheikh et al. proposed to quantify the loss of image information in the distortion process and explore the relationship between image information and visual quality. They modeled natural images in the wavelet domain using *Gaussian scale mixtures* (GSM) (Wang and Bovik 2006) and employed the *natural scene statistics* (NSS) model for IQA (Wang and Bovik 2006). The *visual information fidelity* (VIF) method was derived from the combination of a statistical model for natural scenes, an image distortion model, and a HVS model in an information-theoretic setting. The experimental results demonstrate that it outperformed traditional IQA methods by a sizable margin. Sheikh et al. also furnished a statistical evaluation of recent FR algorithms (Tao et al. 2007b).

Since FR metrics need full information of images on the comparison between corresponding regions of the original image and the degraded image, this requirement makes the metrics less than optimal for some applications, which require images and videos to be broadcasted or transmitted through the data network. For these applications, FR metrics might be impossible or too expensive to allocate the extra bandwidth that is required to send information of images. Therefore, researchers are seeking a rational IQA method that could work without any prior information. NR IQA emerges in demand, it does not require any information of the original image and only uses a distortion analysis of the distorted image to assess its quality. It addresses a fundamental distinction between fidelity and quality, that is, HVS usually does not need any reference to determine the subjective quality of a target image. This kind of metric is most promising in the context of a video broadcast scenario, since the original images or videos are not accessible to terminal users in practice.

NR IQA is a relatively new research direction with a promising future. However, NR is also a difficult task, and most NR metrics are designed for one or a set of predefined specific distortion types and are unlikely to generalize for images with other types of distortions, for example, the blocking effect in JPEG, the ringing and blurring artifacts in JPEG2000. In practical applications, they are useful only when the distortion type of the distorted images are fixed and known. So there is a big gap between NR metrics and real scenarios. Wang and Bovik (2006) proposed a computational and memory efficient quality model for JPEG images. Li (2002) proposed to appraise the image quality by three objective measures: edge-sharpness level, random noise level, and structural noise level, which jointly provide a heuristic approach to characterizing most important aspects of visual quality. Sheikh et al. (2006) use NSS models to blindly measure the quality of the image compressed using the JPEG2000 scheme.

They claim that natural scenes contain nonlinear dependencies that are disturbed by the compression process, and this disturbance can be quantified and related to human perceptual quality.

FR metrics may not be applicable because of the absence of original images. On the other hand, NR or "blind" IQA is such an extremely difficult task that it is impossible to apply a universal NR metric to practical applications. Although there is an urgent demand for NR QA methods that are applicable to a wide variety of distortions, unfortunately no such method has been proposed and extensively tested. RR IQA provides an alternative solution that compromises the FR and NR methods. It only makes use of partial information from the original images to evaluate the perceptual quality of the distorted image. In general, certain features or physical measures are extracted from the original image and then transmitted to the receiver as extra information for evaluating the quality of the image or video.

RR metrics may be less accurate for evaluating image quality than the FR metrics, but they are less complicated and make real-time implementations more affordable. RR metrics are very useful for monitoring quality on transmission network. In such contexts, the features are transmitted with the coded sequence if they correspond to a reasonable overhead. They also can be used to track image quality degradations and control the streaming resources in real-time visual communication systems. Recently, the VQEG (2000a) report has included RR image and video quality assessment as one of its directions for future development. RR methods are useful in a number of applications. Masry et al. (2006) presented a computationally efficient video distortion metric that can operate in FR and RR model as required. This metric is based on a model of the HVS and implemented using the wavelet transform and separable filters. The visual model is parameterized using a set of video frames and the associated quality scores. Wolf and Pinson (2005) presented a new RR video quality monitoring system under low bandwidth. This system utilizes less than 10 kbits/s of reference information from the source video stream.

In the RR model, the key issue is to determine which features have the best ability to capture distortion between the original image and the distorted image. The best features are the ones that are able to produce the highest correlation between objective quality scores and subjective ones. In order to select the best feature to design a reduced description IQA method, Lu et al. (Gao et al. 2008a; Lu et al. 2008; Li et al. 2009) utilized some transforms to extract features for representing images based on HVS. Wang and Bovik (2006) proposed an RR IQA method called the *wavelet-domain natural image statistic metric* (WNISM), which achieves promising performance for image visual perception quality evaluation. The underlying factor in WNISM is that the marginal distribution of wavelet coefficients of a natural image conforms to the generalized Gaussian distribution. Based on this fact, WNISM measures the quality of a distorted image by the fitting error between the wavelet coefficients of the distorted image and the Gaussian distribution of the reference. They use the Kullback–Leibler distance to represent this error, so that only a relatively small number of RR features are needed for the evaluation of image quality.

In recent years, neural computing has emerged as a practical technology, with successful applications in many fields (Petersena et al. 2002). The majority of these applications are concerned with problems in image processing. They can address most of the various steps that are involved in the image processing chain: from the preprocessing to the image understanding level. For example, in the process of image construction, Wang and Wahl trained a Hopfield *artificial neural network* (ANN) for the reconstruction of 2D images from pixel data obtained from projections (Wang and Wahl 1997). In image enhancement, Chandrasekaran et al. (1996) used a novel feed-forward architecture to classify an input window as containing an edge or not. The weights of the network were set manually instead of being obtained from training. In the phase of image understanding, it is considered as part of artificial intelligence or perception, which involves recognition, classification, and relational matching. So it seems that neural networks can be used to advantage in certain problems of image understanding, especially in feature extraction. Feature extraction can be seen as a special kind of data reduction, of which the goal is to find a subset of informative variables based on image data. In image processing, an effective approach to extract feature can be employed to represent image sparsely. A well-known feature-extraction ANN is Oja's neural implementation of a one-dimensional *principal component analysis* (PCA) (Oja 1982). To aim at IQA, feature extraction is probably the most important stage, and effective features can well reflect the quality of digital images and vice versa. So the neural network for feature extraction is used to evaluate image quality in RR mode. In Bishop (1995) and Lampinen and Oja (1998), Callet et al. (2006) use a *convolutional neural network* (CNN) that allows a continuous time scoring of the video to complete the QA in MPEG-2 video. Objective features are extracted on a frame-by-frame basis on both the reference and the distorted sequences. They are derived from a perceptual-based representation and integrated along the temporal axis using a *time-delay neural network* (TDNN). By realizing a nonlinear mapping between nonsubjective features extracted from the video frames and subjective scores, experimental results demonstrated that TDNN can be useful to assess the perceived quality of video sequences. Gastaldo and Zunino (2004) use a *circular back propagation* (CBP) feed-forward network to process objective features extracted from JPEG images and to return the associated quality scores. Gastaldo et al. (2002) feed the CBP network, estimating the corresponding perceived quality; objective features are continuously extracted from compressed video streams on a frame-by-frame basis. The resulting adaptive modeling of subjective perception supports a real-time system for monitoring displayed video quality.

Although the aforementioned RR metrics achieve a good solution for IQA problems, there is still much room to further improve the performance of RR IQA, because the existing methods fail to consider the statistical correlations of transformed coefficients in different subbands and the visual response characteristics of the mammalian cortical simple cells. Moreover, wavelet transforms cannot explicitly extract the image geometric information; for example, lines, curves, and wavelet coefficients are dense for smooth image edge contours.

To target the aforementioned problems, to further improve the performance of RR IQA, and to broaden RR IQA related applications, a novel HVS driven framework is proposed. The new framework is consistent with HVS: MGA decomposes images for feature extraction to mimic the multichannel structure of HVS, CSF reweights MGA decomposed coefficients to mimic the nonlinearities inherent in HVS, and JND produces a noticeable variation in sensory experience. This framework contains a number of different ways for IQA because MGA offers a series of transforms including wavelet (Mallet 1989), contourlet (Do and Vetterli 2005), *wavelet-based contourlet transform* (WBCT) (Eslami and Radha 2004), and *hybrid wavelets and directional filter banks* (HWD) (Eslami and Radha 2005), and different transforms capture different types of image geometric information. Extensive experiments based on the laboratory for image and video engineering (LIVE) database (Sheikh et al. 2003) against subjective MOS (VQEG 2000a) have been conducted to demonstrate the effectiveness of the new framework.

2   Multiscale Geometric Analysis

Multiscale geometric analysis (MGA) (Romberg 2003) is such a framework for optimally representing high-dimensional function. It is developed, enhanced, formed, and perfected in signal processing, computer vision, machine learning, and statistics. MGA can detect, organize, represent, and manipulate data, for example, edges, which nominally span a high-dimensional space but contain important features approximately concentrated on lower-dimensional subsets, for example, curves.

In recent decades, scientists devoted themselves to finding a simple way to present images. With the advances in science, MGA has formed a big family with a large number of members. Wavelet is the first one that can analyze images by multiscale and multidirection transforms, and can recover the original image in a lossless way. Due to their good *nonlinear approximation* (NLA) performance for piecewise smooth functions in one dimensions, wavelets have been successfully applied to many image processing tasks, such as low bit-rate compression and denoising. To utilize wavelets, we can catch point or zero-dimensional discontinuities effectively. However, in two-dimension or other higher dimensions, the wavelet is not able to depict the singulars efficiently and effectively. In essence, wavelets in two-dimension obtained by a tensor-product of one-dimensional wavelets will be good at isolating the discontinuities at edge points, but will not see the smoothness along the contours.

Taking ❷ *Fig. 1* as an example (Do and Vetterli 2005) for approximating the contour efficiently, the wavelet is limited to using brushes of square shapes along the contour, with different sizes corresponding to the multiresolution of wavelets. As the resolution gets finer, the limitation of the wavelet scheme can be clearly seen since it requires many finer dots to capture the contour. The X-let styles, which are the expected transforms, on the other hand, have more freedom in making brush strokes in different directions and rectangular shapes that follow the contour. As hoped, the expected NLA methods are much more efficient than the wavelet.

In addition, wavelets have only three directions and lack the important feature of directionality; hence, they are not efficient in retaining textures and fine details in these applications. There have been several efforts toward developing geometrical image transforms. According to the above analysis, the expected multiscale geometrical image transforms should contain the following features (Do and Vetterli 2005). Multi-resolution, localization, critical sampling, directionality, and anisotropy.

◘ Fig. 1

The wavelet versus other lets approximating the 2D contour.

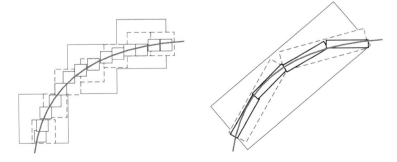

Recently, some more powerful MGA representations were created. In 1998, Cands pioneered a nonadaptive method for high dimension functions representation, named Ridgelet (Cands 1998). And in the same year, Donoho gave a method for constructing the orthonormal ridgelet. The ridgelet can approximate the multivariable functions containing the line-singularity effectively, but for the others with curve-singularity, it performs only as well as the wavelet. In 2000, Pennec and Mallat proposed a new system of representation, called bandelet (Pennec and Mallat 2000). It can represent the images based on the edges and track the geometrically regular directions of images. Thus, if one knew the geometrically regular directions of the images, bandelet would lead to optimal sparse representations for images and have great potential in image compression. In order to achieve optimal approximation behavior in a certain sense for 2D piecewise smooth functions in R^2 where the discontinuity curve is a C^2 function, Cands and Donoho constructed the curvelet (Cands and Donoho 2000) transform. More specifically, an M-term NLA for such piecewise smooth function using curvelets has L^2 square error decaying. An attractive property of the curvelet system is that such correct approximation behavior is simply obtained via thresholding a fixed transform. The key features of the curvelet elements are that they exhibit very high directionality and anisotropy. However, the original construction of the curvelet transform is intended for functions defined in the continuum space; when the critical sampling is desirable, the development of discrete transforms for sampled images remains a challenge.

Based on the key features, namely directionality and anisotropy, which make curvelets an efficient representation for 2D piecewise smooth functions with smooth discontinuity curves, Do and Vetterli proposed a new image transform, contourlet, which is also called the *pyramidal directional filter bank* (PDFB) (Bamberger and Smith 1992). The contourlet first uses the Laplacian pyramid (LP) (Burt and Adelson 1983) to decompose the image and capture point singularity. Then, it combines all the singular points in every direction into one coefficient by directional filter bank. The contourlet provides a multiscale and directional decomposition for images with a small redundancy factor, and a frame expansion for images with frame elements like contour segments, which is the reason that it is named contourlet. The connection between the developed discrete- and continuous-domain constructions is made precise via a new directional multi-resolution analysis, which provides successive refinements at both spatial and directional resolutions. The contourlet transform can be designed to satisfy the anisotropy scaling relation for curves and thus it provides a curvelet-like decomposition for images. The contourlet transform aims to achieve an optimal approximation rate of piecewise smooth functions with discontinuities along twice continuously differentiable curves. Therefore, it captures areas with subsection smooth contours. WBCT and HWD are designed to optimize the representation of image features without redundancy.

For IQA, one needs to find MGA transforms that perform excellently for reference image reconstruction, have perfect perception of orientation, are computationally tractable, and are sparse and effective for image representation. Among all requirements for IQA, the effective representation of visual information is especially important. Natural images are not simple stacks of 1D piecewise smooth scan-lines, and points of discontinuity are typically located along smooth curves owing to smooth boundaries of physical objects. As a result of a separable extension from 1D bases, the 2D wavelet transform is not good at representing visual information of images. Consequently, when we dispose of the image with linear features the wavelet transform is not effective. However, MGA transforms can capture the characteristics of images, for example, lines, curves, cuneiforms, and the contours of object. As mentioned in ❷ *Table 1*, different transforms of MGA capture different features of an image and

◻ Table 1
The performance of PSNR, MSSIM, and WNISM on the LIVE database

Metric	Type	JPEG					JPEG2000				
		CC	ROCC	OR	MAE	RMSE	CC	ROCC	OR	MAE	RMSE
PSNR	FR	0.9229	0.8905	0.1886	7.118	9.154	0.933	0.9041	0.0947	6.407	8.313
MSSIM	FR	0.9674	0.9485	0.04	4.771	5.832	0.949	0.9368	0.0651	5.422	6.709
WNISM	RR	0.9291	0.9069	0.1486	6.0236	8.2446	0.9261	0.9135	0.1183	6.135	7.9127

complement each other. Based on the mentioned requirements for IQA, it is reasonable to consider a wide range of explicit interactions between multiscale methods and geometry, for example, contourlet (Do and Vetterli 2005), WBCT (Eslami and Radha 2004), and HWD (Eslami and Radha 2005).

3   Multiscale Geometric Analysis for Image Quality Assessment

As discussed in ❷ Sect. 2, MGA contains a series of transforms, which can analyze and approximate geometric structure while providing near optimal sparse representations. The image sparse representation means we can represent the image by a small number of components, so small visual changes of the image will affect these components significantly. Therefore, sparse representations can be well utilized for IQA. In this chapter, a novel framework for IQA (Gao et al. 2008b) is developed by applying MGA transforms to decompose images and extract effective features. This framework (Wang and Bovik 2006) quantifies the errors between the distorted and the reference images by mimicking the error sensitivity function in the HVS. The objective of this framework is to provide IQA results, which have good consistency with subjective perception values. ❷ *Figure 2* shows the framework for IQA.

3.1   MGA-Based Feature Extraction

A number of MGA transforms (Do 2001), which are contourlet, WBCT, and HWD, are considered for image decomposition and feature extraction. Moreover, the wavelet is utilized as the baseline for comparison. In this framework, MGA is utilized to decompose images and then extract features to mimic the multichannel structure of HVS. Moreover, there is a wide application of neural computing that can be applied to perform feature extraction for image representation sparsely. Some methods of neural computing are selected to extract features of image data further. Feature extraction (Bishop 1995) is treated as a means for reducing dimensionality of the image data and preserving feature data separability well, for example, feed-forward ANN, *self-organizing feature map* (SOMs) and Hopfield ANN.

3.1.1   Wavelet Transform

The wavelet transform (Mallat 1989) is well known as an approach to analyze signals in both time and frequency domains simultaneously and adaptively. Features are extracted effectively

◘ Fig. 2

MGA-based IQA framework. (Sender side is applied to extract the normalized histogram of the reference image. Receiver side is applied to extract the normalized histogram of the distorted image. Ancillary channel is applied to transmit the extracted normalized histogram.)

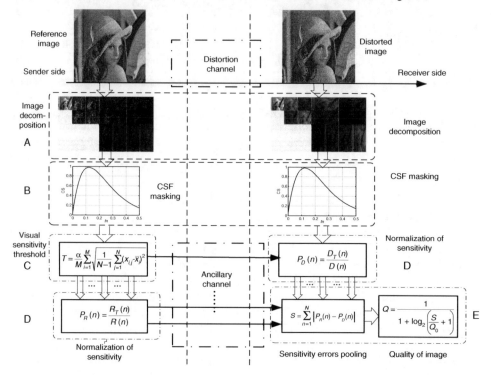

from signals, especially non-stationary signals, by multiscale operation. In this article, three levels of wavelet transform are applied to decompose the image into nine highpass subbands and a lowpass residual subband. Then all the highpass subbands are selected for feature extraction in the proposed framework. ❷ *Figure 3* shows decomposition of the image using wavelet transforms and a set of selected subbands (marked with white dashed boxes and white numerals) are used for feature extraction in the proposed framework.

3.1.2 Contourlet Transform

The contourlet transform (Mallat 1989) can capture the intrinsic geometrical structure that is key in visual information. Contourlet consists of two major stages: the multiscale decomposition and the direction decomposition. At the first stage, it uses a *Laplacian pyramid* (LP) (Bishop 1995) to capture the point discontinuities. For the second stage, it uses *directional filter banks* (DFB) (Burt and Adelson 1983) to link point discontinuities into linear structures. Here, every image is decomposed into three pyramidal levels. Based on the characteristics of DFB for decomposition, half of the directional subbands are selected for the feature extraction. The decomposition of images using the contourlet transform and a set of selected subbands

◘ **Fig. 3**
Wavelet transform-based image decomposition.

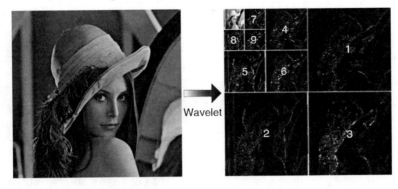

◘ **Fig. 4**
Contourlet-transform-based image decomposition.

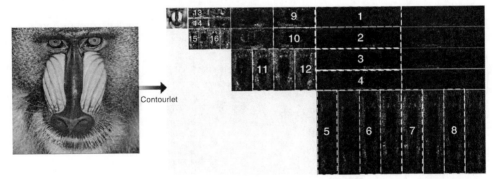

(marked with white dashed boxes and numerals) are used for feature extraction in the proposed framework, as shown in ❷ *Fig. 4.*

3.1.3 WBCT

WBCT (Eslami and Radha 2004) has a construction similar to the contourlet transform, which consists of two filter bank stages. The first stage is subband decomposition by wavelet transform. The second stage of the WBCT is angular decomposition. In this stage, WBCT employs DFB in the contourlet transform. Here, each image is decomposed into two wavelet levels, and the number of DFB decomposition levels at each highpass of the finest and finer scales is equal to 3. The image is decomposed into 48 high-frequency directional subbands and a lowpass residual subband in all. Like the contourlet transform, half of the subbands at each fine scale are selected to extract the features of the image. The decomposition of the image using the WBCT transform and a set of selected subbands (marked with numerals and gray blocks) are applied for feature extraction in the proposed framework, as demonstrated in ❷ *Fig. 5.*

◻ Fig. 5

Wavelet-based contourlet transform (WBCT)-based image decomposition.

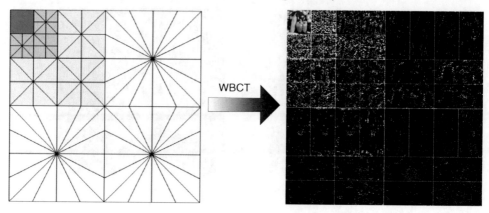

◻ Fig. 6

Hybrid wavelets and directional filter banks (HWD) transform-based image decomposition.

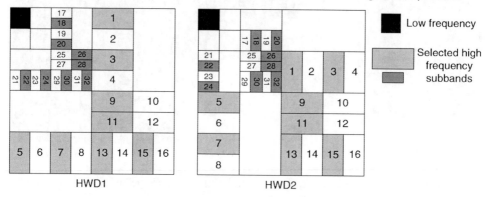

3.1.4 HWD Transform

Like WBCT, the HWD transform (Eslami and Radha 2005) employs wavelets as the multiscale decomposition. The DFB and modified DFB are applied to some of the wavelets subbands. When the HWD transform (HWD1 and HWD2 respectively) is employed in the proposed framework, each image is decomposed into two wavelet levels, and the number of DFB decomposition levels at each highpass of the finest and finer scales is equal to 3. Thus, the image is decomposed into 36 high-frequency directional subbands and a lowpass residual subband in all. Half of the directional subbands at each fine scale (for HWD1 and HWD2, 16 directional subbands in all, respectively) are selected to extract features of the image. The HWD decomposition of the image and a set of selected subbands (marked with numerals and gray blocks) are applied for feature extraction in the proposed frameworks, as demonstrated in ❷ *Fig. 6.*

3.2 CSF Masking

MGA is introduced to decompose images and then extract features to mimic the multichannel structure of HVS; that is, HVS (Wandell 1995) works similar to a filter bank (containing filters with various frequencies). CSF (Miloslavski and Ho 1998) measures how sensitive we are to the various frequencies of visual stimuli; that is, we are unable to recognize a stimuli pattern if its frequency of visual stimuli is too high. For example, given an image consisting of horizontal black and white stripes, we will perceive it as a gray image if stripes are very thin; otherwise, these stripes can be distinguished. Because coefficients in different frequency subbands have different perceptual importance, it is essential to balance the MGA decomposed coefficients via a weighting scheme, CSF masking. In this framework, the CSF masking coefficients are obtained using the *modulation transfer function* (MTF) (Miloslavski and Ho 1998), that is,

$$H(f) = a(b + cf)e^{(-cf)^d} \tag{1}$$

where $f = f_n \times f_s$, the center frequency of the band, is the radial frequency in cycles/degree of the visual angle subtended, f_n is the normalized spatial frequency with units of cycles/pixel, and f_s is the sampling frequency with units of pixels/degree. According to Bamberger and Smith (1992), a, b, c, and d are 2.6, 0.192, 0.114, and 1.1, respectively. The sampling frequency f_s is defined as (Nadenau et al. 2003)

$$f_s = \frac{2vr \cdot \tan(0.5°)}{0.0254} \tag{2}$$

where v is the viewing distance with units of meters and is the resolution power of the display with units of pixels/in. In this framework, v is 0.8 m (about 2–2.5 times the height of the display), the display is 21 in. with a resolution of 1024 × 768. According to the Nyquist sampling theorem, if changes from 0 to $f_s/2$, so f_n changes from 0 to 0.5. Because MGA is utilized to decompose an image into three scales from coarse to fine, we have three normalized spatial frequencies, $f_{n1} = 3/32$, $f_{n2} = 3/16$, $f_{n3} = 3/8$. Weighting factors are identical for coefficients in an identical scale.

For example, if the contourlet transform is utilized to decompose an image, a series of contourlet coefficients $c_{i,j}^k$ is obtained, where k denotes the level index (the scale sequence number) of the contourlet transform, i stands for the serial number of the directional subband index at the kth level, and j represents the coefficient index. By using CSF masking, the coefficient $c_{i,j}^k$ is scaled to $x_{i,j}^k = H(f_k) \cdot c_{i,j}^k$.

3.3 JND Threshold

Because HVS is sensitive to coefficients with larger magnitude, it is valuable to preserve visually sensitive coefficients. JND, a research result in psychophysics, is a suitable means for this. It measures the minimum amount by which stimulus intensity must be changed to produce a noticeable variation in the sensory experience. In our framework, MGA is introduced to decompose an image and highpass subbands contain the primary contours and textures information of the image. CSF masking makes coefficients have similar perceptual importance in different frequency subbands, and then JND is calculated to obtain a threshold to remove visually insensitive coefficients. The number of visually sensitive coefficients reflects the visual quality of the reconstructed images. The lower the JND threshold is, the more coefficients are utilized for image reconstruction and the better the visual quality of the reconstructed image is. Therefore, the normalized histogram reflects the visual quality of an image. Here, the JND threshold is defined as

$$T = \frac{\alpha}{M} \sum_{i=1}^{M} \sqrt{\frac{1}{N_i - 1} \sum_{j=1}^{N_i} (x_{i,j} - \bar{x}_i)^2} \tag{3}$$

where $x_{i,j}$ is the jth coefficient of the ith subband in the finest scale and $\bar{x}_i$ is the mean value of the ith subband coefficients, M is the number of selected subbands in the finest scale, N_i is the number of coefficients of the ith subband, and α is a tuning parameter corresponding to different types of distortion.

Using the JND threshold technique, some parameters are so delicate that they have to be selected using empirical values. Others may be tunable by trial and error or a cross validation process by observing the overall performance of the system. For optimizing the JND threshold technique, some methods of neural computing (Bishop 1995) might be applied to train parameters or variables for autofit thresholds. Neural networks provides a whole family of divergent formalisms to optimize corresponding values for improving the overall performance, for example, the *radial basis function* (RBF) network.

3.4 Normalization of Sensitivity

By using the JND threshold T, we can count the number of visually sensitive coefficients in the nth selected subband and define the value as $C_T(n)$, which means the number of coefficients in the nth selected subband that are larger than T obtained from ❷ Eq. 5. The number of coefficients in the nth selected subband is $C(n)$. Therefore, for a given image, we can obtain the normalized histogram with L bins (L subbands are selected) for representation and the nth entry is given by

$$P(n) = \frac{C_T(n)}{C(n)} \tag{4}$$

3.5 Sensitivity Errors Pooling

Based on ❷ Eq. 4, the normalized histograms can be obtained for both the reference and the distorted images as $P_R(n)$ and $P_D(n)$, respectively. In this framework, the metrics of the distorted image quality can be defined as

$$Q = \frac{1}{1 + \log_2\left(\frac{S}{Q_0} + 1\right)} \tag{5}$$

where $S = \sum_{n=1}^{L} |P_R(n) - P_D(n)|$ is the city-block distance between $P_R(n)$ and $P_D(n)$, and Q_0 is a constant used to control the scale of the distortion measure. In this framework, we set Q_0 as 0.1. The log function is introduced here to reduce the effects of large S and enlarge the effects of small S, so that we can analyze a large scope of S conveniently. There is no particular reason to choose the city-block distance, which can be replaced by others, for example, the Euclidean norm. This is also the case for the base 2 for the logarithm. The entire function preserves the monotonic property of S.

4   Performance Evaluation

In this section, we compare the performance of the proposed framework based on different MGA transforms with standard IQA methods, that is, PSNR, WNISM, and MSSIM, based on the following experiments: the consistency experiment, the cross-image and cross-distortion experiment, and the rationality experiment. At the beginning of this section, the image database for evaluation is first briefly described.

The LIVE database (Sheikh et al. 2003) is widely used to evaluate the image quality measures, in which 29 high-resolution RGB color images are compressed at a range of quality levels using either JPEG or JPEG2000, producing a total of 175 JPEG images and 169 JPEG2000 images. To remove any nonlinearity due to the subjective rating process and to facilitate comparison of the metrics in a common analysis space, following the procedure given in the *video quality experts group* (VQEG) (VQEG 2000) test, variance-weighted regression analysis is used in a fitting procedure to provide a nonlinear mapping between the objective and subjective MOS. After the nonlinear mapping, the following three metrics are used as evaluation criteria (VQEG 2000): Metric 1 is the Pearson linear *correlation coefficient* (CC) between the objective and MOS after the variance-weighted regression analysis, which provides an evaluation of prediction accuracy; Metric 2 is the Spearman *rank-order correlation coefficient* (ROCC) between the objective and subjective scores, which is considered as a measure of the prediction monotonicity; Metric 3 is the *outlier ratio* (OR), the percentage of the number of predictions outside the range of twice the standard deviation of the predictions after the nonlinear mapping, which is a measure of the prediction consistency. In addition, the *mean absolute error* (MAE) and the *root mean square error* (RMSE) of the fitting procedure are calculated after the nonlinear mapping.

In the following parts, we compare the performance of different IQA methods based on the aforementioned image database and metrics.

4.1   The Consistency Experiment

In this subsection, the performance of the proposed IQA framework will be compared with PSNR (Avcibas et al. 2002), WNISM (Wang and Bovik 2006), and the well-known full reference assessment metric, MSSIM (Wang and Bovik 2006). The evaluation results for all IQA methods being compared are given as benchmarks in ❷ *Table 1*. ❷ *Table 2* shows the

🞏 Table 2

The performance of the proposed IQA framework with different MGA transforms on the LIVE database

Metric		JPEG					JPEG2000				
		CC	ROCC	OR	MAE	RMSE		CC	ROCC	OR	MAE	RMSE
Wavelet	1	0.9587	0.9391	0.0914	5.1012	6.5812	6	0.9487	0.9333	0.0473	5.3486	6.8453
Contourlet	3	0.9493	0.9309	0.1029	5.3177	6.6941	2	0.9451	0.9273	0.0710	5.5818	6.9616
WBCT	3	0.9728	0.9527	0.0457	4.1162	5.3750	6	0.9565	0.9390	0.0414	4.9891	6.4718
HWD1	2	0.9704	0.9526	0.0514	4.3305	5.5646	5	0.9540	0.9333	0.0493	5.2857	6.7972
HWD2	3	0.9728	0.9543	0.0400	4.1135	5.3206	5	0.9540	0.9362	0.0473	5.1357	6.6312

evaluation results of the proposed IQA framework with different MGA transforms, e.g., wavelet, contourlet, WBCT, HWD1, and HWD2. ❯ *Figures 7* and ❯ *8* present the scatter plots of MOS versus the predicted score using objective metrics after the nonlinear mapping.

❯ *Table 2* shows that the results of the MGA transforms using the proposed framework provide a higher effectiveness for IQA for JPEG and JPEG2000 images, respectively.

❏ **Fig. 7**

Scatter plots of mean opinion score (MOS) versus different image quality assessment (IQA) methods for JPEG and JPEG2000 images.

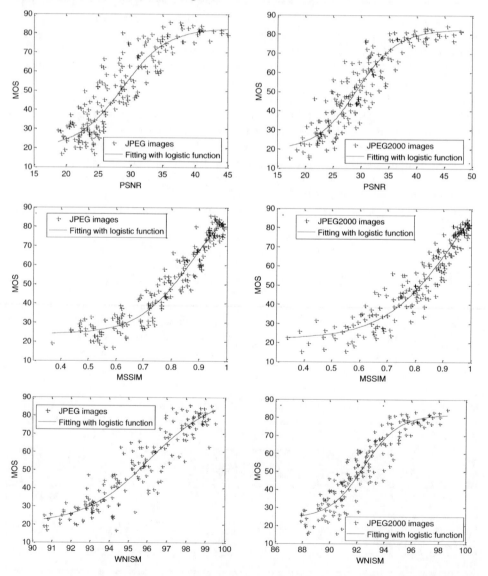

☐ **Fig. 8**

Trend plots of Lena with different distortions using the proposed framework with contourlet.

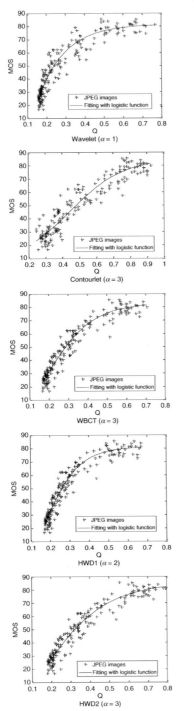
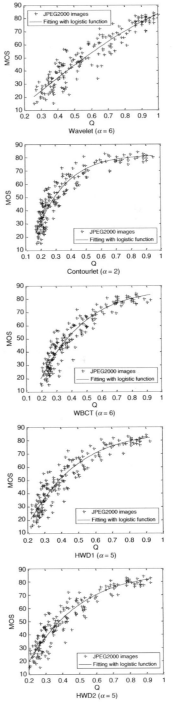

Comparing ❯ *Table 2* with ❯ *Table 1*, one can observe that the metric obtained by the proposed framework provides much better performance than WNISM, including better prediction accuracy (higher CC), better prediction monotonicity (higher ROCC), and better prediction consistency (lower OR, MAE, and RMSE). Particularly when HWD is employed in the proposed framework, better performance will be achieved than with the MSSIM index. The only tuning parameter α in the proposed framework responds to different distortions.

Since the key stage of IQA is how to represent images effectively and efficiently, it is necessary to investigate different transforms. The wavelet is localized in the spatial domain and the frequency domain and can extract local information with high efficiency, so it optimally approximates a target function with 1D singularity. However, the wavelet cannot achieve the sparse representation of edges even if it captures the point singularity effectively. In order to represent edges in an image sparsely, the contourlet approach analyzes the scale and the orientation respectively and reduces the redundancy by approximating images with line-segments similar basis ultimately. To further reduce the redundancy, HWD is proposed to represent images sparsely and precisely. Redundancy is harmful for IQA. If an image is represented by a large number of redundant components, small visual changes will affect the quality of the image slightly. Both transforms are nonredundant, multiscale, and multi-orientation for image decomposition. ❯ *Table 3* shows the proposed IQA framework with HWD works much better than previous standards.

4.2   The Rationality Experiment

To verify the rationality of the proposed framework, contourlet and WBCT are chosen to test the Lena image with different distortions: blurring (with smoothing window of $W*W$), additive Gaussian noise (mean = 0, variance = V), impulsive salt–pepper noise (density = D), and JPEG compression (compression rate = R).

❯ *Figures 9* and ❯ *10* (all images are 8 bits/pixel and resized from 512×512 to 128×128 for visualization) show the relationships between the Lena image with different distortions and the IQA methods prediction trend for the corresponding image.

◘ Table 3

The value of different IQA metrics for images in ❯ *Fig. 11*

Metric	(b)	(c)	(d)
PSNR	24.8022	24.8013	24.8041
MSSIM	0.9895	0.9458	0.6709
Wavelet	1.0000	0.3208	0.2006
Contourlet	1.0000	0.2960	0.2423
WBCT	1.0000	0.2345	0.1929
HWD1	1.0000	0.2740	0.2166
HWD2	1.0000	0.2704	0.2094

■ Fig. 9
■ Fig. 9
Trend plots of Lena with different distortions using the proposed framework with contourlet transform (all images are 8 bits/pixel and resized from 512 × 512 to 128 × 128 for visibility).

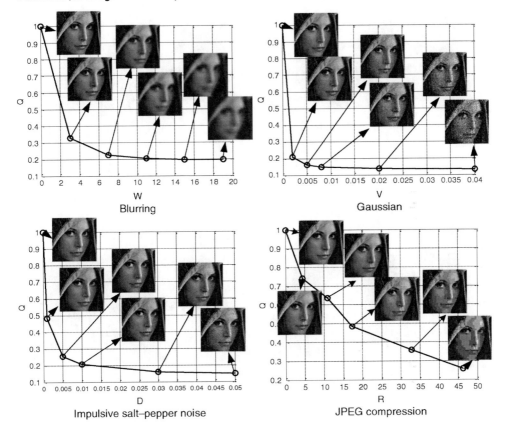

For JPEG compression, we find that Q for IQA drops with increasing intensity of different distortions, which is consistent with the tendency of the decreasing image quality. That is the proposed IQA framework works well for the JPEG distortion. The coding scheme of JPEG (JPEG2000) is based on the discrete cosine transform (discrete wavelet transform). In the JPEG (JPEG2000) coding stage, the lowpass subband is compressed in a high compression rate, and highpass subbands are compressed in a low compression rate to achieve a good compression rate for the whole image while preserving the visual quality of the image. This procedure is similar to the IQA framework; that is, information in the lowpass subband is not considered because most information in the lowpass subband is preserved; and the output value Q is obtained from highpass subbands only because low compression rates are utilized on them and the quality of the reconstructed image is strongly relevant to highpass subbands (the defined JND threshold). Therefore, the proposed scheme adapts well for JPEG and JPEG2000 distortions.

For blurring, additive Gaussian noise distortion, and impulsive salt–pepper noise, Q for IQA drops sharply initially and then slowly because MGA transforms cannot explicitly extract

⬛ Fig. 10

Trend plots of Lena with different distortions using the proposed framework with WBCT transform (all images are 8 bits/pixel and resized from 512 × 512 to 128 × 128 for visibility).

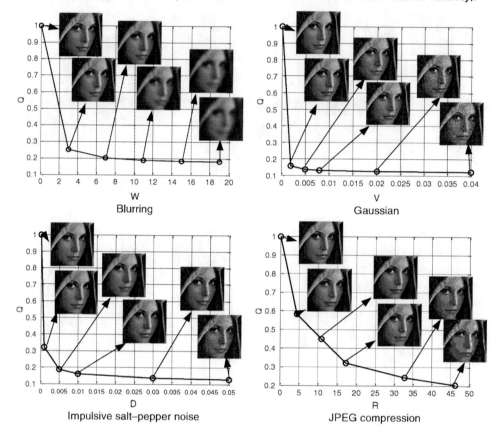

effective information from images with these distortions. However, based on these figures, Q can still reflect the quality of distorted images with blurring, additive Gaussian noise distortion, and impulsive salt–pepper noise, although the performance is not good.

4.3   Sensitivity Test Experiment of the Proposed Framework

Currently, MSE and PSNR are the most commonly used objective quality metrics for images. However, they do not correlate well with the perceived quality measurement. ❷ *Figure 11* shows three degraded "Lena" images with different types of distortion but with the same PSNR; however, their perceived qualities are obviously varied. As shown in ❷ *Table 3*, the proposed framework with different MGA transforms can distinguish them very well. It is noted that Q is insensitive to the changes of intensity, in other words, a slight change of gray value at the image level will not affect the image quality when Q equals 1, so it is consistent with the HVS character that only when the distorted image and the reference image have the same variance of light can the two images have the same visual quality.

◻ **Fig. 11**

Lena image with the same PSNR but different perceived quality. (a) The reference image; (b) mean shift; (c) contrast stretching; (d) JPEG compression.

a
b
c
d

5   Conclusion

This chapter has described a RR IQA framework by incorporating merits of MGA and HVS. In this framework, sparse image representation based on MGA is used to mimic the multichannel structure of HVS, and then CSF is used to balance the magnitude of the coefficients obtained by MGA mimic nonlinearities of HVS, and JND is used to produce a noticeable variation in sensory experience. The quality of a distorted image is measured by comparing the normalized histogram of the distorted image and that of the reference image. Thorough empirical studies show that the novel framework with a suitable image decomposition method performs better than the conventional standard RR IQA method. Since RR methods require full limited information of the reference image, it could be a serious impediment for many applications. It is essential to develop NR image quality metrics that blindly estimate the quality of images. Recently, tensor-based approaches (Tao et al. 2006, 2007a, b) have been demonstrated to be effective for image representation in classification problems, so it is valuable to introduce them for image quality.

Acknowledgments

This research was supported by the National Natural Science Foundation of China (60771068, 60702061, 60832005), the Ph.D. Programs Foundation of the Ministry of Education of China (No. 20090203110002), the Natural Science Basic Research Plan in the Shaanxi Province of

China (2009JM8004), the Open Project Program of the National Laboratory of Pattern Recognition (NLPR) in China, and the National Laboratory of Automatic Target Recognition, Shenzhen University, China.

References

Avcibas I, Sankur B, Sayood K (2002) Statistical evaluation of image quality measures. J Electron Imaging 11(2):206–213

Bamberger RH, Smith JT (1992) A filter bank for the directional decomposition of images: theory and design. IEEE Trans Signal Process 40(4):882–893

Bishop CM (1995) Neural networks for pattern recognition. Oxford University Press, London

Burt PJ, Adelson EH (1983) The Laplacian pyramid as a compact image code. IEEE Trans Commun 31(4):532–540

Callet PL, Christian VG, Barba D (2006) A convolutional neural network approach for objective video quality assessment. IEEE Trans Neural Netw 17(5): 1316–1327

Cands EJ (1998) Ridgelets: theory and applications. PhD thesis, Department of statistics, Stanford University

Cands EJ, Donoho DL (2000) Curvelets – a surprising effective nonadaptive representation for objects with edges. Curves and surfaces, Vanderbilt University Press, Nashville, TN, pp 105–120

Chandrasekaran V, Palaniswami M, Caelli TM (1996) Range image segmentation by dynamic neural network architecture. Pattern Recogn 29(2):315–329

Daly S (1993) The visible difference predictor: an algorithm for the assessment of image fidelity. In: Watson AB (ed) Digital images and human vision, The MIT Press, Cambridge, MA, pp 179–206

Damera-Venkata N, Thomas DK, Wilson SG, Brian LE, Bovik AC (2000) Image quality assessment based on a degradation model. IEEE Trans Image Process 9(4): 636–650

Do MN (2001) Directional multiresolution image representations. PhD thesis, École Polytechnique Fédérale de Lausanne

Do MN, Vetterli M (2005) The contourlet transform: an efficient directional multiresolution image representation. IEEE Trans Image Process 14(12):2091–2106

Eskicioglu AM, Fisher PS (1995) Image quality measures and their performance. IEEE Trans Commun 43 (12):2959–2965

Eslami R, Radha H (2004) Wavelet-based contourlet transform and its application to image coding. In: Proceedings of the IEEE international conference on image processing, Singapore, pp 3189–3192

Eslami R, Radha H (2005) New image transforms using hybrid wavelets and directional filter

banks: analysis and design. In: Proceedings of the IEEE international conference on image processing, Singapore, pp 11–14

Gao XB, Lu W, Li XL, Tao DC (2008a) Wavelet-based contourlet in quality evaluation of digital images. Neurocomputing 72(12):378–385

Gao XB, Lu W, Li XL, Tao DC (2008b) Image quality assessment based on multiscale geometric analysis. Accepted to IEEE Trans Image Process

Gastaldo P, Rovetta S, Zunino R (2002) Objective quality assessment of MPEG-2 video streams by using CBP neural networks. IEEE Trans Neural Netw 13 (4):939–947

Gastaldo P, Zunino R (2004) No-reference quality assessment of JPEG images by using CBP neural networks. In: Proceedings of the IEEE international symposium on circuits and systems, Vancouver, Canada, pp 772–775

Karunasekera SA, Kingsbury NG (1995) A distortion measure for blocking artifacts in images based on human visual sensitivity. IEEE Trans Image Process 4(6):713–724

Lampinen J, Oja E (1998) Neural network systems, techniques and applications. Image processing and pattern recognition, Academic Press, New York

Li X (2002) Blind image quality assessment. In: Proceedings of the IEEE international conference on image processing, Singapore, pp 449–452

Li XL, Tao DC, Gao XB, Lu W (2009) A natural image quality evaluation metric based on HWD. Signal Process 89(4):548–555

Lu W, Gao XB, Tao DC, Li XL (2008) A wavelet-based image quality assessment method. Int J Wavelets Multiresolut Inf Process 6(4):541–551

Lubin J (1995) A visual discrimination mode for image system design and evaluation. In: Peli E (ed) Visual models for target detection and recognition, Singapore, World Scientific, pp 207–220

Mallat S (1989) A theory for multiresolution decomposition: the wavelet representation. IEEE Trans Pattern Anal Mach Intell 11(7):674–693

Mannos JL, Sakrison DJ (1974) The effect of visual fidelity criterion on the encoding of images. IEEE Trans Inf Theory 20(2):525–536

Masry M, Hemami SS, Sermadevi Y (2006) A scalable wavelet-based video distortion metric and applications. IEEE Trans Circuits Syst Video Technol 16(2): 260–273

Miloslavski M, Ho YS (1998) Zerotree wavelet image coding based on the human visual system model. In: Proceedings of the IEEE Asia-Pacific conference on circuits and systems, pp 57–60

Miyahara M, Kotani K, Algazi RV (1998) Objective picture quality scale (PQS) for image coding. IEEE Trans Commun 46(9):1215–1225

Nadenau MJ, Reichel J, Kunt M (2003) Wavelet-based color image compression: exploiting the contrast sensitivity. IEEE Trans Image Process 12(1):58–70

Oja E (1982) A simplified neuron model as a principal component analyzer. J Math Biol 15(3):267–273

Pennec EL, Mallat S (2000) Image compression with geometrical wavelets. In: Proceedings of the IEEE international conference on image processing, Vancouver, Canada, pp 661–664

Petersena E, Ridder DD, Handelsc H (2002) Image processing with neural networks – a review. Pattern Recogn 35:2279C2301

Rec. ITU-R Methodology for the subjective assessment of the quality of television pictures, Recommendation ITU-R Rec. BT. 500-11

Romberg JK (2003) Multiscale geometric image processing. PhD thesis, Rice University

Sheikh HR, Sabir MF, Bovik AC (2006) A statistical evaluation of recent full reference image quality assessment algorithms. IEEE Trans Image Process 15(11):3440–3451

Sheikh HR, Wang Z, Cormack L, Bovik AC (2003) LIVE image quality assessment database. http://live.ece.utexas.edu/research/quality

Tao DC, Li XL, Wu XD, Maybank SJ (2006) Human carrying status in visual surveillance. In: Proceedings of the IEEE international conference on computer vision and pattern recognition, Hong Kong, China, pp 1670–1677

Tao DC, Li XL, Wu XD, Maybank SJ (2007a) General tensor discriminant analysis and Gabor features for gait recognition. IEEE Trans Pattern Anal Mach Intell 29(10):1700–1715

Tao DC, Li XL, Wu XD, Maybank SJ (2007b) Supervised tensor learning. Knowl Inf Syst (Springer) 13(1): 1–42

VQEG (2000a) Final report from the video quality experts group on the validation of objective models of video quality assessment, phase II VQEG, 2003. http://www.vqeg.org

VQEG (2000b) Final report from the video quality experts group on the validation of objective models of video quality assessment. http://www.vqeg.org

Wandell BA (1995) Foundations of vision, 1st edn. Sinauer, Sunderland, MA

Wang Z, Bovik AC (2006) Modern image quality assessment. Morgan and Claypool, New York

Wang YM, Wahl FM (1997) Vector-entropy optimization based neural-network approach to image reconstruction from projections. IEEE Trans Neural Netw 8(5):1008–1014

Watson AB (1993) DCT quantization matrices visually optimized for individual images. In: Proceedings of the SPIE human vision, visual processing and digital display IV. Bellingham, WA, 1913(14):202–216

Wolf S, Pinson MH (2005) Low bandwidth reduced reference video quality monitoring system. In: First international workshop on video processing and quality metrics for consumer electronics. Scottsdale, AZ

12 Nonlinear Process Modelling and Control Using Neurofuzzy Networks

Jie Zhang
School of Chemical Engineering and Advanced Materials, Newcastle University, Newcastle upon Tyne, UK
jie.zhang@newcastle.ac.uk

G. Rozenberg et al. (eds.), *Handbook of Natural Computing*, DOI 10.1007/978-3-540-92910-9_12,
© Springer-Verlag Berlin Heidelberg 2012

Abstract

This chapter presents neurofuzzy networks for nonlinear process modeling and control. The neurofuzzy network uses local linear models to model a nonlinear process and the local linear models are combined using center of gravity (COG) defuzzification. In order to be able to provide accurate long-range predictions, a recurrent neurofuzzy network structure is developed. An advantage of neurofuzzy network models is that they are easy to interpret. Insight about the process characteristics at different operating regions can be easily obtained from the neurofuzzy network parameters. Based on the neurofuzzy network model, nonlinear model predictive controllers can be developed as a nonlinear combination of several local linear model predictive controllers that have analytical solutions. Applications to the modeling and control of a neutralization process and a fed-batch process demonstrate that the proposed recurrent neurofuzzy network is very effective in the modeling and control of nonlinear processes.

1 Introduction

Advanced process control, optimization, and process monitoring require accurate process models. Process models can be broadly divided into two categories: first principles models and empirical models. First principles models are developed based upon process knowledge and, hence, they are generally reliable. However, the development of first principles models is usually time consuming and effort demanding, especially for complex processes. To overcome this difficulty, empirical models based upon process operational data should be utilized.

Neural networks have been shown to possess good function approximation capability (Cybenko 1989; Girosi and Poggio 1990; Park and Sandberg 1991) and have been applied to process modeling by many researchers (Bhat and McAvoy 1990; Bulsari 1995; Morris et al. 1994; Tian et al. 2002; Zhang 2004). A neural network can learn the underlying process model from a set of process operation data. Conventional network modeling typically results in a "black box" model. One potential limitation of conventional neural network models is that they can be difficult to interpret and can lack robustness when applied to unseen data.

One approach to improve model robustness and open up the "black box" models is through the combined use of both process knowledge and process operation data. Process knowledge can, for example, be used to decompose the process operation into a number of local operating regions such that, within each region, a reduced order linear model can be used to approximate the local behavior of the process. In fact, a nonlinear process can always be locally linearized around a particular operating point and the locally linearized model is valid within a region around that operating point. Fuzzy sets provide an appropriate means for defining operating regions since the definition of local operating regions is often vague in nature and there usually exists overlapping among different regions. This leads to the fuzzy modeling approach (Yager and Filev 1994).

One fuzzy modeling approach was developed by Takagi and Sugeno (1985). In this approach, each model input is assigned several fuzzy sets with the corresponding membership functions being defined. Through logical combinations of these fuzzy inputs, the model input space is partitioned into several fuzzy regions. A local linear model is used within each region and the global model output is obtained through the center of gravity (COG) defuzzification which is essentially the interpolation of local model outputs. This modeling approach is very powerful in that it decomposes a complex system into several less complex subsystems. Based on a similar principle, Johansen and Foss (1993) proposed an approach to construct NARMAX (Nonlinear AutoRegressive Moving Average with eXogenous inputs) models using local ARMAX (AutoRegressive Moving Average with eXogenous inputs) models. In their approach, a nonlinear system is decomposed into several operating regimes and, within each regime, a local ARMAX model is developed. A model validate function is defined for each local model and the global NARMAX model is then obtained by interpolating these local ARMAX models.

Fuzzy models can be implemented by using neurofuzzy networks. Neurofuzzy network representations have emerged as a powerful approach to the solution of many engineering problems (Blanco et al. 2001b; Brown and Harris 1994; Harris et al. 1996; Horikawa et al. 1992; Jang 1992; Jang and Sun 1995; Jang et al. 1997; Nie and Linkens 1993; Omlin et al. 1998; Wang 1994; Zhang and Morris 1994; Zhang et al. 1998; Zhang and Morris 1999; Zhang 2005, 2006). Jang (Jang 1992; Jang and Sun 1995; Jang et al. 1997) proposed the ANFIS (adaptive-network-based fuzzy inference system) architecture to represent fuzzy models. Through back propagation training, ANFIS is adapted to refine, or derive, the fuzzy if-then rules using system input–output data. Fuzzy reasoning is capable of handling imprecise and uncertain information while neural networks can be identified using real plant data. Neurofuzzy networks combine the advantages of both fuzzy reasoning and neural networks. Process knowledge can also be embedded into neurofuzzy networks in terms of both fuzzy membership partitions and as reduced order local models. Zhang and Morris propose two types of feed forward neurofuzzy networks for nonlinear processes modeling (Zhang and Morris 1995). Most of the reported neurofuzzy network models are, however, one-step-ahead prediction models in that the current process output is used as a network input to predict the process output at the next sampling time step.

In some process-control applications, such as long-range predictive control and batch process optimal control, models capable of providing accurate multistep-ahead prediction are more appropriate. It is well known that nonlinear multistep-ahead prediction models can be built using dynamic neural networks. Good examples are globally recurrent neural networks (e.g., Su et al. 1992; Werbos 1990), Elman networks (Elman 1990; Scott and Ray 1993), dynamic feed forward networks with filters (Morris et al. 1994), and locally recurrent networks (Frasconi et al. 1992; Tsoi and Back 1994; Zhang et al. 1998). In globally recurrent networks, the network outputs are fed back to the network inputs through time delay units. Su et al. (1992) showed that a feed forward neural network trained for one-step-ahead predictions has difficulties in providing acceptable long-term predictions, unlike a globally recurrent network. In an Elman network, the hidden neuron outputs at the previous time step are fed back to all the hidden neurons. Such a topology is similar to a nonlinear state space representation in dynamic systems. Scott and Ray (1993) demonstrated the performance of an Elman network for nonlinear process modeling. Filter networks incorporate filters of the form $N(s)/D(s)$ into the network interconnections (Morris et al. 1994). In locally recurrent networks, the output of a hidden neuron is fed back to its input through one or several time delay units. Zhang et al. (1998) proposed a sequential orthogonal training strategy which allows hidden neurons to be gradually added to avoid an unnecessarily large network structure.

A type of recurrent neurofuzzy network is proposed by Zhang and Morris (1999) and it allows the construction of a "global" nonlinear multistep-ahead prediction model from the fuzzy conjunction of a number of "local" dynamic models. In this recurrent neurofuzzy network, the network output is fed back to the network input through one or more time delay units. This particular structure ensures that predictions from a recurrent neurofuzzy network are multistep-ahead or long-range predictions. Both process knowledge and process input–output data are used to build multistep-ahead prediction models. Process knowledge is used to initially partition the process nonlinear characteristics into several local operating regions and to aid the initialization of the corresponding network weights. Process input–output data are then used to train the network. Membership functions of the local regimes are identified and local models discovered through network training. In the training of a recurrent neurofuzzy network, the training objective is to minimize the multistep-ahead prediction errors. Therefore, a successfully trained recurrent neurofuzzy network is able to provide good long-range predictions.

A recurrent neurofuzzy network model can be used to develop a novel type of nonlinear model based on a long-range predictive controller. Based upon the local linear models contained in a recurrent neurofuzzy network, local linear model-based predictive controllers can be developed. These local controllers can be combined through COG defuzzification to form a global nonlinear model-based long-range predictive controller. The advantage of this approach is that analytical solutions usually exist for the local model predictive controllers (except for the cases where nonlinear constraints are present), hence avoiding computation that requires numerical optimization procedures and the uncertainty in converging to the global minimum which are typically seen in previously reported nonlinear model-based predictive control strategies. When nonlinear constraints are present, the difficulty in computation and analysis can also be eased through the combination of local controllers. For instance, a nonlinear constraint may never be active in a particular operating region and can therefore be relaxed in that region. Within a particular operating region, a nonlinear constraint may be approximated by a linear constraint. Furthermore, control actions calculated from local linear models in incremental form contain integral actions which can naturally eliminate static control offsets.

This chapter is structured as follows. ❯ Section 2 presents neurofuzzy networks and recurrent neurofuzzy networks for nonlinear process modeling. Applications of the proposed recurrent neurofuzzy networks to the modeling of pH dynamics in a continuous stirred tank reactor (CSTR) are presented in ❯ Sect. 3. A long-range model-based predictive control technique using recurrent neurofuzzy network models is discussed in ❯ Sect. 4. ❯ Section 5 presents the modeling and multi-objective optimization control of a fed-batch process using recurrent neurofuzzy networks. Finally, ❯ Sect. 6 draws some concluding remarks.

2 Neurofuzzy Networks for Nonlinear Process Modeling

2.1 Fuzzy Models

The global operation of a nonlinear process is divided into several local operating regions. Within each local region, R_i, a reduced order linear model in ARX (AutoRegressive with eXogenous inputs) form is used to represent the process behavior. Fuzzy sets are used to define process operating regions such that the fuzzy dynamic model of a nonlinear process can be described in the following way:

R_i: IF operating condition i
THEN

$$\hat{y}_i(t) = \sum_{j=1}^{no} a_{ij} y(t-j) + \sum_{j=1}^{ni} b_{ij} u(t-j)$$

$$(i = 1, 2, \ldots, nr)$$

(1)

The final model output is obtained through COG defuzzification as:

$$\hat{y}(t) = \frac{\sum_{i=1}^{nr} \mu_i \hat{y}_i(t)}{\sum_{i=1}^{nr} \mu_i} \tag{2}$$

In the above model, y is the process output, u is the process input, $\hat{y}_i$ is the prediction of the process output in the ith operating region, nr is the number of fuzzy operating regions, ni and no are the time lags in u and y, respectively, μ_i is the membership function for the ith model, a_{ij} and b_{ij} are the ARX model parameters, and t represents discrete time.

The above model represents a one-step-ahead prediction model in that the process output at time $t-1$, $y(t-1)$ is used to predict the process output at time t, $y(t)$. A long-term prediction fuzzy dynamic model of a nonlinear process can be described as follows:

R_i: IF operating condition i
THEN

$$\hat{y}_i(t) = \sum_{j=1}^{no} a_{ij}\hat{y}(t-j) + \sum_{j=1}^{ni} b_{ij}u(t-j) \tag{3}$$

$$(i = 1, 2, \ldots, nr)$$

The final model output is obtained through COG defuzzification as indicated in ❱ Eq. 2. In the above model, $\hat{y}_i$ is the prediction of the process output in the ith operating region. The above model is a multistep-ahead prediction model since it uses previous model outputs, $\hat{y}(t-j)$, instead of previous process outputs, $y(t-j)$, to predict the present process output.

Operating regions of a process can usually be defined by one or several process variables. A number of fuzzy sets, such as "low," "medium," and "high," can be defined for each of these process variables. An operating region can then be defined through logical combinations of the fuzzy sets of those variables. Suppose that x and y are the process variables used to define the process-operating regions and they are assigned the fuzzy sets: "low," "medium," and "high." The ith operating region can be defined, for example, as

$$x \text{ is high AND } y \text{ is medium}$$

The membership function for this operating region can be obtained in a number of ways. One approach to calculate it is

$$\mu_i = min(\mu_h(x), \mu_m(y)) \tag{4}$$

and another approach is

$$\mu_i = \mu_h(x)\mu_m(y) \tag{5}$$

In the above equations, μ_i is the membership function of the ith operating region, $\mu_h(x)$ is the membership function of x being "high," and $\mu_m(y)$ is the membership function of y being "medium." The second approach allows the calculation of model error gradients which are required in gradient-based network training. For the convenience of using gradient-based network training algorithms, the second approach is adopted here. However, if gradient free training algorithms, such as those based on genetic algorithms, are used, then either approach can be adopted.

The local linear model used in this work can be transformed into incremental form in terms of the model input variables which are used as the manipulated variables in control applications. Consider the following discrete time local linear model:

$$y(t) = a_1 y(t-1) + a_2 y(t-2) + \cdots + a_{no} y(t-no) + b_1 u(t-1) + b_2 u(t-2) \\ + \cdots + b_{ni} u(t-ni) \tag{6}$$

The prediction of y at time $t-1$ is:

$$y(t-1) = a_1 y(t-2) + a_2 y(t-3) + \ldots + a_{no} y(t-no-1) \\ + b_1 u(t-2) + b_2 u(t-3) + \ldots + b_{ni} u(t-ni-1) \tag{7}$$

Subtracting ❷ Eq. 7 from ❷ Eq. 6 and rearranging gives the model in incremental form:

$$y(t) = (a_1 + 1)y(t-1) + (a_2 - a_1)y(t-2) + \cdots + (a_{no} - a_{no-1})y(t-no) \\ - a_{no}y(t-no-1) + b_1 \Delta u(t-1) + b_2 \Delta u(t-2) + \cdots + b_{ni} \Delta u(t-ni) \tag{8}$$

where $\Delta u(t-1) = u(t-1) - u(t-2)$. Incremental models are particularly useful for controller development since control actions calculated from them naturally contain integral actions which eliminate static control offsets.

❷ *Figure 1* illustrates the idea of how membership functions span across the operating regions. Process operating regions are determined by two process variables, x_1 and x_2, each being partitioned into two fuzzy sets: "small" and "large." There exist four operating regions and they are: (1) x_1 is small and x_2 is small; (2) x_1 is small and x_2 is large; (3) x_1 is large and x_2 is small; and (4) x_1 is large and x_2 is large. ❷ *Figure 1* shows the membership functions of the four operating regions. It can be seen that there exists overlapping among operating regions and the transition from one operating region to another can be achieved in a smooth fashion. The fuzzy approach in representing process-operating regions can handle uncertain and imprecise information.

◻ **Fig. 1**
Membership functions for operating regions.

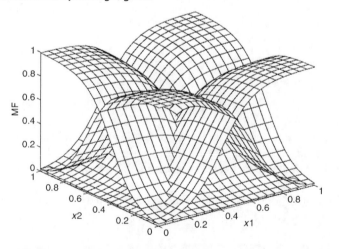

Fuzzy modeling is particularly suitable for processes which can be divided into several operating regions. An example would be some batch processes which have different dynamics at different batch stages. Another example would be processes producing different grades of products. These processes can have different dynamics associated with the production of different grades of products. Inside an operating region, the process dynamics could be represented by a fairly simple model, such as a reduced order linear model. In different operating regions, the local models may be very different from each other. By means of fuzzy modeling, a complex global model can be developed in terms of several fairly simple local models.

2.2 Neurofuzzy Networks

The above fuzzy model can be represented by the neurofuzzy networks shown in ❯ *Figs. 2* and ❯ *3*. ❯ *Figure 2* shows a feed forward neurofuzzy network (Zhang and Morris 1995) while ❯ *Fig. 3* shows a recurrent neurofuzzy network (Zhang and Morris 1999). A recurrent neurofuzzy network is more appropriate for building long-range prediction models (Zhang and Morris 1999). Both types of neurofuzzy networks contain five layers: an input layer, a fuzzification layer, a rule layer, a function layer, and a defuzzification layer. Several different types of neurons are employed in the network. Inputs to the fuzzification layer are process variables used for defining fuzzy operating regions. Each of these variables is transformed into several fuzzy sets in the fuzzification layer where each neuron corresponds to a particular fuzzy set with the actual membership function being given by the neuron output. Three types of neuron activation functions are used and they are: the sigmoidal function, the Gaussian function, and the complement sigmoidal function. ❯ *Figure 4* shows the shapes of the membership functions. The two fuzzy sets at the left and right sides in ❯ *Fig. 4* are represented

■ Fig. 2
The structural approach neurofuzzy network.

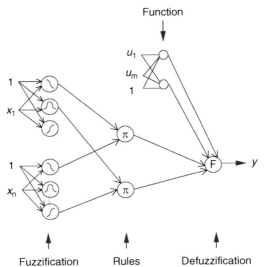

◘ Fig. 3

The structural approach recurrent neurofuzzy network.

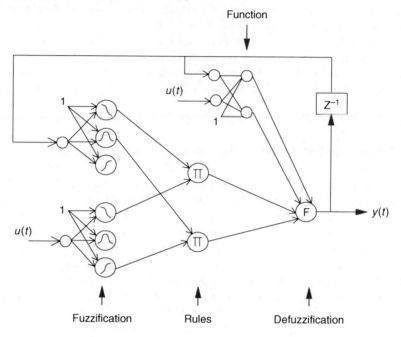

◘ Fig. 4

Shapes of membership functions.

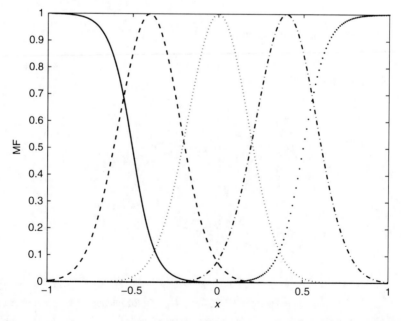

by neurons with the complement sigmoidal activation function and the sigmoidal activation function, respectively. Other fuzzy sets are represented by neurons with the Gaussian activation function. Weights in this layer determine the membership functions of the corresponding fuzzy sets. By changing the weights in this layer, appropriately shaped membership functions can be obtained. Weights in this layer are initialized based on process knowledge and are refined during network training.

Neurons corresponding to the fuzzy set on the right end of ❯ *Fig. 4* use the sigmoid function and their outputs are of the form:

$$y = \frac{1}{1 + e^{-(w_1 x + w_0)}} \tag{9}$$

where y is the neuron output, x is the neuron input, w_1 is the input weight, and w_0 is a bias. Neurons corresponding to the fuzzy set on the left end of ❯ *Fig. 4* use the complement sigmoid function and their outputs are given as:

$$y = 1 - \frac{1}{1 + e^{-(w_1 x + w_0)}} = \frac{e^{-(w_1 x + w_0)}}{1 + e^{-(w_1 x + w_0)}} \tag{10}$$

The activation function used by neurons corresponding to fuzzy sets between the two terminal fuzzy sets in ❯ *Fig. 4* is the Gaussian function. Outputs of such neurons are of the form:

$$y = e^{-(w_1 x + w_0)^2} \tag{11}$$

The shapes and positions of the membership functions are determined by the values of the fuzzification layer weights. By appropriately selecting the fuzzification layer weights, desired membership functions can be obtained.

Each neuron in the rule layer corresponds to a fuzzy operating region of the process being modeled. Its inputs are the fuzzy sets which determine this operating region. Its output is the product of its input and is the membership function of this fuzzy operating region. Neurons in the rule layer implement the fuzzy intersection defined by ❯ Eq. 5. The construction of the rule layer is guided by some knowledge about the process. Knowledge on the number of operating regions and how each region is established is used to set up the rule layer.

Neurons in the function layer implement the local linear models in the fuzzy operating regions. Each neuron corresponds to a particular operating region and is a linear neuron. Its output is a summation of its weighted inputs and a bias which represents the constant term in a local model. Weights in the function layer are the local model parameters. The defuzzification layer performs the COG defuzzification and gives the final network output. Inputs to the defuzzification neuron are membership functions of the fuzzy operating regions and the local model outputs in these regions. The neuron activation function used is given in ❯ Eq. 2. With some processes, it may be possible to define the operating regions by using just one process variable. In such cases, the rule layer in ❯ *Figs. 2* and ❯ *3* can be removed since the fuzzification layer directly gives the fuzzy operating regions.

2.3 Training of Neurofuzzy Networks

Recurrent neurofuzzy networks can be trained using any of the number of training methods, such as the back propagation method (Rumelhart et al. 1986), the conjugate gradient method

(Leonard and Kramer 1990), Levenberg–Marquardt optimization (Marquardt 1963), or methods based on genetic algorithms (Blanco et al. 2001a; Mak et al. 1999). In this study, recurrent neurofuzzy networks are trained using the Levenberg–Marquardt algorithm with regularization. The training objective function can be defined as:

$$J = \frac{1}{N} \sum_{t=1}^{N} (\hat{y}(t) - y(t))^2 + \lambda ||W||^2 \tag{12}$$

where N is the number of data points, $\hat{y}$ is the network prediction, y is the target value, t represents the discrete time, W is a vector of network weights, and λ is the regularization parameter.

The objective of regularized training is to penalize excessively large network weights, which do not contribute significantly to the reduction of model errors, so that the trained neural network has a smooth function surface. An intuitive interpretation of this modified objective function is that a weight that does not influence the first term very much will be kept close to zero by the second term. A weight that is important for model fit will, however, not be affected very much by the second term. The appropriate value of λ is obtained through a cross validation procedure. Data for building a recurrent neurofuzzy network model is partitioned into a training data set and a testing data set. Several values of λ are considered and the one resulting in the smallest error on the testing data is adopted. For linear models, minimization of ❷ Eq. 12 leads to the well-known ridge regression formula. Regularization has been widely used in statistical model building and a variety of techniques have been developed, such as ridge regression (Hoerl and Kennard 1970), principal component regression (Geladi and Kowalski 1986), and partial least squares regression (Geladi and Kowalski 1986).

In the Levenberg–Marquardt training method, network weights are adjusted as follows.

$$\Delta W(k+1) = -\eta \left(\frac{1}{N} \sum_{t=1}^{N} \frac{\partial \hat{y}(t)}{\partial W(k)} \left(\frac{\partial \hat{y}(t)}{\partial W(k)} \right)^T + \delta I \right)^{-1} \frac{\partial J}{\partial W(k)} \tag{13}$$

$$W(k+1) = W(k) + \Delta W(k+1) \tag{14}$$

where $W(k)$ and $W(k)$ are vectors of weights and weight adaptations at training step k, respectively, η is the learning rate, δ is a parameter to control the searching step size. A large value of δ gives a small step in the gradient direction and a small value of δ gives a searching step close to the Gauss–Newton step. Like other types of gradient-based training methods, training can be terminated when the error gradient is less than a prespecified value. Training can also be terminated by a cross validation-based stopping criterion as described below. For different weights, the gradient $\partial J / \partial W$ can be calculated accordingly. Due to the recurrence in network connections, the gradient $\partial J / \partial W$ needs to be calculated in the "back propagation through time" fashion (Werbos 1990).

A cross validation-based "early stopping" mechanism is used to minimize the problem of over-fitting. Using the early stopping mechanism, network training is stopped at a point beyond which over-fitting would obviously occur. This can be explained using ❷ Fig. 5 which represents a typical neural network learning curve. In ❷ Fig. 5, the vertical axis represents the sum of squared errors (SSE), the horizontal axis represents the network training steps, the solid line represents the network error on training data, and the dashed line represents the network error on testing data. During the initial training stage, both training error and testing error decrease quite quickly. As training progresses, the training error will decrease

Fig. 5
Neural network learning curves.

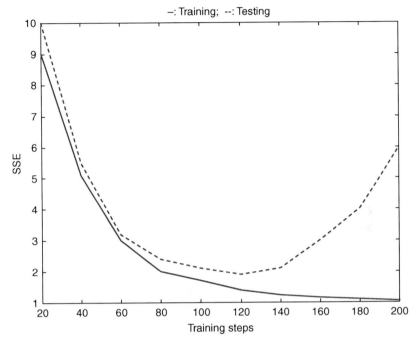

slowly, and the testing error will start to increase, sometimes very quickly, after certain training steps. The appropriate point to stop training is that point at which the testing error is at its minimum. During network training, both training and testing errors are continuously monitored to detect the appropriate stopping point. "Early stopping" has an implicit effect on regularization as shown by Sjoberg et al. (1995).

3 Modeling of a Neutralization Process Using a Recurrent Neurofuzzy Network

A recurrent neurofuzzy network was used to model a neutralization process which was taken from McAvoy et al. (1972). The neutralization process takes place in a CSTR which is shown in ▸ *Fig. 6*. There are two input streams to the CSTR. One is acetic acid of concentration C_1 at a flow rate F_1 and the other sodium hydroxide of concentration C_2 at a flow rate F_2. The mathematical equations of the CSTR can be described as follows by assuming that the tank level is perfectly controlled (McAvoy et al. 1972).

$$V \frac{d\xi}{dt} = F_1 C_1 - (F_1 + F_2)\xi \tag{15}$$

$$V \frac{d\zeta}{dt} = F_2 C_2 - (F_1 + F_2)\zeta \tag{16}$$

Fig. 6

A continuous stirred tank reactor for pH neutralization.

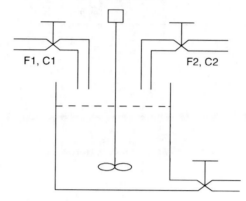

F1, C1   F2, C2

$$[H^+]^3 + (K_a + \zeta)[H^+]^2 + (K_a(\zeta - \xi) - K_w)[H^+] - K_w K_a = 0 \qquad (17)$$

$$pH = -\log_{10}[H^+] \qquad (18)$$

where

$$\xi = [HAC] + [AC^-] \qquad (19)$$

$$\zeta = [Na^+] \qquad (20)$$

❯ *Table 1* gives the meaning and the initial setting of each variable. It is well-known that pH dynamics are highly nonlinear. The steady-state relationship between acid flow rate and pH in the reactor (the titration curve) is plotted in ❯ *Fig. 7*. It can be seen from ❯ *Fig. 7* that the process gain is very high in the medium pH region while being quite low in both low- and high-pH regions. A recurrent neurofuzzy network model was developed to model the nonlinear dynamic relationships between the acetic acid flow rate and the pH in the reactor.

To generate training, testing, and unseen validation data, multilevel random perturbations were added to the flow rate of acetic acid while other inputs to the reactor were kept constant. Three sets of data were generated. One set was used as training data, another set was used as testing data, and the remaining set was used as unseen validation data. The simulated pH values in the reactor were corrupted with random noise in the range $(-0.3, 0.3)$ representing measurement noise. Training, testing, and validation data are plotted in ❯ *Figs. 8–10*, respectively.

A recurrent neurofuzzy network was used to build a multistep-ahead prediction model for the neutralization process. Based on the process characteristics, the nonlinear pH-operating region is divided into three local regions: pH low, pH medium, and pH high. The fuzzification layer weights were initialized based on the titration curve. Weights for the function layer were initialized as random numbers in the range $(-0.2, 0.2)$. Initially, within each local region a second-order linear model was selected. If the identified model is not sufficiently adequate then the local model order and/or the number of fuzzy operating regions are increased.

Table 1
The physical parameters used in simulation

Variable	Meaning	Initial setting
V	Volume of tank	1 l
F_1	Flow rate of acid	0.081 l/min
F_2	Flow rate of base	0.512 l/min
C_1	Concentration of acid in F_1	0.32 moles/l
C_2	Concentration of acid in F_2	0.05005 moles/l
K_a	Acid equilibrium constant	1.8×10^{-5}
K_w	Water equilibrium constant	1.0×10^{-14}

Fig. 7
Titration curve.

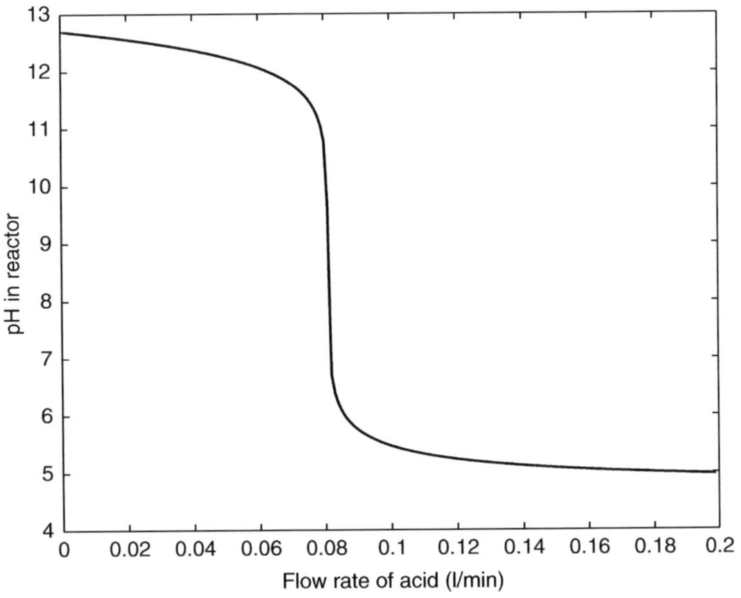

The recurrent neurofuzzy network was trained using the Levenberg–Marquardt training method together with regularization and a cross-validation-based stopping criterion. After network training, the following multistep-ahead prediction fuzzy model was identified:

IF pH low
THEN $\hat{y}(t) = 0.6367\hat{y}(t-1) + 0.2352\hat{y}(t-2) - 3.56u(t-1) - 1.81u(t-2) - 1.214$
IF pH medium
THEN $\hat{y}(t) = 0.3309\hat{y}(t-1) + 0.2798\hat{y}(t-2) - 94.5u(t-1) - 22.53u(t-2) + 13.117$
IF pH high
THEN $\hat{y}(t) = 0.6631\hat{y}(t-1) + 0.2595\hat{y}(t-2) - 6.16u(t-1) - 1.12u(t-2) + 1.244$

□ Fig. 8
Training data.

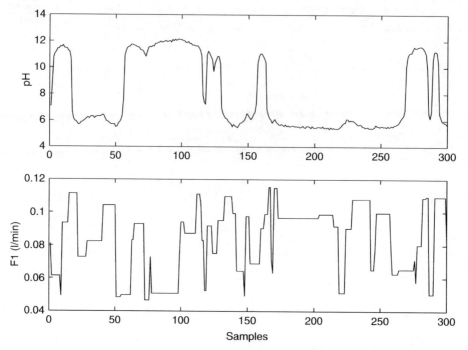

□ Fig. 9
Testing data.

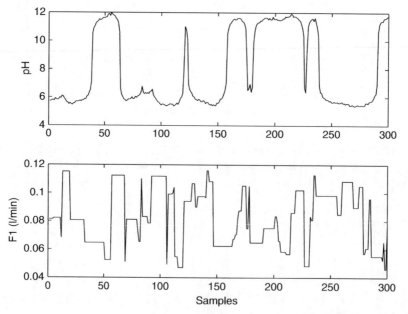

Fig. 10
Validation data.

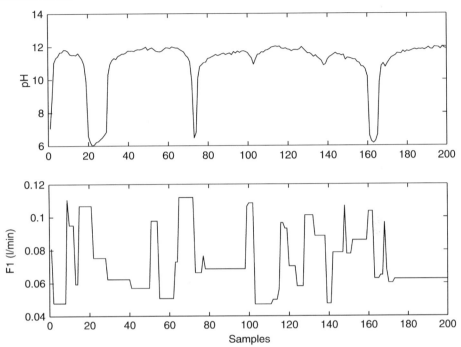

In the above model, $\hat{y}$ is the predicted pH in the reactor, u is the flow rate of acetic acid, and t is the discrete time. The identified membership functions (normalized) represented by the fuzzification layer weights, for the three fuzzy sets: pH low, pH medium, and pH high, are plotted in ❯ *Fig. 11*. It can be seen that the identified partition of process-operating regions agrees with the operating regions indicated by the titration curve. The static process gains in the pH low, pH medium, and pH high regions are easily calculated from the local model parameters as -41.92, -300.61, and -65.12, respectively. Thus, the local model associated with the operating region, pH medium, has much higher gain than the local models for the other two operating regions. This is exactly what one would expect from the titration curve. Neurofuzzy network models can, hence, provide certain insight into the modeled processes and are easy to interpret. Using COG defuzzification, the process gain can be calculated from the local model gains and the membership functions of the local operating regions. The calculated process gain is plotted in ❯ *Fig. 12* and it matches with what is indicated by the titration curve. The above recurrent neurofuzzy network model is much easier to interpret than strictly black box models, with the fuzzification layer weights indicating how the process operation is partitioned into different operating regions and the function layer weights showing the local dynamic behavior of the process. This again is in marked contrast to conventional neural network models where the modeling function of the network weights are difficult, if not impossible, to understand. The local linear dynamic models can also be transformed into frequency-domain transfer function forms and interpreted using pole-zero positions.

⬛ **Fig. 11**

Membership functions learnt by the recurrent neurofuzzy network.

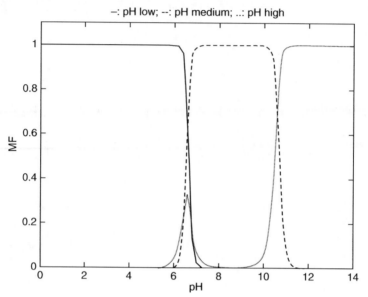

⬛ **Fig. 12**

Identified process gains at different operating conditions.

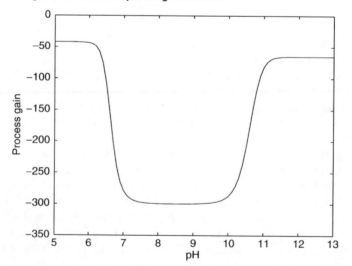

Long-term predictions of pH from the recurrent neurofuzzy network on the unseen validation data are plotted in ❷ *Fig. 13*. It can be seen that the long-term predictions are quite accurate. The SSE of multistep-ahead predictions from the recurrent neurofuzzy network model on training, testing, and validation data are given in ❷ *Table 2*. For the purpose

◘ Fig. 13
Multistep-ahead predictions on validation data.

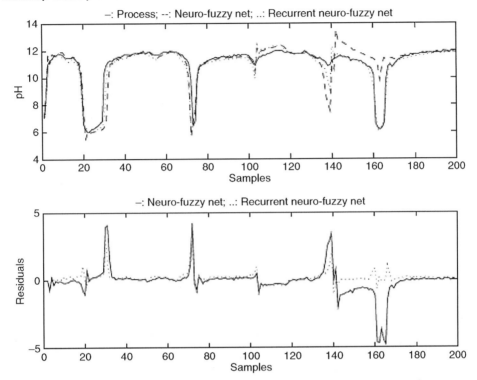

◘ Table 2
SSE for multistep-ahead predictions

	Training	Testing	Validation
Feed forward neurofuzzy net	307	419	211
Recurrent neurofuzzy net	36	37	13

of comparison, a feed forward neurofuzzy network was also used to model the pH dynamics and the following one-step-ahead prediction model was identified.

IF pH low

THEN $\hat{y}(t) = 0.5651y(t-1) + 0.2712y(t-2) - 4.92u(t-1) - 0.52u(t-2) + 1.424$

IF pH medium

THEN $\hat{y}(t) = 0.3011y(t-1) + 0.2289y(t-2) - 100.39u(t-1)$
$- 30.07u(t-2) + 14.695$

IF pH high

THEN $\hat{y}(t) = 0.667y(t-1) + 0.2463y(t-2) - 0.78u(t-1) - 3.87u(t-2) + 1.323$

This model can give accurate one-step-ahead predictions with the SSE on training, testing, and validation data being 18.5, 24.4, and 11.1, respectively. However, if this model is used for multistep-ahead prediction, the prediction accuracy deteriorates significantly as is shown in ❯ *Table 2*. Multistep-ahead predictions of pH from the feed forward neurofuzzy network are plotted in ❯ *Fig. 13*. Both ❯ *Fig. 13* and ❯ *Table 2* show that the recurrent neurofuzzy network can achieve much accurate multistep-ahead predictions than the feed forward neurofuzzy network.

A number of model-validity tests for nonlinear model identification procedures have been developed, for example, the statistical chi-square test (Leontaris and Billings 1987), the Akaike Information Criterion (AIC) (Akaike 1974), the predicted squared error criterion (Barron 1984), and the high order correlation tests (Billings and Voon 1986). The validity of a neurofuzzy network model can be assessed using the following high order correlation tests (Billings and Voon 1986).

$$\Phi_{\varepsilon\varepsilon}(\tau) = E[\varepsilon(t-\tau)\varepsilon(t)] = \delta(\tau) \quad \forall \tau \tag{21}$$

$$\Phi_{u\varepsilon}(\tau) = E[u(t-\tau)\varepsilon(t)] = 0 \quad \forall \tau \tag{22}$$

$$\Phi_{u^2\varepsilon}(\tau) = E\{[u^2(t-\tau) - \overline{u^2(t)}]\varepsilon(t)\} = 0 \quad \forall \tau \tag{23}$$

$$\Phi_{u^2\varepsilon^2}(\tau) = E\{[u^2(t-\tau) - \overline{u^2(t)}]\varepsilon^2(t)\} = 0 \quad \forall \tau \tag{24}$$

$$\Phi_{\varepsilon(\varepsilon u)}(\tau) = E[\varepsilon(t)\varepsilon(t-1-\tau)u(t-1-\tau)] = 0 \quad \forall \tau \tag{25}$$

where ε is the model residual and $\overline{u^2}$ is the time average of u^2. These tests look into the cross correlation amongst model residuals and inputs. Normalization to give all tests a range of plus and minus one and approximate the 95% confidence bounds at $1.96/\sqrt{N}$, N being the number of testing data, make the tests independent of signal amplitudes and easy to interpret. If these correlation tests are satisfied, then the model residuals are a random sequence and are not predictable from the model inputs. ❯ *Figure 14* shows the correlation-based model validation test (Billings and Voon 1986) results for the recurrent neurofuzzy network-based multistep-ahead prediction model. In ❯ *Fig. 14*, plots a–e are $\Phi_{\varepsilon\varepsilon}(\tau)$, $\Phi_{u\varepsilon}(\tau)$, $\Phi_{u^2\varepsilon}(\tau)$, $\Phi_{u^2\varepsilon^2}(\tau)$, and $\Phi_{\varepsilon(\varepsilon u)}(\tau)$, respectively. The dash-dot lines in each plot are the 95% confidence bounds. It can be seen that only a couple of points are slightly outside the 95% confidence bounds and, hence, the model can be regarded as being adequate for long-term predictions.

4 Nonlinear Model-Based Control Through Combination of Local Linear Controllers Based on Neurofuzzy Network Models

The neurofuzzy network topology proposed with its local linear models allow local linear model-based predictive controllers, such as generalized predictive controllers (GPC) (Clarke et al. 1987), to be developed. These local controllers can then be combined in a similar method to that used in the heterogeneous control strategy proposed by Kuipers and Astrom (Kuipers and Astrom 1994). The global controller output is obtained by combining the local model outputs in the format similar to the COG defuzzification. This is illustrated by letting

☐ **Fig. 14**

Model validation tests for the recurrent neurofuzzy network.

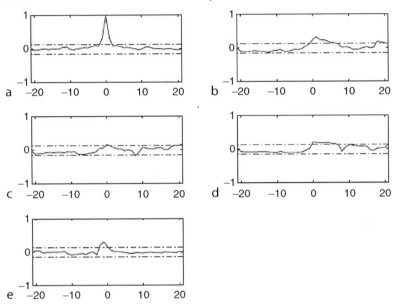

μ_i and u_i be the membership function and output of the local controller for the ith operating region, respectively, and calculating the global controller output, u, as

$$u = \frac{\sum\limits_{i=1}^{nr} \mu_i u_i}{\sum\limits_{i=1}^{nr} \mu_i} \tag{26}$$

The advantage of using local linear models is that a nonlinear model predictive controller can be constructed through several local linear model predictive controllers which usually have analytical solutions. Hence, the control actions of this nonlinear model predictive controller can be obtained analytically avoiding the time-consuming numerical search procedures and the uncertainty in convergence to the global optimum which are typically seen in conventional nonlinear model-based predictive control strategies (Eaton and Rawlings 1990; Hernandez and Arkun 1993; Li and Biegler 1989; Maner and Doyle 1997; Saint-Donat et al. 1991; Sistu et al. 1993). Furthermore, control actions calculated from local linear models in incremental form contain integral actions which can naturally eliminate static control offsets.

Neurofuzzy network-based nonlinear model predictive control strategy is derived as follows. Consider the following local linear model

$$A(q^{-1})y(t) = B(q^{-1})u(t-1) \tag{27}$$

where A and B are polynomials in the backward shift operator q^{-1}.

$$A(q^{-1}) = 1 + a_1 q^{-1} + \ldots + a_{no} q^{-no} \tag{28}$$

$$B(q^{-1}) = 1 + b_1 q^{-1} + \ldots + b_{ni} q^{-ni} \tag{29}$$

To derive a j-step ahead predictor of $y(t + j)$ based on ❷ Eq. 27, consider the following Diophantine equation.

$$1 = E_j(q^{-1})A(q^{-1})\Delta + q^{-j}F_j(q^{-1}) \tag{30}$$

where $E_j(q^{-1})$ and $F_j(q^{-1})$ are polynomials in q^{-1} uniquely defined given $A(q^{-1})$ and the prediction interval j and Δ is the differencing operator, $1 - q^{-1}$. Multiplying ❷ Eq. 9 by $E_j(q^{-1})\Delta q^j$ gives

$$E_j(q^{-1})A(q^{-1})\Delta y(t + j) = E_j(q^{-1})B(q^{-1})\Delta u(t + j - 1) \tag{31}$$

The substitution of $E_j(q^{-1})A(q^{-1})\Delta$ from ❷ Eq. 12 gives

$$y(t + j) = E_j(q^{-1})B(q^{-1})\Delta u(t + j - 1) + F_j(q^{-1})y(t) \tag{32}$$

Therefore, the j-step ahead prediction of y, given measured output data up to time t and any given input $u(t + i)$ for $i > 1$, is

$$\hat{y}(t + j|t) = G_j(q^{-1})\Delta u(t + j - 1) + F_j(q^{-1})y(t) \tag{33}$$

where $G_j(q^{-1}) = E_j(q^{-1})B(q^{-1})$. The Diophantine equation can be recursively solved (Clarke et al. 1987). The following vector equation can be written for a prediction horizon of N.

$$\hat{Y} = GU + F \tag{34}$$

where

$$\hat{Y} = [\hat{y}(t + 1)\hat{y}(t + 2)\ldots\hat{y}(t + N)]^T \tag{35}$$

$$U = [\Delta u(t)\Delta u(t + 1)\ldots\Delta u(t + N - 1)]^T \tag{36}$$

$$F = [f(t + 1)f(t + 2)\ldots f(t + N)]^T \tag{37}$$

The matrix G is lower-triangular of dimension $N \times N$:

$$G = \begin{bmatrix} g_0 & 0 & \cdots & 0 \\ g_1 & g_0 & \cdots & 0 \\ \cdots & \cdots & \cdots & \cdots \\ g_{N-1} & g_{N-2} & \cdots & g_0 \end{bmatrix} \tag{38}$$

Suppose that a future set point or reference sequence $[w(t + j); j = 1, 2, \ldots]$ is available. The control objective function can be written as

$$J = (\hat{Y} - W)^T(\hat{Y} - W) + \lambda U^T U \tag{39}$$

The unconstrained minimization of the above objective function leads to the following optimal control strategy.

$$U = (G^T G + \lambda I)^{-1} G^T (W - F) \tag{40}$$

Since the first element of U is $\Delta u(t)$, the current control action is given by

$$u(t) = u(t-1) + \bar{g}^T(W - F) \tag{41}$$

where $\bar{g}^T$ is the first row of $(G^TG + \lambda I)^{-1}G^T$. The control action obtained in ❷ Eq. 41 hence contains an integral action which provides zero static offset.

A neurofuzzy network-based heterogeneous long-range predictive controller was developed for the pH reactor. This consisted of three local linear GPC controllers based on the following three local linear incremental models transformed from the recurrent neurofuzzy network model:

> IF pH low
> THEN $\hat{y}(t) = 1.6367\hat{y}(t-1) - 0.4015\hat{y}(t-2) - 3.56\Delta u(t-1) - 1.81\Delta u(t-2)$
> IF pH medium
> THEN $\hat{y}(t) = 1.3309\hat{y}(t-1) - 0.0511\hat{y}(t-2) - 94.5\Delta u(t-1) - 22.53\Delta u(t-2)$
> IF pH high
> THEN $\hat{y}(t) = 1.6631\hat{y}(t-1) - 0.4036\hat{y}(t-2) - 6.16\Delta u(t-1) - 1.12\Delta u(t-2)$

The control horizons for the three local GPC controllers were all set to one and prediction horizons were all set to five. A linear GPC controller was also developed using a linear model identified from the data shown in ❷ Figs. 8 and ❷ 9 again with a control horizon of one and a prediction horizon of five. ❷ Figure 15 shows the performance of the neurofuzzy network-based heterogeneous long-range predictive controller and the linear GPC controller

◻ **Fig. 15**
Control performance for set-point tracking.

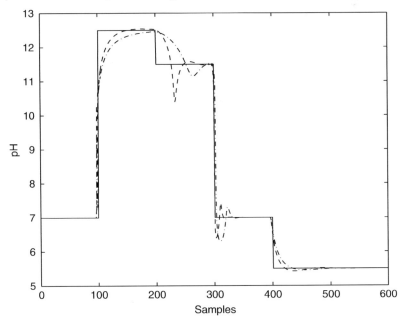

for set-point tracking over a very wide range of process operation. In ❷ *Fig. 15*, the solid line represents the set points, the dashed line represents the responses under the neurofuzzy GPC controller, and the dash-dot line represents the responses under the linear GPC controller. It can be seen that the response under the linear GPC controller is very sluggish in the pH low and pH high regions. However, the response under the neurofuzzy GPC controller is satisfactory in all operating regions. This is due to the fact that the neurofuzzy GPC controller output in a particular operating region is mainly determined by the corresponding local controller which is based on a more appropriate local model for that region.

To examine the performance of the neurofuzzy network-based long-range predictive controller in rejecting disturbances, negative and positive step changes in base concentration (C_2) were added to the process. ❷ *Figure 16* shows the disturbances and the performance of the neurofuzzy network-based predictive controller and the linear GPC controller. In the upper plot of❷ *Fig. 16*, the solid line represents the set points, the dashed line represents the responses under the neurofuzzy GPC controller, and the dash-dot line represents the responses under the linear GPC controller. It can be seen that the neurofuzzy network-based predictive controller performs very well whereas the linear model-based controller performs very poorly.

Although a gain scheduled controller can also be developed to control the neutralization process (Astrom and Wittenmark 1989), the gain scheduling strategy is usually used on controllers with a simple form such as a PID controller where the controller gain is scheduled according to a monitored process variable. Traditional gain-scheduling techniques are not straightforward to be implemented on long-range model-based predictive controllers.

◘ Fig. 16

Control performance for disturbance rejection.

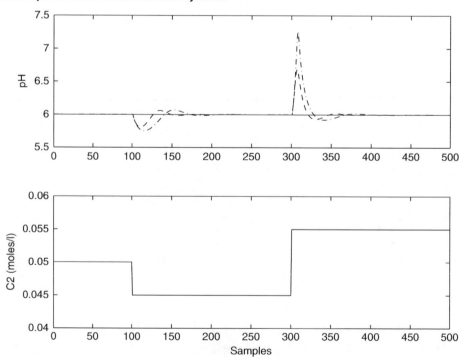

The recurrent neurofuzzy network-based long-range predictive control strategy presented in this section represents a model parameter scheduled long-range predictive control strategy.

5 Modeling and Optimal Control of Batch Processes Using Neurofuzzy Networks

5.1 A Fed-Batch Reactor

The fed-batch reactor is taken from Terwiesch et al. (1998). The following reaction system

$$A + B \xrightarrow{k_1} C$$

$$B + B \xrightarrow{k_2} D$$

is conducted in an isothermal semi-batch reactor. The objective in operating this reactor is, through the addition of reactant B, to convert as much as possible of reactant A to the desired product, C, in a specified time $t_f = 120$ min. It would not be optimal to add all B initially as the second-order side reaction yielding the undesired species D will be favored at high concentration of B. To keep this undesired species low, the reactor is operated in semi-batch mode where B is added in a feed stream with concentration $b_{\text{feed}} = 0.2$. Based on the reaction kinetics and material balances in the reactor, the following mechanistic model can be developed.

$$\frac{d[A]}{dt} = -k_1[A][B] - \frac{[A]}{V}u \tag{42}$$

$$\frac{d[B]}{dt} = -k_1[A][B] - 2k_2[B]^2 + \frac{b_{\text{feed}} - [B]}{V}u \tag{43}$$

$$\frac{d[C]}{dt} = k_1[A][B] - \frac{[C]}{V}u \tag{44}$$

$$\frac{d[D]}{dt} = 2k_2[B]^2 - \frac{[D]}{V}u \tag{45}$$

$$\frac{dV}{dt} = u \tag{46}$$

In the above equations, $[A]$, $[B]$, $[C]$, and $[D]$ denote, respectively, the concentrations of A, B, C, and D, V is the current reaction volume, u is the reactant feed rate, and the reaction rate constants have the nominal value $k_1 = 0.5$ and $k_2 = 0.5$. At the start of reaction, the reactor contains $[A](0) = 0.2$ moles/L of A, no B ($[B](0) = 0$) and is fed to 50% ($V(0) = 0.5$).

5.2 Recurrent Neurofuzzy Network Modeling

In this study, it is assumed that the above mechanistic model is not available (due to, e.g., unknown reaction mechanism or unknown parameters, which are commonly encountered in practice) so a data-based empirical model has to be utilized. Since the main interest in this process is on the end-of-batch product quality, the data-based empirical model should offer accurate long-range predictions. A recurrent neurofuzzy network is used to model the process from the process operation data. ❯ *Figure 17* shows the data from a typical batch run. It can be seen from ❯ *Fig. 17* that the process behaves nonlinearly and the dynamic characteristics vary with the batch operating stages. ❯ *Figure 17* indicates that the reaction volume could be used to divide the process operation into several stages, for example, labeled as "low volume," "medium volume," and "high volume." Four batches of process operation under different feeding policies were simulated to produce the data for recurrent neurofuzzy network modeling. In agile responsive batch processing, it is highly desirable that a model can be developed using data from a limited number of process runs so that the model can be quickly developed and applied to the process. Based on this consideration, the data for developing recurrent neurofuzzy network model is limited to data from just four batch runs. It should be noted here that the model can be retrained when data from more batch runs are available.

After network training, the following model was identified.

IF $V(t)$ is low

THEN
$$y_1(t) = 0.4577y_1(t-1) + 0.5449y_1(t-2) - 0.0339V(t-1)$$
$$+ 0.0064V(t-2) + 0.0264V(t-3)$$
$$y_2(t) = 0.5678y_2(t-1) + 0.4340y_2(t-2) + 0.0066V(t-1)$$
$$+ 0.0077V(t-2) - 0.0141V(t-3)$$

◼ **Fig. 17**
Data from a typical batch run.

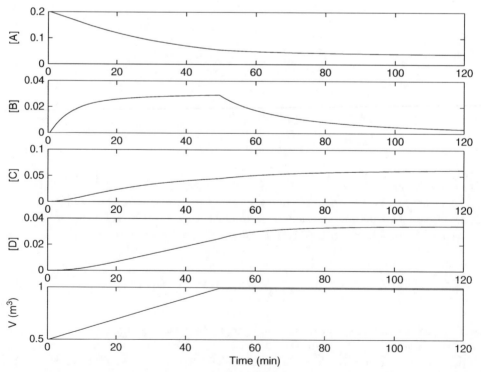

IF $V(t)$ is medium

THEN

$$y_1(t) = 0.8247y_1(t-1) + 0.0683y_1(t-2) - 0.0002V(t-1)$$
$$\quad - 0.0002V(t-2) + 0.0079V(t-3)$$
$$y_2(t) = 0.2323y_2(t-1) + 0.7725y_2(t-2) - 0.0611V(t-1)$$
$$\quad + 0.00002V(t-2) + 0.0608V(t-3)$$

IF $V(t)$ is high

THEN

$$y_1(t) = 0.8553y_1(t-1) + 0.1004y_1(t-2) - 0.0134V(t-1)$$
$$\quad + 0.003V(t-2) + 0.0133V(t-3)$$
$$y_2(t) = -0.9829y_2(t-1) + 1.9827y_2(t-2) - 0.0143V(t-1)$$
$$\quad + 0.0223V(t-2) - 0.008V(t-3)$$

In the above model, y_1 is $[C]$, y_2 is $[D]$, V is the reaction volume, and t is the discrete time. When performing long-range predictions using the above neurofuzzy network model, the future reaction volume $V(t)$ can be simply obtained by integrating the feeding rate (u). The identified membership functions for reaction volume "low," "medium," and "high" are shown in ❷ Fig. 18. The identified recurrent neurofuzzy network model was tested on an additional unseen testing batch, which is not used when developing the recurrent neurofuzzy network model. ❷ Figure 19 shows the long-range predictions of $[C]$ as well as the model prediction

◻ **Fig. 18**
Identified membership functions.

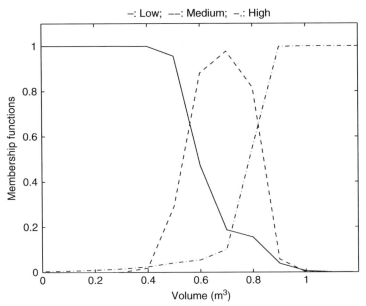

☐ **Fig. 19**

Long-range predictions of [C].

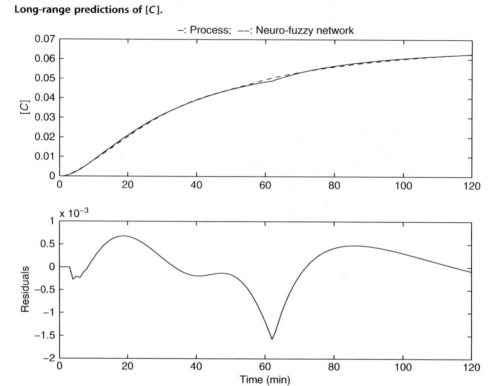

residuals, on the unseen testing batch, whereas ❷ *Fig. 20* shows those for [D]. In the top plots of ❷ *Figs. 19* and ❷ *20*, the solid lines represent the actual values whereas the dashed lines represent the model predictions. The bottom plots of ❷ *Figs. 19* and ❷ *20* show the recurrent neurofuzzy network model prediction residuals. It can be seen from ❷ *Figs. 19* and ❷ *20* that the model predictions can be considered as being accurate.

The above local linear models are easier to interpret than a pure black box model (e.g., a conventional neural network model). From the local model parameters, the local model characteristics, such as gain and time constant, can be calculated. From the identified membership functions, different process-operating regions can be visualized. ❷ *Table 3* shows the calculated static gain of different local models. It can be seen that [C] has much higher gain in the "Volume low" region than other regions, whereas [D] has high gain in the "Volume medium" region. Insights into these local model characteristics could help process operators in understanding the process operation and in judging the optimal control actions. It should be noticed that the static gains shown in ❷ *Table 3* can only give a rough guidance about the operational characteristics in the local regions since a batch process does not have a steady state. The local dynamic gains can be calculated from the local linear models. ❷ *Figure 21* shows the local model dynamic gains for [C] while ❷ *Fig. 22* shows those for [D].

◻ Fig. 20

Long-range predictions of [D].

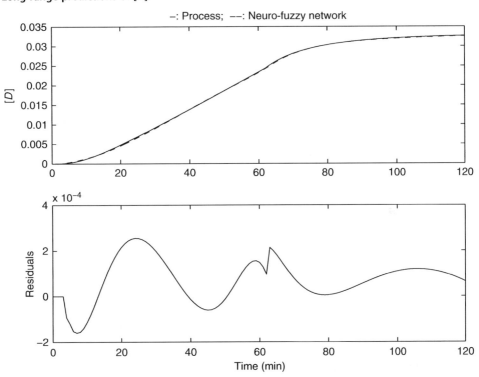

5.3 Multi-objective Optimal Control of the Fed-Batch Reactor

The objectives in operating this process are to maximize the amount of the final product $[C](t_f)V(t_f)$ and minimize the amount of the final undesired species $[D](t_f)V(t_f)$. This can be formulated as a multi-objective optimization problem with the goal-attainment method (Gembicki 1974):

$$\mathbf{F}(U) = \begin{bmatrix} -[C](t_f)V(t_f) \\ [D](t_f)V(t_f) \end{bmatrix} \tag{47}$$

$$\min_{U,\gamma} \gamma$$

$$\text{s.t. } \mathbf{F}(U) - \mathbf{W}\gamma \le \mathbf{GOAL}$$

$$0 \le u_i \le 0.01 \ (i = 1, 2, \ldots, m)$$

$$V(t_f) \le 1$$

where γ is a scalar variable, $\mathbf{W} = [w_1 \ w_2]^\mathrm{T}$ are weighting parameters, $\mathbf{GOAL} = [g_1 \ g_2]^\mathrm{T}$ are the goals that the objectives attain, $U = [u_1, u_2, \ldots, u_m]$ is a sequence of the reactant-feeding rates and V is the reaction volume. In this study, the batch time (120 min) is divided into $m = 10$ segments of equal length and, within the ith segment, the reactant feeding rate is u_i.

◘ **Table 3**

Static gains calculated from the local models

	Volume low	Volume medium	Volume high
$[C]$	0.4231	0.0701	0.0655
$[D]$	−0.1111	0.0583	−0.0064

◘ **Fig. 21**

The dynamic gains for the local models for [C].

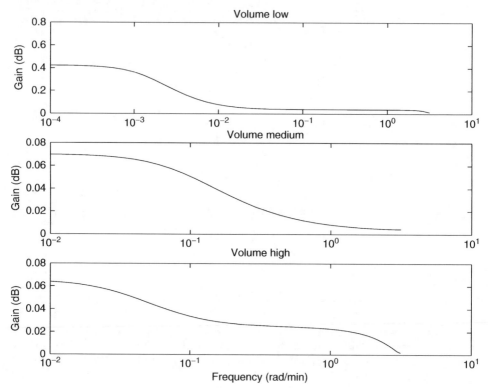

Two cases are studied here: Case 1 – **GOAL** $= [-0.065\ 0.01]^{\mathrm{T}}$ and Case 2 – **GOAL** $= [-0.1\ 0.02]^{\mathrm{T}}$. In both cases the weighting parameters w_1 and w_2 are selected as 1. Case 1 emphasizes on less by-product while Case 2 emphasizes on more product.

The optimization problem was solved using the MATLAB Optimization Toolbox function "attgoal." The calculated optimal reactant feeding policies for both cases are plotted in ❷ *Fig. 23*. It can be seen that the feeding policies are very different for the two cases.

Under these feeding policies, the trajectories of process variables in Cases 1 and 2 are respectively shown in ❷ *Figs. 24* and ❷ *25*. In Case 1 the values of $[C](t_f)$ and $[D](t_f)]$ are 0.0547 and 0.0202, respectively, whereas in Case 2 those are 0.0617 and 0.0349, respectively. It can be seen that these two different feeding policies lead to different operation objectives. Looking back at the local linear models and their gains given in ❷ *Table 3*, the control

◻ Fig. 22

The dynamic gains for the local models for [D].

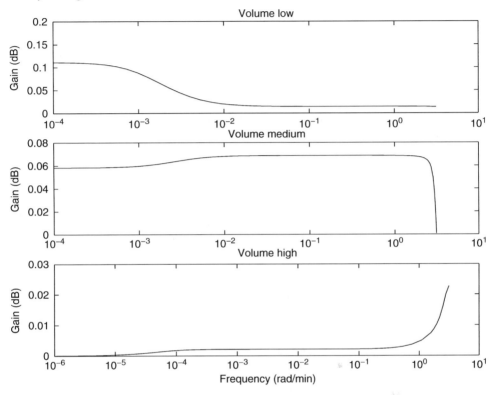

◻ Fig. 23

Optimal reactant feeding policies.

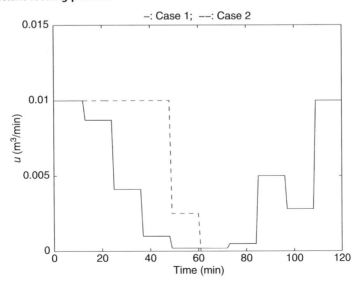

☐ **Fig. 24**

Trajectories of process variables under optimal control in Case 1.

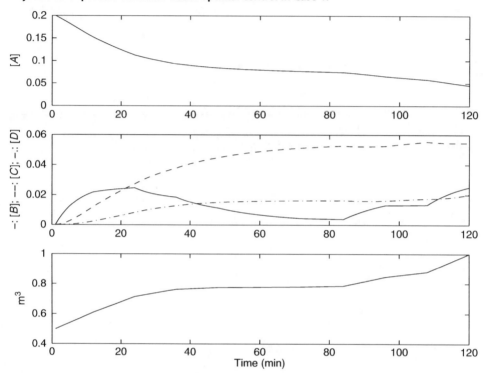

profiles given in ❯ *Fig. 23* are easy to comprehend. Case 1 emphasizes on less by-product and ❯ *Table 3* shows that the gain for by-product is high during the reaction volume medium stage, thus the control profile exhibits low feed rate during the reaction volume medium stage to limit the growth of by-product. Whereas Case 2 emphasizes on more product and ❯ *Table 3* indicates that the gain for product at low reaction volume is high, thus the control profile for Case 1 has maximum feed rate at the beginning to ensure a large amount of product being produced through fast growth of reaction volume during the reaction volume low stage as can be seen in ❯ *Fig. 25*.

Terwiesch et al. (1998) reported a nominal control policy and a minimal risk control policy for this fed-batch process based on the mechanistic model. ❯ *Figure 26* shows these mechanistic model-based optimal control policies whereas ❯ *Figure 27* shows the trajectories of $[C]$ and $[D]$ under these control policies. ❯ *Table 4* gives the values of $[C](t_f)$ and $[D](t_f)$ under different control policies. It can be seen that the recurrent neurofuzzy network model-based optimal control policies give comparable performance as the mechanistic model-based optimal control policies. Results from the recurrent neurofuzzy network model-based optimal control policy (Case 2) is very close to those from mechanistic model-based optimal control policy (nominal). Although the final product concentration $[C](t_f)$ under the recurrent neurofuzzy network model-based optimal control policy (Case 1) is lower than those under the mechanistic model-based optimal control policies, the final concentration of the undesired species $[D](t_f)$ under the recurrent neurofuzzy network model-based optimal control policy (Case 1) is also lower than those under the mechanistic model-based optimal control policies.

□ Fig. 25

Trajectories of process variables under optimal control in Case 2.

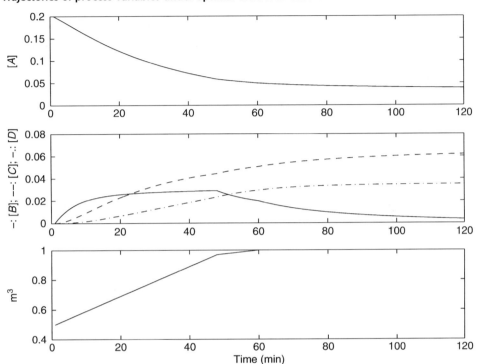

□ Fig. 26

Mechanistic model-based nominal and minimal risk control policies.

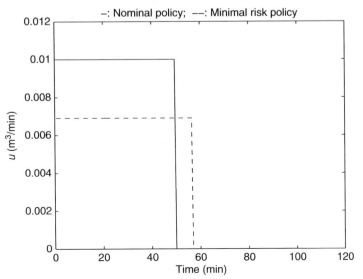

◘ **Fig. 27**

Trajectories of [C] and [D] under the nominal and minimal risk control policies.

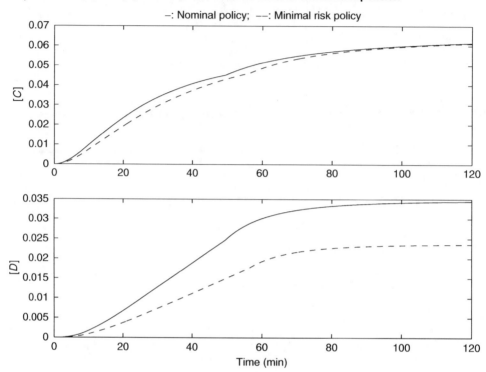

◘ **Table 4**

Values of [C](t_f) and [D](t_f) under different control policies

Control policies	[C](t_f)	[D](t_f)
Recurrent neurofuzzy network – Case 1	0.0547	0.0202
Recurrent neurofuzzy network – Case 2	0.0617	0.0349
Nominal (Terwiesch et al. 1998)	0.0615	0.0345
Minimal risk (Terwiesch et al. 1998)	0.0612	0.0236

6 Conclusions

Recurrent neurofuzzy networks are very effective in building long-range prediction models for nonlinear processes from process operational data. A recurrent neurofuzzy network can offer accurate long-term predictions which are crucial to the success of many advanced process-control techniques such as nonlinear long-range predictive control and optimal batch process control. Neurofuzzy network models are easier to interpret than conventional neural network models. The weights in the fuzzification layer determine the membership functions of the fuzzy operating regions while the weights in the function layer determine the local models in these operating regions. The neurofuzzy network-based approaches are therefore not purely "black box" approaches which are difficult to interpret.

Neurofuzzy network process models provide a basis for building a novel type of nonlinear controller which is composed of several local linear controllers. The global controller output is obtained by combining local controller outputs based upon their membership functions. A neurofuzzy network model-based predictive controller can be obtained by combining several local linear model-based predictive controllers which have analytical solutions. The advantage of this technique is that both the time-consuming numerical optimization methods and the uncertainty in convergence to the global optimum, which are typically seen in conventional nonlinear model-based predictive control, are avoided. Furthermore, control actions obtained, based on local incremental models, contain integration actions which can naturally eliminate static offsets. In most of the reported nonlinear model-based controllers, however, static offsets are usually eliminated using some ad hoc methods, such as by coupling a PI controller.

Applications to the modeling and control of a neutralization process and a fed-batch reactor demonstrate that the recurrent neurofuzzy networks are capable of providing accurate long-range predictions. The developed models are easy to comprehend and local dynamic characteristics of the modeled processes can easily be obtained.

References

Akaike H (1974) A new look at the statistical model identification. IEEE Trans Automatic Control 19:716–723

Astrom KJ, Wittenmark B (1989) Adaptive control, 2nd edn. Addison-Wesley, Reading, MA

Barron AR (1984) Predicted squared error: a criterion for automatic model selection. In: Farlow SJ (ed) Self organising methods. Marcel Dekker, New York, pp 87–103

Bhat NV, McAvoy TJ (1990) Use of neural nets for dynamical modelling and control of chemical process systems. Comput Chem Eng 14:573–583

Billings SA, Voon WSF (1986) Correlation based model validity tests for non-linear models. Int J Control 44:235–244

Blanco A, Delgado M, Pegalajar MC (2001a) A real-coded genetic algorithm for training recurrent neural networks. Neural Netw 14:93–105

Blanco A, Delgado M, Pegalajar MC (2001b) Fuzzy automaton induction using neural networks. Int J Approx Reasoning 27:1–26

Brown M, Harris CJ (1994) Neurofuzzy adaptive modelling and control. Prentice Hall, Hemel Hempstead

Bulsari AB (ed) (1995) Computer-aided chemical engineering. Neural networks for chemical engineers, vol 6. Elsevier, Amsterdam

Clarke DW, Mohtadi C, Tuffs PS (1987) Generalised predictive control, Parts 1 and 2. Automatica 23:859–875

Cybenko G (1989) Approximation by superpositions of a sigmoidal function. Math Control Signal Syst 2:303–314

Eaton JW, Rawlings JB (1990) Feedback control of chemical processes using on-line optimization techniques. Comput Chem Eng 14:469–479

Elman JL (1990) Finding structures in time. Cogn Sci 14:179–211

Frasconi P, Gori M, Soda G (1992) Local feedback multi-layered networks. Neural Comput 4:120–130

Geladi P, Kowalski BR (1986) Partial least-squares regression: a tutorial. Anal Chim Acta 185:1–17

Gembicki FW (1974) Vector optimisation for control with performance and parameter sensitivity indices, Ph.D. Thesis, Case Western Reserve University

Girosi F, Poggio T (1990) Networks and the best approximation property. Biol Cybern 63:169–179

Harris CJ, Brown M, Bossley KM, Mills DJ, Ming F (1996) Advances in neurofuzzy algorithms for real-time modelling and control. Eng Appl Artif Intell 9:1–16

Hernandez E, Arkun Y (1993) Control of nonlinear systems using polynomial ARMA models. AIChE J 39:446–460

Hoerl AE, Kennard RW (1970) Ridge regression: biased estimation for nonorthogonal problems. Technometrics 12:55–67

Horikawa S, Furuhashi T, Uchikawa Y (1992) On fuzzy modeling using fuzzy neural networks with the back-propagation algorithm. IEEE Trans Neural Netw 3:801–806

Jang JSR (1992) Self-learning fuzzy controllers based on temporal back propagation. IEEE Trans Neural Netw 3:714–723

Jang JSR, Sun CT (1995) Neuro-fuzzy modeling and control. Proc IEEE 83:378–406

Jang JSR, Sun CT, Mizutani E (1997) Neuro-fuzzy and soft computing: a computational approach to learning and machine intelligence. Prentice Hall, Englewood Cliffs, NJ

Johansen TA, Foss BA (1993) Constructing NARMAX models using ARMAX models. Int J Control 58:1125–1153

Kuipers B, Astrom K (1994) The composition and validation of heterogeneous control laws. Automatica 30:233–249

Leonard JA, Kramer MA (1990) Improvement of the back-propagation algorithm for training neural networks. Comput Chem Eng 14:337–341

Leontaris IJ, Billings SA (1987) Model selection and validation methods for nonlinear systems. Int J Control 45:311–341

Li WC, Biegler LT (1989) Multistep, Newton-type control strategies for constrained, nonlinear processes. Chem Eng Res Des 67:562–577

Mak MW, Ku KW, Lu YL (1999) On the improvement of the real time recurrent learning algorithm for recurrent neural networks. Neurocomputing 24:13–36

Maner BR, Doyle FJ III (1997) Polymerization reactor control using autoregressive-plus Volterra-based MPC. AIChE J 43:1763–1784

Marquardt D (1963) An algorithm for least squares estimation of nonlinear parameters. SIAM J Appl Math 11:431–441

McAvoy TJ, Hsu E, Lowenthal S (1972) Dynamics of pH in controlled stirred tank reactor. Ind Eng Chem Process Des Dev 11:68–70

Morris AJ, Montague GA, Willis MJ (1994) Artificial neural networks: studies in process modelling and control. Chem Eng Res Des 72:3–19

Nie J, Linkens DA (1993) Learning control using fuzzified self-organising radial basis function networks. IEEE Trans Fuzzy Syst 1:280–287

Omlin CW, Thornber KK, Giles CL (1998) Fuzzy finite-state automata can be deterministically encoded into recurrent neural networks. IEEE Trans Fuzzy Syst 6:76–89

Park J, Sandberg IW (1991) Universal approximation using radial basis function networks. Neural Comput 3:246–257

Rumelhart DE, Hinton GE, Williams RJ (1986) Learning internal representations by error propagation. In: Rumelhart DE, McClelland JL (eds) Parallel distributed processing. MIT Press, Cambridge, MA

Saint-Donat J, Bhat N, McAvoy TJ (1991) Neural net based model predictive control. Int J Control 54:1452–1468

Scott GM, Ray WH (1993) Creating efficient nonlinear network process models that allow model interpretation. J Process Control 3:163–178

Sistu PB, Gopinath RS, Bequette BW (1993) Computational issues in nonlinear predictive control. Comput Chem Eng 17:361–366

Sjoberg J, Zhang Q, Ljung L, Benveniste A, Delyon B, Glorennec P, Hjalmarsson H, Juditsky A (1995) Nonlinear black-box modelling in system identification: a unified overview. Automatica 31: 1691–1724

Su HT, McAvoy TJ, Werbos P (1992) Long term prediction of chemical processes using recurrent neural networks: a parallel training approach. Ind Eng Chem Res 31:1338–1352

Takagi T, Sugeno M (1985) Fuzzy identification of systems and its application to modelling and control. IEEE Trans Syst Man Cybern 15:116–132

Terwiesch P, Ravemark D, Schenker B, Rippin DWT (1998) Semi-batch process optimization under uncertainty: theory and experiments. Comput Chem Eng 22:201–213

Tian Y, Zhang J, Morris AJ (2002) Optimal control of a batch emulsion copolymerisation reactor based on recurrent neural network models. Chem Eng Process 41:531–538

Tsoi AC, Back AD (1994) Locally recurrent globally feedforward networks: a critical review of architectures. IEEE Trans Neural Netw 5:229–239

Wang LX (1994) Adaptive fuzzy systems and control: design and stability analysis. Prentice Hall, Englewood Cliffs, NJ

Werbos PJ (1990) Backpropagation through time: what it does and how to do it. Proc IEEE 78:1550–1560

Yager RR, Filev DP (1994) Essentials of fuzzy modelling and control. Wiley, New York

Zhang J (2004) A reliable neural network model based optimal control strategy for a batch polymerisation reactor. Ind Eng Chem Res 43:1030–1038

Zhang J (2005) Modelling and optimal control of batch processes using recurrent neuro-fuzzy networks. IEEE Trans Fuzzy Syst 13:417–427

Zhang J (2006) Modelling and multi-objective optimal control of batch processes using recurrent neuro-fuzzy networks. Int J Automation Comput 3:1–7

Zhang J, Morris AJ (1994) On-line process fault diagnosis using fuzzy neural networks. Intell Syst Eng 3:37–47

Zhang J, Morris AJ (1995) Fuzzy neural networks for nonlinear systems modelling. IEE Proc, Control Theory Appl 142:551–561

Zhang J, Morris AJ (1999) Recurrent neuro-fuzzy networks for nonlinear process modelling. IEEE Trans Neural Netw 10:313–326

Zhang J, Morris AJ, Martin EB (1998) Long term prediction models based on mixed order locally recurrent neural networks. Comput Chem Eng 22: 1051–1063

13 Independent Component Analysis

Seungjin Choi
Pohang University of Science and Technology, Pohang, South Korea
seungjin@postech.ac.kr

G. Rozenberg et al. (eds.), *Handbook of Natural Computing*, DOI 10.1007/978-3-540-92910-9_13,
© Springer-Verlag Berlin Heidelberg 2012

Abstract

Independent component analysis (ICA) is a statistical method, the goal of which is to decompose multivariate data into a linear sum of non-orthogonal basis vectors with coefficients (encoding variables, latent variables, and hidden variables) being statistically independent. ICA generalizes widely used subspace analysis methods such as principal component analysis (PCA) and factor analysis, allowing latent variables to be non-Gaussian and basis vectors to be non-orthogonal in general. ICA is a density-estimation method where a linear model is learned such that the probability distribution of the observed data is best captured, while factor analysis aims at best modeling the covariance structure of the observed data. We begin with a fundamental theory and present various principles and algorithms for ICA.

1   Introduction

Independent component analysis (ICA) is a widely used multivariate data analysis method that plays an important role in various applications such as pattern recognition, medical image analysis, bioinformatics, digital communications, computational neuroscience, and so on. ICA seeks a decomposition of multivariate data into a linear sum of non-orthogonal basis vectors with coefficients being statistically as independent as possible.

We consider a linear generative model where m-dimensional observed data $x \in \mathbb{R}^m$ is assumed to be generated by a linear combination of n basis vectors $\{a_i \in \mathbb{R}^m\}$,

$$x = a_1 s_1 + a_2 s_2 + \cdots + a_n s_n \tag{1}$$

where $\{s_i \in \mathbb{R}\}$ are *encoding variables*, representing the extent to which each basis vector is used to reconstruct the data vector. Given N samples, the model (❷ Eq. 1) can be written in a compact form

$$X = AS \tag{2}$$

where $X = [x(1), \ldots, x(N)] \in \mathbb{R}^{m \times N}$ is a data matrix, $A = [a_1, \ldots, a_n] \in \mathbb{R}^{m \times n}$ is a basis matrix, and $S = [s(1), \ldots, s(N)] \in \mathbb{R}^{n \times N}$ is an encoding matrix with $s(t) = [s_1(t), \ldots, s_n(t)]^\top$.

Dual interpretation of basis encoding in the model (❷ Eq. 2) is given as follows:

- When columns in X are treated as data points in m-dimensional space, columns in A are considered as *basis vectors* and each column in S is *encoding* that represents the extent to which each basis vector is used to reconstruct data vector.
- Alternatively, when rows in X are data points in N-dimensional space, rows in S correspond to basis vectors and each row in A represents encoding.

A strong application of ICA is a problem of *blind source separation* (BSS), the goal of which is to restore *sources* S (associated with encodings) without the knowledge of A, given the data matrix X. ICA and BSS have often been treated as an identical problem since they are closely related to each other. In BSS, the matrix A is referred to as a *mixing matrix*. In practice, we find a linear transformation W, referred to as a *demixing matrix*, such that the rows of the output matrix

$$Y = WX \tag{3}$$

are statistically independent. Assume that sources (rows of S) are statistically independent. In such a case, it is well known that WA becomes a *transparent transformation* when the rows of Y are statistically independent. The transparent transformation is given by $WA = P\Lambda$, where P is a permutation matrix and Λ is a nonsingular diagonal matrix involving scaling. This transparent transformation reflects two indeterminacies in ICA (Comon 1994): (1) scaling ambiguity; (2) permutation ambiguity. In other words, entries of Y correspond to scaled and permuted entries of S.

Since Jutten and Herault's first solution (Jutten and Herault 1991) to ICA, various methods have been developed so far, including a neural network approach (Cichocki and Unbehauen 1996), information maximization (Bell and Sejnowski 1995), natural gradient (or relative gradient) learning (Amari et al. 1996; Cardoso and Laheld 1996; Amari and Cichocki 1998), maximum likelihood estimation (Pham 1996; MacKay 1996; Pearlmutter and Parra 1997; Cardoso 1997), and nonlinear principal component analysis (PCA) (Karhunen 1996; Oja 1995; Hyvärinen and Oja 1997). Several books on ICA (Lee 1998; Hyvärinen et al. 2001; Haykin 2000; Cichocki and Amari 2002; Stone 1999) are available, serving as a good resource for a thorough review and tutorial on ICA. In addition, tutorial papers on ICA (Hyvärinen 1999; Choi et al. 2005) are useful resources.

This chapter begins with a fundamental idea, emphasizing why independent components are sought. Then, well-known principles to tackle ICA are introduced, leading to an objective function to be optimized. The natural gradient algorithm for ICA is explained. We also elucidate how we incorporate nonstationarity or temporal information into the standard ICA framework.

2 Why Independent Components?

PCA is a popular subspace analysis method that has been used for dimensionality reduction and feature extraction. Given a data matrix $X \in \mathbb{R}^{m \times N}$, the covariance matrix R_{xx} is computed by

$$R_{xx} = \frac{1}{N} XHX^{\top}$$

where $H = I_{N \times N} - \frac{1}{N} 1_N 1_N^{\top}$ is the *centering matrix*, where $I_{N \times N}$ is the $N \times N$ identity matrix and $1_N = [1, \ldots, 1]^{\top} \in \mathbb{R}^N$. The rank-$n$ approximation of the covariance matrix R_{xx} is of the form

$$R_{xx} \approx U\Lambda U^{\top}$$

where $U \in \mathbb{R}^{m \times n}$ contains n eigenvectors associated with n largest eigenvalues of R_{xx} in its columns and the corresponding eigenvalues are in the diagonal entries of Λ (diagonal matrix). Then, principal components $z(t)$ are determined by projecting data points $x(t)$ onto these eigenvectors, leading to

$$z(t) = U^{\top} x(t)$$

or in a compact form

$$Z = U^{\top} X$$

It is well known that rows of Z are uncorrelated with each other.

☐ **Fig. 1**

Two-dimensional data with two main arms are fitted by two different basis vectors: **(a)** principal component analysis (PCA) makes the implicit assumption that the data have a Gaussian distribution and determines the optimal basis vectors that are orthogonal, which are not efficient at representing non-orthogonal distributions; **(b)** independent component analysis (ICA) does not require that the basis vectors be orthogonal and considers non-Gaussian distributions, which is more suitable in fitting more general types of distributions.

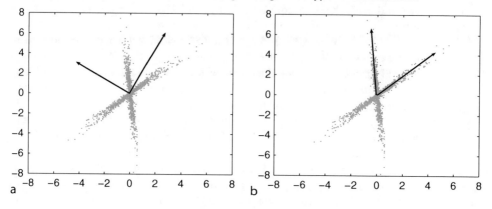

ICA generalizes PCA in the sense that latent variables (components) are non-Gaussian and *A* is allowed to be a non-orthogonal transformation, whereas PCA considers only orthogonal transformation and implicitly assumes Gaussian components. ❷ *Figure 1* shows a simple example, emphasizing the main difference between PCA and ICA.

A core theorem that plays an important role in ICA is presented. It provides a fundamental principle for various unsupervised learning algorithms for ICA and BSS.

Theorem 1 (Skitovich–Darmois) *Let* $\{s_1, s_2, \ldots, s_n\}$ *be a set of independent random variables. Consider two random variables* x_1 *and* x_2 *which are linear combinations of* $\{s_i\}$,

$$y_1 = \alpha_1 s_1 + \cdots + \alpha_n s_n$$
$$y_2 = \beta_1 s_1 + \cdots + \beta_n s_n$$

$$(4)$$

where $\{\alpha_i\}$ *and* $\{\beta_i\}$ *are real constants. If* y_1 *and* y_2 *are statistically independent, then each variable* s_i *for which* $\alpha_i \beta_i \neq 0$ *is Gaussian.*

Consider the linear model (❷ Eq. 2) for $m = n$. Throughout this chapter, the simplest case, where $m = n$ (square mixing), is considered. The global transformation can be defined as $G = WA$, where A is the mixing matrix and W is the demixing matrix. With this definition, the output $y(t)$ can be written as

$$y(t) = Wx(t) = Gs(t)$$

$$(5)$$

It is assumed that both A and W are nonsingular, hence G is nonsingular. Under this assumption, one can easily see that if $\{y_i(t)\}$ are mutually independent non-Gaussian signals, then invoking ❷ Theorem 1, G has the following decomposition

$$G = P\Lambda$$

$$(6)$$

This explains how ICA performs BSS.

3 Principles

The task of ICA is to estimate the mixing matrix A or its inverse $W = A^{-1}$ (referred to as the *demixing matrix*) such that elements of the estimate $y = A^{-1}x = Wx$ are as independent as possible. For the sake of simplicity, the index t is often left out if the time structure does not have to be considered. In this section, four different principles are reviewed: (1) maximum likelihood estimation; (2) mutual information minimization; (3) information maximization; and (4) negentropy maximization.

3.1 Maximum Likelihood Estimation

Suppose that sources s are independent with marginal distributions $q_i(s_i)$

$$q(s) = \prod_{i=1}^{n} q_i(s_i) \tag{7}$$

In the linear model, $x = As$, a single factor in the likelihood function is given by

$$
\begin{aligned}
p(x|A, q) &= \int p(x|s, A)q(s)ds \\
&= \int \prod_{j=1}^{n} \delta\left(x_j - \sum_{i=1}^{n} A_{ji}s_i \right) \prod_{i=1}^{n} q_i(s_i)ds
\end{aligned}
\tag{8}
$$

$$= |\det A|^{-1} \prod_{i=1}^{n} q_i\left(\sum_{j=1}^{n} A_{ij}^{-1} x_j \right) \tag{9}$$

Then, we have

$$p(x|A, q) = |\det A|^{-1} r(A^{-1}x) \tag{10}$$

The log-likelihood is written as

$$\log p(x|A, q) = -\log|\det A| + \log q(A^{-1}x) \tag{11}$$

which can be also written as

$$\log p(x|W, q) = \log|\det W| + \log p(y) \tag{12}$$

where $W = A^{-1}$ and y is the estimate of s with the true distribution $q(\cdot)$ replaced by a hypothesized distribution $p(\cdot)$. Since sources are assumed to be statistically independent, (❯ Eq. 12) is written as

$$\log p(x|W, q) = \log|\det W| + \sum_{i=1}^{n} \log p_i(y_i) \tag{13}$$

The demixing matrix W is determined by

$$\widehat{W} = \arg\max_{W} \left\{ \log|\det W| + \sum_{i=1}^{n} \log p_i(y_i) \right\} \tag{14}$$

It is well known that maximum likelihood estimation is equivalent to Kullback matching where the optimal model is estimated by minimizing the Kullback–Leibler (KL) divergence between the empirical distribution and the model distribution. Consider KL divergence from the empirical distribution $\tilde{p}(x)$ to the model distribution $p_\theta(x) = p(x|A, q)$

$$KL[\tilde{p}(x)||p_\theta(x)] = \int \tilde{p}(x) \log \frac{\tilde{p}(x)}{p_\theta(x)} dx = -H(\tilde{p}) - \int \tilde{p}(x) \log p_\theta(x) dx \qquad (15)$$

where $H(\tilde{p}) = -\int \tilde{p}(x) \log \tilde{p}(x) dx$ is the entropy of $\tilde{p}$. Given a set of data points, $\{x_1, \ldots, x_N\}$ drawn from the underlying distribution $p(x)$, the empirical distribution $\tilde{p}(x)$ puts probability $\frac{1}{N}$ on each data point, leading to

$$\tilde{p}(x) = \frac{1}{N} \sum_{t=1}^{N} \delta(x - x_t) \qquad (16)$$

It follows from (❷ Eq. 15) that

$$\arg\min_{\theta} KL[\tilde{p}(x)||p_\theta(x)] \equiv \arg\max_{\theta} \langle \log p_\theta(x) \rangle_{\tilde{p}} \qquad (17)$$

where $\langle \cdot \rangle_{\tilde{p}}$ represents the expectation with respect to the distribution $\tilde{p}$. Plugging (❷ Eq. 16) into the right-hand side of (❷ Eq. 15), leads to

$$\langle \log p_\theta(x) \rangle_{\tilde{p}} = \frac{1}{N} \int \sum_{t=1}^{N} N\delta(x - x_t) \log p_\theta(x) dx = \frac{1}{N} \sum_{t=1}^{N} \log p_\theta(x_t) \qquad (18)$$

Apart from the scaling factor $\frac{1}{N}$, this is just the log-likelihood function. In other words, maximum likelihood estimation is obtained from the minimization of (❷ Eq. 15).

3.2 Mutual Information Minimization

Mutual information is a measure for statistical independence. Demixing matrix W is learned such that the mutual information of $y = Wx$ is minimized, leading to the following objective function:

$$\begin{aligned} \mathcal{I}_{mi} &= \int p(y) \log \left[\frac{p(y)}{\prod_{i=1}^{n} p_i(y_i)} \right] dy \\ &= -H(y) - \left\langle \sum_{i=1}^{n} \log p_i(y_i) \right\rangle_y \end{aligned} \qquad (19)$$

where $H(\cdot)$ represents the entropy, that is,

$$H(y) = -\int p(y) \log p(y) dy \qquad (20)$$

and $\langle \cdot \rangle_y$ denotes the statistical average with respect to the distribution $p(y)$. Note that $p(y) = \frac{p(x)}{|\det W|}$. Thus, the objective function (❷ Eq. 19) is given by

$$\mathcal{I}_{mi} = -\log |\det W| - \sum_{i=1}^{n} \langle \log p_i(y_i) \rangle \qquad (21)$$

where $\langle \log p(x) \rangle$ is left out since it does not depend on parameters W. For online learning, only instantaneous value is taken into consideration, leading to

$$\mathcal{I}_{mi} = -\log|\det W| - \sum_{i=1}^{n} \log p_i(y_i) \tag{22}$$

3.3 Information Maximization

Infomax (Bell and Sejnowski 1995) involves the maximization of the output entropy $z = g(y)$ where $y = Wx$ and $g(\cdot)$ is a squashing function (e.g., $g_i(y_i) = \frac{1}{1+e^{-y_i}}$). It was shown that Infomax contrast maximization is equivalent to the minimization of KL divergence between the distribution of $y = Wx$ and the distribution $p(s) = \prod_{i=1}^{n} p_i(s_i)$. In fact, Infomax is nothing but mutual information minimization in the CA framework.

The Infomax contrast function is given by

$$\mathcal{I}_I(W) = H(g(Wx)) \tag{23}$$

where $g(y) = [g_1(y_1), \dots, g_n(y_n)]^\top$. If $g_i(\cdot)$ is differentiable, then it is the cumulative distribution function of some probability density function $q_i(\cdot)$,

$$g_i(y_i) = \int_{-\infty}^{y_i} q_i(s_i) ds_i$$

Let us choose a squashing function $g_i(y_i)$

$$g_i(y_i) = \frac{1}{1 + e^{-y_i}} \tag{24}$$

where $g_i(\cdot) : \mathbb{R} \to (0,1)$ is a monotonically increasing function.

Let us consider an n-dimensional random vector $\hat{s}$, the joint distribution of which is factored into the product of marginal distributions:

$$q(\hat{s}) = \prod_{i=1}^{n} q_i(\hat{s}_i) \tag{25}$$

Then $g_i(\hat{s}_i)$ is distributed uniformly on $(0,1)$, since $g_i(\cdot)$ is the cumulative distribution function of $\hat{s}_i$. Define $u = g(\hat{s}) = [g_1(\hat{s}_1), \dots, g_n(\hat{s}_n)]^\top$, which is distributed uniformly on $(0,1)^n$.

Define $v = g(Wx)$. Then, the Infomax contrast function is rewritten as

$$\begin{aligned} \mathcal{I}_I(W) &= H(g(Wx)) \\ &= H(v) \\ &= -\int p(v) \log p(v) dv \\ &= -\int p(v) \log\left(\frac{p(v)}{\prod_{i=1}^{n} 1_{(0,1)}(v_i)}\right) dv \\ &= -KL[v \| u] \\ &= -KL[g(Wx) \| u] \end{aligned} \tag{26}$$

where $1_{(0,1)}(\cdot)$ denotes a uniform distribution on $(0,1)$. Note that KL divergence is invariant under an invertible transformation f,

$$
\begin{aligned}
KL[f(u)\|f(v)] &= KL[u\|v]\\
&= KL[f^{-1}(u)\|f^{-1}(v)]
\end{aligned}
$$

Therefore, we have

$$
\begin{aligned}
\mathscr{J}_I(W) &= -KL[g(Wx)\|u]\\
&= -KL[Wx\|g^{-1}(u)] \qquad\qquad (27)\\
&= -KL[Wx\|\hat{s}]
\end{aligned}
$$

It follows from (❷ Eq. 27) that maximizing $\mathscr{J}_I(W)$ (Infomax principle) is identical to the minimization of the KL divergence between the distribution of the output vector $y = Wx$ and the distribution $\hat{s}$ whose entries are statistically independent. In other words, Infomax is equivalent to mutual information minimization in the framework of ICA.

3.4   Negentropy Maximization

Negative entropy or negentropy is a measure of distance to Gaussianity, yielding a larger value for a random variable whose distribution is far from Gaussian. Negentropy is always nonnegative and vanishes if and only if the random variable is Gaussian. Negentropy is defined as

$$
J(y) = H(y^G) - H(y) \qquad\qquad (28)
$$

where $H(y) = \mathbb{E}\{-\log p(y)\}$ represents the entropy and y^G is a Gaussian random vector whose mean vector and covariance matrix are the same as y. In fact, negentropy is the KL divergence of $p(y^G)$ from $p(y)$, that is

$$
\begin{aligned}
J(y) &= KL\big[p(y)\|p(y^G)\big]\\
&= \int p(y)\log\frac{p(y)}{p(y^G)}\,dy
\end{aligned} \qquad\qquad (29)
$$

leading to (❷ Eq. 28).

Let us discover a relation between negentropy and mutual information. To this end, we consider mutual information $I(y)$:

$$
\begin{aligned}
I(y) &= I(y_1,\ldots,y_n)\\
&= \sum_{i=1}^{n} H(y_i) - H(y)\\
&= \sum_{i=1}^{n} H(y_i^G) - \sum_{i=1}^{n} J(y_i) + J(y) - H(y^G)\\
&= J(y) - \sum_{i=1}^{n} J(y_i) + \frac{1}{2}\log\left[\frac{\prod_{i=1}^{n}[R_{yy}]_{ii}}{\det R_{yy}}\right]
\end{aligned} \qquad\qquad (30)
$$

where $R_{yy} = \mathbb{E}\{yy^\top\}$ and $[R_{yy}]_{ii}$ denotes the ith diagonal entry of R_{yy}.

Assume that y is already whitened (decorrelated), that is, $R_{yy} = I$. Then the sum of marginal negentropies is given by

$$\sum_{i=1}^{n} J(y_i) = J(y) - I(y) + \frac{1}{2}\log\underbrace{\left[\frac{\prod_{i=1}^{n}[R_{yy}]_{ii}}{\det R_{yy}}\right]}_{0}$$

$$= -H(y) - \int p(y)\log p(y^G)dy - I(y)$$

$$= -H(x) - \log|\det W| - I(y) - \int p(y)\log p(y^G)dy$$

(31)

Invoking $R_{yy} = I$, (❷ Eq. 31) becomes

$$\sum_{i=1}^{n} J(y_i) = -I(y) - H(x) - \log|\det W| + \frac{1}{2}\log|\det R_{yy}|$$

(32)

Note that

$$\frac{1}{2}\log|\det R_{yy}| = \frac{1}{2}\log|\det(WR_{xx}W^{\top})|$$

(33)

Therefore, we have

$$\sum_{i=1}^{n} J(y_i) = -I(y)$$

(34)

where irrelevant terms are omitted. It follows from (❷ Eq. 34) that maximizing the sum of marginal negentropies is equivalent to minimizing the mutual information.

4   Natural Gradient Algorithm

In ❷ Sect. 3, four different principles lead to the same objective function

$$\mathscr{J} = -\log|\det W| - \sum_{i=1}^{n}\log p_i(y_i)$$

(35)

That is, ICA boils down to learning W, which minimizes (❷ Eq. 35),

$$\widehat{W} = \arg\min_{W}\left\{-\log|\det W| - \sum_{i=1}^{n}\log p_i(y_i)\right\}$$

(36)

An easy way to solve (❷ Eq. 36) is the gradient descent method, which gives a learning algorithm for W that has the form

$$\Delta W = -\eta\frac{\partial\mathscr{J}}{\partial W}$$
$$= -\eta\{W^{-\top} - \varphi(y)x^{\top}\}$$

(37)

where $\eta > 0$ is the learning rate and $\varphi(y) = [\varphi_1(y_1),\ldots,\varphi_n(y_n)]^{\top}$ is the negative *score function* whose ith element $\varphi_i(y_i)$ is given by

$$\varphi_i(y_i) = -\frac{d\log p_i(y_i)}{dy_i}$$

(38)

A popular ICA algorithm is based on natural gradient (Amari 1998), which is known to be efficient since the steepest descent direction is used when the parameter space is on a Riemannian manifold. The natural gradient ICA algorithm can be derived (Amari et al. 1996). Invoking (❍ Eq. 38), we have

$$d\left\{-\sum_{i=1}^{n}\log q_i(y_i)\right\} = \sum_{i=1}^{n}\varphi_i(y_i)dy_i \tag{39}$$

$$= \varphi^\top(y)dy \tag{40}$$

where $\varphi(y) = [\varphi_1(y_1)\ldots\varphi_n(y_n)]^\top$ and dy is given in terms of dW as

$$dy = dWW^{-1}y \tag{41}$$

Define a modified coefficient differential dV as

$$dV = dWW^{-1} \tag{42}$$

With this definition, we have

$$d\left\{-\sum_{i=1}^{n}\log q_i(y_i)\right\} = \varphi^\top(y)dVy \tag{43}$$

We calculate an infinitesimal increment of $\log|\det W|$, then we have

$$d\{\log|\det W|\} = \mathrm{tr}\{dV\} \tag{44}$$

where $\mathrm{tr}\{\cdot\}$ denotes the trace that adds up all diagonal elements.

Thus, combining ❍ Eqs. 43 and ❍ 44 gives

$$d\mathscr{J} = \varphi^\top(y)dVy - \mathrm{tr}\{dV\} \tag{45}$$

The differential in (❍ Eq. 45) is in terms of the modified coefficient differential matrix dV. Note that dV is a linear combination of the coefficient differentials dW_{ij}. Thus, as long as dW is nonsingular, dV represents a valid search direction to minimize (❍ Eq. 35), because dV spans the same tangent space of matrices as spanned by dW. This leads to a stochastic gradient learning algorithm for V given by

$$\Delta V = -\eta\frac{d\mathscr{J}}{dV}$$
$$= \eta\{I - \varphi(y)y^\top\} \tag{46}$$

Thus the learning algorithm for updating W is described by

$$\Delta W = \eta\Delta VW$$
$$= \eta\{I - \varphi(y)y^\top\}W \tag{47}$$

5 Flexible ICA

The optimal nonlinear function $\varphi_i(y_i)$ is given by (❍ Eq. 38). However, it requires knowledge of the probability distributions of sources that are not available. A variety of hypothesized density models has been used. For example, for super-Gaussian source signals, the unimodal or hyperbolic-Cauchy distribution model (MacKay 1996) leads to the nonlinear function given by

$$\varphi_i(y_i) = \tanh(\beta y_i) \tag{48}$$

Such a sigmoid function was also used by Bell and Sejnowski (1995). For sub-Gaussian source signals, the cubic nonlinear function $\varphi_i(y_i) = y_i^3$ has been a favorite choice. For mixtures of sub- and super-Gaussian source signals, according to the estimated kurtosis of the extracted signals, the nonlinear function can be selected from two different choices (Lee et al. 1999).

The flexible ICA (Choi et al. 2000) incorporates the generalized Gaussian density model into the natural gradient ICA algorithm, so that the parameterized nonlinear function provides flexibility in learning. The *generalized Gaussian* probability distribution is a set of distributions parameterized by a positive real number α, which is usually referred to as the *Gaussian exponent* of the distribution. The Gaussian exponent α controls the "peakiness" of the distribution. The probability density function (PDF) for a generalized Gaussian is described by

$$p(y; \alpha) = \frac{\alpha}{2\lambda \Gamma\left(\frac{1}{\alpha}\right)} e^{-\left|\frac{y}{\lambda}\right|^{\alpha}} \tag{49}$$

where $\Gamma(x)$ is the Gamma function given by

$$\Gamma(x) = \int_0^{\infty} t^{x-1} e^{-t} dt \tag{50}$$

Note that if $\alpha = 1$, the distribution becomes the standard "Laplacian" distribution. If $\alpha = 2$, the distribution is a standard normal distribution (see ❷ *Fig. 2*).

For a generalized Gaussian distribution, the kurtosis can be expressed in terms of the Gaussian exponent, given by

$$\kappa_\alpha = \frac{\Gamma\left(\frac{5}{\alpha}\right)\Gamma\left(\frac{1}{\alpha}\right)}{\Gamma^2\left(\frac{3}{\alpha}\right)} - 3 \tag{51}$$

◻ **Fig. 2**
The generalized Gaussian distribution is plotted for several different values of Gaussian exponent, α = 0.8, 1, 2, 4.

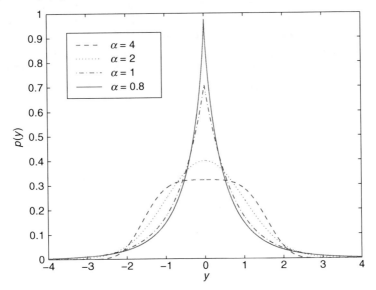

□ Fig. 3

The plot of kurtosis κ_α versus Gaussian exponent α: (a) for a leptokurtic signal; (b) for a platykurtic signal.

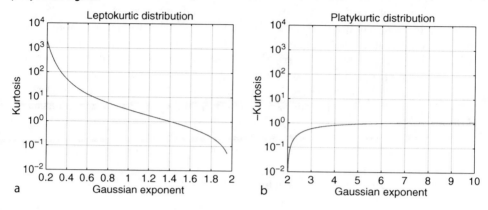

a

b

The plots of kurtosis κ_α versus the Gaussian exponent α for leptokurtic and platykurtic signals are shown in ❷ *Fig. 3*.

From the parameterized generalized Gaussian density model, the nonlinear function in the algorithm (❷ Eq. 47) is given by

$$\varphi_i(y_i) = \frac{d \log p_i(y_i)}{dy_i}$$

$$= |y_i|^{\alpha_i - 1} \operatorname{sgn}(y_i) \tag{52}$$

where $\operatorname{sgn}(y_i)$ is the signum function of y_i.

Note that for $\alpha_i = 1$, $\varphi_i(y_i)$ in (❷ Eq. 38) becomes a signum function (which can also be derived from the Laplacian density model for sources). The signum nonlinear function is favorable for the separation of speech signals since natural speech is often modeled as a Laplacian distribution. Note also that for $\alpha_i = 4$, $\varphi_i(y_i)$ in (❷ Eq. 38) becomes a cubic function, which is known to be a good choice for sub-Gaussian sources.

In order to select a proper value of the Gaussian exponent α_i, we estimate the kurtosis of the output signal y_i and select the corresponding α_i from the relationship in ❷ *Fig. 3*. The kurtosis of y_i, κ_i, can be estimated via the following iterative algorithm:

$$\kappa_i(t+1) = \frac{M_{4i}(t+1)}{M_{2i}^2(t+1)} - 3 \tag{53}$$

where

$$M_{4i}(t+1) = (1-\delta)M_{4i}(t) + \delta|y_i(t)|^4 \tag{54}$$

$$M_{2i}(t+1) = (1-\delta)M_{2i}(t) + \delta|y_i(t)|^2 \tag{55}$$

where δ is a small constant, say 0.01.

In general, the estimated kurtosis of the demixing filter output does not exactly match the kurtosis of the original source. However, it provides an idea about whether the estimated source is a sub-Gaussian signal or a super-Gaussian signal. Moreover, it was shown (Cardoso

1997; Amari and Cardoso 1997) that the performance of the source separation is not degraded even if the hypothesized density does not match the true density. For these reasons, we suggest a practical method where only several different forms of nonlinear functions are used.

6   Differential ICA

In a wide sense, most ICA algorithms based on unsupervised learning belong to the Hebb-type rule or its generalization with adopting nonlinear functions. Motivated by the differential Hebb rule (Kosko 1986) and differential decorrelation (Choi 2002, 2003), we introduce an ICA algorithm employing differential learning and natural gradients, which leads to a differential ICA algorithm. A random walk model is first introduced for latent variables, in order to show that differential learning is interpreted as the maximum likelihood estimation of a linear generative model. Then the detailed derivation of the differential ICA algorithm is presented.

6.1   Random Walk Model for Latent Variables

Given a set of observation data, $\{x(t)\}$, the task of learning the linear generative model (❯ Eq. 1) under the constraint of the latent variables being statistically independent, is a semiparametric estimation problem. The maximum likelihood estimation of basis vectors $\{a_i\}$ involves a probabilistic model for latent variables that are treated as nuisance parameters.

In order to show a link between the differential learning and the maximum likelihood estimation, a random walk model for latent variables $s_i(t)$ is considered, which is a simple Markov chain, that is,

$$s_i(t) = s_i(t-1) + \varepsilon_i(t) \tag{56}$$

where the innovation $\varepsilon_i(t)$ is assumed to have zero mean with a density function $q_i(\varepsilon_i(t))$. In addition, innovation sequences $\{\varepsilon_i(t)\}$ are assumed to be mutually independent white sequences, that is, they are spatially independent and temporally white as well.

Let us consider latent variables $s_i(t)$ over an N-point time block. The vector $\bar{s}_i$ can be defined as

$$\bar{s}_i = [s_i(0), \ldots, s_i(N-1)]^\top \tag{57}$$

Then the joint PDF of $\bar{s}_i$ can be written as

$$\begin{aligned} p_i(\bar{s}_i) &= p_i(s_i(0), \ldots, s_i(N-1)) \\ &= \prod_{t=0}^{N-1} p_i(s_i(t)|s_i(t-1)) \end{aligned} \tag{58}$$

where $s_i(t) = 0$ for $t < 0$ and the statistical independence of the innovation sequences is taken into account.

It follows from the random walk model (❯ Eq. 56) that the conditional probability density of $s_i(t)$ given its past samples can be written as

$$p_i(s_i(t)|s_i(t-1)) = q_i(\varepsilon_i(t)) \tag{59}$$

Combining (❯ Eq. 58) and (❯ Eq. 59) leads to

$$p_i(\bar{s}_i) = \prod_{t=0}^{N-1} q_i(\varepsilon_i(t))$$

$$= \prod_{t=0}^{N-1} q_i(s_i'(t)) \tag{60}$$

where $s_i'(t) = s_i(t) - s_i(t-1)$, which is the first-order approximation of the differentiation.

Taking the statistical independence of the latent variables and (❯ Eq. 60) into account, then the joint density $p(\bar{s}_1, \ldots, \bar{s}_n)$ can be written as

$$p(\bar{s}_1, \ldots, \bar{s}_n) = \prod_{i=1}^{n} p_i(\bar{s}_i)$$

$$= \prod_{t=0}^{N-1} \prod_{i=1}^{n} q_i(s_i'(t)) \tag{61}$$

The factorial model given in (❯ Eq. 61) will be used as an optimization criterion in deriving the differential ICA algorithm.

6.2 Algorithm

Denote a set of observation data by

$$\mathcal{X} = \{\bar{x}_1, \ldots, \bar{x}_n\} \tag{62}$$

where

$$\bar{x}_i = [x_i(0), \ldots, x_i(N-1)]^\top \tag{63}$$

Then the normalized log-likelihood is given by

$$\frac{1}{N} \log p(\mathcal{X}|A) = -\log|\det A| + \frac{1}{N} \log p(\bar{s}_1, \ldots, \bar{s}_n)$$

$$= -\log|\det A| + \frac{1}{N} \sum_{t=0}^{N-1} \sum_{i=1}^{n} \log q_i(s_i'(t)) \tag{64}$$

The inverse of A is denoted by $W = A^{-1}$. The estimate of latent variables is denoted by $y(t) = Wx(t)$. With these defined variables, the objective function (i.e., the negative normalized log-likelihood) is given by

$$\mathcal{I}_{di} = -\frac{1}{N} \log p(\mathcal{X}|A)$$

$$= -\log|\det W| - \frac{1}{N} \sum_{t=0}^{N-1} \sum_{i=1}^{n} \log q_i(y_i'(t)) \tag{65}$$

where s_i is replaced by its estimate y_i and $y_i'(t) = y_i(t) - y_i(t-1)$ (the first-order approximation of the differentiation).

For online learning, the sample average is replaced by the instantaneous value. Hence the online version of the objective function (❯ Eq. 65) is given by

$$\mathcal{I}_{di} = -\log|\det W| - \sum_{i=1}^{n} \log q_i(y_i'(t)) \tag{66}$$

Note that objective function ($\bullet$ Eq. 66) is slightly different from ($\bullet$ Eq. 35) used in the conventional ICA based on the minimization of mutual information or the maximum likelihood estimation.

We derive a natural gradient learning algorithm that finds a minimum of ($\bullet$ Eq. 66). To this end, we follow the way that was discussed in (Amari et al. 1997; Amari 1998; Choi et al. 2000). The total differential $d\,\mathcal{J}_{di}(W)$ due to the change dW is calculated

$$
\begin{aligned}
d\mathcal{J}_{di} &= \mathcal{J}_{di}(W + dW) - \mathcal{J}_{di}(W) \\
&= d\{-\log|\det W|\} + d\left\{-\sum_{i=1}^{n}\log q_i(y_i'(t))\right\}
\end{aligned}
\tag{67}
$$

Define

$$
\varphi_i(y_i') = -\frac{d\log q_i(y_i')}{dy_i'}
\tag{68}
$$

and construct a vector $\varphi(y') = \varphi_1(y_1'),\ldots,\varphi_n(y_n')^{\top}$.

With this definition, we have

$$
\begin{aligned}
d\left\{-\sum_{i=1}^{n}\log q_i(y_i'(t))\right\} &= \sum_{i=1}^{n}\varphi_i(y_i'(t))dy_i'(t) \\
&= \varphi^{\top}(y'(t))dy'(t)
\end{aligned}
\tag{69}
$$

One can easily see that

$$
d\{-\log|\det W|\} = \text{tr}\{dWW^{-1}\}
\tag{70}
$$

Define a modified differential matrix dV by

$$
dV = dWW^{-1}
\tag{71}
$$

Then, with this modified differential matrix, the total differential $d\mathcal{J}_{di}(W)$ is computed as

$$
d\mathcal{J}_{di} = -\text{tr}\{dV\} + \varphi^{\top}(y'(t))dVy'(t)
\tag{72}
$$

A gradient descent learning algorithm for updating V is given by

$$
\begin{aligned}
V(t+1) &= V(t) - \eta_t\frac{d\mathcal{J}_{di}}{dV} \\
&= \eta_t\{I - \varphi(y'(t))y'^{\top}(t)\}
\end{aligned}
\tag{73}
$$

Hence, it follows from the relation ($\bullet$ Eq. 71) that the updating rule for W has the form

$$
W(t+1) = W(t) + \eta_t\{I - \varphi(y'(t))y'^{\top}(t)\}W(t)
\tag{74}
$$

7 Nonstationary Source Separation

So far, we assumed that sources are stationary random processes where the statistics do not vary over time. In this section, we show how the natural gradient ICA algorithm is modified to handle nonstationary sources. As in Matsuoka et al. (1995), the following assumptions are made in this section.

AS1 The mixing matrix A has full column rank.

AS2 Source signals $\{s_i(t)\}$ are statistically independent with zero mean. This implies that the covariance matrix of the source signal vector, $R_s(t) = E\{s(t)s^\top(t)\}$, is a diagonal matrix, that is,

$$R_s(t) = \text{diag}\{r_1(t), \ldots, r_n(t)\} \tag{75}$$

where $r_i(t) = E\{s_i^2(t)\}$ and E denotes the statistical expectation operator.

AS3 $\frac{r_i(t)}{r_j(t)}$ ($i, j = 1, \ldots, n$ and $i \neq j$) are not constant with time.

It should be pointed out that the first two assumptions (AS1, AS2) are common in most existing approaches to source separation, however the third assumption (AS3) is critical in this chapter. For nonstationary sources, the third assumption is satisfied and it allows one to separate linear mixtures of sources via second-order statistics (SOS).

For stationary source separation, the typical cost function is based on the mutual information, which requires knowledge of the underlying distributions of the sources. Since the probability distributions of the sources are not known in advance, most ICA algorithms rely on hypothesized distributions (e.g., see Choi et al. 2000 and references therein). Higher-order statistics (HOS) should be incorporated either explicitly or implicitly.

For nonstationary sources, Matsuoka et al. have shown that the decomposition (❷ Eq. 6) is satisfied if cross-correlations $E\{y_i(t)y_j(t)\}$ ($i, j = 1, \ldots, n$, $i \neq j$) are zeros at any time instant t, provided that the assumptions (AS1)–(AS3) are satisfied. To eliminate cross-correlations, the following cost function was proposed in Matsuoka et al. (1995),

$$\mathscr{I}(W) = \frac{1}{2}\left\{\sum_{i=1}^{n} \log E\{y_i^2(t)\} - \log \det\left(E\{y(t)y^\top(t)\}\right)\right\} \tag{76}$$

where $\det(\cdot)$ denotes the determinant of a matrix. The cost function given in (❷ Eq. 76) is a nonnegative function, which takes minima if and only if $E\{y_i(t)y_j(t)\} = 0$, for $i, j = 1, \ldots, n$, $i \neq j$. This is the direct consequence of Hadamard's inequality, which is summarized below.

Theorem 2 (Hadamard's inequality) *Suppose $K = [k_{ij}]$ is a nonnegative definite symmetric $n \times n$ matrix. Then,*

$$\det(K) \leq \prod_{i=1}^{n} k_{ii} \tag{77}$$

with equality iff $k_{ij} = 0$, for $i \neq j$.

Take the logarithm on both sides of (❷ Eq. 77) to obtain

$$\sum_{i=1}^{n} \log k_{ii} - \log \det(K) \geq 0 \tag{78}$$

Replacing the matrix K by $E\{y(t)y^\top(t)\}$, one can easily see that the cost function (❷ Eq. 76) has the minima iff $E\{y_i(t)y_j(t)\} = 0$, for $i, j = 1, \ldots, n$ and $i \neq j$.

We compute

$$\begin{aligned} d\left\{\log \det(E\{y(t)y^\top(t)\})\right\} &= 2d\{\log \det W\} + d\{\log \det C(t)\} \\ &= 2\text{tr}\left\{W^{-1}dW\right\} + d\{\log \det C(t)\} \end{aligned} \tag{79}$$

Define a modified differential matrix dV as

$$dV = W^{-1}dW \tag{80}$$

Then, we have

$$d\left\{\sum_{i=1}^{n}\log E\{y_i^2(t)\}\right\} = 2E\{y^\top(t)\Lambda^{-1}(t)dVy(t)\} \tag{81}$$

Similarly, we can derive the learning algorithm for W that has the form

$$\begin{aligned}\Delta W(t) &= \eta_t\{I - \Lambda^{-1}(t)y(t)y^\top(t)\}W(t)\\ &= \eta_t\Lambda^{-1}(t)\{\Lambda(t) - y(t)y^\top(t)\}W(t)\end{aligned} \tag{82}$$

8   Spatial, Temporal, and Spatiotemporal ICA

ICA decomposition, $X = AS$, inherently has duality. Considering the data matrix $X \in \mathbb{R}^{m\times N}$ where each of its rows is assumed to be a time course of an attribute, ICA decomposition produces n independent time courses. On the other hand, regarding the data matrix in the form of $X^\top$, ICA decomposition leads to n independent patterns (for instance, images in fMRI or arrays in DNA microarray data).

The standard ICA (where X is considered) is treated as *temporal ICA* (tICA). Its dual decomposition (regarding $X^\top$) is known as *spatial ICA* (sICA). Combining these two ideas leads to *spatiotemporal ICA* (stICA). These variations of ICA were first investigated in Stone et al. (2000). Spatial ICA or spatiotemporal ICA were shown to be useful in fMRI image analysis (Stone et al. 2000) and gene expression data analysis (Liebermeister 2002; Kim and Choi 2005).

Suppose that the singular value decomposition (SVD) of X is given by

$$X = U\,DV^\mathrm{T} = \left(U\,D^{1/2}\right)\left(V\,D^{1/2}\right)^\mathrm{T} = \widetilde{U}\widetilde{V}^\mathrm{T} \tag{83}$$

where $U \in \mathbb{R}^{m\times n}$, $D \in \mathbb{R}^{n\times n}$, and $V \in \mathbb{R}^{N\times n}$ for $n \le \min(m, N)$.

8.1   Temporal ICA

Temporal ICA finds a set of independent time courses and a corresponding set of dual unconstrained spatial patterns. It embodies the assumption that each row vector in $\widetilde{V}^\top$ consists of a linear combination of n independent sequences, that is, $\widetilde{V}^\top = \widetilde{A}_T S_T$, where $S_T \in \mathbb{R}^{n\times N}$ has a set of n independent temporal sequences of length N, and $\widetilde{A}_T \in \mathbb{R}^{n\times n}$ is an associated mixing matrix.

Unmixing by $Y_T = W_T \widetilde{V}^\top$ where $W_T = P\widetilde{A}_T^{-1}$, leads one to recover the n dual patterns A_T associated with the n independent time courses, by calculating $A_T = \widetilde{U}W_T^{-1}$, which is a consequence of $\widetilde{X} = A_T Y_T = \widetilde{U}\widetilde{V}^\top = \widetilde{U}W_T^{-1}Y_T$.

8.2   Spatial ICA

Spatial ICA seeks a set of independent spatial patterns S_S and a corresponding set of dual unconstrained time courses A_S. It embodies the assumption that each row vector in $\widetilde{U}^\top$ is composed of a linear combination of n independent spatial patterns, that is, $\widetilde{U}^\top = \widetilde{A}_S S_S$, where $S_S \in \mathbb{R}^{n\times m}$ contains a set of n independent m-dimensional patterns and $\widetilde{A}_S \in \mathbb{R}^{n\times n}$ is an encoding variable matrix (mixing matrix).

Define $Y_S = W_S \widetilde{U}^\top$ where W_S is a permuted version of $\widetilde{A}_S^{-1}$. With this definition, the n dual time courses $A_S \in \mathbb{R}^{N\times n}$ associated with the n independent patterns are computed as $A_S = \widetilde{V} W_S^{-1}$, since $\widetilde{X}^\top = A_S Y_S = \widetilde{U}\widetilde{V}^T = \widetilde{V} W_S^{-1} Y_S$. Each column vector of A_S corresponds to a temporal mode.

8.3 Spatiotemporal ICA

In linear decomposition sICA enforces independence constraints over space to find a set of independent spatial patterns, whereas tICA embodies independence constraints over time to seek a set of independent time courses. Spatiotemporal ICA finds a linear decomposition by maximizing the degree of independence over space as well as over time, without necessarily producing independence in either space or time. In fact, it allows a trade-off between the independence of arrays and the independence of time courses.

Given $\widetilde{X} = \widetilde{U}\widetilde{V}^T$, stICA finds the following decomposition:

$$\widetilde{X} = S_S^\top \Lambda S_T \tag{84}$$

where $S_S \in \mathbb{R}^{n\times m}$ contains a set of n independent m-dimensional patterns, $S_T \in \mathbb{R}^{n\times N}$ has a set of n independent temporal sequences of length N, and Λ is a diagonal scaling matrix. There exist two $n \times n$ mixing matrices, W_S and W_T, such that $S_S = W_S\widetilde{U}^\top$ and $S_T = W_T\widetilde{V}^\top$. The following relation

$$\begin{aligned}\widetilde{X} &= S_S^\top \Lambda S_T \\ &= \widetilde{U} W_S^\top \Lambda W_T \widetilde{V}^\top \\ &= \widetilde{U}\widetilde{V}^\top\end{aligned} \tag{85}$$

implies that $W_S^\top \Lambda W_T = I$, which leads to

$$W_T = W_S^{-\top}\Lambda^{-1} \tag{86}$$

Linear transforms, W_S and W_T, are found by jointly optimizing objective functions associated with sICA and tICA. That is, the objective function for stICA has the form

$$\mathscr{J}_{\text{stICA}} = \alpha\mathscr{J}_{\text{sICA}} + (1-\alpha)\mathscr{J}_{\text{tICA}} \tag{87}$$

where $\mathscr{J}_{\text{sICA}}$ and $\mathscr{J}_{\text{tICA}}$ could be Infomax criteria or log-likelihood functions and α defines the relative weighting for spatial independence and temporal independence. More details on stICA can be found in Stone et al. (2002).

9 Algebraic Methods for BSS

Up to now, online ICA algorithms in a framework of unsupervised learning have been introduced. In this section, several algebraic methods are explained for BSS where matrix decomposition plays a critical role.

9.1 Fundamental Principle of Algebraic BSS

Algebraic methods for BSS often make use of the eigen-decomposition of correlation matrices or cumulant matrices. Exemplary algebraic methods for BSS include FOBI (Cardoso 1989), AMUSE (Tong et al. 1990), JADE (Cardoso and Souloumiac 1993), SOBI (Belouchrani et al. 1997), and SEONS (Choi et al. 2002). Some of these methods (FOBI and AMUSE) are based on simultaneous diagonalization of two symmetric matrices. Methods such as JADE, SOBI, and SEONS make use of joint approximate diagonalization of multiple matrices (more than two). The following theorem provides a fundamental principle for algebraic BSS, justifying why simultaneous diagonalization of two symmetric data matrices (one of them is assumed to be positive definite) provides a solution to BSS.

Theorem 3 *Let $\Lambda_1, D_1 \in \mathbb{R}^{n \times n}$ be diagonal matrices with positive diagonal entries and $\Lambda_2, D_2 \in \mathbb{R}^{n \times n}$ be diagonal matrices with nonzero diagonal entries. Suppose that $G \in \mathbb{R}^{n \times n}$ satisfies the following decompositions:*

$$D_1 = G\Lambda_1 G^\top \tag{88}$$

$$D_2 = G\Lambda_2 G^\top \tag{89}$$

Then the matrix G is the generalized permutation matrix, i.e., $G = P\Lambda$ if $D_1^{-1}D_2$ and $\Lambda_1^{-1}\Lambda_2$ have distinct diagonal entries.

Proof It follows from (❷ Eq. 88) that there exists an orthogonal matrix Q such that

$$\left(G\Lambda_1^{\frac{1}{2}}\right) = \left(D_1^{\frac{1}{2}}\right)Q \tag{90}$$

Hence,

$$G = D_1^{\frac{1}{2}} Q \Lambda_1^{-\frac{1}{2}} \tag{91}$$

Substitute (❷ Eq. 91) into (❷ Eq. 89) to obtain

$$D_1^{-1}D_2 = Q\Lambda_1^{-1}\Lambda_2 Q^\top \tag{92}$$

Since the right-hand side of (❷ Eq. 92) is the eigen-decomposition of the left-hand side of (❷ Eq. 92), the diagonal elements of $D_1^{-1}D_2$ and $\Lambda_1^{-1}\Lambda_2$ are the same. From the assumption that the diagonal elements of $D_1^{-1}D_2$ and $\Lambda_1^{-1}\Lambda_2$ are distinct, the orthogonal matrix Q must have the form $Q = P\Psi$, where Ψ is a diagonal matrix whose diagonal elements are either $+1$ or -1. Hence, we have

$$\begin{aligned} G &= D_1^{\frac{1}{2}} P\Psi\Lambda_1^{-\frac{1}{2}} \\ &= PP^\top D_1^{\frac{1}{2}} P\Psi\Lambda_1^{-\frac{1}{2}} \\ &= P\Lambda \end{aligned} \tag{93}$$

where

$$\Lambda = P^\top D_1^{\frac{1}{2}} P\Psi\Lambda_1^{-\frac{1}{2}}$$

which completes the proof.

9.2 AMUSE

As an example of ❿ Theorem 3, we briefly explain AMUSE (Tong et al. 1990), where a BSS solution is determined by simultaneously diagonalizing the equal-time correlation matrix of x (t) and a time-delayed correlation matrix of $x(t)$.

It is assumed that sources $\{s_i(t)\}$ (entries of $s(t)$) are uncorrelated stochastic processes with zero mean, that is,

$$\mathbb{E}\{s_i(t)s_j(t-\tau)\} = \delta_{ij}\gamma_i(\tau) \tag{94}$$

where δ_{ij} is the Kronecker delta and $\gamma_i(\tau)$ are distinct for $i = 1,\ldots,n$, given τ. In other words, the equal-time correlation matrix of the source, $R_{ss}(0) = \mathbb{E}\{s(t)s^\top(t)\}$, is a diagonal matrix with distinct diagonal entries. Moreover, a time-delayed correlation matrix of the source, $R_{ss}(\tau) = \mathbb{E}\{s(t)s^\top(t-\tau)\}$, is diagonal as well, with distinct nonzero diagonal entries.

It follows from (❿ Eq. 2) that the correlation matrices of the observation vector $x(t)$ satisfy

$$R_{xx}(0) = AR_{ss}(0)A^\top \tag{95}$$

$$R_{xx}(\tau) = AR_{ss}(\tau)A^\top \tag{96}$$

for some nonzero time-lag τ and both $R_{ss}(0)$ and $R_{ss}(\tau)$ are diagonal matrices since sources are assumed to be spatially uncorrelated.

Invoking ❿ Theorem 3, one can easily see that the inverse of the mixing matrix, A^{-1}, can be identified up to its rescaled and permuted version by the simultaneous diagonalization of $R_{xx}(0)$ and $R_{xx}(\tau)$, provided that $R_{ss}^{-1}(0)R_{ss}(\tau)$ has distinct diagonal elements. In other words, a linear transformation W is determined such that $R_{yy}(0)$ and $R_{yy}(\tau)$ of the output $y(t) = Wx(t)$ are simultaneously diagonalized:

$$R_{yy}(0) = (WA)R_{ss}(0)(WA)^\top$$

$$R_{yy}(\tau) = (WA)R_{ss}(\tau)(WA)^\top$$

It follows from ❿ Theorem 3 that WA becomes the transparent transformation.

9.3 Simultaneous Diagonalization

We explain how two symmetric matrices are simultaneously diagonalized by a linear transformation. More details on simultaneous diagonalization can be found in Fukunaga (1990). Simultaneous diagonalization consists of two steps (whitening followed by a unitary transformation):

1. First, the matrix $R_{xx}(0)$ is whitened by

$$z(t) = D_1^{-\frac{1}{2}}U_1^\top x(t) \tag{97}$$

where D_1 and U_1 are the eigenvalue and eigenvector matrices of $R_{xx}(0)$,

$$R_{xx}(0) = U_1 D_1 U_1^\top \tag{98}$$

Then, we have

$$R_{zz}(0) = D_1^{-\frac{1}{2}}U_1^\top R_{xx}(0)U_1 D_1^{-\frac{1}{2}} = I_m$$

$$R_{zz}(\tau) = D_1^{-\frac{1}{2}}U_1^\top R_{xx}(\tau)U_1 D_1^{-\frac{1}{2}}$$

2. Second, a unitary transformation is applied to diagonalize the matrix $R_{zz}(\tau)$. The eigen-decomposition of $R_{zz}(\tau)$ has the form

$$R_{zz}(\tau) = U_2 D_2 U_2^\top \tag{99}$$

Then, $y(t) = U_2^\top z(t)$ satisfies

$$R_{yy}(0) = U_2^\top R_{zz}(0) U_2 = I_m$$
$$R_{yy}(\tau) = U_2^\top R_{zz}(\tau) U_2 = D_2$$

Thus, both matrices $R_{xx}(0)$ and $R_{xx}(\tau)$ are simultaneously diagonalized by a linear transform $W = U_2^\top D_1^{-\frac{1}{2}} U_1^\top$. It follows from ❷ Theorem 3 that $W = U_2^\top D_1^{-\frac{1}{2}} U_1^\top$ is a valid demixing matrix if all the diagonal elements of D_2 are distinct.

9.4 Generalized Eigenvalue Problem

The simultaneous diagonalization of two symmetric matrices can be carried out without going through two-step procedures. From the discussion in ❷ Sect. 9.3, we have

$$W R_{xx}(0) W^\top = I_n \tag{100}$$

$$W R_{xx}(\tau) W^\top = D_2 \tag{101}$$

The linear transformation W, which satisfies (❷ Eq. 100) and (❷ Eq. 101) is the eigenvector matrix of $R_{xx}^{-1}(0) R_{xx}(\tau)$ (Fukunaga 1990). In other words, the matrix W is the generalized eigenvector matrix of the pencil $R_{xx}(\tau) - \lambda R_{xx}(0)$ (Molgedey and Schuster 1994).

Recently Chang et al. (2000) proposed the matrix pencil method for BSS where they exploited $R_{xx}(\tau_1)$ and $R_{xx}(\tau_2)$ for $\tau_1 \neq \tau_2 \neq 0$. Since the noise vector was assumed to be temporally white, two matrices $R_{xx}(\tau_1)$ and $R_{xx}(\tau_2)$ are not theoretically affected by the noise vector, that is,

$$R_{xx}(\tau_1) = A R_{ss}(\tau_1) A^\top \tag{102}$$

$$R_{xx}(\tau_2) = A R_{ss}(\tau_2) A^\top \tag{103}$$

Thus, it is obvious that we can find an estimate of the demixing matrix that is not sensitive to the white noise. A similar idea was also exploited in Choi and Cichocki (2000a, b).

In general, the generalized eigenvalue decomposition requires the symmetric-definite pencil (one matrix is symmetric and the other is symmetric and positive definite). However, $R_{xx}(\tau_2) - \lambda R_{xx}(\tau_1)$ is not symmetric-definite, which might cause a numerical instability problem that results in complex-valued eigenvectors.

The set of all matrices of the form $R_1 - \lambda R_2$ with $\lambda \in \mathbb{R}$ is said to be a *pencil*. Frequently we encounter the case where R_1 is symmetric and R_2 is symmetric and positive definite. Pencils of this variety are referred to as *symmetric-definite pencils* (Golub and Loan 1993).

Theorem 4 (Golub and Loan 1993, p. 468) *If $R_1 - \lambda R_2$ is symmetric-definite, then there exists a nonsingular matrix $U = [u_1, \ldots, u_n]$ such that*

$$U^\top R_1 U = \text{diag}\{\gamma_1(\tau_1), \ldots, \gamma_n(\tau_1)\} \tag{104}$$

$$U^\top R_2 U = \text{diag}\{\gamma_1(\tau_2), \ldots, \gamma_n(\tau_2)\} \tag{105}$$

Moreover $R_1 u_i = \lambda_i R_2 u_i$ for $i = 1, \ldots, n$, and $\lambda_i = \frac{\gamma_i(\tau_1)}{\gamma_i(\tau_2)}$.

It is apparent from ❯ Theorem 4 that R_1 should be symmetric and R_2 should be symmetric and positive definite so that the generalized eigenvector U can be a valid solution if $\{\lambda_i\}$ are distinct.

10 Software

A variety of ICA software is available. ICA Central (URL:http://www.tsi.enst.fr/icacentral/) was created in 1999 to promote research on ICA and BSS by means of public mailing lists, a repository of data sets, a repository of ICA/BSS algorithms, and so on. ICA Central might be the first place where you can find data sets and ICA algorithms. In addition, here are some widely used software packages.

- *ICALAB Toolboxes* (http://www.bsp.brain.riken.go.jp/ICALAB/): ICALAB is an ICA Matlab software toolbox developed in the Laboratory for Advanced Brain Signal Processing in the RIKEN Brain Science Institute, Japan. It consists of two independent packages, including ICALAB for signal processing and ICALAB for image processing, and each package contains a variety of algorithms.
- *FastICA* (http://www.cis.hut.fi/projects/ica/fastica/): This is the FastICA Matlab package that implements fast fixed-point algorithms for non-Gaussianity maximization (Hyvärinen et al. 2001). It was developed in the Helsinki University of Technology, Finland, and other environments (R, C++, Physon) are also available.
- *Infomax ICA* (http://www.cnl.salk.edu/~tewon/ica_cnl.html): Matlab and C codes for Bell and Sejnowski's Infomax algorithm (Bell and Sejnowski 1995) and extended Infomax (Lee 1998) where a parametric density model is incorporated into Infomax to handle both super-Gaussian and sub-Gaussian sources.
- *EEGLAB* (http://sccn.ucsd.edu/eeglab/): EEGLAB is an interactive Matlab toolbox for processing continuous and event-related EEG, MEG, and other electrophysiological data using ICA, time/frequency analysis, artifact rejection, and several modes of data visualization.
- *ICA: DTU Toolbox* (http://isp.imm.dtu.dk/toolbox/ica/): ICA: DTU Toolbox is a collection of ICA algorithms that includes: (1) icaML, which is an efficient implementation of Infomax; (2) icaMF, which is an iterative algorithm that offers a variety of possible source priors and mixing matrix constraints (e.g., positivity) and can also handle over- and under-complete mixing; and (3) icaMS, which is a "one shot" fast algorithm that requires time correlation between samples.

11 Further Issues

- *Overcomplete representation*: Overcomplete representation enforces the latent space dimension n to be greater than the data dimension m in the linear model (❯ Eq. 1). Sparseness constraints on latent variables are necessary to learn fruitful representation (Lewicki and Sejnowski 2000).
- *Bayesian ICA*: Bayesian ICA incorporates uncertainty and prior distributions of latent variables in the model (❯ Eq. 1). Independent factor analysis (Attias 1999) is pioneering work along this direction. The EM algorithm for ICA was developed in Welling and

Weber (2001) and a full Bayesian ICA (also known as ensemble learning) was developed in Miskin and MacKay (2001).

- *Kernel ICA*: Kernel methods were introduced to consider statistical independence in reproducing kernel Hilbert space (Bach and Jordan 2002), developing kernel ICA.
- *Nonnegative ICA*: Nonnegativity constraints were imposed on latent variables, yielding nonnegative ICA (Plumbley 2003). Rectified Gaussian priors can also be used in Bayesian ICA to handle nonnegative latent variables.
- *Sparseness*: Sparseness is another important characteristic of sources, besides independence. Sparse component analysis is studied in Lee et al. (2006).
- *Beyond ICA*: Independent subspace analysis (Hyvärinen and Hoyer 2000) and tree-dependent component analysis (Bach and Jordan 2003) generalizes ICA, allowing intra-dependence structures in feature subspaces or clusters.

12   Summary

ICA has been successfully applied to various applications of machine learning, pattern recognition, and signal processing. A brief overview of ICA has been presented, starting from fundamental principles on learning a linear latent variable model for parsimonious representation. Natural gradient ICA algorithms were derived in the framework of maximum likelihood estimation, mutual information minimization, Infomax, and negentropy maximization. We have explained flexible ICA where generalized Gaussian density was adopted such that a flexible nonlinear function was incorporated into the natural gradient ICA algorithm. Equivariant nonstationary source separation was presented in the framework of natural gradient as well. Differential learning was also adopted to incorporate a temporal structure of sources. We also presented a core idea and various methods for algebraic source separation. Various software packages for ICA were introduced, for easy application of ICA. Further issues were also briefly mentioned so that readers can follow the status of ICA.

References

Amari S (1998) Natural gradient works efficiently in learning. Neural Comput 10(2):251–276

Amari S, Cardoso JF (1997) Blind source separation: semiparametric statistical approach. IEEE Trans Signal Process 45:2692–2700

Amari S, Chen TP, Cichocki A (1997) Stability analysis of learning algorithms for blind source separation. Neural Networks 10(8):1345–1351

Amari S, Cichocki A (1998) Adaptive blind signal processing – neural network approaches. Proc IEEE (Special Issue on Blind Identification and Estimation) 86(10):2026–2048

Amari S, Cichocki A, Yang HH (1996) A new learning algorithm for blind signal separation. In: Touretzky DS, Mozer MC, Hasselmo ME (eds) Advances in neural information processing systems (NIPS), vol 8. MIT Press, Cambridge, pp 757–763

Attias H (1999) Independent factor analysis. Neural Comput 11:803–851

Bach F, Jordan MI (2002) Kernel independent component analysis. JMLR 3:1–48

Bach FR, Jordan MI (2003) Beyond independent components: trees and clusters. JMLR 4:1205–1233

Bell A, Sejnowski T (1995) An information maximisation approach to blind separation and blind deconvolution. Neural Comput 7:1129–1159

Belouchrani A, Abed-Merain K, Cardoso JF, Moulines E (1997) A blind source separation technique using second order statistics. IEEE Trans Signal Process 45:434–444

Cardoso JF (1989) Source separation using higher-order moments. In: Proceedings of the IEEE international conference on acoustics, speech, and signal processing (ICASSP), Paris, France, 23–26 May 1989

Cardoso JF (1997) Infomax and maximum likelihood for source separation. IEEE Signal Process Lett 4(4): 112–114

Cardoso JF, Laheld BH (1996) Equivariant adaptive source separation. IEEE Trans Signal Process 44 (12):3017–3030

Cardoso JF, Souloumiac A (1993) Blind beamforming for non Gaussian signals. IEE Proc-F 140(6):362–370

Chang C, Ding Z, Yau SF, Chan FHY (2000) A matrix-pencil approach to blind separation of colored nonstationary signals. IEEE Trans Signal Process 48(3): 900–907

Choi S (2002) Adaptive differential decorrelation: a natural gradient algorithm. In: Proceedings of the international conference on artificial neural networks (ICANN), Madrid, Spain. Lecture notes in computer science, vol 2415. Springer, Berlin, pp 1168–1173

Choi S (2003) Differential learning and random walk model. In: Proceedings of the IEEE international conference on acoustics, speech, and signal processing (ICASSP), IEEE, Hong Kong, pp 724–727

Choi S, Cichocki A (2000a) Blind separation of nonstationary and temporally correlated sources from noisy mixtures. In: Proceedings of IEEE workshop on neural networks for signal processing, IEEE, Sydney, Australia. pp 405–414

Choi S, Cichocki A (2000b) Blind separation of nonstationary sources in noisy mixtures. Electron Lett 36(9):848–849

Choi S, Cichocki A, Amari S (2000) Flexible independent component analysis. J VLSI Signal Process 26(1/2): 25–38

Choi S, Cichocki A, Belouchrani A (2002) Second order nonstationary source separation. J VLSI Signal Process 32:93–104

Choi S, Cichocki A, Park HM, Lee SY (2005) Blind source separation and independent component analysis: a review. Neural Inf Process Lett Rev 6(1): 1–57

Cichocki A, Amari S (2002) Adaptive blind signal and image processing: learning algorithms and applications. Wiley, Chichester

Cichocki A, Unbehauen R (1996) Robust neural networks with on-line learning for blind identification and blind separation of sources. IEEE Trans Circ Syst Fund Theor Appl 43:894–906

Comon P (1994) Independent component analysis, a new concept? Signal Process 36(3):287–314

Fukunaga K (1990) An introduction to statistical pattern recognition. Academic, New York

Golub GH, Loan CFV (1993) Matrix computations, 2nd edn. Johns Hopkins, Baltimore

Haykin S (2000) Unsupervised adaptive filtering: blind source separation. Prentice-Hall

Hyvärinen A (1999) Survey on independent component analysis. Neural Comput Surv 2:94–128

Hyvärinen A, Hoyer P (2000) Emergence of phase- and shift-invariant features by decomposition of natural images into independent feature subspaces. Neural Comput 12(7):1705–1720

Hyvärinen A, Karhunen J, Oja E (2001) Independent component analysis. Wiley, New York

Hyvärinen A, Oja E (1997) A fast fixed-point algorithm for independent component analysis. Neural Comput 9:1483–1492

Jutten C, Herault J (1991) Blind separation of sources, part I: an adaptive algorithm based on neuromimetic architecture. Signal Process 24:1–10

Karhunen J (1996) Neural approaches to independent component analysis and source separation. In: Proceedings of the European symposium on artificial neural networks (ESANN), Bruges, Belgium, pp 249–266

Kim S, Choi S (2005) Independent arrays or independent time courses for gene expression data. In: Proceedings of the IEEE international symposium on circuits and systems (ISCAS), Kobe, Japan, 23–26 May 2005

Kosko B (1986) Differential Hebbian learning. In: Proceedings of American Institute of Physics: neural networks for computing, Snowbird. American Institute of Physics, Woodbury, pp 277–282

Lee TW (1998) Independent component analysis: theory and applications. Kluwer

Lee TW, Girolami M, Sejnowski T (1999) Independent component analysis using an extended infomax algorithm for mixed sub-Gaussian and super-Gaussian sources. Neural Comput 11(2):609–633

Lewicki MS, Sejnowski T (2000) Learning overcomplete representation. Neural Comput 12(2):337–365

Li Y, Cichocki A, Amari S (2006) Blind estimation of channel parameters and source components for EEG signals: a sparse factorization approach. IEEE Trans Neural Networ 17(2):419–431

Liebermeister W (2002) Linear modes of gene expression determined by independent component analysis. Bioinformatics 18(1):51–60

MacKay DJC (1996) Maximum likelihood and covariant algorithms for independent component analysis. Technical Report Draft 3.7, University of Cambridge, Cavendish Laboratory

Matsuoka K, Ohya M, Kawamoto M (1995) A neural net for blind separation of nonstationary signals. Neural Networks 8(3):411–419

Miskin JW, MacKay DJC (2001) Ensemble learning for blind source separation. In: Roberts S, Everson R (eds) Independent component analysis: principles and practice. Cambridge University Press, Cambridge, UK, pp 209–233

Molgedey L, Schuster HG (1994) Separation of a mixture of independent signals using time delayed correlations. Phys Rev Lett 72:3634–3637

Oja E (1995) The nonlinear PCA learning rule and signal separation – mathematical analysis. Technical Report A26, Helsinki University of Technology, Laboratory of Computer and Information Science

Pearlmutter B, Parra L (1997) Maximum likelihood blind source separation: a context-sensitive generalization of ICA. In: Mozer MC, Jordan MI, Petsche T (eds) Advances in neural information processing systems (NIPS), vol 9. MIT Press, Cambridge, pp 613–619

Pham DT (1996) Blind separation of instantaneous mixtures of sources via an independent component analysis. IEEE Trans Signal Process 44(11):2768–2779

Plumbley MD (2003) Algorithms for nonnegative independent component analysis. IEEE Trans Neural Network 14(3):534–543

Stone JV (2004) Independent component analysis: a tutorial introduction. MIT Press, Cambridge

Stone JV, Porrill J, Porter NR, Wilkinson IW (2002) Spatiotemporal independent component analysis of event-related fMRI data using skewed probability density functions. NeuroImage 15(2):407–421

Tong L, Soon VC, Huang YF, Liu R (1990) AMUSE: a new blind identification algorithm. In: Proceedings of the IEEE international symposium on circuits and systems (ISCAS), IEEE, New Orleans, pp 1784–1787

Welling M, Weber M (2001) A constrained EM algorithm for independent component analysis. Neural Comput 13:677–689

14 Neural Networks for Time-Series Forecasting

G. Peter Zhang
Department of Managerial Sciences, Georgia State University,
Atlanta, GA, USA
gpzhang@gsu.edu

G. Rozenberg et al. (eds.), *Handbook of Natural Computing*, DOI 10.1007/978-3-540-92910-9_14,
© Springer-Verlag Berlin Heidelberg 2012

Abstract

Neural networks has become an important method for time series forecasting. There is increasing interest in using neural networks to model and forecast time series. This chapter provides a review of some recent developments in time series forecasting with neural networks, a brief description of neural networks, their advantages over traditional forecasting models, and some recent applications. Several important data and modeling issues for time series forecasting are highlighted. In addition, recent developments in several methodological areas such as seasonal time series modeling, multi-period forecasting, and the ensemble method are reviewed.

1   Introduction

Time series forecasting is an active research area that has received a considerable amount of attention in the literature. Using the time series approach to forecasting, forecasters collect and analyze historical observations to determine a model to capture the underlying data-generating process. Then the model is extrapolated to forecast future values. This approach is useful for applications in many domains such as business, economics, industry, engineering, and science. Much effort has been devoted, over the past three decades, to the development and improvement of time series forecasting models.

There has been an increasing interest in using neural networks to model and forecast time series. Neural networks have been found to be a viable contender when compared to various traditional time series models (Zhang et al. 1998; Balkin and Ord 2000; Jain and Kumar 2007). Lapedes and Farber (1987) report the first attempt to model nonlinear time series with neural networks. De Groot and Wurtz (1991) present a detailed analysis of univariate time series forecasting using feedforward neural networks for two benchmark nonlinear time series. Chakraborty et al. (1992) conduct an empirical study on multivariate time series forecasting with neural networks. Atiya et al. (1999) present a case study of multistep river flow forecasting. Poli and Jones (1994) propose a stochastic neural net model based on the Kalman filter for nonlinear time series prediction. Weigend et al. (1990, 1992) and Cottrell et al. (1995) address the issue of network structure for forecasting real-world time series. Berardi and Zhang (2003) investigate the bias and variance issue in the time series forecasting context. Liang (2005) proposes a Bayesian neural network for time series analysis. In addition, results from several large forecasting competitions (Balkin and Ord 2000; Weigend and Gershenfeld 1994) suggest that neural networks can be a very useful addition to the time series forecasting toolbox.

Time series forecasting has been dominated by linear methods for decades. Linear methods are easy to develop and implement and they are also relatively simple to understand and interpret. However, it is important to understand the limitation of the linear models. They are not able to capture nonlinear relationships in the data. In addition, the approximation of linear models to complicated nonlinear relationships is not always satisfactory as evidenced by the well-known M-competition where a majority of commonly used linear methods were tested with more than 1,000 real time series data (Makridakis et al. 1982). The results clearly show that no single model is the best and the best performer is dependent on the data and other conditions. One explanation of the mixed findings is the failure of linear models to account for a varying degree of nonlinearity that is common in real-world problems.

That is, because of the inherent nonlinear characteristics in the data, no single linear model is able to approximate all types of nonlinear data structure equally well.

Neural networks provide a promising alternative tool for forecasters. The inherently nonlinear structure of neural networks is particularly useful for capturing the complex underlying relationship in many real-world problems. Neural networks are perhaps more versatile methods for forecasting applications in that, not only can they find nonlinear structures in a problem, they can also model linear processes. For example, the capability of neural networks in modeling linear time series has been studied and reported by several researchers (Hwang 2001; Medeiros and Pedreira 2001; Zhang 2001).

In addition to the nonlinear modeling capability, neural networks have several other features that make them valuable for time series forecasting. First, neural networks are data-driven nonparametric methods that do not require many restrictive assumptions on the underlying stochastic process from which data are generated. As such, they are less susceptible to the model misspecification problem than parametric methods. This "learning from data" feature is highly desirable in various forecasting situations where time series data are usually easy to collect but the underlying data-generating mechanism is not known. Second, neural networks have been shown to have universal functional approximating capability in that they can accurately approximate many types of complex functional relationships. This is an important and powerful characteristic as a time series model aims to accurately capture the functional relationship between the variable to be forecast and its historical observations. The combination of the above mentioned characteristics makes neural networks a quite general and flexible tool for forecasting.

Research efforts on neural networks for time series forecasting are considerable and numerous applications of neural networks for forecasting have been reported. Adya and Collopy (1998) and Zhang et al. (1998) reviewed the relevant literature in forecasting with neural networks. There has been a significant advance in research in this area since then. The purpose of this chapter is to summarize some of the important recent work with a focus on time series forecasting using neural networks.

2   Neural Networks

Neural networks are computing models for information processing. They are useful for identifying the fundamental functional relationship or pattern in the data. Although many types of neural network models have been developed to solve different problems, the most widely used model by far for time series forecasting has been the feedforward neural network. ❯ *Figure 1* is a popular one-output feedforward neural network model. It is composed of several layers of basic processing units called neurons or nodes. Here, the network model has one input layer, one hidden layer, and one output layer. The nodes in the input layer are used to receive information from the data. For a time series forecasting problem, past lagged observations $(y_t, y_{t-1}, \ldots, y_{t-p})$ are used as inputs. The hidden layer is composed of nodes that are connected to both the input and the output layer and is the most important part of a network. With nonlinear transfer functions, hidden nodes can process the information received by input nodes. The output from the network is used to predict the future value(s) of a time series. If the focus is on one-step-ahead forecasting, then only one output node is needed. If multistep-ahead forecasting is needed, then multiple nodes may be employed in the output layer. In a feedforward network, information is one directional. That is, it goes through the input nodes to hidden nodes and to output nodes, and there is no feedback from the

◘ **Fig. 1**

A typical feedforward neural network for time series forecasting.

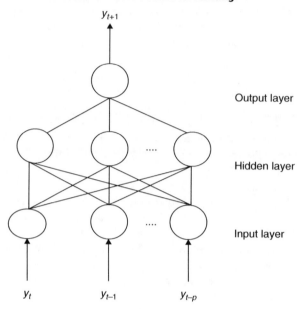

network output. The feedforward neural network illustrated in ❯ *Fig. 1* is functionally equivalent to a nonlinear autoregressive model

$$y_{t+1} = f(y_t, y_{t-1}, \ldots, y_{t-p}) + \varepsilon_{t+1}$$

where y_t is the observed time series value for variable y at time t and ε_{t+1} is the error term at time $t+1$. This model suggests that a future time series value, y_{t+1}, is an autoregressive function of its past observations, y_t, y_{t-1}, ..., y_{t-p} plus a random error. Because any time series forecasting model assumes that there is a functional relationship between the future value and the past observations, neural networks can be useful in identifying this relationship.

In developing a feedforward neural network model for forecasting tasks, specifying its architecture in terms of the number of input, hidden, and output neurons is an important yet nontrivial task. Most neural network applications use one output neuron for both one-step-ahead and multistep-ahead forecasting. However, as argued by Zhang et al. (1998), it may be beneficial to employ multiple output neurons for direct multistep-ahead forecasting. The input neurons or variables are very important in any modeling endeavor and especially important for neural network modeling because the success of a neural network depends, to a large extent, on the patterns represented by the input variables. For a time series forecasting problem, one needs to identify how many and what past lagged observations should be used as the inputs. Finally, the number of hidden nodes is usually unknown before building a neural network model and must be chosen during the model-building process. This parameter is useful for capturing the nonlinear relationship between input and output variables.

Before a neural network can be used for forecasting, it must be trained. Neural network training refers to the estimation of connection weights. Although the estimation process is similar to that in the regression modeling where one minimizes the sum of squared errors, the neural network training process is more difficult and complicated due to the nature of nonlinear optimization process. There are many training algorithms developed in the literature and the most influential one is the backpropagation algorithm by Werbos (1974) and Rumelhart et al. (1986). The basic idea of backpropagation training is to use a gradient-descent approach to adjust and determine weights such that an overall error function such as the sum of squared errors can be minimized.

In addition to the most popular feedforward neural networks, other types of neural networks can also be used for time series forecasting purposes. For example, Barreto (2008) provides a review of time series forecasting using the self-organizing map. Recurrent neural networks (Connor et al. 1994; Kuan and Liu 1995; Kermanshahi 1998; Vermaak and Botha 1998; Parlos et al. 2000; Mandic and Chambers 2001; Huskent and Stagge 2003; Ghiassi et al. 2005; Cai et al. 2007; Jain and Kumar 2007; Menezes and Barreto 2008) that explicitly account for the dynamic nonlinear pattern are good alternatives to feedforward networks for certain time series forecasting problems. In a recurrent neural network, there are cycles or feedback connections among neurons. Outputs from a recurrent network can be directly fed back to inputs, generating dynamic feedbacks on errors of past patterns. In this sense, recurrent networks can model richer dynamics than feedforward networks just like linear autoregressive and moving average (ARMA) models that have certain advantages over autoregressive (AR) models. However, much less attention has been paid to the research and applications of recurrent networks, and the superiority of recurrent networks over feedforward networks has not been established. The practical difficulty of using recurrent neural networks may lie in the facts that (1) recurrent networks can assume very different architectures and it may be difficult to specify appropriate model structures to experiment with, and (2) it is more difficult to train recurrent networks due to the unstable nature of training algorithms.

For an in-depth coverage of many aspects of networks, readers are referred to a number of excellent books including Smith (1993), Bishop (1995), and Ripley (1996). For neural networks for forecasting research and applications, readers may consult Azoff (1994), Weigend and Gershenfeld (1994), Gately (1996), Zhang et al. (1998), and Zhang (2004).

3   Applications in Time Series Forecasting

Time series are data collected over time and this is one of the most commonly available forms of data in many forecasting applications. Therefore, it is not surprising that the use of neural networks for time series forecasting has received great attention in many different fields. Given that forecasting problems arise in so many different disciplines and the literature on forecasting with neural networks is scattered in so many diverse fields, it is difficult to cover all neural network applications in time series forecasting problems in this review. ❷ *Table 1* provides a sample of recent time series forecasting applications reported in the literature since 2005. For other forecasting applications of neural networks, readers are referred to several survey articles such as Dougherty (1995) for transportation modeling and forecasting, Wong and Selvi (1998) and Fadlalla and Lin (2001) for financial applications, Krycha and Wagner (1999) for management science applications, Vellido et al. (1999) and Wong et al. (2000) for business applications, Maier and Dandy (2000) for water resource forecasting, and Hippert et al. (2001) for short-term load forecasting.

As can be seen from ❷ *Table 1*, a wide range of time series forecasting problems have been solved by neural networks. Some of these application areas include environment (air pollutant concentration, carbon monoxide concentration, drought, and ozone level), business and

■ Table 1

Some recent neural network applications in time series forecasting

Forecasting problems	Studies
Air pollutant concentration	Gautama et al. (2008)
Carbon monoxide concentration	Chelani and Devotta (2007)
Demand	Aburto and Weber (2007)
Drought	Mishra and Desai (2006)
Electrical consumption	Azadeh et al. (2007)
Electricity load	Hippert et al. (2005), Xiao et al. (2009)
Electricity price	Pino et al. (2008)
Energy demand	Abdel-Aal (2008)
Exchange rate	Zhang and Wan (2007)
Food grain price	Zou et al. (2007)
Food product sales	Doganis et al. (2006)
Gold price changes	Parisi et al. (2008)
Inflation	Nakamura (2005)
Inventory	Doganis et al. (2008)
Macroeconomic time series	Teräsvirta et al. (2005)
Stock index option price	Wang (2009)
Stock returns volatility	Bodyanskiy and Popov (2006)
Tourism demand	Palmer et al. (2006), Chu (2008)
Traffic flow	Jiang and Adeli (2005)
Ozone level	Coman et al. (2008)
River flow	Jain and Kumar (2007)
Wind speed	Cadenas and Rivera (2009)

finance (product sales, demand, inventory, stock market movement and risk, exchange rate, futures trading, commodity and option price), tourism and transportation (tourist volume and traffic flow), engineering (wind speed and river flow) and energy (electrical consumption and energy demand). Again this is only a relatively small sample of the application areas to which neural networks have been applied. One can find many more application areas of neural networks in time series analysis and forecasting.

4   Neural Network Modeling Issues

Developing a neural network model for a time series forecasting application is not a trivial task. Although many software packages exist to ease users' effort in building a neural network model, it is critical for forecasters to understand many important issues around the model-building process. It is important to point out that building a successful neural network is a combination of art and science, and software alone is not sufficient to solve all problems in the process. It is a pitfall to blindly throw data into a software package and then hope it will automatically give a satisfactory forecast.

Neural network modeling issues include the choice of network type and architecture, the training algorithm, as well as model validation, evaluation, and selection. Some of these can be solved during the model-building process while others must be carefully considered and planned before actual modeling starts.

4.1 Data Issues

Regarding the data issues, the major decisions a neural network forecaster must make include data preparation, data cleaning, data splitting, and input variable selection. Neural networks are data-driven techniques. Therefore, data preparation is a very critical step in building a successful neural network model. Without a good, adequate, and representative data set, it is impossible to develop a useful predictive model. The reliability of neural network models often depends, to a large degree, on the quality of data.

There are several practical issues around the data requirement for a neural network model. The first is the size of the sample used to build a neural network. While there is no specific rule that can be followed for all situations, the advantage of having a large sample should be clear because not only do neural networks typically have a large number of parameters to estimate, but also it is often necessary to split data into several portions to avoid overfitting, to select the model, and to perform model evaluation and comparison. A larger sample provides a better chance for neural networks to adequately capture the underlying data-generating process. Although large samples do not always give superior performance over small samples, forecasters should strive to get as large a size as they can. In time series forecasting problems, Box and Jenkins (1976) have suggested that at least 50, or better 100, observations are necessary to build linear autoregressive integrated moving average (ARIMA) models. Therefore, for nonlinear modeling, a larger sample size should be more desirable. In fact, using the longest time series available for developing forecasting models is a time-tested principle in forecasting (Armstrong 2001). Of course, if observations in the time series are not homogeneous or the underlying data-generating process changes over time, then larger sample sizes may not help and can even hurt the performance of neural networks.

The second issue is the data splitting. Typically, for neural network applications, all available data are divided into an in-sample and an out-of-sample. The in-sample data are used for model fitting and selection, while the out-of-sample is used to evaluate the predictive ability of the model. The in-sample data sometimes are further split into a training sample and a validation sample. This division of data means that the true size of the sample used in model building is smaller than the initial sample size. Although there is no consensus on how to split the data, the general practice is to allocate more data for model building and selection. That is, most studies in the literature use convenient ratios of splitting for in-sample and out-of-sample, such as 70:30%, 80:20%, and 90:10%. It is important to note that in data splitting the issue is not about what proportion of data should be allocated in each sample but about sufficient data points in each sample to ensure adequate learning, validation, and testing. When the size of the available data set is large, different splitting strategies may not have a major impact. But it is quite different when the sample size is small. According to Chatfield (2001), forecasting analysts typically retain about 10% of the data as a hold-out sample. Granger (1993) suggests that for nonlinear modeling, at least 20% of the data should be held back for an out-of-sample evaluation. Hoptroff (1993) recommends that at least ten data points should be in the test sample, while Ashley (2003) suggests that a much larger out-of-sample size is necessary in order to achieve statistically significant improvement for forecasting problems.

In addition, data splitting generally should be done randomly to make sure each subsample is representative of the population. However, time series data are difficult or impossible to split randomly because of the desire to keep the autocorrelation structure of the time series observations. For many time series problems, data splitting is typically done at researchers' discretion. However, it is important to make sure that each portion of the sample is characteristic of the true data-generating process. LeBaron and Weigend (1998) evaluate the effect of data splitting on time series forecasting and find that data splitting can cause more sample variation which in turn causes the variability of forecast performance. They caution the pitfall of ignoring variability across the splits and drawing too strong conclusions from such splits.

Data preprocessing is another issue that is often recommended to highlight important relationships or to create more uniform data to facilitate neural network learning, meet algorithm requirements, and avoid computation problems. Azoff (1994) summarizes four methods typically used for input data normalization. They are along-channel normalization, across-channel normalization, mixed-channel normalization, and external normalization. However, the necessity and effect of data normalization on network learning and forecasting are still not universally agreed upon. For example, in modeling and forecasting seasonal time series, some researchers (Gorr 1994) believe that data preprocessing is not necessary because the neural network is a universal approximator and is able to capture all of the underlying patterns well. Empirical studies (Nelson et al. 1999), however, find that pre-deseasonalization of the data is critical in improving forecasting performance. Zhang and Qi (2002, 2005) further demonstrate that for time series containing both trend and seasonal variations, preprocessing the data by both detrending and deseasonalization should be the most appropriate way to build neural networks for best forecasting performance.

4.2 Network Design

Neural network design and architecture selection are important yet difficult tasks. Not only are there many ways to build a neural network model and a large number of choices to be made during the model building and selection process, but also numerous parameters and issues have to be estimated and experimented with before a satisfactory model may emerge. Adding to the difficulty is the lack of standards in the process. Numerous rules of thumb are available but not all of them can be applied blindly to a new situation. In building an appropriate model for the forecasting task at hand, some experiments are usually necessary. Therefore, a good experimental design is needed. For discussions of many aspects of modeling issues, readers may consult Kaastra and Boyd (1996), Zhang et al. (1998), Coakley and Brown (1999), and Remus and O'Connor (2001).

A feedforward neural network is characterized by its architecture determined by the number of layers, the number of nodes in each layer, the transfer or activation function used in each layer, as well as how the nodes in each layer connect to nodes in adjacent layers. Although partial connections between nodes in adjacent layers and direct connection from input layer to output layer are possible, the most commonly used neural network is the fully connected one in which each node in one layer is fully connected only to all nodes in the adjacent layers.

The size of the output layer is usually determined by the nature of the problem. For example, in most time series forecasting problems, one output node is naturally used for one-step-ahead forecasting, although one output node can also be employed for multistep-ahead forecasting, in which case iterative forecasting mode must be used. That is, forecasts for more than twosteps ahead in the time horizon must be based on earlier forecasts. This may not be effective for multistep forecasting as pointed out by Zhang et al. (1998) which is in line with Chatfield (2001) who discusses the potential benefits of using different forecasting models for different lead times. Therefore, for multistep forecasting, one may either use multiple output nodes or develop multiple neural networks each for one particular step forecasting.

The number of input nodes is perhaps the most important parameter for designing an effective neural network forecaster. For causal forecasting problems, it corresponds to the number of independent or predictor variables that forecasters believe are important in predicting the dependent variable. For univariate time series forecasting problems, it is the number of past lagged observations. Determining an appropriate set of input variables is vital for neural networks to capture the essential underlying relationship that can be used for successful forecasting. How many and what variables to use in the input layer will directly affect the performance of neural networks in both in-sample fitting and out-of-sample forecasting, resulting in the under-learning or over-fitting phenomenon. Empirical results (Lennon et al. 2001; Zhang et al. 2001; Zhang 2001) also suggest that the input layer is more important than the hidden layer in time series forecasting problems. Therefore, considerable attention should be given to input variable selection especially for time series forecasting. Balkin and Ord (2000) select the ordered variables (lags) sequentially by using a linear model and a forward stepwise regression procedure. Medeiros et al. (2006) also use a linear variable selection approach to choosing the input variables.

Although there is substantial flexibility in choosing the number of hidden layers and the number of hidden nodes in each layer, most forecasting applications use only one hidden layer and a small number of hidden nodes. In practice, the number of hidden nodes is often determined by experimenting with a number of choices and then selecting using the cross-validation approach or the performance on the validation set. Although the number of hidden nodes is an important factor, a number of studies have shown that the forecasting performance of neural networks is not very sensitive to this parameter (Bakirtzis et al. 1996; Khotanzad et al. 1997; Zhang et al. 2001). Medeiros et al. (2006) propose a statistical approach to selecting the number of hidden nodes by sequentially applying the Lagrange multiplier type tests.

Once a particular neural network architecture is determined, it must be trained so that the parameters of the network can be estimated from the data. To be effective in performing this task, a good training algorithm is needed. Training a neural network can be treated as a nonlinear mathematical optimization problem and different solution approaches or algorithms can have quite different effects on the training result. As a result, training with different algorithms and repeating with multiple random initial weights can be helpful in getting better solutions to the neural network training problem. In addition to the popular basic back-propagation training algorithm, users should be aware of many other (sometimes more effective) algorithms. These include the so-called second-order approaches such as conjugate gradient descent, quasi-Newton, and Levenberg–Marquardt (Bishop 1995).

4.3 Model Selection and Evaluation

Using the selection of a neural network model is typically done using the cross-validation process. That is, the in-sample data is split into a training set and a validation set. The network parameters are estimated with the training sample, while the performance of the model is evaluated with the validation sample. The best model selected is the one that has the best performance on the validation sample. Of course, in choosing competing models, one must also apply the principle of parsimony. That is, a simpler model that has about the same performance as a more complex model should be preferred.

Model selection can also be done with solely the in-sample data. In this regard, several in-sample selection criteria are used to modify the total error function to include a penalty term that penalizes for the complexity of the model. In-sample model selection approaches are typically based on some information-based criteria such as Akaike's information criterion (AIC) and Bayesian (BIC) or Schwarz information criterion (SIC). However, it is important to note the limitation of these criteria as empirically demonstrated by Swanson and White (1995) and Qi and Zhang (2001). Egrioglu et al. (2008) propose a weighted information criterion to select the model. Other in-sample approaches are based on pruning methods such as node and weight pruning (Reed 1993) as well as constructive methods such as the upstart and cascade correlation approaches (Fahlman and Lebiere 1990; Frean 1990).

After the modeling process, the finally selected model must be evaluated using data not used in the model-building stage. In addition, as neural networks are often used as a nonlinear alternative to traditional statistical models, the performance of neural networks needs to be compared to that of statistical methods. As Adya and Collopy (1998) point out, "if such a comparison is not conducted it is difficult to argue that the study has taught one much about the value of neural networks." They further propose three evaluation criteria to objectively evaluate the performance of a neural network: (1) comparing it to well-accepted (traditional) models; (2) using true out-of-sample data and (3) ensuring enough sample size in the out-of-sample data (40 for classification problems and 75 for time series problems). It is important to note that the test sample served as the out-of-sample should not in any way be used in the model-building process. If the cross-validation is used for model selection and experimentation, the performance on the validation sample should not be treated as the true performance of the model.

5   Methodological Issues

5.1   Modeling Trend and Seasonal Time Series

Many business and economic time series exhibit both seasonal and trend variations. Seasonality is a periodic and recurrent pattern caused by factors such as weather, holidays, repeating promotions, as well as the behavior of economic agents (Hylleberg 1992). Because of the frequent occurrence of these time series in practice, how to model and forecast seasonal and trend time series has long been a major research topic that has significant practical implications.

Traditional analyses of time series are mainly concerned with modeling the autocorrelation structure of a time series, and typically require that the data under study be stationary. Trend and seasonality in time series violate the condition of stationarity. Thus, the removal of the trend and seasonality is often desired in time series analysis and forecasting. For example, the well-known Box–Jenkins approach to time series modeling relies entirely on the stationarity assumption. The classic decomposition technique decomposes a time series into trend, seasonal factor, and irregular components. The trend and seasonality are often estimated and removed from the data first before other components are estimated. Seasonal ARIMA models also require that the data be seasonally differenced to achieve stationarity condition (Box and Jenkins 1976).

However, seasonal adjustment is not without controversy. Ghysels et al. (1996) suggest that seasonal adjustment might lead to undesirable nonlinear properties in univariate time series. Ittig (1997) also questions the traditional method for generating seasonal indexes and proposed a nonlinear method to estimate the seasonal factors. More importantly, some empirical studies find that seasonal fluctuations are not always constant over time and at least in some time series, seasonal components and nonseasonal components are not independent, and thus not separable (Hylleberg 1994).

In the neural network literature, Gorr (1994) points out that neural networks should be able to simultaneously detect both the nonlinear trend and the seasonality in the data. Sharda and Patil (1992) examine 88 seasonal time series from the M-competition and find that neural networks can model seasonality effectively and pre-deseasonalizing the data is not necessary. Franses and Draisma (1997) find that neural networks can also detect possible changing seasonal patterns. Hamzacebi (2008) proposes a neural network with seasonal lag to directly model seasonality. Farway and Chatfield (1995), however, find mixed results with the direct neural network approach. Kolarik and Rudorfer (1994) report similar findings. Based on a study of 68 time series from the M-competition, Nelson et al. (1999) find that neural networks trained on deseasonalized data forecast significantly better than those trained on seasonally non-adjusted data. Hansen and Nelson (2003) find that the combination of transformation, feature extraction, and neural networks through stacked generalization gives more accurate forecasts than classical decomposition or ARIMA models.

Due to the controversies around how to use neural networks to best model trend and/or seasonal time series, several researchers have systematically studied the issue recently. Qi and Zhang (2008) investigate the issue of how to best use neural networks to model trend time series. With a simulation study, they address the question: what is the most effective way to model and forecast trend time series with neural networks? A variety of different underlying data-generating processes are considered that have different trend mechanisms. Results show that directly modeling the trend component is not a good strategy and differencing the data first is the overall most effective approach in modeling trend time series. Zhang and Qi (2005) further look into the issue of how to best model time series with both trend and seasonal components. Using both simulated and real time series, they find that preprocessing the data by both detrending and deseasonalization is the most appropriate way to build neural networks for best forecasting performance. This finding is supported by Zhang and Kline (2007) who empirically examine the issue of data preprocessing and model selection using a large data set of 756 quarterly time series from the M3 forecasting competition.

5.2 Multi-period Forecasting

One of the methodological issues that has received limited attention in the time series forecasting as well as the neural network literature is multi-period (or multistep) forecasting. A forecaster facing a multiple-period forecasting problem typically has a choice between the iterated method – using a general single-step model to iteratively generate forecasts, and the direct method – using a tailored model that directly forecasts the future value for each forecast horizon. Which method the forecaster should use is an important research and practical question.

Theoretically, the direct method should be more appealing because it is less sensitive to model misspecification (Chevillon and Hendry 2005). Several empirical studies on the relative performance of iterated vs. direct forecasts, however, yield mixed findings. Findley (1985) find some improvement in forecasting accuracy using the direct method. Based on several simulated autoregressive and moving average time series, Stoica and Nehorai (1989) find no significant differences between the two methods. Bhansali (1997), however, shows that the iterated method has a clear advantage over the direct method for finite autoregressive processes. Kang (2003) uses univariate autoregressive models to forecast nine US economic time series and found inconclusive results regarding which multistep method is preferred. Ang et al. (2006) find that iterated forecasts of US GDP growth outperformed direct forecasts at least during the 1990s. The most comprehensive empirical study to date was undertaken by Marcellino et al. (2006) who compared the relative performance of the iterated vs. direct forecasts with several univariate and multivariate autoregressive (AR) models. Using 170 US monthly macroeconomic time series spanning 1959 to 2002, they find that iterated forecasts are generally better than direct forecasts under several different scenarios. In addition, they showed that direct forecasts are increasingly less accurate as the forecast horizon increases.

For nonlinear models, multi-period forecasting receives little attention in the literature because of the analytical challenges and the computational difficulties (De Gooijer and Kumar 1992). Lin and Granger (1994) recommend the strategy of fitting a new model for each forecast horizon in nonlinear forecasting. Tiao and Tsay (1994) find that the direct forecasting method can dramatically outperform the iterated method, especially for long-term forecasts. In the neural network literature, mixed findings have been reported (Zhang et al. 1998). Although Zhang et al. (1998) argue for using the direct method, Weigend et al. (1992) and Hill et al. (1996) find that the direct method performed much worse than the iterated method. Kline (2004) specifically addresses the issue of the relative performance of iterated vs. direct methods. He proposes three methods – iterated, direct, and joint (using one neural network model to predict all forecast horizons simultaneously) – and compares them using a subset of quarterly series from the M3 competition. He finds that the direct method significantly outperformed the iterated method. In a recent study, Hamzacebi et al. (2009) also reports better results achieved by using the direct method.

5.3   Ensemble Models

One of the major developments in neural network time series forecasting is model combining or ensemble modeling. The basic idea of this multi-model approach is the use of each component model's unique capability to better capture different patterns in the data. Both theoretical and empirical findings have suggested that combining different models can be an effective way to improve the predictive performance of each individual model, especially when the models in the ensemble are quite different. Although a majority of the neural ensemble literature is focused on pattern classification problems, a number of combining schemes have been proposed for time series forecasting problems. For example, Pelikan et al. (1992) and Ginzburg and Horn (1994) combine several feedforward neural networks for time series forecasting. Wedding and Cios (1996) describe a combining methodology using radial basis function networks and the Box–Jenkins models. Goh et al. (2003) use an ensemble of boosted Elman networks for predicting drug dissolution profiles. Medeiros and Veiga (2000) consider a hybrid time series forecasting system with neural networks used to control the time-varying parameters of a smooth transition autoregressive model. Armano et al. (2005) use a combined genetic-neural model to forecast stock indexes. Zhang (2003) proposes a hybrid neural-ARIMA model for time series forecasting. Aslanargun et al. (2007) use a similar hybrid model to forecast tourist arrivals and find improved results. Liu and Yao (1999) develop a simultaneous training system for negatively correlated networks to overcome the limitation of sequential or independent training methods. Khashei et al. (2008) propose a hybrid artificial neural network and fuzzy regression model for financial time series forecasting. Freitas and Rodrigues (2006) discuss the different ways of combining Gaussian radial basis function networks. They also propose a prefiltering methodology to address the problem caused by nonstationary time series. Wichard and Ogorzalek (2007) use several different model architectures with an iterated prediction procedure to select the final ensemble members.

An ensemble can be formed by multiple network architectures, the same architecture trained with different algorithms, different initial random weights, or even different methods. The component networks can also be developed by training with different data such as the resampling data or with different inputs. In general, as discussed in Sharkey (1996) and Sharkey and Sharkey (1997), the neural ensemble formed by varying the training data typically has more component diversity than that trained on different starting points, with a different number of hidden nodes, or using different algorithms. Thus, this approach is the most commonly used in the literature. There are many different ways to alter training data including cross-validation, bootstrapping, using different data sources or different preprocessing techniques, as well as a combination of the above techniques (Sharkey 1996). Zhang and Berardi (2001) propose two data-splitting schemes to form multiple sub-time-series upon which ensemble networks are built. They find that the ensemble achieves significant improvement in forecasting performance. Zhang (2007b) proposes a neural ensemble model based on the idea of adding noises to the input data and forming different training sets with the jittered input data.

6   Conclusions

Neural networks have become an important tool for time series forecasting. They have many desired features that are quite suitable for practical applications. This chapter provides a general overview of the neural networks for time series forecasting problems. Successful application areas of neural networks as well as critical modeling issues are reviewed. It should be emphasized that each time series forecasting situation requires a careful study of the problem characteristics, careful examination of the data characteristics, prudent design of the modeling strategy, and full consideration of modeling issues. Many rules of thumb in neural networks may not be useful for a new application although good forecasting principles and established guidelines should be followed. Researchers need to be aware of many pitfalls that could arise in using neural networks in their research and applications (Zhang 2007a).

It is important to recognize that although some of the modeling issues are unique to neural networks, some are general issues for any forecasting method. Therefore, good forecasting practice and principles should be followed. It is beneficial to consult Armstrong (2001) which provides a good source of information on useful principles for forecasting model building, evaluation, and use.

Neural networks have achieved great successes in the field of time series forecasting. It is, however, important to note that they may not always yield better results than traditional methods for every forecasting task under all circumstances. Therefore, researchers should not focus only on neural networks and completely ignore the traditional methods in their forecasting applications. A number of forecasting competitions suggest that no single method including neural networks is universally the best for all types of problem in every situation. Thus it may be beneficial to combine different models (e.g., combine neural networks and statistical models) in improving forecasting performance. Indeed, efforts to find better ways to use neural networks for time series forecasting should continue.

References

Abdel-Aal RE (2008) Univariate modeling and forecasting of monthly energy demand time series using abductive and neural networks. Comput Ind Eng 54:903–917

Aburto L, Weber R (2007) Improved supply chain management based on hybrid demand forecasts. Appl Soft Comput 7(1):136–144

Adya M, Collopy F (1998) How effective are neural networks at forecasting and prediction? A review and evaluation. J Forecasting 17:481–495

Ang A, Piazzesi M, Wei M (2006) What does the yield curve tell us about GDP growth? J Econometrics 131:359–403

Armano G, Marchesi M, Murru A (2005) A hybrid genetic-neural architecture for stock indexes forecasting. Info Sci 170(1):3–33

Armstrong JS (2001) Principles of forecasting: A handbook for researchers and practitioners. Kluwer, Boston, MA

Ashley R (2003) Statistically significant forecasting improvements: how much out-of-sample data is likely necessary? Int J Forecasting 19(2):229–239

Aslanargun A, Mammadov M, Yazici B, Yolacan S (2007) Comparison of ARIMA, neural networks and hybrid models in time series: tourist arrival forecasting. J Stat Comput Simulation 77(1):29–53

Atiya AF, El-Shoura SM, Shaheen SI, El-Sherif MS (1999) A comparison between neural-network forecasting techniques-case study: river flow forecasting. IEEE Trans Neural Netw 10(2):402–409

Azadeh A, Ghaderi SF, Sohrabkhani S (2007) Forecasting electrical consumption by integration of neural network, time series and ANOVA. Appl Math Comput 186:1753–1761

Azoff EM (1994) Neural network time series forecasting of financial markets. Wiley, Chichester, UK

Bakirtzis AG, Petridis V, Kiartzis SJ, Alexiadis MC, Maissis AH (1996) A neural network short term load forecasting model for the Greek power system. IEEE Trans Power Syst 11(2):858–863

Balkin DS, Ord KJ (2000) Automatic neural network modeling for univariate time series. Int J Forecasting 16:509–515

Barreto GA (2008) Time series prediction with the self-organizing map: a review. Stud Comput Intell 77:135–158

Berardi LV, Zhang PG (2003) An empirical investigation of bias and variance in time series forecasting: modeling considerations and error evaluation. IEEE Trans Neural Netw 14(3):668–679

Bhansali RJ (1997) Direct autoregressive predictions for multistep prediction: order selection and performance relative to the plug in predictors. Stat Sin 7:425–449

Bishop M (1995) Neural networks for pattern recognition. Oxford University Press, Oxford

Bodyanskiy Y, Popov S (2006) Neural network approach to forecasting of quasiperiodic financial time series. Eur J Oper Res 175:1357–1366

Box GEP, Jenkins G (1976) Time series analysis: forecasting and control. Holden-Day, San Francisco, CA

Cadenas E, Rivera W (2009) Short term wind speed forecasting in La Venta, Oaxaca, México, using artificial neural networks. Renewable Energy 34(1):274–278

Cai X, Zhang N, Venayagamoorthy GK, Wunsch DC (2007) Time series prediction with recurrent neural networks trained by a hybrid PSO-EA algorithm. Neurocomputing 70:2342–2353

Chakraborty K, Mehrotra K, Mohan KC, Ranka S (1992) Forecasting the behavior of multivariate time series using neural networks. Neural Netw 5:961–970

Chatfield C (2001) Time-series forecasting. Chapman & Hall/CRC, Boca Raton, FL

Chelani AB, Devotta S (2007) Prediction of ambient carbon monoxide concentration using nonlinear time series analysis technique. Transportation Res Part D 12:596–600

Chevillon G, Hendry DF (2005) Non-parametric direct multi-step estimation for forecasting economic processes. Int J Forecasting 21:201–218

Chu FL (2008) Analyzing and forecasting tourism demand with ARAR algorithm. Tourism Manag 29(6):1185–1196

Coakley JR, Brown CE (1999) Artificial neural networks in accounting and finance: modeling issues. Int J Intell Syst Acc Finance Manag 9:119–144

Coman A, Ionescu A, Candau Y (2008) Hourly ozone prediction for a 24-h horizon using neural networks. Environ Model Software 23(12):1407–1421

Connor JT, Martin RD, Atlas LE (1994) Recurrent neural networks and robust time series prediction. IEEE Trans Neural Netw 51(2):240–254

Cottrell M, Girard B, Girard Y, Mangeas M, Muller C (1995) Neural modeling for time series: a statistical stepwise method for weight elimination. IEEE Trans Neural Netw 6(6):1355–1364

De Gooijer JG, Kumar K (1992) Some recent developments in non-linear time series modeling, testing, and forecasting. Int J Forecasting 8:135–156

De Groot C, Wurtz D (1991) Analysis of univariate time series with connectionist nets: a case study of two classical examples. Neurocomputing 3:177–192

Doganis P, Aggelogiannaki E, Patrinos P, Sarimveis H (2006) Time series sales forecasting for short shelf-life food products based on artificial neural networks and evolutionary computing. J Food Eng 75:196–204

Doganis P, Aggelogiannaki E, Sarimveis H (2008) A combined model predictive control and time series forecasting framework for production-inventory systems. Int J Prod Res 46(24):6841–6853

Dougherty M (1995) A review of neural networks applied to transport. Transportation Res Part C 3(4): 247–260

Egrioglu E, Aladag CAH, Gunay S (2008) A new model selection strategy in artificial neural networks. Appl Math Comput 195:591–597

Fadlalla A, Lin CH (2001) An analysis of the applications of neural networks in finance. Interfaces 31(4): 112–122

Fahlman S, Lebiere C (1990) The cascade-correlation learning architecture. In: Touretzky D (ed) Advances in neural information processing systems, vol 2. Morgan Kaufmann, Los Altos, CA, pp 524–532

Farway J, Chatfield C (1995) Time series forecasting with neural networks: a comparative study using the airline data. Appl Stat 47:231–250

Findley DF (1985) Model selection for multi-step-ahead forecasting. In: Baker HA, Young PC (eds) Proceedings of the seventh symposium on identification and system parameter estimation. Pergamon, Oxford, New York, pp 1039–1044

Franses PH, Draisma G (1997) Recognizing changing seasonal patterns using artificial neural networks. J Econometrics 81:273–280

Frean M (1990) The Upstart algorithm: A method for constructing and training feed-forward networks. Neural Comput 2:198–209

Freitas and Rodrigues (2006) Model combination in neural-based forecasting. Eur J Oper Res 173: 801–814

Gately E (1996) Neural networks for financial forecasting. Wiley, New York

Gautama AK, Chelanib AB, Jaina VK, Devotta S (2008) A new scheme to predict chaotic time series of air pollutant concentrations using artificial neural network and nearest neighbor searching. Atmospheric Environ 42:4409–4417

Ghiassi M, Saidane H, Zimbra DK (2005) A dynamic artificial neural network model for forecasting time series events. Int J Forecasting 21(2):341–362

Ghysels E, Granger CWJ, Siklos PL (1996) Is seasonal adjustment a linear or nonlinear data filtering process? J Bus Econ Stat 14:374–386

Ginzburg I, Horn D (1994) Combined neural networks for time series analysis. Adv Neural Info Process Syst 6:224–231

Goh YW, Lim PC, Peh KK (2003) Predicting drug dissolution profiles with an ensemble of boosted neural networks: A time series approach. IEEE Trans Neural Netw 14(2):459–463

Gorr L (1994) Research prospective on neural network forecasting. Int J Forecasting 10:1–4

Granger CWJ (1993) Strategies for modelling nonlinear time-series relationships. Econ Rec 69(206):233–238

Hansen JV, Nelson RD (2003) Forecasting and recombining time-series components by using neural networks. J Oper Res Soc 54(3):307–317

Hamzacebi C (2008) Improving artificial neural networks' performance in seasonal time series forecasting. Inf Sci 178:4550–4559

Hamzacebi C, Akay D, Kutay F (2009) Comparison of direct and iterative artificial neural network forecast approaches in multi-periodic time series forecasting. Expert Syst Appl 36(2):3839–3844

Hill T, O'Connor M, Remus W (1996) Neural network models for time series forecasts. Manag Sci 42: 1082–1092

Hippert HS, Pedreira CE, Souza RC (2001) Neural networks for short-term load forecasting: a review and evaluation. IEEE Trans Power Syst 16(1):44–55

Hippert HS, Bunn DW, Souza RC (2005) Large neural networks for electricity load forecasting: are they overfitted? Int J Forecasting 21(3):425–434

Hoptroff RG (1993) The principles and practice of time series forecasting and business modeling using neural networks. Neural Comput Appl 1:59–66

Huskent M, Stagge P (2003) Recurrent neural networks for time series classification. Neurocomputing 50:223–235

Hwang HB (2001) Insights into neural-network forecasting of time series corresponding to ARMA (p,q) structures. Omega 29:273–289

Hylleberg S (1992) General introduction. In: Hylleberg S (ed) Modelling seasonality. Oxford University Press, Oxford, pp 3–14

Hylleberg S (1994) Modelling seasonal variation. In: Hargreaves CP (ed) Nonstationary time series analysis and cointegration. Oxford University Press, Oxford, pp 153–178

Ittig PT (1997) A seasonal index for business. Decis Sci 28(2):335–355

Jain A, Kumar AM (2007) Hybrid neural network models for hydrologic time series forecasting. Appl Soft Comput 7:585–592

Jiang X, Adeli H (2005) Dynamic wavelet neural network model for traffic flow forecasting, J Transportation Eng 131(10):771–779

Kang I-B (2003) Multi-period forecasting using different models for different horizons: an application to U.S. economic time series data. Int J Forecasting 19:387–400

Kaastra I, Boyd M (1996) Designing a neural network for forecasting financial and economic time series. Neurocomputing 10:215–236

Kermanshahi B (1998) Recurrent neural network for forecasting next 10 years loads of nine Japanese utilities. Neurocomoputing 23:125–133

Khashei M, Hejazi SR, Bijari M (2008) A new hybrid artificial neural networks and fuzzy regression model for time series forecasting. Fuzzy Sets Syst 159:769–786

Khotanzad A, Afkhami-Rohani R, Lu TL, Abaye A, Davis M, Maratukulam DJ (1997) ANNSTLF—A neural-network-based electric load forecasting system. IEEE Trans Neural Netw 8(4):835–846

Kline DM (2004) Methods for multi-step time series forecasting with neural networks. In: Zhang GP (ed) Neural networks for business forecasting. Idea Group, Hershey, PA, pp 226–250

Kolarik T, Rudorfer G (1994) Time series forecasting using neural networks. APL Quote Quad 25:86–94

Krycha KA, Wagner U (1999) Applications of artificial neural networks in management science: a survey. J Retailing Consum Serv 6:185–203

Kuan C-M, Liu T (1995) Forecasting exchange rates using feedforward and recurrent neural networks. J Appl Economet 10:347–364

Lapedes A, Farber R (1987) Nonlinear signal processing using neural networks: Prediction and system modeling. Technical Report LA-UR-87-2662, Los Alamos National Laboratory, Los Alamos, NM

LeBaron B, Weigend AS (1998) A bootstrap evaluation of the effect of data splitting on financial time series. IEEE Trans Neural Netw 9(1):213–220

Lennon B, Montague GA, Frith AM, Gent C, Bevan V (2001) Industrial applications of neural networks— An investigation. J Process Control 11:497–507

Liang F (2005) Bayesian neural networks for nonlinear time series forecasting. Stat Comput 15:13–29

Lin J-L, Granger CWJ (1994) Forecasting from nonlinear models in practice. J Forecasting 13:1–9

Liu Y, Yao X (1999) Ensemble learning via negative correlation. Neural Netw 12:1399–1404

Maier HR, Dandy GC (2000) Neural networks for the prediction and forecasting of water resource variables: a review of modeling issues and applications. Environ Model Software 15:101–124

Makridakis S, Anderson A, Carbone R, Fildes R, Hibdon M, Lewandowski R, Newton J, Parzen E, Winkler R (1982) The accuracy of extrapolation (time series) methods: results of a forecasting competition. J Forecasting 1(2):111–153

Mandic D, Chambers J (2001) Recurrent neural networks for prediction: learning algorithms, architectures and stability. Wiley, Chichester, UK

Marcellino M, Stock JH, Watson MW (2006) A comparison of direct and iterated multistep AR methods for forecasting macroeconomic time series. J Economet 135:499–526

Medeiros MC, Pedreira CE (2001) What are the effects of forecasting linear time series with neural networks? Eng Intell Syst 4:237–424

Medeiros MC, Veiga A (2000) A hybrid linear-neural model for time series forecasting. IEEE Trans Neural Netw 11(6):1402–1412

Medeiros MC, Terasvirta T, Rech G (2006) Building neural network models for time series: a statistical approach. J Forecasting 25:49–75

Menezes JMP, Barreto GA (2008) Long-term time series prediction with the NARX network: an empirical evaluation. Neurocomputing 71:3335–3343

Mishra AK, Desai VR (2006) Drought forecasting using feed-forward recursive neural network. Ecol Model 198:127–138

Nakamura E (2005) Inflation forecasting using a neural network. Econ Lett 86:373–378

Nelson M, Hill T, Remus T, O'Connor M (1999) Time series forecasting using neural networks: should the data be deseasonalized first? J Forecasting 18:359–367

Palmer A, Montano JJ, Sese A (2006) Designing an artificial neural network for forecasting tourism time series. Tourism Manag 27:781–790

Parisi A, Parisi F, Díaz D (2008) Forecasting gold price changes: rolling and recursive neural network models. J Multinational Financial Manag 18(5):477–487

Parlos AG, Rais OT, Atiya AF (2000) Multi-step-ahead prediction using dynamic recurrent neural networks. Neural Netw 13:765–786

Pelikan E, de Groot C, Wurtz D (1992) Power consumption in West-Bohemia: improved forecasts with decorrelating connectionist networks. Neural Netw World 2(6):701–712

Pino P, Parreno J, Gomez A, Priore P (2008) Forecasting next-day price of electricity in the Spanish energy market using artificial neural networks. Eng Appl Artif Intell 21:53–62

Poli I, Jones DR (1994) A neural net model for prediction. J Am Stat Assoc 89:117–121

Qi M, Zhang GP (2001) An investigation of model selection criteria for neural network time series forecasting. Eur J Oper Res 132:666–680

Qi M, Zhang GP (2008) Trend time-series modeling and forecasting with neural networks. IEEE Trans Neural Netw 19(5):808–816

Reed R (1993) Pruning algorithms—A survey. IEEE Trans Neural Netw 4(5):740–747

Remus W, O'Connor M (2001) Neural networks for time series forecasting. In: Armstrong JS (ed) Principles of forecasting: a handbook for researchers and practitioners. Kluwer, Norwell, MA, pp 245–256

Ripley BD (1996) Pattern recognition and neural networks. Cambridge University Press, Cambridge

Rumelhart DE, McClelland JL, PDP Research Group (1986) Parallel distributed processing: explorations in the microstructure of cognition, Foundations, vol 1. MIT Press, Cambridge, MA

Sharda R, Patil RB (1992) Connectionist approach to time series prediction: an empirical test. J Intell Manufacturing 3:317–323

Sharkey CJ (1996) On combining artificial neural nets. Connect Sci 8:299–314

Sharkey CJ, Sharkey EN (1997) Combining diverse neural nets. Knowledge Eng Rev 12(3):231–247

Smith M (1993) Neural networks for statistical modeling. Van Nostrand Reinhold, New York

Stoica P, Nehorai A (1989) On multi-step prediction errors methods for time series models. J Forecasting 13:109–131

Swanson NR, White H (1995) A model-selection approach to assessing the information in the term structure using linear models and artificial neural networks. J Bus Econ Stat 13:265–275

Tiao GC, Tsay RS (1994) Some advances in non-linear and adaptive modeling in time-series. J Forecasting 13:109–131

Teräsvirta T, van Dijk D, Medeiros MC (2005) Linear models, smooth transition autoregressions, and neural networks for forecasting macroeconomic time series: a re-examination. Int J Forecasting 21(4):755–774

Vellido A, Lisboa PJG, Vaughan J (1999) Neural networks in business: a survey of applications (1992–1998). Expert Syst Appl 17:51–70

Vermaak J, Botha EC (1998) Recurrent neural networks for short-term load forecasting. IEEE Trans Power Syst 13(1):126–132

Wang Y-H (2009) Nonlinear neural network forecasting model for stock index option price: hybrid GJR-GARCH approach. Expert Syst Appl 36(1):564–570

Wedding K II, Cios JK (1996) Time series forecasting by combining RBF networks, certainty factors, and the Box-Jenkins model. Neurocomputing 10:149–168

Weigend AS, Gershenfeld NA (1994) Time series prediction: forecasting the future and understanding the past. Addison-Wesley, Reading, MA

Weigend AS, Huberman BA, Rumelhart DE (1990) Predicting the future: a connectionist approach. Int J Neural Syst 1:193–209

Weigend AS, Huberman BA, Rumelhart DE (1992) Predicting sunspots and exchange rates with connectionist networks. In: Casdagli M, Eubank S (eds) Nonlinear modeling and forecasting. Addison-Wesley, Redwood City, CA, pp 395–432

Werbos P (1974) Beyond regression: new tools for prediction and analysis in the behavioral sciences. Ph.D. thesis, Harvard University

Wichard J, Ogorzalek M (2007) Time series prediction with ensemble models applied to the CATS benchmark. Neurocomputing 70:2371–2378

Wong BK, Selvi Y (1998) Neural network applications in finance: a review and analysis of literature (1990–1996). Inf Manag 34:129–139

Wong BK, Lai VS, Lam J (2000) A bibliography of neural network business applications research: 1994–1998. Comput Oper Res 27:1045–1076

Xiao Z, Ye SJ, Zhong B, Sun CX (2009) BP neural network with rough set for short term load forecasting. Expert Syst Appl 36(1):273–279

Zhang G, Patuwo EP, Hu MY (1998) Forecasting with artificial neural networks: the state of the art. Int J Forecasting 14:35–62

Zhang GP (2001) An investigation of neural networks for linear time-series forecasting. Comput Oper Res 28:1183–1202

Zhang GP (2003) Time series forecasting using a hybrid ARIMA and neural network model. Neurocomputing 50:159–175

Zhang GP (2004) Neural networks in business forecasting. Idea Group, Hershey, PA

Zhang GP (2007a) Avoiding pitfalls in neural network research. IEEE Trans Syst Man and Cybern 37:3–16

Zhang GP (2007b) A neural network ensemble method with jittered training data for time series forecasting. Inf Sci 177:5329–5346

Zhang GP, Berardi LV (2001) Time series forecasting with neural network ensembles: An application for exchange rate prediction. J Oper Res Soc 52(6):652–664

Zhang GP, Kline D (2007) Quarterly time-series forecasting with neural networks. IEEE Trans Neural Netw 18(6):1800–1814

Zhang GP, Qi M (2002) Predicting consumer retail sales using neural networks. In: Smith K, Gupta J (eds) Neural networks in business: techniques and applications. Idea Group, Hershey, PA, pp 26–40

Zhang GP, Qi M (2005) Neural network forecasting for seasonal and trend time series. Eur J Oper Res 160(2):501–514

Zhang GP, Patuwo EP, Hu MY (2001) A simulation study of artificial neural networks for nonlinear time series forecasting. Comput Oper Res 28:381–396

Zhang YQ, Wan X (2007) Statistical fuzzy interval neural networks for currency exchange rate time series prediction. Appl Soft Comput 7:1149–1156

Zou HF, Xia GP, Yang FT, Wang HY (2007) An investigation and comparison of artificial neural network and time series models for Chinese food grain price forecasting. Neurocomputing 70:2913–2923

15 SVM Tutorial — Classification, Regression and Ranking

Hwanjo Yu[1] · Sungchul Kim[2]
[1]Data Mining Lab, Department of Computer Science and Engineering, Pohang University of Science and Technology, Pohang, South Korea
hwanjoyu@postech.ac.kr
[2]Data Mining Lab, Department of Computer Science and Engineering, Pohang University of Science and Technology, Pohang, South Korea
subright@postech.ac.kr

G. Rozenberg et al. (eds.), *Handbook of Natural Computing*, DOI 10.1007/978-3-540-92910-9_15,
© Springer-Verlag Berlin Heidelberg 2012

Abstract

Support vector machines (SVMs) have been extensively researched in the data mining and machine learning communities for the last decade, and applied in various domains. They represent a set of supervised learning techniques that create a function from training data, which usually consists of pairs of an input object, typically vectors, and a desired output. SVMs learn a function that generates the desired output given the input, and the learned function can be used to predict the output of a new object. They belong to a family of generalized linear classifier where the classification (or boundary) function is a hyperplane in the feature space. This chapter introduces the basic concepts and techniques of SVMs for learning classification, regression, and ranking functions.

1 Introduction

Support vector machines are typically used for learning classification, regression, or ranking functions, for which they are called classifying SVM, support vector regression (SVR), or ranking SVM (RankSVM), respectively. Two special properties of SVMs are that they achieve high generalization by maximizing the margin, and they support an efficient learning of nonlinear functions using the kernel trick. This chapter introduces these general concepts and techniques of SVMs for learning classification, regression, and ranking functions. In particular, the SVMs for binary classification are first presented in ❷ Sect. 2, SVR in ❷ Sect. 3, ranking SVM in ❷ Sect. 4, and another recently developed method for learning ranking SVM called Ranking Vector Machine (RVM) in ❷ Sect. 5.

2 SVM Classification

SVMs were initially developed for classification (Burges 1998) and have been extended for regression (Smola and Schölkopf 1998) and preference (or rank) learning (Herbrich et al. 2000; Yu 2005). The initial form of SVMs is a binary classifier where the output of the learned function is either positive or negative. A multiclass classification can be implemented by combining multiple binary classifiers using the pairwise coupling method (Hastie and Tibshirani 1998; Friedman 1998). This section explains the motivation and formalization of SVM as a binary classifier, and the two key properties – margin maximization and the kernel trick.

Binary SVMs are classifiers that discriminate data points of two categories. Each data object (or data point) is represented by an n-dimensional vector. Each of these data points belongs to only one of two classes. A *linear* classifier separates them with a hyperplane. For example, ❷ *Fig. 1* shows two groups of data and separating hyperplanes that are lines in a two-dimensional space. There are many linear classifiers that correctly classify (or divide) the two groups of data such as L1, L2, and L3 in ❷ *Fig. 1*. In order to achieve maximum separation between the two classes, the SVM picks the hyperplane that has the largest margin. The margin is the summation of the shortest distance from the separating hyperplane to the nearest data point of both categories. Such a hyperplane is likely to generalize better, meaning that the hyperplane can correctly classify "unseen" or testing data points.

SVMs do the mapping from the input space to the feature space to support nonlinear classification problems. The kernel trick is helpful for doing this by allowing the absence of the

□ Fig. 1
Linear classifiers (hyperplane) in two-dimensional spaces.

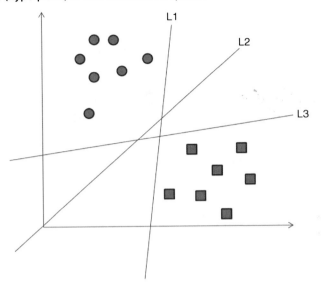

exact formulation of the mapping function which could introduce a case of the curse of dimensionality problem. This makes a linear classification in the new space (or the feature space) equivalent to nonlinear classification in the original space (or the input space). SVMs do this by mapping input vectors to a higher dimensional space (or feature space) where a maximal separating hyperplane is constructed.

2.1   Hard-Margin SVM Classification

To understand how SVMs compute the hyperplane of maximal margin and support nonlinear classification, we first explain the hard-margin SVM where the training data is free of noise and can be correctly classified by a linear function.

The data points D in ❷ *Fig. 1* (or training set) can be expressed mathematically as follows:

$$D = \{(\mathbf{x}_1, y_1), (\mathbf{x}_2, y_2), \ldots, (\mathbf{x}_m, y_m)\} \tag{1}$$

where $\mathbf{x}_i$ is an n-dimensional real vector, y_i is either 1 or -1 denoting the class to which the point $\mathbf{x}_i$ belongs. The SVM classification function $F(\mathbf{x})$ takes the form

$$F(\mathbf{x}) = \mathbf{w} \cdot \mathbf{x} - b \tag{2}$$

$\mathbf{w}$ is the weight vector and b is the bias, which will be computed by the SVM in the training process.

First, to correctly classify the training set, $F(\cdot)$ (or $\mathbf{w}$ and b) must return positive numbers for positive data points and negative numbers otherwise, that is, for every point $\mathbf{x}_i$ in D,

$$\mathbf{w} \cdot \mathbf{x}_i - b > 0 \quad \text{if } y_i = 1, \text{ and}$$
$$\mathbf{w} \cdot \mathbf{x}_i - b < 0 \quad \text{if } y_i = -1$$

Fig. 2

SVM classification function: the hyperplane maximizing the margin in a two-dimensional space.

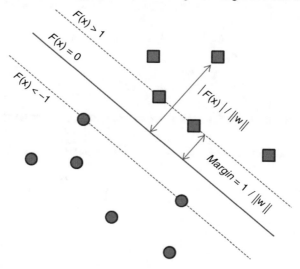

These conditions can be revised into

$$y_i(\mathbf{w} \cdot \mathbf{x}_i - b) > 0, \forall(\mathbf{x}_i, y_i) \in D \tag{3}$$

If there exists such a linear function F that correctly classifies every point in D or satisfies ❷ Eq. 3, D is called *linearly separable*.

Second, F (or the hyperplane) needs to maximize the *margin*. The margin is the distance from the hyperplane to the closest data points. An example of such a hyperplane is illustrated in ❷ *Fig. 2*. To achieve this, ❷ Eq. 3 is revised into the following:

$$y_i(\mathbf{w} \cdot \mathbf{x}_i - b) \geq 1, \forall(\mathbf{x}_i, y_i) \in D \tag{4}$$

Note that ❷ Eq. 4 includes the equality sign, and the right side becomes 1 instead of 0. If D is linearly separable, or every point in D satisfies ❷ Eq. 3, then there exists such an F that satisfies ❷ Eq. 4. This is because, if there exist such $\mathbf{w}$ and b that satisfy ❷ Eq. 3, they can always be rescaled to satisfy ❷ Eq. 4.

The distance from the hyperplane to a vector $\mathbf{x}_i$ is formulated as $\frac{|F(\mathbf{x}_i)|}{||\mathbf{w}||}$. Thus, the margin becomes

$$margin = \frac{1}{||\mathbf{w}||} \tag{5}$$

because when $\mathbf{x}_i$ are the closest vectors, $F(\mathbf{x})$ will return to 1 according to ❷ Eq. 4. The closest vectors that satisfy ❷ Eq. 4 with the equality sign are called *support vectors*.

Maximizing the margin becomes minimizing $||\mathbf{w}||$. Thus, the training problem in an SVM becomes a constrained optimization problem as follows:

$$\text{minimize:} \quad Q(\mathbf{w}) = \frac{1}{2}||\mathbf{w}||^2 \tag{6}$$

$$\text{subject to:} \ y_i(\mathbf{w} \cdot \mathbf{x}_i - b) \geq 1, \forall(\mathbf{x}_i, y_i) \in D \tag{7}$$

The factor of $\frac{1}{2}$ is used for mathematical convenience.

2.1.1 Solving the Constrained Optimization Problem

The constrained optimization problem ❷ Eqs. 6 and ❷ 7 is called a *primal problem*. It is characterized as follows:

- The objective function (❷ Eq. 6) is a *convex* function of **w**.
- The constraints are *linear* in **w**.

Accordingly, we may solve the constrained optimization problem using the method of Lagrange multipliers (Bertsekas 1995). First, we construct the Lagrange function

$$J(\mathbf{w}, b, \alpha) = \frac{1}{2}\mathbf{w} \cdot \mathbf{w} - \sum_{i=1}^{m} \alpha_i \{y_i(\mathbf{w} \cdot \mathbf{x}_i - b) - 1\} \tag{8}$$

where the auxiliary nonnegative variables α are called Lagrange multipliers. The solution to the constrained optimization problem is determined by the saddle point of the Lagrange function $J(\mathbf{w}, b, \alpha)$, which has to be minimized with respect to **w** and b; it also has to be maximized with respect to α. Thus, differentiating $J(\mathbf{w}, b, \alpha)$ with respect to **w** and b and setting the results equal to zero, we get the following two conditions of optimality:

$$\text{Condition 1:} \frac{\partial J(\mathbf{w}, b, \alpha)}{\partial \mathbf{w}} = 0 \tag{9}$$

$$\text{Condition 2:} \frac{\partial J(\mathbf{w}, b, \alpha)}{\partial b} = 0 \tag{10}$$

After rearrangement of terms, Condition 1 yields

$$\mathbf{w} = \sum_{i=1}^{m} \alpha_i y_i, \mathbf{x}_i \tag{11}$$

and Condition 2 yields

$$\sum_{i=1}^{m} \alpha_i y_i = 0 \tag{12}$$

The solution vector **w** is defined in terms of an expansion that involves the m training examples.

As noted earlier, the primal problem deals with a convex cost function and linear constraints. Given such a constrained optimization problem, it is possible to construct another problem called the *dual problem*. The dual problem has the same optimal value as the primal problem, but with the Lagrange multipliers providing the optimal solution.

To postulate the dual problem for the primal problem, ❷ Eq. 8 is first expanded term by term, as follows:

$$J(\mathbf{w}, b, \alpha) = \frac{1}{2}\mathbf{w} \cdot \mathbf{w} - \sum_{i=1}^{m} \alpha_i y_i \mathbf{w} \cdot \mathbf{x}_i - b \sum_{i=1}^{m} \alpha_i y_i + \sum_{i=1}^{m} \alpha_i \tag{13}$$

The third term on the right-hand side of ❷ Eq. 13 is zero by virtue of the optimality condition of ❷ Eq. 12. Furthermore, from ❷ Eq. 11, we have

$$\mathbf{w} \cdot \mathbf{w} = \sum_{i=1}^{m} \alpha_i y_i \mathbf{w} \cdot \mathbf{x} = \sum_{i=1}^{m} \sum_{j=1}^{m} \alpha_i \alpha_j y_i y_j \mathbf{x}_i \mathbf{x}_j \tag{14}$$

Accordingly, setting the objective function $J(\mathbf{w}, b, \alpha) = Q(\alpha)$, ❷ Eq. 13 can be formulated as

$$Q(\alpha) = \sum_{i=1}^{m} \alpha_i - \frac{1}{2} \sum_{i=1}^{m} \sum_{j=1}^{m} \alpha_i \alpha_j y_i y_j \mathbf{x}_i \cdot \mathbf{x}_j \tag{15}$$

where the α_i are nonnegative.

The dual problem can be now stated as follows:

$$\text{maximize: } Q(\alpha) = \sum_{i} \alpha_i - \frac{1}{2} \sum_{i} \sum_{j} \alpha_i \alpha_j y_i y_j \mathbf{x}_i \mathbf{x}_j \tag{16}$$

$$\text{subject to: } \sum_{i} \alpha_i y_i = 0 \tag{17}$$

$$\alpha \geq 0 \tag{18}$$

Note that the dual problem is cast entirely in terms of the training data. Moreover, the function $Q(\alpha)$ to be maximized depends only on the input patterns in the form of a set of dot products $\{\mathbf{x}_i \cdot \mathbf{x}_j\}_{(i,j)=1}^{m}$.

Having determined the optimum Lagrange multipliers, denoted by α_i^*, the optimum weight vector $\mathbf{w}^*$ may be computed using ❷ Eq. 11 and so can be written as

$$\mathbf{w}^* = \sum_{i} \alpha_i^* y_i \mathbf{x}_i \tag{19}$$

Note that according to the property of Kuhn–Tucker conditions of optimization theory, the solution of the dual problem α_i^* must satisfy the following condition:

$$\alpha_i^* \{y_i(\mathbf{w}^* \cdot \mathbf{x}_i - b) - 1\} = 0 \quad \text{for } i = 1, 2, \ldots, m \tag{20}$$

and either α_i^* or its corresponding constraint $\{y_i(\mathbf{w}^* \cdot \mathbf{x}_i - b) - 1\}$ must be nonzero. This condition implies that only when $\mathbf{x}_i$ is a support vector or $y_i(\mathbf{w}^* \cdot \mathbf{x}_i - b) = 1$, its corresponding coefficient α_i will be nonzero (or nonnegative from ❷ Eq. 18). In other words, the $\mathbf{x}_i$ whose corresponding coefficients α_i are zero will not affect the optimum weight vector $\mathbf{w}^*$ due to ❷ Eq. 19. Thus, the optimum weight vector $\mathbf{w}^*$ will only depend on the support vectors whose coefficients are nonnegative.

Once the nonnegative α_i^* and their corresponding support vectors are computed, we can compute the bias b using a positive support vector $\mathbf{x}_i$ from the following equation:

$$b^* = 1 - \mathbf{w}^* \cdot \mathbf{x}_i \tag{21}$$

The classification of ❷ Eq. 2 now becomes

$$F(\mathbf{x}) = \sum_{i} \alpha_i y_i \mathbf{x}_i \cdot \mathbf{x} - b \tag{22}$$

2.2 Soft-Margin SVM Classification

The discussion so far has focused on linearly separable cases. However, the optimization problem ❷ Eqs. 6 and ❷ 7 will not have a solution if D is not linearly separable. To deal with such cases, a *soft margin* SVM allows mislabeled data points while still maximizing the margin. The method introduces slack variables, ξ_i, which measure the degree of misclassification. The following is the optimization problem for a soft margin SVM.

$$\text{minimize:} \quad Q_1(w, b, \xi_i) = \frac{1}{2}||\mathbf{w}||^2 + C\sum_i \xi_i \tag{23}$$

$$\text{subject to:} \quad y_i(\mathbf{w} \cdot \mathbf{x}_i - b) \geq 1 - \xi_i, \qquad \forall (\mathbf{x}_i, y_i) \in D \tag{24}$$

$$\xi_i \geq 0 \tag{25}$$

Due to the ξ_i in ❷ Eq. 24, data points are allowed to be misclassified, and the amount of misclassification will be minimized while maximizing the margin according to the objective function (❷ Eq. 23). C is a parameter that determines the trade-off between the margin size and the amount of error in training.

Similarly to the case of hard-margin SVM, this primal form can be transformed to the following dual form using the Lagrange multipliers:

$$\text{maximize:} \quad Q_2(\alpha) = \sum_i \alpha_i - \sum_i \sum_j \alpha_i \alpha_j y_i y_j \mathbf{x}_i \mathbf{x}_j \tag{26}$$

$$\text{subject to:} \quad \sum_i \alpha_i y_i = 0 \tag{27}$$

$$C \geq \alpha \geq 0 \tag{28}$$

Note that neither the slack variables ξ_i nor their Lagrange multipliers appear in the dual problem. The dual problem for the case of nonseparable patterns is thus similar to that for the simple case of linearly separable patterns except for a minor but important difference. The objective function $Q(\alpha)$ to be maximized is the same in both cases. The nonseparable case differs from the separable case in that the constraint $\alpha_i \geq 0$ is replaced with the more stringent constraint $C \geq \alpha_i \geq 0$. Except for this modification, the constrained optimization for the nonseparable case and computations of the optimum values of the weight vector $\mathbf{w}$ and bias b proceed in the same way as in the linearly separable case.

Just as for the hard-margin SVM, α constitutes a dual representation for the weight vector such that

$$\mathbf{w}^* = \sum_{i=1}^{m_s} \alpha_i^* y_i \mathbf{x}_i \tag{29}$$

where m_s is the number of support vectors whose corresponding coefficient $\alpha_i > 0$. The determination of the optimum values of the bias also follows a procedure similar to that described before. Once α and b are computed, the function ❷ Eq. 22 is used to classify new objects.

Relationships among α, ξ, and C can be further disclosed using the Kuhn–Tucker conditions that are defined by

$$\alpha_i \{ y_i(\mathbf{w} \cdot \mathbf{x}_i - b) - 1 + \xi_i \} = 0, \quad i = 1, 2, \ldots, m \tag{30}$$

and

$$\mu_i \xi_i = 0, \quad i = 1, 2, \ldots, m \tag{31}$$

❷ Eq. 30 is a rewrite of ❷ Eq. 20 except for the replacement of the unity term $(1 - \xi_i)$. As for ❷ Eq. 31, the μ_i are Lagrange multipliers that have been introduced to enforce the nonnegativity of the slack variables ξ_i for all i. At the saddle point, the derivative of the

◘ Fig. 3

Graphical relationships among α_i, ξ_i, and C.

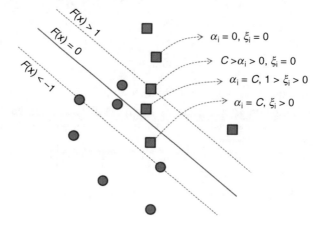

Lagrange function for the primal problem with respect to the slack variable ξ_i is zero, the evaluation of which yields

$$\alpha_i + \mu_i = C \tag{32}$$

By combining ❯ Eqs. 31 and ❯ 32, we see that

$$\xi_i = 0 \quad \text{if } \alpha_i < C, \text{ and} \tag{33}$$

$$\xi_i \geq 0 \quad \text{if } \alpha_i = C \tag{34}$$

We can graphically display the relationships among α_i, ξ_i, and C in ❯ Fig. 3.

Data points outside the margin will have $\alpha = 0$ and $\xi = 0$ and those on the margin line will have $C > \alpha > 0$ and still $\xi = 0$. Data points within the margin will have $\alpha = C$. Among them, those correctly classified will have $1 > \xi > 0$ and misclassified points will have $\xi > 1$.

2.3   Kernel Trick for Nonlinear Classification

If the training data is not linearly separable, there is no straight hyperplane that can separate the classes. In order to learn a nonlinear function in that case, linear SVMs must be extended to nonlinear SVMs for the classification of nonlinearly separable data. The process of finding classification functions using nonlinear SVMs consists of two steps. First, the input vectors are transformed into high-dimensional feature vectors where the training data can be linearly separated. Then, SVMs are used to find the hyperplane of maximal margin in the new feature space. The separating hyperplane becomes a linear function in the transformed feature space but a nonlinear function in the original input space.

Let **x** be a vector in the n-dimensional input space and $\varphi(\cdot)$ be a nonlinear mapping function from the input space to the high-dimensional feature space. The hyperplane representing the decision boundary in the feature space is defined as follows:

$$\mathbf{w} \cdot \varphi(\mathbf{x}) - b = 0 \tag{35}$$

where $\mathbf{w}$ denotes a weight vector that can map the training data in the high-dimensional feature space to the output space, and b is the bias. Using the $\varphi(\cdot)$ function, the weight becomes

$$\mathbf{w} = \sum \alpha_i y_i \varphi(\mathbf{x}_i) \tag{36}$$

The decision function of ❸ Eq. 22 becomes

$$F(\mathbf{x}) = \sum_i^m \alpha_i y_i \varphi(\mathbf{x}_i) \cdot \varphi(\mathbf{x}) - b \tag{37}$$

Furthermore, the dual problem of the soft-margin SVM (❸ Eq. 26) can be rewritten using the mapping function on the data vectors as follows:

$$Q(\alpha) = \sum_i \alpha_i - \frac{1}{2} \sum_i \sum_j \alpha_i \alpha_j y_i y_j \varphi(\mathbf{x}_i) \cdot \varphi(\mathbf{x}_j) \tag{38}$$

holding the same constraints.

Note that the feature mapping functions in the optimization problem and also in the classifying function always appear as dot products, for example, $\varphi(\mathbf{x}_i) \cdot \varphi(\mathbf{x}_j)$. $\varphi(\mathbf{x}_i) \cdot \varphi(\mathbf{x}_j)$ is the inner product between pairs of vectors in the transformed feature space. Computing the inner product in the transformed feature space seems to be quite complex and suffers from the curse of dimensionality problem. To avoid this problem, the kernel trick is used. The kernel trick replaces the inner product in the feature space with a kernel function K in the original input space as follows.

$$K(\mathbf{u}, \mathbf{v}) = \varphi(\mathbf{u}) \cdot \varphi(\mathbf{v}) \tag{39}$$

Mercer's theorem proves that a kernel function K is valid if and only if the following conditions are satisfied, for any function $\psi(\mathbf{x})$ (refer to Christianini and Shawe-Taylor (2000) for the proof in detail):

$$\int K(\mathbf{u}, \mathbf{v}) \psi(\mathbf{u}) \psi(\mathbf{v}) dx dy \le 0$$
$$\text{where } \int \psi(x)^2 dx \le 0 \tag{40}$$

Mercer's theorem ensures that the kernel function can be always expressed as the inner product between pairs of input vectors in some high-dimensional space, thus the inner product can be calculated using the kernel function only with input vectors in the original space without transforming the input vectors into the high-dimensional feature vectors.

The dual problem is now defined using the kernel function as follows:

$$\text{maximize: } Q_2(\alpha) = \sum_i \alpha_i - \sum_i \sum_j \alpha_i \alpha_j y_i y_j K(\mathbf{x}_i, \mathbf{x}_j) \tag{41}$$

$$\text{subject to: } \sum_i \alpha_i y_i = 0 \tag{42}$$

$$C \ge \alpha \ge 0 \tag{43}$$

The classification function becomes

$$F(\mathbf{x}) = \sum_i \alpha_i y_i K(\mathbf{x}_i, \mathbf{x}) - b \tag{44}$$

Since $K(\cdot)$ is computed in the input space, no feature transformation will actually be done or no $\varphi(\cdot)$ will be computed, and thus the weight vector $\mathbf{w} = \sum \alpha_i y_i \varphi(\mathbf{x})$ will not be computed either in nonlinear SVMs.

The following are popularly used kernel functions:

- Polynomial: $K(a, b) = (a \cdot b + 1)^d$
- Radial Basis Function (RBF): $K(a, b) = \exp(-\gamma \|a - b\|^2)$
- Sigmoid: $K(a, b) = \tanh(\kappa a \cdot b + c)$

Note that the kernel function is a kind of similarity function between two vectors where the function output is maximized when the two vectors become equivalent. Because of this, SVM can learn a function from any shapes of data beyond vectors (such as trees or graphs) as long as we can compute a similarity function between any pair of data objects. Further discussions on the properties of these kernel functions are out of the scope of this chapter. Instead, we will give an example of using the polynomial kernel for learning an XOR function in the following section.

2.3.1 Example: XOR Problem

To illustrate the procedure of training a nonlinear SVM function, assume that the training set of ❯ *Table 1* is given.

❯ *Figure 4* plots the training points in the 2D input space. There is no linear function that can separate the training points.

To proceed let

$$K(\mathbf{x}, \mathbf{x}_i) = (1 + \mathbf{x} \cdot \mathbf{x}_i)^2 \tag{45}$$

If we denote $\mathbf{x} = (x_1, x_2)$ and $\mathbf{x}_i = (x_{i1}, x_{i2})$, the kernel function is expressed in terms of monomials of various orders as follows.

$$K(\mathbf{x}, \mathbf{x}_i) = 1 + x_1^2 x_{i1}^2 + 2x_1 x_2 x_{i1} x_{i2} + x_2^2 x_{i2}^2 + 2x_1 x_{i1} + 2x_2 x_{i2} \tag{46}$$

The image of the input vector $\mathbf{x}$ induced in the feature space is therefore deduced to be

$$\varphi(\mathbf{x}) = (1, x_1^2, \sqrt{2}x_1 x_2, x_2^2, \sqrt{2}x_1, \sqrt{2}x_2) \tag{47}$$

◻ Table 1

XOR problem

Input vector x	Desired output y
$(-1, -1)$	-1
$(-1, +1)$	$+1$
$(+1, -1)$	$+1$
$(+1, +1)$	-1

Fig. 4
XOR problem.

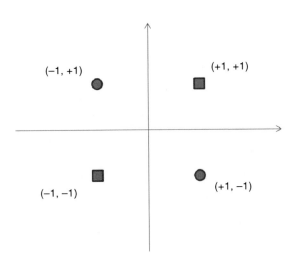

Based on this mapping function, the objective function for the dual form can be derived from ❷ Eq. 41 as follows.

$$Q(\alpha) = \alpha_1 + \alpha_2 + \alpha_3 + \alpha_4$$
$$- \frac{1}{2}(9\alpha_1^2 - 2\alpha_1\alpha_2 - 2\alpha_1\alpha_3 + 2\alpha_1\alpha_4 \tag{48}$$
$$+ 9\alpha_2^2 + 2\alpha_2\alpha_3 - 2\alpha_2\alpha_4 + 9\alpha_3^2 - 2\alpha_3\alpha_4 + \alpha_4^2)$$

Optimizing $Q(\alpha)$ with respect to the Lagrange multipliers yields the following set of simultaneous equations:

$$9\alpha_1 - \alpha_2 - \alpha_3 + \alpha_4 = 1$$
$$-\alpha_1 + 9\alpha_2 + \alpha_3 - \alpha_4 = 1$$
$$-\alpha_1 + \alpha_2 + 9\alpha_3 - \alpha_4 = 1$$
$$\alpha_1 - \alpha_2 - \alpha_3 + 9\alpha_4 = 1$$

Hence, the optimal values of the Lagrange multipliers are

$$\alpha_1 = \alpha_2 = \alpha_3 = \alpha_4 = \frac{1}{8}$$

This result denotes that all four input vectors are support vectors and the optimum value of $Q(\alpha)$ is

$$Q(\alpha) = \frac{1}{4}$$

and

$$\frac{1}{2}\|w\|^2 = \frac{1}{4}, \quad \text{or} \quad \|w\| = \frac{1}{\sqrt{2}}$$

From ❷ Eq. 36, we find that the optimum weight vector is

$$w = \frac{1}{8}[-\varphi(\mathbf{x}_1) + \varphi(\mathbf{x}_2) + \varphi(\mathbf{x}_3) - \varphi(\mathbf{x}_4)]$$

$$= \frac{1}{8}\left[-\begin{bmatrix} 1 \\ 1 \\ \sqrt{2} \\ 1 \\ -\sqrt{2} \\ -\sqrt{2} \end{bmatrix} + \begin{bmatrix} 1 \\ 1 \\ -\sqrt{2} \\ 1 \\ -\sqrt{2} \\ \sqrt{2} \end{bmatrix} + \begin{bmatrix} 1 \\ 1 \\ -\sqrt{2} \\ 1 \\ \sqrt{2} \\ -\sqrt{2} \end{bmatrix} - \begin{bmatrix} 1 \\ 1 \\ \sqrt{2} \\ 1 \\ \sqrt{2} \\ \sqrt{2} \end{bmatrix}\right] = \begin{bmatrix} 0 \\ 0 \\ -\frac{1}{\sqrt{2}} \\ 0 \\ 0 \\ 0 \end{bmatrix} \qquad (49)$$

The bias b is 0 because the first element of w is 0. The optimal hyperplane becomes

$$w \cdot \varphi(\mathbf{x}) = \begin{bmatrix} 0 & 0 & \frac{-1}{\sqrt{2}} & 0 & 0 & 0 \end{bmatrix} \begin{bmatrix} 1 \\ x_1^2 \\ \sqrt{2}x_1x_2 \\ x_2^2 \\ \sqrt{2}x_1 \\ \sqrt{2}_2 \end{bmatrix} = 0 \qquad (50)$$

which reduces to

$$-x_1x_2 = 0 \qquad (51)$$

this is the optimal hyperplane, the solution of the XOR problem. It makes the output $y = 1$ for both input points $x_1 = x_2 = 1$ and $x_1 = x_2 = -1$, and $y = -1$ for both input points $x_1 = 1, x_2 = -1$ or $x_1 = -1, x_2 = 1$. ❷ *Figure 5* represents the four points in the transformed feature space.

❏ **Fig. 5**

The four data points of the XOR problem in the transformed feature space.

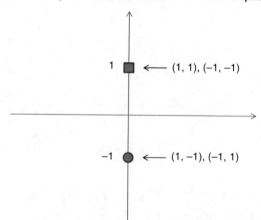

3   SVM Regression

SVM regression (SVR) is a method to estimate a function that maps from an input object to a real number based on training data. Similarly to the classifying SVM, SVR has the same properties of the margin maximization and kernel trick for nonlinear mapping.

A training set for regression is represented as follows.

$$D = \{(\mathbf{x}_1, y_1), (\mathbf{x}_2, y_2), \ldots, (\mathbf{x}_m, y_m)\} \tag{52}$$

where $\mathbf{x}_i$ is a n-dimensional vector, y is the real number for each $\mathbf{x}_i$. The SVR function $F(\mathbf{x}_i)$ makes a mapping from an input vector $\mathbf{x}_i$ to the target y_i and takes the form:

$$F(\mathbf{x}) \Longrightarrow \mathbf{w} \cdot \mathbf{x} - b \tag{53}$$

where $\mathbf{w}$ is the weight vector and b is the bias. The goal is to estimate the parameters ($\mathbf{w}$ and b) of the function that give the best fit of the data. An SVR function $F(\mathbf{x})$ approximates all pairs $(\mathbf{x}_i, y_i)$ while maintaining the differences between estimated values and real values under ε precision. That is, for every input vector $\mathbf{x}$ in D,

$$y_i - \mathbf{w} \cdot \mathbf{x}_i - b \leq \varepsilon \tag{54}$$

$$\mathbf{w} \cdot \mathbf{x}_i + b - y_i \leq \varepsilon \tag{55}$$

The margin is

$$margin = \frac{1}{||\mathbf{w}||} \tag{56}$$

By minimizing $||\mathbf{w}||^2$ to maximize the margin, the training in SVR becomes a constrained optimization problem, as follows:

$$\text{minimize: } L(\mathbf{w}) = \frac{1}{2}||\mathbf{w}||^2 \tag{57}$$

$$\text{subject to: } y_i - \mathbf{w} \cdot \mathbf{x_i} - b \leq \varepsilon \tag{58}$$

$$\mathbf{w} \cdot \mathbf{x_i} + b - y_i \leq \varepsilon \tag{59}$$

The solution of this problem does not allow any errors. To allow some errors to deal with noise in the training data, the soft margin SVR uses slack variables ξ and $\hat{\xi}$. Then the optimization problem can be revised as follows:

$$\text{minimize: } L(\mathbf{w}, \xi) = \frac{1}{2}||\mathbf{w}||^2 + C\sum_i (\xi_i^2, \hat{\xi}_i^2), C > 0 \tag{60}$$

$$\text{subject to: } y_i - \mathbf{w} \cdot \mathbf{x_i} - b \leq \varepsilon + \xi_i, \quad \forall (\mathbf{x}_i, y_i) \in D \tag{61}$$

$$\mathbf{w} \cdot \mathbf{x_i} + b - y_i \leq \varepsilon + \hat{\xi}_i, \quad \forall (\mathbf{x}_i, y_i) \in D \tag{62}$$

$$\xi, \hat{\xi}_i \geq 0 \tag{63}$$

The constant $C > 0$ is the trade-off parameter between the margin size and the number of errors. The slack variables ξ and $\hat{\xi}$ deal with infeasible constraints of the optimization problem by imposing a penalty on excess deviations that are larger than ε.

To solve the optimization problem ❷ Eq. 60, we can construct a Lagrange function from the objective function with Lagrange multipliers as follows:

$$\text{minimize: } L = \frac{1}{2}\|\mathbf{w}\|^2 + C\sum_i (\xi_i + \hat{\xi}_i) - \sum_i (\eta_i \xi_i + \hat{\eta}_i \hat{\xi}_i)$$

$$- \sum_i \alpha_i (\varepsilon + \eta_i - y_i + \mathbf{w} \cdot \mathbf{x}_i + b)$$

$$- \sum_i \hat{\alpha}_i (\varepsilon + \hat{\eta}_i + y_i - \mathbf{w} \cdot \mathbf{x}_i - b)$$

(64)

$$\text{subject to: } \eta, \hat{\eta}_i \geq 0$$

(65)

$$\alpha, \hat{\alpha}_i \geq 0$$

(66)

where $\eta_i, \hat{\eta}_i, \alpha,$ and $\hat{\alpha}_i$ are the Lagrange multipliers that satisfy positive constraints. The following is the process to find the saddle point by using the partial derivatives of L with respect to each Lagrangian multiplier for minimizing the function L:

$$\frac{\partial L}{\partial b} = \sum_i (\alpha_i - \hat{\alpha}_i) = 0$$

(67)

$$\frac{\partial L}{\partial \mathbf{w}} = \mathbf{w} - \Sigma(\alpha_i - \hat{\alpha}_i)\mathbf{x}_i = 0, \mathbf{w} = \sum_i (\alpha_i - \hat{\alpha}_i)\mathbf{x}_i$$

(68)

$$\frac{\partial L}{\partial \hat{\xi}_i} = C - \hat{\alpha}_i - \hat{\eta}_i = 0, \hat{\eta}_i = C - \hat{\alpha}_i$$

(69)

The optimization problem with inequality constraints can be changed to the following dual optimization problem by substituting ❷ Eqs. 67, ❷ 68, and ❷ 69 into ❷ 64.

$$\text{maximize: } L(\alpha) = \sum_i y_i(\alpha_i - \hat{\alpha}_i) - \varepsilon \sum_i (\alpha_i + \hat{\alpha}_i)$$

(70)

$$- \frac{1}{2}\sum_i \sum_j (\alpha_i - \hat{\alpha}_i)(\alpha_i - \hat{\alpha}_i)\mathbf{x}_i \mathbf{x}_j$$

(71)

$$\text{subject to: } \sum_i (\alpha_i - \hat{\alpha}_i) = 0$$

(72)

$$0 \leq \alpha, \hat{\alpha} \leq C$$

(73)

The dual variables $\eta, \hat{\eta}_i$ are eliminated in revising ❷ Eq. 64 into ❷ Eq. 70. ❷ Eqs. 68 and ❷ 69 can be rewritten as follows:

$$w = \sum_i (\alpha_i - \hat{\alpha}_i)\mathbf{x}_i$$

(74)

$$\eta_i = C - \alpha_i$$

(75)

$$\hat{\eta}_i = C - \hat{\alpha}_i$$

(76)

where w is represented by a linear combination of the training vectors x_i. Accordingly, the SVR function $F(x)$ becomes the following function:

$$F(x) \Longrightarrow \sum_i (\alpha_i - \hat{\alpha}_i) x_i x + b \qquad (77)$$

❯ Eq. 77 can map the training vectors to target real values which allow some errors but it cannot handle the nonlinear SVR case. The same kernel trick can be applied by replacing the inner product of two vectors x_i, x_j with a kernel function $K(x_i, x_j)$. The transformed feature space is usually high dimensional, and the SVR function in this space becomes nonlinear in the original input space. Using the kernel function K, the inner product in the transformed feature space can be computed as fast as the inner product $x_i \cdot x_j$ in the original input space. The same kernel functions introduced in ❯ Sect. 2.3 can be applied here.

Once we replace the original inner product with a kernel function K, the remaining process for solving the optimization problem is very similar to that for the linear SVR. The linear optimization function can be changed by using the kernel function as follows:

$$\text{maximize:} \quad L(\alpha) = \sum_i y_i(\alpha_i - \hat{\alpha}_i) - \varepsilon \sum_i (\alpha_i + \hat{\alpha}_i)$$
$$-\frac{1}{2} \sum_i \sum_j (\alpha_i - \hat{\alpha}_i)(\alpha_i - \hat{\alpha}_i) K(x_i, x_j) \qquad (78)$$

$$\text{subject to:} \quad \sum_i (\alpha_i - \hat{\alpha}_i) = 0 \qquad (79)$$

$$\hat{\alpha}_i \geq 0, \alpha_i \geq 0 \qquad (80)$$

$$0 \leq \alpha, \hat{\alpha} \leq C \qquad (81)$$

Finally, the SVR function $F(x)$ becomes the following using the kernel function:

$$F(x) \Longrightarrow \sum_i (\hat{\alpha}_i - \alpha_i) K(x_i, x) + b \qquad (82)$$

4 SVM Ranking

Ranking SVM, learning a ranking (or preference) function, has resulted in various applications in information retrieval (Herbrich et al. 2000; Joachims 2002; Yu et al. 2007). The task of learning *ranking* functions is distinguished from that of learning *classification* functions as follows:

1. While a training set in a classification is a set of data objects and their class labels, in *ranking* a training set is an ordering of data. Let "A is preferred to B" be specified as "$A \succ B$." A training set for ranking SVM is denoted as $R = \{(x_1, y_i), \ldots, (x_m, y_m)\}$ where y_i is the ranking of x_i, that is, $y_i < y_j$ if $x_i \succ x_j$.
2. Unlike a classification function, which outputs a distinct class for a data object, a ranking function outputs a *score* for each data object, from which a *global ordering* of data is constructed. That is, the target function $F(x_i)$ outputs a score such that $F(x_i) > F(x_j)$ for any $x_i \succ x_j$.

If not stated, R is assumed to be strict ordering, which means that for all pairs $\mathbf{x}_i$ and $\mathbf{x}_j$ in a set D, either $\mathbf{x}_i \succ_R \mathbf{x}_j$ or $\mathbf{x}_i \prec_R \mathbf{x}_j$. However, it can be straightforwardly generalized to weak orderings. Let R^* be the optimal ranking of data in which the data is ordered perfectly according to the user's preference. A ranking function F is typically evaluated by how closely its ordering R^F approximates R^*.

Using the techniques of SVM, a global ranking function F can be learned from an ordering R. For now, assume F is a *linear* ranking function such that

$$\forall\{(\mathbf{x}_i, \mathbf{x}_j) : y_i < y_j \in R\} : F(\mathbf{x}_i) > F(\mathbf{x}_j) \Longleftrightarrow \mathbf{w} \cdot \mathbf{x}_i > \mathbf{w} \cdot \mathbf{x}_j \tag{83}$$

A weight vector $\mathbf{w}$ is adjusted by a learning algorithm. We say that an ordering R is *linearly rankable* if there exists a function F (represented by a weight vector $\mathbf{w}$) that satisfies ❷ Eq. 83 for all $\{(\mathbf{x}_i, \mathbf{x}_j) : y_i < y_j \in R\}$.

The goal is to learn F that is concordant with the ordering R and also generalize well beyond R. That is to find the weight vector $\mathbf{w}$ such that $\mathbf{w} \cdot \mathbf{x}_i > \mathbf{w} \cdot \mathbf{x}_j$ for most data pairs $\{(\mathbf{x}_i, \mathbf{x}_j) : y_i < y_j \in R\}$.

Though this problem is known to be NP-hard (Cohen et al. 1998), the solution can be approximated using SVM techniques by introducing (nonnegative) slack variables ξ_{ij} and minimizing the upper bound $\sum \xi_{ij}$ as follows (Herbrich et al. 2000):

$$\text{minimize: } L_1(\mathbf{w}, \xi_{ij}) = \frac{1}{2}\mathbf{w} \cdot \mathbf{w} + C\sum \xi_{ij} \tag{84}$$

$$\text{subject to: } \forall\{(\mathbf{x}_i, \mathbf{x}_j) : y_i < y_j \in R\} : \mathbf{w} \cdot \mathbf{x}_i \geq \mathbf{w} \cdot \mathbf{x}_j + 1 - \xi_{ij} \tag{85}$$

$$\forall(i,j) : \xi_{ij} \geq 0 \tag{86}$$

By the constraint (❷ Eq. 85) and by minimizing the upper bound $\sum \xi_{ij}$ in (❷ Eq. 84), the above optimization problem satisfies orderings on the training set R with minimal error. By minimizing $\mathbf{w} \cdot \mathbf{w}$ or by maximizing the margin ($= \frac{1}{\|\mathbf{w}\|}$), it tries to maximize the generalization of the ranking function. We will explain how maximizing the margin corresponds to increasing the generalization of *ranking* in ❷ Sect. 4.1. C is the soft margin parameter that controls the trade-off between the margin size and the training error.

By rearranging the constraint (❷ Eq. 85) we get

$$\mathbf{w}(\mathbf{x}_i - \mathbf{x}_j) \geq 1 - \xi_{ij} \tag{87}$$

The optimization problem becomes equivalent to that of *classifying SVM* on *pairwise difference vectors* $(\mathbf{x}_i - \mathbf{x}_j)$. Thus, we can extend an existing SVM implementation to solve the problem.

Note that the support vectors are the data pairs $(\mathbf{x}_i^s, \mathbf{x}_j^s)$ such that constraint (❷ Eq. 87) is satisfied with the *equality* sign, that is, $\mathbf{w}(\mathbf{x}_i^s - \mathbf{x}_j^s) = 1 - \xi_{ij}$. Unbounded support vectors are the ones on the margin (i.e., their slack variables $\xi_{ij} = 0$), and bounded support vectors are the ones within the margin (i.e., $1 > \xi_{ij} > 0$) or misranked (i.e., $\xi_{ij} > 1$). As done in the classifying SVM, a function F in ranking SVM is also expressed only by the support vectors.

Similarly to the classifying SVM, the primal problem of ranking SVM can be transformed to the following dual problem using the Lagrange multipliers:

$$\text{maximize: } L_2(\alpha) = \sum_{ij} \alpha_{ij} - \sum_{ij}\sum_{uv} \alpha_{ij}\alpha_{uv}K(\mathbf{x}_i - \mathbf{x}_j, \mathbf{x}_u - \mathbf{x}_v) \tag{88}$$

$$\text{subject to: } C \geq \alpha \geq 0 \tag{89}$$

Once transformed to the dual, the kernel trick can be applied to support the nonlinear ranking function. $K(\cdot)$ is a kernel function. α_{ij} is a coefficient for pairwise difference vectors $(\mathbf{x}_i - \mathbf{x}_j)$. Note that the kernel function is computed for $P^2(\sim m^4)$ times where P is the number of data pairs and m is the number of data points in the training set, thus solving the ranking SVM takes $O(m^4)$ at least. Fast training algorithms for ranking SVM have been proposed (Joachims 2006) but they are limited to linear kernels.

Once α is computed, $\mathbf{w}$ can be written in terms of the pairwise difference vectors and their coefficients such that

$$\mathbf{w} = \sum_{ij} \alpha_{ij}(\mathbf{x}_i - \mathbf{x}_j) \tag{90}$$

The ranking function F on a new vector $\mathbf{z}$ can be computed using the kernel function replacing the dot product as follows:

$$F(\mathbf{z}) = \mathbf{w} \cdot \mathbf{z} = \sum_{ij} \alpha_{ij}(\mathbf{x}_i - \mathbf{x}_j) \cdot \mathbf{z} = \sum_{ij} \alpha_{ij} K(\mathbf{x}_i - \mathbf{x}_j, \mathbf{z}). \tag{91}$$

4.1 Margin-Maximization in Ranking SVM

We now explain the margin-maximization of the ranking SVM, to reason about how the ranking SVM generates a ranking function of *high generalization*. Some essential properties of ranking SVM are first established. For convenience of explanation, it is assumed that a training set R is linearly rankable and thus we use hard-margin SVM, that is, $\xi_{ij} = 0$ for all (i, j) in the objective (❯ Eq. 84) and the constraints (❯ Eq. 85).

In the ranking formulation, from ❯ Eq. 83, the linear ranking function $F_\mathbf{w}$ projects data vectors onto a weight vector $\mathbf{w}$. For instance, ❯ *Fig. 6* illustrates linear projections of four vectors $\{\mathbf{x}_1, \mathbf{x}_2, \mathbf{x}_3, \mathbf{x}_4\}$ onto two different weight vectors $\mathbf{w}_1$ and $\mathbf{w}_2$, respectively, in a two-dimensional space. Both $F_{\mathbf{x}_1}$ and $F_{\mathbf{x}_2}$ make the same ordering R for the four vectors, that is, $\mathbf{x}_1 >_R \mathbf{x}_2 >_R \mathbf{x}_3 >_R \mathbf{x}_4$. The ranking difference of two vectors $(\mathbf{x}_i, \mathbf{x}_j)$ according to a ranking

◘ **Fig. 6**
Linear projection of four data points.

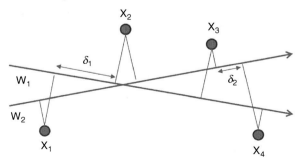

function $F_{\mathbf{w}}$ is denoted by the geometric distance of the two vectors projected onto $\mathbf{w}$, that is, formulated as $\frac{\mathbf{w}(\mathbf{x}_i - \mathbf{x}_j)}{||\mathbf{w}||}$.

Corollary 1 *Suppose $F_{\mathbf{w}}$ is a ranking function computed by the hard-margin ranking SVM on an ordering R. Then, the support vectors of $F_{\mathbf{w}}$ represent the data pairs that are closest to each other when projected to $\mathbf{w}$, thus closest in ranking.*

Proof The support vectors are the data pairs $(\mathbf{x}_i^s, \mathbf{x}_j^s)$ such that $\mathbf{w}(\mathbf{x}_i^s - \mathbf{x}_j^s) = 1$ in constraint (❷ Eq. 87), which is the smallest possible value for all data pairs $\forall (\mathbf{x}_i, \mathbf{x}_j) \in R$. Thus, its ranking difference according to $F_{\mathbf{w}}\left(= \frac{\mathbf{w}(\mathbf{x}_i^s - \mathbf{x}_j^s)}{||\mathbf{w}||}\right)$ is also the smallest among them (Vapnik 1998).

Corollary 2 *The ranking function F, generated by the hard-margin ranking SVM, maximizes the minimal difference of any data pairs in ranking.*

Proof By minimizing $\mathbf{w} \cdot \mathbf{w}$, the ranking SVM maximizes the margin $\delta = \frac{1}{||\mathbf{w}||} = \frac{\mathbf{w}(\mathbf{x}_i^s - \mathbf{x}_j^s)}{||\mathbf{w}||}$ where $(\mathbf{x}_i^s, \mathbf{x}_j^s)$ are the support vectors that denote, from the proof of Corollary 1, the minimal difference of any data pairs in the ranking.

The soft margin SVM allows bounded support vectors whose $\xi_{ij} > 0$ as well as unbounded support vectors whose $\xi_{ij} = 0$, in order to deal with noise and allow small errors for the R which is not completely linearly rankable. However, the objective function in (❷ Eq. 84) also minimizes the number of the slacks and thus the amount of error, and the support vectors are the close data pairs in the ranking. Thus, maximizing the margin generates the effect of maximizing the differences of close data pairs in the ranking.

From Corollaries 1 and 2, we observe that ranking SVM improves the generalization performance by maximizing the minimal ranking difference. For example, consider the two linear ranking functions $F_{\mathbf{w}_1}$ and $F_{\mathbf{w}_2}$ in ❷ *Fig. 6*. Although the two weight vectors $\mathbf{w}_1$ and $\mathbf{w}_2$ make the same ordering, intuitively $\mathbf{w}_1$ generalizes better than $\mathbf{w}_2$ because the distance between the closest vectors on $\mathbf{w}_1$ (i.e., δ_1) is larger than that on $\mathbf{w}_2$ (i.e., δ_2). The SVM computes the weight vector $\mathbf{w}$ that maximizes the differences of close data pairs in the ranking. Ranking SVMs find a ranking function of high generalization in this way.

5 Ranking Vector Machine: An Efficient Method for Learning the 1-Norm Ranking SVM

This section presents another rank learning method, *RVM*, a revised 1-norm ranking SVM that is better for feature selection and more scalable to large datasets than the standard ranking SVM.

We first develop a 1-norm ranking SVM, a ranking SVM that is based on the *1-norm* objective function. (The standard ranking SVM is based on the 2-norm objective function.) The 1-norm ranking SVM learns a function with much fewer support vectors than the standard SVM. Thereby, its testing time is much faster than 2-norm SVMs and provides better feature selection properties. (The function of the 1-norm SVM is likely to utilize fewer features by using fewer support vectors (Fung and Mangasarian 2004).) Feature selection is also important in ranking. Ranking functions are relevance or preference functions in document or data retrieval. Identifying key features increases the interpretability of the function. Feature selection for nonlinear kernels is especially challenging, and the fewer the number of support vectors, the more efficiently feature selection can be done (Guyon and Elisseeff 2003; Mangasarian and Wild 1998; Cao et al. 2007; Yu et al. 2003; Cho et al. 2008).

We next present an RVM that revises the 1-norm ranking SVM for fast training. The RVM trains much faster than standard SVMs while not compromising the accuracy when the training set is relatively large. *The key idea of a RVM is to express the ranking function with "ranking vectors" instead of support vectors.* Support vectors in ranking SVMs are pairwise difference vectors of the closest pairs as discussed in ❷ Sect. 4. Thus, the training requires investigating every *data pair* as potential candidates of support vectors, and the number of data pairs are quadratic to the size of training set. On the other hand, the ranking function of the RVM utilizes each training data object instead of data pairs. Thus, the number of variables for optimization is substantially reduced in the RVM.

5.1   1-Norm Ranking SVM

The goal of 1-norm ranking SVM is the same as that of the standard ranking SVM, that is, to learn F that satisfies ❷ Eq. 83 for most $\{(\mathbf{x}_i, \mathbf{x}_j) : y_i < y_j \in R\}$ and generalize well beyond the training set. In the 1-norm ranking SVM, ❷ Eq. 83 can be expressed using the F of ❷ Eq. 91 as follows:

$$F(\mathbf{x}_u) > F(\mathbf{x}_v) \Longrightarrow \sum_{ij}^{P} \alpha_{ij}(\mathbf{x}_i - \mathbf{x}_j) \cdot \mathbf{x}_u > \sum_{ij}^{P} \alpha_{ij}(\mathbf{x}_i - \mathbf{x}_j) \cdot \mathbf{x}_v \qquad (92)$$

$$\Longrightarrow \sum_{ij}^{P} \alpha_{ij}(\mathbf{x}_i - \mathbf{x}_j) \cdot (\mathbf{x}_u - \mathbf{x}_v) > 0 \qquad (93)$$

Then, replacing the inner product with a kernel function, the 1-norm ranking SVM is formulated as

$$\textit{minimize:} \quad L(\alpha, \xi) = \sum_{ij}^{P} \alpha_{ij} + C \sum_{ij}^{P} \xi_{ij} \qquad (94)$$

$$s.t.: \sum_{ij}^{P} \alpha_{ij} K(\mathbf{x}_i - \mathbf{x}_j, \mathbf{x}_u - \mathbf{x}_v) \geq 1 - \xi_{uv}, \ \forall\{(u, v) : y_u < y_v \in R\} \qquad (95)$$

$$\alpha \geq 0, \ \xi \geq 0 \qquad (96)$$

While the standard ranking SVM suppresses the weight $\mathbf{w}$ to improve the generalization performance, the 1-norm ranking suppresses α in the objective function. Since the weight is expressed by the sum of the coefficient times pairwise ranking difference vectors, suppressing the coefficient α corresponds to suppressing the weight $\mathbf{w}$ in the standard SVM. (Mangasarian proves it in Mangasarian (2000).) C is a user parameter controlling the trade-off between the margin size and the amount of the error, ξ, and K is the kernel function. P is the number of pairwise difference vectors ($\sim m^2$).

The training of the 1-norm ranking SVM becomes a linear programming (LP) problem, thus solvable by LP algorithms such as the Simplex and Interior Point method (Mangasarian

2000, 2006; Fung and Mangasarian 2004). Just as for the standard ranking SVM, K needs to be computed P^2 ($\sim m^4$) times, and there are P number of constraints (❷ Eq. 95) and α to compute. Once α is computed, F is computed using the same ranking function as the standard ranking SVM, that is, ❷ Eq. 91.

The accuracies of the 1-norm ranking SVM and standard ranking SVM are comparable, and both methods need to compute the kernel function $O(m^4)$ times. In practice, the training of the standard SVM is more efficient because fast decomposition algorithms have been developed such as sequential minimal optimization (SMO) (Platt 1998), while the 1-norm ranking SVM uses common LP solvers.

It is shown that *1-norm SVMs use much fewer support vectors than standard 2-norm SVMs*, that is, the number of positive coefficients (i.e., $\alpha > 0$) after training is much fewer in the 1-norm SVMs than in the standard 2-norm SVMs (Mangasarian 2006; Fung and Mangasarian 2004). This is because, unlike the standard 2-norm SVM, the support vectors in the 1-norm SVM are not bounded to those close to the boundary in classification or the minimal ranking difference vectors in ranking. Thus, the testing involves much fewer kernel evaluations, and it is more robust when the training set contains noisy features (Zhu et al. 2003).

5.2   Ranking Vector Machine

Although the 1-norm ranking SVM has merits over the standard ranking SVM in terms of the testing efficiency and feature selection, its training complexity is very high with respect to the number of data points. In this section, we present an RVM that revises the 1-norm ranking SVM to reduce the training time substantially. The RVM significantly reduces the number of variables in the optimization problem while not compromising the accuracy. *The key idea of RVM is to express the ranking function with "ranking vectors" instead of support vectors.* The support vectors in ranking SVMs are chosen from pairwise difference vectors, and the number of pairwise difference vectors are quadratic to the size of the training set. On the other hand, the ranking vectors are chosen from the training vectors, thus the number of variables to optimize is substantially reduced.

To theoretically justify this approach, we first present the representer theorem.

Theorem 1   (Representer Theorem (Schölkopf et al. 2001)).
Denote by Ω: $[0, \infty) \to \mathcal{R}$ a strictly monotonic increasing function, by $\mathcal{X}$ a set, and by $c : (\mathcal{X} \times \mathcal{R}^2)^m \to \mathcal{R} \cup \{\infty\}$ an arbitrary loss function. Then each minimizer $F \in \mathcal{H}$ of the regularized risk

$$c((x_1, y_1, F(x_1)), \ldots, (x_m, y_m, F(x_m))) + \Omega(\|F\|_{\mathcal{H}}) \tag{97}$$

admits a representation of the form

$$F(x) = \sum_{i=1}^{m} \alpha_i K(x_i, x) \tag{98}$$

The proof of the theorem is presented in Schölkopf et al. (2001).

Note that, in the theorem, the loss function c is *arbitrary* allowing *coupling* between data points (x_i, y_i), and the regularizer Ω has to be monotonic.

Given such a loss function and regularizer, the representer theorem states that although we might be trying to solve the optimization problem in an infinite-dimensional space $\mathcal{H}$, containing linear combinations of kernels centered on *arbitrary* points of $\mathscr{X}$, the solution lies in the span of m particular kernels – those centered on the *training* points (Schölkopf et al. 2001).

Based on the theorem, our ranking function F can be defined as ❷ Eq. 98, which is based on the *training points* rather than arbitrary points (or pairwise difference vectors). Function (❷ Eq. 98) is similar to function (❷ Eq. 91) except that, unlike the latter using pairwise difference vectors $(\mathbf{x}_i - \mathbf{x}_j)$ and their coefficients (α_{ij}), the former utilizes the training vectors $(\mathbf{x}_i)$ and their coefficients (α_i). With this function, ❷ Eq. 92 becomes the following.

$$F(\mathbf{x}_u) > F(\mathbf{x}_v) \Longrightarrow \sum_i^m \alpha_i K(\mathbf{x}_i, \mathbf{x}_u) > \sum_i^m \alpha_i K(\mathbf{x}_i, \mathbf{x}_v) \tag{99}$$

$$\Longrightarrow \sum_i^m \alpha_i (K(\mathbf{x}_i, \mathbf{x}_u) - K(\mathbf{x}_i, \mathbf{x}_v)) > 0. \tag{100}$$

Thus, we set the loss function c as follows.

$$c = \sum_{\forall\{(u,v):y_u<y_v\in R\}} \left(1 - \sum_i^m \alpha_i (K(\mathbf{x}_i, \mathbf{x}_u) - K(\mathbf{x}_i, \mathbf{x}_v))\right) \tag{101}$$

The loss function utilizes couples of data points penalizing misranked pairs, that is, it returns higher values as the number of misranked pairs increases. Thus, the loss function is order sensitive, and it is an instance of the function class c in ❷ Eq. 97.

We set the regularizer $\Omega(\|f\|_{\mathscr{H}}) = \sum_i^m \alpha_i$ $(\alpha_i \geq 0)$, which is strictly monotonically increasing. Let P be the number of pairs $(u, v) \in R$ such that $y_u < y_v$, and let $\xi_{uv} = 1 - \sum_i^m \alpha_i(K(\mathbf{x}_i, \mathbf{x}_u) - K(\mathbf{x}_i, \mathbf{x}_v))$. Then, the RVM is formulated as follows.

$$\text{minimize: } L(\alpha, \xi) = \sum_i^m \alpha_i + C \sum_{ij}^P \xi_{ij} \tag{102}$$

$$\text{s.t.: } \sum_i^m \alpha_i(K(\mathbf{x}_i, \mathbf{x}_u) - K(\mathbf{x}_i, \mathbf{x}_v)) \geq 1 - \xi_{uv}, \forall\{(u,v) : y_u < y_v \in R\} \tag{103}$$

$$\alpha, \xi \geq 0 \tag{104}$$

The solution of the optimization problem lies in the span of kernels centered on the training points (i.e., ❷ Eq. 98) as suggested in the representer theorem. Just as the 1-norm ranking SVM, the RVM suppresses α to improve the generalization, and forces ❷ Eq. 100 by constraint (❷ Eq. 103). Note that there are only m number of α_i in the RVM. Thus, the kernel function is evaluated $O(m^3)$ times while the standard ranking SVM computes it $O(m^4)$ times.

Another rationale of RVM or a rationale for using training vectors instead of pairwise difference vectors in the ranking function is that the support vectors in the 1-norm ranking SVM are not the closest pairwise difference vectors, thus expressing the ranking function with pairwise difference vectors becomes not as beneficial in the 1-norm ranking SVM. To explain this further, consider *classifying* SVMs. Unlike the 2-norm (classifying) SVM, the support vectors in the 1-norm (classifying) SVM are *not* limited to those close to the decision boundary. This makes it possible that the 1-norm (classifying) SVM can express the similar boundary function with fewer support vectors. Directly extended from the 2-norm (classifying) SVM, the 2-norm *ranking* SVM improves the generalization by maximizing the closest pairwise ranking difference that corresponds to the margin in the 2-norm (classifying) SVM as discussed in ❷ Sect. 4. Thus, the 2-norm ranking SVM expresses the function with the closest pairwise difference vectors (i.e., the support vectors). However, the 1-norm ranking SVM improves the generalization by suppressing the coefficients α just as the 1-norm (classifying) SVM. Thus, the support vectors in the 1-norm ranking SVM are not the closest pairwise difference vectors any more, and thus expressing the ranking function with pairwise difference vectors becomes not as beneficial in the 1-norm ranking SVM.

5.3 Experiment

This section evaluates the RVM on synthetic datasets (❷ Sect. 5.3.1) and a real-world dataset (❷ Sect. 5.3.2). The RVM is compared with the state-of-the-art ranking SVM provided in SVM-light. Experiment results show that the RVM trains substantially faster than the SVM-light for nonlinear kernels while their accuracies are comparable. More importantly, the number of ranking vectors in the RVM is multiple orders of magnitude smaller than the number of support vectors in the SVM-light. Experiments are performed on a Windows XP Professional machine with a Pentium IV 2.8 GHz and 1 GB of RAM. We implemented the RVM using C and used CPLEX (http://www.ilog.com/products/cplex/) for the LP solver. The source codes are freely available at http://iis.postech.ac.kr/rvm (Yu et al. 2008).

Evaluation Metric: MAP (mean average precision) is used to measure the ranking quality when there are only two classes of ranking (Yan et al. 2003), and NDCG is used to evaluate ranking performance for IR applications when there are multiple levels of ranking (Baeza-Yates and Ribeiro-Neto 1999; Burges et al. 2004; Cao et al. 2006; Xu and Li 2007). Kendall's τ is used when there is a global ordering of data and the training data is a subset of it. Ranking SVMs as well as the RVM minimize the amount of error or mis-ranking, which corresponds to optimizing the Kendall's τ (Joachims 2002; Yu 2005). Thus, we use the Kendall's τ to compare their accuracy.

Kendall's τ computes the overall accuracy by comparing the similarity of two orderings R^* and R^F. (R^F is the ordering of D according to the learned function F.) The Kendall's τ is defined based on the number of concordant pairs and discordant pairs. If R^* and R^F agree on how they order a pair, x_i and x_j, the pair is concordant, otherwise it is discordant. The accuracy of function F is defined as the number of concordant pairs between R^* and R^F per the total number of pairs in D as follows.

$$F(R^*, R^F) = \frac{\text{\# of concordant pairs}}{\binom{|R|}{2}}$$

For example, suppose R^* and R^F order five points $x_1, \ldots, x_5$ as follow:

$$(x_1, x_2, x_3, x_4, x_5) \in R^*$$
$$(x_3, x_2, x_1, x_4, x_5) \in R^F$$

Then, the accuracy of F is 0.7, as the number of discordant pairs is 3, i.e., $\{x_1, x_2\}$, $\{x_1, x_3\}$, $\{x_2, x_3\}$ while all remaining seven pairs are concordant.

5.3.1 Experiments on Synthetic Datasets

Below is the description of the experiments on synthetic datasets.

1. We randomly generated a training and a testing dataset D_{train} and D_{test}, respectively, where D_{train} contains m_{train} (= 40, 80, 120, 160, 200) data points of n (e.g., 5) dimensions (i.e., m_{train}-by-n matrix), and D_{test} contains m_{test} (= 50) data points of n dimensions (i.e., m_{test}-by-n matrix). Each element in the matrices is a random number between zero and one. (We only did experiments on the dataset of up to 200 objects for performance reasons. Ranking SVMs run intolerably slow on datasets larger than 200.)
2. A global ranking function F^* is randomly generated, by randomly generating the weight vector $\mathbf{w}$ in $F^*(\mathbf{x}) = \mathbf{w} \cdot \mathbf{x}$ for linear, and in $F^*(\mathbf{x}) = \exp\left(-||\mathbf{w} - \mathbf{x}||\right)^2$ for RBF function.

◻ **Fig. 7**
Accuracy: (a) Linear (b) RBF.

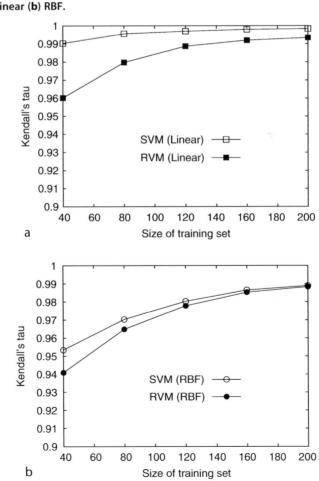

3. D_{train} and D_{test} are ranked according to F^*, which forms the global ordering R^*_{train} and R^*_{test} on the training and testing data.

4. We train a function F from R^*_{train}, and test the accuracy of F on R^*_{test}.

We tuned the soft margin parameter C by trying $C = 10^{-5}$, 10^{-5}, ..., 10^5, and used the highest accuracy for comparison. For the linear and RBF functions, we used linear and RBF kernels accordingly. This entire process is repeated 30 times to get the mean accuracy.

Accuracy: ❯ *Figure 7* compares the accuracies of the RVM and the ranking SVM from the SVM-light. The ranking SVM outperforms RVM when the size of the dataset is small, but their difference becomes trivial as the size of the dataset increases. This phenomenon can be explained by the fact that when the training size is too small, the number of potential ranking vectors becomes too small to draw an accurate ranking function whereas the number of potential support vectors is still large. However, as the size of the training set increases, RVM

◧ **Fig. 8**
Training time: (a) Linear Kernel (b) RBF Kernel.

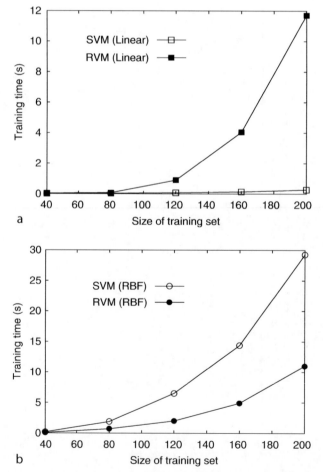

becomes as accurate as the ranking SVM because the number of potential ranking vectors becomes large as well.

Training Time: ❷ *Figure 8* compares the training time of the RVM and the SVM-light. While the SVM-light trains much faster than RVM for linear kernel (SVM-light is specially optimized for linear kernels), *the RVM trains significantly faster than the SVM-light for RBF kernels.*

Number of Support (or Ranking) Vectors: ❷ *Figure 9* compares the number of support (or ranking) vectors used in the function of the RVM and the SVM-light. The RVM's model uses a significantly smaller number of support vectors than the SVM-light.

Sensitivity to Noise: In this experiment, the sensitivity of each method is compared to noise. Noise is inserted by switching the orders of some data pairs in R^*_{train}. We set the size of the training set $m_{\text{train}} = 100$ and the dimension $n = 5$. After R^*_{train} is made from a random function F^*, k vectors are randomly picked from the R^*_{train} and switched with their adjacent vectors in

■ **Fig. 9**
Number of support (or ranking) vectors: (a) Linear Kernel (b) RBF Kernel.

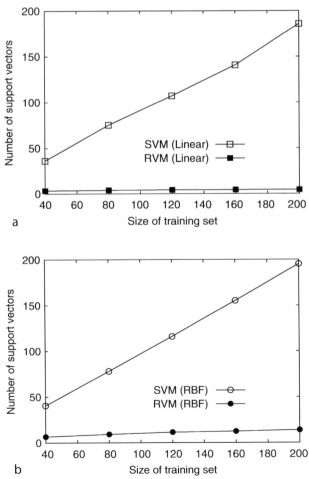

■ **Fig. 10**
Sensitivity to noise (m_{train} = 100): (a) Linear (b) RBF.

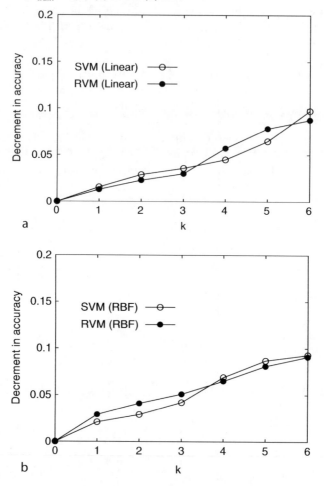

the ordering to implant noise in the training set. ● *Figure 10* shows the decrements of the accuracies as the number of misorderings increases in the training set. Their accuracies are moderately decreasing as the noise increases in the training set, and their sensitivities to noise are comparable.

5.3.2   Experiment on Real Dataset

In this section, we experiment using the OHSUMED dataset obtained from LETOR, the site containing benchmark datasets for ranking (LETOR). OHSUMED is a collection of documents and queries on medicine, consisting of 348,566 references and 106 queries. There are in total 16,140 query-document pairs upon which relevance judgments are made. In this dataset, the relevance judgments have three levels: "definitely relevant," "partially relevant," and

◘ **Table 2**

Experiment results: accuracy (Acc), training time (Time), and number of support or ranking vectors (#SV or #RV)

		Query 1 $\|D\| = 134$			Query 2 $\|D\| = 128$			Query 3 $\|D\| = 182$		
		Acc	Time	#SV or #RV	Acc	Time	#SV or #RV	Acc	Time	#SV or #RV
RVM	Linear	**0.5484**	**0.23**	**1.4**	**0.6730**	**0.41**	**3.83**	0.6611	1.94	1.99
	RBF	0.5055	0.85	4.3	0.6637	0.41	2.83	**0.6723**	**4.71**	**1**
SVM	Linear	**0.5634**	**1.83**	**92**	0.6723	1.03	101.66	0.6588	4.24	156.55
	RBF	0.5490	3.05	92	**0.6762**	**3.50**	**102**	**0.6710**	**55.08**	**156.66**

"irrelevant." The OHSUMED dataset in LETOR extracts 25 features. We report our experiments on the first three queries and their documents. We compare the performance of RVM and SVM-light on them. We tuned the parameters threefold using cross-validation by trying C and $\gamma = 10^{-6}, 10^{-5}, \ldots, 10^6$ for the linear and RBF kernels and compared the highest performances. The training time is measured for training the model with the tuned parameters. The whole process was repeated three times and the mean values reported.

❯ *Table 2* shows the results. The accuracies of the SVM and RVM are comparable overall; SVM shows a little higher accuracy than RVM for query 1, but for the other queries their accuracy differences are not statistically significant. More importantly, *the number of ranking vectors in RVM is significantly smaller than that of support vectors in SVM.* For example, for query 3, an RVM having just one ranking vector outperformed an SVM with over 150 support vectors. *The training time of the RVM is significantly shorter than that of SVM-light.*

References

Baeza-Yates R, Ribeiro-Neto B (eds) (1999) Modern information retrieval. ACM Press, New York

Bertsekas DP (1995) Nonlinear programming. Athena Scientific, Belmont, MA

Burges C, Shaked T, Renshaw E, Lazier A, Deeds M, Hamilton N, Hullender G (2004) Learning to rank using gradient descent. In: Proceedings of the international conference on machine learning (ICML'04), Oregon State University, Corvallis, OR, USA

Burges CJC (1998) A tutorial on support vector machines for pattern recognition. Data Mining Knowl Discov 2:121–167

Cao B, Shen D, Sun JT, Yang Q, Chen Z (2007) Feature selection in a kernel space. In: Proceedings of the international conference on machine learning (ICML'07), Oregon State University, Corvallis, OR, USA

Cao Y, Xu J, Liu TY, Li H, Huang Y, Hon HW (2006) Adapting ranking SVM to document retrieval. In: Proceedings of the ACM SIGIR international conference on information retrieval (SIGIR'06), New York

Cho B, Yu H, Lee J, Chee Y, Kim I (2008) Nonlinear support vector machine visualization for risk factor analysis using nomograms and localized radial basis function kernels. IEEE Trans Inf Technol Biomed 12(2)

Christianini N, Shawe-Taylor J (2000) An introduction to support vector machines and other kernel-based learning methods. Cambridge University Press, Cambridge, UK

Cohen WW, Schapire RE, Singer Y (1998) Learning to order things. In: Proceedings of the advances in neural information processing systems (NIPS'98), Cambridge, MA

Friedman H (1998) Another approach to polychotomous classification. Tech. rep., Stanford University, Department of Statistics, Stanford, CA 10:1895–1924

Fung G, Mangasarian OL (2004) A feature selection Newton method for support vector machine classification. Comput Optim Appl 28:185–202

Guyon I, Elisseeff A (2003) An introduction to variable and feature selection. J Mach Learn Res 3:1157–1182

Hastie T, Tibshirani R (1998) Classification by pairwise coupling. In: Advances in neural information processing systems. MIT Press, Cambridge, MA

Herbrich R, Graepel T, Obermayer K (eds) (2000) Large margin rank boundaries for ordinal regression. MIT Press, Cambridge, MA

Joachims T (2002) Optimizing search engines using clickthrough data. In: Proceedings of the ACM SIGKDD international conference on knowledge discovery and data mining (KDD'02), Paris, France

Joachims T (2006) Training linear SVMs in linear time. In: Proceedings of the ACM SIGKDD international conference on knowledge discovery and data mining (KDD'06) Philadelphia, PA, USA

Liu T-Y (2009) Learning to rank for information retrieval. Found Trends Inf Retr 3(3):225–331

Mangasarian OL (2000) Generalized support vector machines. MIT Press, Cambridge, MA

Mangasarian OL (2006) Exact 1-norm support vector machines via unconstrained convex differentiable minimization. J Mach Learn Res 7:1517–1530

Mangasarian OL, Wild EW (1998) Feature selection for nonlinear kernel support vector machines. Tech. rep., University of Wisconsin, Madison

Platt J (1998) Fast training of support vector machines using sequential minimal optimization. In: Schölkopf B, Burges CJC (eds) Advances in kernel methods: support vector machines, MIT Press, Cambridge, MA

Schölkopf B, Herbrich R, Smola AJ, Williamson RC (2001) A generalized representer theorem. In: Proceedings of COLT, Amsterdam, The Netherlands

Smola AJ, Schölkopf B (1998) A tutorial on support vector regression. Tech. rep., NeuroCOLT2 Technical Report NC2-TR-1998-030

Vapnik V (1998) Statistical learning theory. John Wiley and Sons, New York

Xu J, Li H (2007) Ada Rank: a boosting algorithm for information retrieval. In: Proceedings of the ACM SIGIR international conference on information retrieval (SIGIR'07), New York

Yan L, Dodier R, Mozer MC, Wolniewicz R (2003) Optimizing classifier performance via the Wilcoxon-Mann-Whitney statistics. In: Proceedings of the international conference on machine learning (ICML'03), Washington, DC

Yu H (2005) SVM selective sampling for ranking with application to data retrieval. In: Proceedings of the international conference on knowledge discovery and data mining (KDD'05), Chicago, IL

Yu H, Hwang SW, Chang KCC (2007) Enabling soft queries for data retrieval. Inf Syst 32:560–574

Yu H, Kim Y, Hwang SW (2008) RVM: An efficient method for learning ranking SVM. Tech. rep., Department of Computer Science and Engineering, Pohang University of Science and Technology (POSTECH), Pohang, Korea, http://iis.hwanjoyu.org/rvm

Yu H, Yang J, Wang W, Han J (2003) Discovering compact and highly discriminative features or feature combinations of drug activities using support vector machines. In: IEEE computer society bioinformatics conference (CSB'03), Stanford, CA, pp 220–228

Zhu J, Rosset S, Hastie T, Tibshriani R (2003) 1-norm support vector machines. In: Proceedings of the advances in neural information processing systems (NIPS'00) Berlin, Germany

16 Fast Construction of Single-Hidden-Layer Feedforward Networks

Kang Li[1] · Guang-Bin Huang[2] · Shuzhi Sam Ge[3]
[1]School of Electronics, Electrical Engineering and Computer Science, Queen's University, Belfast, UK
k.li@ee.qub.ac.uk
[2]School of Electrical and Electronic Engineering, Nanyang Technological University, Singapore
egbhuang@ntu.edu.sg
[3]Social Robotics Lab, Interactive Digital Media Institute, The National University of Singapore, Singapore
elegesz@nus.edu.sg

G. Rozenberg et al. (eds.), *Handbook of Natural Computing*, DOI 10.1007/978-3-540-92910-9_16,
© Springer-Verlag Berlin Heidelberg 2012

Abstract

In this chapter, two major issues are addressed: (i) how to obtain a more compact network architecture and (ii) how to reduce the overall computational complexity. An integrated analytic framework is introduced for the fast construction of single-hidden-layer feedforward networks (SLFNs) with two sequential phases. The first phase of the algorithm focuses on the computational efficiency for fast computation of the unknown parameters and fast selection of the hidden nodes. The second phase focuses on improving the performance of the network obtained in the first phase. The proposed algorithm is evaluated on several benchmark problems.

1   Introduction

The single-hidden-layer feedforward neural network (SLFN) represents a large class of flexible and efficient structures in the neural network literature. Due to their excellent capabilities, SLFNs have been widely used in many areas such as pattern recognition, bioinformatics, signal processing, time series prediction, nonlinear system modeling and control, etc. (Hong et al. 2008).

An SLFN is a linear combination of some basis functions that are usually nonlinear functions of the inputs. Many SLFNs with different types of basis functions have been proposed and intensively researched in the last three decades. Generally speaking, there are two mainstream architectures for SLFNs (Huang et al. 2006a, b; Peng and Irwin 2008): (i) SLFNs with additive nodes, and (ii) SLFNs with radial basis function (RBF) nodes.

An SLFN with n additive nodes can be formulated by

$$y(\mathbf{x}) = \sum_{i=1}^{n} \theta_i g(\mathbf{w}_i^\mathsf{T} \mathbf{x} + b_i) \tag{1}$$

where $\mathbf{x} \in \mathfrak{R}^d$ is the input vector, $\mathbf{w}_i \in \mathfrak{R}^d$ is the input weight vector connecting the input layer to the ith hidden node, $b_i \in \mathfrak{R}$ is the bias of the ith hidden node, θ_i is the output weight connecting the ith hidden node to the output node, and g is the hidden node activation function.

The RBF networks belong to a specific class of SLFNs that use RBF nodes in the hidden layer, and each RBF node has its own center and impact factor with its output being some radially symmetric function of the distance between the input and the center (Peng et al. 2007; Li et al. 2009):

$$y(\mathbf{x}) = \sum_{i=1}^{n} \theta_i g(\gamma_i \|\mathbf{x} - \mathbf{c}_i\|) \tag{2}$$

where $\mathbf{x} \in \mathfrak{R}^d$ is again the neural input vector, $\mathbf{c}_i \in \mathfrak{R}^d$ and $\gamma_i \in \mathfrak{R}$ are the center and impact factors of the ith RBF node, and θ_i is the weight connecting the ith RBF hidden node to the output node.

In neural network applications, learning is a critical phase. For SLFNs with additive nodes, the early training algorithms make use of the first derivative information, that is, backpropagation with adaptive learning rate and momentum (BPAM), conjugate gradient, QuickProp, etc. (Bishop 1995). Advanced training algorithms like the Levenberg–Marquardt (LM) algorithm, which make use of the second derivative information, have proven to be more efficient and have been widely used in applications (Marquardt 1963; Hagan and Menhaj 1994).

For the training of SLFNs with RBF kernels, the conventional approach takes a two-stage procedure, that is, unsupervised learning of both the centers and impact factors for the RBF nodes and supervised learning of the linear output weights. With respect to the center location, clustering techniques have been proposed (Sutanto et al. 1997; Musavi et al. 1992). Once the centers and the impact factors are determined, the linear output weights can be obtained using Cholesky decomposition, orthogonal least squares or singular value decomposition (Press et al. 1992; Lawson and Hanson 1974; Serre 2002; Chen et al. 1991).

In contrast to the conventional two-stage learning procedure, supervised learning methods aim to optimize all the network parameters. To improve the convergence, various techniques have been introduced. For example, hybrid algorithms combine the gradient-based search for the nonlinear parameters (the impact factors and centers) of the RBF nodes and the least-squares estimation of the linear output weights (McLoone et al. 1998; Peng et al. 2003; Panchapakesan et al. 2002). Second-order algorithms have also been proposed, which use an additional adaptive momentum term to the Levenberg–Marquardt (LM) algorithm in order to maintain the conjugacy between successive minimization directions, resulting in good convergence for some well-known hard problems (Ampazis and Perantonis 2002).

In addition to the network learning, to control the network complexity of SLFNs is another important issue, and a number of additive (or constructive, or growing) methods, and subtractive (or destructive, or pruning) methods have been proposed (Mao and Huang 2005; Chng et al. 1996; Miller 1990; Platt 1991; Kadirkamanathan and Niranjan 1993; Yingwei et al. 1997; Huang et al. 2005). In particular, it has been shown that one can add nodes to the network one by one until a stopping criterion is satisfied, thus obtaining more compact networks (Chen et al. 1991; Chng et al. 1996). The techniques they used are subset selection algorithms, where a small number of nodes are selected from a large pool of candidates based on the orthogonal least squares (OLS) and its variants or the fast recursive algorithm (Chen et al. 1991; Mao and Huang 2005; Korenberg 1998; Zhang and Billings 1996; Chen and Wigger 1995; Li et al. 2005). The elegance of the method is that the net contribution of the newly added node can be explicitly identified without solving the whole least-square problem, leading to significantly reduced computational complexity.

Generally speaking, conventional network training and selection methods for SLFNs can be extremely time consuming if both issues have to be considered simultaneously (Peng et al. 2006). Unlike the conventional learning schemes, a more general network learning concept, namely, the extreme learning machine (ELM), was proposed for SLFNs (Huang et al. 2006b, c; Liang et al. 2006a). For ELM, all the nonlinear parameters in the hidden nodes, including both the input weights and hidden nodes' biases for SLFNs with additive nodes or both the centers and impact factors for SLFNs with RBF kernels, are randomly chosen independently of the training data. Thus the training is transformed to a standard least-squares problem, leading to a significant improvement of the generalization performance at a fast learning speed. The output weights in the SLFNs are linearly analytically determined (Huang et al. 2006b, c; Liang et al. 2006a). It has been proven in an incremental method that the SLFNs with randomly generated hidden nodes, independent of the training data, according to any continuous sampling distribution, can approximate any continuous target functions (Huang et al. 2006a). The activation functions for additive nodes in such SLFNs can be any bounded piecewise continuous function and the activation functions for RBF nodes can be any integrable piecewise continuous function. Further, they extended ELM to a much wider type of hidden nodes, including fuzzy rules as well as additive nodes, etc. (Huang and Chen 2007). The ELM has also been applied in some other areas like biomedical engineering

(Xu et al. 2005), bioinformatics (Handoko et al. 2006; Zhang et al. 2007; Wang and Huang 2005) human–computer interfaces (HCI) (Liang et al. 2006b), text classification (Liu et al. 2005), terrain reconstruction (Yeu et al. 2006) and communication channel equalization (Li et al. 2005).

This chapter addresses two critical issues arising from a wide range of applications for SLFNs with random hidden nodes: (i) how to obtain a more compact network architecture; (ii) how to reduce the overall computational complexity. These are the two bottlenecks restricting the application of the SLFNs to some large databases or complex problems. In this chapter, it will be shown that the above two objectives can be achieved using an integrated framework. A detailed performance evaluation will be performed on the benchmark problems in regression, time series prediction, and nonlinear system modeling and identification.

2   Review of SLFNs with Random Nodes

In this section, SLFNs with fully randomly generated hidden nodes are briefly reviewed. In particular, the ELM concept (Huang et al. 2006b, c; Huang and Chen 2007) is discussed. To begin with, a unified description is introduced.

The output of a generalized SLFN with n hidden nodes can be represented by

$$y(\mathbf{x}) = \sum_{i=1}^{n} \theta_i g(\omega_i; \mathbf{x}) \tag{3}$$

where ω is the vector of nonlinear parameters of the hidden nodes and θ_i is the weight connecting the ith hidden node to the output node. $g(\omega_i; \mathbf{x})$ is the output of the ith hidden node with respect to the input vector $\mathbf{x}$. Obviously, according to formulas (❷ 1), (❷ 2), and (❷ 3), for the SLFN with additive hidden nodes, $g(\omega_i; \mathbf{x}) = g(\mathbf{w}_i^T \mathbf{x} + b_i)$, and for the SLFN with RBF hidden nodes, $g(\omega_i; \mathbf{x}) = g(\gamma_i \|\mathbf{x} - \mathbf{c}_i\|)$.

Now suppose N samples $\{(\mathbf{x}_i, t_i)\}_{i=1}^{N}$ are used to train the network (❷ Eq. 3). Here, $\mathbf{x}_i \in \mathfrak{R}^d$ is the input vector and $t_i \in \mathfrak{R}$ is a target (it should be noted that all the analysis in this chapter can be linearly extended to SLFNs with multi-output nodes). If an SLFN with n hidden nodes can approximate these N samples with zero error, it implies that there exist θ_i and ω_i such that

$$y(\mathbf{x}_j) = \sum_{i=1}^{n} \theta_i g(\omega_i; \mathbf{x}_j) = t_j, j = 1, \ldots, N \tag{4}$$

❷ Equation 4 can be rewritten compactly as:

$$\mathbf{y} = \mathbf{P}\Theta = \mathbf{t} \tag{5}$$

where

$$\mathbf{P}(\omega_1, \ldots, \omega_n, \mathbf{x}_1, \ldots, \mathbf{x}_n) = [\mathbf{p}_1, \mathbf{p}_2, \ldots, \mathbf{p}_n] = \begin{bmatrix} g(\omega_1, \mathbf{x}_1) & \cdots & g(\omega_n, \mathbf{x}_1) \\ \vdots & \cdots & \vdots \\ g(\omega_1, \mathbf{x}_N) & \cdots & g(\omega_n, \mathbf{x}_N) \end{bmatrix} \tag{6}$$

$$\Theta = [\theta_1, \ldots, \theta_n]^T, \mathbf{t} = [t_1, \ldots, t_N]^T, \mathbf{y} = [y(\mathbf{x}_1), \ldots, y(\mathbf{x}_N)]^T$$

$\mathbf{P}$ is the output matrix of the hidden layer; the ith column of $\mathbf{P}$ is the ith hidden node's output vector corresponding to inputs $\mathbf{x}_1, \mathbf{x}_2, \ldots, \mathbf{x}_N$ and the jth row of $\mathbf{P}$ is the output vector of the hidden layer with respect to input $\mathbf{x}_j$. In the regression context, $\mathbf{P}$ is also called the regression matrix, and the ith column of $\mathbf{P}$ corresponds to the ith regressor term (hidden node).

It has been proven that the universal approximation capability of SLFNs with random hidden nodes generated according to any continuous sampling distribution can approximate any continuous target function (Huang et al. 2006a; Huang and Chen 2007). Therefore, the hidden node parameters ω_i of SLFNs *need not be tuned during training and may simply be assigned with random values* (Huang et al. 2006a, b; Huang and Chen 2007). ❷ Equation 5 then becomes a linear system and the output weights Θ are estimated as:

$$\Theta = \mathbf{P}^{\dagger}\mathbf{P} \tag{7}$$

where $\mathbf{P}^{\dagger}$ is the Moore–Penrose generalized inverse (Rao and Mitra 1971) of the hidden layer output matrix $\mathbf{P}$.

The above introduction reveals the three issues in the ELM that have to be dealt with in applications: (i) the ELM usually requires a larger number of hidden nodes, thus a compact network implementation is demanded; (ii) a large network may lead to the singularity problem in the matrix computation, and controlling the network complexity can also reduce or eliminate this problem; and (iii) a large network generates a high computational overhead in deriving the output weights and so a more efficient algorithm is needed. In fact, these three problems are also closely coupled. If the performance of the network can be improved, the number of required hidden nodes and the corresponding number of output weights can be significantly reduced, leading to an overall reduction of the computational complexity. Moreover, the improved network performance will eventually reduce or eliminate the singularity problem. Once $\mathbf{P}^{\mathrm{T}}\mathbf{P}$ becomes nonsingular, a number of efficient algorithms can be used for fast computation of the output weights. The basic idea in this chapter is to introduce an integrated framework, to further select the hidden nodes from the randomly generated hidden node pools in ELM, in order to further improve the network performance and to efficiently calculate the output weights as well. To propose this framework, the problem of fast construction of SLFNs will be formulated first.

3   Problem Formulation for Fast Construction of SLFNs

In this chapter, the network construction is carried out in two phases. The first phase is carried out by selecting the hidden nodes one by one from a pool of candidates, which are randomly generated at the beginning of the construction process, and then compute the linear output weights. Two scenarios are considered:

1. **Case 1**: The number of candidates exactly matches the desired number of hidden nodes for the SLFNs. In this case, the issue in this chapter is to efficiently compute the output weights and to increase the numerical stability. In the original ELM algorithm (Huang et al. 2006b), this is achieved via ❷ Eq. 7 which is practically implemented using singular value decomposition. In this chapter, in order to further reduce the computational complexity and to increase the numerical stability, the selection process is still embedded in the algorithm, that is, the output weights are calculated only after all the hidden nodes are ordered according to their contribution to the cost function.

2. **Case 2**: The number of candidates is much bigger than required, and only a small subset of hidden nodes is selected. This is a standard subset selection problem, and the issue is to both reduce the computational complexity in the selection process and in the computation of output weights. For this subset selection problem, the OLS method and its variants (Chen et al. 1991; Chng et al. 1996; Korenberg 1988; Zhang and Billings 1996; Chen and

Wigger 1995; Zhu and Billings 1996; Gomm and Yu 2000) are perhaps among the most well-known approaches for the model selection.

Therefore, a selection process needs to be implemented in the above two scenarios. This selection problem can be described as follows.

Suppose that a set of M randomly generated nodes serves as a pool of candidates from which n hidden nodes will be selected, and a set of N samples $(\mathbf{x}_i, t_i)$ is used to train the network. This leads to the following regression matrix of candidate nodes:

$$\left.\begin{array}{l} \Phi = [\phi_1, \phi_2, \ldots, \phi_M] \\ \phi_i = [\phi_i(1), \phi_i(2), \cdots, \phi_i(N)]^{\mathrm{T}}, i = 1, 2, \ldots, M \end{array}\right\} \quad (8)$$

where ϕ_i is the output vector of the ith candidate node corresponding to the N training samples.

Now, the problem is to select, say, n hidden nodes, the corresponding regressor terms are denoted as $\mathbf{p}_1, \mathbf{p}_2, \ldots, \mathbf{p}_n$, and $\mathbf{p}_i \in \{\phi_1, \cdots, \phi_M\}, i = 1, \ldots, n$ forming the regression matrix of the SLFN hidden layer:

$$\mathbf{P} = [\mathbf{p}_1, \mathbf{p}_2, \ldots, \mathbf{p}_n] \quad (9)$$

producing the following SLFN outputs:

$$\mathbf{y} = \mathbf{P}\Theta + \mathbf{e} = \mathbf{t} \quad (10)$$

where $\mathbf{e} = [e_1, \ldots, e_N]^{\mathrm{T}} \in \mathfrak{R}^N$ is the modeling error vector.

Now, the objective of network training is to minimize the error with respect to the cost function

$$J = \|\mathbf{e}\| = \|\mathbf{y} - \mathbf{P}\theta\| \quad (11)$$

where $\|\cdot\|$ denotes the two-norm of a vector.

If $\mathbf{P}$ is of full column rank, the optimum estimation of the output weights is

$$\Theta = [\mathbf{P}^{\mathrm{T}}\mathbf{P}]^{-1}\mathbf{P}^{\mathrm{T}}\mathbf{y} \quad (12)$$

Theoretically, there are $M!/n!/(M-n)!$ possible combinations of n hidden nodes out of the total M candidates. Obviously, obtaining the optimal subset can be very expensive or impossible if M is a very large number, and part of this is also referred to as the curse of dimensionality in the literature. Practically, this problem can be partially solved using fast forward network growing or backward pruning methods. In the following section, an efficient approach will be employed.

4 Fast Network Construction Approach for SLFNs – Phase 1: Subset Selection

Before introducing the detailed scheme a core algorithm will be presented.

4.1 Fast Selection of One Hidden Node

The core idea of the fast network construction is to select hidden nodes for SLFNs one by one from a pool of candidates, each time the reduction of the cost function is maximized.

This procedure is iterated until n hidden nodes are selected. Therefore, the major objective in this section is to propose a recursive algorithm to select the hidden nodes.

To begin with, suppose that k hidden nodes have been selected, producing the following regression matrix

$$\mathbf{P_k} = [\mathbf{p}_1, \mathbf{p}_2, \ldots, \mathbf{p}_k] \tag{13}$$

The corresponding cost function is given by

$$J(\mathbf{P}_k) = \mathbf{y}^T\mathbf{y} - \mathbf{y}^T\mathbf{P}_k(\mathbf{P}_k^T\mathbf{P}_k)^{-1}\mathbf{P}_k^T\mathbf{y} \tag{14}$$

If $\mathbf{P}_k$ is of full column rank, $(\mathbf{P}_k^T\mathbf{P}_k)$ in ❷ Eq. 14 is symmetric and positive definite.

Define

$$\mathbf{W} = \mathbf{P}_k^T\mathbf{P}_k = [w_{i,j}]_{k \times k} \tag{15}$$

then $\mathbf{W}$ can be decomposed as

$$\mathbf{W} = \mathbf{P}_k^T\mathbf{P}_k = \tilde{\mathbf{A}}^T\mathbf{D}\tilde{\mathbf{A}} \tag{16}$$

where $\mathbf{D} = \mathrm{diag}(d_1, \ldots, d_k)$ is a diagonal matrix and $\tilde{\mathbf{A}} = [\tilde{a}_{i,j}]_{k \times k}$ is a unity upper triangular matrix.

Define

$$\mathbf{A} = \mathbf{D}\tilde{\mathbf{A}} = [a_{i,j}]_{k \times k}, \, a_{i,j} = \begin{cases} 0, & j < i \\ d_i\tilde{a}_{i,j}, & j \geq i \end{cases} \tag{17}$$

Obviously, $\tilde{a}_{i,i} = 1$, therefore $d_i = a_{i,i}, i = 1, \ldots, k$.

According to ❷ Eq. 16, it can be derived that

$$a_{i,j} = w_{i,j} - \sum_{s=1}^{i-1} a_{s,i}a_{s,j}/a_{s,s}, i = 1, \ldots, k, j = i, \ldots, k \tag{18}$$

Define

$$\mathbf{a}_y = \mathbf{A}\Theta = \mathbf{D}\tilde{\mathbf{A}}\Theta = [a_{1,y}, \ldots, a_{k,y}]^T \tag{19}$$

and

$$\mathbf{w}_y = \mathbf{P}_k^T\mathbf{y} = [w_{1,y}, \ldots, w_{k,y}]^T \tag{20}$$

Left-multiplying both sides of ❷ Eq. 12 with $\mathbf{W}$ for $\mathbf{P} = \mathbf{P}_k$ and substituting ❷ Eq. 16, we have

$$\tilde{\mathbf{A}}^T\mathbf{D}\tilde{\mathbf{A}}\theta = \mathbf{P}_k^T\mathbf{y} \text{ or } \tilde{\mathbf{A}}^T\mathbf{a}_y = \mathbf{w}_y \tag{21}$$

Noting that $\tilde{\mathbf{A}}$ is a unity upper triangular matrix, and noting the relationship of $\tilde{\mathbf{A}}$ with $\mathbf{A}$ defined in ❷ Eq. 17, $\mathbf{a}_y$ in ❷ Eq. 21 could be computed as

$$a_{i,y} = w_{i,y} - \sum_{i=1}^{k-1} a_{s,i}a_{s,y}/a_{s,s}, i = 1, \ldots, k \tag{22}$$

Substituting ❯ Eq. 16 into ❯ Eq. 14 and noting ❯ Eq. 21 or

$$\mathbf{a}_y = \left(\tilde{\mathbf{A}}^{\mathrm{T}}\right)^{-1}\mathbf{w}_y = \left(\tilde{\mathbf{A}}^{\mathrm{T}}\right)^{-1}\mathbf{P}_k^{\mathrm{T}}\mathbf{y} \tag{23}$$

we have

$$J(\mathbf{P}_k) = \mathbf{y}^{\mathrm{T}}\mathbf{y} - \mathbf{a}_y^{\mathrm{T}}\mathbf{D}^{-1}\mathbf{a}_y = \mathbf{y}^{\mathrm{T}}\mathbf{y} - \sum_{i=1}^{k} a_{i,y}^2 / a_{i,i} \tag{24}$$

Now, suppose that one more hidden node is selected with the corresponding regressor term $\mathbf{p}_{k+1}$, the cost function becomes

$$J(\mathbf{P}_{k+1}) = \mathbf{y}^{\mathrm{T}}\mathbf{y} - \sum_{i=1}^{k+1} a_{i,y}^2 / a_{i,i} \tag{25}$$

where $\mathbf{P}_{k+1} = [\mathbf{P}_k \ \mathbf{p}_{k+1}]$.

The net reduction of the cost function due to adding one more hidden node is given by

$$\Delta J_{k+1}(\mathbf{p}_{k+1}) = J(\mathbf{P}_k) - J(\mathbf{P}_{k+1}) = a_{k+1,y}^2 / a_{k+1,k+1} \tag{26}$$

where $a_{k+1,y}$, $a_{k+1,k+1}$ are computed using ❯ Eqs. 18 and ❯ 22 as k increases by 1.

According to ❯ Eqs. 18 and ❯ 22, it can be seen that elements $a_{i,j}$ and $a_{i,y}$ for $i = 1, \ldots, k$ and $j = i, \ldots, k$, which correspond to the previously selected hidden nodes $\mathbf{p}_1, \mathbf{p}_2, \ldots, \mathbf{p}_m$, do not change as a new hidden node is added into the SLFN. Therefore, the selection of the next hidden node is formulated as the following optimization problem according to ❯ Eq. 26

$$\min\{J([\mathbf{P}_k, \phi])\} = J(\mathbf{P}_k) - \max\{\Delta J_{k+1}(\phi)\}$$
$$\text{s.t. } \phi \in \{\phi_1, \ldots, \phi_M\}, \phi \notin \{\mathbf{p}_1, \ldots, \mathbf{p}_k\} \tag{27}$$

where $\{\phi_1, \ldots, \phi_M\}$ is the candidate node pool.

According to ❯ Eq. 27, the contributions of all remaining candidate nodes in $\Phi = \{\phi_1, \ldots, \phi_M\}$ need to be calculated using ❯ Eq. 26. To achieve this, the dimension of $\mathbf{A}$, $\mathbf{a}_y$ defined above will be augmented to store the information of all remaining candidate nodes in Φ. To achieve this, redefine

$$\Phi = [\mathbf{P}_k, \mathbf{C}_{M-k}]$$
$$\mathbf{C}_{M-k} = [\phi_{k+1}, \ldots, \phi_M] \tag{28}$$

where the first k regressor terms in Φ (i.e., $\mathbf{P}_k$) correspond to the selected k nodes, the remaining $M - k$ terms $\mathbf{C}_{M-k} = [\phi_{k+1}, \ldots, \phi_M]$ are candidates, forming the candidate pool $\mathbf{C}_{M-k}$.

Now, for an SLFN with k hidden nodes, augment the $k \times k$ matrices $\tilde{\mathbf{A}}$ and $\mathbf{A}$, and the $k \times 1$ vector $\mathbf{a}_y$ to $k \times M$ and $M \times 1$, respectively, in order to store information about all the remaining candidates in $\mathbf{C}_{M-k}$ for selection of the next hidden node.

Based on ❯ Eq. 18, $\mathbf{A}$ is redefined as

$$\mathbf{A} = [a_{i,j}]_{k \times M}$$
$$a_{i,j} = \begin{cases} 0, & j < i \\ w_{i,j} - \sum_{s=1}^{i-1} \frac{a_{s,i}a_{s,j}}{a_{s,s}}, & i \le j \le M \end{cases} \tag{29}$$

where

$$w_{i,j} = \begin{cases} \mathbf{p}_i^T \mathbf{p}_j, & j \leq k \\ \mathbf{p}_i^T \phi_j, & j > k \end{cases} \tag{30}$$

Based on ❷ Eq. 17, $\tilde{\mathbf{A}}$ is redefined as

$$\tilde{\mathbf{A}} = [\tilde{a}_{i,j}]_{k \times M}, \tilde{a}_{i,j} = a_{i,j}/a_{i,i} \tag{31}$$

and based on ❷ Eq. 23, vector $\mathbf{a}_y$ is redefined as

$$\mathbf{a}_y = [a_{i,y}]_{M \times 1}$$

$$a_{i,y} = \begin{cases} \mathbf{y}^T \mathbf{p}_i - \sum\limits_{s=1}^{i-1} a_{s,i} a_{s,y}/a_{s,s}, & i \leq k \\ \mathbf{y}^T \phi_i - \sum\limits_{s=1}^{k} a_{s,i} a_{s,y}/a_{s,s}, & i > k \end{cases} \tag{32}$$

In addition, one more $M \times 1$ vector $\mathbf{b}$ is defined as

$$\mathbf{b} = [b_i]_{M \times 1}$$

$$b_i = \begin{cases} \mathbf{p}_i^T \mathbf{p}_i - \sum\limits_{s=1}^{i-1} a_{s,i} a_{s,i}/a_{s,s}, & i \leq k \\ \phi_i^T \phi_i - \sum\limits_{s=1}^{k} a_{s,i} a_{s,i}/a_{s,s}, & i > k \end{cases} \tag{33}$$

Then

$$b_i = a_{i,i}, i = 1, \ldots, k \tag{34}$$

Thus, based on ❷ Eq. 26, the contribution of each of the candidates in $\mathbf{C}_{M-k}$ to the cost function can be computed as follows:

$$\Delta J_{k+1}(\phi_i) = a_{i,y}^2/b_i, i = k+1, \ldots, M \tag{35}$$

and that from $\mathbf{C}_{M-k}$ which gives the maximum contribution is then selected as the $(k+1)$th hidden node.

For example, assume

$$j = \arg \max_{k < i \leq M} \{\Delta J_{k+1}(\phi_i)\} \tag{36}$$

then ϕ_j is selected as the $(k+1)$th hidden node, and re-denoted as $\mathbf{p}_{k+1} = \phi_j$, and the selected hidden node regression matrix becomes $\mathbf{P}_{k+1} = [\mathbf{P}_k \ \mathbf{p}_{k+1}]$, while the candidate pool is reduced in size and becomes $\mathbf{C}_{M-k-1}$, and the remaining candidates in $\mathbf{C}_{M-k-1}$ are re-indexed as $\phi_{k+2}, \ldots, \phi_M$. Thus, the full regression matrix Φ becomes $\Phi = [\mathbf{P}_{k+1} \ \mathbf{C}_{M-k-1}]$. This change leads to some changes in matrices $\mathbf{A}$ and $\tilde{\mathbf{A}}$.

In detail, due to the interchange of ϕ_{k+1} and ϕ_j in Φ, columns $k+1$ and j of $\mathbf{A}$ and $\tilde{\mathbf{A}}$ should also be interchanged, that is

$$\begin{cases} \bar{a}_{i,k+1} = a_{i,j}, \bar{a}_{i,j} = a_{i,k+1} \\ \bar{\tilde{a}}_{i,k+1} = \tilde{a}_{i,j}, \bar{\tilde{a}}_{i,j} = \tilde{a}_{i,k+1} \end{cases}, i = 1, \ldots, k \tag{37}$$

where $\bar{a}_{i,k+1}$ and $\bar{a}_{i,j}$ denote the ith elements in columns $k+1$ and j of the updated matrices $\mathbf{A}$ and $\tilde{\mathbf{A}}$, respectively. To denote an updated element in these matrices, a bar is applied to the element hereafter.

Similarly, the $(k+1)$th and jth elements in vectors $\mathbf{a}_y$ and $\mathbf{b}$ should also be interchanged, that is,

$$\begin{cases} \bar{a}_{k+1,y} = a_{j,y}, \bar{a}_{j,y} = a_{k+1,y} \\ \bar{b}_{k+1} = b_j, \bar{b}_j = b_{k+1} \end{cases} \tag{38}$$

In addition, as the $(k+1)$th hidden node is now selected, a new row (the $(k+1)$th row) will be appended to matrices $\mathbf{A}$ and $\tilde{\mathbf{A}}$ (❯ Eqs. 29–31), that is

$$\begin{cases} a_{k+1,i} = \mathbf{p}_{k+1}^{\mathrm{T}}\phi_i - \sum_{s=1}^{k} \tilde{a}_{s,k+1}a_{s,i} &, i = k+1, \ldots, M \\ \tilde{a}_{k+1,i} = a_{k+1,i}/a_{k+1,k+1} \end{cases} \tag{39}$$

Finally, the $(k+2)$th to the last elements of both vectors $\mathbf{a}_y$ and b need to be updated if the $(k+2)$th hidden node will be selected from the remaining candidates in $\mathbf{C}_{M-k-1}$. This can be done as follows, according to their definitions (❯ 32) and (❯ 33), respectively

$$\begin{cases} \bar{a}_{i,y} = a_{i,y} - \tilde{a}_{k+1,i}a_{k+1,y} &, i = k+1, \ldots, M \\ \bar{b}_i = b_i - \tilde{a}_{k+1,i}a_{k+1,i} \end{cases} \tag{40}$$

This section has provided a framework to iteratively select the $(k+1)$th hidden node for a pool of candidates for $k = 0, 1, \ldots, (n-1)$. In the following sections, an integrated framework will be proposed for both the selection of hidden nodes and the computation of output weights.

4.2 Fast Construction of SLFNs – Phase 1

4.2.1 Algorithm: Fast construction algorithm for SLFNs – Phase 1 (FCA-I)

1. Initialization phase:
   Collect N data samples, randomly generate M candidate hidden nodes, producing the full regression matrix $\boldsymbol{\Phi} = [\phi_1, \phi_2, \ldots, \phi_M]$. Let the cost function $J = \mathbf{y}^{\mathrm{T}}\mathbf{y}$, and calculate $a_{i,y} = \mathbf{y}^{\mathrm{T}}\phi_i$, $b_i = \phi_i^{\mathrm{T}}\phi_i$ for $i = 1, \ldots, M$. Finally, calculate $\mathbf{A}$ and $\tilde{\mathbf{A}}$ according to ❯ Eqs. 29 and ❯ 31. Let $k = 0$.

2. Construction phase:
   (a) Calculate $\Delta J_{k+1}(\phi_i)$ using ❯ Eq. 35 for $i = k+1, \ldots, M$.
   (b) Based on the values of $\Delta J_{k+1}(\phi_i)$, $i = k+1, \ldots, M$, search for the hidden node which gives the maximum contribution to the cost function, assuming $\phi_j = \arg\max\{\Delta J_{k+1}(\phi_i), k < i \leq M\}$, then update the cost function as $J = J - \Delta J_{k+1}(\phi_j)$.
   (c) Reorder columns in $\mathbf{A}$ and $\tilde{\mathbf{A}}$ correspondingly, according to ❯ Eq. 37 and also update these two matrices according to ❯ Eq. 39. Finally, rearrange and update elements of $\mathbf{a}_y$ and $\mathbf{b}$ according to ❯ Eq. 38 and ❯ Eq. 40.

(d) If k is less than the desired number (say n) of hidden nodes or the cost function (the training error) is higher than expected, let $k \leftarrow k + 1$ and go to ❷ step 2a for the next round of hidden node selection.

3. Once the selection procedure is terminated, the output weights can be immediately computed from the upper triangular equation (❷ 19) using backward substitution:

$$\theta_k = a_{k,y}/a_{k,k} - \sum_{i=k+1}^{n} \tilde{a}_{k,i}\theta_i, \quad k = n, \ldots, 1 \tag{41}$$

where $\theta_n = a_{n,y}/a_{n,n}$.

Remark Some other stopping criteria can also be used in ❷ step 2d, for example, the construction phase may stop when Akaike's information criterion (AIC) begins to increase (Akaike 1974).

5   Fast Construction of SLFNs – Phase 2: Fine Tuning

The above phase 1 fast construction of SLFNs is by nature a variant of the forward subset select algorithm, which selects one hidden node at a time by maximizing the reduction of error. The main advantage is its computational efficiency; however, the disadvantage is that the resultant network is not necessarily compact due to the fact that the later introduced regressor terms (neurons) may affect the contribution of previously selected regressor terms (nodes). Then, the previously selected regressor terms may become insignificant due to the late introduced regressor terms (Peng et al. 2006; Li et al. 2006).

To improve the performance of the network obtained in phase 1, this section will introduce a method for fine-tuning the network structure. This is done by reassessing all selected hidden nodes within the same analytic framework developed in phase 1, and any insignificant regressor terms will be removed and replaced within the same analytic framework, leading to improved network performance without introducing a significant amount of computation effort.

5.1   Reevaluation of a Selected Hidden Node

The core issue in the phase 2 network construction is to reassess the contribution of all selected hidden nodes and all candidate nodes. In this section, algorithms will be introduced to reassess any of the hidden nodes selected in the phase 1 fast SLFN construction procedure.

Suppose a hidden node (out of all selected n nodes), say $\mathbf{p}_i$, $1 \leq i < n$, is to be reevaluated. Its contribution $\Delta J_n(\mathbf{p}_i)$ to the cost function needs to be recalculated and then compared with the candidate hidden nodes. Denote the maximum candidate contribution as $\Delta J_n(\phi_j)$. If $\Delta J_n(\mathbf{p}_i) < \Delta J_n(\phi_j)$, $\mathbf{p}_i$ is said to be insignificant and will be replaced with ϕ_j, and $\mathbf{p}_i$ will be put back into the candidate pool. This procedure will further reduce the cost function by $\Delta J_n(\phi_j) - \Delta J_n(\mathbf{p}_i)$, thus the network performance is further improved.

To reevaluate a selected hidden node, say $\mathbf{p}_i$, its contribution to the cost function and that of the candidates $\phi_{n+1}, \ldots, \phi_M$ needs to be recalculated. To achieve efficient computation, the regression context has to be reconstructed. From the phase 1 fast SLFN construction, it can be seen that the proper regression context should include $\mathbf{A}$, $\tilde{\mathbf{A}}$, $\mathbf{a}_y$, and $\mathbf{b}$, which are used to compute the contribution of a hidden node. The algorithm to reconstruct the regression context is as follows.

The first step in the regression context reconstruction is to move the selected hidden node of interest, say $\mathbf{p}_i$, $i \in \{1, \ldots, n-1\}$, to the nth position in the selected regression matrix $\mathbf{P}_n$, as if it was the last selected hidden node. This can be achieved by a series of interchanges between two adjacent regressor terms $\mathbf{p}_k$ and $\mathbf{p}_{k+1}$ for $k = i, \ldots, n-1$. And each time two adjacent regressor terms are interchanged in their position, the regression context, including $\mathbf{A}$, $\tilde{\mathbf{A}}$, $\mathbf{a}_y$, and $\mathbf{b}$, will be reconstructed once.

To begin with, consider two adjacent regressor terms $\mathbf{p}_k$ and $\mathbf{p}_{k+1}$ for $k = i, i+1, \ldots, n-1$ are interchanged. Then the n regressor terms in the new selected order becomes $\mathbf{p}_1, \cdots, \mathbf{p}_{k-1}, \hat{\mathbf{p}}_k, \hat{\mathbf{p}}_{k+1}, \mathbf{p}_{k+1}, \cdots, \mathbf{p}_n$, where $\hat{\mathbf{p}}_k = \mathbf{p}_{k+1}$ and $\hat{\mathbf{p}}_{k+1} = \mathbf{p}_k$. This shift of positions for the two adjacent regressor terms leads to the changes in $\mathbf{A}$, $\tilde{\mathbf{A}}$, $\mathbf{a}_y$, and $\mathbf{b}$.

For $\mathbf{A}$, according to ❷ Eq. 17, from rows 1 to $k-1$, we have

$$\left.\begin{aligned}
\hat{w}_{i,k} &= \mathbf{p}_i^T \hat{p}_k = \mathbf{p}_i^T \mathbf{p}_{k+1} = w_{i,k+1} \\
\hat{w}_{i,k+1} &= \mathbf{p}_i^T \hat{p}_{k+1} = \mathbf{p}_i^T \mathbf{p}_k = w_{i,k} \\
\hat{w}_{i,j} &= w_{i,j}, i = 1, \ldots, k-1; j = i, \ldots, k-1, k+2, \ldots, M
\end{aligned}\right\} \quad (42)$$

therefore

$$\left.\begin{aligned}
\hat{a}_{i,k} &= a_{i,k+1}, \hat{a}_{i,k+1} = a_{i,k} \\
\hat{a}_{i,j} &= a_{i,j}, i = 1, \ldots, k-1; j = i, \ldots, k-1, k+2, \ldots, M
\end{aligned}\right\} \quad (43)$$

According to ❷ Eqs. 42 and ❷ 43, it can be found that only columns k and $k+1$ are interchanged in $\mathbf{A}$ for rows 1 to $k-1$, and the rest are unchanged.

Noting that $\hat{w}_{k,k+1} = \hat{p}_k^T \hat{p}_{k+1} = \mathbf{p}_{k+1}^T \mathbf{p}_k = w_{k+1,k}$, for the kth row of $\mathbf{A}$, it holds that

$$\hat{a}_{k,k+1} = a_{k,k+1} \quad (44)$$

Noting that $\hat{w}_{k,k} = \hat{p}_k^T \hat{p}_k = \mathbf{p}_{k+1}^T \mathbf{p}_{k+1} = w_{k+1,k+1}$,

$$\hat{a}_{k,k} = \hat{w}_{k,k} - \sum_{s=1}^{k-1} \frac{\hat{a}_{s,k} \hat{a}_{s,k}}{\hat{a}_{s,s}} = a_{k+1,k+1} + a_{k,k+1} a_{k,k+1} / a_{k,k} \quad (45)$$

While for other elements of the kth row in $\mathbf{A}$ it holds that

$$\hat{a}_{k,j} = \hat{w}_{k,j} - \sum_{s=1}^{k-1} \frac{\hat{a}_{s,k} \hat{a}_{s,j}}{\hat{a}_{s,s}} = a_{k,j} + a_{k,k+1} a_{k,j} / a_{k,k}, j = k+2, \ldots, n \quad (46)$$

Also, because $\hat{w}_{k+1,k+1} = \hat{p}_{k+1}^T \hat{p}_{k+1} = \mathbf{p}_k^T \mathbf{p}_k = w_{k,k}$, for the $(k+1)$th row of $\mathbf{A}$, it holds that

$$\hat{a}_{k+1,k+1} = \hat{w}_{k+1,k+1} - \sum_{s=1}^{k} \hat{a}_{s,k+1} \hat{a}_{s,k+1} / \hat{a}_{s,s} = a_{k,k} - (a_{k,k+1})^2 / \hat{a}_{k,k} \quad (47)$$

and for $j = k+2, \cdots, M$ it holds that

$$\begin{aligned}
\hat{a}_{k+1,j} &= \hat{w}_{k+1,j} - \sum_{s=1}^{k} \hat{a}_{s,k+1} \hat{a}_{s,j} / \hat{a}_{s,s} \\
&= a_{k,j} - a_{k,k+1} \hat{a}_{k,j} / \hat{a}_{k,k}, j = k+2, \ldots, M
\end{aligned} \quad (48)$$

Since $\hat{w}_{i,j} = \hat{\mathbf{p}}_i^T \hat{\mathbf{p}}_j = \mathbf{p}_i^T \mathbf{p}_j = w_{i,j}$, $i, j > k + 1$, for the elements of the $(k + 2)$th row in $\mathbf{A}$, it holds that

$$
\begin{aligned}
\hat{a}_{k+2,j} &= \hat{w}_{k+2,j} - \sum_{s=1}^{k+1} \hat{a}_{s,k+2} \hat{a}_{s,j} / \hat{a}_{s,s} \\
&= w_{k+2,j} - \sum_{s=1}^{k-1} \frac{a_{s,k+2} a_{s,j}}{a_{s,s}} - \frac{\hat{a}_{k,k+2} \hat{a}_{k,j}}{\hat{a}_{k,k}} - \frac{\hat{a}_{k+1,k+2} \hat{a}_{k+1,j}}{\hat{a}_{k+1,k+1}}
\end{aligned}
\tag{49}
$$

From the kth and $(k + 1)$th row of $\mathbf{A}$, it could be derived that

$$
\frac{\hat{a}_{k,k+2} \hat{a}_{k,j}}{\hat{a}_{k,k}} - \frac{\hat{a}_{k+1,k+2} \hat{a}_{k+1,j}}{\hat{a}_{k+1,k+1}} = \frac{a_{k,k+2} a_{k,j}}{a_{k,k}} - \frac{a_{k+1,k+2} a_{k+1,j}}{a_{k+1,k+1}}
\tag{50}
$$

which implies that the $(k + 2)$th row of $\mathbf{A}$ has no change. Furthermore, it could be derived that rows $(x + 2)$ to n of $\mathbf{A}$ have no change.

Similarly, it could be derived that for vector $\mathbf{a}_y$ only two elements are changed as follows

$$
\left.
\begin{aligned}
\hat{a}_{k,y} &= a_{k+1,y} + a_{k,k+1} a_{k,y} / a_{k,k} \\
\hat{a}_{k+1,y} &= a_{k,y} - a_{k,k+1} \hat{a}_{k,j} / \hat{a}_{k,k}
\end{aligned}
\right\}
\tag{51}
$$

For vector $\mathbf{b}$, only the kth and the $(k + 1)$th elements are changed, that is

$$
\hat{b}_k = \hat{a}_{k,k}, \hat{b}_{k+1} = \hat{a}_{k+1,k+1}
\tag{52}
$$

To summarize, if two adjacent regressor terms $\mathbf{p}_k$ and $\mathbf{p}_{k+1}$ are interchanged in position, the regression context of $\mathbf{A}$, $\mathbf{a}_y$ and $\mathbf{b}$ can be reconstructed as shown in ❷ Eq. 53, where the changed elements (with hat) are highlighted.

$$
\mathbf{A} \Rightarrow
\begin{bmatrix}
a_{1,1} & \cdots & \hat{a}_{1,k} & \hat{a}_{1,k+1} & \cdots & a_{1,k} & a_{1,k+1} & \cdots & a_{1,M} \\
 & \ddots & \vdots & \vdots & & \vdots & \vdots & & \vdots \\
 & & \hat{a}_{k,k} & \hat{a}_{k,k+1} & \cdots & \hat{a}_{k,k} & \hat{a}_{k,k+1} & \cdots & \hat{a}_{k,M} \\
 & & & \hat{a}_{k+1,k+1} & \cdots & \hat{a}_{k+1,k} & \hat{a}_{k+1,k+1} & \cdots & \hat{a}_{k+1,M} \\
 & \mathbf{0} & & & \ddots & \vdots & \vdots & & \vdots \\
 & & & & & a_{k,k} & a_{k,k+1} & \cdots & a_{k,M}
\end{bmatrix},
$$

$$
\mathbf{a}_y \Rightarrow
\begin{bmatrix}
a_{1,y} \\
\vdots \\
\hat{a}_{k,y} \\
\hat{a}_{k+1,y} \\
\vdots \\
a_{M,y}
\end{bmatrix}
, \mathbf{b} \Rightarrow
\begin{bmatrix}
b_1 \\
\vdots \\
\hat{a}_k \\
\hat{a}_{k+1} \\
\vdots \\
b_M
\end{bmatrix}
\tag{53}
$$

The above analysis shows how the regression context is changed as two adjacent regressor terms interchange their positions in the selected regression matrix $\mathbf{P}_n$. Through a series of such interchanges between two adjacent regressor terms, $\mathbf{p}_i$ can be moved to the nth position in the selected regression matrix $\mathbf{P}_n$. Then the contribution of $\mathbf{p}_i$ of interest can be computed as

$$\Delta J_n(\mathbf{p}_i) = \Delta J_n(\hat{\mathbf{p}}_n) = a_{n,y}^2/a_{n,n} \tag{54}$$

where $\mathbf{p}_i = \hat{\mathbf{p}}_n$.

5.2 Reevaluation of a Candidate Hidden Node

The main objective in the phase 2 fast SLFN construction is to reevaluate the contribution of a selected hidden node, and this contribution is then compared with that of the candidate nodes. If its contribution is less than a candidate node, it will be replaced. The contribution of a candidate node to the cost function is calculated as if the last selected hidden node is pruned, and the following quantities will have to be computed:

$$\left.\begin{array}{l} a_{j,y}^{(-i)} = a_{j,y} + a_{n,j}a_{n,y}/a_{n,n} \\ b_j^{(-i)} = b_j + (a_{n,n+1})^2/a_{n,n} \end{array}\right\}, j = n+1,\ldots,M \tag{55}$$

These quantities are in fact the values of the corresponding elements in vector $\mathbf{a}_y$ and $\mathbf{b}$ when the regressor term $\mathbf{p}_i$ is removed from the model. Then, the contribution of all candidates can be computed as

$$\Delta J_n(\phi_j) = (a_{j,y}^{(-i)})^2/b_j^{(-i)}, j = n+1,\ldots,M \tag{56}$$

Now, the next step is to identify the maximum contribution among all the candidates

$$\Delta J_n(\phi_s) = \max_{j=n+1,\ldots,M}\{\Delta J_n(\phi_j)\} \tag{57}$$

If $\Delta J_n(\phi_s) > \Delta J_n(\mathbf{p}_i)$, then $\mathbf{p}_i$ should be replaced by ϕ_s, while $\mathbf{p}_i$ should be put back into the candidate pool, taking the position of ϕ_s in the candidate pool. In this case, the regression context needs to be updated again to reflect the interchange of ϕ_s and $\mathbf{p}_i$.

For matrix $\mathbf{A}$, only the nth and sth column starting from row 1 to row $n-1$ need to be interchanged, noting that $\mathbf{p}_i$ has already been moved to the nth position in the regression matrix $\mathbf{P}_n$. For the nth row of $\mathbf{A}$, the elements are calculated as

$$\left.\begin{array}{l} \hat{a}_{n,n} = b_s^{(-i)}, \hat{a}_{n,s} = a_{n,s}, w_{n,k} = \hat{\mathbf{p}}_n^T\phi_k \\ \hat{a}_{n,k} = w_{n,k} - \sum_{j=1}^{n-1}\frac{a_{j,n}a_{j,k}}{a_{j,j}} \\ k = n+1,\ldots,M, k \neq s \end{array}\right\} \tag{58}$$

where $\hat{\mathbf{p}}_n$ is the $\mathbf{p}_i$ which has now been moved to the nth position. In the meantime, the nth to Mth elements of vector $\mathbf{a}_y$ and $\mathbf{b}$ are updated, respectively, as

$$\left.\begin{array}{l} \hat{a}_{n,y} = a_{n,y}^{(-i)}, a_{s,y} = a_{n,y} - \hat{a}_{n,s}\hat{a}_{n,y}/\hat{a}_{n,n}, \\ a_{j,y} = a_{j,y}^{(-i)} - \frac{\hat{a}_{n,j}\hat{a}_{n,y}}{\hat{a}_{n,n}}, j = n+1,\ldots,M, j \neq s \end{array}\right\} \tag{59}$$

and

$$
\left.
\begin{aligned}
\hat{b}_n &= b_s^{(-i)}, \hat{b}_s = a_{n,n} - \left(\hat{a}_{n,n+1}\right)^2/\hat{a}_{n,n} \\
\hat{b}_j &= b_j^{(-i)} - \left(\hat{a}_{n,n+1}\right)^2/\hat{a}_{n,n}, j = n+1,\ldots,M, j \neq s
\end{aligned}
\right\}
\tag{60}
$$

It should be noted that if **A** is updated, the corresponding element of $\tilde{\mathbf{A}}$ should be recalculated as well according to its definition.

5.3 Fast Construction of SLFNs: Phase 2

5.3.1 Algorithm: Fast Construction Algorithm for SLFNs – Phase 2 (FCA-II)

1. Let $i = n - 1$, go to ❯ step 2 to perform a check loop.
2. For $j = i, \ldots, n - 1$, exchange regressor terms $\mathbf{p}_j$ and $\mathbf{p}_{j+1}$ to move the xth regressor term to the nth position. Elements of **A**, $\mathbf{a}_y$, and **b** are updated correspondingly for each regressor term interchange as ❯ Eqs. 42–53.
3. Compute the contribution of $\mathbf{p}_i$ once it has been moved to the nth position in the regression matrix $\mathbf{P}_n$ according to ❯ Eq. 54 and those of all the $M - n$ candidates according to ❯ Eq. 56.
4. Identify the candidate ϕ_s that has the maximum contributions. If $\Delta J_n(\mathbf{p}_i) \geq \Delta J_n(\phi_s)$ let $i \leftarrow i - 1$; otherwise select ϕ_s as the new nth regressor by interchanging $\mathbf{p}_i$ with ϕ_j and modify the regression context as ❯ Eqs. 58–60, and then let $i = k - 1$.
5. If $i = 0$, all the selected regressors are reevaluated (in this case, the cost function converges to a minimum), and are terminated from this reviewing procedure. Otherwise, continue the reevaluation procedure from ❯ step 2. Note that as each insignificant regressor is being replaced, the cost function is further reduced by $\Delta J_k(\phi_s) - \Delta J_k(\mathbf{p}_i)$. The monotonic decrease will make the cost function converge to a minimum, hence improving the model performance.

6 Performance Evaluation

6.1 Benchmark Problems

The performance of the proposed fast construction algorithm (FCA) for SLFNs was evaluated using the benchmark problems described in ❯ Table 1, which includes two regression applications (California Housing and Abalone) (Blake and Merz 1998), one time series prediction (Mackey and Glass 1977), and one nonlinear system modeling (Piroddi and Spinelli 2003).

1. **California Housing**: California Housing is a dataset obtained from the StatLib repository. There are 20,640 observations for predicting the price of houses in California. Information on the variables was collected using all the block groups in California from the 1990 census.
2. **Abalone**: The abalone problem is to estimate the age of abalone from physical measurements (Blake and Merz 1998).

☐ **Table 1**

Specification of benchmark datasets

Dataset	# Attributes	# Training data	# Testing data
CalHousing	8	8,000	12,640
Abalone	8	3,000	1,177
Mackey–Glass	4	20,000	500
Ident	6	20,000	500

3. **Mackey–Glass**: The need for time series prediction arises in many real-world problems such as detecting arrhythmia in heartbeats, stock market indices, etc. One of the classical benchmark problems in the literature is the chaotic Mackey–Glass differential delay equation given by Mackey and Glass (1977):

$$\frac{dx(t)}{dt} = \frac{ax(t-\tau)}{1 + x^{10}(t-\tau)} - bx(t) \tag{61}$$

for $a = 0.2, b = 0.1$, and $\tau = 17$. Integrating the equation over the time interval $[t, t + \Delta t]$ by the trapezoidal rule yields:

$$x(t + \Delta t) = \frac{2 - b\Delta t}{2 + \Delta t}x(t) + \frac{a\Delta t}{2 + b\Delta t}\left[\frac{x(t + \Delta t - \tau)}{1 + x^{10}(t + \Delta t - \tau)} + \frac{x(t-\tau)}{1 + x^{10}(t-\tau)}\right] \tag{62}$$

The time series is generated under the condition $x(t - \tau) = 0.3$ for $0 \le t \le \tau$ and predicted with $\upsilon = 50$ sample steps ahead using the four past samples: $s_{n-\upsilon}, s_{n-\upsilon-6}, s_{n-\upsilon-12}$, and $s_{n-\upsilon-18}$. Hence, the nth input–output instance is:

$$\begin{aligned}\mathbf{x}_n &= \left[s_{n-\upsilon}, s_{n-\upsilon-6}, s_{n-\upsilon-12}, s_{n-\upsilon-18}\right]^{\mathrm{T}} \\ y_n &= s_n\end{aligned} \tag{63}$$

In this simulation, $\Delta t = 1$, and the training samples were from $t = 1$ to 20,000 and the testing samples were from $t = 20,001$ to $t = 20,500$.

4. **Ident**: Modeling and identification of nonlinear dynamic systems has been intensively researched in recent decades due to its wide applications in almost all engineering sectors. This chapter considers the following artificial nonlinear discrete system (Piroddi and Spinelli 2003):

$$\begin{aligned}y(t) = {} & 0.8y(t-1) + u(t-1) - 0.3u(t-2) \\ & - 0.4u(t-3) + 0.25u(t-1)u(t-2) \\ & - 0.2u(t-2)u(t-3) - 0.3u(t-1)^3 \\ & + 0.24u(t-2)^3 + \epsilon(t)\end{aligned} \tag{64}$$

where $\epsilon(t)$ is a random noise within the range $[0, 0.02]$.

Data points numbering 20,500 were generated in this case where $u(t)$ was uniformly distributed within $[-1, 1]$. The first 20,000 samples were used for network training and the other 500 points for testing.

In this example, the inputs of the SLFN are $y(t-1), y(t-2)$, $y(t-3), u(t-1), u(t-2), u(t-3)$, and the output is $y(t)$. In this chapter, a slightly large number of training samples were generated in the last two examples, with the aim to test the computational demand of the algorithms when the size of the problem becomes large.

The first three problems were used in the literature on the performance evaluation of ELM (see Huang et al. 2006a, b); therefore, the results published in these papers can be directly used for the comparison purpose.

Following the same procedures as in Huang et al. (2006a, b), both the Gaussian RBF activation function and the sigmoidal additive activation function were used in the simulations. In the simulations, all the inputs and outputs were normalized so that they fall within the range [0, 1]. For SLFNs with additive hidden nodes, the input weights and biases were randomly chosen from the range [−1, 1]. For SLFNs with RBF hidden nodes, the centers were

◘ **Table 2**
Comparison for FCA-I and ELM on the regression problems

Data sets	Algorithms	Training time (s)	RMSE		# Nodes
			Training	Testing	
CalHousing	FCA-I(Sigmoid)	0.0094	0.1452	0.1457	10
	ELM(Sigmoid)	0.0361	0.1449	0.1452	
	FCA-I(RBF)	0.0124	0.1528	0.1521	
	ELM(RBF)	0.0343	0.1567	0.1566	
	FCA-I(Sigmoid)	0.0500	0.1399	0.1410	20
	ELM(Sigmoid)	0.0984	0.1403	0.1413	
	FCA-I(RBF)	0.0500	0.1402	0.1411	
	ELM(RBF)	0.0950	0.1404	0.1415	
	FCA-I(Sigmoid)	0.1189	0.1357	0.1371	30
	ELM(Sigmoid)	0.2015	0.1355	0.1373	
	FCA-I(RBF)	0.1219	0.1351	0.1364	
	ELM(RBF)	0.2002	0.1366	0.1377	
Abalone	FCA-I(Sigmoid)	0.0000	0.0903	0.0907	5
	ELM(Sigmoid)	0.0016	0.0903	0.0890	
	FCA-I(RBF)	0.0000	0.1052	0.1031	
	ELM(RBF)	0.0032	0.0956	0.0935	
	FCA-I(Sigmoid)	0.0064	0.0793	0.0818	10
	ELM(Sigmoid)	0.0154	0.0788	0.0811	
	FCA-I(RBF)	0.0031	0.0841	0.0858	
	ELM(RBF)	0.0063	0.0855	0.0857	
	FCA-I(Sigmoid)	0.0125	0.0757	0.0796	20
	ELM(Sigmoid)	0.0157	0.0758	0.0801	
	FCA-I(RBF)	0.0124	0.0766	0.0797	
	ELM(RBF)	0.0158	0.0765	0.0798	

randomly chosen from the range $[-1, 1]$, and the impact factors were randomly chosen from the range $(0, 0.5)$. Each of the test cases reported in this chapter were performed 10 times, and the results reported are the mean values. All the simulations in this chapter were performed in the MATLAB 6.0 environment running on an ordinary laptop with 1.66 GHZ CPU.

6.2   Performance Evaluation of FCA-I with ELM

The ELM has been compared with those methods in previous works (Huang et al. 2006a, b), and it has been shown that the ELM outperforms those methods in terms of computational complexity and network performance, for details please refer to Huang et al. (2006a, b).

■ Table 3

Comparison for FCA-I and ELM on the time-series and system identification problems

Data sets	Algorithms	Training time (s)	RMSE		# Nodes
			Training	Testing	
Ident	**FCA-I(Sigmoid)**	0.0157	0.1058	0.1117	10
	ELM(Sigmoid)	0.0469	0.1258	0.1332	
	FCA-I(RBF)	0.0188	0.2697	0.2765	
	ELM(RBF)	0.0437	0.2642	0.2695	
	FCA-I(Sigmoid)	0.0638	0.0675	0.0701	20
	ELM(Sigmoid)	0.1250	0.0718	0.0743	
	FCA-I(RBF)	0.0626	0.1455	0.1512	
	ELM(RBF)	0.1250	0.1400	0.1455	
	FCA-I(Sigmoid)	0.1438	0.0553	0.0561	30
	ELM(Sigmoid)	0.2626	0.0561	0.0571	
	FCA-I(RBF)	0.1440	0.1114	0.1152	
	ELM(RBF)	0.2625	0.0989	0.1020	
Mackey–Glass	**FCA-I(Sigmoid)**	0.0155	0.0092	0.0089	10
	ELM(Sigmoid)	0.0406	0.0100	0.0097	
	FCA-I(RBF)	0.0187	0.0690	0.0656	
	ELM(RBF)	0.0437	0.0501	0.0478	
	FCA-I(Sigmoid)	0.0653	0.0045	0.0042	20
	ELM(Sigmoid)	0.1265	0.0046	0.0043	
	FCA-I(RBF)	0.0626	0.0118	0.0111	
	ELM(RBF)	0.1250	0.0104	0.0099	
	FCA-I(Sigmoid)	0.1451	0.0025	0.0024	30
	ELM(Sigmoid)	0.2564	0.0025	0.0023	
	FCA-I(RBF)	0.1423	0.0035	0.0033	
	ELM(RBF)	0.2532	0.0039	0.0038	

Therefore, here the performance evaluation was only made between FCA-I and ELM on the four benchmark problems. For all the problems studied here, 10 trials were performed for each algorithm, and in each trial the hidden nodes were randomly generated. It should be noted that in order to compare the performance of FCA-I with ELM, the number of candidates for SLFNs should be equal to the desired number of hidden nodes, that is $M = n$. This implies that the FCA-I proposed in the previous section will not select the hidden nodes from a large pool of candidates. In this case, FCA-I will perform the same task as ELM, that is to compute the linear output weights for fixed SLFNs with random nodes.

❯ *Table 2* summarizes the results (mean values over the 10 trials) for the first two regression problems. ❯ *Table 2* reveals that: (1) the training and testing performance of both the FCA-I and ELM are similar in almost all cases for the two regression problems; (2) the FCA-I performed consistently faster than ELM in all cases; (3) for these two regression

◻ Table 4

Comparison of FCA-I and OLS on the regression problems

Data sets	Algorithms	Training time (s)	RMSE		# Nodes out of 100
			Training	Testing	
CalHousing	FCA-I(Sigmoid)	0.2376	0.1401	0.1413	10
	OLS(Sigmoid)	1.8467	0.1411	0.1423	
	FCA-I(RBF)	0.2364	0.1401	0.1411	
	OLS(RBF)	1.8781	0.1396	0.1403	
	FCA-I(Sigmoid)	0.4515	0.1334	0.1350	20
	OLS(Sigmoid)	3.49840	0.13438	0.13619	
	FCA-I(RBF)	0.4469	0.1332	0.1341	
	OLS(RBF)	3.5391	0.1332	0.1345	
	FCA-I(Sigmoid)	0.6425	0.1305	0.1322	30
	OLS(Sigmoid)	5.00010	0.13084	0.13271	
	FCA-I(RBF)	0.6407	0.1304	0.1322	
	OLS(RBF)	5.0046	0.1295	0.1307	
Abalone	FCA-I(Sigmoid)	0.0316	0.0794	0.0820	5
	OLS(Sigmoid)	0.2002	0.0798	0.0819	
	FCA-I(RBF)	0.0313	0.0823	0.0822	
	OLS(RBF)	0.1970	0.0805	0.0827	
	FCA-I(Sigmoid)	0.0689	0.0764	0.0797	10
	OLS(Sigmoid)	0.3797	0.0765	0.0794	
	FCA-I(RBF)	0.0659	0.0765	0.0789	
	OLS(RBF)	0.3876	0.0760	0.0778	
	FCA-I(Sigmoid)	0.1250	0.0745	0.0813	20
	OLS(Sigmoid)	0.7313	0.0745	0.0798	
	FCA-I(RBF)	0.1234	0.0745	0.0786	
	OLS(RBF)	0.7330	0.0745	0.0783	

16 Fast Construction of Single-Hidden-Layer Feedforward Networks

problems, the network size can be quite small, and further increases of the number of hidden nodes did not have significant impact on the network performance (training and testing errors).

❯ *Table 3* lists the results for the last two problems, that is, the time series prediction and nonlinear system modeling. Several observations can be identified from ❯ *Table 3*: (1) SLFNs with additive hidden nodes perform much better than SLFNs with RBF nodes in these two problems; (2) the training and testing performance of both the FCA-I and ELM again are quite similar; (3) the FCA-I performs consistently faster than the ELM in all cases, as observed in the first two regression problems; (4) the number of hidden nodes for SLFNs does have significant impact on the network performance. It implies that these two problems are more complex than the previous two cases, and therefore require SLFNs of bigger number of random hidden nodes.

■ Table 5
Comparison of FCA-I and OLS on the time-series and system identification problems

Data sets	Algorithms	Training time (s)	RMSE		# Nodes out of 100
			Training	Testing	
Ident	FCA-I(Sigmoid)	0.2986	0.0659	0.0645	10
	OLS(Sigmoid)	2.3312	0.0610	0.0621	
	FCA-I(RBF)	0.2984	0.1128	0.1159	
	OLS(RBF)	2.3453	0.1155	0.1200	
	FCA-I(Sigmoid)	0.5659	0.0478	0.0477	20
	OLS(Sigmoid)	4.4079	0.0477	0.0484	
	FCA-I(RBF)	0.5655	0.0781	0.0803	
	OLS(RBF)	4.4327	0.0746	0.0759	
	FCA-I(Sigmoid)	0.8063	0.0410	0.0414	30
	OLS(Sigmoid)	6.2485	0.0409	0.0412	
	FCA-I(RBF)	0.8062	0.0582	0.0588	
	OLS(RBF)	6.2847	0.0594	0.0600	
Mackey–Glass	FCA-I(Sigmoid)	0.2985	0.0070	0.0066	10
	OLS(Sigmoid)	2.3470	0.0066	0.0064	
	FCA-I(RBF)	0.2969	0.0111	0.0105	
	OLS(RBF)	2.3580	0.0138	0.0132	
	FCA-I(Sigmoid)	0.5704	0.0030	0.0028	20
	OLS(Sigmoid)	4.4406	0.0028	0.0026	
	FCA-I(RBF)	0.5703	0.0042	0.0039	
	OLS(RBF)	4.4546	0.0050	0.0048	
	FCA-I(Sigmoid)	0.8079	0.0018	0.0018	30
	OLS(Sigmoid)	6.2860	0.0019	0.0018	
	FCA-I(RBF)	0.8091	0.0025	0.0025	
	OLS(RBF)	6.2891	0.0026	0.0025	

To conclude, both ❷ *Tables 2* and ❷ *3* have shown that the FCA-I performs consistently faster than the ELM for fixed SLFNs with random nodes.

6.3 Performance Evaluation of FCA-I with OLS

As discussed in previous sections, a compact network is always desirable for many real-life applications. Therefore, to select a subset of nodes from a large pool of randomly generated candidates is one of the most popular approaches. For this subset selection problem, OLS (Chen and Wigger 1995; Chen et al. 1989) has been intensively used in the literature. In this study, both FCA-I and OLS are applied to the selection of hidden nodes for the four problems.

❏ Table 6
Comparison of FCA-(I+II) and FCA-I only on the regression problems

Data sets	Algorithms	RMSE		# Nodes out of 100
		Training	Testing	
CalHousing	FCA-I(Sigmoid)	0.1401	0.1413	10
	FCA-(I+II)(Sigmoid)	0.1380	0.1393	
	FCA-I(RBF)	0.1401	0.1411	
	FCA-(I+II)(RBF)	0.1379	0.1390	
	FCA-I(Sigmoid)	0.1334	0.1350	20
	FCA-(I+II)(Sigmoid)	0.1313	0.1330	
	FCA-I(RBF)	0.1332	0.1341	
	FCA-(I+II)(RBF)	0.1315	0.1329	
	FCA-I(Sigmoid)	0.1305	0.1322	30
	FCA-(I+II)(Sigmoid)	0.1285	0.1308	
	FCA-I(RBF)	0.1304	0.1322	
	FCA-(I+II)(RBF)	0.1280	0.1295	
Abalone	FCA-I(Sigmoid)	0.0794	0.0820	5
	FCA-(I+II)(Sigmoid)	0.0781	0.0815	
	FCA-I(RBF)	0.0823	0.0822	
	FCA-(I+II)(RBF)	0.0794	0.0809	
	FCA-I(Sigmoid)	0.0764	0.0797	10
	FCA-(I+II)(Sigmoid)	0.0757	0.0791	
	FCA-I(RBF)	0.0765	0.0789	
	FCA-(I+II)(RBF)	0.0756	0.0783	
	FCA-I(Sigmoid)	0.0745	0.0813	20
	FCA-(I+II)(Sigmoid)	0.0738	0.0865	
	FCA-I(RBF)	0.0745	0.0786	
	FCA-(I+II)(RBF)	0.0738	0.0781	

For all four problems, the number of candidate nodes is set to be 100, that is, the hidden nodes are selected from a pool of 100 randomly generated candidates.

❯ *Table 4* summarizes the results for the first two regression problems. ❯ *Table 4* shows that: (1) the training and testing performance of both the FCA-I and OLS are similar in all cases for the two regression problems; (2) the FCA-I performs significantly faster than OLS in all cases; (3) the network performance is not significantly improved as the number of hidden nodes increases.

❯ *Table 5* lists the results for the last two problems, that is, the time series prediction and nonlinear system modeling. It is observed that: (1) SLFNs with additive nodes perform much better than SLFNs with RBF nodes in these two cases; (2) the training and testing performance of both the FCA-I and OLS again are quite similar; (3) FCA-I performs significantly faster than OLS in all cases, this observation agrees with the previous findings for the first two regression problems; (4) the network performance improves as the number of hidden nodes increases.

◼ Table 7

Comparison of FCA-(I+II) and FCA-I only on the time-series and system identification problems

Data sets	Algorithms	RMSE		# Nodes out of 100
		Training	Testing	
Ident	FCA-I(Sigmoid)	0.0645	0.0659	10
	FCA-(I+II)(Sigmoid)	0.0534	0.0539	
	FCA-I(RBF)	0.1128	0.1159	
	FCA-(I+II)(RBF)	0.0973	0.0994	
	FCA-I(Sigmoid)	0.0478	0.0477	20
	FCA-(I+II)(Sigmoid)	0.0428	0.0429	
	FCA-I(RBF)	0.0781	0.0803	
	FCA-(I+II)(RBF)	0.0623	0.0641	
	FCA-I(Sigmoid)	0.0410	0.0414	30
	FCA-(I+II)(Sigmoid)	0.0330	0.0335	
	FCA-I(RBF)	0.0582	0.0588	
	FCA-(I+II)(RBF)	0.0499	0.0513	
Mackey–Glass	FCA-I(Sigmoid)	0.0070	0.0066	10
	FCA-(I+II)(Sigmoid)	0.0047	0.0044	
	FCA-I(RBF)	0.0111	0.0105	
	FCA-(I+II)(RBF)	0.0096	0.0091	
	FCA-I(Sigmoid)	0.0030	0.0028	20
	FCA-(I+II)(Sigmoid)	0.0023	0.0022	
	FCA-I(RBF)	0.0042	0.0039	
	FCA-(I+II)(RBF)	0.0032	0.0030	
	FCA-I(Sigmoid)	0.0018	0.0018	30
	FCA-(I+II)(Sigmoid)	0.0015	0.0014	
	FCA-I(RBF)	0.0025	0.0025	
	FCA-(I+II)(RBF)	0.0020	0.0020	

To conclude, both ❷ *Tables 4* and ❷ 5 have shown that the FCA-I performs consistently faster than the OLS for subset selection problems.

6.4   Performance Evaluation of FCA-II with FCA-I

While the previous sections on performance evaluation focus on the computational complexity of FCA-I with OLS and ELM, this section focuses on the efficacy of FCA-II on improving the network performance of the SLFNs with random nodes. It has been shown in the previous sections that FCA-I, OLS, and ELM produce the network of similar generalization performance, except for their difference in the computational complexity. This section will compare the network compactness produced by FCA (FCA-I + FCA-II) with FCA-I only. ❷ *Tables 6* and ❷ 7 show the performance of the two algorithms. Again, each test case in the tables was performed 10 times, and these values are the mean values of the 10 tests. It reveals that phase II fast network construction (FCA-II) can improve the compactness of the network produced in phase I when the network size is small. Among these four benchmark problems, the improvements on regression problems are less significant while more significant on the time series and nonlinear dynamic systems modeling.

7   Conclusion

In this chapter, an FCA has been proposed for a class of SLFNs with fully random hidden nodes. An integrated framework has been developed through a two-phase network construction procedure with improved computational efficiency and network performance. Detailed performance evaluation has been performed on four benchmark problems ranging from regression, to time series prediction, to nonlinear system identification. For the first phase fast construction algorithm, FCA-I, the results show that it can consistently achieve faster computation of the linear weights than the ELM algorithm for fixed SLFNs, and consistently perform faster selection of hidden nodes than the OLS method. It has also been shown that for SLFNs with random hidden nodes, hidden node selection can significantly improve the performance of compact networks. Finally, it has been shown that the second phase (fine-tuning phase) fast construction algorithm, FCA-II, can significantly improve the network performance on time series and nonlinear system modeling problems. We believe that the fast construction approach proposed in this chapter can be used in generalized SLFNs with non-neural-alike hidden nodes proposed in Huang and Chen (2007), which is worth investigating in the future.

Acknowledgment

K. Li would like to acknowledge the helpful comments from Lei Chen of the National University of Singapore. He would also like to acknowledge the support of the International Exchange program of Queen's University Belfast.

References

Akaike H (1974) New look at the statistical model identification. IEEE Trans Automat Cont AC-19(1): 716–723

Ampazis N, Perantonis SJ (2002) Two highly efficient second-order algorithms for training feedforward networks. IEEE Trans Neural Netw 13(3): 1064–1074

Bishop CM (1995) Neural networks for pattern recognition. Clarendon Press, Oxford

Blake C, Merz C (1998) UCI repository of machine learning databases. In: Department of Information and Computer Sciences, University of California, Irvine, CA. http://www.ics.uci.edu/~mlearn/MLRepository.html

Chen S, Wigger J (1995) Fast orthogonal least squares algorithm for efficient subset model selection. IEEE Trans Signal Process 43(7):1713–1715

Chen S, Billings SA, Luo W (1989) Orthogonal least squares methods and their application to non-linear system identification. Int J Control 50(5):1873–1896

Chen S, Cowan CFN, Grant PM (1991) Orthogonal least squares learning algorithm for radial basis functions. IEEE Trans Neural Netw 2:302–309

Chng ES, Chen S, Mulgrew B (1996) Gradient radial basis function networks for nonlinear and nonstationary time series prediction. IEEE Trans Neural Netw 7(1):190–194

Gomm JB, Yu DL (March 2000) Selecting radial basis function network centers with recursive orthogonal least squares training. IEEE Trans Neural Netw 11(2):306–314

Hagan MT, Menhaj MB (1994) Training feedforward networks with the Marquardt algorithm. IEEE Trans Neural Netw 5(6):989–993

Handoko SD, Keong KC, Soon OY, Zhang GL, Brusic V (2006) Extreme learning machine for predicting HLA-peptide binding. Lect Notes Comput Sci 3973:716–721

Hong X, Mitchell RJ, Chen S, Harris CJ, Li K, Irwin G (2008) Model selection approaches for nonlinear system identification: a review. Int J Syst Sci 39(10):925–946

Huang G-B, Chen L (2007) Convex incremental extreme learning machine. Neurocomputing 70 (16–18):3056–3062

Huang G-B, Saratchandran P, Sundararajan N (2005) A generalized growing and pruning RBF (GGAP-RBF) neural network for function approximation. IEEE Trans Neural Netw 16(1):57–67

Huang G-B, Chen L, Siew C-K (2006a) Universal approximation using incremental constructive feedforward networks with random hidden nodes. IEEE Trans Neural Netw 17(4):879–892

Huang G-B, Zhu Q-Y, Siew C-K (2006b) Extreme learning machine: theory and applications. Neurocomputing 70:489–501

Huang G-B, Zhu Q-Y, Mao KZ, Siew C-K, Saratchandran P, Sundararajan N (2006c) Can threshold networks be trained directly? IEEE Trans Circuits Syst II 53(3):187–191

Kadirkamanathan V, Niranjan M (1993) A function estimation approach to sequential learning with neural networks. Neural Comput 5:954–975

Korenberg MJ (1988) Identifying nonlinear difference equation and functional expansion representations: the fast orthogonal algorithm. Ann Biomed Eng 16:123–142

Lawson L, Hanson RJ (1974) Solving least squares problem. Prentice-Hall, Englewood Cliffs, NJ

Li K, Peng J, Irwin GW (2005) A fast nonlinear model identification method. IEEE Trans Automa Cont 50(8):1211–1216

Li K, Peng J, Bai EW (2006) A two-stage algorithm for identification of nonlinear dynamic systems. Automatica 42(7):1189–1197

Li K, Peng J, Bai E (2009) Two-stage mixed discrete-continuous identification of radial basis function (RBF) neural models for nonlinear systems. IEEE Trans Circuits Syst I Regular Pap 56(3):630–643

Li M-B, Huang G-B, Saratchandran P, Sundararajan N (2005) Fully complex extreme learning machine. Neurocomputing 68:306–314

Liang N-Y, Huang G-B, Saratchandran P, Sundararajan N (2006a) A fast and accurate on-line sequential learning algorithm for feedforward networks. IEEE Trans Neural Netw 17(6):1411–1423

Liang N-Y, Saratchandran P, Huang G-B, Sundararajan N (2006b) Classification of mental tasks from EEG signals using extreme learning machine. Int J Neural Syst 16(1):29–38

Liu Y, Loh HT, Tor SB (2005) Comparison of extreme learning machine with support vector machine for text classification. Lect Notes Comput Sci 3533:390–399

Mackey MC, Glass L (1977) Oscillation and chaos in physiological control systems. Science 197:287–289

Mao KZ, Huang G-B (2005) Neuron selection for RBF neural network classifier based on data structure preserving criterion. IEEE Trans Neural Netw 16(6):1531–1540

Marquardt D (1963) An algorithm for least-squares estimation of nonlinear parameters. SIAM J Appl Math 11:431–441

McLoone S, Brown MD, Irwin GW, Lightbody G (1998) A hybrid linear/nonlinear training algorithm for feedforward neural networks. IEEE Trans Neural Netw 9:669–684

Miller AJ (1990) Subset selection in regression. Chapman & Hall, London

Musavi M, Ahmed W, Chan K, Faris K, Hummels D (1992) On training of radial basis function classifiers. Neural Netw 5:595–603

Panchapakesan C, Palaniswami M, Ralph D, Manzie C (2002) Effects of moving the centers in an RBF network. IEEE Trans Neural Netw 13:1299–1307

Peng H, Ozaki T, Haggan-Ozaki V, Toyoda Y (2003) A parameter optimization method for radial basis function type models. IEEE Trans Neural Netw 14:432–438

Peng J, Li K, Huang DS (2006) A hybrid forward algorithm for RBF neural network construction. IEEE Trans Neural Netw 17(11):1439–1451

Peng J, Li K, Irwin GW (2007) A novel continuous forward algorithm for RBF neural modelling. IEEE Trans Automat Cont 52(1):117–122

Peng J, Li K, Irwin GW (2008) A new Jacobian matrix for optimal learning of single-layer neural nets. IEEE Trans Neural Netw 19(1):119–129

Piroddi L, Spinelli W (2003) An identification algorithm for polynomial NARX models based on simulation error minimization. Int J Cont 76:1767–1781

Platt J (1991) A resource-allocating network for function interpolation. Neural Comput 3:213–225

Press WH, Teukolsky SA, Vetterling WT, Flannery BP (1992) Numerical recipes in C: the art of scientific computing. Cambridge University Press, Cambridge

Rao CR, Mitra SK (1971) Generalized inverse of matrices and its applications. Wiley, New York

Serre D (2002) Matrices: theory and applications. Springer, New York

Sutanto EL, Mason JD, Warwick K (1997) Mean-tracking clustering algorithm for radial basis function centre selection. Int J Control 67:961–977

Wang D, Huang G-B (2005) Protein sequence classification using extreme learning machine. In: Proceedings of international joint conference on neural networks (IJCNN2005), Montreal, Canada, 31 July – 4 August 2005

Xu J-X, Wang W, Goh JCH, Lee G (2005) Internal model approach for gait modeling and classification. In: the 27th annual international conference of the IEEE, Engineering in Medicine and Biology Society (EMBS), Shanghai, China, 1–4 September 2005

Yeu C-WT, Lim M-H, Huang G-B, Agarwal A, Ong Y-S (2006) A new machine learning paradigm for terrain reconstruction. IEEE Geosci Rem Sens Lett 3(3):382–386

Yingwei L, Sundararajan N, Saratchandran P (1997) A sequential learning scheme for function approximation using minimal radial basis function (RBF) neural networks. Neural Comput 9:461–478

Zhang GL, Billings SA (1996) Radial basis function network configuration using mutual information and the orthogonal least squares algorithm. Neural Netw 9:1619–1637

Zhang R, Huang G-B, Sundararajan N, Saratchandran P (2007) Multi-category classification using an extreme learning machine for microarray gene expression cancer diagnosis. IEEE/ACM Trans Comput Biol Bioinform 4(3):485–495

Zhu QM, Billings SA (1996) Fast orthogonal identification of nonlinear stochastic models and radial basis function neural networks. Int J Control 64(5):871–886

17 Modeling Biological Neural Networks

Joaquin J. Torres[1] · *Pablo Varona*[2]
[1]Institute "Carlos I" for Theoretical and Computational Physics and Department of Electromagnetism and Matter Physics, Facultad de Ciencias, Universidad de Granada, Spain
jtorres@ugr.es
[2]Departamento de Ingeniería Informática, Universidad Autónoma de Madrid, Spain
pablo.varona@uam.es

G. Rozenberg et al. (eds.), *Handbook of Natural Computing*, DOI 10.1007/978-3-540-92910-9_17,
© Springer-Verlag Berlin Heidelberg 2012

Abstract

In recent years, many new experimental studies on emerging phenomena in neural systems have been reported. The high efficiency of living neural systems to encode, process, and learn information has stimulated an increased interest among theoreticians in developing mathematical approaches and tools to model biological neural networks. In this chapter we review some of the most popular models of neurons and neural networks that help us understand how living systems perform information processing. Beyond the fundamental goal of understanding the function of the nervous system, the lessons learned from these models can also be used to build bio-inspired paradigms of artificial intelligence and robotics.

1   Introduction

Neural systems, including those of small insects and invertebrates, are the most efficient and adaptable information-processing devices even when compared to state-of-the-art human technology. There are many computing strategies that can be learned from living neural networks. Computational neuroscience studies the nervous system from the point of view of its functionality and relies both on experimental data and on theoretical models of individual neurons and networks. Experimental data alone is not enough to address information processing because neural systems are only partially observable with the current electrophysiological and imaging techniques. Models provide a mechanism to address the role of all variables that are thought to be involved in a particular aspect of the system under study. In many cases, the models can provide new insight and hypotheses that can be tested experimentally. Computational neuroscience involves multidisciplinary research and contributes to a better understanding of the nervous system (including several brain diseases), to the development of prosthetic devices and brain–machine interfaces, and to the design of novel paradigms of artificial intelligence and robotics.

When modeling biological neural media, one must take into account different components that extend over several temporal and spatial scales and involve individual cells, synapses, and structural topology. This chapter is organized as follows. In the first section we review the theoretical descriptions of single neurons, from detailed biophysical paradigms to simplified models. In the second section we review how single neurons can receive external signals through different kinds of synapses. In the third section we describe small circuit paradigms. Finally in the fourth section we describe models of large neural networks.

2   The Neuron Level

Informational processes in the nervous system take place at subcellular, cellular, network, and system levels. The timescales of these processes can be very different, and there is a cumulative interaction between all temporal and spatial scales. In this context it is not straightforward to select a single unit of information processing to describe neural activity. A widespread view, which is followed here, considers neurons as the basic units of information processing in the brain.

During the last six decades, phenomenological models have been developed to understand individual neuron behavior and their network dynamics. How detailed does the description of a neuron have to be to build a model that is biologically realistic and computationally tractable?

In this section we will try to answer this question by reviewing some of the most popular neuronal models. For a more detailed description of these and other models see, for example, Koch and Segev (1998), Koch (1999), Gerstner and Kistler (2002), and Izhikevich (2004).

Neuron models can be assigned into classes depending on the general goal of the modeling. If one wishes to understand, for example, how information is generated and propagated in a single neuron due to a complex synaptic bombardment, a detailed conductance-based model may be required. This model will have to describe the distribution of active properties in the membrane throughout the morphology and the propagation of the electrical activity along different compartments. On the other hand, if one wants to describe emerging phenomena in a network due to the synchronization of many units, it may be reasonable enough to use a simplified model that reproduces the main features of neuronal activity related to this overall goal. Here, modeling approaches are divided into two classes: biophysical models and simplified models. In many cases, neural models built on simplified paradigms can lead to more detailed biophysical models based on the same dynamical principles but implemented with more biophysically realistic mechanisms. By the same token, a realistic model can also be simplified by keeping only those features that allow us to reproduce the behavior that is important for the modeling goal. A good indication that the level of the description was chosen wisely comes if the model can reproduce with the same parameters the main bifurcations observed in the experiments.

2.1 Biophysical Models

2.1.1 Hodgkin–Huxley Model

Neurons talk to each other in the language of action potentials. An action potential, also called a spike, is a rapid change in voltage across the membrane. Biophysical models describe the generation of action potentials in terms of the electrical properties of the membranes. Detailed conductance-based neuron models take into account the ionic currents flowing across the membrane (Koch 1999) and describe their active (voltage dependent) and passive (capacitive and voltage independent) electrical properties (see ❷ *Fig. 1*). Neural membranes contain

❑ **Fig. 1**
Electrical circuit that represents a single neuron as an isopotential compartment with active and passive electrical properties. The circuit on the right includes the description of two active ionic channels (Na⁺ and K⁺) and one passive leakage current.

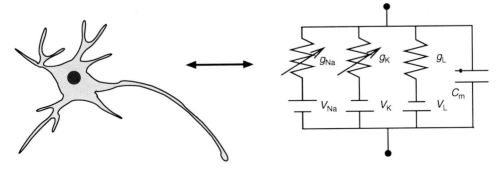

several types of voltage-dependent sodium, potassium, and calcium channels. The dynamics of some of these channels can also depend on the concentration of specific ions. In addition, there is a leakage current that is often considered to be independent of the voltage level. The flow of all these currents results in changes in the voltage across the membrane. The probability that a type of ionic channel is open depends nonlinearly on the membrane voltage and on the current state of the channel. This was first modeled by Hodgkin and Huxley (H–H) (1952).

Hodgkin and Huxley expressed the nonlinear conductance dependencies of the ionic channels in a set of coupled nonlinear differential equations that describe the electrical activity of the cell as

$$C_m \frac{dV(t)}{dt} = g_L[V_L - V(t)] + g_{Na}m(t)^3 h(t)[V_{Na} - V(t)] + g_K n(t)^4[V_K - V(t)] + I$$
$$\frac{dm(t)}{dt} = \frac{m_\infty(V(t)) - m(t)}{\tau_m(V(t))}$$
$$\frac{dh(t)}{dt} = \frac{h_\infty(V(t)) - h(t)}{\tau_h(V(t))} \tag{1}$$
$$\frac{dn(t)}{dt} = \frac{n_\infty(V(t)) - n(t)}{\tau_n(V(t))}$$

where $V(t)$ is the membrane potential, and $m(t)$, $h(t)$, and $n(t)$ represent empirical variables describing the activation and inactivation of the ionic conductances. In this model, there are two active channels, a Na^+ channel and a K^+ channel, and a passive leakage channel L. I is an external current that acts as a stimulus to the model; C_m is the capacitance of the membrane; g_L, g_{Na}, and g_K are the maximum conductances of the ionic channels; and V_L, V_{Na}, V_K are the reversal potentials for the different ionic channels. The steady-state values of the active conductance variables m_∞, h_∞, n_∞ and the time constants τ_m, τ_h, τ_n have a nonlinear voltage dependence, typically through sigmoidal or exponential functions, whose parameters can be estimated from experimental data.

The Hodgkin–Huxley model is one of the most successful phenomenological descriptions of neuronal activity, and can be enhanced by the addition of other voltage and ionic-concentration currents. These active membrane conductances can enable neurons to generate different spike patterns that are commonly observed in a variety of motor neural circuits and brain regions. The biophysical mechanisms of spike generation enable individual neurons to encode different stimulus features into distinct spike patterns. Spikes and bursts of spikes of different durations code for different stimulus features, which can be quantified without *a priori* assumptions about those features (Kepecs and Lisman 2003). The different timescales and the nonlinear interaction between the different variables can also account for different coding mechanisms, preferred input/output relations, and even the creation of new information (Latorre et al. 2006; Rabinovich et al. 2006).

2.1.2 Multicompartmental Models

The H–H ODEs (❷ Eq. 1) represent point neurons. As mentioned above, some modeling goals may require additional descriptions of the morphology, and thus a multicompartmental description of the neuron is needed. In this case, two adjacent isopotential compartments

◘ Fig. 2

Multicompartment models are built out of several isopotential compartments linked by axial resistances between them.

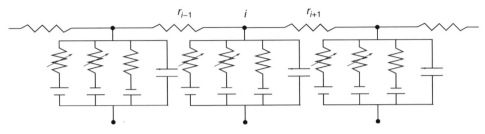

i and j can be linked through axial resistances $r_{i,j}$, and the membrane potential of the ith compartment can be described with the following equation:

$$C_i \frac{\mathrm{d}V_i}{\mathrm{d}t} = g_{\mathrm{L}}(V_{\mathrm{L}_i} - V_i) + \frac{(V_{i+1} - V_i)}{r_{i,i+1}} + \frac{(V_{i-1} - V_i)}{r_{i,i-1}} + I_{\mathrm{act},i}$$

where $I_{\mathrm{act},i}$ is the sum of all active currents in the ith compartment and $r_{i,j}$ is the axial resistance between compartments i and j (see ❷ *Fig.* 2). Following this strategy, one can build complex morphology patterns for single neuron models (Koch and Segev 1998; Varona et al. 2000).

2.2 Simplified Models

There is a large list of models derived from the H–H model, which in general try to simplify the complexity of the H–H equations by reducing the number of variables or their associated nonlinearities. In many cases, these simplifications allow the use of theoretical analysis that adds to the insight provided by the numerical simulations of the models. The most abstract models do not keep any biophysical parameter and just consider neurons as basic integrators or pure linear threshold gates. One of the main motivations to simplify the H–H model is to reduce the underlying computations and, as a result, to allow for large-scale network simulations.

2.2.1 FitzHugh–Nagumo Model

The FitzHugh–Nagumo model describes the excitation properties of a neuron without a detailed description of the nonlinearities associated with the electrochemical properties of the ionic channels. This model is often written as

$$\frac{\mathrm{d}V(t)}{\mathrm{d}t} = V(t) - cV^3(t) - W(t) + I$$

$$\frac{\mathrm{d}W(t)}{\mathrm{d}t} = V(t) + bW(t) - a$$

where $V(t)$ is the membrane potential, $W(t)$ describes the dynamics of fast currents, and I is an external current. The parameter values a, b, and c are constants chosen to generate spiking activity with different features. This reduced model with only polynomial nonlinearities accounts for the description of oscillatory spiking neural dynamics including bistability (Nagumo et al. 1962; FitzHugh 1961).

2.2.2 Morris–Lecar Model

The Morris–Lecar model (1981) maintains the description of the nonlinearities associated with two ionic channels while reducing the number of dimensions by considering an instantaneously responding voltage-sensitive Ca^{2+} conductance:

$$\frac{dV(t)}{dt} = g_L[V_L - V(t)] + n(t)g_n[V_n - V(t)] + g_m\, m_\infty(V(t))[V_m - V(t)] + I$$

$$\frac{dn(t)}{dt} = \lambda(V(t))[n_\infty(V(t)) - n(t)]$$

$$m_\infty(V) = \frac{1}{2}\left(1 + \tanh\frac{V - V_m}{V_m^0}\right)$$

$$n_\infty(V) = \frac{1}{2}\left(1 + \tanh\frac{V - V_n}{V_n^0}\right)$$

$$\lambda(V) = \phi_n \cosh\frac{V - V_n}{2\,V_n^0}$$

where $V(t)$ is the membrane potential, $n(t)$ describes the recovery activity of the K^+ current, and I is an external current. This simplified model, which reduces the number of dynamical variables of the H–H model, allows the use of phase plane methods for bifurcation analysis. The Morris–Lecar type models are widely used to study spike rate and precise timing codes (Gutkin 1998).

2.2.3 Hindmarsh–Rose Model

The H–R model is a simplified three-dimensional model that uses a polynomial approximation to the right-hand side of a Hodgkin–Huxley model (Hindmarsh and Rose 1984)

$$\frac{dV(t)}{dt} = W(t) + aV(t)^2 - bV(t)^3 - Z(t) + I$$

$$\frac{dW(t)}{dt} = C - dV(t)^2 - W(t)$$

$$\frac{dZ(t)}{dt} = r[s(V(t) - V_0) - Z(t)]$$

where $V(t)$ is the membrane potential, $W(t)$ describes fast currents, $Z(t)$ describes slow currents, and I is an external current. This simple model can display complex spiking and spiking-bursting regimes. H–R type models have been successfully used to build artificial electronic neurons that can interact bidirectionally with living neurons to recover a central pattern generator rhythm (Szucs et al. 2000; Pinto et al. 2000).

2.2.4 Integrate and Fire Models

Integrate and fire models reduce the computational cost of calculating the nonlinearities associated with the action potential by describing the dynamics of subthreshold activity and imposing the spike when a threshold is reached. The first integrate and fire model was introduced by Lapicque in 1907 (Lapicque 1907), before the mechanisms responsible for action potential generation were known. Many of these types of models can be described as

$$\frac{dV(t)}{dt} = \begin{cases} \dfrac{-V(t) + R_m I}{\tau} & 0 < V(t) < \theta \\ \dfrac{V(t_0^+)}{\tau} = 0 & V(t_0^-) = \theta \end{cases}$$

where $V(t)$ is the neuron membrane potential, θ is the threshold for spike generation, R_m is the membrane resistance, $\tau \equiv R_m \times C_m$, and I is an external stimulus current. A spike occurs when the neuron reaches the threshold θ in $V(t)$ after which the cell is reset to the resting state.

2.2.5 Map Models

Discrete time maps are simple phenomenological models of spiking and bursting neurons in the form of difference equations that are computationally very fast, and thus adequate to model large populations of neurons (Cazelles et al. 2001; Rulkov 2001). An example of such a map is

$$V(t+1) = \frac{\alpha}{1 + V(t)^2} + W(t) + I$$
$$W(t+1) = W(t) - \sigma V(t) - \beta$$

where $V(t)$ represents the spiking activity and $W(t)$ represents a slow variable. The parameters of the map allow us to tune the type of spiking activity that this model can produce (Rulkov 2002).

2.2.6 McCulloch and Pitts Model

The McCulloch–Pitts paradigm was the first computational model for an artificial neuron (McCulloch and Pitts 1943). This model neglects the relative timing of neural spikes and is also known as a linear threshold device model that can be described as

$$V(t+1) = \Theta(I(t) - \theta)$$
$$\Theta(X) = 1, \quad X > 0$$
$$= 0, \quad X \leq 0$$

where $V(t)$ is the variable that represents the activity of a neuron that receives a total input $I(t)$ from other neurons, and θ is the neuron threshold for firing. This model is also used in computational neuroscience and has proved to be very adequate to address the role of connection topologies in different neural systems (Nowotny and Huerta 2003; Garcia-Sanchez and Huerta 2003).

❯ *Figure 3* represents several time series of spiking activity from some of the models described in this section under the stimulus of a constant current injection. Of course, constant currents are not a usual stimulus to a neuron. Stimuli arrive to these cells in the form of synapses, which will be the subject of the next section.

■ **Fig. 3**

Time series of spiking activity from several widely used neuronal models. From top to bottom: Hodgkin–Huxley model (H–H), FitzHugh–Nagumo model (F–N), Hindmarsh–Rose (H–R) model (in a chaotic spiking-bursting regime), and integrate and fire model (I–F).

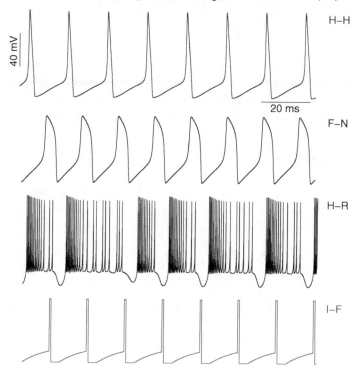

3 The Synapse Level

In the previous section we briefly reviewed several mathematical techniques used by neuroscientists to model the electrical activity of neurons. These techniques model, at different levels of description, the so-called neuronal signal, action potential, or neuronal *spike*. In an actual neuron, an action potential is generated when the sum of all involved membrane ionic currents is enough to overcome an excitation threshold. When this occurs, it is said that the cell is excitable. Once the spike has been generated, it propagates through the axon. At the end of the axon, the signal can be transmitted to other neurons through complex inter-neuron structures called synapses.

In order to build biologically inspired neural network models, it is fundamental to understand how neurons communicate through the synapses. To go further, let us first briefly review the different types of synaptic connections that can be found in living neural systems.

Traditionally and depending on the position at which the neural signal generated in the *presynaptic* neuron is transmitted to the *postsynaptic* neuron, the synapses have been denoted as (see ❷ *Fig. 4* for an illustration) *axodendritic* (when the transmission occurs from the axon of the presynaptic neuron to the dendrites of the postsynaptic neuron), *axoaxonic* (between the two axons, as in several types of "inhibitory" synapses), and *axosomatic* (between the axon

◘ **Fig. 4**

Classification of synapses depending on their relative location along the cell body of both presynaptic and postsynaptic neurons.

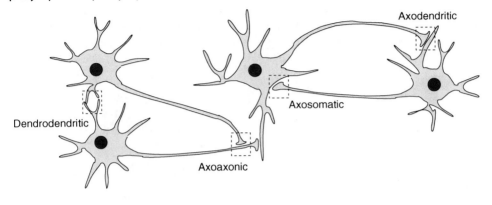

of the presynaptic neuron and the soma of the postsynaptic neuron). In recent years, many other types of connections between different parts of the neurons involved in the synapse have been discovered, including *dendrodendritic, dendro-axonic,* and *dendro-somatic synapses* (see for instance Pinault et al. 1997). In many models of neural networks built with point neurons, all these types of synapses play the same role, and a frequent assumption is that neurons i and j are connected through a synapse w_{ij}, without the specification of the precise spatial location at which it occurs.

3.1 Electrical Versus Chemical Synapses

More relevant for a realistic modeling of neural networks is the classification of synapses depending on the mechanism that the cell uses for the transmission of the signal. The synapse can be either *electrical* or *chemical.* In an electrical synapse, the separation between the presynaptic and postsynaptic membranes is about 3.5 nm (Kandel et al. 2000) and they are joined by specific protein structures called *gap junctions* (see ❷ *Fig. 5*), which are specialized ionic channels that connect the cytoplasm of both cells (Kandel et al. 2000). When a neural signal (action potential) comes to the gap junction, it depolarizes (or hyperpolarizes) the membrane, which induces the opening of the channels and the diffusion of ions through them from one neuron to another. In a chemical synapse, however, the separation is larger than in the electrical synapses, namely, 20–40 nm (Kandel et al. 2000), so there is no physical contact among the cells. In this case the transmission of the signal is actively mediated by the release, in this space, of some chemical messengers called *neurotransmitters*, which are generated at the end of the axon of the presynaptic neuron (see ❷ *Fig. 5*). Signal transmission in a chemical synapse occurs as follows: when a presynaptic action potential arrives at the synapse, it produces the opening of some neurotransmitter vesicles near the membrane, the so-called *ready releasable pool* of vesicles. This causes the release and diffusion of a large number of neurotransmitters in the inter-synaptic space. The neurotransmitters have affinity for certain molecular receptors in the postsynaptic membrane and, after binding to them, they induce the opening of specific ionic channels. This produces the depolarization (or hyperpolarization) of the postsynaptic

Scheme showing the differences during signal transmission between electrical synapses (panel **a**) and chemical synapses (panel **b**). Signal transmission in electrical synapses is a passive mechanism in which ions can quickly diffuse between presynaptic and postsynaptic neurons. In chemical synapses, signal transmission occurs as the result of different biophysical mechanisms, which occur sequentially in different timescales, and which generate an excitatory (or inhibitory) postsynaptic potential, namely, EPSP (or IPSP for inhibitory synapses), during a time delay of τ_{in}.

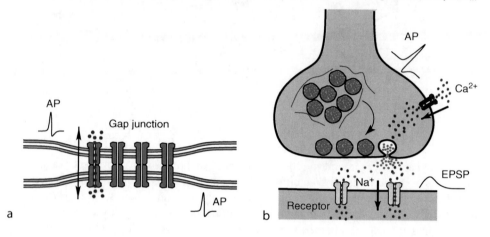

membrane. Finally, the released vesicles are replaced from others that are in a *reserve pool* of vesicles, located relatively far from the membrane (see ❯ *Fig. 5*). In both cases, the communication among neurons is the result of a flux of ions through the postsynaptic membrane constituting a synaptic current, namely I_{syn}. Theoretically, one can see the effect of such a current on the excitability of the membrane of the postsynaptic neuron i using, for instance, a standard Hodgkin–Huxley formalism (see ❯ Sect. 1):

$$C_m \frac{dV_i(t)}{dt} = -\left[\sum_\kappa I_{i,\kappa}(t) + I_i^{syn}(t)\right] \qquad (2)$$

where C_m is the capacitance of the membrane, I_κ represents the different ionic currents defining the excitability of the neuron membrane (see ❯ Sect. 1), and $I_i^{syn}(t) = \sum_j I_{ij}^{syn}(t)$ is the total synaptic current received by the postsynaptic neuron i. Using Ohm's Law, one has for an electrical synapse

$$I_{ij}^{syn}(t) = \bar{G}_{ij} \times [V_i(t) - V_j(t)] \qquad (3)$$

where $\bar{G}_{ij} =$ constant is the maximum synaptic conductance. That is, ❯ Eq. 3 takes into account the passive diffusion of ions through the synapse. For a chemical synapse, however, one has

$$I_{ij}^{syn} = G_{ij}(t) \times (V_i - E_j^{syn}) \qquad (4)$$

with E_j^{syn} being the synaptic reversal potential, that is, the voltage value at which the synaptic current cancels. Moreover, $G_{ij}(t)$ now depends on time following a complex dynamics that considers all the biophysical processes affecting the release and trafficking of neurotransmitters, their posterior binding to the synaptic receptors, and the permeability of ionic channels, in the postsynaptic membrane, to the passage of ions through them, as explained above. The shape of $G_{ij}(t)$ is usually modeled using an α function, namely,

$$G_{ij}(t) = \bar{G}_{ij} \times \Theta(t - t_j^{sp}) \times \alpha(t - t_j^{sp}) \qquad (5)$$

with $\alpha(t) = t \times e^{-t/\tau_{in}}$ (see ❷ Fig. 6), where t_j^{sp} is the time at which the presynaptic spike occurs, τ_{in} is the time constant for the synaptic transmission, and $\Theta(x)$ is the Heaviside step function (Also the α-function can be modeled as the difference of two exponentials, namely, $\alpha(t) = e^{-t/\tau_1} - e^{-t/\tau_2}$ with $\tau_1 > \tau_2$.) A computationally efficient alternative to the α-function is discussed in Destexhe et al. (1994).

These two different processes for electrical and chemical synapses make their functionality strongly different, in such a way that electrical synapses are faster (a small fraction of a millisecond) and the signal can be transmitted in both directions. In the chemical synapses, however, the transmission of the signal is slower (about 1–5 ms) and they are intrinsically asymmetric because information is transmitted from the presynaptic neuron to the postsynaptic neuron. (This fact has strong implications under a theoretical point of view, if one wants to build and study neural networks models that include symmetric or asymmetric synaptic connections. Large neural networks with symmetric connections can be studied using standard equilibrium statistical physics techniques, which is not possible for asymmetric

☐ **Fig. 6**
Temporal behavior of an α function $\alpha(t)$ (as defined in the text), which traditionally has been used to model an evoked postsynaptic response in a single chemical synapse after the arrival of a presynaptic action potential. The function has an initially fast increase, which accounts for the effect of the neurotransmitter binding to the postsynaptic receptors, and an exponential decay with a relatively slow time constant τ_{in}, which considers the typical time delay observed in chemical synapses. Note that $\alpha(t)$ can be used to model both excitatory and inhibitory synapses, the main functional differences being due to the particular value of the synaptic reversal potential E_j^{syn}, which is around zero for excitatory synapses and strongly hyperpolarized for inhibitory synapses (see main text).

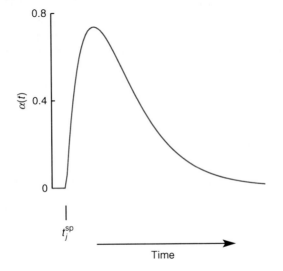

connections; see, for instance, Peretto (1992)). Concerning their functionality, electrical synapses are important to enhance fast synchronization of a large group of electrically connected neurons or to transmit a metabolic signal among neurons (because different types of ions or other molecules can diffuse through gap-junctions). The transmission of signals on chemical synapses, however, involves more complex nonlinear processes that act like gain functions, which amplify the effect of the weak electrical presynaptic currents, depolarizing relatively large extensions of the postsynaptic membrane.

3.2 Excitatory and Inhibitory Synapses

The synapses can also be classified into *excitatory*, when the release of neurotransmitters produces a depolarization of the postsynaptic membrane potential (typically due to the influx of Na^+ or Ca^{2+} ions through the postsynaptic ionic channels) or *inhibitory*, when there is a hyperpolarization of the postsynaptic membrane potential (i.e., there is a flux of K^+ ions outside the cell, or an influx of Cl^- ions from the extracellular medium) (see ❷ *Fig. 7*). For these different effects, not only the type of neurotransmitter but also the type of the postsynaptic receptor is important. The most conventional neurotransmitter associated with an excitatory postsynaptic response is glutamate, which activates both the *fast* AMPA/kainate and the relatively *slow* NMDA excitatory receptors (Koch and Segev 1998). The resulting generated synaptic current is produced by the influx of Na^+ or Ca^{2+} ions (i.e., is an inward current) and it has associated an approximate reversal potential of $E^{syn} = 0$ mV. (Note that the synaptic reversal potential is smaller than, for instance, the Na^+ equilibrium potential, which is about +55 mV. This is because, in excitatory synapses, the synaptic channels are permeable for both Na^+ and K^+ ions, which induces a flux of both ions in opposite directions (inward for Na^+ and outward for k^+) resulting in a net synaptic current that cancels at the reversal potential of $E^{syn} = 0$ mV (Kandel et al. 2000).) Other neurotransmitters associated with excitatory responses are aspartate, acetylcholine, and dopamine. On the other hand, gamma-aminobutyric acid (GABA) is the most common neurotransmitter involved in inhibitory synapses in the central nervous system and, similarly to glutamate, it activates a *fast* inhibitory receptor, or GABAa, and a relatively *slow* inhibitory receptor named GABAb (Koch and Segev 1998; Kandel et al. 2000). In the spinal cord, however, the most common neurotransmitter associated with inhibitory responses is glycine. The activation of the GABAa receptor produces a hyperpolarized synaptic current due to the influx of Cl^- ions (inward current) and has $E^{syn} \approx E_{Cl} = -70$ mV. The activation of the GABAb receptor, however, produces the efflux of K^+ ions from the cell to the extracellular medium (outward current) and has a reversal potential of approximately $E^{syn} = E_K = -95$ mV.

3.3 Dynamic Synapses

When modeling neural systems, synapses have been traditionally considered as *static* identities with the only possible modification of their maximum conductances ($\bar{G}_{ij}$) due to learning (Hopfield 1982), this being a *slow* process that takes place in a temporal scale of seconds, minutes, or years. Recently, it has been reported in actual neural media that the synaptic strength (measured as the amplitude of the generated postsynaptic response) also varies in short timescales producing a fluctuating postsynaptic response (Dobrunz and Stevens 1997),

◻ **Fig. 7**

Differences between excitatory (panel **a**) and inhibitory (panel **b**) chemical synapses. In excitatory synapses, an excitatory postsynaptic ionic current (EPSC) is generated, which can be inward (due to the influx of Na^+ ions as plotted in the panel) or outward (due to the efflux of K^+ ions through the channel) and which induces an excitatory or depolarized postsynaptic membrane potential (EPSP). In inhibitory synapses, an inhibitory postsynaptic current (IPSC) is generated, usually due to the influx of Cl^- ions through the channel (i.e., it is an inward current) and it produces a hyperpolarization of the postsynaptic membrane potential (IPSP).

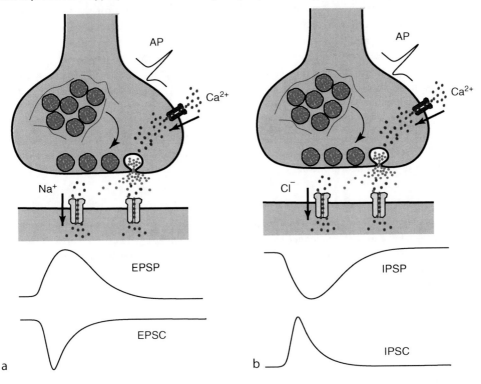

which can be depressed (synaptic depression) or enhanced (synaptic facilitation) depending on the presynaptic activity (Abbott et al. 1997; Markram et al. 1998; Tsodyks et al. 1998). This defines the so-called *dynamic synapses*. In recent years, several studies have shown that activity-dependent processes such as synaptic depression and facilitation have a key role in information processing in living neural systems. For instance, it is well known that short-term depression plays an important role in several emerging phenomena in the brain, such as selective attention (Buia and Tiesinga 2005; McAdams and Maunsell 1999) and cortical gain control (Abbott et al. 1997), and it is responsible for the complex switching behavior between activity patterns observed in neural network models with depressing and facilitating synapses (Pantic et al. 2002; Cortes et al. 2006).

From a biophysical point of view, synaptic depression is a consequence of the fact that the number of available neurotransmitters in the synaptic buttons is limited in the ready releasable pool. Therefore, the neuron needs some time to recover these synaptic resources in order to transmit the next incoming spike. As a consequence, the dynamics of the synapse is affected by an activity-dependent mechanism that produces nonlinear effects in the postsynaptic response (see ❷ *Fig. 8*).

On the other hand, synaptic facilitation is produced by the influx of calcium ions into the neuron near the synapse through voltage-sensitive channels after the arrival of each presynaptic spike. These ions bind to some acceptor that gates and facilitates the neurotransmitter vesicle depletion in such a way that the postsynaptic response increases for successive spikes (Kamiya and Zucker 1994) due mainly to residual cytosolic calcium (see ❷ *Fig. 9*). Facilitation, therefore, increases the synaptic strength for high-frequency presynaptic stimuli. However, given a train of presynaptic spikes, a strong facilitating effect for the first incoming spikes produces an even stronger depressing effect for the last spikes in the train. The phenomenon of synaptic facilitation has been reported to be relevant for synchrony and selective attention (Buia and Tiesinga 2005), and in the detection of bursts of action potentials (Matveev and Wang 2000; Destexhe and Marder 2004). More recently, it has been proved that synaptic facilitation had a positive effect, for instance, in the efficient transmission of temporal correlations between spike trains arriving from different synapses (Mejias and Torres 2008). This feature, known as spike coincidence detection, has been measured in vivo in cortical neurons and is related to some dynamical processes that affect neuron firing thresholds (Azouz and Gray 2000), so that it seems to be an important mechanism for efficient transmission of information in actual neural media.

◼ Fig. 8

Simulated train of EPSPs generated in a single synapse after the arrival of a train of presynaptic spikes at frequency of 5 Hz (*top panels*) and 15 Hz (*bottom panels*), when the synapse is *static* (a), and *dynamic* with only a depressing mechanism (b) and with depressing and facilitating mechanisms (c). The synapse parameters were $U_{SE} = 0.1$ and $\tau_{rec} = \tau_{fac} = 0$ for panel (a), $U_{SE} = 0.3$, $\tau_{rec} = 600$ ms, $\tau_{fac} = 0$ for panel (b), and $U_{SE} = 0.02$, $\tau_{rec} = 600$ ms, $\tau_{fac} = 6$ s for panel (c).

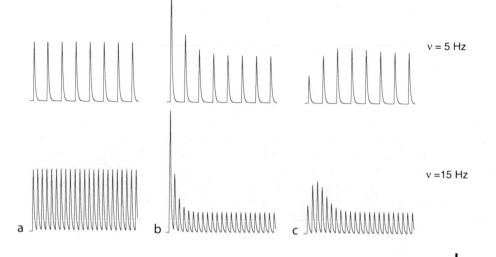

☐ **Fig. 9**
Biophysical mechanism for facilitation: After a first presynaptic spike that induces a given EPSP, the existence of a certain amount of residual-free cytosolic calcium increases the neurotransmitter probability release for a second spike that enlarges the secondary EPSP.

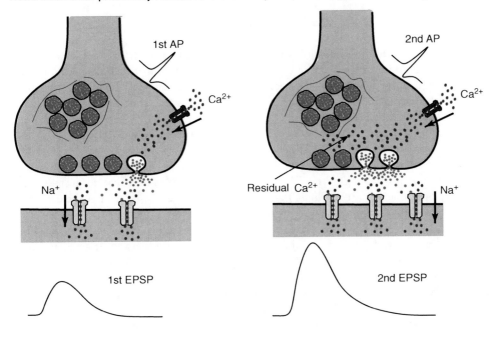

One of the most important theoretical approaches to modeling dynamic synapses is the phenomenological model presented in Tsodyks and Markram (1997), which considers that the state of a synapse between a presynaptic neuron j and a postsynaptic neuron i is governed by the system of equations

$$\frac{dx_j(t)}{dt} = \frac{z_j(t)}{\tau_{rec}} - U_j(t)x_j(t)\delta(t - t_j^{sp})$$
$$\frac{dy_j(t)}{dt} = -\frac{y_j(t)}{\tau_{in}} + U_j(t)x_j(t)\delta(t - t_j^{sp})$$
$$\frac{dz_j(t)}{dt} = \frac{y_j(t)}{\tau_{in}} - \frac{z_j(t)}{\tau_{rec}}$$

(6)

where $x_j(t)$, $y_j(t)$, $z_j(t)$ are the fractions of neurotransmitters in recovered, active, and inactive states, respectively. (Neurotransmitters in a recovered state are those that are in vesicles refilling the ready releasable pool, active neurotransmitters are those that have been released and can produce a postsynaptic response, and inactive neurotransmitters are the remaining.) Here, τ_{in} and τ_{rec} are the inactivation and recovery time constants, respectively. Depressing synapses are obtained, for instance, for $U_j(t) = U_{SE}$ constant, which represents the maximum number of neurotransmitters that can be released (activated) after the arrival of each presynaptic spike. The delta functions appearing in ❷ Eq. 6 consider that a spike arrives to the synapse at a fixed time $t = t_j^{sp}$. Typical values of these parameters in cortical depressing synapses are $\tau_{in} = 3$ ms, $\tau_{rec} = 800$ ms, and $U_{SE} = 0.5$ (Tsodyks and Markram 1997). The synaptic facilitation mechanism can be introduced assuming that $U_j(t)$ has its own dynamics related with the release of calcium from intracellular stores and the influx of calcium from the extracellular medium each time a spike arrives (see for instance ❷ *Fig. 5*), which in this phenomenological model is $U_j(t) \equiv u_j(t)(1 - U_{SE}) + U_{SE}$ with

$$\frac{du_j(t)}{dt} = \frac{u_j(t)}{\tau_{fac}} + U_{SE}[1 - u_j(t)]\delta(t - t_j^{sp}) \tag{7}$$

Here, $u_j(t)$ is a dynamical variable that takes into account the influx of calcium ions into the presynaptic neuron near the synapse through voltage-sensitive ion channels (Bertram et al. 1996). As we described previously, these ions can usually bind to some acceptor that gates and facilitates the release of neurotransmitters. A typical value for the facilitation time constant is $\tau_{fac} = 530$ ms (Markram et al. 1998). The variable $U_j(t)$ represents then the maximum fraction of neurotransmitters that can be activated, both by the arrival of a presynaptic spike (U_{SE}) and by means of facilitating mechanisms (i.e., $u_j(t)(1 - U_{SE})$).

One can use then $y(t)$ instead of the "static" $\alpha(t)$ function in ❷ Eqs. 4 and ❷ 5 to model the corresponding synaptic current I_{ij}^{syn}. However, one can also assume that, for instance, the excitatory postsynaptic current (EPSC) generated in a particular synapse after the arrival of a presynaptic spike is proportional to the fraction of neurotransmitters that are in the active state, that is, $I_{ij}^{syn} = -A_{SE} \times y_j(t)$ (the minus indicates that the current is inward in this particular case), where A_{SE} is the maximum postsynaptic current that can be generated and, typically, it takes a value $A_{SE} \approx 42.5$ pA for cortical neurons. The previous assumption for I_{ij}^{syn} remains valid as long as the corresponding generated postsynaptic potential is below the threshold for the generation of a postsynaptic spike, which is about $V_{th} = 13$ mV. In this situation, the postsynaptic potential varies slowly compared to the variation in the synaptic conductance (which is determined by $y_j(t)$) and, therefore, it can be considered almost constant during the generation of the synaptic current.

4 The Network Level

4.1 Modeling Small Networks

Once one knows how to model individual neurons and synapses one is ready to build models of neural networks. The relationship between the dynamics of individual neurons and the activity of a large network in the nervous system is complex and difficult to assess experimentally. In most cases, there is a lack of knowledge about the details of the connections and the properties of the individual cells. In large-scale models, most of the relevant parameters can only be estimated. By studying small circuits of invertebrates first, one can have a closer relationship between a model and the living neural network, and thus a more detailed understanding of the basic principles underlying neural dynamics. Another advantage in this case is that the predictions of the models can be more easily tested in systems where neurons and connections are accessible and identifiable.

4.1.1 Central Pattern Generators

Central pattern generators (CPGs) are small neural circuits that produce stereotyped cyclic outputs without rhythmic sensory or central input (Marder and Calabrese 1996; Stein et al.

1997; Selverston 2005). CPGs are oscillators that underlie the production of motor commands for muscles that execute rhythmic motion involved in locomotion (Getting et al. 1989), respiration (Ramirez et al. 2004), heart beat (Cymbalyuk et al. 2002), etc. CPG circuits produce signals in the form of spatiotemporal patterns with specific phase lags between the activity of different motoneurons that innervate muscles.

Because of their relative simplicity, the network architecture and the main features of CPG neurons and synapses are known much better than any other brain circuits. Examples of typical invertebrate CPG networks are shown in ❷ *Fig. 10*. Common to many CPG circuits are network topologies built with electrical and mutual inhibitory connections. The cells of these circuits typically display spiking–bursting activity. The characteristics of the spatiotemporal patterns generated by the CPG, such as burst frequency, phase, duration of the slow wave, etc., are determined by the intrinsic properties of each individual neuron, the properties of the synapses, and the architecture of the circuit. Regulation mechanisms at all levels contribute to network stability and the generation of robust and flexible motor patterns.

◻ **Fig. 10**
Examples of CPG topologies in invertebrates: (a) the gastric and pyloric CPGs in crustacea (modified from Selverston and Moulins (1987)), (b) the feeding CPG in *Planorbis* (modified from Arshavsky et al. (1985)), (c) the swimming CPG in *Tritonia* (modified from Getting (1989)), (d) the swimming CPG in *Clione* (modified from Arshavsky et al. (1991)). Dots represent inhibitory synapses while arrows represent excitatory synapses. Electrotonic gap junctions are represented by resistors. Note that all topologies are non-open, that is, each neuron receives feedback from at least one other neuron of the CPG (Huerta et al. 2001). As individual neurons are identifiable from one preparation to another in these small circuits, the connectivity diagrams can be fully established.

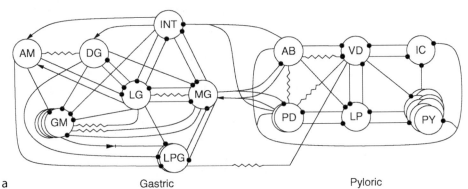

a　　　　　　Gastric　　　　　　　　　Pyloric

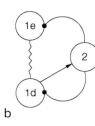 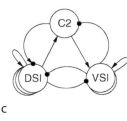 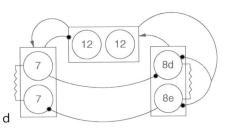

b　　　　　c　　　　　d

Although CPGs autonomously establish rhythmic firing patterns, they also receive supervision signals from higher centers and from local reflex pathways. These inputs allow the animal to constantly adapt its behavior to the environment by producing different stable spatiotemporal patterns (Selverston and Moulins 1987; Rabinovich et al. 2006). Thus, CPGs produce robust rhythms but at the same time are flexible to incoming signals. There are many experimental results that show that CPGs can be reconfigured by neuromodulatory substances in the blood or released synaptically. These substances induce changes both in the connectivity (by altering synaptic strength) and in the intrinsic properties of the neurons (by affecting the voltage-dependent conductances) (Simmers and Moulins 1988; Marder and Bucher 2007).

Many modeling studies have contributed to understanding the interaction between intrinsic and network properties in CPGs as illustrated here with a few examples. Many of these modeling efforts use the pyloric CPG of crustaceans (Prinz 2006), one of the best known neural circuits. This CPG is composed of 14 neurons (see ❷ *Fig. 10a*) that produce a spiking–bursting activity that keeps a characteristic phase relationship between the bursts of other members due, mainly, to the inhibitory connection topology of the CPG (Selverston and Moulins 1987). The electrical activity of the neurons in this CPG is propagated to the muscles where the spikes are integrated to exert force on different places of the pylorus. The main task of the pyloric CPG is to transport shredded food from the stomach to the digestive system.

A remarkable property of the motoneurons of the pyloric CPG is that, when they are isolated from other cells in the circuit, they display highly irregular, and, in fact, chaotic spiking–bursting activity (Abarbanel et al. 1996; Elson et al. 1998, 1999; Selverston et at. 2000). Models of the pyloric CPG have been able to explain how the mutual inhibitory connections and the slow subcellular dynamics are responsible for the regularity of the rhythms (Varona et al. 2001a, b) (see ❷ *Fig. 11*). The dynamical richness of the individual neurons contributes to a fast response in the form of a new CPG rhythm when incoming signals or neuromodulatory input arrives to the circuit. Models have also suggested that non-open connection topologies are widespread in many different CPGS to produce regular rhythms while preserving the effect of individual rich dynamics to maximize the gain of information (Huerta et al. 2001; Stiesberg et al. 2007).

Recent experimental findings have revealed important new phenomena in the dynamics of CPG circuits. One of them is related to the presence of robust and cell-type-specific intraburst firing patterns in the bursting neurons of the pyloric CPG (Szucs et al. 2003).

❏ **Fig. 11**
Reduced pyloric CPG model and the regular triphasic rhythm produced by this circuit. Individual neurons are modeled with H–H chaotic dynamics. Neurons within this circuit produce neural signatures in the form of cell-specific interspike intervals (Szucs et al. 2003; Latorre et al. 2006).

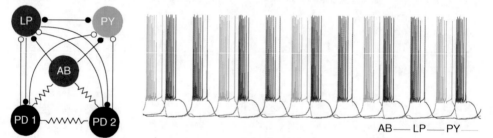

These patterns, termed as interspike interval signatures or neural signatures, depend on the intrinsic biophysical properties of the bursting neurons as well as on the synaptic connectivity of the surrounding network, and can influence the overall activity of the system. The existence of these precise and reproducible intra-burst spiking patterns might indicate that they contribute to specific computational tasks. Modeling studies have shown that H–H models can react distinctly to the neural signatures and point them out as elements of a multicoding mechanism for spiking–bursting neurons (Latorre et al. 2006). Neural signatures coexist with the generation of robust rhythms (see ❷ *Fig. 11*) and thus the bursting signal can contain information about the who and the what of the information. Beyond the context of the CPGs, and taking into account that neural signatures have also been found in vertebrate neurons (Zeck and Masland 2007), these results suggest the design of artificial neural network paradigms with units that use multicoding strategies and local information discrimination for information processing.

Another phenomenon that can provide inspiration for bio-inspired CPG control in robotics is related to the existence of dynamical invariants (such as the maintenance of the ratio between the change in burst periods and the change in phase lags) when the CPG suffers perturbations (Reyes et al. 2003). These dynamical invariants are built through the cellular and synaptic contributions to network properties, and can account for a wide range of efficient rhythms in response to external stimuli for many types of motor tasks.

4.1.2 Small Sensory Networks

Simple circuits of sensory systems have also contributed to a better understanding of intrinsic sensory network dynamics and its role in the sensory–motor transformation. Some of the clearest examples are found in invertebrate systems in which the different processing stages are more directly accessible (Lewis and Kristan 1998; Levi and Camhi 2000; Laurent et al. 2001; Beenhakker and Nusbaum 2004). In some cases, modeling results have provided striking hypotheses about small sensory networks that then have been tested experimentally. One of these examples is related to the dual role of the gravimetric sensory network of the mollusk *Clione*. This model will also serve to illustrate the use of rate models as another type of simplified paradigms for network activity.

Clione is a marine mollusk that has to maintain a continuous motor activity in order to keep its preferred head-up orientation. The motor activity is controlled by the wing central pattern generator and the tail motoneurons that use the signals from its gravity sensory organs, the statocysts (Panchin et al. 1995). A statocyst is a small sphere in which the statolith, a stone-like structure, moves according to the gravitational field. The statolith exerts pressure on the statocyst receptor neurons (SRCs) that are located in the statocyst internal wall. When excited, these mechanoreceptors send signals to the neural systems responsible for wing beating and tail orientation. The SRCs also receive input from a pair of cerebral hunting interneurons (H) (Panchin et al. 1995), which are activated in the presence of the prey by signals from the chemoreceptors. *Clione* does not have a visual system and although its chemosensors can detect the presence of prey, they are not direction sensitive. When hunting behavior is triggered, *Clione* loops and turns in a complex trajectory trying to locate its prey.

A six-receptor neural network model with synaptic inhibitions was built to describe a single statocyst (Varona et al. 2002) (see ❷ *Fig. 12*). The model of the statocyst sensory network suggested the hypothesis that the SRC sensory network has a dual role and participates not

■ **Fig. 12**

Panel a: Lotka–Volterra rate model of the gravimetric sensory network of the mollusk *Clione*. Panels b and c: Time series of rate activity for each neuron (colors correspond to the different neurons in panel a, units are dimensionless). This network displays irregular winnerless competition (WLC) dynamics with activation phase locks (black arrows) for those neurons that are active in a particular time window.

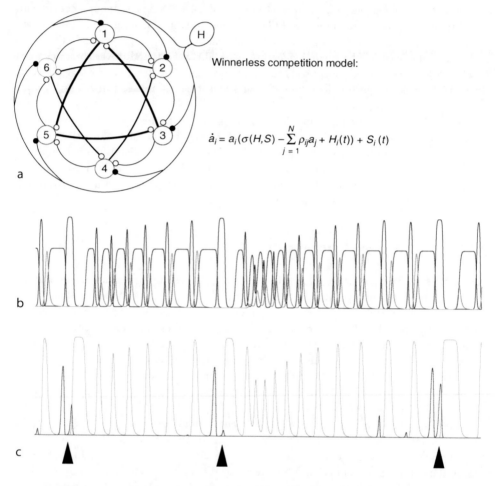

Winnerless competition model:

$$\dot{a}_i = a_i \left(\sigma(H,S) - \sum_{j=1}^{N} \rho_{ij} a_j + H_i(t) \right) + S_i(t)$$

a

b

c

only in keeping a head-up position during routine swimming, but also in the organization of the complex hunting behavior of *Clione* (Varona et al. 2002).

The model of a statocyst receptor network consists of six neurons (see ❷ *Fig. 12*, panel a). Each receptor neuron follows a Lotka–Volterra type dynamics given by the equation shown in panel a, where $a_i > 0$ represents the instantaneous spiking rate of the neurons, $H_i(t)$ represents the excitatory stimulus from the cerebral hunting interneuron to neuron i, and $S_i(t)$ represents the action of the statolith on the receptor that is pressing. When there is no stimulus from the hunting neuron ($H_i = 0$, $\forall i$) or the statolith ($S_i = 0$, $\forall i$), then $\sigma(H, S) = -1$ and all neurons are silent; $\sigma(H, S) = 1$ when the hunting neuron is active and/or the statolith is pressing one of the receptors. $\rho_{i,j}$ is the connection matrix that follows the diagram illustrated in the figure.

When there is no activation of the sensory neurons by the hunting neuron, the effect of the statolith ($S_i \neq 0$) in the model is to induce a higher rate of activity on one of the neurons (the neuron i where it rests for a big enough S_i value). We assume that this higher rate of activity affects the behavior of the motoneurons to organize the head-up position. The other neurons are either silent or have a lower rate of activity and we can suppose that they do not influence the posture of *Clione*. The model displays a winner-take-all situation for normal swimming.

When the hunting neuron is active, a completely different behavior arises. We assume that the action of the hunting neuron overrides the effect of the statolith and thus $S_i \approx 0, \forall i$. The network with a stimuli from the hunting neuron establishes a winnerless competition (WLC) among the SRCs. As a result of this competition, the SRCs display a highly irregular, and in fact chaotic (Varona et al. 2002) switching activity that can drive the complex loops and turns of *Clione*'s hunting search. ❷ *Figure 12* shows an illustration of the nonsteady switching activity of the receptors. An interesting phenomenon can be seen in this multifunctional network. Although the timing of each activity is irregular, the sequence of the switching among the SRC activity (when it is present) is the same at all times. Arrows in ❷ *Fig. 12* point out this fact for neuron 4 (cf. panels b and c). The activation phase lock among the statocyst receptor neurons emerges in spite of the highly irregular timing of the switching dynamics. The complex switching activity among the receptors keeps a coordinated relation in the activation of the signals. This is a desirable property if the statocyst receptor signals are active participants in the control of *Clione*'s tail and wing motion during hunting behavior.

The Lotka–Volterra rate formalism allows us to establish the conditions for stability of the sequential switching. The predictions of this model have been demonstrated experimentally (Levi et al. 2004, 2005) and indeed the gravimetric network has a distinct role during routine swimming and during hunting. During normal swimming, only the neurons that are excited by the statolith are active, which leads to a winner-take-all mode as a result of inhibitory connections in the network. However, during hunting search behavior, the hunting neuron triggers WLC between all statocyst neurons whose signals participate in the generation of a complex motion that the animal uses to scan the space until it finds its prey.

Beyond this simple network, WLC dynamics has been proposed as a general paradigm to represent transient dynamics in many types of neural functions (Rabinovich et al. 2006): from sensory encoding to decision making in cognitive tasks (Rabinovich et al. 2008). The mathematical image of WLC is a stable heteroclinic channel, that is a set of trajectories in the vicinity of a heteroclinic skeleton that consists of saddles and unstable separatrices that connect their surroundings (Rabinovich et al. 2008). This image satisfies the competing requirements of stability and flexibility, which is required in most descriptions of neural activity.

4.2 Modeling Large Networks

Since the pioneering work of McCulloch and Pitts (1943), neural networks models have been widely used to artificially mimic (with more or less success) some features of the complex behavior of different neural systems. This is due to the fact that, as a first approximation, neural systems and even the whole brain can be viewed as large networks constituted by nodes connected by edges or *synapses*, where the nodes are excitable. This means that each element has a threshold and a refractory time between consecutive responses, a behavior that in some cases impedes thermal equilibrium and induces complex emergent phenomenology including, for instance, learning and memory recall, pattern formation, and chaotic neural activity. In recent decades, many different theoretical approaches and computational techniques (with different levels of description for neuron, synapses, and network topology) have been developed to study extended neural systems (see for instance, Brette et al. (2007) for a recent review about the modeling of networks of spiking neurons and Swiercz et al. (2006) for a recent model of synaptic plasticity).

One of the most successful models of large neural networks was due to Amari (1972) and Hopfield (1982) who separately defined the so-called Amari–Hopfield attractor neural network (ANN). This is a neural network where the edges are fixed in time and weighted according to a prescription (Hebb 1949), which in a sense stores or memorizes information (during a slow learning process) from a set of given patterns of activity. As a consequence, these patterns become attractors of the phase space dynamics and one may interpret that the system shows retrieval of the stored patterns. This process, known as *associative memory*, can be considered as a suitable representation of the memory and recall processes observed in the brain. Actual neural systems do much more than just recalling a memory and staying there, however. That is, one should expect dynamic instabilities or other destabilizing mechanisms on the attractor that allow for recall of new memories, for instance. This expectation is reinforced by recent experiments suggesting that synapses can also undergo rapid changes with time, as we explained in a previous section, which may both determine brain complex tasks (Pantic et al. 2002; Abbott et al. 1997; Tsodyks et al. 1998) and induce irregular and perhaps chaotic activity (Barrie et al. 1996; Korn and Faure 2003). In order to include these synaptic features in neural networks models, we first mathematically define a large neural network model.

4.2.1 Mathematical Definition of a Neural Network

In a general form, a large neural network can be defined by a set of N nodes or *neurons* (placed in some specific positions x_i of a d-dimensional vector space) whose state is represented by some variable s_i. Neurons are connected to each other by edges or *synapses* whose state, *strength*, or *weight* is represented by a real variable w_{ij} (see ❷ *Fig. 13*). The map of connections defines the *topological* structure of the network. Following the standard ANN model introduced above, we assume that w_{ij} fixed and containing information from a set of M

❏ Fig. 13

(a) Scheme of an artificial neural network with arbitrary network topology and symmetric connections $w_{ij} = w_{ji}$ and (b) analogy between the variables defining the dynamic of a neural network and those related with the dynamics of actual neural systems.

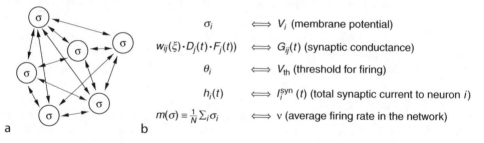

	σ_i	$\Longleftrightarrow$ V_i (membrane potential)
$w_{ij}(\xi) \cdot D_j(t) \cdot F_j(t))$		$\Longleftrightarrow$ $G_{ij}(t)$ (synaptic conductance)
	θ_i	$\Longleftrightarrow$ V_{th} (threshold for firing)
	$h_i(t)$	$\Longleftrightarrow$ $I_i^{syn}(t)$ (total synaptic current to neuron i)
$m(\sigma) \equiv \frac{1}{N}\Sigma_i \sigma_i$		$\Longleftrightarrow$ ν (average firing rate in the network)

a   b

patterns of activity $\xi^v = \{\xi_i^v, i = 1, \ldots, N\}, v = 1, \ldots, M$ are previously stored in a slow learning process, that is, $w_{ij} = w_{ij}(\xi)$.

In many neural networks models in the literature, neurons have been considered as binary elements with their state being $s_i = +1, -1$ (in the so-called $-1, +1$ code) or $\sigma_i = \frac{1+s_i}{2} = 0, 1$ (in the 0, 1 code). In biologically motivated neural networks models, each one of these two possible values represents a neuron in a silent state (below the threshold for firing) or a neuron that fires an action potential. Several recent studies have already shown that binary neurons may capture the essence of cooperation in many more complex settings (see, for instance, Pantic et al. (2002) in the case of integrate and fire neuron models of pyramidal cells).

In the following, the more "realistic" 0, 1 code shall be taken into consideration. Each node σ_i then receives information from its neighbors (the neurons that are directly connected to it), which is measured in terms of the so-called *local field*

$$h_i(\sigma, t) = \sum_{j \neq i} w_{ij}(\xi)\sigma_j(t) \tag{8}$$

In the absence of other external factors, when this field is larger than some threshold θ_i the neuron i fires, so that one can update the current state of a particular neuron (assuming discrete time update) as

$$\sigma_i(t+1) = \Theta[h_i(\sigma, t) - \theta_i] \tag{9}$$

where $\Theta(X)$ is the Heaviside step function. If the network is in the presence of different sources of noise (such as the stochastic neurotransmitter release or the stochasticity associated with the opening and closing of different ionic channels responsible for the excitability of the neuron), which can affect the dynamics of the neurons, one can substitute the dynamics (❯ Eq. 9) by the most realistic stochastic dynamics

$$\text{Prob}[\sigma_i(t+1) = +1] = \mathscr{S}[2\beta\{h_i(\sigma, t) - \theta_i\}] \tag{10}$$

where $\mathscr{S}(X)$ is a nonlinear sigmoidal function, such as, for instance, $\mathscr{S}(X) = \frac{1}{2}[1 + \tanh(X)]$. The dynamical rule (❯ 10) can be updated in parallel and synchronously for all neurons in the network, or sequentially at random, one neuron per time step (although one can also think of other types of updating). In the last case and for fixed symmetric weights $w_{ij} = w_{ji}$, the steady state is characterized by a Gibbs equilibrium probability distribution $P(\sigma) \propto e^{-\beta\mathscr{H}(\sigma)}$ with $\mathscr{H}(\sigma) = -\frac{1}{2}\sum_i h_i(\sigma)$.

From this very general and simple picture, one can easily find a simple analogy between the variables defining the dynamics of artificial neural networks and the variables involved in the dynamics of actual extended neural systems, as shown in ❯ Fig. 13b. Note that in this scheme, extra variables $D_j(t)$ and $F_j(t)$ modulating the maximum conductances are introduced due to learning $w_{ij}(\xi)$, and these could define, for instance, the dynamics of the depressing and facilitating synaptic mechanisms introduced in ❯ Sect. 3.

4.2.2 Neural Networks with Dynamic Synapses

The stochastic dynamics (❯ 10) can be derived from a more general mathematical formalism, which assumes that both neurons and synapses can suffer stochastic changes in different timescales due to the presence of one (or some times more) thermal bath at a given temperature $T \equiv \beta^{-1}$. The dynamics of the system is then represented by a general *master equation* (for more details, see, for instance, Torres et al. (1997, 2007a) and Cortes et al. (2006)). Some particular cases of this general model, with a biological motivation, account for fast synaptic changes coupled with neuron activity (one can think, for instance, of the stochasticity of the opening and closing of the neurotransmitter vesicles, the stochasticity of the postsynaptic receptor, which in turn has several sources, for example, variations of the glutamate concentration in the synaptic cleft, and differences in the power released from different locations on the active zone of the synapses (Franks et al. 2003)). As a consequence, the resulting model exhibits much more varied and intriguing behavior than the standard static Amari–Hopfield model. For example, the network exhibits high sensitivity to external stimuli and, in some conditions, chaotic jumping among the stored memories, which allows for better exploring of the stored information.

For a better understanding of how these instabilities can emerge in the memory attractors and the resulting network behavior, an example will be described. In particular, we will study (in a biologically motivated neural network) the consequences of introducing a realistic dynamics for synaptic transmission, such as that described by the set of equations (❯ 6 and ❯ 7) in the stochastic dynamics of neurons described by ❯ Eq. 10.

To go further it is necessary to propose some simplifications. First, the dynamics (❯ Eq. 10) assumes a discrete time evolution of neuron states, so one can consider that the typical temporal scale for neuron dynamics in ❯ Eq. 10 is larger than the absolute refractory period in actual neurons, which is about (5 ms). On the other hand, one has that the synaptic delay $\tau_{in} = 2$ ms $\ll \tau_{rec}, \tau_{fac}$, so $y(t) = y(t^{sp})e^{-(t-t^{sp})/\tau_{in}} \approx y(t^{sp}) = U(t^{sp})x(t^{sp})$. This allows us to consider a discrete dynamics for synapse states and to reduce the dimensionality of the system ❯ Eqs. 6 and ❯ 7 from four to two equations (note that $x + y + z = 1$, because they are molar fractions), which is important when one has to simulate large networks. One then has that the generated synaptic current in a single synapse is $I_{ij}^{syn}(t) \propto U_j(t)x_j(t)$. Therefore, the total synaptic current that arrives at a particular neuron i (using the analogy described in ❯ Fig. 13) can be written as

$$I_i^{syn}(t) = h_i(t) = \sum_j w_{ij}(\xi)D_j(t)F_j(t)\sigma_j(t) \tag{11}$$

where $D_j(t) \equiv x_j(t)$ and $F_j(t) \equiv U_j(t)/U_{SE}$ represent the depressing and facilitating mechanisms, respectively, affecting the maximum conductances $w_{ij}(\xi)$, with

$$x_j(t+1) = x_j(t) + \frac{1 - x_j(t)}{\tau_{rec}} - U_j(t)x_j(t)\sigma_j(t)$$
$$U_j(t+1) = U_j(t) + \frac{U_{SE} - U_j(t)}{\tau_{fac}} + U_{SE}[1 - U_j(t)]\sigma_j(t) \tag{12}$$

Here, the binary character of the neuron variables σ_i is used to substitute the delta functions $\delta(t - t_j^{sp})$ in ❯ Eqs. 6 and ❯ 7 by $\sigma_j(t)$. The model described by ❯ Eqs. 10–12 can be considered as the more general model of a biologically inspired ANN, which includes dynamic synapses with depressing and facilitating mechanisms. It reduces to the Amari–Hopfield model for static synapses when $x_i = F_i = 1$ (that is $\tau_{rec}, \tau_{fac} \ll 1$); $\theta_i = \frac{1}{2}\sum_{j\neq i} w_{ij}(\xi)$ and $w_{ij}(\xi)$ are given by the standard *covariance learning rule*, namely,

$$w_{ij}(\xi) = \frac{1}{Nf(1-f)}\sum_{\mu=1}^{M}(\xi_i^\mu - f)(\xi_j^\mu - f) \tag{13}$$

where $\xi^\mu = \{\xi_i^\mu = 0, 1, i = 1\ldots, N\}$ are M random patterns of neuron activity with mean $\langle \xi_i^\mu \rangle = f$. To measure the degree of correlation between the current network state and one of these stored patterns it is useful to introduce the overlap function

$$m^\mu(\sigma) \equiv \frac{1}{Nf(1-f)} \sum_i (\xi^\mu - f)\sigma_i \qquad (14)$$

Assuming a synchronous and parallel updating of neurons states (e.g., a little dynamics) using ❷ Eq. 10 and random unbiased patterns ($f = 1/2$), the previous model can be studied within the standard mean-field approach $\sigma_i \rightarrow \langle \sigma_i \rangle$, where $\langle \cdot \rangle$ means an average over the probability distribution $P_t(\sigma)$. In the simplest case of only a single pattern, or for a finite number of patterns, that is, when the loading parameter $\alpha \equiv M/N \rightarrow 0$ in the thermodynamic limit ($N \rightarrow \infty$), the dynamics of the network is driven by a $6M$-dimensional discrete map (see Torres et al. (2007b) for details)

$$\mathbf{v}_{t+1} = \mathcal{F}(\mathbf{v}_t) \qquad (15)$$

where $\mathcal{F}(X)$ is a complex multidimensional nonlinear function of

$$\mathbf{v}_t = \{m_+^\mu(t), m_-^\mu(t), x_+^\mu(t), x_-^\mu(t), u_+^\mu(t), u_-^\mu(t)\} \quad \mu = 1, \ldots, M \qquad (16)$$

This vector represents a set of order parameters defined as averages of the microscopic dynamical variables over the sites that are active and inert, respectively, in a given pattern μ, that is

$$c_+^\mu(t) \equiv \frac{1}{Nf} \sum_{i \in Act(\mu)} c_i(t), \quad c_-^\mu(t) \equiv \frac{1}{N(1-f)} \sum_{i \notin Act(\mu)} c_i(t) \qquad (17)$$

with c_i being $\sigma_i(t)$, $x_i(t)$, and $u_i(t)$, respectively.

One can easily compute the fixed point solutions of the dynamics (❷ Eq. 15) and study their local stability. Similarly to the Amari–Hopfield standard model one finds that the stored memories ξ^μ are stable attractors in some regions of the space of relevant parameters, such as T, U_{SE}, τ_{rec}, and τ_{fac}. Their stability and the error in the memory recall process, however, is highly influenced by the particular value of these parameters (see for instance ❷ Fig. 14a, b). In particular, there are some critical values for which the memories destabilize, and an oscillatory regime, in which the activity of the network is visiting different memories, can emerge. ❷ Figure 14c shows an example of such oscillatory behavior, and a typical phase diagram with different dynamical phases in the (τ_{rec}, τ_{fac}) space is shown in ❷ Fig. 14d. The critical values for the destabilization of memories depend on the relative competition between depression and facilitation, which are two processes with, a priori, different effects over the attractors (synaptic depression tries to put the activity of the network out of attractors whereas facilitation induces a faster access of the activity to the attractors). The emergence of oscillations between different patterns of activity induced by activity–dependent synaptic processes has been recently related to the appearance of oscillations between "up and down" activity states observed in cortical areas of living animals (Holcman and Tsodyks 2006).

The instability in the memories induced by an activity-dependent synaptic process, such as depression and facilitation, enhances the sensitivity of the network to external stimuli. An example of the latter is shown in ❷ Fig. 15. In this simulation, an external varying stimulus is presented to the network during a temporal window at which τ_{fac} (for the dynamic synapse case) is continuously increased. The stimulus consists of short pulses of amplitude δ,

□ **Fig. 14**

Panels **a** and **b** show the steady state or fixed point memory attractors (as in the standard Amari–Hopfield model), described in terms of the overlap $m \equiv m^1$, which can be reached in the dynamics of the general model of ANN with dynamic synapses. Data points correspond to simulations of ❯ Eqs. 6 and ❯ 7 for different values τ_{rec} and τ_{fac}, and lines are the corresponding mean field steady-state curves. There are *critical* values of the synaptic parameters for which these fixed point memory solutions disappear and oscillatory attractors such as that shown in panel **c** can emerge. In Panel **d** a typical phase diagram in the space (τ_{rec}, τ_{fac}) for $T = U_{SE} = 0.1$ is presented, which shows the critical lines at which the different emergent phases appear.

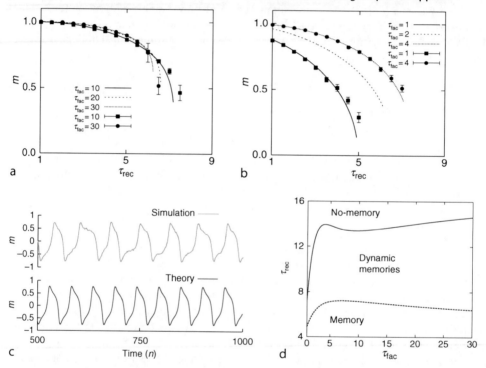

presented at a given frequency such that the activity of the network is already in a memory attractor. The sign of the stimulus changes in such a way that every time it is presented it tries to drive the activity of the network out of the attractor. The analysis of this simulation shows that dynamic synapses allow for a better response to the stimulus even for $\delta \ll 1$ (weak signals). This behavior can be useful for living animals to respond efficiently to a continuously varying environment.

The phenomenology described above is general and it has been also observed in more realistic situations, such as for instance, in networks of integrate and fire neurons, with other type of stored patterns (e.g., more overlapping patterns), for a large number of stored patterns, and for a continuous dynamics (see Pantic et al. (2002)).

Another important issue in the study of ANN is the maximum number of memories M_{max}, per neuron, that the network is able to store and recall without having significant errors, which defines the critical (or maximum) storage capacity of the system. This capacity is

☐ **Fig. 15**

Comparison of the responses to a varying external stimulus in an ANN with dynamic (left) and static (right) synapses, for increasing values of the τ_{fac}. Only in the case of dynamic synapses, can the system optimally respond even for a very weak stimulus.

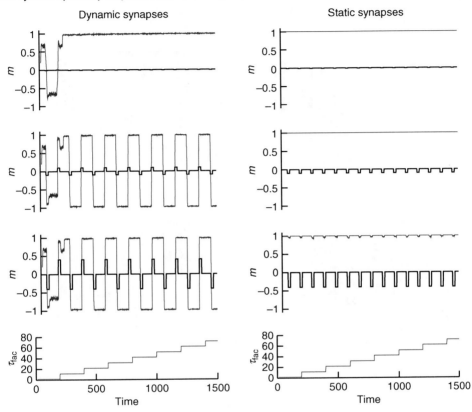

commonly denoted by the ratio $\alpha_c = M_{max}/N$. It is well known that the critical storage capacity is affected by certain considerations about actual neural systems, such as constraints in the range of values of the synaptic strength (Fusi and Abbott 2007), the mean activity level of the stored patterns (Tsodyks and Feigelman 1988; Amit and Tsodyks 1991), or the topology of the network (McGraw and Menzinger 2003; Torres et al. 2004). Recently the effect of activity-dependent synaptic processes such as depression and facilitation on storage capacity has been reported (Mejias and Torres 2009). The main results can be summarized in ❷ *Fig. 16*. This shows that synaptic facilitation improves the storage capacity with respect to the case of depressing synapses, for a certain range of the synaptic parameters. Moreover, if the level of depression is not very large, facilitation can increase the critical storage capacity, reaching in some cases the value obtained with static synapses (which is the maximum that one can obtain considering a Hebbian learning rule with unbiased random patterns in a fully connected network). These results suggest that a certain level of facilitation in the synapses might be positive for an efficient memory retrieval, while the function of strongly depressed synapses could be more oriented to other tasks concerning, for instance, the dynamical processing of data, as explained above.

☐ **Fig. 16**

Maximum storage capacity α_c in a "realistic" model of ANN with dynamic synapses computed within a general mean-field approach and for different sets of the synapse parameters (with $\tau_{rec} = 2$ in the top panels and $U_{SE} = 0.02$ in the bottom panels). Contour lines and different levels of gray define the regions in the space of relevant parameters with similar storage capacity.

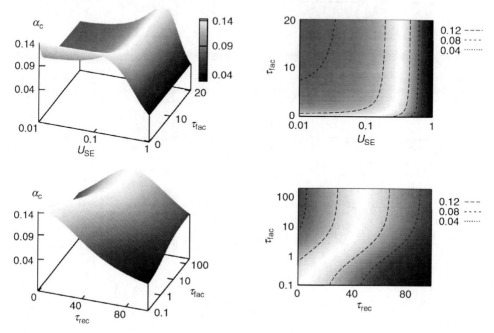

4.2.3 Network Topology

Large networks, such as those described above, are usually modeled using fully connected networks. This is convenient because it allows for simple analytical treatments such as mean-field approaches, as seen previously. Extended neural systems, however, present a map of connectivity that can largely differ from such a simple picture and which, in general, are very difficult to determine. An important question to address then is how the network structure can influence the functionality of the system. There have been many theoretical studies focused on this particularly interesting issue. For instance, it has been recently reported that ANN with activity-dependent synapses work optimally for particular network topologies where the level of "heterogeneity" (measured as the number of different connectivity degrees) in the network is maximal (Johnson et al. 2008). Moreover, it is reasonable to think that during development a given neural system can self-organize to reach an optimal topological structure that allows for a better performance of the system during the realization of a particular task (Beggs and Plenz 2003). Finally, recent studies have also shown the complex interplay between network topology and neuron activity to explain the proper or improper functioning of a given neural system, which can be related to brain diseases such as, for instance, epilepsy (Dyhrfjeld-Johnsen et al. 2007; Roxin et al. 2004).

5   Summary

In this chapter we reviewed some theoretical approaches currently used in the modeling of biological neural networks from single cells to large ensembles of neural networks. Models are useful mathematical tools to understand complex emergent behavior observed in living neural media, which arises from different phenomena occurring at subcellular, cellular, network, and system levels, and involving different spatial and temporal scales.

In many cases, models contribute to explain specific aspects of a particular neural system, but there are also efforts to unveil general principles underlying information processing in the brain (Ashby 1960; Rosenblatt 1962; Freeman 1972; von der Malsburg 1981; Hopfield 1982; Rabinovich et al. 2006). In computational neuroscience, both bottom–up and top–down approaches are needed. Often models are accused of being able to fit any particular behavior through the *ad hoc* tuning of their frequently large parameter space. In fact, biology also provides different ways to achieve similar neural dynamics, and self-regulation mechanisms rely on this. It is important to emphasize that the progress in computational neuroscience research will go hand in hand with the development of new experimental protocols to observe and stimulate neural activity, which will constrain the parameter space and validate model hypotheses.

Computational neuroscience is a relatively young discipline that has a lot of work ahead in the study of universal paradigms that neural systems use for coding, decoding, and to achieve robust and flexible information processing. Artificial intelligence will benefit from the work of computational neuroscience, whose results also contribute to bio-inspired robotics, and the design of novel brain–machine interfaces for medical and industrial applications.

References

Abarbanel HDI, Huerta R, Rabinovich MI, Rulkov NF, Rowat P, Selverston AI (1996) Synchronized action of synaptically coupled chaotic model neurons. Neural Comp 8(8):1567–1602

Abbott LF, Valera JA, Sen K, Nelson SB (1997) Synaptic depression and cortical gain control. Science 275(5297):220–224

Amari S (1972) Characteristics of random nets of analog neuron-like elements. IEEE Trans Syst Man Cybern 2:643–657

Amit D, Tsodyks M (1991) Quantitative study of attractor neural networks retrieving at low spike rates: II. low-rate retrieval in symmetric networks. Network: Comput Neural Syst 2:275–294

Arshavsky YI, Beloozerova IN, Orlovsky GN, Panchin YV, Pavlova GA (1985) Control of locomotion in marine mollusc Clione limacina. i. efferent activity during actual and fictitious swimming. Exp Brain Res 58(2):255–293

Arshavsky YI, Grillner S, Orlovsky GN, Panchin YV (1991) Central generators and the spatiotemporal pattern of movements. In: Fagard J, Wolff PH (eds) The development of timing control. Elsevier, Amsterdam, pp 93–115

Ashby WR (1960) Design for a brain, 2nd edn. Wiley, New York

Azouz R, Gray CM (2000) Dynamic spike threshold reveals a mechanism for synaptic coincidence detection in cortical neurons in vivo. Proc Natl Acad Sci USA 97(14):8110–8115

Barrie JM, Freeman WJ, Lenhart MD (1996) Spatiotemporal analysis of prepyriform, visual, auditory, and somesthetic surface EEGs in trained rabbits. J Neurophysiol 76(1):520–539

Beenhakker MP, Nusbaum MP (2004) Mechanosensory activation of a motor circuit by coactivation of two projection neurons. J Neurosci 24(30):6741–6750

Beggs JM, Plenz D (2003) Neuronal avalanches in neocortical circuits. J Neurosci 23(35):11167–11177

Bertram R, Sherman A, Stanley EF (1996) Single-domain/bound calcium hypothesis of transmitter release and facilitation. J Neurophysiol 75(5):1919–1931

Brette R et al (2007) Simulation of networks of spiking neurons: a review of tools and strategies. J Comp Neurosci 23(3):349–398

Buia CI, Tiesinga PHE (2005) Rapid temporal modulation of synchrony in cortical interneuron networks with synaptic plasticity. Computational

Neuroscience: Trends in Research 2005. Neurocomputing 6566:809–815

Cazelles B, Courbage M, Rabinovich MI (2001) Anti-phase regularization of coupled chaotic maps modelling bursting neurons. Europhys Lett 56(4):504–509

Cortes JM, Torres JJ, Marro J, Garrido PL, Kappen HJ (2006) Effects of fast presynaptic noise in attractor neural networks. Neural Comp 18(3):614–633

Cymbalyuk G, Gaudry O, Masino M, Calabrese R (2002) Bursting in leech heart interneurons: cell-autonomous and network-based mechanisms. J Neurosci 22(24):10580–10592

Destexhe A, Marder E (2004) Plasticity in single neuron and circuit computations. Nature 431 (7010):789–795

Destexhe A, Mainen ZF, Sejnowski TJ (1994) An efficient method for computing synaptic conductances based on a kinetic model of receptor binding. Neural Comp 6(1):14–18

Dobrunz LE, Stevens CF (1997) Heterogeneity of release probability, facilitation, and depletion at central synapses. Neuron 18(6):995–1008

Dyhrfjeld-Johnsen J, Santhakumar V, Morgan RJ, Huerta R, Tsimring L, Soltesz I (2007) Topological determinants of epileptogenesis in large-scale structural and functional models of the dentate gyrus derived from experimental data. J Neurophysiol 97(2):1566–1587

Elson R, Selverston AI, Huerta R, Rabinovich MI, Abarbanel HDI (1998) Synchronous behavior of two coupled biological neurons 81(25):5692–5695

Elson R, Huerta R, Abarbanel HDI, Rabinovich MI, Selverston AI (1999) Dynamic control of irregular bursting in an identified neuron of an oscillatory circuit. Phys Rev Lett. J Neurophysiol 82(1):115–122

FitzHugh R (1961) Impulses and physiological states in theoretical models of nerve membrane. Biophys J 1(6):445–466

Franks KM, Stevens CF, Sejnowski TJ (2003) Independent sources of quantal variability at single glutamatergic synapses. J Neurosci 23(8):3186–3195

Freeman W (1972) Progress in theoretical biology, vol 2. Academic, New York

Fusi S, Abbott L (2007) Limits on the memory storage capacity of bounded synapses. Nat Neurosci 10(4):485–493

Garcia-Sanchez M, Huerta R (2003) Design parameters of the fan-out phase of sensory systems. J Comp Neurosci 15(1):5–17

Gerstner W, Kistler W (2002) Spiking neuron models. Cambridge University Press, Cambridge

Getting PA (1989) Emerging principles governing the operation of neural networks. Ann Rev Neurosci 12:185–204

Gutkin BS, Ermentrout GB (1998) Dynamics of membrane excitability determine interspike interval variability: a link between spike generation mechanisms and cortical spike train statistics. Neural Comp 10(5):1047–1065

Hebb DO (1949) The organization of behavior. Wiley, New York

Hindmarsh JL, Rose RM (1984) A model of neuronal bursting using three coupled first order differential equations. Proc R Soc Lond B 221(1222):87–102

Hodgkin AL, Huxley AF (1952) A quantitative description of membrane current and its application to conduction and excitation in nerve. J Physiol 117(4):500–544

Holcman D, Tsodyks M (2006) The emergence of up and down states in cortical networks. PLoS Comput Biol 2(3):174–181

Hopfield JJ (1982) Neural networks and physical systems with emergent collective computational abilities. Proc Natl Acad Sci USA 79(8):2554–2558

Huerta R, Varona P, Rabinovich MI, Abarbanel HDI (2001) Topology selection by chaotic neurons of a pyloric central pattern generator. Biol Cybern 84(1):L1–L8

Izhikevich E (2004) Which model to use for cortical spiking neurons? IEEE Trans Neural Netw 15(5):1063–1070

Johnson S, Marro J, Torres JJ (2008) Functional optimization in complex excitable networks. EPL 83:46006 (1–6)

Kamiya H, Zucker RS (1994) Residual Ca^{2+} and short-term synaptic plasticity. Nature 371(6498):603–606

Kandel ER, Schwartz JH, Jessell TM (2000) Principles of neural science, 4th edn. McGraw-Hill, New York, pp 10–11

Kepecs A, Lisman J (2003) Information encoding and computation with spikes and bursts. Network: Comput Neural Syst 14(1):103–118

Koch C (1999) Biophysics of computation. Oxford University Press, New York

Koch C, Segev I (1998) Methods in neuronal modeling: from ions to networks, 2nd edn. MIT Press, London, pp 6–17

Korn H, Faure P (2003) Is there chaos in the brain? II. experimental evidence and related models. C R Biol 326(9):787–840

Lapicque L (1907) Recherches quantitatives sur l'excitation électrique des nerfs traitée comme une polarisation. J Physiol Pathol Gen 9:620–635

Latorre R, Rodriguez FB, Varona P (2006) Neural signatures: multiple coding in spiking bursting cells. Biol Cybern 95(2):169–183

Laurent G, Stopfer M, Friedrich RW, Rabinovich MI, Volkovskii A, Abarbanel HDO (2001) Odor encoding as an active, dynamical process: experiments, computation, and theory. Annu Rev Neurosci 24:263–297

Levi R, Camhi JM (2000) Population vector coding by the giant interneurons of the cockroach. J Neurosci 20(10):3822–3829

Levi R, Varona P, Arshavsky YI, Rabinovich MI, Selverston AI (2004) Dual sensorymotor function for a molluskan statocyst network. J Neurophysiol 91(1):336–345

Levi R, Varona P, Arshavsky YI, Rabinovich MI, Selverston AI (2005) The role of sensory network dynamics in generating a motor program. J Neurosci 25(42):9807–9815

Lewis JE, Kristan WB (1998) Quantitative analysis of a directed behavior in the medicinal leech: implications for organizing motor output. J Neurosci 18(4):1571–1582

Marder E, Bucher D (2007) Understanding circuit dynamics using the stomatogastric nervous system of lobsters and crabs. Annu Rev Physiol 69:291–316

Marder E, Calabrese RL (1996) Principles of rhythmic motor pattern generation. Physiol Rev 76(3): 687–717

Markram H, Wang Y, Tsodyks M (1998) Differential signaling via the same axon of neocortical pyramidal neurons. Proc Natl Acad Sci USA 95(9):5323–5328

Matveev V, Wang XJ (2000) Differential short-term synaptic plasticity and transmission of complex spike trains: to depress or to facilitate? Cerebral Cortex 10 (11):1143–1153

McAdams C, Maunsell J (1999) Effects of attention on orientation-tuning functions of single neurons in macaque cortical area V4. J Neurosci 19(1):431–441

McCulloch WS, Pitts WH (1943) A logical calculus of ideas immanent in nervous activity. Bull Math Biophys 5:115–133

McGraw PM, Menzinger M (2003) Topology and computational performance of attractor neural networks. Phys Rev E 68(4 Pt 2):047102

Mejias JF, Torres JJ (2008) The role of synaptic facilitation in spike coincidence detection. J Comp Neurosci 24(2):222–234

Mejias JF, Torres JJ (2009) Maximum memory capacity on neural networks with short-term synaptic depression and facilitation. Neural Comput 21(3): 851–871

Morris C, Lecar H (1981) Voltage oscillations in the barnacle giant muscle fiber. Biophys J 35(1):193–213

Nagumo J, Arimoto S, Yoshizawa S (1962) An active pulse transmission line simulating nerve axon. Proc IRE 50:2061–2070

Nowotny T, Huerta R (2003) Explaining synchrony in feed-forward networks: are McCulloch-Pitts neurons good enough? Biol Cybern 89(4):237–241

Panchin YV, Arshavsky YI, Deliagina TG, Popova LB, Orlovsky GN (1995) Control of locomotion in marine mollusk Clione limacina. ix. neuronal mechanisms of spatial orientation. J Neurophysiol 75(5):1924–1936

Pantic L, Torres JJ, Kappen HJ, Gielen SCA (2002) Associative memory with dynamic synapses. Neural Comp 14(12):2903–2923

Peretto P (1992) An introduction to the modeling of neural networks. Cambridge University Press, Cambridge

Pinault D, Smith Y, Deschnes M (1997) Dendrodendritic and axoaxonic synapses in the thalamic reticular nucleus of the adult rat. J Neurosci 17(9):3215–3233

Pinto RD, Varona P, Volkovskii AR, Szucs A, Abarbanel HD, Rabinovich MI (2000) Synchronous behavior of two coupled electronic neurons. Phys Rev E 62(2 Pt B):2644–2656

Prinz A (2006) Insights from models of rhythmic motor systems. Curr Opin Neurobiol 16(6):615–620

Rabinovich MI, Varona P, Selverston AI, Abarbanel HDI (2006) Dynamical principles in neuroscience. Rev Mod Phys 78:1213–1265

Rabinovich MI, Huerta R, Varona P, Afraimovich VS (2008) Transient cognitive dynamics, metastability, and decision making. PLoS Comput Biol 4(5): e1000072

Ramirez J, Tryba A, Pena F (2004) Pacemaker neurons and neuronal networks: an integrative view. Curr Opin Neurobiol 6(6):665–674

Reyes MB, Huerta R, Rabinovich MI, Selverston AI (2003) Artificial synaptic modification reveals a dynamical invariant in the pyloric CPG. Eur J Appl Physiol 102(6):667–675

Rosenblatt F (1962) Principles of neurodynamics: perceptions and the theory of brain mechanisms. Spartan Books, New York

Roxin A, Riecke H, Solla SA (2004) Self-sustained activity in a small-world network of excitable neurons. Phys Rev Lett 92(19):198101

Rulkov NF (2001) Regularization of synchronized chaotic bursts. Phys Rev Lett 86(1):183–186

Rulkov NF (2002) Modeling of spiking-bursting neural behavior using two-dimensional map. Phys Rev E 65(4 Pt 1):041922

Selverston A (2005) A neural infrastructure for rhythmic motor patterns. Cell Mol Neurobiol 25(2): 223–244

Selverston AI, Moulins M (eds) (1987) The crustacean stomatogastric system. Springer, Berlin

Selverston AI, Rabinovich MI, Abarbanel HDI, Elson R, Szcs A, Pinto RD, Huerta R, Varona P (2000) Reliable circuits from irregular neurons: a dynamical approach to understanding central pattern generators. J Physiol (Paris) 94(5–6):357–374

Simmers AJ, Moulins M (1988) A disynaptic sensorimotor pathway in the lobster stomatogastric system. J Neurophysiol 59(3):740–756

Stein SG, Grillner S, Selverston AI, Douglas GS (eds) (1997) Neurons, networks, and motor behavior. MIT Press, Cambridge

Stiesberg GR, Reyes MB, Varona P, Pinto RD, Huerta R (2007) Connection topology selection in central pattern generators by maximizing the gain of information. Neural Comp 19(4):974–993

Swiercz W, Cios KJ, Staley K, Kurgan L, Accurso F, Sagel S (2006) A new synaptic plasticity rule for networks of spiking neurons. IEEE Trans Neural Netw 17(1): 94–105

Szucs A, Varona P, Volkovskii AR, Abarbanel HD, Rabinovich MI, Selverston AI (2000) Interacting biological and electronic neurons generate realistic oscillatory rhythms. Neuroreport 11(11):563–569

Szucs A, Pinto RD, Rabinovich MI, Abarbanel HD, Selverston AI (2003) Synaptic modulation of the interspike interval signatures of bursting pyloric neurons. J Neurophysiol 89(3):1363–1377

Torres JJ, Garrido PL, Marro J (1997) Neural networks with fast time-variation of synapses. J Phys A: Math Gen 30:7801–7816

Torres JJ, Munoz MA, Marro J, Garrido PL (2004) Influence of topology on the performance of a neural network. Computational Neuroscience: Trends in Research. 58–60:229–234

Torres JJ, Cortes JM, Marro J (2007a) Information processing with unstable memories. Neurocomputing AIP Conf Proc 887:115–128

Torres JJ, Cortes JM, Marro J, Kappen HJ (2007b) Competition between synaptic depression and facilitation in attractor neural networks. Neural Comp 19(10):2739–2755

Tsodyks M, Feigelman M (1988) The enhanced storage capacity in neural networks with low activity level. Europhys Lett 6:101–105

Tsodyks MV, Markram H (1997) The neural code between neocortical pyramidal neurons depends on neurotransmitter release probability. Proc Natl Acad Sci USA 94(2):719–723

Tsodyks MV, Pawelzik K, Markram H (1998) Neural networks with dynamic synapses. Neural Comp 10(4):821–835

Varona P, Ibarz JM, López L, Herreras O (2000) Macroscopic and subcellular factors shaping population spikes. J Neurophysiol 83(4):2192–2208

Varona P, Torres JJ, Huerta R, Abarbanel HDI, Rabinovich MI (2001a) Regularization mechanisms of spiking-bursting neurons. Neural Netw 14(6–7): 865–875

Varona P, Torres JJ, Abarbanel HDI, Rabinovich MI, Elson RC (2001b) Dynamics of two electrically coupled chaotic neurons: experimental observations and model analysis. Biol Cybern 84(2): 91–101

Varona P, Rabinovich MI, Selverston AI, Arshavsky YI (2002) Winnerless competition between sensory neurons generates chaos: a possible mechanism for molluscan hunting behavior. Chaos 12(3): 672–677

von der Malsburg C (1981) The correlation theory of brain function. MPI Biophysical Chemistry, Internal Report

Zeck GM, Masland RH (2007) Spike train signatures of retinal ganglion cell types. Eur J Neurosci 26(2): 367–380

18 Neural Networks in Bioinformatics

Ke Chen[1] · *Lukasz A. Kurgan*[2]
[1]Department of Electrical and Computer Engineering, University of
Alberta, Edmonton, AB, Canada
kchen1@ece.ualberta.ca
[2]Department of Electrical and Computer Engineering, University of
Alberta, Edmonton, AB, Canada
lkurgan@ece.ualberta.ca

G. Rozenberg et al. (eds.), *Handbook of Natural Computing*, DOI 10.1007/978-3-540-92910-9_18,
© Springer-Verlag Berlin Heidelberg 2012

Abstract

Over the last two decades, neural networks (NNs) gradually became one of the indispensable tools in bioinformatics. This was fueled by the development and rapid growth of numerous biological databases that store data concerning DNA and RNA sequences, protein sequences and structures, and other macromolecular structures. The size and complexity of these data require the use of advanced computational tools. Computational analysis of these databases aims at exposing hidden information that provides insights which help with understanding the underlying biological principles. The most commonly explored capability of neural networks that is exploited in the context of bioinformatics is prediction. This is due to the existence of a large body of raw data and the availability of a limited amount of data that are annotated and can be used to derive the prediction model. In this chapter we discuss and summarize applications of neural networks in bioinformatics, with a particular focus on applications in protein bioinformatics. We summarize the most often used neural network architectures, and discuss several specific applications including prediction of protein secondary structure, solvent accessibility, and binding residues.

1   Introduction

The term "bioinformatics" was coined relatively recently, that is, it did not appear in the literature until 1991 (Boguski 1998). However, the first studies that concerned the field of bioinformatics appeared already in the 1960s when the first protein and nucleic acid sequence database was established. The National Institutes of Health (NIH) defines bioinformatics as "research, development, or application of computational tools and approaches for expanding the use of biological, medical, behavioral, or health data, including those to acquire, store, organize, archive, analyze, or visualize such data" (NIH Working Definition of Bioinformatics and Computational Biology 2000). We note that bioinformatics is usually constrained to molecular genetics and genomics. In a review by Luscombe et al. (2001), this term is defined as "conceptualizing biology in terms of macromolecules (in the sense of physical chemistry) and then applying 'informatics' techniques (derived from disciplines such as applied math, computer science, and statistics) to understand and organize the information associated with these molecules, on a large-scale." The key observations concerning the above definition are that bioinformatics research is interdisciplinary, that is, it requires knowledge of physics, biochemistry, and informatics, and that it concerns large-scale analysis, that is, only scalable computational methods can be used. Since bioinformatics spans a wide variety of research areas, that is, sequence analysis, genome annotation, evolutionary biology, etc., we are not able to discuss all these research topics. Instead, we concentrate on the approaches concerning protein bioinformatics, that is, the scope of this chapter is limited to the application of bioinformatics in protein-related topics.

The last two decades observed an increased interest in the application of machine learning techniques, and particularly artificial neural networks (NNs), in protein bioinformatics. The most common application of the NNs is prediction; we assume that prediction concerns targets that are both discrete and real valued. The popularity of NNs stems from two key advantages that distinguish them from many other machine-learning methods. First, after the NN model is trained, the use of the model to perform prediction is very efficient, that is, computations are fast. This allows for a high throughput prediction of massive amounts of data, which is an inherent feature of a significant majority of bioinformatics projects. Second, NN-based models provide high-quality results for many prediction tasks, for example, the leading methods in protein secondary structure prediction and protein solvent accessibility prediction are based on NNs. These successful applications raised the profile of NNs, which are currently being applied in dozens of other prediction tasks.

First, we introduce the relevant biological background. Next, we summarize the most popular NN architectures that are applied in protein bioinformatics and the key prediction methods that utilize NNs. Finally, we provide a more detailed analysis of NN-based solutions for the prediction of protein secondary structure, solvent accessibility, and binding residues.

2 Biological Background

Proteins are essential elements of virtually all living organisms. They participate in every process within cells. For instance, some proteins serve as enzymes that catalyze biochemical reactions which are vital to metabolism. Proteins are also important in cell signaling, immune responses, cell adhesion, and the cell cycle, to name just a few of their functions. They are large polymeric organic molecules which are composed of amino acids (also called residues). Amino acid (AA) is a small molecule that includes an amino ($-NH_2$) (except the proline amino acid) and a carboxyl ($-COOH$) group that are linked to a carbon atom. The AA formula, $NH_2CHRCOOH$, where N, H, C, and O are the nitrogen, hydrogen, carbon, and oxygen atoms, respectively, also incorporates R which denotes an organic substituent (so-called side chain), see ❷ Fig. 1 (Panel A). There are a total of 20 AAs that make up all proteins. They all share the same $NH_2CHCOOH$ group and have different R-group. The side chains determine physiochemical properties, such as charge, weight, and hydrophobicity, of

❑ Fig. 1
Panel A shows the chemical structure of AAs; the side chain (R-group) differentiates the structure of different AAs. Panel B shows a protein chain (linear sequence) composed of AAs where each circle represents one AA.

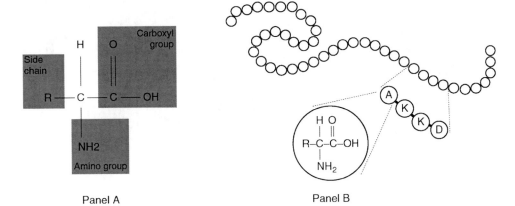

Panel A   Panel B

individual AAs. AAs are abbreviated using either three-letter or one-letter encoding. In the one-letter case, the 20 AAs are encoded as A, C, D, E, F, G, H, I, K, L, M, N, P, Q, R, S, T, V, W, and Y. The AAs are joined together in a linear form through chemical bonds between the carboxyl group and the amino groups of two adjacent AAs, see ❷ *Fig. 1* (Panel B).

The protein structure is defined at four levels, which include the primary, secondary, tertiary, and quaternary structures, see ❷ *Fig. 2*:

- *Primary structure* is the linear order of AAs, also called amino acid sequence. The AA sequence is translated from genes (DNA).
- *Secondary structure* is defined as regular and repetitive spatially local structural patterns in the protein structure. The secondary structure is stabilized by hydrogen bonds. The most common secondary structures are helix, strand, and coil. Secondary structures are present in different regions of the same and different protein molecules.

◻ **Fig. 2**

Examples of different levels of protein structure. Panels A, B, and C show the structure of a globular domain of the human prion protein. Panel A shows the AA sequence (using one-letter encoding) and the corresponding secondary structure for each AA (using encoding in which H, E, and C stand for helix, strand, and coil, respectively) and also a graphical format where the horizontal line denotes the coil, waves denotes helices, and arrows denotes strands. Panel B shows a cartoon representation of the spatial arrangement of the secondary structures. Helices are shown in red, strands in yellow, and coils as black lines. Panel C shows the tertiary structure of the protein in which individual atoms are represented using spheres. Panel D shows the quaternary structure of a microtubule, which is assembled from α-tubulin (dark gray spheres) and β-tubulin (light gray spheres) proteins. The tubulin is assembled into a hollow cylindrical shape.

```
LGGYMLGSAMSRPIIHFGSDYEDRYYRENMHRYPNQVYYRPCDEYSNQNNFVHDCVNITI    (AA sequence)
CCCCEECCCCCCCCCCCCCHHHHHHHHHHHHHHCCCCCEECCCCCCCCCCCHHHHHHHHHH    (Sec. structure)
```

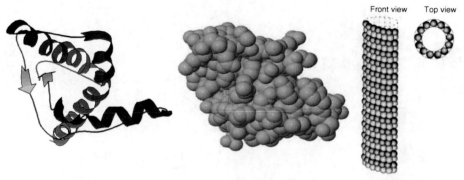

```
KQHTVTTTTKGENFTETDVKMMERVVEQMCITQYERCSQAYYQR    (Continued AA sequence)
HHHHHHHHHHHCCCCCHHHHHHHHHHHHHHHHHHHHHHHHHHCCC    (Continued sec. structure)
```

Panel A

Front view   Top view

Panel B   Panel C   Panel D

- *Tertiary structure* defines the overall three-dimensional shape of a single protein molecule. It concerns the spatial arrangement of the secondary structures and is represented by the coordinates of all atoms in the protein. It is generally believed that the tertiary structure of a given protein is defined by its primary sequence and that each protein has a unique tertiary structure.
- *Quaternary structure* is the arrangement of multiple protein structures in a multi-subunit complex. The individual proteins are assembled into a larger molecule usually with a given geometrical shape, for example, protofilament or a spherical shape. For instance, a microtubule is the assembly of α-tubulin and β-tubulin proteins which takes the form of a hollow cylindrical filament.

While as of January 2009 the primary structure is known for over 6.4 million nonredundant proteins, the corresponding structure is known for only about 55,000 proteins. We emphasize that knowledge of the structure is of pivotal importance for learning and manipulating a protein's function, which for instance is exploited in modern drug design. The significant and widening gap between the set of known protein sequences and protein structures motivates the development of machine learning models that use the known structures to predict structures for the unsolved sequences.

3   Neural Network Architectures in Protein Bioinformatics

Although more than a dozen NN architectures have been developed and adopted, one of the first and simplest architectures, the feedforward neural network (FNN), is the most frequently applied in protein bioinformatics. Besides FNN, the recurrent neural network (RNN) and the radial basis function neural network (RBF) architectures also found several applications in the prediction of bioinformatics data.

A common feature of all prediction applications in protein bioinformatics is the necessity to convert the input (biological) data into the data that can be processed by the NN. This usually involves encoding of the biological data into a fixed-size feature vector. For instance, the primary protein structure is represented as a variable length string of characters with an alphabet of 20 letters (AAs), see ❷ *Fig. 2* (Panel A). This sequence is converted into a vector of numerical features that constitutes the input to the NN. For instance, the vector could include 20 counts of the occurrence of the 20 amino acids in the sequence. The following discussion assumes that the input data are already encoded into the feature vector.

3.1   Feedforward Neural Networks

The FNN architecture usually consists of three layers, an input layer, a hidden layer, and an output layer. The input layer accepts the input feature vector and the output layer generates the predictions. The hidden layer is responsible for capturing the prediction model. Each layer consists of a number of nodes and each node in a given layer connects with every other node in the following layer, see ❷ *Fig. 3*. The connections are associated with weights v_{ij} and w_{ij} between the ith node in one layer and the jth node in the next layer. The nodes process the input values, which are computed as the weighted sum of values passed from the previous layer, using activation functions. The two most frequently used activation functions are:

◘ Fig. 3

Architecture of FNN. The input layer contains _n_ nodes (which equals the number of features in the input feature vector), the hidden layer contains _m_ nodes and the output layer contains _k_ nodes. The weight between the _i_th node of the input layer and the _j_th node of the hidden layer is denoted by _v$_{ij}$_. The weight between the _i_th node of the hidden layer and the _j_th node of the output layer is denoted by _w$_{ij}$_.

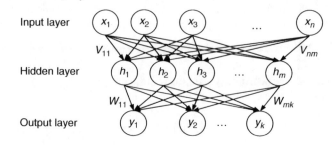

$$\phi(v_i) = \tanh(v_i)$$

$$\phi(v_i) = (1 + e^{-v_i})^{-1}$$

where v_i is the weighted sum of the inputs. The values of the former hyperbolic tangent function range between -1 and 1, while the values of the latter logistic function range between 0 and 1. Some applications also utilize radial basis activation functions.

Learning using the FNN-based prediction model is performed by adjusting the connection weight values to minimize the prediction error on training data. For a given input feature vector $\{x_i\}$, the observations (the prediction outcomes) are denoted as $\{y_i\}$. The goal of the FNN is to find a function $f\colon X{\rightarrow}Y$, which describes the relation between inputs X and observations Y. The merit of function f is measured with a cost function $C = E[(f(x_i) - y_i)^2]$. For a training dataset with n samples, the cost function is

$$C = \sum_{i=1}^{n} \frac{[f(x_i) - y_i]^2}{n}$$

Based on the amount of error associated with the outputs of the network in comparison with the expected result (cost function), the adjustment of the connection weights is carried out using a backpropagation algorithm. The n input feature vectors are fed multiple times (each presentation of the entire training dataset is called an epoch) until the weight values do not change or a desired value of the cost function is obtained.

FNN is the most widely applied among the NN architectures in protein bioinformatics. The applications include:

- Prediction of the secondary structure of protein (Jones 1999; Rost et al. 1994; Dor and Zhou 2007a; Hung and Samudrala 2003; Petersen et al. 2000; Qian and Sejnowski 1988). The aim of these methods is to predict the secondary structure state (helix, strand, or coil) for every AA in the input protein sequence.
- Prediction of relatively solvent accessibility of protein residues (Rost and Sander 1994; Garg et al. 2005; Adamczak et al. 2005; Ahmad et al. 2003; Dor and Zhou 2007b; Pollastri

et al. 2002a). The solvent accessibility is defined as a fraction of a surface area of a given AA that is accessible to the solvent. The AAs with high solvent accessibility are usually on the protein surface.

- Prediction of binding residues (Jeong et al. 2004; Ahmad and Sarai 2005; Zhou and Shan 2001; Ofran and Rost 2007). The binding residues are those AAs in a given protein that are involved in interactions with another molecule. The interactions could concern other proteins, DNA, RNA, ions, etc., and they usually implement protein functions.
- Prediction of transmembrane regions (Gromiha et al. 2005; Natt et al. 2004; Jacoboni et al. 2001). Some proteins are embedded into cell membranes and they serve as pumps, channels, receptors, and energy transducers for the cell. The goal of this prediction method is to find which AAs in the input protein sequence are embedded into the membrane.
- Prediction of subcellular location of proteins (Zou et al. 2007; Cai et al. 2002; Reinhardt and Hubbard 1998; Emanuelsson et al. 2000). These methods predict the location of the proteins inside a cell. The locations include cytoplasm, cytoskeleton, endoplasmic reticulum, Golgi apparatus, mitochondrion, nucleus, etc.

3.2 Recurrent Neural Networks

A recurrent neural network (RNN) is a modification to the FNN architecture. In this case, a "context" layer is added, and this layer retains information across observations. In each iteration, a new feature vector is fed into the input layer. The previous contents of the hidden layer are copied to the context layer and then fed back into the hidden layer in the next iteration, see ❷ *Fig. 4*.

When an input feature vector is fed into the input layer, the RNN processes are as follows:

1. Copy the input vector values into the input nodes.
2. Compute hidden node activations using net input from input nodes and from the nodes in the context layer.
3. Compute output node activations.
4. Compute the new weight values using the backpropagation algorithm.
5. Copy new hidden node weights to the context layer.

❑ **Fig. 4**
Architecture of RNN. Like FNN, RNN also contains an input layer, a hidden layer, and an output layer. An additional context layer is connected to the hidden layer.

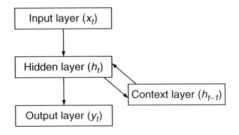

Since the trainable weights, that is, weights between the input and hidden layers and between the hidden and output layers, are feedforward only, the standard backpropagation algorithm is applied to learn the weight values. The weights between the context and the hidden layers play a special role in the computation of the cost function. The error values they receive come from the hidden nodes and so they depend on the error at the hidden nodes at the tth iteration. During the training of the RNN model we consider a gradient of a cost (error) function which is determined by the activations at both the present and the previous iterations.

The RNN architecture was successfully applied in the prediction of:

- Beta-turns (Kirschner and Frishman 2008). Beta turns are the most frequent subtypes of coils, which are one of the secondary protein structures.
- Secondary structure of proteins (Chen and Chaudhari 2007).
- Continuous B-cell epitopes (Saha and Raghava 2006). B-cell epitopes are the antigenic regions of proteins recognized by the binding sites of immunoglobulin molecules. They play an important role in the development of synthetic vaccines and in disease diagnosis. The goal of this prediction method is to find AAs that correspond to the epitopes.
- Number of residue contacts (Pollastri et al. 2002b), which is defined as the number of contacts a given AA makes in the three-dimensional protein molecule. The knowledge of the contacts helps in learning the tertiary protein structure.

3.3 Radial Basis Function Neural Networks

Radial basis function (RBF) neural networks also incorporate three layers: an input layer, a hidden layer with a nonlinear RBF activation function, and a linear output layer, see ❯ *Fig. 5.* During the process of training the RBF model:

1. The input vectors are mapped onto each RBF in the hidden layer. The RBF is usually implemented as a Gaussian function. The Gaussian functions are parameterized, that is, values of the center and spread are established, using the training dataset. The commonly

◻ **Fig. 5**
Architecture of RBF neural network. The network is fully connected between the input and the hidden layers (each node in the input layer is connected with all nodes in the hidden layer), and all the weights are usually assumed to be equal to 1. The nodes in the hidden layer are fully connected with a single node in the output layer, and the weight values are optimized to minimize the cost function.

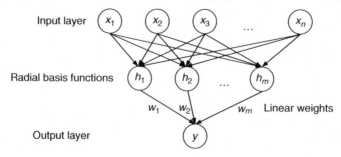

used methods include K-means clustering or, alternatively, a random subset of the training vectors can be used as the centers.

2. In regression problems (the prediction outcomes are real values), the output layer is a linear combination of values produced by the hidden layer, which corresponds to the mean predicted output. In prediction problems (the prediction outcomes are discrete), the output layer is usually implemented using a sigmoid function of a linear combination of hidden layer values, representing a posterior probability.

RBF networks are faster to train when compared with FNN and RNN. They also have an advantage of not suffering from local minima in the same way as FNN, that is, the FNN may not be able to find globally best solution, but it may get stuck in a local minimum of the cost function. This is because the only parameters that are adjusted in the learning process of the RBF network are associated with the linear mapping from the hidden layer to the output layer. The linearity ensures that the error surface is quadratic and therefore it has a single, usually relatively easy to find, minimum. At the same time, the quality of the prediction is usually higher when using a properly designed and trained FNN.

RBF networks were utilized in several applications that include:

- Prediction of inter-residue contact maps (Zhang and Huang 2004). The contact maps include binary entries that define whether a given AA is or is not in contact with any other AA in the tertiary structure. The knowledge of contacts helps in the reconstruction of the tertiary protein structure.
- Prediction of protease cleavage sites (Yang and Thomson 2005). Protease cleavage is performed by enzymes, which are proteins that catalyze biological reactions. Knowledge of how a given protease cleaves the proteins is important for designing effective inhibitors to treat some diseases. This prediction method aims at finding AAs in the protein sequence that are involved in this interaction.
- Prediction of targets for protein-targeting compounds (Niwa 2004). This method aims at the prediction of biological targets (proteins) that interact with given chemical compounds. This has applications in drug design where large libraries of chemical compounds are screened to find compounds that interact with a given protein and which, as a result, modify (say, inhibit) the protein's function.

One of the important parameters in the design of any of the three above mentioned architectures, that is, FNN, RNN, and RBFNN, is the number of nodes. The number of input nodes usually equals the number of input features. Most commonly, there is only one output node that corresponds to the predicted outcome, although in some cases NNs are used to generate multiple outcomes simultaneously, that is, prediction of the protein secondary structure requires three outcomes. The number of nodes in the hidden layer is chosen by the designer of the network. This number depends on the application and the desired quality of the prediction.

4 Applications of Neural Networks in Protein Bioinformatics

NNs are used in a variety of protein bioinformatics applications. They can be categorized into:

- Prediction of protein structure including secondary structure and secondary structure content, contact maps, structural contacts, boundaries of structural domains, specific types of local structures like beta-turns, etc.

- Prediction of binding sites and ligands, which includes prediction of binding residues and prediction of various properties of the binding ligands.
- Prediction of protein properties such as physicochemical proteins, localization in the host organism, etc.

Specific applications include prediction of a number of residue contacts (Pollastri et al. 2002), protein contact maps (Zhang and Huang 2004), helix-helix (Fuchs et al. 2009) and disulfide contacts (Martelli et al. 2004), beta and gamma turns (Kirschner and Frishman 2008; Kaur and Raghava 2003, 2004), secondary structure (Jones 1999; Rost et al. 1994; Dor and Zhou 2007a; Hung and Samudrala 2003; Petersen et al. 2000; Qian and Sejnowski 1988; Chen and Chaudhari 2007), domain boundaries (Ye et al. 2008), transmembrane regions (Gromiha et al. 2005; Natt et al. 2004; Jacoboni et al. 2001), binding sites and functional sites (Jeong et al. 2004; Ahmad and Sarai 2005; Zhou and Shan 2001; Ofran and Rost 2007; Yang and Thomson 2005; Lin et al. 2005; Lundegaard et al. 2008; Blom et al. 1996; Ingrell et al. 2007), residue solvent accessibility (Rost and Sander 1994; Garg et al. 2005; Adamczak et al. 2005; Ahmad et al. 2003; Dor and Zhou 2007b; Pollastri et al. 2002), subcellular location (Zou et al. 2007; Cai et al. 2002; Reinhardt and Hubbard 1998; Emanuelsson et al. 2000), secondary structure content (Muskal and Kim 1992; Cai et al. 2003; Ruan et al. 2005), backbone torsion angles (Xue et al. 2008; Kuang et al. 2004), protein structural class (Chandonia and Karplus 1995; Cai and Zhou 2000), signal peptides (Plewczynski et al. 2008; Sidhu and Yang 2006), continuous B-cell epitopes (Saha and Raghava 2006), binding affinities, toxicity, and pharmacokinetic parameters of organic compounds (Vedani and Dobler 2000), biological targets of chemical compounds (Niwa 2004), and prediction of spectral properties of green fluorescent proteins (Nantasenamat et al. 2007). We observe a growing interest in applying NNs in this domain, see ❯ *Fig. 6.* The NN-based applications in protein bioinformatics were published in a number of

■ **Fig. 6**
Number of journal publications (*y*-axis) concerning the applications of NNs in protein bioinformatics in the last two decades. The included publications do not constitute an exhaustive list of corresponding studies, but rather they provide a set of the most significant and representative developments.

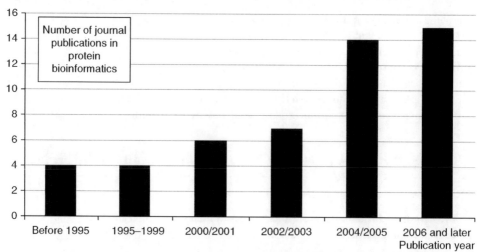

high-impact scientific journals such as (in alphabetical order) Bioinformatics; BMC Bioinformatics; Gene; IEEE/ACM Transactions on Computational Biology and Bioinformatics; Journal of Computational Chemistry; Journal of Computer-Aided Molecular Design; Journal of Medicinal Chemistry; Journal of Molecular Biology; Nucleic Acids Research; PLoS Computational Biology; Protein Science; Proteins; and Proteomics. The number, scope, and quality of the above venues strongly indicate the important role of this research.

The prediction of the secondary structure, residue solvent accessibility, and binding sites attracted the most attention in the context of the NN-based solutions. Therefore, the following sections concentrate on these three topics.

4.1   Prediction of Protein Secondary Structure with Neural Networks

Protein secondary structure is defined as a regular and repetitive spatially local structural pattern in protein structures. Several methods are used to define the protein secondary structure from a protein's three-dimensional structure. The most commonly used method is the Dictionary of Protein Secondary Structure (DSSP) (Kabsch and Sander 1983), which assigns eight types of secondary structures based on hydrogen-bonding patterns. These types include 3/10 helix, alpha helix, pi helix, extended strand in parallel and/or antiparallel β-sheet conformation, isolated β-bridge, hydrogen bonded turn, bend, and coil. The eight-state secondary structure is often aggregated into a three-state secondary structure. The first three types are combined into the helix state, the following two types into the strand state, and the last three types into the coil state. Most of the existing computational methods predict the three-state secondary structure instead of the eight-state structure. The main goal of these methods is to obtain the secondary structure using only AA sequences of the protein as the input. ❯ *Figure 2* shows the AA sequence, the corresponding secondary structure for each AA, the spatial arrangement of the secondary structure, and the overall three-dimensional structure of the human prion protein. This protein is associated with several prion diseases such as fatal familial insomnia and Creutzfeldt–Jakob disease.

The first study concerning the prediction of protein secondary structure using an NN appeared in 1988 (Qian and Sejnowski 1988). This model is a typical three-layer FNN in which the input layer contains $13 \times 21 = 273$ nodes representing a stretch of 13 continuous AAs in the sequence, and the output layer contains three nodes representing the three secondary structure states. Each AA in the sequence is encoded using 21 binary features indicating the type of the AA at a given position in the sequence. This early method was trained using a very small dataset of 106 protein sequences, which limited its quality.

One of the most successful and commonly used models for the prediction of protein secondary structure, named PSIPRED, was proposed by Jones in 1999 (Jones 1999). It is a two-stage NN that takes a position-specific scoring matrix (PSSM), which is generated from the protein sequence using the PSI-BLAST (Position Specific Iterated Basic Local Alignment Search Tool) algorithm (Altschul et al. 1997) as the input. The architecture of PSIPRED is summarized in ❯ *Fig. 7*.

In the first stage, the input protein sequence is represented by the PSSM using a window of size 15 which is centered over the predicted AA. PSSM includes 20 dimensions for each AA, which correspond to substitution scores for each of the 20 AAs. The scores quantify which AAs are likely to be present/absent at a given position in the sequence in a set of known sequences that are similar to the sequence being predicted. This is based on the assumption that

■ **Fig. 7**

Architecture of PSIPRED algorithm. The algorithm is two-stage and includes two 3-layer FNNs, where the output of the first stage network feeds into the input to the second stage network. In the first stage, a window of 15 positions over the PSSM profile generated by the PSI-BLAST program from the input protein sequence is used. Each position in the input is represented by a vector of 21 values (the ith AA in the window is represented as $m_{i,1}m_{i,2} \ldots m_{i,21}$). The 21×15 values are fed into the input layer. The output layer in the first stage NN contains three nodes that represent the probabilities of forming helix, strand, and coil structures (the predicted probabilities for the central AA in the window are represented as $y_{8,1}$, $y_{8,2}$, and $y_{8,3}$). These probabilities, using a window of 15 positions, are fed into the second-stage NN. The output from the second-stage NN is the final prediction that represents the probabilities of three types of the secondary structure: $z_{8,1}$, $z_{8,2}$, and $z_{8,3}$.

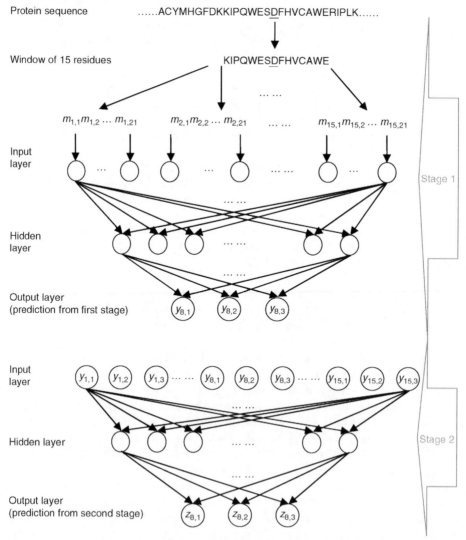

similarity in the sequence often implies similarity in the structure. The positive scores indicate that a given AA substitution occurs more frequently than expected by chance, while negative scores indicate that the substitution occurs less frequently than expected. The 20 scores from the PSSM together with a single feature that indicates the terminus of the sequence are fed into the input layer of the first-stage NN. As a result, the input layer contains $15 \times 21 = 315$ nodes. The hidden layer contains 75 nodes and the output layer contains three nodes which indicate the probabilities of the three secondary structure states.

In the second stage, the predicted probabilities of the secondary structures from the first stage for a window of 15 AAs centered over the position being predicted are fed into the input layer. The second layer exploits the fact that secondary structures form segments in the protein sequence, see ❷ *Fig. 2a,* and thus information about the structure of the AAs in the adjacent positions in the sequence is helpful to determine the structure of a given AA. The input layer contains $4 \times 15 = 60$ nodes (the value indicating the terminus of the sequence is also included), the hidden layer contains 60 nodes, and the output layer contains three nodes. The PSIPRED method can be accessed, as a web server, at http://bioinf.cs.ucl.ac.uk/psipred/. Interested users can also download a stand-alone version of this popular prediction method.

One of the recently proposed NN-based methods performs the prediction of the secondary structure using a cascaded bidirectional recurrent neural network (BRNN) (Chen and Chaudhari 2007). Similar to the PSIPRED design, the first BRNN (sequence-to-structure BRNN) predicts the secondary structure based on the input AA sequences. The second BRNN (structure-to-structure BRNN) refines the raw predictions from the first BRNN. The learning algorithm used to develop this method is based on the backpropagation.

The last two decades observed the development of several methods based on NN for the prediction of protein secondary structure. The performance of the methods mainly depends on the representation of the protein sequence and the size of the training dataset. Since the beta-sheets (strands adjacent in the tertiary structure) are established between AAs that are far away in the sequence, the window-based methods (including all present methods for the prediction of protein secondary structure) are inherently incapable of grasping the long-range interactions, which results in a relatively poor result for strands.

4.2 Prediction of Binding Sites with Neural Networks

A protein performs its function through interactions with other molecules, called ligands, which include another protein, DNA, RNA, small organic compounds, or metal ions. Knowledge of the binding sites, which are defined as the AAs that directly interact with the other molecules, is crucial to understand the protein's function. More specifically, an AA is a part of the binding site if the distance from at least one atom of this AA to any atom of the ligand is less than a cutoff threshold D. The values of D vary in different studies and they usually range between 3.5 and 6 Å (Zhou and Shan 2001; Ofran and Rost 2007; Ahmad et al. 2004; Kuznetsov et al. 2006).

In one of the recent works by Jeong and colleagues, the FNN architecture is used for the prediction of RNA-binding sites (Jeong et al. 2004). Each AA in the input protein sequence is encoded by a vector of 24 values, of which 20 values indicate the AA type (using binary encoding), one value represents the terminus of the sequence, and the remaining three values correspond to the probabilities of three types of the secondary structure predicted using the PHD program. Using a window of size of 41 residues, the corresponding design includes

$24 \times 41 = 984$ input nodes. The hidden layer includes 30 nodes and the output layer consists of a single node that provides the prediction.

In another recent study by Ahmad and Sarai, a three-layer FNN is utilized for the prediction of DNA-binding sites (Ahmad and Sarai 2005). The architecture of this model is relatively simple, that is, 100 nodes in the input layer, 2 nodes in the hidden layer, and 1 node in the output layer. The input layer receives values from the PSSM computed over the input protein sequence with the window size of 5, which results in 100 features per AA.

Prediction of protein–protein interaction sites uses designs that are similar to the designs utilized in the prediction of DNA/RNA-binding sites. Zhou and Shan proposed a three-layer FNN to predict the protein–protein interaction sites (Zhou and Shan 2001). In their design, the PSSM and solvent-accessible area generated by the DSSP program (Kabsch and Sander 1983) for the predicted AAs and the 19 spatially nearest neighboring surface AAs make up the input. As a result, the input layer contains $21 \times 20 = 420$ nodes. The hidden layer includes 75 nodes. This method predicts the protein–protein interaction sites from the protein's three-dimensional structure, since this information is necessary to compute the relative solvent accessibility values and to find the 19 spatially nearest AAs. In a recent study by Ofran and Rost, a classical FNN model is used for the prediction of protein–protein interaction sites from the protein sequence (Ofran and Rost 2007). In this case, AAs in the input protein sequence are represented using PSMM, predicted values of solvent accessibility, predicted secondary structure state, and a conservation score. The window size is set to include eight AAs surrounding the position that is being predicted, and the above-mentioned information concerning these nine amino acids is fed into the input layer.

NNs were also applied for prediction of metal-binding sites (Lin et al. 2005), binding sites for a specific protein family, that is, the binding sites of MHC I (Lundegaard et al. 2008), and prediction of functional sites, that is, the cleavage sites (Blom et al. 1996) and phosphorylation sites (Ingrell et al. 2007).

4.3   Prediction of Relative Solvent Accessibility with Neural Networks

Relative solvent accessibility (RSA) reflects the percentage of the surface area of a given AA in the protein sequence that is accessible to the solvent. RSA value, which is normalized to the [0, 1] interval, is defined as the ratio between the solvent accessible surface area (ASA) of an AA within a three-dimensional structure and the ASA of its extended tripeptide (Ala-X-Ala) conformation:

$$RSA = \frac{RSA \text{ in 3D structure}}{RSA \text{ in extended tripeptide conformation}}$$

The first study that concerned prediction of RSA from the protein sequence was published in 1994 by Rost and Sander (Rost and Sander 1994). In this work, the AA is encoded by the percentage of the occurrence of each AA type at this position in the sequence in multiple sequence alignment, which is similar to the values provided in the PSSM matrix. The input to the two-layer FNN is based on a window of size 9 which is centered on the AA that is being predicted and is used, which results in 9×20 features, together with the AA composition of the entire protein sequence, length of the sequence (using four values), and distance of the window from two termini of the sequence (using four values for each terminus). As a result, the network has a total of $180 + 20 + 4 + 8 = 212$ nodes in the input layer. The output layer contains one node representing the predicted RSA value and no hidden layer is used in this model.

The past decade observed the development of several NN-based methods for the prediction of RSA values (Garg et al. 2005; Adamczak et al. 2005; Ahmad et al. 2003; Dor and Zhou 2007b; Pollastri et al. 2002). These methods share similar architectures and therefore we discuss one representative model proposed by Garg et al. (2005). This method is a two-stage design in which both stages are implemented using FNNs. Two sources of information are used to generate inputs for the NNs from the protein sequence, the PSSM profile, and the secondary structure predicted with the PSIPRED algorithm. The input features are extracted using a window of size 11 centered on the AA that is being predicted. The values from PSSM in the window are fed into the first NN. This results in the input layer with $11 \times 21 = 231$ nodes. The hidden layer contains ten nodes and the output layer has one node. In the second stage NN, the predicted RSA values of the AAs in the window and the predicted probabilities of the three secondary structure types predicted by PSIPRED in the same window are fed into the input layer. This results in $11 \times 4 = 44$ nodes in the input layer. The hidden layer includes ten nodes and the single node in the output layer corresponds to the final prediction. The architecture of this method is shown in ❷ *Fig. 8*. The second layer exploits the observation that information about the secondary structure and solvent accessibility of the AAs in the adjacent positions in the sequence is useful in determining the solvent accessibility of a given AA. We observe that a similar design is used to implement the PSIPRED method.

5 Summary

We summarized the applications of neural networks (NN) in bioinformatics, with a particular focus on protein bioinformatics. We show that numerous applications that aim at predictions of a variety of protein-related information, such as structure, binding sites, and localization, are designed and implemented using NNs. The most popular architecture used in these methods is a simple three-layer feedforward NN, although other architectures such as RBF and recurrent NNs are also applied. Some of the protein bioinformatics applications use multilayered designs in which two (or more) NNs are used in tandem. We show that the popularity of the NN-based designs has been growing over the last decade. Three applications that enjoy the most widespread use are discussed in greater detail. They include protein secondary structure prediction, prediction of binding sites, and prediction of relative solvent accessibility. We contrast and analyze the architectures of the corresponding NN models. We conclude that the extent and quality of the applications that are based on NNs indicate that this methodology provides sound and valuable results for the bioinformatics community.

We acknowledge several other useful resources that discuss the applications of a broader range of machine learning techniques in bioinformatics. Although none of these contributions is solely devoted to NNs, some of them discuss NNs together with other similar techniques. A survey by Narayanan and colleagues discusses applications of classification methods (nearest neighbor and decision trees), NNs, and genetic algorithms in bioinformatics (Narayanan et al. 2002). Another survey contribution by Kapetanovic and coworkers concerns clustering and classification algorithms, including NNs and support vector machines (Kapetanovic et al. 2004). The most recent review by Fogel discusses a host of computational intelligence techniques, including NNs, fuzzy systems, and evolutionary computation, in the context of

■ **Fig. 8**

Architecture of the model proposed in Garg et al. (2005) for the prediction of relative solvent accessibility. The method includes two stages implemented using FNNs. In the first stage, a window of 11 AAs is used, and each AA is represented by a vector of 21 values. The vector is taken from the PSSM profile generated by the PSI-BLAST algorithm. The output layer of the first stage generates one value that represents the predicted RSA value, which is further refined using the second stage. The predicted RSA values and the secondary structure probabilities predicted using PSIPRED of the AAs in the window are fed into the second-stage FNN. The output from the second-stage NN constitutes the final predicted RSA value.

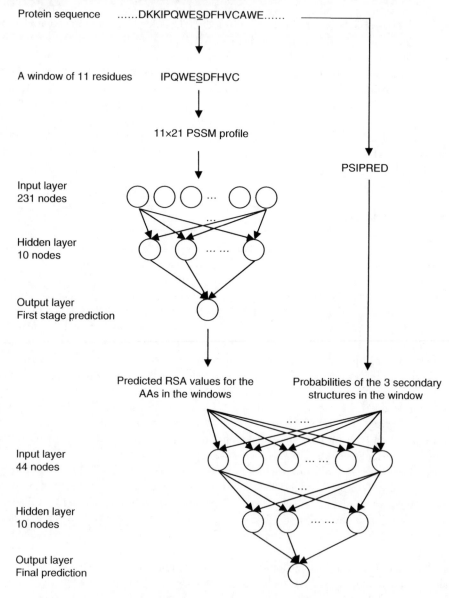

bioinformatics (Fogel 2008). Several other surveys that do not treat NNs but which focus on the use of other related techniques in bioinformatics were published in recent years. They include a paper by Byvatov and Schneider (2003) that concerns applications of support vector machines; a contribution by Saeys et al. (2007) that discusses feature selection methods; a survey concerning Bayesian networks by Wilkinson (2007); a review of supervised classification, clustering, and probabilistic graphical models by Larranaga et al. (2006); and a recent contribution by Miller et al. (2008) that focuses on clustering.

References

Adamczak R, Porollo A, Meller J (2005) Combining prediction of secondary structure and solvent accessibility in proteins. Proteins 59:467–475

Ahmad S, Gromiha MM, Sarai A (2003) Real value prediction of solvent accessibility from amino acid sequence. Proteins 50:629–635

Ahmad S, Gromiha MM, Sarai A (2004) Analysis and prediction of DNA-binding proteins and their binding residues based on composition, sequence and structural information. Bioinformatics 20:477–486

Ahmad S, Sarai A (2005) PSSM-based prediction of DNA binding sites in proteins. BMC Bioinformatics 6:33

Altschul SF, Madden TL, Schaffer AA, Zhang JH, Zhang Z, Miller W, Lipman DJ (1997) Gapped BLAST and PSI-BLAST: a new generation of protein database search programs. Nucleic Acids Res 17:3389–3402

Blom N, Hansen J, Blaas D, Brunak S (1996) Cleavage site analysis in picornaviral polyproteins: discovering cellular targets by neural networks. Protein Sci 5:2203–2216

Boguski MS (1998) Bioinformatics – a new era. Trends Guide Bioinformatics (Suppl S):1–3

Byvatov E, Schneider G (2003) Support vector machine applications in bioinformatics. Appl Bioinformatics 2(2):67–77

Cai YD, Zhou GP (2000) Prediction of protein structural classes by neural network. Biochimie 82:783–785

Cai YD, Liu XJ, Chou KC (2002) Artificial neural network model for predicting protein subcellular location. Comput Chem 26:179–182

Cai YD, Liu XJ, Chou KC (2003) Prediction of protein secondary structure content by artificial neural network. J Comput Chem 24:727–731

Chandonia JM, Karplus M (1995) Neural networks for secondary structure and structural class predictions. Protein Sci 4:275–285

Chen J, Chaudhari N (2007) Cascaded bidirectional recurrent neural networks for protein secondary structure prediction. IEEE/ACM Trans Comput Biol Bioinform 4:572–582

Dor O, Zhou Y (2007a) Achieving 80% ten-fold cross-validated accuracy for secondary structure prediction by large-scale training. Proteins 66:838–845

Dor O, Zhou Y (2007b) Real-SPINE: an integrated system of neural networks for real-value prediction of protein structural properties. Proteins 68:76–81

Emanuelsson O, Nielsen H, Brunak S, von Heijne G (2000) Predicting subcellular localization of proteins based on their N-terminal amino acid sequence. J Mol Biol 300:1005–1016

Fogel GB (2008) Computational intelligence approaches for pattern discovery in biological systems. Brief Bioinform 9(4):307–316

Fuchs A, Kirschner A, Frishman D (2009) Prediction of helix-helix contacts and interacting helices in polytopic membrane proteins using neural networks. Proteins 74:857–871

Garg A, Kaur H, Raghava GP (2005) Real value prediction of solvent accessibility in proteins using multiple sequence alignment and secondary structure. Proteins 61:318–324

Gromiha MM, Ahmad S, Suwa M (2005) TMBETA-NET: discrimination and prediction of membrane spanning beta-strands in outer membrane proteins. Nucleic Acids Res 33:W164–167

Hung LH, Samudrala R (2003) PROTINFO: secondary and tertiary protein structure prediction. Nucleic Acids Res 31:3296–3299

Ingrell CR, Miller ML, Jensen ON, Blom N (2007) NetPhosYeast: prediction of protein phosphorylation sites in yeast. Bioinformatics 23:895–897

Jacoboni I, Martelli PL, Fariselli P, De Pinto V, Casadio R (2001) Prediction of the transmembrane regions of beta-barrel membrane proteins with a neural network-based predictor. Protein Sci 10:779–787

Jeong E, Chung IF, Miyano S (2004) A neural network method for identification of RNA-interacting residues in protein. Genome Inform 15:105–116

Jones DT (1999) Protein secondary structure prediction based on position-specific scoring matrices. J Mol Biol 292:195–202

Kabsch W, Sander C (1983) Dictionary of protein secondary structure: pattern recognition of hydrogen-bonded and geometrical features. Biopolymers 22:2577–2637

Kapetanovic IM, Rosenfeld S, Izmirlian G (2004) Overview of commonly used bioinformatics methods and their applications. Ann NY Acad Sci 1020:10–21

Kaur H, Raghava GP (2003) A neural-network based method for prediction of gamma-turns in proteins from multiple sequence alignment. Protein Sci 12:923–929

Kaur H, Raghava GP (2004) A neural network method for prediction of beta-turn types in proteins using evolutionary information. Bioinformatics 20:2751–2758

Kirschner A, Frishman D (2008) Prediction of beta-turns and beta-turn types by a novel bidirectional Elman-type recurrent neural network with multiple output layers (MOLEBRNN). Gene 422:22–29

Kuang R, Leslie CS, Yang AS (2004) Protein backbone angle prediction with machine learning approaches. Bioinformatics 20:1612–1621

Kuznetsov IB, Gou Z, Li R, Hwang S (2006) Using evolutionary and structural information to predict DNA-binding sites on DNA-binding proteins. Proteins 64:19–27

Larranaga P, Calvo B, Santana R, Bielza C, Galdiano J, Inza I, Lozano JA, Armananzas R, Santafe G, Perez A, Robles V (2006) Machine learning in bioinformatics. Brief Bioinformatics 7(1):86–112

Lin CT, Lin KL, Yang CH, Chung IF, Huang CD, Yang YS (2005) Protein metal binding residue prediction based on neural networks. Int J Neural Syst 15:71–84

Lundegaard C, Lamberth K, Harndahl M, Buus S, Lund O, Nielsen M (2008) NetMHC-3.0: accurate web accessible predictions of human, mouse and monkey MHC class I affinities for peptides of length 8–11. Nucleic Acids Res 36:W509–512

Luscombe NM, Greenbaum D, Gerstein M (2001) What is bioinformatics? A proposed definition and overview of the field. Methods Inf Med 40:346–358

Martelli PL, Fariselli P, Casadio R (2004) Prediction of disulfide-bonded cysteines in proteomes with a hidden neural network. Proteomics 4:1665–1671

Miller DJ, Wang Y, Kesidis G (2008) Emergent unsupervised clustering paradigms with potential application to bioinformatics. Front Biosci 13:677–690

Muskal SM, Kim SH (1992) Predicting protein secondary structure content. A tandem neural network approach. J Mol Biol 225:713–727

Nantasenamat C, Isarankura-Na-Ayudhya C, Tansila N, Naenna T, Prachayasittikul V (2007) Prediction of GFP spectral properties using artificial neural network. J Comput Chem 28:1275–1289

Narayanan A, Keedwell EC, Olsson B (2002) Artificial intelligence techniques for bioinformatics. Appl Bioinformatics 1(4):191–222

Natt NK, Kaur H, Raghava GP (2004) Prediction of transmembrane regions of beta-barrel proteins using ANN- and SVM-based methods. Proteins 56:11–18

NIH Working Definition of Bioinformatics and Computational Biology (2000) BISTIC Definition Committee, http://www.bisti.nih.gov/

Niwa T (2004) Prediction of biological targets using probabilistic neural networks and atom-type descriptors. J Med Chem 47:2645–2650

Ofran Y, Rost B (2007) Protein-protein interaction hotspots carved into sequences. PLoS Comput Biol 3:e119

Petersen TN, Lundegaard C, Nielsen M, Bohr H, Bohr J, Brunak S, Gippert GP, Lund O (2000) Prediction of protein secondary structure at 80% accuracy. Proteins 41:17–20

Plewczynski D, Slabinski L, Ginalski K, Rychlewski L (2008) Prediction of signal peptides in protein sequences by neural networks. Acta Biochim Pol 55:261–267

Pollastri G, Baldi P, Fariselli P, Casadio R (2002a) Prediction of coordination number and relative solvent accessibility in proteins. Proteins 47:142–153

Pollastri G, Baldi P, Fariselli P, Casadio R (2002b) Prediction of coordination number and relative solvent accessibility in proteins. Proteins 47:142–153

Qian N, Sejnowski TJ (1988) Predicting the secondary structure of globular proteins using neural network models. J Mol Biol 202:865–884

Reinhardt A, Hubbard T (1998) Using neural networks for prediction of the subcellular location of proteins. Nucleic Acids Res 26:2230–2236

Rost B, Sander C (1994) Conservation and prediction of solvent accessibility in protein families. Proteins 20:216–226

Rost B, Sander C, Schneider R (1994) PHD – an automatic mail server for protein secondary structure prediction. Comput Appl Biosci 10:53–60

Ruan J, Wang K, Yang J, Kurgan LA, Cios KJ (2005) Highly accurate and consistent method for prediction of helix and strand content from primary protein sequences. Artif Intell Med 35:19–35

Saeys Y, Inza I, Larrañaga P (2007) A review of feature selection techniques in bioinformatics. Bioinformatics 23(19):2507–2517

Saha S, Raghava GP (2006) Prediction of continuous B-cell epitopes in an antigen using recurrent neural network. Proteins 65:40–48

Sidhu A, Yang ZR (2006) Prediction of signal peptides using bio-basis function neural networks and decision trees. Appl Bioinformatics 5:13–19

Vedani A, Dobler M (2000) Multi-dimensional QSAR in drug research. Predicting binding affinities, toxicity and pharmacokinetic parameters. Prog Drug Res 55:105–135

Wilkinson DJ (2007) Bayesian methods in bioinformatics and computational systems biology. Brief Bioinformatics 8(2):109–116

Xue B, Dor O, Faraggi E, Zhou Y (2008) Real-value prediction of backbone torsion angles. Proteins 72:427–433

Yang ZR, Thomson R (2005) Bio-basis function neural network for prediction of protease cleavage sites in proteins. IEEE Trans Neural Netw 16:263–274

Ye L, Liu T, Wu Z, Zhou R (2008) Sequence-based protein domain boundary prediction using BP neural network with various property profiles. Proteins 71:300–307

Zhang GZ, Huang DS (2004) Prediction of inter-residue contacts map based on genetic algorithm optimized radial basis function neural network and binary input encoding scheme. J Comput Aided Mol Des 18:797–810

Zhou HX, Shan Y (2001) Prediction of protein interaction sites from sequence profile and residue neighbor list. Proteins 44:336–343

Zou L, Wang Z, Huang J (2007) Prediction of subcellular localization of eukaryotic proteins using position-specific profiles and neural network with weighted inputs. J Genet Genomics 34:1080–1087

19 Self-organizing Maps

Marc M. Van Hulle
Laboratorium voor Neurofysiologie, K.U. Leuven, Belgium
marc@neuro.kuleuven.be

G. Rozenberg et al. (eds.), *Handbook of Natural Computing*, DOI 10.1007/978-3-540-92910-9_19,
© Springer-Verlag Berlin Heidelberg 2012

Abstract

A topographic map is a two-dimensional, nonlinear approximation of a potentially high-dimensional data manifold, which makes it an appealing instrument for visualizing and exploring high-dimensional data. The self-organizing map (SOM) is the most widely used algorithm, and it has led to thousands of applications in very diverse areas. In this chapter we introduce the SOM algorithm, discuss its properties and applications, and also discuss some of its extensions and new types of topographic map formation, such as those that can be used for processing categorical data, time series, and tree-structured data.

1 Introduction

One of the most prominent features of the mammalian brain is the *topographical* organization of its sensory cortex: neighboring nerve cells (neurons) can be driven by stimuli originating from neighboring positions in the sensory input space, and neighboring neurons in a given brain area project to neighboring neurons in the next area. In other words, the connections establish a so-called *neighborhood-preserving* or *topology-preserving* map, or *topographic map* for short. In the visual cortex, this is called a *retinotopic* map; in the somatosensory cortex, a *somatotopic* map (a map of the body surface); and in the auditory cortex, a *tonotopic map* (of the spectrum of possible sounds).

The study of topographic map formation, from a theoretical perspective, started with basically two types of self-organizing processes, gradient-based learning and competitive learning, and two types of network architectures (❷ *Fig. 1*) (for a review, see Van Hulle [2000]). In the first architecture, which is commonly referred to as the Willshaw–von der

❑ Fig. 1

(a) Willshaw–von der Malsburg model. Two isomorphic, rectangular lattices of neurons are shown: one represents the input layer and the other the output layer. Neurons are represented by circles: filled circles denote active neurons ("winning" neurons); open circles denote inactive neurons. As a result of the weighted connections from the input to the output layer, the output neurons receive different inputs from the input layer. Two input neurons are labeled (i,j) as well as their corresponding output layer neurons (i′,j′). Neurons i and i′ are the only active neurons in their respective layers. (b) Kohonen model. The common input all neurons receive is directly represented in the input space, v ∈ V ⊆ ℝ^d. The "winning" neuron is labeled as i*: its weight (vector) is the one that best matches the current input (vector).

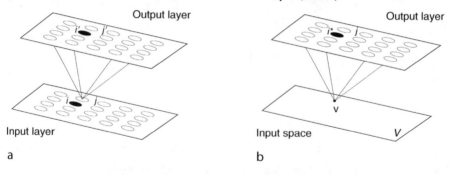

a   b

Malsburg model (Willshaw and von der Malsburg 1976), there are two sets of neurons, arranged in two (one- or) two-dimensional layers or *lattices* (❷ *Fig. 1a*). (A lattice is an undirected graph in which every non-border vertex has the same, fixed number of incident edges, and which usually appears in the form of an array with a rectangular or simplex topology.) Topographic map formation is concerned with learning a mapping for which neighboring neurons in the input lattice are connected to neighboring neurons in the output lattice.

The second architecture is far more studied, and is also the topic of this chapter. Now we have continuously valued inputs taken from the input space $\mathbb{R}^d$, or the data manifold $V \subseteq \mathbb{R}^d$, which need not be rectangular or have the same dimensionality as the lattice to which it projects (❷ *Fig. 1b*). To every neuron i of the lattice A corresponds a reference position in the input space, called the weight vector $\mathbf{w}_i = [w_{ij}] \in \mathbb{R}^d$. All neurons receive the same input vector $\mathbf{v} = [v_1, \ldots, v_d] \in V$. Topographic map formation is concerned with learning a map $\mathscr{V}_A$ of the data manifold V (gray shaded area in ❷ *Fig. 2*), in such a way that neighboring lattice neurons, i, j, with lattice positions $\mathbf{r}_i, \mathbf{r}_j$, code for neighboring positions, $\mathbf{w}_i, \mathbf{w}_j$, in the input space (cf., the inverse mapping, Ψ). The forward mapping, Φ, from the input space to the lattice is not necessarily topology-preserving – neighboring weights do not necessarily correspond to neighboring lattice neurons – even after learning the map, due to the possible mismatch in dimensionalities of the input space and the lattice (see, e.g., ❷ *Fig. 3*). In practice, the map is represented in the input space in terms of neuron weights that are connected by straight lines, if the corresponding neurons are the nearest neighbors in the lattice (e.g., see the left panel of ❷ *Fig. 2* or ❷ *Fig. 3*). When the map is topology preserving, it can be used for visualizing the data distribution by projecting the original data points onto the map. The advantage of having a flexible map, compared to, for example, a plane specified by principal components analysis (PCA), is demonstrated in ❷ *Fig. 4*. We observe that the three classes are better separated with a topographic map than with PCA. The most popular learning algorithm

◻ **Fig. 2**
Topographic mapping in the Kohonen architecture. In the *left panel*, the topology-preserving map $\mathscr{V}_A$ of the data manifold $V \subseteq \mathbb{R}^d$ (gray-shaded area) is shown. The neuron weights, $\mathbf{w}_i, \mathbf{w}_j$, are connected by a straight line since the corresponding neurons i, j in the lattice A (*right panel*), with lattice coordinates $\mathbf{r}_i, \mathbf{r}_j$, are nearest neighbors. The forward mapping Φ is from the input space to the lattice; the backward mapping Ψ is from the lattice to the input space. The learning algorithm tries to make neighboring lattice neurons, i, j, code for neighboring positions, $\mathbf{w}_i, \mathbf{w}_j$, in the input space.

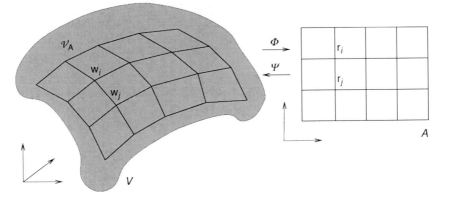

◘ Fig. 3

Example of a one-dimensional lattice consisting of four neurons i,j,k,l in a two-dimensional rectangular space. The distance between the weight vectors of neurons i,j, d_{ij} is larger than between that of neurons i,l, d_{il}. This means that, at least in this example, neighboring neuron weights do not necessarily correspond to neighboring neurons in the lattice.

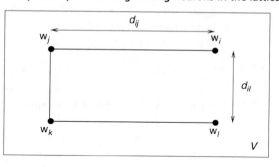

◘ Fig. 4

Oil flow data set visualized using PCA (*left panel*) and a topographic map (*right panel*). The latter was obtained with the generative topographic map (GTM) algorithm (Bishop et al. 1996, 1998). Since the GTM performs a nonlinear mapping, it is better able to separate the three types of flow configurations: laminar (*red crosses*), homogeneous (*blue plusses*), and annular (*green circles*). (Bishop 2006, reprinted with permission.)

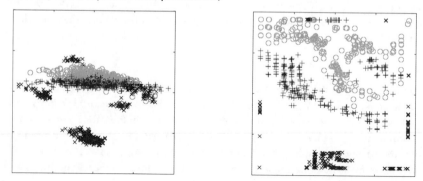

for this architecture is the self-organizing map (SOM) algorithm developed by Teuvo Kohonen (1982, 1984), whence this architecture is often referred to as Kohonen's model.

1.1 Chapter Overview

We start with the basic version of the SOM algorithm where we discuss the two stages of which it consists: the competitive and the cooperative ones. The discussion then moves on to the topographic ordering properties of the algorithm: how it unfolds and develops topographically ordered maps, whether there exists a mathematical proof of ordering, and whether topological defects in the map could still occur after the learning process has ended. We also discuss the convergence properties of the algorithm, and in what sense the converged weights are modeling the input density (is the weight density a linear function of the input density?).

We then discuss applications of the SOM algorithm, thousands of which have been reported in the open literature. Rather than attempting an extensive overview, the applications are grouped into three areas: vector quantization, regression, and clustering. The latter is the most important one since it is a direct consequence of the data visualization and exploration capabilities of the topographic map. A number of important applications are highlighted, such as WEBSOM (self-organizing maps for internet exploration) (Kaski et al. 1998) for organizing large document collections; PicSOM (Laaksonen et al. 2002) for content-based image retrieval; and the emergent self-organizing maps (ESOM) (Ultsch and Mörchen 2005), for which the MusicMiner (Risi et al. 2007) is considered, for organizing large collections of music, and an application for classifying police reports of criminal incidents.

Later, an overview of a number of extensions of the SOM algorithm is given. The motivation behind these is to improve the original algorithm, or to extend its range of applications, or to develop new ways to perform topographic map formation.

Three important extensions of the SOM algorithm are then given in detail. First, we discuss the growing topographic map algorithms. These algorithms consider maps with a dynamically defined topology so as to better capture the fine structure of the input distribution. Second, since many input sources have a temporal characteristic, which is not captured by the original SOM algorithm, several algorithms have been developed based on a recurrent processing of time signals (recurrent topographic maps). It is a heavily researched area since some of these algorithms are capable of processing tree-structured data. Third, another topic of current research is the kernel topographic map, which is in line with the "kernelization" trend of mapping data into a feature space. Rather than Voronoi regions, the neurons are equipped with overlapping activation regions, in the form of kernel functions, such as Gaussians. Important future developments are expected for these topographic maps, such as the visualization and clustering of structure-based molecule descriptions, and other biochemical applications.

Finally, a conclusion to the chapter is formulated.

2   SOM Algorithm

The SOM algorithm distinguishes two stages: the *competitive* stage and the *cooperative* stage. In the first stage, the best matching neuron is selected, that is, the "winner," and in the second stage, the weights of the winner are adapted as well as those of its immediate lattice neighbors. Only the minimum Euclidean distance version of the SOM algorithm is considered (also the dot product version exists, see Kohonen 1995).

2.1   Competitive Stage

For each input $\mathbf{v} \in V$, the neuron with the smallest Euclidean distance ("winner-takes-all," WTA) is selected, which we call the "winner":

$$i^* = \arg\min_i \|\mathbf{w}_i - \mathbf{v}\| \tag{1}$$

By virtue of the minimum Euclidean distance rule, we obtain a Voronoi tessellation of the input space: to each neuron corresponds a region in the input space, the boundaries of which are perpendicular bisector planes of lines joining pairs of weight vectors (the gray-shaded area in ❯ *Fig. 5* is the Voronoi region of neuron *j*). Remember that the neuron weights are connected by straight lines (links or edges): they indicate which neurons are nearest neighbors in the lattice. These links are important for verifying whether the map is topology preserving.

2.2 Cooperative Stage

It is now crucial to the formation of topographically ordered maps that the neuron weights are not modified independently of each other, but as topologically related subsets on which similar kinds of weight updates are performed. During learning, not only the weight vector of the winning neuron is updated, but also those of its lattice neighbors, which end up responding to similar inputs. This is achieved with the neighborhood function, which is centered at the winning neuron, and decreases with the lattice distance to the winning neuron. (Besides the neighborhood function, also the neighborhood set exists, consisting of all neurons to be updated in a given radius from the winning neuron [see Kohonen, (1995)].

The weight update rule in incremental mode is given by (with incremental mode we mean that the weights are updated each time an input vector is presented, contrasted with batch mode where the weights are only updated after the presentation of the full training set ["batch"]):

$$\Delta \mathbf{w}_i = \eta \, \Lambda(i, i^*, \sigma_\Lambda(t)) \, (\mathbf{v} - \mathbf{w}_i), \quad \forall i \in A \qquad (2)$$

with Λ the neighborhood function, that is, a scalar-valued function of the lattice coordinates of neurons i and i^*, $\mathbf{r}_i$ and $\mathbf{r}_i^*$, mostly a Gaussian:

$$\Lambda(i, i^*) = \exp\left(-\frac{\|\mathbf{r}_i - \mathbf{r}_{i^*}\|^2}{2\sigma_\Lambda^2}\right) \qquad (3)$$

with range σ_Λ (i.e., the standard deviation). (we further drop the parameter $\sigma_\Lambda(t)$ from the neighborhood function to simplify the notation.) The positions $\mathbf{r}_i$ are usually taken to be

◻ **Fig. 5**
Definition of quantization region in the self-organizing map (SOM). Portion of a lattice (*thick lines*) plotted in terms of the weight vectors of neurons *a*, ..., *k*, in the two-dimensional input space, that is, w$_a$, ..., w$_k$.

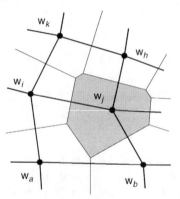

the nodes of a discrete lattice with a regular topology, usually a two-dimensional square or rectangular lattice. An example of the effect of the neighborhood function in the weight updates is shown in ❷ *Fig. 6* for a 4 × 4 lattice. The parameter σ_Λ, and usually also the learning rate η, are gradually decreased over time. When the neighborhood range vanishes, the previous learning rule reverts to standard unsupervised competitive learning (UCL) (note that the latter is unable to form topology-preserving maps, pointing to the importance of the neighborhood function).

As an example, a 10 × 10 square lattice is trained with the SOM algorithm on a uniform square distribution $[-1,1]^2$, using a Gaussian neighborhood function of which the range $\sigma_\Lambda(t)$ is decreased as follows:

$$\sigma_\Lambda(t) = \sigma_{\Lambda 0} \exp\left(-2\sigma_{\Lambda 0}\, \frac{t}{t_{\max}}\right) \tag{4}$$

with t the present time step, $t_{\max}$ the maximum number of time steps, and $\sigma_{\Lambda 0}$ the range We take spanned by the neighborhood function at $t = 0$. We take $t_{\max} = 100,000$ and $\sigma_{\Lambda 0} = 5$, and the learning rate $\eta = 0.01$. The initial weights (i.e., for $t = 0$) are chosen randomly from the same square distribution. Snapshots of the evolution of the lattice are shown in ❷ *Fig. 7*. We observe that the lattice initially tangles, then contracts, unfolds, and expands so as to span the input distribution. This two-phased convergence process is an important property of the SOM algorithm, and it has been thoroughly studied from a mathematical viewpoint in the following terms: (1) the topographic ordering of the weights and, thus, the formation of

◻ Fig. 6
The effect of the neighborhood function in the SOM algorithm. Starting from a perfect arrangement of the weights of a square lattice (*full lines*), the weights nearest to the current input (indicated with the cross) receive the largest updates, those further away smaller updates, resulting in the updated lattice (*dashed lines*).

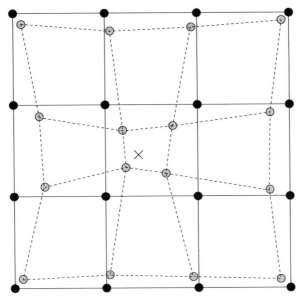

□ **Fig. 7**

Evolution of a 10 × 10 lattice with a rectangular topology as a function of time. The outer squares outline the uniform input distribution. The values given below the squares represent time.

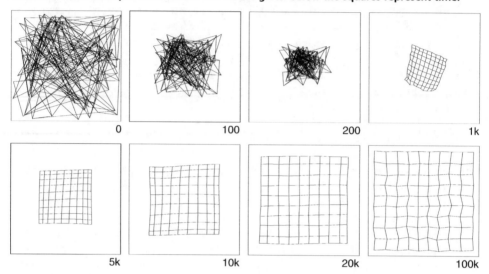

topology-preserving mappings and (2) the convergence of these weights (energy function minimization). Both topics will be discussed next. Finally, the astute reader would have noticed that at the end of the learning phase, the lattice is smooth, but then suddenly becomes more erratic. This is an example of a phase transition, and it has been widely studied for the SOM algorithm (see Der and Herrmann (1993)).

Finally, since the speed of the convergence depends on the learning rate, a version without one has also been developed, called batch map (Kohonen 1995):

$$\mathbf{w}_i = \frac{\sum_\mu \Lambda(i^*, i)\mathbf{v}^\mu}{\sum_\mu \Lambda(i^*, i)}, \quad \forall i \tag{5}$$

and it leads to a faster convergence of the map.

2.3 Topographic Ordering

In the example of ❷ *Fig. 7*, a two-dimensional square lattice has been used for mapping a two-dimensional uniform, square distribution. One can also use the same lattice for mapping a non-square distribution, so that there is a topological mismatch, for example, a circular and an L-shaped distribution (❷ *Fig. 8a, b*). The same lattice and simulation setup is used as before, but now only the final results are shown. Consider first the circular distribution: the weight distribution is now somewhat nonuniform. For the L-shaped distribution, one sees that there are several neurons outside the support of the distribution, and some of them even have a zero (or very low) probability of being active: hence, they are often called "dead" units. It is hard to find a better solution for these neurons without clustering them near the inside corner of the L-shape.

□ Fig. 8

Mapping of a 10 × 10 neuron lattice onto a circular (**a**) and an L-shaped (**b**) uniform distribution, and a 40-neuron one-dimensional lattice onto a square uniform distribution (**c**).

a   b   c

We can also explore the effect of a mismatch in lattice dimensionality. For example, we can develop a one-dimensional lattice ("chain") in the same two-dimensional square distribution as before. (Note that it is now impossible to preserve all of the topology.) We see that the chain tries to fill the available space as much as possible (❷ *Fig. 8c*): the resulting map approximates the so-called space-filling *Peano curve*. (A Peano curve is an infinitely and recursively convoluted fractal curve which represents the continuous mapping of, for example, a one-dimensional interval onto a two-dimensional surface; Kohonen (1995) pp. 81, 87.)

2.3.1 Proofs or Ordering

It is clear that the neighborhood function plays a crucial role in the formation of topographically ordered weights. Although this may seem evident, the ordering itself is very difficult to describe (and prove!) in mathematical terms. The mathematical treatments that have been considered are, strictly speaking, only valid for one-dimensional lattices developed in one-dimensional spaces. Cottrell and Fort (1987) presented a mathematical stringent (but quite long) proof of the ordering process for the one-dimensional case. For a shorter constructive proof, one can refer to Kohonen (1995, pp. 100–105); for an earlier version, see Kohonen (1984, pp. 151–154). The results of Kohonen (1984) and Cottrell and Fort (1987) were extended by Erwin and coworkers (1992) to the more general case of a monotonically decreasing neighborhood function. However, the same authors also state that a strict proof of convergence is unlikely to be found for the higher-than-one-dimensional case.

2.3.2 Topological Defects

As said before, the neighborhood function plays an important role in producing topographically ordered lattices; however, this does not imply that one is guaranteed to obtain it. Indeed, if one decreases the neighborhood range too fast, then there could be topological defects (Geszti 1990; Heskes and Kappen 1993). These defects are difficult to iron out, if at all, when the neighborhood range vanishes. In the case of a chain, one can obtain a so-called *kink* (❷ *Fig. 9*).

Consider, as a simulation example, a rectangular lattice sized $N = 24 \times 24$ neurons with the input samples taken randomly from a two-dimensional uniform distribution, $p(\mathbf{v})$, within

◻ Fig. 9

Example of a topological defect ("kink") in a chain consisting of 4 neurons i,j,k,l in a two-dimensional rectangular space.

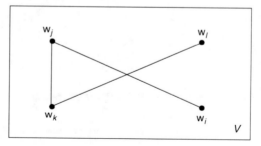

◻ Fig. 10

Example of the formation of a topological defect called "twist" in a 24 × 24 lattice. The evolution is shown for different time instances (values below the squares).

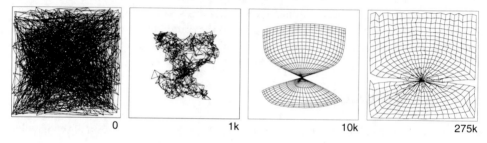

the square $[0,1]^2$. The initial weight vectors are randomly drawn from this distribution. Now incremental learning is performed and the range is decreased as follows:

$$\sigma_\Lambda(t) = \sigma_{\Lambda 0} \; \exp\left(-2 \; \sigma_{\Lambda 0} \frac{t}{t_{\max}}\right) \tag{6}$$

but now with t the present time step and $t_{\max} = 275,000$. For the learning rate, $\eta = 0.015$ is taken. The evolution is shown in ❯ *Fig. 10*. The neighborhood range was too rapidly decreased since the lattice is twisted and, even if one continues the simulation, with zero neighborhood range, the *twist* will not be removed.

2.4 Weight Convergence, Energy Function

Usually, in neural network learning algorithms, the weight update rule performs gradient descent on an energy function E (also termed error, cost, or distortion objective function):

$$\Delta w_{ij} \propto -\frac{\partial E}{\partial w_{ij}} \tag{7}$$

so that convergence to a local minimum in E can be easily shown, for example, in an average sense. However, contrary to intuition, the SOM algorithm *does not* perform gradient descent on an energy function as long as the neighborhood range has not vanished (when $\sigma_\Lambda = 0$, an energy function exists: one is minimizing the mean squared error (MSE) due to the quantization of the input space into Voronoi regions). Hence, strictly speaking, one cannot judge the degree of optimality achieved by the algorithm during learning.

In defense of the lack of an energy function, Kohonen emphasizes that there is no theoretical reason why the SOM algorithm *should* ensue from such a function (Kohonen 1995, p. 122) since (1) the SOM algorithm is aimed at developing (specific) topological relations between clusters in the input space and (2) it yields an approximative solution of an energy function (by virtue of its connection with the Robbins–Munro stochastic approximation technique), so that convergence is not a problem in practice.

Besides this, a lot of effort has been devoted to developing an energy function that is minimized during topographic map formation (Tolat 1990; Kohonen 1991; Luttrell 1991; Heskes and Kappen 1993; Erwin et al. 1992). Both Luttrell (1991) and Heskes and Kappen (1993) were able to show the existence of an energy function by modifying the definition of the winner. Heskes and Kappen first ascribed a *local error*, e_i, to each neuron, i, at time t:

$$e_i(\mathbf{W}, \mathbf{v}, t) = \frac{1}{2} \sum_{j \in A} \Lambda(i, j) \|\mathbf{v} - \mathbf{w}_j\|^2 \tag{8}$$

with $\mathbf{W} = [\mathbf{w}_i]$ the vector of all neuron weights, and then defined the winner as the neuron for which the local error is minimal:

$$i^* = \arg \min_i \sum_j \Lambda(i, j) \|v - w_j\|^2 \tag{9}$$

This was also the equation introduced by Luttrell (1991), but with $\sum_j \Lambda(i,j) = 1$. The actual weight update rule remains the same as in the original SOM algorithm and, thus, still considers several neurons around the winner. The solution proposed by Kohonen (1991) takes a different starting point, but it leads to a more complicated learning rule for the winning neuron (for the minimum Euclidean distance SOM algorithm, see Kohonen (1995, pp. 122–124)).

2.4.1 Weight Density

The next question is, "In what sense are the converged weights modeling the input probability density?" What is the distribution of the weights versus that of the inputs? Contrary to what was originally assumed (Kohonen 1984), the weight density at convergence, also termed the (inverse of the) magnification factor, is not a linear function of the input density $p(\mathbf{v})$.

Ritter and Schulten (1986) have shown that, for a one-dimensional lattice, developed in a one-dimensional input space:

$$p(w_i) \propto p(v)^{\frac{2}{3}} \tag{10}$$

in the limit of an infinite density of neighbor neurons (continuum approximation). Furthermore, when a discrete lattice is used, the continuum approach undergoes a correction, for example, for a neighborhood set with σ_Λ neurons around each winner neuron ($\Lambda(i, i^*) = 1$ *iff* $|r_i - r_{i^*}| \le \sigma_\Lambda$):

$$p(w_i) \propto p^\alpha \tag{11}$$

with:

$$\alpha = \frac{2}{3} - \frac{1}{3\sigma_A^2 + 3(\sigma_A + 1)^2}$$

For a discrete lattice of N neurons, it is expected that for $N \to \infty$ and for minimum mean-squared error (MSE) quantization, in d-dimensional space, the weight density will be proportional to (Kohonen 1995):

$$p(\mathbf{w}_i) \propto p^{\frac{1}{1+\frac{d}{d}}}(\mathbf{v}) \tag{12}$$

or that in the one-dimensional case:

$$p(w_i) \propto p(v)^{\frac{1}{3}} \tag{13}$$

The connection between the continuum and the discrete approach was established for the one-dimensional case by Ritter (1991), for a discrete lattice of N neurons, with $N \to \infty$, and for a neighborhood set with σ_A neurons around each "winner" neuron.

Finally, regardless of the effect of the neighborhood function or set, it is clear that the SOM algorithm tends to undersample high probability regions and oversample low probability ones. This affects the separability of clusters: for example; when the clusters overlap, the cluster boundary will be more difficult to delineate in the overlap region than for a mapping which has a linear weight distribution (Van Hulle 2000).

3 Applications of SOM

The graphical map displays generated by the SOM algorithm are easily understood, even by nonexperts in data analysis and statistics. The SOM algorithm has led to literally thousands of applications in areas ranging from automatic speech recognition, condition monitoring of plants and processes, cloud classification, and microarray data analysis, to document and image organization and retrieval (for an overview, see Centre (2003 http://www.cis.hut.fi/research/som-bibl/)). The converged neuron weights yield a model of the training set in three ways: vector quantization, regression, and clustering.

3.1 Vector Quantization

The training samples are modeled in such a manner that the average discrepancy between the data points and the neuron weights is minimized. In other words, the neuron weight vectors should "optimally" quantize the input space from which the training samples are drawn, just like one would desire for an adaptive vector quantizer (Gersho and Gray 1991). Indeed, in standard unsupervised competitive learning (UCL), and also the SOM algorithm, when the neighborhood has vanished ("zero-order" topology), the weight updates amount to centroid estimation (usually the mean of the samples which activate the corresponding neuron) and minimum Euclidean distance classification (Voronoi tessellation), and we attempt to minimize the mean squared error due to quantization, or some other quantization metric that one wishes to use. In fact, there exists an intimate connection between the batch version of the UCL rule and the zero-order topology SOM algorithm, on the one hand, and the generalized Lloyd algorithm for building vector quantizers, on the other hand (Luttrell 1989, 1990) (for a review, see Van Hulle (2000)). Luttrell showed that the neighborhood function can be considered as a probability density function of which the range is chosen to capture the noise process responsible for the distortion of the quantizer's output code (i.e., the index of the winning neuron), for example, due to noise in the communication channel. Luttrell adopted for the noise process a zero-mean Gaussian, so that there is theoretical justification for choosing a Gaussian neighborhood function in the SOM algorithm.

3.2 Regression

One can also interpret the map as a case of non-parametric regression: no prior knowledge is assumed about the nature or shape of the function to be regressed. Non-parametric regression is perhaps the first successful statistical application of the SOM algorithm (Ritter et al. 1992; Mulier and Cherkassky 1995; Kohonen 1995): the converged topographic map is intended to capture the principal dimensions (principal curves and principal manifolds) of the input space. The individual neurons represent the "knots" that join piecewise smooth functions, such as splines, which act as interpolating functions for generating values at intermediate positions. Furthermore, the lattice coordinate system can be regarded as an (approximate) global coordinate system of the data manifold (❍ *Fig. 2*).

3.3 Clustering

The most widely used application of the topographic map is clustering, that is, the partitioning of the data set into subsets of "similar" data, without using prior knowledge about these subsets. One of the first demonstrations of clustering was by Ritter and Kohonen (1989). They had a list of 16 animals (birds, predators, and preys) and 13 binary attributes for each one of them (e.g., large size or not, hair or not, two legged or not, can fly or not, etc.). After training a 10×10 lattice of neurons with these vectors (supplemented with the 1-out-of-16 animal code vector, thus, in total, a 29-dimensional binary vector), and labeling the winning neuron for each animal code vector, a natural clustering of birds, predators, and prey appeared in the map. The authors called this the "semantic map."

In the previous application, the clusters and their boundaries were defined by the user. In order to visualize clusters more directly, one needs an additional technique. One can compute the mean Euclidean distance between a neuron's weight vector and the weight vectors of its nearest neighbors in the lattice. The maximum and minimum of the distances found for all neurons in the lattice is then used for scaling these distances between 0 and 1; the lattice then becomes a gray scale image with white pixels corresponding to, for example, 0 and black pixels to 1. This is called the U-matrix (Ultsch and Siemon 1990), for which several extensions have been developed to remedy the oversampling of low probability regions (possibly transition regions between clusters) (Ultsch and Mörchen 2005).

An important example is WEBSOM (Kaski et al. 1998). Here, the SOM is used for organizing document collections ("document map"). Each document is represented as a vector of keyword occurrences. Similar documents then become grouped into the same cluster. After training, the user can zoom into the map to inspect the clusters. The map is manually or automatically labeled with keywords (e.g., from a man-made list) in such a way that, at each zoom level, the same density of keywords is shown (so as not to clutter the map with text).

WEBSOM has also been applied to visualizing clusters in patents based on keyword occurrences in patent abstracts (Kohonen et al. 1999).

An example of a content-based image retrieval system is the PicSOM (Laaksonen et al. 2002). Here, low-level features (color, shape, texture, etc.) of each image are considered. A separate two-dimensional SOM is trained for each low-level feature (in fact, a hierarchical SOM). In order to be able to retrieve one particular image from the database, one is faced with a semantic gap: how well do the low-level features correlate with the image contents? To bridge this gap, *relevance feedback* is used: the user is shown a number of images and he/she has to decide which ones are relevant and which ones are not (close or not to the image the user is interested in). Based on the trained PicSOM, the next series of images shown are then supposed to be more relevant, and so on.

For high-dimensional data visualization, a special class of topographic maps called emergent self-organizing maps (ESOM) (Ultsch and Mörchen 2005) can be considered. According to Ultsch, emergence is the ability of a system to produce a phenomenon on a new, higher level. In order to achieve emergence, the existence and cooperation of a large number of elementary processes is necessary. An emergent SOM differs from the traditional SOM in that a very large number of neurons (at least a few thousand) are used (even larger than the number of data points). The ESOM software is publicly available from http://databionic-esom.sourceforge.net/.

As an example, Ultsch and coworkers developed the *MusicMiner* for organizing large collections of music (Risi et al. 2007). Hereto, low-level audio features were extracted from raw audio data, and static and temporal statistics were used for aggregating these low-level features into higher-level ones. A supervised feature selection was performed to come up with a nonredundant set of features. Based on the latter, an ESOM was trained for clustering and visualizing collections of music. In this way, consistent clusters were discovered that correspond to music genres (❷ *Fig. 11*). The user can then navigate the sound space and interact with the maps to discover new songs that correspond to his/her taste.

◘ Fig. 11

Organizing large collections of music by means of an ESOM trained on high-level audio features (*MusicMiner*). Shown is the map with several music genres labeled. (Risi et al. 2007, reprinted with permission.)

As an example an ESOM application in the realm of text mining, the case of police reports of criminal incidents is considered. When a victim of a violent incident makes a statement to the police, the police officer has to judge whether, for example, domestic violence is involved. However, not all cases are correctly recognized and are, thus, wrongly assigned the label "nondomestic violence." Because it is very time consuming to classify cases and to verify whether or not the performed classifications are correct, text mining techniques and a reliable classifier are needed. Such an automated triage system would result in major cost and time savings. A collection of terms (thesaurus), obtained by a frequency analysis of key words, was constructed, consisting of 123 terms. A 50 × 82 toroidal lattice (in a toroidal lattice, the vertical and horizontal coordinates are circular) was used, and trained with the ESOM algorithm on 4,814 reports of the year 2007; the validation set consisted of 4,738 cases (of the year 2006). The outcome is shown in ❷ *Fig. 12* (Poelmans 2008, unpublished results).

4   Extensions of SOM

Although the original SOM algorithm has all the necessary ingredients for developing topographic maps, many adapted versions have emerged over the years (for references, see Kohonen (1995) and http://www.cis.hut.fi/research/som-bibl/ which contains over 7,000 articles). For some of these, the underlying motivation was to improve the original algorithm, or to extend its range of applications, while for others the SOM algorithm has served as a source of inspiration for developing new ways to perform topographic map formation.

One motivation was spurred by the need to develop a learning rule that performs gradient descent on an energy function (as discussed above). Another was to remedy the occurrence of

◻ **Fig. 12**
Toroidal lattice trained with the ESOM algorithm showing the distribution of domestic violence cases (*red squares*) and nondomestic violence cases (*green squares*). The background represents the sum of the distances between the weight vector of each neuron and those of its nearest neighbors in the lattice, normalized by the largest occurring sum (*white*) (i.e., the U-matrix). It is observed that the domestic violence cases appear in one large cluster and a few smaller clusters, mostly corresponding to violence in homosexual relationships. (Courtesy of Jonas Poelmans.)

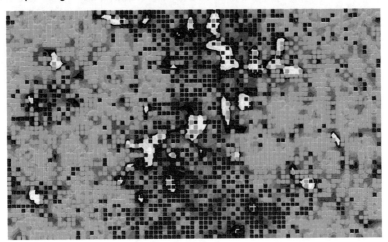

dead units, since they do not contribute to the representation of the input space (or the data manifold). Several researchers were inspired by Grossberg's idea (1976) of adding a "conscience" to frequently winning neurons to feel "guilty" and to reduce their winning rates. The same heuristic idea has also been adopted in combination with topographic map formation (DeSieno 1988; Van den Bout and Miller 1989; Ahalt et al. 1990). Others exploit measures based on the local distortion error to equilibrate the neurons' "conscience" (Kim and Ra 1995; Chinrungrueng and Séquin 1995; Ueda and Nakano 1993). A combination of the two conscience approaches is the learning scheme introduced by Bauer and coworkers (1996).

A different strategy is to apply a competitive learning rule that minimizes the mean absolute error (MAE) between the input samples **v** and the N weight vectors (also called the *Minkowski metric* of power one) (Kohonen 1995, pp. 120, 121) (see also Lin et al. (1997)). Instead of minimizing a (modified) distortion criterion, a more natural approach is to optimize an information-theoretic criterion directly. Linsker was among the first to explore this idea in the context of topographic map formation. He proposed a principle of *maximum information preservation* (Linsker 1988) – *infomax* for short – according to which a processing stage has the property that the output signals will optimally discriminate, in an information-theoretic sense, among possible sets of input signals applied to that stage. In his 1989 article, he devised a learning rule for topographic map formation in a probabilistic WTA network by maximizing the average mutual information between the output and the signal part of the input, which was corrupted by noise (Linsker 1989). Another algorithm is the vectorial boundary adaptation rule (VBAR) which considers the region spanned by a quadrilateral (four neurons forming a square region in the lattice) as the quantization region (Van Hulle 1997a, b), and which is able to achieve an equiprobabilistic map, that is, a map for which every neuron has the same chance to be active (and, therefore, maximizes the information-theoretic entropy).

Another evolution is the growing topographic map algorithms. In contrast to the original SOM algorithm, its growing map variants have a dynamically defined topology, and they are believed to better capture the fine structure of the input distribution. They will be discussed in the next section.

Many input sources have a temporal characteristic, which is not captured by the original SOM algorithm. Several algorithms have been developed based on a recurrent processing of time signals and a recurrent winning neuron computation. Also tree structured data can be represented with such topographic maps. Recurrent topographic maps will be discussed in this chapter.

Another important evolution is the kernel-based topographic maps: rather than Voronoi regions, the neurons are equipped with overlapping activation regions, usually in the form of kernel functions, such as Gaussians (❷ *Fig. 19*). Also for this case, several algorithms have been developed, and a number of them will be discussed in this chapter.

5   Growing Topographic Maps

In order to overcome the topology mismatches that occur with the original SOM algorithm, as well as to achieve an optimal use of the neurons (cf., dead units), the geometry of the lattice has to match that of the data manifold it is intended to represent. For that purpose, several so-called growing (incremental or structure-adaptive) self-organizing map algorithms have been developed. What they share is that the lattices are gradually built up and, hence, do not have a predefined structure (i.e., number of neurons and possibly also lattice dimensionality)

(❯ *Fig. 14*). The lattice is generated by a successive insertion (and possibly an occasional deletion) of neurons and connections between them. Some of these algorithms can even guarantee that the lattice is free of topological defects (e.g., since the lattice is a subgraph of a Delaunay triangularization, see further). The major algorithms for growing self-organizing maps will be briefly reviewed. The algorithms are structurally not very different; the main difference is with the constraints imposed on the lattice topology (fixed or variable lattice dimensionality). The properties common to these algorithms are first listed, using the format suggested by Fritzke (1996).

- The network is an undirected graph (lattice) consisting of a number of nodes (neurons) and links or edges connecting them.
- Each neuron, i, has a weight vector $\mathbf{w}_i$ in the input space V.
- The weight vectors are updated by moving the winning neuron i^*, and its topological neighbors, toward the input $\mathbf{v} \in V$:

$$\Delta \mathbf{w}_{i^*} = \eta_{i^*}(\mathbf{v} - \mathbf{w}_{i^*}) \tag{14}$$

$$\Delta \mathbf{w}_i = \eta_i(\mathbf{v} - \mathbf{w}_i), \quad \forall i \in \mathcal{N}_{i^*} \tag{15}$$

with $\mathcal{N}_{i^*}$ the set of direct topological neighbors of neuron i^* (neighborhood set), and with η_{i^*} and η_i the learning rates, $\eta_{i^*} \eta_i$.
- At each time step, the local error at the winning neuron, i^*, is accumulated:

$$\Delta E_{i^*} = (\text{error measure}) \tag{16}$$

The error term is coming from a particular area around $\mathbf{w}_{i^*}$, and is likely to be reduced by inserting new neurons in that area. A central property of these algorithms is the possibility to choose an arbitrary error measure as the basis for insertion. This extends their applications from unsupervised learning ones, such as data visualization, combinatorial optimization, and clustering analysis, to supervised learning ones, such as classification and regression. For example, for vector quantization, $\Delta E_{i^*} = \|\mathbf{v} - \mathbf{w}_{i^*}\|^2$. For classification, the obvious choice is the classification error. All models reviewed here can, in principle, be used for supervised learning applications by associating output values to the neurons, for example, through kernels such as radial basis functions. This makes most sense for the algorithms that adapt their dimensionality to the data.
- The accumulated error of each neuron is used to determine (after a fixed number of time steps) where to insert new neurons in the lattice. After an insertion, the error information is locally redistributed, which increases the probability that the next insertion will be somewhere else. The local error acts as a kind of memory where much error has occurred; the exponential decay of the error stresses more the recently accumulated error.
- All parameters of the algorithm stay constant over time.

5.1 Competitive Hebbian Learning and Neural Gas

Historically, the first algorithm to develop topologies was introduced by Martinetz and Schulten, and it is a combination of two methods: competitive Hebbian learning (CHL) (Martinetz 1993) and the neural gas (NG) (Martinetz and Schulten 1991).

The principle behind CHL is simple: for each input, create a link between the winning neuron and the second winning neuron (i.e., with the second smallest Euclidean distance to the input), if that link does not already exist. Only weight vectors lying in the data manifold develop links between them (thus, nonzero input density regions). The resulting graph is a subgraph of the (induced) Delaunay triangularization (❷ *Fig. 13*), and it has been shown to optimally preserve topology in a very general sense.

In order to position the weight vectors in the input space, Martinetz and Schulten (1991) have proposed a particular kind of vector quantization method, called neural gas (NG). The main principle of NG is for each input $\mathbf{v}$ update the k nearest-neighbor neuron weight vectors, with k decreasing over time until only the winning neuron's weight vector is updated. Hence, one has a neighborhood function but now in input space. The learning rate also follows a decay schedule. Note that the NG by itself does not delete or insert any neurons. The NG requires finetuning of the rate at which the neighborhood shrinks to achieve a smooth convergence and proper modeling of the data manifold.

The combination of CHL and NG is an effective method for topology learning. The evolution of the lattice is shown in ❷ *Fig. 14* for a data manifold that consists of three-, two-, and one-dimensional subspaces (Martinetz and Schulten 1991). We see that the lattice has successfully filled and adapted its dimensionality to the different subspaces. For this reason, visualization is only possible for low-dimensional input spaces (hence, it is not suited for data visualization purposes where a mapping from a potentially high-dimensional input space to a low-dimensional lattice is desired). A problem with the algorithm is that one needs to decide *a priori* the number of neurons, as required by the NG algorithm (Fritzke 1996): depending on the complexity of the data manifold, very different numbers may be appropriate. This problem is overcome in the growing neural gas (GNG) (Fritzke 1995a) (see ❷ Sect. 5.2).

❑ **Fig. 13**

Left panel: Delaunay triangularization. The neuron weight positions are indicated with open circles; the thick lines connect the nearest neighbor weights. The borders of the Voronoi polygons, corresponding to the weights, are indicated with thin lines. *Right panel:* Induced Delaunay triangularization. The induced triangularization is obtained by masking the original triangularization with the input data distribution (two disconnected *gray* shaded regions). (Fritzke 1995a, reprinted with permission.)

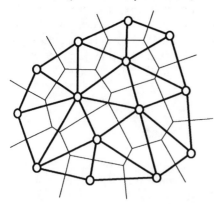
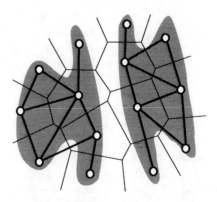

◘ Fig. 14

Neural gas algorithm, combined with competitive Hebbian learning, applied to a data manifold consisting of a right parallelepiped, a rectangle, and a circle connecting a line. The dots indicate the positions of the neuron weights. Lines connecting neuron weights indicate lattice edges. Shown are the initial result (*top left*), and further the lattice after 5,000, 10,000, 15,000, 25,000, and 40,000 time steps (*top-down* the first column, then *top-down* the second column). (Martinetz and Schulten 1991, reprinted with permission.)

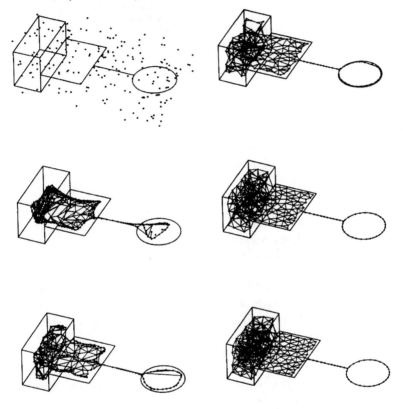

5.2 Growing Neural Gas

Contrary to CHL/NG, the growing neural gas (GNG) poses no explicit constraints on the lattice. The lattice is generated, and constantly updated, by the competitive Hebbian learning technique (CHL, see above; Martinetz 1993). The algorithm starts with two randomly placed and connected neurons (**❷** *Fig. 15*, left panel). Unlike the CHL/NG algorithm, after a fixed number, λ, of time steps, the neuron, i, with the largest accumulated error is determined and a new neuron inserted between i and one of its neighbors. Hence, the GNG algorithm exploits the topology to position new neurons between existing ones, whereas in the CHL/NG the topology is not influenced by the NG algorithm. Error variables are locally redistributed and another λ time step is performed. The lattice generated is a subgraph of a Delaunay triangularization, and can have different dimensionalities in different regions of the data manifold. The end-result is very similar to CHL/NG (**❷** *Fig. 15*, right panel).

□ **Fig. 15**

Growing neural gas algorithm applied to the same data configuration as in ❯ *Fig. 14*. Initial lattice (*left panel*) and lattice after 20,000 time steps (*right panel*). Note that the last one is not necessarily the final result because the algorithm could run indefinitely. (Fritzke 1995a, reprinted with permission.)

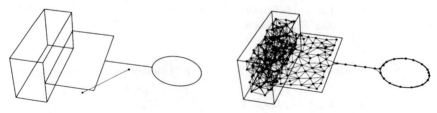

5.3 Growing Cell Structures

In the growing cell structures (GCS) algorithm (Fritzke 1994), the model consists of hypertetrahedrons (or simplices) of a dimensionality chosen in advance (hence, the lattice dimensionality is fixed). Note that a d_A-dimensional hypertetrahedron has $d_A + 1$ vertices, with d_A the lattice dimensionality, and $d_A \leq d$, with d the input space dimensionality. Examples for $d = 1, 2,$ and 3 are a line, a triangle, and a tetrahedron, respectively.

The model is initialized with exactly one hypertetrahedron. Always after a prespecified number of time steps, the neuron, i, with the maximum accumulated error is determined and a new neuron is inserted by splitting the longest of the edges emanating from i. Additional edges are inserted to rebuild the structure in such a way that it consists only of d_A-dimensional hypertetrahedrons: Let the edge which is split connect neurons i and j, then the newly inserted neuron should be connected to i and j and with all *common* topological neighbors of i and j.

Since the GCS algorithm assumes a fixed dimensionality for the lattice, it can be used for generating a dimensionality reducing mapping from the input space to the lattice space, which is useful for data visualization purposes.

5.4 Growing Grid

In the growing grid algorithm (GG) (Fritzke 1995b), the lattice is a rectangular grid of a certain dimensionality d_A. The starting configuration is a d_A-dimensional hypercube, for example, a 2×2 lattice for $d_A = 2$, a $2 \times 2 \times 2$ lattice for $d_A = 3$, and so on. To keep this structure consistent, it is necessary to always insert complete (hyper-)rows and (hyper-)columns. Since the lattice dimensionality is fixed, and possibly much smaller than the input space dimensionality, the GG is useful for data visualization.

Apart from these differences, the algorithm is very similar to the ones described above. After λ time steps, the neuron with the largest accumulated error is determined, and the longest edge emanating from it is identified, and a new complete hyper-row or -column is inserted such that the edge is split.

5.5   Other Algorithms

There exists a wealth of other algorithms, such as the dynamic cell structures (DCS) (Bruske and Sommer 1995), which is similar to the GNG; the growing self-organizing map (GSOM, also called hypercubical SOM) (Bauer and Villmann 1997), which has some similarities to GG but it adapts the lattice dimensionality; incremental grid growing (IGG), which introduces new neurons at the lattice border and adds/removes connections based on the similarities of the connected neurons' weight vectors (Blackmore and Miikkulainen 1993); and one that is also called the growing self-organizing map (GSOM) (Alahakoon et al. 2000), which also adds new neurons at the lattice border, similar to IGG, but does not delete neurons, and which contains a spread factor to let the user control the spread of the lattice.

In order to study and exploit hierarchical relations in the data, hierarchical versions of some of these algorithms have been developed. For example, the growing hierarchical self-organizing map (GHSOM) (Rauber et al. 2002) develops lattices at each level of the hierarchy using the GG algorithm (insertion of columns or rows). The orientation in space of each lattice is similar to that of the parent lattice, which facilitates the interpretation of the hierarchy, and which is achieved through a careful initialization of each lattice. Another example is adaptive hierarchical incremental grid growing (AHIGG) (Merkl et al. 2003) of which the hierarchy consists of lattices trained with the IGG algorithm, and for which new units at a higher level are introduced when the local (quantization) error of a neuron is too large.

6   Recurrent Topographic Maps

6.1   Time Series

Many data sources such as speech have a temporal characteristic (e.g., a correlation structure) that cannot be sufficiently captured when ignoring the order in which the data points arrive, as in the original SOM algorithm. Several self-organizing map algorithms have been developed for dealing with sequential data, such as those using:

- fixed-length windows, for example, the time-delayed SOM (Kangas 1990), among others (Martinetz et al. 1993; Simon et al. 2003; Vesanto 1997);
- specific sequence metrics (Kohonen 1997; Somervuo 2004);
- statistical modeling incorporating appropriate generative models for sequences (Bishop et al. 1997; Tiňo et al. 2004);
- mapping of temporal dependencies to spatial correlation, for example, as in traveling wave signals or potentially trained, temporally activated lateral interactions (Euliano and Principe 1999; Schulz and Reggia 2004; Wiemer 2003);
- recurrent processing of time signals and recurrent winning neuron computation based on the current input and the previous map activation, such as with the temporal Kohonen map (TKM) (Chappell and Taylor 1993), the recurrent SOM (RSOM) (Koskela et al. 1998), the recursive SOM (RecSOM) (Voegtlin 2002), the SOM for structured data (SOMSD) (Hagenbuchner et al. 2003), and the merge SOM (MSOM) (Strickert and Hammer 2005).

Several of these algorithms proposed recently, which shows the increased interest in representing time series with topographic maps. For some of these algorithms, also tree structured data can be represented (see later). We focus on the recurrent processing of time signals, and we briefly describe the models listed. A more detailed overview can be found elsewhere (Barreto and Araújo 2001; Hammer et al. 2005). The recurrent algorithms essentially differ in the context, that is, the way by which sequences are internally represented.

6.1.1 Overview of Algorithms

The TKM extends the SOM algorithm with recurrent self-connections of the neurons, such that they act as leaky integrators (❷ *Fig. 16a*). Given a sequence $[\mathbf{v}_1, \ldots, \mathbf{v}_t]$, $\mathbf{v}_j \in \mathbb{R}^d$, $\forall j$, the integrated distance ID_i of neuron i with weight vector $\mathbf{w}_i \in \mathbb{R}^d$ is:

$$ID_i(t) = \alpha \|\mathbf{v}_t - \mathbf{w}_i\|^2 + (1-\alpha)ID_i(t-1) \tag{17}$$

with $\alpha \in (0,1)$ a constant determining the strength of the context information, and with $ID_i(0) \overset{\Delta}{=} 0$. The winning neuron is selected as $i^*(t) = \mathrm{argmin}_i ID_i(t)$, after which the network is updated as in the SOM algorithm. ❷ Equation 17 has the form of a leaky integrator, integrating previous distances of neuron i, given the sequence.

The RSOM uses, in essence, the same dynamics; however, it integrates over the directions of the individual weight components:

$$ID_{ij}(t) = \alpha(v_{jt} - \mathbf{w}_i) + (1-\alpha)ID_{ij}(t-1) \tag{18}$$

so that the winner is then the neuron for which $\|[ID_{ij}(t)]\|^2$ is the smallest. It is clear that this algorithm stores more information than the TKM. However, both the TKM and the RSOM compute only a leaky average of the time series and they do not use any explicit context.

The RecSOM is an algorithm for sequence prediction. A given sequence is recursively processed based on the already computed context. Hereto, each neuron i is equipped with a

❑ **Fig. 16**
Schematic representation of four recurrent SOM algorithms: TKM (a), RecSOM (b), SOMSD (c), and MSOM (d). Recurrent connections indicate leaky integration; *double circles* indicate the neuron's weight- and context vectors; i^* and the *filled circles* indicate the winning neuron; (t) and (t − 1) represent the current and the previous time steps, respectively.

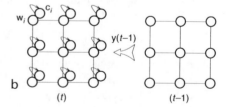

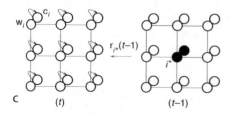
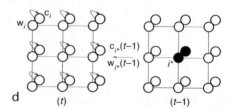

weight and, additionally, a context vector $\mathbf{c}_i \in \mathbb{R}^N$ that stores an activation profile of the whole map, indicating in which context the weight vector should arise (❷ *Fig. 16b*). The integrated distance is defined as:

$$ID_i(t) = \alpha \|\mathbf{v}_t - \mathbf{w}_i\|^2 + \beta \|\mathbf{y}(t-1) - \mathbf{c}_i\|^2 \tag{19}$$

with $\mathbf{y}(t-1) = [\exp(-ID_1(t-1)), \ldots, \exp(-ID_N(t-1))]$, $\alpha, \beta > 0$ constants to control the respective contributions from pattern and context matching, and with $ID_i(0) \stackrel{\Delta}{=} 0$. The winner is defined as the neuron for which the integrated distance is minimal. The equation contains the exponential function in order to avoid numerical explosion: otherwise, the activation, ID_i, could become too large because the distances with respect to the contexts of all N neurons could accumulate. Learning is performed on the weights as well as the contexts, in the usual way (thus, involving a neighborhood function centered around the winner): the weights are adapted toward the current input sequences, the contexts toward the recursively computed contexts $\mathbf{y}$.

The SOMSD was developed for processing labeled trees with fixed fan-out k. The limiting case of $k = 1$ covers sequences. We further restrict ourselves to sequences. Each neuron has, besides a weight, a context vector $\mathbf{c}_i \in \mathbb{R}^{d_A}$, with d_A the dimensionality of the lattice. The winning neuron i^* for a training input at time t is defined as (❷ *Fig. 16c*):

$$i^* = \arg\min \alpha \|\mathbf{v}_t - \mathbf{w}_i\|^2 + (1-\alpha)\|r_{i^*}(t-1) - \mathbf{c}_i\|^2 \tag{20}$$

with r_{i^*} the lattice coordinate of the winning neuron. The weights $\mathbf{w}_i$ are moved in the direction of the current input, as usual (i.e., with a neighborhood), and the contexts $\mathbf{c}_i$ in the direction of the lattice coordinates with the winning neuron of the previous time step (also with a neighborhood).

The MSOM algorithm accounts for the temporal context by an explicit vector attached to each neuron that stores the preferred context of that neuron (❷ *Fig. 16d*). The MSOM characterizes the context by a "merging" of the weight and the context of the winner in the previous time step (whence the algorithm's name: merge SOM). The integrated distance is defined as:

$$ID_i(t) = \alpha \|\mathbf{w}_i - \mathbf{v}_t\|^2 + (1-\alpha)\|\mathbf{c}_i - \mathbf{C}_t\|^2 \tag{21}$$

with $\mathbf{c}_i \in \mathbb{R}^d$, and with $\mathbf{C}_t$ the expected (merged) weight/context vector, that is, the context of the previous winner:

$$\mathbf{C}_t = \gamma \mathbf{c}_{i^*}(t-1) + (1-\gamma)\mathbf{w}_{i^*}(t-1) \tag{22}$$

with $\mathbf{C}_0 \stackrel{\Delta}{=} 0$. Updating of $\mathbf{w}_i$ and $\mathbf{c}_i$ are then done in the usual SOM way, thus, with a neighborhood function centered around the winner. The parameter, α, is controlled so as to maximize the entropy of the neural activity.

6.1.2 Comparison of Algorithms

Hammer and coworkers (2004) pointed out that several of the mentioned recurrent self-organizing map algorithms share their principled dynamics, but differ in their internal representations of context. In all cases, the context is extracted as the relevant part of the activation of the map in the previous time step. The notion of "relevance" thus differs between the algorithms (see also Hammer et al. 2005). The recurrent self-organizing algorithms can be divided into two categories: the representation of the context in the data space, such as for the TKM and MSOM, and in a space that is related to the neurons, as for SOMSD and RecSOM. In the first case, the storage capacity is restricted by the input dimensionality. In the latter case, it can be enlarged simply by adding more neurons to the lattice. Furthermore, there are essential differences in the dynamics of the algorithms. The TKM does not converge to the optimal weights; RSOM does it but the parameter α occurs both in the encoding formula and in the dynamics. In the MSOM algorithm, they can be controlled separately. Finally, the algorithms differ in memory and computational complexity (RecSOM is quite demanding, SOMSD is fast, and MSOM is somewhere in the middle), the possibility to apply different lattice types (such as hyperbolic lattices; Ritter 1998), and their capacities (MSOM and SOMSD achieve the capacity of finite state automata, but TKM and RSOM have smaller capacities; RecSOM is more complex to judge).

As an example, Voegtlin (2002) used the Mackey–Glass time series, a well-known one-dimensional time-delay differential equation, for comparing different algorithms:

$$\frac{dv}{dt} = bv(t) + \frac{av(t-\tau)}{1 + v(t-\tau)^{10}} \tag{23}$$

which for $\tau > 16.8$ generates a chaotic time series. Voegtlin used $a = 0.2$, $b = -0.1$, and $\tau = 17$. A sequence of values is plotted in ❷ *Fig. 17*, starting from uniform input conditions. For training, the series is sampled every three time units. This example was also taken up by Hammer et al. (2004) for comparing their MSOM. Several 10×10 maps were trained using 150,000 iterations; note that the input dimensionality $d = 1$ in all cases. ❷ *Figure 18* shows the quantization error plotted as a function of the index of the past input (index = 0 means the present). The error is expressed in terms of the average standard deviation of the given sequence and the winning neuron's receptive field over a window of 30 time steps (i.e., delay vector). We observe large fluctuations for the SOM, which is due to the temporal regularity of the series and the absence of any temporal coding by the SOM algorithm. We also observe that the RSOM algorithm is not really better than the SOM algorithm. On the contrary, the RecSOM, SOMSD, and MSOM algorithms (the MSOM was trained with a neural gas neighborhood function, for details see Strickert and Hammer (2003a)) display a slow increase in error as a function of the past, but with a better performance for the MSOM algorithm.

■ Fig. 17

Excerpt from the Mackey–Glass chaotic time series. (Strickert and Hammer 2003b, reprinted with permission.)

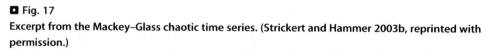
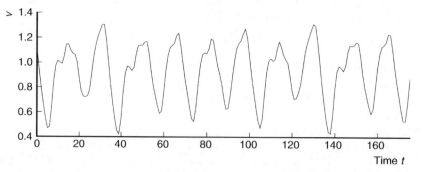

◘ **Fig. 18**

Temporal quantization error of different algorithms for the Mackey–Glass time series plotted as a function of the past (index = 0 is present). (Strickert and Hammer 2003b, reprinted with permission.)

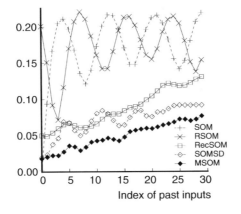

6.2 Tree Structures

Binary trees, and also trees with limited fan-out k, have been successfully processed with the SOMSD and the MSOM by extending the neuron's single context vector to several context vectors (one for each subtree). Starting from the leafs of a tree, the integrated distance ID of a tree with a given label and the k subtrees can be determined, and the context defined. The usual learning can then be applied to the weights and contexts. As a result of learning, a topographic mapping of trees according to their structure and labels arises. Up to now, only preliminary results of the capacities of these algorithms for tree structures have been obtained.

7 Kernel Topographic Maps

Rather than developing topographic maps with disjoint and uniform activation regions (Voronoi tessellation), such as in the case of the SOM algorithm (❷ *Fig. 5*), and its adapted versions, algorithms have been introduced that can accommodate neurons with overlapping activation regions, usually in the form of kernel functions, such as Gaussians (❷ *Fig. 19*). For these *kernel-based topographic maps*, or *kernel topographic maps*, as they are called (they are also sometimes called *probabilistic topographic maps* since they model the input density with a kernel mixture), several learning principles have been proposed (for a review, see Van Hulle (2009)). One motivation to use kernels is to improve, besides the biological relevance, the density estimation properties of topographic maps. In this way, we can combine the unique visualization properties of topographic maps with an improved modeling of clusters in the data. Usually, homoscedastic (equal variance) Gaussian kernels are used, but heteroscedastic (differing variances) Gaussian kernels and other kernel types have also been adopted. The following sections will review the kernel-based topographic map formation algorithms and mention a number of applications. The diversity in algorithms reflects the differences in strategies behind them. As a result, these algorithms have their specific strengths (and weaknesses) and, thus, their own application types.

□ Fig. 19

Kernel-based topographic map. Example of a 2 × 2 map (cf. rectangle with thick lines in V-space) for which each neuron has a Gaussian kernel as output function. Normally, a more condensed representation is used where, for each neuron, a circle is drawn with center the neuron weight vector, and radius the kernel range.

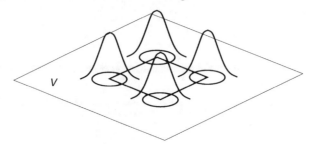

7.1   SOM Algorithm Revisited

The starting point is again Kohonen's SOM algorithm. To every neuron a homoscedastic Gaussian kernel is associated with the center corresponding to the neuron's weight vector. Kostiainen and Lampinen (2002) showed that the SOM algorithm can be seen as the equivalent of a maximum likelihood procedure applied to a homoscedastic Gaussian mixture density model, but with the exception that a winner neuron (and, thus, kernel) is selected (the definition of the "winner" i^* [❷ Eq. 1] is equivalent to looking for the Gaussian kernel with the largest output). The position of the winner's kernel is then updated, and possibly also those of other kernels, given the neighborhood function. In a traditional maximum likelihood procedure, there are no winners, and all kernels are updated (Redner and Walker 1984). This means that, for example, for a vanishing neighborhood range, a Gaussian kernel's center is only updated when that neuron is the winner, hence, contrary to the classical case of Gaussian mixture density modeling, the tails of the Gaussian kernels do not lead to center updates (they disappear "under" other kernels), which implies that the kernel radii will be underestimated.

7.2   Elastic Net

Durbin and Willshaw's (1987) elastic net can be considered as one of the first accounts on kernel-based topographic maps. The elastic net was used for solving the traveling salesman problem (TSP). In the TSP, the objective is to find the shortest, closed tour that visits each city once and that returns to its starting point (e.g., the right panel in ❷ *Fig. 20*). When we represent the location of each city by a point $\mathbf{v}^\mu$ in the two-dimensional input space $V \subseteq \mathbb{R}^2$, and a tour by a sequence of N neurons – which comprise a ring or closed chain A – then a solution to the TSP can be envisaged as a mapping from V-space onto the neurons of the chain. Evidently, we expect the neuron weights to coincide with the input points ("cities") at convergence.

The algorithm of the elastic net can be written as follows (in our format):

$$\Delta \mathbf{w}_i = 2\eta \left(\sum_\mu \Lambda^\mu(i)(\mathbf{v}^\mu - \mathbf{w}_i) + \kappa(\mathbf{w}_{i+1} - 2\mathbf{w}_i + \mathbf{w}_{i-1}) \right), \quad \forall i \qquad (24)$$

◘ **Fig. 20**

One-dimensional topographic map used for solving the traveling salesman problem. The lattice has a ring topology (closed chain); the points represent cities and are chosen randomly from the input distribution demarcated by the square box. The evolution of the lattice is shown for three time instants, at $t = 0$ (initialization); 7,000; and 10,000 (from left to right). The weights of the lattice at $t = 0$ form a circle positioned at the center of mass of the input distribution. (Reprinted from Ritter and Schulten (1988), ©1988 IEEE.)

where each weight, $\mathbf{w}_i$, represents a point on the elastic net. The first term on the right-hand side is a force that drags each point $\mathbf{w}_i$ on the chain A toward the cities $\mathbf{v}^\mu$, and the second term is an elastic force that tends to keep neighboring points on the chain close to each other (and thus tends to minimize the overall tour length). The function $\Lambda^\mu(i)$ is a *normalized* Gaussian:

$$\Lambda^\mu(i) = \frac{\exp(-\|\mathbf{v}^\mu - \mathbf{w}_i\|^2/2\sigma_\Lambda^2)}{\sum_j \exp(-\|\mathbf{v}^\mu - \mathbf{w}_j\|^2/2\sigma_\Lambda^2)} \tag{25}$$

with $\mathbf{w}_i$ the center of the Gaussian and σ_Λ its range, which is gradually decreased over time (as well as η, and also κ). By virtue of this kernel, the elastic net can be viewed as a homoscedastic Gaussian mixture density model, fitted to the data points by a penalized maximum likelihood term (for a formal account, see Durbin et al. (1989)). The elastic net algorithm looks similar to Kohonen's SOM algorithm except that $\Lambda(i,j)$ has been replaced by $\Lambda^\mu(i)$, and that a second term is added. Interestingly, the SOM algorithm can be used for solving the TSP even without the second term (Ritter et al. 1992), provided we take more neurons in our chain than cities, and that we initialize the weights on a circle (a so-called N-gon) positioned at the center of mass of the input distribution. An example of the convergence process for a 30-city case using a $N = 100$ neuron chain is shown in ❷ *Fig. 20*.

The elastic net has been used for finding trajectories of charged particles with multiple scattering in high-energy physics experiments (Gorbunov and Kisel 2006), and for predicting the protein folding structure (Ball et al. 2002). Furthermore, it has been used for clustering applications (Rose et al. 1993). Finally, since it also has a close relationship with "snakes" in computer vision (Kass et al. 1987) (for the connection, see Abrantes and Marques (1995)), the elastic net has also been used for extracting the shape of a closed object from a digital image, such as finding the lung boundaries from magnetic resonance images (Gilson et al. 1997).

7.3 Generative Topographic Map

The generative topographic map (GTM) algorithm (Bishop et al. 1996, 1998) develops a topographic map that attempts to find a representation for the input distribution $p(\mathbf{v})$,

19

$\mathbf{v} = [v_1, \ldots, v_d]$, $\mathbf{v} \in V$, in terms of a number L of latent variables $\mathbf{x} = [x_1, \ldots, x_L]$. This is achieved by considering a nonlinear transformation $\mathbf{y}(\mathbf{x}, \mathbf{W})$, governed by a set of parameters $\mathbf{W}$, which maps points in the latent variable space to the input space, much the same way as the lattice nodes in the SOM relate to positions in V-space (inverse mapping Ψ in ❷ *Fig. 2*). If one defines a probability distribution, $p(\mathbf{x})$, on the latent variable space, then this will induce a corresponding distribution, $p(\mathbf{y}|\mathbf{W})$, in the input space.

As a specific form of $p(\mathbf{x})$, Bishop and coworkers take a discrete distribution consisting of a sum of delta functions located at the N nodes of a regular lattice:

$$p(\mathbf{x}) = \frac{1}{N} \sum_{i=1}^{N} \delta(\mathbf{x} - \mathbf{x}_i) \tag{26}$$

The dimensionality, L, of the latent variable space is typically less than the dimensionality, d, of the input space so that the transformation, $\mathbf{y}$, specifies an L-dimensional manifold in V-space. Since $L < d$, the distribution in V-space is confined to this manifold and, hence, is singular. In order to avoid this, Bishop and coworkers introduced a noise model in V-space, namely, a set of radially symmetric Gaussian kernels centered at the positions of the lattice nodes in V-space. The probability distribution in V-space can then be written as follows:

$$p(\mathbf{v}|\mathbf{W}, \sigma) = \frac{1}{N} \sum_{i=1}^{N} p(\mathbf{v}|\mathbf{x}_i, \mathbf{W}, \sigma) \tag{27}$$

which is a homoscedastic Gaussian mixture model. In fact, this distribution is a *constrained* Gaussian mixture model since the centers of the Gaussians cannot move independently from each other but are related through the transformation, $\mathbf{y}$. Moreover, when the transformation is smooth and continuous, the centers of the Gaussians will be topographically ordered by construction. Hence, the topographic nature of the map is an intrinsic feature of the latent variable model and is not dependent on the details of the learning process. Finally, the parameters $\mathbf{W}$ and σ are determined by maximizing the log-likelihood:

$$\ln \mathscr{L}(\mathbf{W}, \sigma) = \ln \prod_{\mu=1}^{M} p(\mathbf{v}^{\mu}|\mathbf{W}, \sigma) \tag{28}$$

and which can be achieved through the use of an expectation-maximization (EM) procedure (Dempster et al. 1977). Because a single two-dimensional visualization plot may not be sufficient to capture all of the interesting aspects of complex data sets, a hierarchical version of the GTM has also been developed (Tiňo and Nabney 2002).

The GTM has been applied to visualize oil flows along multiphase pipelines, where the phases are oil, water, and gas, and the flows can be one of the three types: stratified, homogeneous, and annular (Bishop et al. 1996) (❷ *Fig. 4*, right panel). It has been applied to visualize electropalatographic (EPG) data for investigating the activity of the tongue in normal and pathological speech (Carreira-Perpiñán and Renals 1998) (❷ *Fig. 21*). It has also been applied to the classification of in vivo magnetic resonance spectra of controls and Parkinson's patients (Axelson et al. 2002), to word grouping in document data sets (using the newsgroup data set benchmark) and the exploratory analysis of web navigation sequences (Kabán 2005), and to spatiotemporal clustering of transition states of a typhoon from image sequences of cloud patterns (Kitamoto 2002). In another application, the GTM is used for microarray data analysis (gene expression data) with the purpose of finding low-confidence value genes (D'Alimonte et al. 2005).

□ Fig. 21
Visualization of the trajectory in a 20 × 20 GTM lattice of the activity of the tongue
(electropalatographic (EPG) data) of speaker RK for the utterance fragment "I prefer *Kant to
Hobbes for a good bedtime book.*" (Carreira-Perpiñán and Renals 1998, reprinted with
permission.)

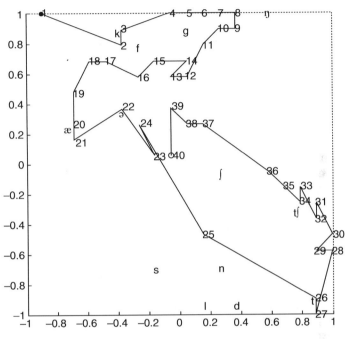

7.4 Regularized Gaussian Mixture Modeling

Heskes (2001) was able to show the direct correspondence between minimum distortion
topographic map formation and maximum likelihood Gaussian mixture density modeling for
the homoscedastic case. The starting point was the traditional distortion (vector quantization)
formulation of the self-organizing map:

$$F_{\text{quantization}} = \sum_{\mu} \sum_{i} P(i|\mathbf{v}^{\mu}) \sum_{j} \Lambda(i.j) \frac{1}{2} \|\mathbf{v}^{\mu} - \mathbf{w}_{j}\|^{2} \qquad (29)$$

with $P(i|\mathbf{v}^{\mu})$ the probability that input $\mathbf{v}^{\mu}$ is assigned to neuron i with weight $\mathbf{w}_i$ (i.e., the
posterior probability, and with $\sum_i P(i|\mathbf{v}^{\mu}) = 1$ and $P(i|\mathbf{v}^{\mu}) \geq 0$). Even if one assigns $\mathbf{v}^{\mu}$ to neuron
i, there exists a confusion probability $\Lambda(i,j)$ that $\mathbf{v}^{\mu}$ is assigned to neuron j. An annealed
version of the self-organizing map is obtained if one adds an entropy term:

$$F_{\text{entropy}} = \sum_{\mu} \sum_{i} P(i|\mathbf{v}^{\mu}) \log\left(\frac{P(i|\mathbf{v}^{\mu})}{Q_i}\right) \qquad (30)$$

with Q_i the prior probability (the usual choice is $Q_i = \frac{1}{N}$, with N the number of neurons in the lattice. The final (free) energy is now:

$$F = \beta F_{\text{quantization}} + F_{\text{entropy}} \quad (31)$$

with β playing the role of an inverse temperature. This formulation is very convenient for an EM procedure. The expectation step leads to:

$$P(i|\mathbf{v}^{\mu}) = \frac{Q_i \exp\left(-\frac{\beta}{2}\sum_j \Lambda(i,j)\,\|\mathbf{v}^{\mu} - \mathbf{w}_j\|\right)}{\sum_s Q_s \exp\left(-\frac{\beta}{2}\sum_j \Lambda(s,j)\,\|\mathbf{v}^{\mu} - \mathbf{w}_j\|\right)} \quad (32)$$

and the maximization step to:

$$\mathbf{w}_i = \frac{\sum_{\mu}\sum_j P(j|\mathbf{v}^{\mu})\Lambda(j,i)\mathbf{v}^{\mu}}{\sum_{\mu}\sum_j P(j|\mathbf{v}^{\mu})\Lambda(j,i)} \quad (33)$$

which is also the result reached by Graepel and coworkers (1998) for the soft topographic vector quantization (STVQ) algorithm (see the next section). Plugging (❷ Eq. 32) into (❷ Eq. 31) leads to an error function, which allows for the connection with a maximum likelihood procedure, for a mixture of homoscedastic Gaussians, when the neighborhood range vanishes ($\Lambda(i,j) = \delta_{ij}$). When the neighborhood is present, Heskes showed that this leads to a term added to the original likelihood.

As an application, Heskes considers market basket analysis. Given are a list of transactions corresponding to the joint set of products purchased by a customer at a given time. The goal of the analysis is to map the products onto a two-dimensional map (lattice) such that neighboring products are "similar." Similar products should have similar conditional probabilities of buying other products. In another application, he considers the case of transactions in a supermarket. The products are summarized in product groups, and the co-occurrence frequencies are given. The result is a two-dimensional density map showing clusters of products that belong together, for example, a large cluster of household products (❷ *Fig. 22*).

7.5 Soft Topographic Vector Quantization

Another approach that considers topographic map formation as an optimization problem is the one introduced by Klaus Obermayer and coworkers (Graepel et al. 1997,1998). They start from the following cost function:

$$E(\mathbf{W}) = \frac{1}{2}\sum_{\mu}\sum_i c_{\mu,i}\sum_j \Lambda(i,j)\,\|\mathbf{v}^{\mu} - \mathbf{w}_j\|^2 \quad (34)$$

with $c_{\mu,i} \in \{0, 1\}$ and for which $c_{\mu,i} = 1$ if $\mathbf{v}^{\mu}$ is assigned to neuron i, else $c_{\mu,i} = 0$ ($\sum_i c_{\mu,i} = 1$); the neighborhood function obeys $\sum_j \Lambda(i,j) = 1$. The $\mathbf{w}_i$, $\forall i$, for which this function is minimal, are the optimal ones. However, the optimization is a difficult task, because it depends both on binary and continuous variables and has many local minima. To avoid this, a technique known as deterministic annealing is applied: the optimization is done on a smooth function parametrized by a parameter β, the so-called free energy. When β is small, the function is smooth and only one global minimum remains; when large, more of the structure of the original cost

◻ **Fig. 22**

Visualization of market basket data in which 199 product groups are clustered based on their co-occurrence frequencies with other products (Heskes (2001), ©2001 IEEE).

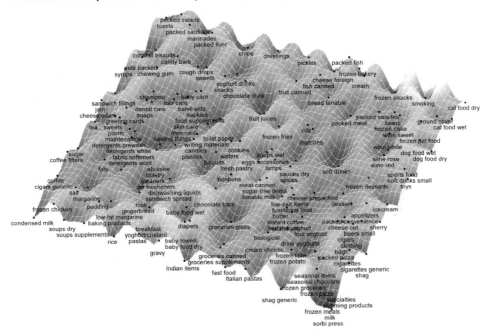

function is reflected in the free energy. One starts with a low value of β and attempts to keep track of the minimum through higher values of β.

The application of the principle of maximum entropy yields the free energy (Graepel et al. 1997):

$$F = -\frac{1}{\beta}\log\sum_{c_{\mu,i}}\exp(-\beta E) \tag{35}$$

which leads to probabilistic assignments of inputs $\mathbf{v}^{\mu}$ to neurons, $P(i|\mathbf{v}^{\mu})$, $\forall i$, that is, the posterior probabilities, and which are given by:

$$P(i|\mathbf{v}^{\mu}) = \frac{\exp(-\frac{\beta}{2}\sum_{j}\Lambda(i,j)\|\mathbf{v}^{\mu}-\mathbf{w}_{j}\|^{2})}{\sum_{s}\exp(-\frac{\beta}{2}\sum_{j}\Lambda(s,j)\|\mathbf{v}^{\mu}-\mathbf{w}_{j}\|^{2})} \tag{36}$$

The fixed point rule for the kernel centers is then:

$$\mathbf{w}_{i} = \frac{\sum_{\mu}\sum_{j}P(j|\mathbf{v}^{\mu})\Lambda(j,i)\mathbf{v}^{\mu}}{\sum_{\mu}\sum_{j}P(j|\mathbf{v}^{\mu})\Lambda(j,i)}, \quad \forall i \tag{37}$$

The updates are done through an EM scheme. We observe that the latter equation is identical to Heskes' rule for regularized Gaussian mixture modeling (❷ Eq. 33).

The STVQ has been generalized to the soft topographic mapping for proximity data (STMP), which can be used for clustering categorical data, given a matrix of pairwise proximities or dissimilarities, which is one of the first accounts of this nature in the topographic map literature. A candidate application is the DNA microarray data sets where the data can be described by matrices with the columns representing tissue samples and the rows genes, and the entries in the matrix correspond to the strength of the gene expression. In Seo and Obermayer (2004), a modified version of the STMP is used for clustering documents ("document map").

7.6   Heteroscedastic Gaussian Kernel Topographic Map Formation

In the literature, only few approaches exist that consider heteroscedastic kernels, perhaps because the kernel radius in the homoscedastic case is often used in an annealing schedule, as shown above in the STVQ and the elastic net algorithms. When using heteroscedastic kernels, a better density estimate is expected. Several algorithms for heteroscedastic kernels have been developed (for a review, see Van Hulle (2009)). We briefly mention a few here.

Bearing in mind what was said earlier about the SOM in connection to Gaussian mixture modeling, one can extend the original batch map, ❯ Eq. 5, to the heteroscedastic case (Van Hulle 2009):

$$\mathbf{w}_i = \frac{\sum_\mu \Lambda(i^*, i)\mathbf{v}^\mu}{\sum_\mu \Lambda(i^*, i)}$$

$$\sigma_i^2 = \frac{\sum_\mu \Lambda(i^*, i)\|\mathbf{v} - \mathbf{w}_i\|^2/d}{\sum_\mu \Lambda(i^*, i)}, \quad \forall i \tag{38}$$

with $i^* = \mathrm{argmax}_i K_i$ (which is no longer equivalent to $i^* = \mathrm{argmin}_i\|\mathbf{v} - \mathbf{w}_i\|$, but which is required since we now have heteroscedastic kernels), that is, an *activity*-based definition of "winner-takes-all," rather than a minimum *Euclidean distance*-based one. Notice again that, by the definition of the winner, the tails of the kernels are cut off, since the kernels overlap.

Recently we introduced (Van Hulle 2005a) a learning algorithm for kernel-based topographic map formation of heteroscedastic Gaussian mixtures that allows for a unified account of distortion error (vector quantization), log-likelihood and Kullback–Leibler divergence, and that generalizes Heskes' (2001) algorithm to the heteroscedastic case.

There is also the heuristic approach suggested by Yin and Allinson (2001), which is minimizing the Kullback–Leibler divergence, based on an idea introduced by Benaim and Tomasini (1991), for the homoscedastic case. Albeit these authors only suggested an incremental, gradient-based learning procedure (thus, with a learning rate), we can cast their format into a fixed point learning scheme:

$$\mathbf{w}_i = \frac{\sum_\mu \Lambda(i^*, i)P(i|\mathbf{v}^\mu)\mathbf{v}^\mu}{\sum_\mu \Lambda(i^*, i)P(i|\mathbf{v}^\mu)}$$

$$\sigma_i^2 = \frac{\sum_\mu \Lambda(i^*, i)P(i|\mathbf{v}^\mu)\|\mathbf{v}^\mu - \mathbf{w}_i\|^2/d}{\sum_\mu \Lambda(i^*, i)P(i|\mathbf{v}^\mu)} \tag{39}$$

with the winner neuron defined as $i^* = \mathrm{argmax}_i P(i|\mathbf{v}^\mu)$, thus, the neuron with the largest posterior probability.

In a still different approach, an input to lattice transformation Φ is considered that admits a kernel function, a Gaussian (Van Hulle 2002a) $\langle\Phi(\mathbf{v}), \Phi(\mathbf{w}_i)\rangle = K(\mathbf{v}, \mathbf{w}_i, \sigma_i)$:

$$K(\mathbf{v}, \mathbf{w}_i, \sigma_i) = \exp\left(-\frac{\|\mathbf{v} - \mathbf{w}_i\|^2}{2\sigma_i^2}\right) \tag{40}$$

When performing topographic map formation, we require that the weight vectors are updated, so as to minimize the expected value of the squared Euclidean distance $\|\mathbf{v} - \mathbf{w}_i\|^2$ and, hence, following our transformation Φ, we instead wish to minimize $\|\Phi(\mathbf{v}) - \Phi(\mathbf{w}_i)\|^2$, which we will achieve by performing gradient descent with respect to $\mathbf{w}_i$. The leads to the following fixed point rules to which we have added a neighborhood function:

$$
\mathbf{w}_i = \frac{\sum_\mu \Lambda(i, i^*) K(\mathbf{v}^\mu, \mathbf{w}_i, \sigma_i) \mathbf{v}^\mu}{\sum_\mu \Lambda(i, i^*) K(\mathbf{v}^\mu, \mathbf{w}_i, \sigma_i)}
$$
$$
\sigma_i^2 = \frac{1}{\rho d} \frac{\sum_\mu \Lambda(i, i^*) K(\mathbf{v}^\mu, \mathbf{w}_i, \sigma_i) \|\mathbf{v}^\mu - \mathbf{w}_i\|^2}{\sum_\mu \Lambda(i, i^*) K(\mathbf{v}^\mu, \mathbf{w}_i, \sigma_i)}
$$

(41)

with ρ a scale factor (a constant) designed to relax the local Gaussian (and d large) assumption in practice, and with $i^* = \arg\max_{\forall i \in A} K(\mathbf{v}, \mathbf{w}_i, \sigma_i)$.

Rather than having a real-valued neural activation, one could also threshold the kernel into a binary variable: in the kernel-based maximum entropy rule (kMER), a neuron, i, is activated by input $\mathbf{v}$ when $\|\mathbf{w}_i - \mathbf{v}\| < \sigma_i$, where σ_i is the kernel radius of neuron i, and which defines a hyperspherical activation region, S_i (Van Hulle 1998). The membership function, $\mathbb{1}_i(\mathbf{v})$, equals unity when neuron i is activated by $\mathbf{v}$, else it is zero. When there are no neurons active for a given input, the neuron that is positioned closest to that input is defined active. The incremental learning rules for the weights and radii of neuron i are as follows:

$$
\Delta\mathbf{w}_i = \eta \sum_j \Lambda(i, j) \Xi_i(\mathbf{v}) \text{sign}(\mathbf{v} - \mathbf{w}_i)
$$
$$
\Delta\sigma_i = \eta\left(\frac{\rho_r}{N}(1 - \mathbb{1}_i(\mathbf{v})) - \mathbb{1}_i(\mathbf{v})\right)
$$

(42)

with sign(.), the sign function taken component-wise, η the learning rate, $\Xi_i(\mathbf{v}) = \frac{\mathbb{1}_i}{\sum_j \mathbb{1}_j}$ a fuzzy membership function, and $\rho_r = \frac{\rho N}{N - \rho}$. It can be shown that the kernel ranges converge to the case where the average probabilities become equal, $\langle \mathbb{1}_i \rangle = \frac{\rho}{N}, \forall i$. By virtue of the latter, kMER is said to generate an equiprobabilistic topographic map (which avoids dead units). The algorithm has been considered for a wide range of applications, such as shape clustering (Van Hulle and Gautama 2004), music signal clustering (Van Hulle 2000), and the linking of patent and scientific publication databases (Deleus and Van Hulle 2001). More recently, a fixed point version, called batch map kMER, was introduced (Gautama and Van Hulle 2006), and applied to handwritten numerals clustering.

7.7 Kernels Other Than Gaussians

In principle, kernels other than Gaussians could be used in topographic map formation. For example, Heskes pointed out that his regularized mixture modeling approach could, in principle, accommodate any kernel of the exponential family, such as the gamma, multinomial, and the Poisson distributions (Heskes 2001).

In another case, the kernel is considered for which the differential entropy of the kernel output will be maximal given a Gaussian input, that is, the incomplete gamma distribution kernel (Van Hulle 2002b).

Another type of kernel is the Edgeworth-expanded Gaussian kernel, which consists of a Gaussian kernel multiplied by a series of Hermite polynomials of increasing order, and of which the coefficients are specified by (the second- and higher-order) cumulants (Van Hulle 2005b).

In still another case, a mixture of Bernoulli distributions is taken (Verbeek et al. 2005) for the specific purpose to better encode binary data (e.g., word occurrence in a document). This also leads to an EM algorithm for updating the posteriors, as well as the expected joint log-likelihood with respect to the parameters of the Bernoulli distributions. However, as the posteriors become quite peaked for higher dimensions, for visualization purposes, a power function of them was chosen. Several applications have been demonstrated, including word grouping in document data sets (newsgroup data set) and credit data analysis (from the University of California, Irvine (UCI) machine learning repository http://archive.ics.uci.edu/ml/).

7.8   Future Developments

An expected development is to go beyond the limitation of the current kernel-based topographic maps that the inputs need to be vectors (we already saw the extension toward categorical data). But in the area of structural pattern recognition, more powerful data structures can be processed, such as strings, trees, and graphs. The SOM algorithm has already been extended toward strings (Kohonen and Somervuo 1998) and graphs, which include strings and trees (Günter and Bunke 2002; Seo and Obermayer 2004; Steil and Sperduti 2007) (see also the SOMSD and the MSOM algorithms above). However, new types of *kernels* for strings, trees, and graphs have been suggested in the support vector machine literature (thus, outside the topographic map literature) (for reviews, see Shawe-Taylor and Cristianini (2004) and Jin et al. (2005)). The integration of these new types of kernels into kernel-based topographic maps is yet to be done, but could turn out to be a promising evolution for biochemical applications, such as visualizing and clustering sets of structure-based molecule descriptions.

8   Conclusion

In this chapter we introduced the self-organizing map (SOM) algorithm, discussed its properties, limitations, and types of application, and reviewed a number of extensions and other types of topographic map formation algorithms, such as the growing, the recurrent, and the kernel topographic maps. We also indicated how recent developments in topographic maps enable us to consider categorical data, time series, and tree-structured data, widening further the application field toward microarray data analysis, document analysis and retrieval, exploratory analysis of web navigation sequences, and the visualization of protein structures and long DNA sequences.

Acknowledgments

The author is supported by the Excellence Financing program (EF 2005) and the CREA Financing program (CREA/07/027) of K.U.Leuven, the Belgian Fund for Scientific Research – Flanders (G.0234.04 and G.0588.09), the Flemish Regional Ministry of Education (Belgium) (GOA 2000/11), the Belgian Science Policy (IUAP P6/29), and the European Commission (NEST-2003-012963, STREP-2002-016276, IST-2004-027017, and ICT-2007-217077).

References

Abrantes AJ, Marques JS (1995) Unified approach to snakes, elastic nets, and Kohonen maps. In: Proceedings of the IEEE International Conference on Acoustics, Speech and Signal Processing (ICASSP'95). IEEE Computer Society, Los Alamitos, CA, vol 5, pp 3427–3430

Ahalt SC, Krishnamurthy AK, Chen P, Melton DE (1990) Competitive learning algorithms for vector quantization. Neural Netw 3:277–290

Alahakoon D, Halgamuge SK, Srinivasan B (2000) Dynamic self organising maps with controlled growth for knowledge discovery. IEEE Trans Neural Netw (Special issue on knowledge discovery and data mining) 11(3):601–614

Axelson D, Bakken IJ, Gribbestad IS, Ehrnholm B, Nilsen G, Aasly J (2002) Applications of neural network analyses to in vivo ^{1}H magnetic resonance spectroscopy of Parkinson disease patients. J Magn Reson Imaging 16(1):13–20

Ball KD, Erman B, Dill KA (2002) The elastic net algorithm and protein structure prediction. J Comput Chem 23(1):77–83

Barreto G, Araújo A (2001) Time in self-organizing maps: an overview of models. Int J Comput Res 10(2):139–179

Bauer H-U, Villmann T (1997) Growing a hypercubical output space in a self-organizing feature map. IEEE Trans Neural Netw 8(2):218–226

Bauer H-U, Der R, Herrmann M (1996) Controlling the magnification factor of self-organizing feature maps. Neural Comput 8:757–771

Benaim M, Tomasini L (1991) Competitive and self-organizing algorithms based on the minimization of an information criterion. In: Proceedings of 1991 international conference in artificial neural networks (ICANN'91). Espoo, Finland. Elsevier Science Publishers, North-Holland, pp 391–396

Bishop CM (2006) Pattern recognition and machine learning. Springer, New York

Bishop CM, Svensén M, Williams CKI (1996) GTM: a principled alternative to the self-organizing map. In: Proceedings 1996 International Conference on Artificial Neural Networks (ICANN'96). Bochum, Germany, 16–19 July 1996. Lecture notes in computer science, vol 1112. Springer, pp 165–170

Bishop CM, Hinton GE, and Strachan IGD (1997) In: Proceedings IEE fifth international conference on artificial neural networks. Cambridge UK, 7–9 July 1997, pp 111–116

Bishop CM, Svensén M, Williams CKI (1998) GTM: the generative topographic mapping. Neural Comput 10:215–234

Blackmore J, Miikkulainen R (1993) Incremental grid growing: encoding high-dimensional structure into a two-dimensional feature map. In: Proceedings of IEEE international conference on neural networks. San Francisco, CA. IEEE Press, Piscataway, NJ, vol 1, pp 450–455

Bruske J, Sommer G (1995) Dynamic cell structure learns perfectly topology preserving map. Neural Comput 7(4):845–865

Carreira-Perpiñán MÁ, Renals S (1998) Dimensionality reduction of electropalatographic data using latent variable models. Speech Commun 26(4):259–282

Centre NNR (2003) Bibliography on the Self-Organizing Map (SOM) and Learning Vector Quantization (LVQ), Helsinki University of Technology. http://liinwww.ira.uka.de/bibliography/Neural/SOM.LVQ.html

Chappell G, Taylor J (1993) The temporal Kohonen map. Neural Netw 6:441–445

Chinrungrueng C, Séquin CH (1995) Optimal adaptive k-means algorithm with dynamic adjustment of learning rate. IEEE Trans Neural Netw 6:157–169

Cottrell M, Fort JC (1987) Etude d'un processus d'auto-organization. Ann Inst Henri Poincaré 23:1–20

D'Alimonte D, Lowe D, Nabney IT, Sivaraksa M (2005) Visualising uncertain data. In: Proceedings European conference on emergent aspects in clinical data analysis (EACDA2005). Pisa, Italy, 28–30 September 2005. http://ciml.di.unipi.it/EACDA2005/papers.html

Deleus FF, Van Hulle MM (2001) Science and technology interactions discovered with a new topographic map-based visualization tool. In: Proceedings of 7th ACM SIGKDD international conference on knowledge discovery in data mining. San Francisco, 26–29 August 2001. ACM Press, New York, pp 42–50

Dempster AP, Laird NM, Rubin DB (1977) Maximum likelihood for incomplete data via the EM algorithm. J R Stat Soc B 39:1–38

Der R, Herrmann M (1993) Phase transitions in self-organizing feature maps. In: Proceedings of 1993 international conference on artificial neuron networks (ICANN'93). Amsterdam, The Netherlands, 13–16 September 1993, Springer, New York, pp 597–600

DeSieno D (1988) Adding a conscience to competitive learning. In: Proceedings of IEEE international conference on neural networks. San Diego, CA, IEEE Press, New York, vol I, pp 117–124

Durbin R, Willshaw D (1987) An analogue approach to the travelling salesman problem using an elastic net method. Nature 326:689–691

Durbin R, Szeliski R, Yuille AL (1989) An analysis of the elastic net approach to the traveling salesman problem. Neural Comput 1:348–358

Erwin E, Obermayer K, Schulten K (1992) Self-organizing maps: ordering, convergence properties and energy functions. Biol Cybern 67:47–55

Euliano NR, Principe JC (1999). A spatiotemporal memory based on SOMs with activity diffusion. In: Oja E, Kaski S (eds) Kohonen maps. Elsevier, Amsterdam, The Netherlands, pp 253–266

Fritzke B (1994) Growing cell structures – a self-organizing network for unsupervised and supervised learning. Neural Netw 7(9):1441–1460

Fritzke B (1995a) A growing neural gas network learns topologies. In: Tesauro G, Touretzky DS, Leen TK (eds) Advances in neural information proceedings systems 7 (NIPS 1994). MIT Press, Cambridge, MA, pp 625–632

Fritzke B (1995b) Growing grid – a self-organizing network with constant neighborhood range and adaptation strength. Neural Process Lett 2(5):9–13

Fritzke B (1996) Growing self-organizing networks – why? In: European symposium on artificial neural networks (ESANN96). Bruges, Belgium, 1996. D Facto Publications, Brussels, Belgium, pp 61–72

Gautama T, Van Hulle MM (2006) Batch map extensions of the kernel-based maximum entropy learning rule. IEEE Trans Neural Netw 16(2):529–532

Gersho A, Gray RM (1991) Vector quantization and signal compression. Kluwer, Boston, MA/Dordrecht

Geszti T (1990) Physical models of neural networks. World Scientific Press, Singapore

Gilson SJ, Middleton I, Damper RI (1997) A localised elastic net technique for lung boundary extraction from magnetic resonance images. In: Proceedings of fifth international conference on artificial neural networks. Cambridge, UK, 7–9 July 1997. Mascarenhas Publishing, Stevenage, UK, pp 199–204

Gorbunov S, Kisel I (2006) Elastic net for stand-alone RICH ring finding. Nucl Instrum Methods Phys Res A 559:139–142

Graepel T, Burger M, Obermayer K (1997) Phase transitions in stochastic self-organizing maps. Phys Rev E 56(4):3876–3890

Graepel T, Burger M, Obermayer K (1998) Self-organizing maps: generalizations and new optimization techniques. Neurocomputing 21:173–190

Grossberg S (1976) Adaptive pattern classification and universal recoding: I. Parallel development and coding of neural feature detectors. Biol Cybern 23:121–134

Günter S, Bunke H (2002) Self-organizing map for clustering in the graph domain, Pattern Recog Lett 23:415–417

Hagenbuchner M, Sperduti A, Tsoi AC (2003) A self-organizing map for adaptive processing of structured data. IEEE Trans Neural Netw 14(3):491–505

Hammer B, Micheli A, Strickert M, Sperduti A (2004) A general framework for unsupervised processing of structured data. Neurocomputing 57:3–35

Hammer B, Micheli A, Neubauer N, Sperduti A, Strickert M (2005) Self organizing maps for time series. In: Proceedings of WSOM 2005. Paris, France, 5–8 September 2005, pp 115–122

Heskes T (2001) Self-organizing maps, vector quantization, and mixture modeling. IEEE Trans Neural Netw 12(6):1299–1305

Heskes TM, Kappen B (1993) Error potentials for self-organization. In: Proceedings of IEEE international conference on neural networks. San Francisco, CA. IEEE Press, Piscataway, NJ, pp 1219–1223

Jin B, Zhang Y-Q, Wang B (2005) Evolutionary granular kernel trees and applications in drug activity comparisons, In: Proceedings of the 2005 IEEE Symposium on Computational Intelligence in Bioinformatics and Computational Biology (CIBCB'05). San Diego, CA, 14–15 November 2005, IEEE Press, Piscataway, NY, pp 1–6

Kabán A (2005) A scalable generative topographic mapping for sparse data sequences. In: Proceedings of the International Conference on Information Technology: Coding and Computing (ITCC'05). Las Vegas, NV, 4–6 April 2005. IEEE Computer Society, vol 1, pp 51–56

Kass M, Witkin A, Terzopoulos D (1987) Active contour models. Int J Comput Vis 1(4):321–331

Kangas J (1990) Time-delayed self-organizing maps. In: Proceedings IEEE/INNS international Joint Conference on neural networks 1990. San Diego, CA, IEEE, New York, vol 2, pp 331–336

Kaski S, Honkela T, Lagus K, Kohonen T (1998) WEBSOM – self-organizing maps of document collections. Neurocomputing 21:101–117

Kim YK, Ra JB (1995) Adaptive learning method in self-organizing map for edge preserving vector quantization. IEEE Trans Neural Netw 6:278–280

Kitamoto A (2002) Evolution map: modeling state transition of typhoon image sequences by spatio-temporal clustering. Lect Notes Comput Sci 2534/2002: 283–290

Kohonen T (1982) Self-organized formation of topologically correct feature maps. Biol Cybern 43:59–69

Kohonen T (1984) Self-organization and associative memory. Springer, Heidelberg

Kohonen T (1991) Self-organizing maps: optimization approaches. In: Kohonen T, Mäkisara K, Simula O, Kangas J (eds) Artificial neural networks. North-Holland, Amsterdam, pp 981–990

Kohonen T (1995) Self-organizing maps, 2nd edn. Springer, Heidelberg

Kohonen T (1997) Self-organizing maps. Springer

Kohonen T, Somervuo P (1998) Self-organizing maps on symbol strings. Neurocomputing 21:19–30

Kohonen T, Kaski S, Salojärvi J, Honkela J, Paatero V, Saarela A (1999) Self organization of a massive document collection. IEEE Trans Neural Netw 11(3):574–585

Koskela T, Varsta M, Heikkonen J, Kaski K (1998) Recurrent SOM with local linear models in time series prediction. In: Verleysen M (ed) Proceedings of 6th European symposium on artificial neural networks (ESANN 1998). Bruges, Belgium, April 22–24, 1998. D-Facto, Brussels, Belgium, pp 167–172

Kostiainen T, Lampinen J (2002) Generative probability density model in the self-organizing map. In: Seiffert U, Jain L (eds) Self-organizing neural networks: Recent advances and applications. Physica-Verlag, Heidelberg, pp 75–94

Laaksonen J, Koskela M, Oja E (2002) PicSOM–self-organizing image retrieval with MPEG-7 content descriptors. IEEE Trans Neural Netw 13(4):841–853

Lin JK, Grier DG, Cowan JD (1997) Faithful representation of separable distributions. Neural Comput 9:1305–1320

Linsker R (1988) Self-organization in a perceptual network. Computer 21:105–117

Linsker R (1989) How to generate ordered maps by maximizing the mutual information between input and output signals. Neural Comput 1:402–411

Luttrell SP (1989) Self-organization: a derivation from first principles of a class of learning algorithms. In: Proceedings IEEE international joint conference on neural networks (IJCNN89). Washington, DC, Part I, IEEE Press, Piscataway, NJ, pp 495–498

Luttrell SP (1990) Derivation of a class of training algorithms. IEEE Trans Neural Netw 1:229–232

Luttrell SP (1991) Code vector density in topographic mappings: scalar case. IEEE Trans Neural Netw 2:427–436

Martinetz TM (1993) Competitive Hebbian learning rule forms perfectly topology preserving maps. In: Proceedings of international conference on artificial neural networks (ICANN93). Amsterdam, The Netherlands, 13–16 September 1993. Springer, London, pp 427–434

Martinetz T, Schulten K (1991) A "neural-gas" network learns topologies. In: Kohonen T, Mäkisara K, Simula O, Kangas J (eds) Proceedings of International Conference on Artificial Neural Networks (ICANN-91). Espoo, Finland, 24–28 June 1991, vol I, North-Holland, Amsterdam, The Netherlands, pp 397–402

Martinetz T, Berkovich S, Schulten K (1993) "Neural-gas" network for vector quantization and its application to time-series prediction. IEEE Trans Neural Netw 4(4):558–569

Merkl D, He S, Dittenbach M, Rauber A (2003) Adaptive hierarchical incremental grid growing: an architecture for high-dimensional data visualization. In: Proceedings of 4th workshop on self-organizing maps (WSOM03). Kitakyushu, Japan, 11–14 September 2003

Mulier F, Cherkassky V (1995) Self-organization as an iterative kernel smoothing process. Neural Comput 7:1165–1177

Rauber A, Merkl D, Dittenbach M (2002) The growing hierarchical self-organizing map: exploratory analysis of high-dimensional data. IEEE Trans Neural Netw 13(6):1331–1341

Redner RA, Walker HF (1984) Mixture densities, maximum likelihood and the EM algorithm. SIAM Rev 26(2):195–239

Risi S, Mörchen F, Ultsch A, Lewark P (2007) Visual mining in music collections with emergent SOM. In: Proceedings of workshop on self-organizing maps (WSOM '07). Bielefeld, Germany, September 3–6, 2007, ISBN: 978-3-00-022473-7, CD ROM, available online at http://biecoll.ub.uni-bielefeld.de

Ritter H (1991) Asymptotic level density for a class of vector quantization processes. IEEE Trans Neural Netw 2(1):173–175

Ritter H (1998) Self-organizing maps in non-Euclidean spaces. In: Oja E, Kaski S (eds) Kohonen maps. Elsevier, Amsterdam, pp 97–108

Ritter H, Kohonen T (1989) Self-organizing semantic maps. Biol Cybern 61:241–254

Ritter H, Schulten K (1986) On the stationary state of Kohonen's self-organizing sensory mapping. Biol Cybern 54:99–106

Ritter H, Schulten K (1988) Kohonen's self-organizing maps: exploring their computational capabilities, In: Proceedings of IEEE international conference on neural networks (ICNN). San Diego, CA. IEEE, New York, vol I, pp 109–116

Ritter H, Martinetz T, Schulten K (1992) Neural computation and self-organizing maps: an introduction. Addison-Wesley, Reading, MA

Rose K, Gurewitz E, Fox GC (1993) Constrained clustering as an optimization method. IEEE Trans Pattern Anal Mach Intell 15(8):785–794

Schulz R, Reggia JA (2004) Temporally asymmetric learning supports sequence processing in multi-winner self-organizing maps. Neural Comput 16(3):535–561

Seo S, Obermayer K (2004) Self-organizing maps and clustering methods for matrix data. Neural Netw 17(8–9):1211–1229

Shawe-Taylor J, Cristianini N (2004) Kernel methods in computational biology. MIT Press, Cambridge, MA

Simon G, Lendasse A, Cottrell M, Fort J-C, Verleysen M (2003) Double SOM for long-term time series prediction. In: Proceedings of the workshop on self-organizing maps (WSOM 2003). Hibikino, Japan, September 11–14, 2003, pp 35–40

Somervuo PJ (2004) Online algorithm for the self-organizing map of symbol strings. Neural Netw 17(8–9):1231–1240

Steil JJ, Sperduti A (2007) Indices to evaluate self-organizing maps for structures. In: Proceedings of the workshop on self-organizing maps (WSOM07) Bielefeld, Germany, 3–6 September 2007. CD ROM, 2007, available online at http://biecoll.ub.uni-bielefeld.de

Strickert M, Hammer B (2003a) Unsupervised recursive sequence processing, In: Verleysen M (ed) European Symposium on Artificial Neural Networks (ESANN 2003). Bruges, Belgium, 23–25 April 2003. D-Side Publications, Evere, Belgium, pp 27–32

Strickert M, Hammer B (2003b) Neural gas for sequences. In: Proceedings of the workshop on self-organizing maps (WSOM'03). Hibikino, Japan, September 2003, pp 53–57

Strickert M, Hammer B (2005) Merge SOM for temporal data. Neurocomputing 64:39–72

Tiňo P, Nabney I (2002) Hierarchical GTM: constructing localized non-linear projection manifolds in a principled way. IEEE Trans Pattern Anal Mach Intell 24(5):639–656

Tiňo P, Kabán A, Sun Y (2004) A generative probabilistic approach to visualizing sets of symbolic sequences. In: Kohavi R, Gehrke J, DuMouchel W, Ghosh J (eds) Proceedings of the tenth ACM SIGKDD international conference on knowledge discovery and data mining (KDD-2004), Seattle, WA, 22–25 August 2004. ACM Press, New York, pp 701–706

Tolat V (1990) An analysis of Kohonen's self-organizing maps using a system of energy functions. Biol Cybern 64:155–164

Ultsch A, Siemon HP (1990) Kohonen's self organizing feature maps for exploratory data analysis. In: Proceedings international neural networks. Kluwer, Paris, pp 305–308

Ultsch A, Mörchen F (2005) ESOM-Maps: Tools for clustering, visualization, and classification with emergent SOM. Technical Report No. 46, Department of Mathematics and Computer Science, University of Marburg, Germany

Ueda N, Nakano R (1993) A new learning approach based on equidistortion principle for optimal vector quantizer design. In: Proceedings of IEEE NNSP93, Linthicum Heights, MD. IEEE, Piscataway, NJ, pp 362–371

Van den Bout DE, Miller TK III (1989) TInMANN: the integer Markovian artificial neural network. In: Proceedings of international joint conference on neural networks (IJCNN89). Washington, DC, 18–22 June 1989, Erlbaum, Englewood Chifts, NJ, pp II205–II211

Van Hulle MM (1997a) Topology-preserving map formation achieved with a purely local unsupervised competitive learning rule. Neural Netw 10(3):431–446

Van Hulle MM (1997b) Nonparametric density estimation and regression achieved with topographic maps maximizing the information-theoretic entropy of their outputs. Biol Cybern 77:49–61

Van Hulle MM (1998) Kernel-based equiprobabilistic topographic map formation. Neural Comput 10(7):1847–1871

Van Hulle MM (2000) Faithful representations and topographic maps: from distortion- to information-based self-organization. Wiley, New York

Van Hulle MM (2002a) Kernel-based topographic map formation by local density modeling. Neural Comput 14(7):1561–1573

Van Hulle MM (2002b) Joint entropy maximization in kernel-based topographic maps. Neural Comput 14(8):1887–1906

Van Hulle MM (2005a) Maximum likelihood topographic map formation. Neural Comput 17(3):503–513

Van Hulle MM (2005b) Edgeworth-expanded topographic map formation. In: Proceedings of workshop on self-organizing maps (WSOM05). Paris, France, 5–8 September 2005, pp 719–724

Van Hulle MM (2009) Kernel-based topographic maps: theory and applications. In: Wah BW (ed) Encyclopedia of computer science and engineering. Wiley, Hoboken, vol 3, pp 1633–1650

Van Hulle MM, Gautama T (2004) Optimal smoothing of kernel-based topographic maps with application to density-based clustering of shapes. J VLSI Signal Proces Syst Signal, Image, Video Technol 37:211–222

Verbeek JJ, Vlassis N, Kröse BJA (2005) Self-organizing mixture models. Neurocomputing 63:99–123

Vesanto J (1997) Using the SOM and local models in time-series prediction. In: Proceedings of workshop on self-organizing maps (WSOM 1997). Helsinki, Finland, 4–6 June 1997. Helsinki University of Technology, Espoo, Finland, pp 209–214

Voegtlin T (2002) Recursive self-organizing maps. Neural Netw 15(8–9):979–992

Wiemer JC (2003) The time-organized map algorithm: extending the self-organizing map to spatiotemporal signals. Neural Comput 15(5):1143–1171

Willshaw DJ, von der Malsburg C (1976) How patterned neural connections can be set up by self-organization. Proc Roy Soc Lond B 194:431–445

Yin H, Allinson NM (2001) Self-organizing mixture networks for probability density estimation. IEEE Trans Neural Netw 12:405–411

Printed by Publishers' Graphics LLC